PRACTITIONER'S GUIDE FOR STATISTICS AND LEAN SIX SIGMA FOR PROCESS IMPROVEMENTS

PRACTITIONER'S GUIDE FOR STATISTICS AND LEAN SIX SIGMA FOR PROCESS IMPROVEMENTS

Mikel J. Harry
Six Sigma Management Institute
Scottsdale, AZ

Prem S. Mann
Department of Economics
Eastern Connecticut State University
Willimantic, CT

Ofelia de Hodgins
International Institute for Learning, Inc.
New York, NY

Chris Lacke
Mathematics Department
Rowan University
Glassboro, NJ

Richard Hulbert
Bank of New York Mellon
New York, NY

A JOHN WILEY & SONS, INC., PUBLICATION

Published by John Wiley & Sons, Inc., Hoboken, New Jersey

Published simultaneously in Canada

For general information on our other products and services or for technical support, please contact our Customer Care Department within the United States at (800) 762-2974, outside the United States at (317) 572-3993 or fax (317) 572-4002.

Wiley also publishes its books in a variety of electronic formats. Some content that appears in print may not be available in electronic formats. For more information about Wiley products, visit our web site at www.wiley.com.

Library of Congress Cataloging-in-Publication Data:

Practitioner's guide for statistics and lean six sigma for process improvements / Mikel J. Harry ... [et al.].
    p. cm.
  Includes index.
  ISBN 978-0-470-11494-0 (cloth)
  1. Process control–Statistical methods. 2. Six sigma (Quality control standard) 3. Statistics–Data processing. I. Harry, Mikel J.
  TS156.8.P72 2009
  658.5072'7--dc22
                             2009034002

Printed in the United States of America
10 9 8 7 6 5 4 3 2 1

CONTENTS

PREFACE

Top management is supposed to be a tree full of owls - hooting when management heads into the wrong part of the forest

—Robert Townsend, *business writer*.

INTRODUCTION

At times the practitioner might wonder why many people go into Lean Six Sigma and statistics. Is their DNA different? What do practitioners of Lean Six Sigma do, practice, or believe in? At a time when success depends on satisfying the need for developing quantitative skills and reasoning abilities, the mystique surrounding Lean Six Sigma persists. *Practitioner's Guide to Statistics and Lean Six Sigma for Process Improvements* shows some of the payoffs for working with the fundamental concepts of statistics combined with the methodology of Lean Six Sigma. The areas of discussion are real. Most of the situations that the practitioner encounters in the book are based on hands-on experience. The goal is for the practitioner to enjoy solving business problems and have fun while doing it. The reader is not expected to become a statistician; it is understood that the practitioner might not remember in a year or so most of the techniques presented here. However, this is not the goal; what we would like to see is that you solve some of your business's most important problems and gain appreciation for the how, the when, and the why. Lean Six Sigma is not about complex formulas and equations; it is about graphical, visual, and simple analytical methods. It is about looking — with a method — at a business problem and its processes and then creating graphical representations to work with in order to solve those business problems and improve the business processes for cost-effectiveness. This is what counts.

We begin the preface of this first edition of *Practitioner's Guide to Statistics and Lean Six Sigma for Process Improvements* with a question that most Lean Six Sigma practitioners have been asking. Just what are statistics and Lean Six Sigma used for? For application purposes, *Practitioner's Guide to Statistics and Lean Six Sigma for Process Improvements* is our response to this question, and an attempt to fill this need. This book represents a combined effort to create excitement, interest, curiosity, and desire in the nonspecialist for cost reduction and value-centric thinking that is based on the fact that errors or mistakes in process execution are directly linked to many of the economic fundamentals that underlie a business. Of course, such fundamentals include things like cycle time, work-in-process, labor cost, and nonvalue activities that consume time and resources, to mention only a few. In this book, we attempt to create a strong desire for reading *Practitioner's Guide to Statistics and Lean Six Sigma for Process Improvements*, by knowing that each of our readers will understand that reducing defects and defect opportunities directly reduces cycles-time, work-in-process, labor, scrap, rework, and equipment utilization. Our efforts have been guided by Thomas Jefferson's notion that an enlightened citizenry, in which practitioners, having acquired a broad knowledge of topics, can then exercise sound judgment in making business and economic decisions. Managers who have learned this material should agree that many of their past management practices and quality programs might need some updating.

The book is designed to introduce Lean Six Sigma practices and statistics. It is also intended to serve as a reference for the Lean Six Sigma community. We envisioned that our readers come from multiple communities — executive, management, and Lean Six Sigma practitioners — and possess diverse education, work experience, and skills. This aim required that the material be presented at several different levels of understanding. Recognizing that practitioners learn in a variety of ways, we have provided alternative approaches to help readers visualize and interpret common types of business problems. Thus, this book contains more than 1000 figures to explain arithmetic, algebraic, statistics, and Lean Six Sigma concepts and results. In writing this book, a struggle with the traditional teaching method in which authors present a problem and immediately solve it was encountered. *Practitioner's Guide to Statistics and Lean Six Sigma for Process Improvements* believes that an important part of problem solving is to encourage reflection by the practitioner before action. Thus, examples covering fundamental concepts offer hints, rather than firm directions, to guide the reader's choice of tools and methods.

The chapters of the book are divided into two different groups— applications and theory. Applications are emphasized in Chapters 1–9, 15, 21, 22, and 23, while theory is emphasized in Chapters 10–14, 16–20 and 24. The chapter sequence 1–24 is a natural one for mastering all of the material. On the other hand, someone who has interest in the material might begin with the summary or overview sections of the first six chapters.

Objective

The *Practitioner's Guide to Statistics and Lean Six Sigma for Process Improvements* is for today's world, where a vast number of individuals and organizations must interact within a rapidly changing, global economy. Perhaps Francis Gouillart and James Kelley said it best in their 1995 McGraw-Hill book, *Transforming the Organization*:

> We once assumed that corporate evolution consists of long periods of stasis, punctuated by periodic adaptations, but the pace of change is too fast for that now. Now, the corporation needs to adapt every day and this is why today's corporations of all types and sizes are looking for top-line growth, reduction in cost structure, and increased capacity – without major capital investment . . .

> Businesses today are looking for improvements in customer satisfaction, investor relations, process quality, sound supply chain, and supplier quality, not to mention the need for gains in organizational capability, cost effectiveness and capacity to remain competitive. These companies must deliver additional value to the bottom line while concurrently improving customer satisfaction.

Another simple example of this mission was apparent during the preparation of this book. The authors are a diverse group that includes men and women, native-born and immigrant members. They were educated at different colleges and universities and pursued diverse careers. In addition, they live or work in a half-dozen states; thus thousands of email messages and telephone calls were required to coordinate the preparation of this book. Now, replace "authors" with "work force," and the practitioner obtains a realistic description of the role of diversity in today's working environment.

Lean Six Sigma is about extraordinary sense, not common sense. Throughout this book, practitioners will encounter the latest ideas and principles of Lean Six Sigma; many of these differ from what managers learned in school or during their professional careers. For example, in science and industry, probability and risk models have become the cluster of designing operations and decision making. To assist in achieving these goals, *Practitioner's Guide to Statistics and Lean Six Sigma for Process Improvements* has stressed the connections among statistics, Six Sigma principles, and Monte Carlo simulations. Naturally, the pursuit of deeper insights into these relationships depends on understanding the basic statistical principles and Lean Six Sigma methodology. No list of publications can capture the continuous sense of invention and innovation in this field; therefore, the goal of this book is to be flexible and simple in its presentation of the contemporary state of Lean Six Sigma and how it impacts our daily lives.

Approach

Since its inception, Six Sigma has expanded far beyond its initial objective of using statistics to improve the quality of manufacturing products. The Six Sigma name was derived from this initial objective.

Critics, unaware of this Six Sigma future-extendibility, compared Six Sigma with *total quality management* (TQM). When significantly higher quality standards are demanded from processes and products, employees must individually and collectively examine the ways in which their work is accomplished. Although narrowing acceptable process variation by a factor of less than two appears to be a minor change, the number of defects resulting from a modified process is reduced by orders of magnitude. Because Six Sigma is about innovating new ways for doing things, not just fixing them after they fail, senior management participation is required to achieve the potential gains from Six Sigma.

Over time, the effectiveness of new and existing analysis methods and tools used to achieve these aggressive quality targets has improved. *Practitioner's Guide to Statistics and Lean Six Sigma for Process Improvements* recognizes that these same techniques may be applied to other processes within a business, to other types of businesses, and perhaps even to managing an entire corporation. The reader should understand that the terms *Lean Sigma* and *Lean Six Sigma* are used interchangeably to describe this ever-broadening capability.

A fundamental concept of Lean Six Sigma is its use of metrics to determine success. Lean Six Sigma has been successfully implemented at many major corporations because it uses business strategies that increase market share and reduce costs. Every process, every number, every metric, every operational definition, and, in general, every step in creating the final product, service, or transaction is consistently questioned along with the integrity of the data. Only by measuring the organization's processes and analyzing the measurements can the organization improve its processes. It is virtually impossible to evaluate the health of an organization or to know where an organization is going without measurements.

In 1883, Lord Kelvin said: "When you can measure what you are speaking about, and express it in numbers, you know something about it; but when you cannot measure it, when you cannot express it in numbers, your knowledge is of a meager and unsatisfactory kind."

On the basis of this statement, it is widely recognized that Lean Six Sigma assists the bottom line in simple tasks such as

Accepting what we do not know

Not doing what we do not know

Not knowing until it is measured

Not measuring what we do not value

Not valuing what we do not measure

Many success stories linking quality directly to business performance are included in the book. In addition, many of the concepts introduced by Lean Six Sigma now appear in quality documents. See, for example, the following quote from the *IBM Quality Policy*.

Product leadership is one of the goals. An essential element of product leadership is quality: the quality of offering which are defect free. Organizations have to do things right the first time. Each stage in the process must produce a defect-free output. Quality is everyone's job. Each function, each individual including all of the suppliers must assume the responsibility for a defect-free operation.

PHILOSOPHY OF THE BOOK

As important as the specific recommendations contained in the *Practitioner's Guide* is the philosophy that learning Lean Six Sigma and the fundamentals of statistics is a task that can be easily accomplished by almost anyone who possesses adequate curiosity, perseverance, will power, and discipline. When readers find themselves asking why processes using these methods are so cost-effective, they are on their way to becoming practitioners of Lean Six Sigma. Practitioners will be encouraged to apply the techniques and ideas introduced here to business problems, root cause analysis, decision making, economic events, controversies, and other real-life situations. To this end, the authors have chosen topics and examples from real-life experiences. Our enthusiasm may mask potential risks of incorrectly applying the Lean Six Sigma tools. There is no substitute for wisdom gained through education and experience. Consider the Law of the Instrument— Give a hammer to a small child and he/she will soon conclude that everything needs pounding. This is also sometimes stated as *If you give a child a hammer, everything looks like a nail*; and *when you really want to drive a nail, everything starts to look like a hammer*. It doesn't matter if it's a rock or a wrench, a blender or a board or a baguette. When someone *really, really* wants to drive that nail, everything suddenly looks like a hammer. In this metaphor it is observed that many variables are correlated but some have no cause-and-effect relationship. The practitioner must therefore rely on a set of Lean Six Sigma methods to identify the critical few causes and leave the remaining trivial many for subsequent analysis.

Software Tools

MINITAB® examples are included throughout the book chapters. References are made to MINITAB, release 15 for Windows® 98, Windows Vista, ME, Windows NT™4, 2000, XP Home or Professional. MINITAB is used by customers from a broad range of industries such as aerospace, financial services, manufacturing, oil and gas, pharmaceuticals, and utilities. For more information on MINITAB, see www.minitab.com.

JMP® Student Edition for Windows XP, Windows Vista, and Mac OS X 10.5 or higher is frequently referenced in the book. JMP is used by customers from a broad range of industries such as aerospace, financial services, manufacturing, oil and gas, pharmaceuticals, and utilities. For more information on the full version of JMP, see www.jmp.com.

Crystal Ball®, from Oracle, examples are included in Chapter 23. Crystal Ball is a leading spreadsheet-based software suite for predictive modeling, forecasting, Monte Carlo simulation and optimization. Crystal Ball is used by customers from a broad range of industries such as aerospace, financial services, manufacturing, oil and gas, pharmaceuticals, and utilities. With Crystal Ball, the practitioner can make tactical decisions to reach objectives and gain a competitive edge under even the most uncertain market conditions. For more information on Crystal Ball, see www.oracle.com/crystalball/index.html.

MindPro®, from the Six Sigma Management Institute, enables the rapid deployment and application of Lean Six Sigma knowledge. This video, which is based on the knowledge system, not only provides a platform for digital and blended training but also includes an integrated *Best Practice Coach* for on-the-job applications. MindPro also includes a *Body of Knowledge* that enables an enterprise to link video coaching and training directly into their company documents, critical processes and existing training materials. For more information on MindPro, see www.sixsigma-mindpro.com.

ACKNOWLEDGMENTS

The main objective of this book is to encourage the use of statistics and Lean Six Sigma in the management of business, service, and manufacturing processes. We believe that logic and ordered reasoning lead to clearer problem definition, superior problem-solving ability, greater self-confidence, and improved management. The concepts and ideas included in the book come from many sources over an extended period of time. Thus, the cited references are generally those of more readily available books and journals rather than the original sources.

During the writing of *Practioner's Guide*, we have received helpful comments and suggestions from many persons.

In particular, Mikel Harry would like to acknowledge the late Bill Smith, Motorola Inc. Although Bill is no longer among us in this world, his central role in the creation and initial development of Six Sigma will always be recognized and admired. Bill was a true pioneer in the field of quality management. Dr. Harry would also like to acknowledge his bride-to-be, Sandy Penwell. Her dedication and support is inspiring and deeply appreciated.

Ofelia de Hodgins wants to thank Dr. Martin Garry Hodgins from the University of North Carolina, Charlotte (UNCC), for his willing support and his major contributions to creating the many figures of the Lean Six Sigma chapters of this book; his dedication is deeply appreciated. In addition, Ofelia wishes to acknowledge the many friends and colleagues who helped mold and shape the manuscript into a viable book. In particular, Ms. de Hodgins appreciates the efforts of - in alphabetical order - Tracee Lee Beebe, Marketing Director of Conversion Service International; Ginger Evans, CEO and President of Dynamic Training; Connie Januzzi from the Six Sigma Management Institute (SSMI); Claudia Nierman, graphic arts designer; and Leonardo Nierman, whose work of art and sculptures are shown in most of the museums of the world. In particular, Ofelia de Hodgins wants to thank Dr. Owen Hill of IBM Corporation and SUNY-New Paltz for long hours of impeccable editing of the Lean Six Sigma chapters. Mr. Nierman's and Dr. Hill's comments and suggestions were insightful and always well taken. Ms. de Hodgins wants to personally thank her triplet-children Garry II, Lizzie, and Cody for their patience, understanding, and consistent support during the writing of this book.

Prem Mann would like to take this opportunity to thank his family for their support and understanding while he was busy working on this project. Chris Lacke would like to thank his wife Cheryl, his kids, Bailey, Maddie, and Nolan, and his parents, Jay and June for all their support and understanding.

R. L. Hulbert wants to thank those who provided support during this process. Special thanks to Joan Fyfe for her inspiration and guidance and Bill Fyfe for friendship above and beyond.

The authors would like to express their utmost gratitude and appreciation to Steve Quigley (Associate Publisher, John Wiley & Sons, Inc.,) and Jacqueline Palmieri (Editorial Program Coordinator, Mathematics & Statistics, John Wiley & Sons, Inc.,). Without their help, push, support, and understanding, this project would never have been completed. Authors would also like to thank Kellsee Chu of John Wiley & Sons, Inc., and Sanchari Sil of Thomson Digital for handling the publication of this book in a very efficient manner.

The efforts and inputs of all were critical in the goal of creating excitement around solving business problems and having fun while doing it.

There is a website for this book that contains helpful and supportive information and manuals. The web address for this site is: ftp://ftp.wiley.com/public/sci_tech_med/guide_statistics. Readers are encouraged to visit this site.

1 Principles of Six Sigma

1.1 OVERVIEW

A Closer Link to Executive Thinking. Within a business, Six Sigma has different meanings for different groups; this difference depends on one's level and respective job role in an organization. At the *enterprise level*, Six Sigma is often deployed as a strategic business initiative. In this context, it focuses on making significant improvements in areas such as business growth, capacity, investor relationships, and customer satisfaction. At the *operations level*, Six Sigma is tactical in nature and is most often directed toward improving delivery time, cost of poor quality (COPQ), defects per unit (DPU), and a host of other critical measures of operational effectiveness and efficiency. At the *process level*, Six Sigma is used to reduce process variability. Reducing variability minimizes the number of defects, shortens process cycle times, and decreases direct costs. At this level, the motto is simple—if you make an improvement, then on a timely basis the gains should be verifiable. In this context, the elimination of a defect, mistake, fault, or error within the "system" must directly translate into a measurable benefit such as reduced headcount, less material, and lower overhead cost. In other words, Six Sigma is a strategic and tactical system for managing total business enterprises. From this perspective, Six Sigma has the capacity and capability to deliver customer and provider satisfaction, which are key ingredients for business success. In short, Six Sigma epitomizes the ideals of business success and optimizes the control function of an enterprise. In its most elemental form, Six Sigma represents 3.4 defects per million opportunities for defect. This perspective of Six Sigma is related to a single opportunity for defect for a single critical-to-quality (CTQ) characteristic. The fundamental idea of Six Sigma is that as performance is improved, quality, capacity, cycle time, inventory levels, and other key factors are also improved. Thus, when these factors are improved, both the provider and the customer experience greater satisfaction in performing business transactions.

1.2 SIX SIGMA ESSENTIALS

The abatement of business risk is essential to Six Sigma. In this text we will explore many of the key concepts underpinning a new definition of quality. We will describe quality as the state in which value is realized for the customer and provider in every aspect of the business relationship. Simply stated, performance meets expectations; in doing so, we will demonstrate that Six Sigma is far more than a simple quality target. The fundamental tenets of Six Sigma are as follows:

> *Thinking Six Sigma.* Explore the big ideas that power the realization of breakthrough performance and then gain insight into how these ideas create value for any type of enterprise.

Practitioner's Guide for Statistics and Lean Six Sigma for Process Improvements. By Mikel J. Harry, Prem S. Mann,
Ofelia C. de Hodgins, Christopher J. Lacke, and Richard Hulbert

Applying Six Sigma. Profile the vital improvement tools that Six Sigma players use when executing their application projects and then grasp how such "mind tools" can be used to solve virtually any problem.

Targeting Six Sigma. Identify, scale, define, empower, and execute Six Sigma projects that achieve higher organizational goals and then learn how to track and validate progress in these projects.

Leading Six Sigma. Identify, select, and train Six Sigma leaders at all levels of an organization without backfilling the vacated positions, and then effectively motivate and retain those leaders to continually achieve forward momentum.

Enabling Six Sigma. Understand the information and reporting needs that underlie the global deployment of Six Sigma and then effectively integrate and fulfill those needs at the local level of the enterprise.

Deploying Six Sigma. Study the essential guidelines for scaling and creating a global Six Sigma deployment plan that will create a critical mass of focused management activity and then put momentum behind that mass at the local level of an enterprise.

Initializing Six Sigma. Interrogate the top programs that support the rapid initialization, deployment, and implementation of Six Sigma and then fully leverage such programs at all levels of an enterprise.

1.2.1   Driving Need

From a layperson's viewpoint, the world appears to be quite predictable at times but unpredictable at other times, although it appears that little has changed. From this perspective, we can appreciate why the human species has been obsessed with the idea of control. Perhaps this driving need for understanding and repeatability of results led to the idea of science. In some cases, we seek to enhance our ability to replicate some object, situation, or phenomenon. We recognize that to replicate a successful business transaction, the provider must establish processes that are capable of yielding high-quality outcomes that are both efficient and effective in terms of cost and time. Today, many businesses deploy such processes both vertically and horizontally throughout their organizations, often interfacing with their customers and suppliers. To this end, the provider must minimize average transactional costs, as well as that of time, while concurrently seeking to maximize quality and volume. At least so goes the theory in the executive mind. Senior executives are always on the lookout for innovative ways to reduce their cost and expand market share. Doing so creates value for all stakeholders. To achieve this, business leaders aspire to increase their organization's capability and capacity with minimal resource investment. They understand that quality and customer satisfaction must be continually improved.

1.2.2   Customer Focus

What is the nature of a customer–provider relationship? Most businesses claim to be customer-focused; however, these same businesses show little or no evidence to support this claim. An organization that is honestly committed to customer satisfaction will implement multiple customer feedback channels and a structured methodology for integrating data into their service delivery processes. The dictionary defines the term *customer* as a person who buys something. Simply stated, this implies that the customer is a person (or perhaps an organization) that receives some form of value in exchange for another form of value, held or originated (fully or partially), by the provider. Obviously, the customer and provider both seek to maximize their respective benefits. The dictionary definition provides us with a fairly large keyhole for viewing satisfaction. Such a state of being is related to the idea of conducting a successful business transaction. The customer has a sense of the extent to which her/his standards have been met by a business transaction, as does the provider. In other words, they both seek a quality transaction. Essentially, customer and provider expectations form the basis on which the idea of quality is based. It is precisely this interaction between the customer and provider that governs the "quality" of the business relationship. For every aspect of producing and consuming a product or service, there are rightful levels of expectation that can be identified and improved only through careful and detailed analysis. A company can meet or exceed these expectations only by deploying performance metrics to guide and manage each key aspect of the business relationship, whether that aspect is making a product on an assembly line or broadcasting a news show.

Here quality is not an absolute standard but rather a relative measure of the gap between rightful expectation and actual performance. As the gap diminishes, the quality of interaction improves in all aspects of the business relationship. In this sense, the Six Sigma definition of quality serves as a management framework for focusing a business and launching actions that yield consistent and dramatic results for the customer and provider. It is a divergence from tradition in that the pursuit of value entitlement, not blind conformance to standards, drives the business relationship.

1.2.3   Core Beliefs

What is the fuel that propels the success of six sigma?   We must all remember that Six Sigma is the epitome and embodiment of "hope" that fuels the collective will. It is hope that moves people to align their values, aims, and goals in a common direction. This is what leaders do; where it exists they sustain it; where it does not exist, they create it. Leaders create and energize hope by realizing visible and measurable success, not just one project at a time, but by achieving many simultaneous successes. Hope drives the human spirit to accomplish great things, and thus hope is the muscle of leadership. Without hope, leaders have nothing to sell. Without something to sell, they are just another player on the field of mediocrity. The collective "shock and awe" of Six Sigma projects is one way to ignite the stove of executive hope. It is the sudden, collective, decisive, and repeatable successes of Six Sigma that cause employees to believe their company is the best. When this attitude pervades an organization, it becomes boundaryless. As this occurs, innovation takes hold. Essentially, the Six Sigma initiative was designed to raise the bar so high that employees would be forced to individually and collectively reexamine the way in which work was done, not just tweak the existing work processes. Given this inaugural aim, it should be apparent that Six Sigma is about innovating new ways of doing things, not just making incremental gains to existing processes.

How is Six Sigma superior to other improvement programs?   Simply stated, Six Sigma has produced astounding economic benefits that have hit the proverbial bottom line of many fine corporations in a verifiable and consistent way, year after year. We have a saying in Six Sigma work, "let the data do the talking." In this spirit, the financial performance achieved by Six Sigma says it all, not to mention the quantum gains in customer satisfaction. Unlike the philosophical and prophetic nature of *total quality management* (TQM), Six Sigma is a repeatable management process based on the idea of measurement. It is a goal-driven, result-oriented, fact-based management system based on scientific principles. Thus, Six Sigma requires that any type or form of business improvement must be verifiable through measurement in everything that a company does or seeks to do everyday in every way. Today, few corporate executives believe that TQM is a viable system of business management. The conclusions were quite apparent in a 1996 study, "Measuring performance after meeting award criteria," published in *Quality Progress* magazine, that TQM practices had less impact than most thought. After examining data from Baldrige and state quality award winners, applicants and nonapplicants, the study's authors concluded that they could not conclusively determine whether quality award-winning companies perform better than others.

Even before this, TQM skepticism was already building. Consider the April 1994 article, "Is TQM dead?" featured in *Quality Digest* magazine. Editor Scott Madison Patton cited study after study that brought the viability of TQM into serious question. Only 20% of *Fortune 500* companies are satisfied with the results of their TQM processes, according to a 1992 Rath & Strong survey. Florida Power & Light remains the only US company to have won Japan's coveted Deming Prize. Its winning strategy was largely dismantled after complaints of excessive bureaucracy and red tape. Patton continued, stating that a survey of 300 electronics companies by the American Electronics Association found that 73% had quality programs in place, but of these, 63% said they had failed to improve quality by even as much as 10%. A study of 30 quality programs by McKinsey & Co. found that two-thirds of them had stalled or fallen short of yielding real improvements. Unlike TQM, Six Sigma is a management tool that astute leaders can employ to masterfully intertwine their personal destiny with that of the corporation. Only when this happens does the potential for business magic begin to surface. Not the smoke-and-mirrors variety of magic, but the kind of real stuff from which dreams are made.

The creation of operational magic is what sound business is all about. When this magic begins to unfold, good leaders suddenly edge toward greatness, followers begin to consciously work smarter and harder, and the world (at large) takes notice. This is the magic of Six Sigma; it can transform good corporations into exemplars. As great leaders wield the power of Six Sigma and begin to leverage the tools of breakthrough, they cross the threshold of destiny. At this point the corporation, its employees, shareholders, and all of those so connected prosper. Employees across the corporation rise to the challenge. When this occurs, an unstoppable revolution begins.

What are the boundaries of Six Sigma? First, a host of well-respected global corporations have directly experienced the promised benefits of Six Sigma. Most of these highly diverse corporations have carefully documented their Six Sigma journey and published their notable successes. Many of their senior executives are outspoken on the merits of Six Sigma and corroborate its power as a management system. In fact, Mr. Jack Welch (former CEO of General Electric) stated that Six Sigma was the most significant undertaking in GE history. He also said that Six Sigma, as a management tool, reaches the control function of a corporation. These facts speak volumes about the power and reach of Six Sigma. At any level of an enterprise, a mix of resources are required to fix a defect or error. If the given defect is eliminated (or otherwise prevented by process or product design), then the improvement is verifiable. In other words, the improvement is real. If an improvement is verifiable, we should see a corresponding savings in labor, material, and/or overhead. If such a savings does not materialize or cannot be verified, then the improvement was not real. Such is the way of Six Sigma. In a nutshell, Six Sigma is about the creation of global value, whereas TQM was generally limited to local quality improvements. In this sense, Six Sigma is about sudden and quantum breakthroughs in business performance (vertically and horizontally), whereas TQM was concerned mostly with achieving gradual improvements in product defect rates (at the local level of an enterprise). Six Sigma is a top–down business imperative (based on cascading performance expectations), whereas TQM is a bottom–up quality program (based on disconnected quality improvements). Again, it is easy to see that TQM is mostly constrained to the business of quality, whereas Six Sigma is concerned with the quality of business.

1.2.4 Deterministic Reasoning

The Nature of Determinism and How It Fits with Six Sigma. In a purely deterministic world we recognize that at any moment in time, a change in some object, event, or phenomenon is dependent on a change in one or more of its underlying determinants. In its simplest form, we can express this idea by the relation $Y = f(X)$, where Y is the dependent (outcome) variable, X is the independent (causal) variable, and f is the function that defines the relationship between Y and X. For example, a sonic boom (Y) is a function of an airplane's speed (X). Physics and mathematics are based on this simple concept. In the real world, life is not that simple; we recognize that other factors exert an influence on Y. Coming back to our analogy, in addition to speed, most of us realize that the existence and strength of a sonic boom depends on other contributing factors such as altitude, wing design, and atmospheric conditions.

Simply knowing the airplane's speed is not enough, by itself, to fully explain the boom effect. Hence, we must extend the relationship $Y = f(X)$ in such a way that $Y = f(x_1, \ldots, x_n)$, where $x_1, \ldots, x_n$ becomes a set of determinants. The world of Six Sigma has been built around this simple belief; without the idea of determinism, the practice of Six Sigma could not exist.

1.2.5   Leverage Principle

What is the leverage principle, and how is it used in the practice of Six Sigma? Not all variables are created equal; some exert more influence than others. This idea is exemplified by the relation (transfer function) $Y = f(x_1, x_2, \ldots, x_n)$. Given such a deterministic model, we understand that a certain amount of change in a particular x may not induce or otherwise cause the same amount of change in Y as some other X. Thus, it can be said that every X may have a different influence on Y. Those x's (x values) that exert a large influence are said to have leverage. Similar to a lever that is used to move a large rock, an "X" with leverage can cause a significant change in Y when compared to the influence of other x's within the same set of causative factors. Those x's that exert a disproportionately large amount of influence on Y are often called the "vital few" variables. These vital few variables stand in stark contrast to the less influential factors, often called the "trivial many" variables.

1.3   QUALITY DEFINITION

Quality and How It Fits in Six Sigma. In general, quality is defined as conformance to standards; however, some authors define quality as a subjective term for which each person has his/her own definition, based on the perceived degree to which the product or service meets customer's expectations. Quality has no specific meaning unless related to a specific function or object. Quality is a perceptual, conditional, and subjective attribute. On the other hand, the Six Sigma view of quality differs from this perception–definition. For Six Sigma, quality is a state in which value entitlement is realized for the customer and provider in every aspect of the relationship. Therefore, the central question for the practitioners regarding quality is: Is my organization in the business of quality? Or, is my organization in the quality of the business? A guide for the practitioner is to determine whether his/her organization is in the business of quality. If the organization is in the business of quality, the key determinant for customer need is utility and the key determinant for the provider is to achieve. If the organization is in the quality of the business the key determinant for customer need is value and the key determinant for the provider is to create. The importance of quality improvement is that it starts with finding the root cause of a defect; this includes problem analysis and problem solving. On the other hand, it is known that Six Sigma is a business improvement method that searches for the causes of the defect and then eliminates these causes by focusing on outputs according to the customer's key needs. From this, we could say that a phase of quality is the quality of the product or the quality of service. Another phase of quality is *product or service offering*—when an organization makes the product or service within the options that customers want, the customers will buy it. Another phase is the *quality of cost conformance*—if the organization meets customer's price better than other offerings, customers will buy it from the organization. Another phase of quality is *provider cost reduction* and therefore, improvement of business margins. Yet, another phase of quality is the *quality of creating value*—if an organization creates value, customers will buy the product or service.

Plainly stated, the Six Sigma definition of quality is a state in which value entitlement is realized for the customer and provider in every aspect of the business relationship. By this definition, quality is not exclusively related to a thing, but rather to a state applicable to all aspects of the business relationship; there exists a perceived state of quality as well as a real state. Thus, quality now encompasses the idea of value and a rightful expectation of entitlement. This definition of quality has spread around the world as Six Sigma quality professionals have become more sophisticated about how they measure and improve what companies do and how companies do it. Specifically, they have expanded the quality envelope to include not only the utilitarian aspects of products and services but also their availability and worth aspects. The value aspects of a product or service should exclude variation, should exclude defects, and should include time reduction. The customer and provider have every right to expect that form, fit, and function requirements for a deliverable (e.g., product, service) process will be created and provided in a quality way. Even when this sounds redundant, customers can reasonably expect that deliverables will be presented to them on a timely basis and in the quantity ordered. Finally, the overall worth of each transaction should always be present in the minds of both customer and provider. Inherently, customer and provider must find economic, social, intellectual, and/or psychological value in each transaction. Since these forms of value are shaped by laws, rules, and regulations and translated by the processes of dictation and negotiation into performance requirements, it is obvious that quality begins with the first customer contact and continues throughout the lifecycle of the business process relationship. In business terms, this is known as delivery.

Because the customer has the rightful expectation that the utility of a product or service will be present at the time of purchase, on delivery, and during its defined lifetime, the provider seeks to reduce the cycle time of its systems, operations, and processes, thereby reducing the time between order entry and shipment. The provider expects to make full use of the capacity and capability inherent to the systems, operations, and processes that realize the product or service. The same could be said for the provider's stakeholders, as they can rightfully expect that their capital investments will yield an attractive return. This same line of reasoning holds throughout the value chain (vertically and horizontally). In addition to utility, availability aspects of value entitlement are part of the total business relationship. Finally, as quality practice becomes integrated with the financial domain of business, Six Sigma metrics and improvement efforts become focused on the "worth" aspects of products and services. Simply stated, the customer can rightfully expect to minimize his/her economic outlay while the producer is entitled to make the highest possible profit. Obviously, this can occur only when the previously mentioned quality aspects achieve their respective levels of entitlement. In summary, and from this perspective, it is clear that quality professionals should focus on the quality of the business (quality of cost) and not on the business of quality (cost of quality). In other words, it is more important to ensure that every dollar spent on the business produces a dollar's worth of value for the customer and provider, than it is to measure how much money is lost as a result of poor quality that is a posteriori focus. The practitioner should again recognize that the idea of worth embodies many aspects, such as emotional and intellectual fulfillment to the customer and provider in maintaining the business relationship. Whereas the customer feels good or satisfied after purchasing a product or service, the provider has these same needs while producing a product, delivering a service, or performing a transaction. In addition, the customer has a rightful worth expectation related to various societal considerations such as the product will not degrade the environment, will not be a hazard to her/his health, or cause harm to others. As may be apparent, the notion of value entitlement turns the classic idea of quality that is conformance to standards, instead of a singular focus on conformance to standards (customer satisfaction). Six Sigma companies strive to determine the customer's and the company's rightful level of expectation in all aspects of the business relationship. Such expectations, some set by the laws of physical possibility and some negotiated, form the quality standards for that relationship. Given this, the inherent meaning of quality constantly changes as a function of the interaction between customer expectations and provider capabilities.

Node Levels.   These are the basic units of data structures. Information is contained in data structures such as trees and linked lists. Each node contains data and links. Chains of interlinked nodes create large complex data structures. The node level is similar to the product or service level. While a product or service design has certain performance specifications, real or theoretical levels of process capability and capacity are required to achieve the database designers' aims. Only when the two are properly identified and integrated can we measure and then reduce the gap between rightful expectation (e.g., design specifications) and actual performance (e.g., process capability). As these gaps are diminished, the quality of interplay improves, thereby cascading to the system level and ultimately to the customer. Thus, we turn to Six Sigma to qualify our designs through the prudent and scientific process of measurement, analysis, and optimization of producibility. As may be intuitively reasoned, the vast majority of ideas, methods, tools, and practices related to the field of producibility analysis can be effectively used by all vertical and horizontal functions within the organization. Only in this manner can we begin to design in quality. By the simple domino effect, we break through to a new level of performance and begin to create the quality of business rather than just using the philosophy of quality in products.

1.4   VALUE CREATION

Value Creation and How It Aligns with the Aims of Six Sigma.   From the customer's viewpoint, Six Sigma translates into higher-quality products and services, delivered on time, at the lowest possible cost. In this sense, Six Sigma offers tremendous value. Through Six Sigma, a customer has better access to products and services, enjoys a significantly higher level of product and service utility, and pays less for these benefits. So, in the customer's head and heart, Six Sigma is about the creation of value. This makes Six Sigma a value proposition for both the customer and the provider. In its first generation, (1984–1994), Six Sigma was initiated at Motorola. At that time, Six Sigma focused primarily on the reduction of defects through quality improvements. It was widely believed that the reduction of defects would naturally and invariably offer economic benefits even though such benefits could not be readily verified by commonly accepted accounting conventions. During the second generation, (1994–2001), Six Sigma was initiated at Asea Brown Boveri (within their large-scale transformer business). Here, Six Sigma was directed toward achieving verifiable cost reductions. Thus, Six Sigma focused on cost reduction, and that became the battle cry for quality improvement. Essentially, this reversed the emphasis of the previous Six Sigma generation. Obviously it worked as evidenced by the wide adoption of Six Sigma since that time. Interestingly, the third generation of Six Sigma (2001–present) is focused on the creation of value. This generation is based on a careful examination of the lessons learned from the application of Six Sigma by more than 30 global organizations over the previous twenty years.

1.4.1   Value

Business improvement is much like filling a bucket with water; of course, you must stop the leaks, but you may also have to use a larger bucket, increase the flow rate, or some combination thereof. In other words, value creation moves the focus of quality closer to the customer (and the provider). Such a shift in emphasis is augmented by the concurrent and increased focus on the key dimensions of quality—utility, availability, and worth. In this context, we can describe value by the simple equation.

$$V = \frac{U \times A}{C}$$

where V = value of cost product or service, event or activity; U = utility of the process that delivers the product/service in terms of form, fit, and function; A = access by the customer to the product/service in terms of volume, timing, and location; and C = cost.

Hence, value is "bang for the buck" from the customer's perspective. Value (per se) is realized when the need for low-cost utility is fully satisfied. However, to create such value, we must employ the four phases of Lean Sigma—innovation, configuration, realization, and attenuation. Low-cost solutions should be developed to meet the customer's utilitarian and access needs. To meet these needs, we must first propose a plan/design to deliver the needed product or service. Then, we must realize the physical form of those ways and means, and finally, we must reduce the value gaps that will usually emerge over time. To implement the Lean Sigma strategy, we employ certain tools and methods. In the course of the innovation phase, we apply these methods and techniques associated with *marketing for Six Sigma* (MFSS). During the configuration phase, we seek to leverage the practices of *design for Six Sigma* (DFSS). During the realization and attenuation phases, we use the tools commonly associated with *processing for Six Sigma* (PFSS). Underlying the Lean Sigma strategy are the principles and practices associated with *leading for Six Sigma* (LFSS).

Through these programs, an organization becomes empowered and is able to upgrade its knowledge base. From this perspective, we realize that the progressive application of Six Sigma tools across the total product lifecycle can have a profound impact on value. In fact, Six Sigma not only preserves value but also seeks to create value. Thus we can define quality as a state in which the customer and provider realize full value entitlement in every aspect of the business relationship.

1.5   BUSINESS, OPERATIONS, PROCESS, AND INDIVIDUAL (BOPI) GOALS

The Vertical Goals of Six Sigma and How They Are Connected to the Bottom Line.   In a strict sense, the goal of Six Sigma is to realize no more that 3.4 adverse consequences per million risk exposures. However, this goal is merely a "model" of what an organization should strive for at the molecular level of its endeavors (e.g., at the opportunity level of a product, process or service). In practice, the Six Sigma goal varies from enterprise to enterprise, as well as within and across organizational functions of each enterprise. This third-generation Six Sigma goal is applicable to each of the four basic tiers of an enterprise—*business, operations, process* and *individual* (BOPI). Thus, we have the BOPI acronym. However, it should be noted that each BOPI level has different yet interrelated Six Sigma goals. By level, the BOPI goals are as follows:

Business Level.   Achieve best-in-class performance for each critical-to-business (CTB) characteristic over a 5-year period. CTBs are those business measures that are considered to be the most important for the organization's economic survival. For example, CTBs include, but are not limited to market share, return on net assets, and customer satisfaction.

Operations Level.   Realize an annualized 78% baseline improvement over a 5-year period for each critical-to-value (CTV) characteristic that links to a business goal. In particular, CTVs are the operational metrics that relate directly to customer satisfaction and market performance. CTVs include, but are not limited to total defects per unit, late deliveries, and warranty returns.

Process Level.   Realize no more than 3.4 defects per million opportunities for each critical-to-quality characteristic (CTQ) associated with the organization's key products, services, and transactions. CTQs are the conditions that must be "right" when to the product or service is being produced. For example, CTQs include, but are not limited to, part thickness, absolute weight, reaction speed, material strength, and telephone hold time.

Individual Level.   Achieve a level of capability equivalent to Cp = 2.0 and Cpk = 1.5 (please see Section 1.2.6) for every critical-to-process (CTP) characteristic that underpins each core process. From this perspective, CTPs are located at the key "adjustment points" within a process. Achieving and sustaining optimal control over the CTPs is essential to effectively and efficiently improve CTQs. For example, CTPs for a machine include but are not be limited to temperature, squeeze pressure, and cutting speed. To translate these idealized goals into the real world, let us consider a corporation's Six Sigma aims. We assume that executive management has issued a directive to improve the company's profit margin by $X\%$ over some defined period of time (say, Z years). In turn, this business goal has been allocated (i.e., flowed down) to the various operations of the business. At this level, the apportioned business goal is translated into expected improvements in certain operational performance metrics such as customer satisfaction, total defects per unit, on-time deliveries, product yield, and service performance. Thus, by closing the "performance gaps" at the operations level, the "economic gaps" can be reduced or eliminated at the business level. Next, these operational goals must be translated into process level objectives. Thus by focusing on critical processes, higher-order goals become actionable. For example, suppose that one of the operational goals is to improve total defects per unit (DPU). To accomplish this objective, the average capability of one or more critical processes must be improved. To illustrate, it might be necessary to improve the *ABC* and *XYZ* process capability from 4σ to 5σ (over the defined period of time to reduce DPU by $X\%$). To achieve this aim, the *ABC* and *XYZ* processes must be fully characterized and subsequently optimized. Naturally, some knobs may have great leverage while others may have little effect on the processes. Once the "vital few knobs" have been identified, each must be fine-tuned to achieve the overall process improvement. In other words, each key process knob must be centered on its ideal operating condition. In addition, the time-related variations around the ideal centering condition must be minimized. Thus, we can now better understand the linkages across the hierarchy of an organization (BOPI). To achieve the global aims of Six Sigma in body and spirit, measurable and verifiable linkages between the business, operations, process, and individual levels of a company must be made.

1.5.1 Differences between Product and Process Capability from a Six Sigma Perspective

Process Capability. This is defined as the spread within which almost all of the parts or values within a data distribution will fall. It is usually described as being within plus ($+$) or minus ($-$) three standard deviations ($\pm 3\sigma$) or six standard deviations (6σ) from the mean value of the dataset. This baseline definition enables us to compare the process capability with the process or specification tolerance. Many quality practitioners treat product and process capability as if they were interchangeable concepts. Presumably, this convention has evolved over time because of the naturally strong correlation between the two. Although related, these two concepts should not be viewed in the same light. At the risk of leading a dead horse to water, we should recognize two things.

Specified Tolerance and Distance to Nearest Specification (DNS). These values are commonly translated into other capability measures. Capability ratios are a simple way of expressing the relationship between the *voice of the process* and *voice of the customer* (VOC). The result is generally expressed as either a dimensionless number or a percentage. The capability ratio is calculated by dividing the difference of the specification limits [*upper specification limit* (USL) and *lower specification limit* (LSL)] by the spread of the data under the area of the normal distribution curve (6σ). The *capability ratio* is denoted by the symbol C_p

Therefore, C_p is expressed as

$$C_p = \frac{\text{USL}-\text{LSL}}{6\sigma}$$

Another quantity, denoted by C_{pk}, is also called the *capability ratio*. This is defined as the distance to nearest specification (in sigma units) divided by 3:

$$C_{pk} = \frac{\text{DNS}}{3}$$

This ratio characterizes the process centering relative to specifications. Some authors do not distinguish between this measure of process centering and the distance between the specification limits.

1.6 UNDERPINNING ECONOMICS

Cost-of-Poor-Quality (COPQ) and How It Should Be Used in a Six Sigma Program. Today, corporations of all types and sizes are looking for top-line growth, reduction in cost structure, and increased capacity, without capital investment. In addition, these organizations are looking for improvements in customer satisfaction, investor relations, supplier quality, not to mention the need for gains in organizational capability and capacity. It can be said that the ability of senior management to yield an additional 5–6% return is no longer a stretch goal; it is fundamental to business growth. Of course, these imperatives are just one of the tasks that senior executives must pursue on an ongoing basis. In short, a business enterprise must be able to create and sustain value. Few companies are aware that the typical corporation leaves the equivalent of 15–25% of every sales dollar on the table and that about 5–6% of sales can be recovered in hard savings through the use of Six Sigma. Such hidden losses result from the untapped capability and dormant capacity in everyday activities. These organizations are usually not aware that achieving a Six Sigma performance level delivers a $1800\times$ improvement over the global 4σ average. They fail to recognize that achieving even a 0.5σ improvement in their operation could create a huge competitive advantage. Unfortunately, these beliefs will directly impact the future of their corporation. Most corporate accounting systems are unable to capture a large part of the true cost of poor quality (COPQ). For example, lost customer opportunities (due to quality issues) are real; however, the organization's financial systems are not designed to detect or report this economic impact. Thus, many executives falsely believe what appears on the accounting dashboard that indicates that the COPQ is about 1% or 2% of sales. In short, this level of COPQ does not trigger any management alarms; thus, the business continues its operations as usual, blinded to the true cost of quality. Even more tragic, the senior leadership team falsely believes that their overall quality level is excellent because their COPQ is low. The best wakeup call for senior leadership teams is "sigma benchmarking." For example, such benchmarking has revealed that the average global corporation operates at roughly a 4σ level. This translates to about 6810 defects per million opportunities (on average). Even more shocking is that the average COPQ for a typical company is 25% of sales. But when we consider a Six Sigma company that produces only 3.4 defects per million opportunities, we realize that this organization is about 1800 times better than its average competitor. This means that a Six Sigma business will experience a COPQ of about one percent of sales. One must wonder how a 4σ company can ever hope to compete with a Six Sigma company? The answer is simple—a 4σ business cannot compete with a Six Sigma business. Of course, everyone thinks their company is the best in existence. We have all heard executives, managers, and employees boast that *nobody does it better than we do*. Around the globe, senior management teams base decisions on the belief that their organization is best in class. Many also falsely believe that they are already at or near the Six Sigma benchmark. When confronted with the results of a sigma benchmarking study, they initially reject the conclusion that they are performing far below a superior level. They often say, we are not that bad. But sooner or later, most organizations come to the realization that they are only average after completion of the benchmarking (test and inspection) phase. On the upside, it is at this point that the senior management team understands their true status and sounds the call to action.

1.6.1 Sigma Benchmarking

The best wakeup call for senior leadership teams is sigma benchmarking. Such benchmarking has revealed that the average global corporation achieves about 4σ. This translates to about 6810 defects per million opportunities (on average). Even more shocking is that the COPQ of average companies is about 25–30% of sales. When we consider a Six Sigma company that produces only 3.4 defects per million opportunities (on average). A Six Sigma organization is about 1800 times better than its average counterpart (4σ). This means that a Six Sigma (6σ) business will experience a COPQ equal to $\cong 1\%$ of sales. From this point of view, how can a 4σ company ever hope to compete with a 6σ company? The answer is simple—a 4σ business cannot compete with a 6σ business.

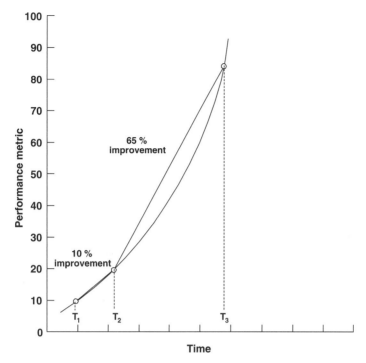

Figure 1.1 Graphical representation of a breakthrough improvement (65%) from T_2 to T_3 (from 10% to 55%).

1.6.2 Breakthrough Goals

Breakthrough goals are sudden departures from historical trends. For example, a change is considered to be small when it does not result in a large jump/change. Thus a change from 10% to 55% (from T_1 to T_3 as the one illustrated in Fig. 1.1) would represent a breakthrough jump while a change from 10% to 25% (from T_1 to T_2 as the one illustrated in Fig. 1.1) would not be a breakthrough because the processes could be easily tweaked to achieve this change. However, a large change cannot occur by simply tweaking the process.

1.6.3 Performance Benchmark

It is reasonable to assume that everyone believes their company to be the best. We all have heard our executives, managers, and employees say: "Nobody does it better than we do." All around the world, senior management teams believe that their organization or corporation is best in class. Many believe that they are already at, or near the Six Sigma benchmark. However, when comparing the empirical evidence with the Six Sigma benchmark chart, their first reaction is to reject the data. However, sooner or later most organizations or corporations come to the realization that they are only average. On the upside, it is at this point that senior management becomes enlightened and sounds the call to *action*.

1.7 PERFORMANCE METRICS

What are "performance metrics," and how are they used by Six Sigma? At a global level, the reader should recognize that many well-respected organizations consider the use of metrics to be the foundation of success. Within these organizations, it is well understood that periodic reviews of key business indices or metrics causes employees to focus on those indices that are essential to the business. Obviously, if the employees are not focused, improvement will be more difficult. These metrics include, but are not limited to, cycle time, field performance, and delivery delinquencies. With such data, a standardized system of feedback and control can be implemented and institutionalized. To ensure consistent and uniform use of performance metrics, many companies prepare and distribute a *metrics manual*, similar to the Performance Metrics Manual published by the Government Electronics Group of Motorola Inc., Scottsdale, Arizona.[1] During the implementation of Six Sigma, we often believe that there are no readily available information or numbers to estimate our existing level of process performance and we believe that considerable effort will be required to measure and identify trends, strengths, and areas for improvement. However, companies measure certain activities and record the data that could provide a valuable starting point for a capability evaluation.

There are many performance metrics available to define and communicate the capability of processes. Each has a unique function, and each individual or organization has its favorites. Specific combinations are useful, and many times several metrics are used together, or in a specific order to gain a deeper knowledge of a process. The appropriate metric to use in a particular situation is a function of the type of data involved, i.e., discrete or continuous, and whether the capability being evaluated is short-term or long-term. It is difficult to compare various metrics because they are defined differently. With a mix of various metrics, it is impossible to say which process is better. To overcome this difficulty, one should be aware that the sigma value (z) is the only metric that applies in all cases allowing direct comparison of one process with another, or one product with another, regardless of their unique characteristics or complexities.

1.8   PROCESS

Almost everything done by a company or organization involves a process. A *process* is any activity or group of activities that takes an input, adds value through these activities, and provides an output to an internal or external customer. Companies and organizations, regardless of size, use thousands of processes every day to create their products, deliver their services, or complete transactions.

1.8.1   Process Models

When at least 80% of the product or service value is derived from machinery, we classify it as an industrial process. It does not include shipping, distribution, or billing activities. When 80% or more of a process depends on human activity, we classify it as a commercial process. A commercial process, such as ordering materials, payroll, or processing customer orders, may support industrial processes or stand on its own as a separate and unique business. The profitability of banks, insurance companies, brokerage firms, credit card financial organizations, and similar organizations depend primarily on their business quality. On the other hand, manufacturing companies prosper only when the quality of their industrial and commercial processes meets or exceeds customer expectations.

1.9   DESIGN COMPLEXITY

Design Complexity and Its Importance to the Vision of Six Sigma.   The basic concept of complexity has a long history and still bewilders the human species today. The dictionary defines complexity as that which is made up of "many elaborately interrelated or interconnected parts, so that much study is needed to understand or operate it." Needless to say, it is likely that each of us have felt this way about something at one time or another. From a theoretical perspective, it is doubtful that we (as humans) will ever be able to fully grasp the limit of its meaning, much like the idea of infinity. Fortunately for those of us involved in Six Sigma work, the idea of complexity takes on a slightly different meaning. From this perspective, we define the complexity of a product or service as the simple sum of all its features (i.e., characteristics), from the interactive system level through the discrete node level. From a practical viewpoint we can ask "How complicated is this unit?" By a simple but perhaps time consuming tally, we could answer this question, particularly so in this age of computers. However, we shall make the definition even simpler; complexity is the aggregate quantity of all independent critical-to-quality (CTQ) characteristics that are assignable to a unit of product or service. Inferentially, we can say that a set of critical features is generally proportional to the larger parent set of all possible features. Given this, we do not have to count all things, just those that are "critical to mission success." Thus, the operational notion of unit complexity is given as the summation of CTQ's, where a CTQ can be generally thought of as a unique "circumstance" related to the "unit" that maintains enough value (of a positive or negative nature) that we seek to periodically assess its content.

1.10   NATURE AND PURPOSE OF SIX SIGMA

Six Sigma is a strategic and tactical system for managing the total business enterprise. From this perspective, Six Sigma has the ability to concurrently deliver customer and provider satisfaction, the key ingredients of business success. In short, this view of Six Sigma epitomizes the ideals of business success and optimizes the control function of an enterprise. Remember that in its most elemental form, Six Sigma represents 3.4 defects per million opportunities for defect. This perspective of Six Sigma is related to a single opportunity for defect or a single critical-to-quality (CTQ) characteristic.

1.10.1   Not Just Defect Reduction

Given that Six Sigma is primarily a business initiative that contains business metrics and is not just another quality program, we can now begin to see that it is more closely aligned with risk abatement than with defect reduction. By focusing on the sources of risk commonly associated with the operation of systems and processes, we can reduce the extent of risk that customers are exposed to when they purchase products and/or services. At the same time, the provider of such products and services benefits from the reduction of process and operational risks. This is how we concurrently realize customer and provider satisfaction. In other words, when businesses apply Six Sigma to reduce exposure to risk, customers increase their confidence of achieving entitlement performance in everything they do.

1.11   NEEDS THAT UNDERLIE SIX SIGMA

What are the cultural challenges associated with Six Sigma?   In the book *Organizational Behavior* by Robert Kreitner and Angelo Kinicki, the culture of an organization is defined as the set of shared, taken-for-granted implicit assumptions that determine how a group perceives, thinks about, and reacts to its various environments. On the basis of this definition, an organization's observable artifacts, espoused values, and basic assumptions characterize its culture. In turn, these continually shape the organization's design and reward systems in a dynamic, ongoing manner. In a domino-like way, the latter two factors mold certain group and social processes, such as decision making, patterns of socialization, group dynamics, communication, and leadership. Of course, such processes determine work attitude, job satisfaction, and motivation. Ultimately, an organization's effectiveness (and its ability to innovate) can be derived by careful analysis of the aforementioned factors. In fact, research has shown that

- Organizational culture is strongly correlated with employee attitudes and behavior.
- The extent of congruence between an individual's values and the organization's values is associated with such things as organizational commitment, job satisfaction, and turnover.
- Organizational culture cannot be used to predict a company's financial performance.
- Business mergers frequently fail from incompatible cultures.

Building on this knowledge and research, we can identify several tried-and-proven ways to create or otherwise embed a certain culture into an organization. Kreitner and Kinicki argue that these ways and means include

- Formal statements of organizational philosophy, mission, vision, and values
- Materials used for recruiting, selection, and socialization
- Design of physical space, work environments, and buildings
- Slogans, acronyms, and sayings
- Deliberate role modeling, training programs, teaching and coaching by managers and supervisors
- Explicit rewards, status symbols (e.g., titles), and promotion criteria
- Stories, legends, or myths about key people and events
- Organizational activities, processes, or outcomes that leaders pay attention to, measure, and control
- Leader reactions to critical incidents and organization crises
- Workflow and organizational structure
- Organizational systems and procedures
- Organizational goals
- Recruitment, selection, and development criteria
- Promotion, layoff, and retirement standards

Is Six Sigma a culture? The answer is a resounding: "*YES*" Why? Consider how Six Sigma is a fact-based, metric-driven, problem-solving, project-oriented initiative. Furthermore, Six Sigma ties performance to pay, has a formal vision, constitutes a value, has defined roles, is deployed by way of knowledge transfer (training), and involves coaching by internal leaders. As is widely known, Six Sigma has an established workflow called DMAIC (define–measure–analyze–improve—control). Finally, we recognize that Six Sigma has its own work design and structure. For these reasons, and many others, we conclude that Six Sigma can significantly influence the existing corporate culture; it can even be a culture, in and of itself.

1.11.1 Looking Across the Organization

It is apparent that the term Six Sigma can carry different (but related) meanings across the vertical and horizontal planes of an organization. Let us now explore Six Sigma across these planes. To unify these divergent views of what Six Sigma really is, we must examine its inner core. The fundamental tenets of Six Sigma are rooted in the abatement of risk in everything a business does or delivers. At a basic level, we recognize that Six Sigma embodies the idea that all defects or errors represent risk but not all forms of risk can be characterized in terms of defects.

1.11.2   Processing for Six Sigma

The first vehicle, *processing for Six Sigma* (PFSS), focuses on reducing risks of systems and processes regardless of their basic nature (industrial or commercial) or vertical strata (business, operations, process). More specifically, PFSS is concerned with reducing the extent of risk exposure inherent in the operation of an existing system or process throughout its useful lifespan. Inversely, PFSS is employed to increase confidence that a system or process will achieve operational entitlement in the short term and Six Sigma performance in the long term. PFSS exercises the breakthrough strategy to the extent that each value risk point critical to the operation of a system or process experiences no more than 3.4 risk exposures per million opportunities for such exposure. We must recognize that this goal exists in light of uncertainties that cannot be feasibly managed. This goal is often translated into the language of quality, 3.4 nonconformances per million opportunities for nonconformance. Still, another way of stating this goal would be to say that Six Sigma is equivalent to one defect (error) per 294,118 CTQs.

1.11.3   Designing for Six Sigma

The second vehicle, *designing for Six Sigma* (DFSS), focuses on reducing the various forms of risk attributable to the design of a product, system, or process, regardless of its nature (industrial or commercial). DFSS is concerned with two equally important and often related aims: (1) it is concerned with reducing the extent of risk exposure inherent in the functional performance and physical attributes of a design (customer satisfaction issues), and (2) it is concerned with reducing the risks associated with the business and operational viability of a design (provider satisfaction issues). Inversely, DFSS can increase our confidence that a product, system or process design will perform to its entitlement level in the presence of uncertainties that cannot be feasibly managed. The goal of DFSS is simple: to exercise the Breakthrough Strategy such that each critical value risk point associated with the functional properties and processability of a design are limited to no more than 3.4 risk exposures per million opportunities for such risk.

1.11.4   Managing for Six Sigma

The third vehicle, *managing for Six Sigma* (MFSS), is the underlying foundation for delivering the breakthrough Six Sigma initiative. It is concerned with the creation, installation, initialization and utilization of the deployment plans, reporting systems, and implementation processes that support PFSS and DFSS. The ultimate goal of MFSS is to attain best in class business performance by improving the operational capability of an organization at an annualized rate of approximately 78% (Fig. 1.1). We can also view MFSS as the unifying component of Six Sigma that encompasses the aims of PFSS and DFSS for the purpose of synergistically maximizing the confidence in realizing value entitlement for both customer and provider in every aspect of the business relationship. Embodied in the new definition of quality, MFSS is the primary agent to effectively and efficiently plan, organize, initialize, execute, and sustain breakthrough improvement for a corporation, business unit, or facility.

1.11.5   Risk Orientation

Six Sigma is about the abatement of risk in all its forms. Although Six Sigma has been driven by this idea from its inception, many quality professionals have inadvertently constrained its full power and potential through the lack of knowledge about its developmental evolution. From this perspective, risk orientation represents a higher level of Six Sigma thinking and practice that is more closely aligned with the language of business (risk, opportunity, time, cost), than with the language of quality (defects, errors). As the number of quality professionals with this expanded orientation reaches critical mass, the conventional interpretation will give way to new ideas. New ideas will stimulate innovation. And innovation will lead to business success. The conventional approach used to encourage executives to think in terms of defects and to speak the language of quality is over; it is time for the world of quality to think in terms of risk and speak the language of business. As this happens, business leaders will view quality professionals in a more opportunistic manner.

1.12   WHY FOCUSING ON THE CUSTOMER IS ESSENTIAL TO SIX SIGMA

A focus on quality is essential to evaluate how well the provider's capability (and capacity) matches customer requirements. Measurement of the scope and depth of this match is critical to producibility analysis. As a result, this writer often refers to producibility analysis as confidence analysis, encouraging a less restrictive definition of the term. Regardless of terminology, each of these shares the same goal and provides the same ways and means to achieve that goal. The reader is strongly encouraged to extend this discussion beyond the matching of product requirements to manufacturing capability, although this is a major factor.

It is equally important, if not more so, to apply the scientific methods of Six Sigma to sales, accounting, legal, engineering, production control, purchasing, and other areas where the idea of confidence engineering applies. Each of these areas implements designs, directly or indirectly, deliberately or unintentionally by careful thought or by historical precedent. While each must consider customer needs/requirements and judiciously match capabilities to these requirements if some form of satisfaction is to result. They must design and produce documents, proposals, or processes. In short, the same ideas, methods, tools and practices that apply to the industrial base of the economy also apply to the commercial base (Fig. 1.2). This presents an overly simplistic view from the provider's perspective. Since customers and providers are both concerned with the minimization of transactional cost and time, there will be competing goals for quality and volume. In management terminology, this commonality becomes visible as quality and volume are translated into process capability and capacity. However, the idea of variation sometimes escapes executives who are often consumed with averages, medians, percents, and baselines. They thus limit their statistical insights to processes and products and fail to recognize their ever-widening range of applications.

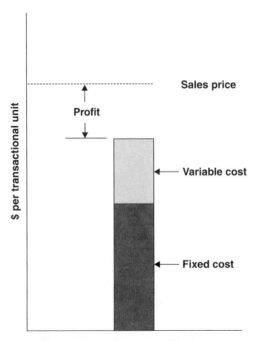

Figure 1.2   Provider's view of business.

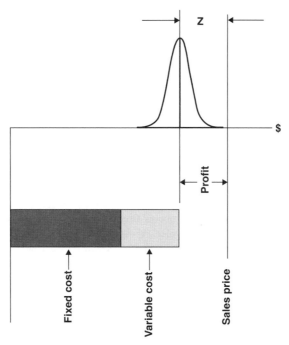

Figure 1.3   The Six Sigma view of business.

For example, if we consider the idea of variable cost as presented in Fig. 1.3, it occurs to this researcher that the term variable cost means just that, its variable. However, the bar chart limit of variable cost is displayed as a point estimate. Is this terminal point of the variable cost represent a central value, or does it represent the $+3\sigma$ limit of its corresponding distribution? Let us assume that it is the average. We also may assume that the underlying distribution of this variable is relatively independent and normal. We recognize its potential covariance with fixed cost but will ignore such second-order details at this time. From our discussion concerning the distribution of variable cost, it should be apparent that we could compute the probability of exceeding the sales price. In other words, we are free to estimate sales and discover the likelihood that the total cost of any given transactional unit will exceed the sales price per unit. Given this, we would then know the confidence of profit by virtue of $1 - R$, where R is the risk. In fact, we could compute this confidence for any given level of profit. Figure 1.3 provides a graphical representation to better facilitate the understanding of this concept. Also, we could approach the problem backward from a statistical perspective. This can be readily accomplished by first establishing a desired sales price and confidence level for the variable called *profit*. Next, we would easily back compute the specific variable cost goals for $\bar{f}$ and $\bar{v}$. Although this is just a simple example, it certainly pragmatically illustrates how we could better study and improve business quality.

However, in many cases the organization's existing accounting system is not properly configured to smoothly facilitate the application of such Six Sigma methods. Often it is easier to just keep doing it the way we always have. Interestingly, it is usually organizations of this type that also expound on their desire to become world-class. Perhaps they simply do not recognize the paradoxical nature of their thinking. The practitioner can now understand why 4σ companies tend to remain 4σ. If a company is to grasp the "golden Six Sigma ring," they must understand that Six Sigma products and services cannot be realized with 4σ business systems supported by 4σ thinking. To do Six Sigma, businesses must think Six Sigma. In this sense, the goal of Six Sigma is just a very small piece of the puzzle. It is important to emphasize that the real leverage of Six Sigma is in the *thinking* piece. Only when we think differently will we *do* differently, and only when businesses think and do things differently only then they will we be able to alter their historical business continuum.

From this discussion, it is easy to synthesize the key points and realize what customers and providers exchange is the value through and the need–do interaction. Simply stated, the idea of business becomes fully realized only when the need–do interactions are fully optimized. Only when such optimization is realized, can we say that the business relationship was abundantly successful. To realize this entitlement level of business, the practitioner must think about the quality of the business processes. Of course, this leads to the idea of breakthrough improvement in process capability and capacity, not just in physical processes, but in mental processes as well; one drives the other. Given this understanding, the practitioner may characterize the idea of business by the relation

$$C_N \xleftrightarrow{\text{exchange}} P_D$$

where the symbolic notation represents the iterative and often interactive operation of those processes necessary to the realization and exchange of value, C_N denotes the many needs of a customer, and P_D is the myriad capabilities and inherent capacity offered by the provider. This bidirectional view of business necessarily broadens the scope of quality from a singular state (customer satisfaction) to a two-dimensional state (customer and provider satisfaction) that matches the very purpose of the business relationship, the optimal exchange of value between the provider and the consumer. Through a more traditional lens, the historical idea of quality was focused solely on the customer. This definition of quality demands that both the customer and provider be recognized as contributory factors/agents within a larger state of quality.

1.13 SUCCESS FACTORS

In the beginning, Six Sigma was a tough sell. One must realize that in 1987 Motorola promised to make a $1800\times$ improvement in 5 years. At that time, an army of corporations that had already experienced any benefits of Six Sigma did not exist. In the 1980s organizations had only theories, beliefs, and faith. Of course, there were many times that organizations felt like throwing in the towel and calling it quits, but somehow these organizations persevered through the natural criticisms, relentless naysayers, and many other obstacles. Perhaps their biggest discovery was that organizations don't have to achieve Six Sigma to beat the competition hands-down. Many times, a half-sigma gain (or less) provided the market momentum necessary to capture more business. In addition, it has been discovered that there is no "point of diminishing return" where quality improvement is concerned (as the world previously believed). In addition, businesses have learned that a progressively higher level of quality demands more and more innovation. It forced them to become more innovative in all of their functions. In a nutshell, these organizations ultimately discovered that Six Sigma brought out the best in all of them. Businesses must always remember that the senior executives within a corporation are the ones who render judgment about the success or failure of an initiative. Simply stated, senior executives are in general concerned with one matter, to demonstrate results that are overtly visible and economically aligned with larger aims, not hidden somewhere in the bowels of the organization under the guise of cost avoidance or future realization. This sheds light on why the quality function in most corporations has remained largely disconnected from the sanctum of managerial (financial) control. While many quality professionals have been overly focused on incremental, perennial, long-term improvement, the senior executives of a corporation still remain focused on the *here* and *now*. Most of the world's greatest business leaders desire changes in orders of magnitude not in small fractions or small percentages. Yesterday's world is simply not the world we live in today. Perhaps Francis Gouillart and James Kelley said it best in their 1995 McGraw-Hill book entitled *Transforming the Organization*. It was here that they wrote: "We once assumed that corporate evolution consists of long periods of stasis, punctuated by periodic adaptations, but the pace of change is too fast for that now. In today's market, companies need to adapt every day." Tom Peters called it "perpetual revolution." Joseph Schumpeter called it "creative destruction." Peter Drucker called it "destabilizing the organization." Alvin Toffler called it the "flex firm." They were all referring to an organizational mentality that refuses to tolerate business as usual. For years, these visionaries have foreshadowed that leading corporations will reinvent themselves again and again. In the words of Tom Peters: "Improvement will not do… only revolution will." This means, in the words of Peter Drucker, systematically abandoning the established, the customary, the familiar, and the comfortable. The practitioner must recognize that without revolutionary thinking, corporations simply cease to grow, cease to prosper, and cease to survive, as this is the law of Darwinian economics. The only way they can survive is through a high-quality, short-cycled process that can produce cash. In other words, corporations rapidly evolve as a result of judicious planning, careful design and coordinated execution in the interest of quantum, fast-track improvement. At the same time, such a short-cycled orientation exists in symbiotic fashion with a viable long-term vision. This holds true, and must be brought into the forefront of consciousness at all levels, across all boundaries and within each and every compartment of a corporation. To unlock the momentum required to drive quantum improvement, an organization must approach Six Sigma in a holistic manner. This means that the initial focus of a Six Sigma curriculum in the short term should be directed toward achieving dramatic and visible financial results. It is when the cash register starts to ring, so to speak, that management sits up and notices and it is only then that curriculum development specialists can start to mold and evolve the content of Six Sigma training in the direction of meeting longer-term needs.

1.14   SOFTWARE APPLICATIONS

1.14.1   Explore Excel

Microsoft Excel®, a Microsoft Office™ product, will provide an elementary view of metrics for Six Sigma by utilizing the graphs and chart features. Charts are visually appealing and make it easy for users to see comparisons, patterns, and trends in data. For instance, rather than having to analyze several columns of worksheet numbers, you can see at a glance whether sales are falling or rising over quarterly periods, or how the actual sales compare to the projected sales.

Excel does have limitations. For example, Excel can only manage 65,534 rows of data. Thus, Minitab is the proper tool for addressing problems requiring large amounts of data.

Visit www.microsoft.com/excel for additional information and customer stories.

1.14.2   Explore MINITAB

MINITAB™ Statistical Software is the ideal package for Six Sigma and other quality improvement projects. From statistical process control (SPC) to design of experiments (DOE), it offers the methods required to implement every phase of your quality project, along with features like Stat Guide and Report Pad that help you understand and communicate your results. No package is more accurate, reliable, or easier to use. In addition to statistical power Minitab offers many exciting features such as

- A powerful graphics engine that delivers engaging results that offer tremendous insight into your data
- An effortless method to create, edit, and update graphs
- The ability to customize your menus and toolbars so you can conveniently access the methods you use most

Visit www.minitab.com for additional information and customer stories.

1.14.3   Explore JMP

JMP®, desktop statistical discovery software from SAS, uses a structured, problem-centered approach for exploring and analyzing data on Windows, Macintosh, and Linux. The intelligent interface guides users to the right analyses. JMP automatically displays graphs with statistics enabling users to visualize and uncover data patterns. JMP has comprehensive statistics software that combines interactive data mining, design of experiments, and statistical quality control in a single package. In addition to its statistical capabilities, JMP also contains

- Interactive graphs and data tables, allowing the user to identify data points of interest by selecting specific data table rows or by clicking on regions of interest in graphs.
- The ability to save output in a variety of useful forms, such as a Microsoft Word document, HTML, and a number of graphics files, including JPG.
- The ability to customize output preferences for repeated use.
- A scripting language suitable for the creation of specialized programs and GUIs.

Visit www.jmp.com for additional information and customer stories.

GLOSSARY

applying Six Sigma   Profiling the vital improvement tools that the Six Sigma practitioners must be able to utilize when executing their applications projects and then grasp how such mind tools can be used to solve virtually any problem.

business, operations, process, individual (BOPI)   The business "goal" for the four basic tiers of an enterprise.

characteristic   Definable or measurable feature of a process, product, or variable.

control specifications   Specifications for the product being manufactured or service delivered.

cost of poor quality (COPQ)   COPQ consists of those costs that are realized as a result of producing defective material.

C_p Process capability: a simple and straightforward indicator of process capability.

C_{pk} Process capability index: an adjustment of C_p for the effect of noncentered distribution.

critical to quality (CTQ)   CTQs are the key measurable characteristics of a product or process whose performance standards or specification limits must be met in order to satisfy the customer. They align improvement or design efforts with customer requirements.

data   Factual information used as a basis for reasoning, discussion, or calculation. Often refers to quantitative information.

deploying Six Sigma   Studying the essential guidelines for scaling and creating a global Six Sigma deployment plan that will originate a critical mass of focused management activity and then learning how to put momentum behind that mass at the local level of an enterprise.

defect   Output that doesn't meet specification(s).

defective   Unit that contains at least one defect.

defective parts per million (Dppm) or parts per million (ppm)   Metric that estimates the number of "escapes" reaching the customer.

defects per million opportunities (DPMO)   Compares dissimilar processes and products. This metric is essentially DPU normalized by process complexity.

design for Six Sigma (DFSS)   A data-driven quality strategy for designing products and processes, and an integral part of a Six Sigma quality initiative. DFSS consists of five interconnected phases: define, measure, analyze, design, and verify.

DMAIC   Acronym for define–measure–analyze–improve—control; refers to a data-driven quality strategy for improving processes.

DPU   Average number of defects per unit.

enabling Six Sigma Understand the information and reporting needs that underlie the global deployment of Six Sigma and then learning how to effectively integrate and fulfill those needs at the local level of an enterprise.

final test yield (FTY) Counts defectives at final test.

initializing Six Sigma Interrogation the of top programs that support the rapid initialization, deployment, and implementation of Six Sigma and then learning how to fully leverage such programs at all levels of an enterprise.

leading for Six Sigma (LFSS) Understanding how to identify, select, and train Six Sigma leaders at all levels of an organization without backfilling the vacated positions and then learning how to effectively motivate and retain those leaders so as to continually enrich forward momentum.

managing for Six Sigma Methology for delivering outstanding business results.

P_p Process performance: a simple and straightforward indicator of process performance; the index used to show long-term potential.

marketing for Six Sigma Customer-facing business units applying the concepts of Six Sigma to quantify and boost results.

P_{pk} Process performance index; adjustment of P_p for the effect of noncentered distribution.

processing for Six Sigma (PFSS) A strategy concerned with reducing the extent of risk exposure inherent to the operation of an existing system or process throughout its useful lifespan.

RTY (rolled throughput yield) Counts defects throughout the process.

Six Sigma A fact-based, goal-driven, results-oriented, strategic, and tactical system of management for the total business enterprise.

statistical process control (SPC) Statistical process control is the application of statistical methods to identify and control the special cause of variation in a process.

statistical process monitoring Every new point is statistically compared with previous points as well as with the distribution as a whole in order to assess likely considerations of process control (i.e., control, shifts, and trends). Forms with zones and rules are created and used to simplify plotting, monitoring, and decisionmaking at the operator level.

targeting Six Sigma Discovering how to best identify, scale, define, empower, and execute Six Sigma projects that are progressively connected to higher organizational goals and then learning how such projects can be effectively tracked, validated, and closed.

total quality management (TQM) A conceptual and philosophical context that requires management and human resources commitment to adopt a perpetual improvement philosophy, through succinct management of all processes, practices, and systems throughout the organization to achieve effectiveness in the organizational performance and fulfilling or exceeding the community expectations.

total defects per unit (TDPU) The total number of defects observed when sampling a population.

REFERENCES

1. Government Electronics Group, *Performance Metrics Manual.*
2. R. Kreitner and A. Kinick, *Organizational Behavior*, 6th ed., 2006.
3. S. M. Patton, "Is TQM dead," *Quality Digest Magazine*, (April 1994).
4. F. Gouillart and J. Kelley, *Transforming the Organization*, McGraw-Hill, 1995.
5. F. Gouillart and J. Kelley, "Measuring performance after meeting award criteria," *Quality Progress Magazine* (1996).

2 Six Sigma Installation

2.1 OVERVIEW

Love of the Journey. A wagon train needed a leader—someone who was a master horseman, trail tracker, and navigator. He had to be logistically minded, be tactically astute, possess extreme emotional fortitude, and have a superior ability to read, motivate, and deal with people. The trail boss was highly adaptive and capable of improvising plans on the fly. He had to be such a person, because the lives of his companions were dependent on his leadership. When Indians threatened a wagon train, he stopped and circled the wagons into a defensive position. He had already insisted that everyone carry a weapon. During an attack he encouraged them to use these guns; then he instilled a sense of hope and survival by implementing battle tactics. When the battle was over, he tended to the wounded, loaded them into wagons, and moved on. In short, he applied his many talents to ensure their survival as they continued their trek. Despite the loss of people, livestock, and equipment, the trail boss returned east to form and lead another caravan. While the members of the wagon train made the trip once, the trail boss made the trip over and over again. Perhaps he was driven by the thrill of victory (and the rewards) or wished to demonstrate his lack of fear. I believe this was how Don Linsenmann[1] felt as he led one DuPont business group through the corporate wilderness, only to return time and time again to lead others. There were 18 in all, and each journey held its own story of promise, peril, agony, and victory.

I believe that is how Dr. Harry and his Six Sigma Academy partner, Mr. Richard Schroeder, view their careers—guiding corporation after corporation, giving up weekends, vacations, and time with their children. Why did they do it? After 18 years as a trail boss, Dr. Harry said he did it because it was the right thing to do, the best thing to do, and the only thing to do.

Today the Six Sigma trail is well marked. It's not as primitive and treacherous as it was in the past. First, there was uncharted territory, then a trail, then a railroad, and now a superhighway. So what excuse do corporations have today for not making the Six Sigma journey and for not achieving dramatically improved performance? Extraordinary senior Champions like Don Linsenmann[1] of DuPont and Gary Reiner[2] of GE have paved the way. A corporation deploying Six Sigma today assumes much less risk than in the past. When Dr. Harry thinks about whether he would retrace his career if he had it to do over, he doesn't immediately come forth with a resounding yea or nay. He paid a heavy personal toll by not participating in many of the life experiences a credit card commercial would call "priceless." Still, if he could turn the clock back, he would do it all over again because "that's just me." As with all passionate leaders, Dr. Harry is simply wired to be ambitious and feels dead if he isn't changing something that most people think can't be changed. Today, there is no lack of Six Sigma knowledge and know-how. A lack of resources is not an excuse because a good leader can always find the necessary resources. It's all about adaptation and improvisation, two skills required for survival. A Six Sigma champion successfully leads one business unit across the deserts, plains, and mountains who will surely return to lead another and another. But if this initial venture fails, this "champion" will become known as the person who caused the initiative to falter.

Practitioner's Guide for Statistics and Lean Six Sigma for Process Improvements. By Mikel J. Harry, Prem S. Mann, Ofelia C. de Hodgins, Christopher J. Lacke, and Richard Hulbert
Copyright © 2010 John Wiley & Sons, Inc.

An individual is not destined at birth to be a leader. It is a skill learned over time and motivated by unrest. Can all people lead? No. You can teach certain elements of leadership, but you cannot teach how to synergistically bring these elements together. Successful Six Sigma practice requires intensive training, technical knowledge, personal skills, and application leadership. When an organization embarks on the Six Sigma journey, it still retains the same old beliefs and practices that have made it what it's always been.

2.2 SIX SIGMA LEADERSHIP—THE FUEL OF SIX SIGMA

We must remember that Six Sigma is the embodiment of hope, and that hope fuels the collective will of the adapting organization. It is hope that causes individuals to align their personal values, aims, and goals with those of the organization. This is what leaders do; they create hope where there is little or none and then forge it into a collective will capable of creating major change.

The collective "shock and awe" of Six Sigma projects is one way to ignite hope. It is the sudden, collective, decisive, and repeatable successes of Six Sigma that cause employees to believe their company is the best. When this attitude pervades an organization, it becomes boundaryless and innovation takes hold. Essentially, a Six Sigma initiative is designed to raise the bar so high that employees will be forced to individually and collectively reexamine how work is done. Given this inaugural aim, it should be apparent that Six Sigma is about innovating, is about new ways of doing things, not just fixing them for incremental gain.

The Way of the Magician. Many claim that Six Sigma is just smoke and mirrors, a sales job for senior management and a snow job for the rest. Well they are right. How else can you get an initiative moving when, at first, there is no initial momentum or substance behind it? As a Six Sigma practitioner, you better have some smoke and mirrors. You must know how to manipulate or clear away the smoke of ambiguity that has obscured the true performance of the company. This smoke also hides the details in which improvements lay hidden. Smoke can also be used to intentionally hide the old ways of conducting business. The senior practitioner is both a strategic and tactical smoke manager. That means that he/she must employ smoke and mirrors where necessary to support the vision needed for breakthrough change. You also must use mirrors to shed light on old problems that have gone unnoticed or have been ignored. Mirrors show us how to improve our business and measure one's own performance. Only with mirrors can you reveal the values and behaviors that can revitalize an organization. Management, operational and process habits are changed by these early Six Sigma successes. By hiding counterproductive values and exposing uncomfortable facts, a leader can reconfigure the operating environment of a business.

Leadership and Six Sigma. Six Sigma is more about leadership than about knowledge, structures, systems, methods, and tools. While these are critical aspects of Six Sigma, leadership provides the energy and impetus required to set a corporation on a new performance trajectory. A company with strong leadership and little knowledge of Six Sigma is better off than a company with enormous Six Sigma knowledge and weak leaders. Yes, it is better to have a critical mass of strong leaders than to have Six Sigma; but a corporation is better off when it possesses both.

Bob Galvin[3] provides us with additional insights into the value of leadership. Here's what he/she had to say, because a leader is human and fallible, his and her leadership is in one sense finite—constrained by mortality and human imperfections. In another sense the leader's influence is almost limitless. He/she can spread hope, lend courage, kindle confidence, impart knowledge, give heart, instill spirit, elevate standards, display vision, set direction, and call for action today, each day, and each tomorrow. The frequency with which one can perform these leadership functions seems without measure. A practitioner's effectiveness and personal resources, rather than attenuating with use, amplify and then reuse and extend the skills. Like a tree whose shadow falls where the tree is not, the consequence of the leader's act radiates beyond the fondest perception. Again we see the paradox of the leader—a finite person with an apparent infinite influence. Galvin describes the talents required to lead the Six Sigma effort in a large corporation. He initially defines Six Sigma deployment CTQ (critical to quality characteristic). We call this leadership, and he articulates the essence and idea of leadership in a way that can greatly benefit all who would lead. Under this leadership style, the leader and those being led both learn how to become better leaders. Such leadership generates an ever-improving institution able to face an ever-increasing number of critical choices. The leader then willingly shares these challenges across the organization. Leadership, like quality, has a binary outcome. It is either successful or not. Yet leadership is more of an art while quality is more of a science. Leaders simply deliver expected results on time.

Leaders versus Managers

What is the difference between a leader and a manager? We can compare the traits for those we would call managers to those of the trail-boss described earlier. These are people who are goal-oriented, politically astute, budget-minded, empathetic, and always diplomatic. They hold down the fort once it has been conquered and maintain the trail after it has been forged. They create policies and procedures, and ensure compliance to them. They don't carry informal forms of power too well, and they value the peace that comes with predictability. They aren't risk-takers, and they shy away from those who might actually rock the boat in favor of those who just talk about it. Most of them are people dedicated only to preserving the organization, rather than moving it ahead. On the other hand, a leader acts with speed and nerve. When leaders launch a process, they move fast. Managers, especially the bureaucratic ones, while frequently surrounded by quivering lieutenants, struggle against the perceived threat. In contrast leaders face the world, kindle confidence and handle each crisis anchored in reality. Dr. Owen Hill, former IBM executive, always uses the county jail story (Christopher Kiernan story) as an example of management at its worst! The manager, Owen says, paints and repairs the jail, supervises his staff, feeds the prisoners, provides heat during cold days, and allows no one to escape; yet little is done to make the prisoner's stay beneficial to him or to society. That requires a leader. Now the question is this: who would you want to shepherd your business through uncharted territory, a leader or a manager?

For many corporations, the implementation of Six Sigma forces the organization into uncharted territory because it requires quantum change for which there is no marked trail or map. The organization (a series of wagon trains) and its constituency are simply seeking a better life for themselves, their customers, and their shareholders. If you are a CEO, would you choose your best manager to lead the effort, or would you choose your best leader? If you want to succeed, you will choose your best leader.

We should mention that while rank is important in formal organizational structures, it is unimportant in the firestorm of rapid change brought about by Six Sigma. Rank provides a sense of order and identity during stable periods, but it does not guide people through intense trials and tribulations. In times of instability and grave uncertainty, a corporation needs someone who can engender faith, trust, and belief required to overcome the challenge at hand. Rank is just a placeholder; it has no innate or assumed ability to lead in times of crisis, or even in times of moderate instability.

The Antileaders. We must remember that the bell curve applies to the strength of leadership across on organization's management. Only a small fraction of people out on the right edge of the curve will be outstanding leaders. Inversely, another small group on the far left may oppose change as vehemently as leaders promote positive change. They are the anti-leaders who have the strength of a leader but who expend that energy in a negative direction. They are counter-change, counter-aim, counter-leaders - and they are the ones who must be defeated. Their base of power must be drained and their aims must be blurred. You cannot move an entire corporation by concentrating on the bulk of employees in the middle of the bell curve. Well, you can, but it requires enormous effort. Instead, the best strategy is to identify the antileaders and convert them through superior logic, tactics, and demonstration projects. If you accomplish this, then those in the middle will naturally establish a new center of performance. You decisively move the entire organization and generate breakthrough by reducing the number and effectiveness of the naysayers while emphasizing actions of those who cause positive change.

Yet the leader does have a responsibility to keep the troops fed, focused, and motivated. Sometimes, the people in a wagon train would become demoralized by severe conditions and events. If the source of demoralization was from lack of food, then the trail boss had to pull a rabbit (or perhaps a buffalo) out of his hat—literally. Wherever the crisis falls on Maslow's hierarchy,[4] the leader is responsible for restoring faith and hope among the wagon train members. Leaders must bridge the past to the future. What and where are the past behaviors that embody the past restrain a corporation from seizing its future? What are these impediments and where are the people who can embody change and lead a corporation toward a better future?

Leaders. In the final analysis, a Six Sigma practitioner is a leader who forces others to confront their faulty beliefs and values. The effective leader then provides the vision, direction, inspiration and support to encourage the adoption of new beliefs and values. A leader must perform the usual planning, budgeting, organizing, staffing, directing, coaching, controlling, and reporting to reach a new value system. Leaders create magic; *magic does not create leaders*. Those who employ leaders should use a common selection criterion—can the individual do magic? Can the candidate manufacture smoke and wield mirrors until the deployment-and-implementation machine is designed, produced, and enabled? Then, once enabled, can he/she decisively penetrate the new frontier and bring home the gold, time and time again? First impressions are everything. That's when the practitioner has to strike with his/her magic act. Knowing that it takes time to generate results, a practitioner must position Six Sigma in a positive light from the beginning. You are asking people to jump on board with a leap of faith. Initially, you are selling an idea that has no form or substance. All you have is a little bit of magic. You're out in the wilderness without a map, compass, or supplies. In effect, you have become part of the wilderness food chain.

That's pretty much the case for the one who, employing Six Sigma, leads a company to greatness. You must rely on successful Six Sigma case studies and testimonials from others who have "been there and done that." Your job is to explain Six Sigma and keep it alive until tangible results are achieved, and it becomes largely self-sustaining. Within a few months, top management will see real returns. People can't get enough of it; they want more and the impact from Six Sigma grows rapidly. There are three main pieces of magic in the act we call Six Sigma. The first creates the need for Six Sigma and redirects a business toward the path to breakthrough. Only senior leadership can do this. In the second, senior management must select a powerful and credible leader to become a senior Champion. He/she must possess integrity, audacity, hardness, and intelligence because the champion will personally own the idea of business quality. The final piece of magic is to put decisionmaking power and control of resources into the hands of that leader.

The easiest of these is to recognize the need for Six Sigma. The hardest is selecting and empowering the right people to lead Six Sigma implementation. That requires both a commitment and an investment. In this sense, those who enlist leaders are by definition equal opportunity employers. There is no discrimination based on education, experience, or physical stature. Leaders come in black, white, tall, short, thin, overweight, male, female, bald, hairy, funny, boring, moody, scrappy, polished, single, married, divorced, gay, introverted or extroverted. But no matter who they are, leaders have to be magicians and masters of smoke and mirrors. This is the only dimension on which a corporation must discriminate when installing its Six Sigma leaders.

There was a sign on the desk of a Marine colonel that said, "I am they. The buck stops right here." The leader must assume this aura of responsibility during trying times and give credit to others when things go well.

Lead by Example. David Leighton[5] former operations officer of HSBC Mexico led in this style. He created synergy between company objectives and the management system called Six Sigma. Although thousands have played their roles very well, it was he who generated the magnetism and force required to move this large Latin American organization toward a new future. Leighton was the organizational magician who pulled the proverbial rabbit out of the corporate hat.

2.3 DEPLOYMENT PLANNING

Five major tasks must be accomplished to successfully deploy Six Sigma:

1. Encourage executive management to own Six Sigma and link it with their business strategy.
2. Define stretch goals and quality metrics.
3. Establish common approaches for issues such as training and selection of people.
4. Capture and share knowledge.
5. Deploy methods to build and maintain momentum.

While there is no one method for deploying Six Sigma, some good practices have evolved over time that we should understand and use wherever it make sense. Each deployment may differ from standard practice as we align deployment plan with company style, organization, industry, or capability. The material outlined in this section is meant to spark thinking as each group produces a deployment manual adapted to their unique situation. To enable the sharing of knowledge, each organization must ensure that deployment approaches, roles, standards, management processes, and metrics are relatively consistent across the organization. This clear definition of roles and responsibilities is vital to any deployment of Six Sigma. Figure 2.1 defines deployment roles for seven employee categories.

2.3.1 Executive Management

Executive management owns and drives Six Sigma. A senior management leader or leaders of Six Sigma is chosen from this group to provide day-to-day top-level management of Six Sigma deployment. Executive Management establishes the corporate-level-goals and sets time and performance expectations. They establish the initial focus of the Six Sigma initiative. The initial focus of Six

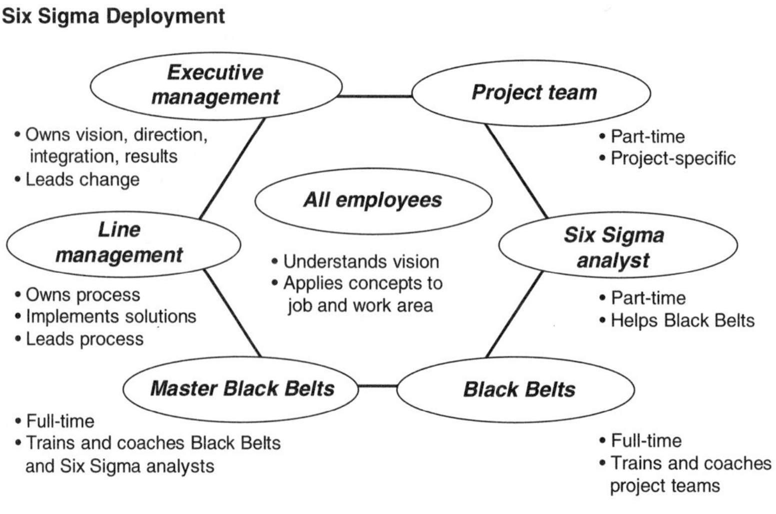

Six Sigma Deployment

Figure 2.1 Illustrative example of Six Sigma roles for seven employee categories.

Sigma can significantly improve the likelihood of success. Remember that nothing is more disappointing or frustrating than a hard restart.

2.3.2 Six Sigma Champion

What is a Six Sigma champion, and how is this role defined? Six Sigma is a fact-driven system of business management that is organized and led by a very special group of leaders known as "Six Sigma champions." They have been formally empowered by their respective senior management to deploy and implement Six Sigma. They must know how to successfully initiate, plan, lead, and sustain the global Six Sigma initiative. They ensure the successful implementation of Six Sigma by identifying and orchestrating highly focused application projects. Each project must be carefully designed to generate verifiable and lasting benefits that aggregate to increased business value. When properly supported, applied and sustained, it has repeatedly been demonstrated that Six Sigma champions can deliver exceptionally high returns that directly impact the bottom line. These benefits are most often realized on a project-by-project basis through the efforts of Six Sigma Black Belts, Green Belts, and White Belts. See Sections 2.3.4.–2.3.7 for descriptions of responsibilities and skills of these highly trained individuals who are collectively referred to as "X-Belts." Champions are the leaders who plan, orchestrate, and direct the X-Belts. They make significant contributions to realizing operational objectives and ensure that X-Belt projects are consistent with the strategic goals of an enterprise. So when anyone says Six Sigma is about smoke and mirrors, Dr. Harry agrees and adds that Jack Germaine[6] at Motorola, Gary Reiner[7] at GE, and Rich Schroeder[8] at AlliedSignal were all magicians.

The Nature of a Champion. Don Linsenmann of DuPont is a magician, too, albeit one of a different kind. He has a strong technical background, and so he is comfortable with those aspects of Six Sigma. Linsenmann also has a very strong business leadership base as a senior executive, and he understands people and how they tick. He also understands the inner workings of DuPont and has an uncanny ability to connect the equations of Six Sigma with real products made by real people in real places—all the way down to the customer's smile. If he were a tailor, he could see the threads that make up the pattern, know how the shirt was made, how the sleeve was attached, how the pockets were formed and how the buttons were sewn on the shirt. When it comes to DuPont, Linsenmann sees the relationships, and how the whole hangs together. This is what gives him his power as a leader. Linsenmann accepts Six Sigma as a deeply held belief, not just a professional obligation. You should have seen him when the first results from Black Belts started rolling in at DuPont. He was no longer Don the businessman or Don the Six Sigma man; he was Don the leader. All of the great Six Sigma champions have shared this experience. That's what occurs to those who master the magic, move the mirrors, distribute the smoke, and guide the path of a corporation. They are communicators. They make the unbelievable become believable and rational. They are con men, master mechanics, part psychologists, and big-time politicians. They know when to throw a spear and when to hand out an olive branch. They know which battles to pick first, and which ones to avoid altogether.

The best leaders understand the connection between strategies, tactics, and tools, and they appreciate that they cannot perform all of the required tasks personally. So they surround themselves with people who are as strong as or stronger than they are. In this manner, a senior champion builds a house, not from deadwood washed up on the corporate beach, but from the strongest trees in the forest. They admire risk-takers and loathe moss-backed generals. Thus, a CEO must begin by selecting the strongest possible executive to fill the role of senior champion.

A senior champion must be a fast tracker—one of the birds that does not fly with the flock – but a bird of a different feather. When finding this person, the CEO should say, "here, you have a shot at the golden corporate ring. If you succeed, you will have it. But if you fail, you will be ostracized and doomed to career stagnation. Come all ye who are called, step forth recognizing that you have all to gain and all to lose. Your number one job priority is to keep the idea of Six Sigma alive." That's one piece of advice Dr. Harry always remembers from Bob Galvin. Keep the idea alive; do not let it die. That's the precept guiding the heart of a Champion as he/she spreads the word about Six Sigma, plants its seeds, waters them, harvests the results and repeats the cycle again and again.

Champions are the fundamental guiding lights. They make Six Sigma credible. They spread, foster, validate, grow, and do what is necessary to install change where change is not welcomed. In other words, a champion works against a stacked deck. That's why total quality management (TQM) fell short of expectations. Either its leaders were not strong enough, or they were not given the resources to create the infrastructure and thrust required for business breakthrough. Its leaders often responded to the demands of those above them, even though they had been charged with the task of changing organizational values. The senior champion designs the course of Six Sigma development, and of the corporation. At the same time, the global leader provides key people with the knowledge they need to leverage those enablers through application projects, each connected to larger business goals and objectives. This is what Six Sigma is about. It's not just about enablers, methods and tools—it is the "what's" and even the "how's" of Six Sigma. At this point in time, most progressive corporations have acquired a good working knowledge of these fundamentals. What most haven't acquired and internalized is the idea of Six Sigma leadership. The difference between a smashing success and a failure is not determined by the choices of statistical tools, software, or methods but it is about the what's, the why's and how's of Six Sigma.

2.3.3   Line Management

Line managers play a pivotal role because they own the processes of the business and, therefore, must ensure that process improvements are realized and maintained. Line managers guide the implementation of Six Sigma and ensure the application of statistics to exploit new opportunities at the process level, they aggregate gains made at the operations level and report performance. They also typically manage Black Belts, and understand the challenges facing them.

2.3.4   Master Black Belts

These are the full-time trainers of your organization, the coaches for Black Belts. Through mentoring, coaching, defining objectives, and leading the way, they develop, define, and carry to completion successful Six Sigma projects. Some Master Black Belt responsibilities are listed below:

- Manage the project selection process
- Ensure that projects begin and remain aligned with organizational CTQs (critical-to-quality requirements).
- Clarify project components if the project team is confused
- Mediate conflicts within project teams as necessary
- Monitor project teams to review measures and data collection plan
- Counsel project teams when there is poor identification of key measures and an inadequate data collection plan
- Facilitate team building, focus on conflict resolution and decisionmaking if project team members dispute data or other process steps
- Conduct project critiques with senior managers, team, and champion after project reviews
- Lead Six Sigma projects that may be more substantial in scope than those lead by Black Belts
- Work in tandem with the Six Sigma champion to
  Create a commitment to reach Six Sigma Project goals
  Tear down roadblocks to the Six Sigma process

Encourage follow-up and monitor project activities

Help transition project ownership to the process owner on completion of the control phase

Recognize progress, promote success, and celebrate accomplishments.

2.3.5 Black Belts

Every completed Black Belt project yields benefits to the corporation from immediate "hard" cost savings validated by accounting. In the early nineties, savings per Black Belt project were about $150,000. In the early 2000s, the average Black Belt project returned more than $250,000 or about $1 million in realized hard benefits per Black Belt per year. Today, the average Black Belt project returns about $350,000 or about $1.4 million per Black Belt per year. For example, if an organization of 100,000 employees selects (2000) future executive leaders and trains them as Black Belts, they can impact the bottom line by about $2.8 billion. This assumes that when key people are installed as Black Belts, their previous positions are not backfilled. Initially, many executives do not believe that their corporation can afford to take approximately 2% of its workforce off current jobs that do not have hard financial targets, reassign them as Six Sigma Black Belts, and yield more than $1.4 million per year per full-time person. A company should select the "right talent" to ensure leadership success. They should select for Black Belt training only those people who are assessed to be good potential leaders. Black Belts should be the best people in a company. Many companies therefore select employees who have been identified as having high potential and sometimes even management trainees as Black Belt candidates. In general, a Black Belt candidate should be

- A self-starter who can perform well with minimum supervision
- Able to work with, lead, train, and mentor team members
- An effective communicator at all levels
- Able to effectively work at multiple levels within the organization
- Computer-literate and competent in mathematics and elementary statistics
- Energetic, enthusiastic, and passionate about excellence.

In addition to the knowledge and skills that Black Belts acquire when working as process improvement experts and change agents, they must also possess competencies in such areas as

- Development and implementation of strategy
- Financial planning and controlling
- International organization development
- Performance management
- Marketing and sales effectiveness

2.3.6 Green Belts

These are part time resources that help Black Belts complete projects and extend the reach of Black Belts. A Green Belt should be a trainee in the Six Sigma methodology and a member of a Six Sigma process improvement team. Some of the broad functions of a Green Belt are

- Introduce the Six Sigma methodology and tools to project team members
- Lead the project team while Six Sigma methodology is applied during Six Sigma projects

Table 2.1 lists some of the basic capabilities and responsibilities that a Six Sigma Green Belt must possess.

TABLE 2.1   Green Belt Roles and Responsibilities

Description	Capabilities	Responsibilities
Part-time project responsibility	Experience in leading a team	Responsible for progress of the project
Apply the Six Sigma methodology to the project	Experience in Six Sigma—at least 2 weeks of training	Responsible for progress of the Six Sigma project
Support the efforts of the business area by spreading the use of the Six Sigma methodology	Experience in Six Sigma—at least 2 weeks of training	Should have a link to the business-leader
Focused on 1–2 Six Sigma projects	Skilled in project management	Leadership in planning team work
Direction and leadership should come from Black Belt	Able to apply basic improvement tools	Should have a link to the business leader
	Good problem-solving skills	
	Good people skills	
	At least 50% resourced	

2.3.7   White Belts

These are the project-specific, part-time resources that provide process and cross-functional knowledge to a team. They are frequently drawn from the area responsible for sustaining the project gains. While all employees need to understand the vision of Six Sigma and eventually be able to apply Six Sigma tools, White Belts may perform distinctive roles during initialization, implementation, and deployment. Six Sigma White Belts are expected to maintain primary and secondary metrics. Project participation of organization's Six Sigma White Belts is in general not expected. White Belts receive 2–4 h of awareness training through classroom or e-learning training. Some organizations skip the Six Sigma Green Belt training and substitute it for Six Sigma Yellow Belt training that is at a lower level of the Green Belt training and therefore takes less time than Green Belt training. Figure 2.2 is a simplified view of the Six Sigma method.

Section Definitions
- *Primary Metric.* Called *process metrics*, this is the metric that practitioners care about and can influence. This metric is almost always a direct output characteristic of a process. It is a measure of a process and not a measure of a high-level business objective. Primary metrics are usually process defects, process cycle time and process consumption[9]

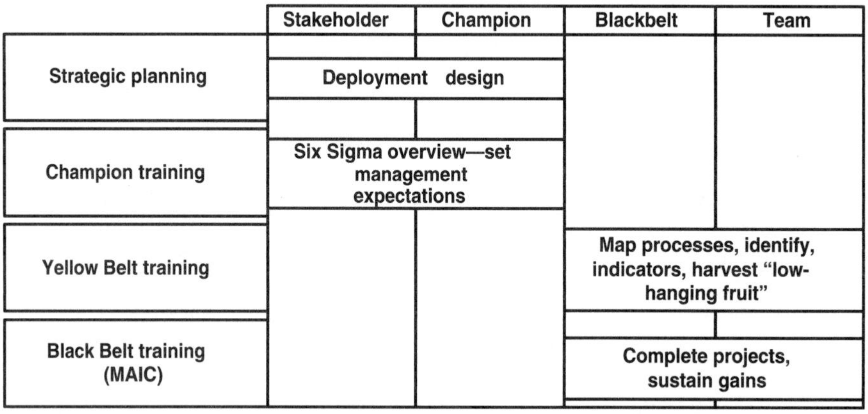

	Stakeholder	Champion	Blackbelt	Team
Strategic planning	Deployment design			
Champion training	Six Sigma overview—set management expectations			
Yellow Belt training			Map processes, identify, indicators, harvest "low-hanging fruit"	
Black Belt training (MAIC)			Complete projects, sustain gains	

Figure 2.2   Simple view of the Six Sigma method.

- *Secondary Metric*. Called *consequential metric*, this could be both business and process metric. Measures anything that goes wrong as a result of improving the primary metric. There can be multiple consequential metrics in a project of improving one process or one primary metric.[6]

2.3.8 Six Sigma Roadmap

The Six Sigma five-phase methodology (DMAIC) can be viewed as a problem-solving approach for product, service, or process improvement. Most of organizations start by using the DMAIC and latter moving to design for Six Sigma (see Chapter 23). For some, this DMAIC methodology may be linear and explicitly defined; however, an iterative approach may be needed, especially for Green and Black Belts new to the tools and techniques of DMAIC. For example, the practitioner may find himself/ herself in a situation in which, on analyzing data that takes place in the *analyze* phase, the gathered data are not enough to find the root cause of the problem. At this point, it is necessary to move back to the *measure* phase. In addition, it is important that the Green Belt (GB) and Black Belt (BB) know enough of the tools and statistical techniques to figure out which ones are useful for each phase, in which ones may be used for more than one phase since the appropriate application of tools becomes critical for effectiveness and correctness. A simple roadmap is provided in Figure 2.3 to guide the practitioner. The practitioner must bear in mind that the table is flexible and is not written in stone.

As many authors have observed, the effectiveness of the Six Sigma DMAIC methodology lies mostly in the structure, rigor, and discipline. In addition, a problem frequently encountered when implementing Six Sigma is that all activities may seem to qualify as Six Sigma projects. Therefore, organizations should develop a process to classify and prioritize projects according to their own organizational strategic goals.

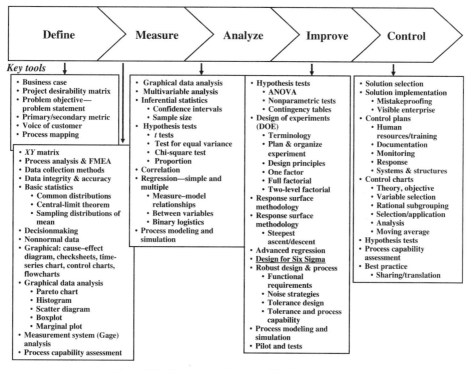

Figure 2.3 Illustration of a simple Six Sigma roadmap.

2.3.9   Characteristics of Effective Metrics

As Lord Kelvin has stated, "Measurements get attention."

One difference between the Eastern and Western cultures is in their treatment of measurements. Westerners have a stronger preoccupation with measurements that may flow from as far back as 1891 when the British physicist Lord Kelvin said, "When you can measure what you are speaking about, and express it in numbers, you know something about it; but when you cannot express it in numbers, your knowledge is of a meager and unsatisfactory kind." Herein is an indication of the role of measurement in the era of total quality management versus the role of measurement in the era of Six Sigma. Quite interestingly, when Dr. Harry was developing Six Sigma and the breakthrough strategy at Motorola, he conducted a comprehensive benchmarking study of existing problem-solving methodologies.[10] He found that none had a separate step or phase devoted solely to measurement. This "mission critical" task was often subordinated to a lower level of the organization or neglected altogether in virtually all of the quality improvement strategies and business-related initiatives of the day. It appeared that many of those in the American quality improvement programs had taken their lead from Japan—quite literally taking measurement for granted rather than treating it as critical. By treating measurement as an "underlying assumption," they significantly diminished its perceived and real importance. Ironically, Black Belts engaged in the rigorous process of measure, analyze, improve, and control (the core steps of the breakthrough strategy) spent most of their time in the measurement phase. Experienced Black Belts were the first to point out that "M" is the most difficult and time-consuming step of the breakthrough strategy. Why? Measurements are critical to achieving quantifiable change and financial benefits. Simply stated, if it can't be measured, it can't be improved. At a global level, the Six Sigma practitioner should recognize that many well-respected organizations consider the use of metrics to be the foundation for achieving success. Within these organizations, it is well understood that periodic reviews of key business indices or metrics cause employees to focus on metrics essential to the business. Metrics include, but are not limited to cycle-time, field performance, and delivery delinquencies. These data make it possible to implement a standardized system of feedback and control. To ensure consistent and uniform use of performance metrics, many companies prepare and distribute a metrics manual.

2.3.10   The Role of Metrics

Only through proper questions can proper actions and results ensue. It is true that the world is changing too quickly to avoid errors and mistakes. There is no grace period for mistakes anymore. One faulty product, one improperly launched service and the competition makes you pay. Today, it is imperative that we ask the right questions at the right time. It is a well-known axiom that you can't improve what you don't measure. Less well understood that business metrics are more than its scale and a standard. They also indicate what companies truly value and thus shape the behavior and culture of the organization.

2.3.11   Six Sigma Performance Metrics

For a metric to be powerful enough to initiate and sustain change, it must possess all these six components. Only then can a company say that the measurement is truly valued. For example, most companies truly value profit. They have a scale of measure, dollars. They have a profit goal. They attach variable pay to achieving the goal. They regularly report progress throughout the organization—and they can pool the goal to the highest level of the enterprise.

Let's take a closer look at business metrics. By doing so, we will see how Six Sigma metrics pull rather than push the business through change and improvement. We will see that, to truly initiate and sustain change, performance metrics at every level of the organization must meet the following six criteria:

1. A metric must have a scale, such as frequency or rate of occurrence, units produced correctly over time, number of defects or dollars. To be effective, the measurement scale must be meaningful, valid, and reliable.

2. The metric must have a standard or goal. If we are measuring injuries, then the goal could be to have no more than 2 missed days of work per year per employee due to injury on the job. Alternatively the goal might be to have an injury rate of no more than 0.05% each month for the entire workforce. In other words, 99.95% of the workforce in any given month has not experienced an injury that caused an absence of more than 1 day from work.

3. Compensation and other forms of recognition must be related to the performance goal for the metric. If, for example we measure safety, then we should also monetarily reward and recognize employees for improving safety. While many companies have a scale of measure and a goal, they do not reward or recognize those who contribute to achieving this goal. Although they may say they value customer satisfaction, they do not compensate employees for improving customer satisfaction.

4. A metric should be reviewed on a regular basis throughout the organization. An organization should distribute performance data to all executives, managers, and employees who can impact the metric.

5. A metric should be capable of being pooled horizontally and vertically within the enterprise. This means that the metric should have meaning and impact across various functions and levels of the organization. High-level metrics can be cascaded down through the business, operations and process levels of the enterprise as a family of supporting measures. In turn, they are monitored and reported up, down and across the organization as needed.

6. A metric must be highly correlated with one or more of the 12 criteria for performance metrics at the business, operations, and/or process level of the organization. See Section 2.3.13 for a list of the 12 criteria.

For a metric to initiate and sustain change, it must meet the above six criteria. Only then can a company say that the measurement is truly valued.

2.3.12   Profit and Measurement

What about ethics? Most companies say they value ethics. But do they have a scale of measure? Do they have a goal? Are employees rewarded for ethical behavior? If the six criteria are not met, a company does not truly value what they purport to value. Since behavior is shaped by values, to influence behavior you must develop strong metrics that meet all six criteria. In this way, you ensure that employees at all levels of the organization are working toward common goals. Imagine that you start with a clean sheet of paper and define metrics and goals for each factorial combination of quality. Then you attach compensation to these metrics and consistently communicate the resulting performance throughout the organization. Such visibility and reward, in turn, would drive the behavior necessary to realize the goals. If people cannot reach the goals using available resources and toolsets, they will naturally reach beyond themselves to acquire new tools, resources and methods. If common sense and classic total quality management (TQM) do not work, then they will reach out and draw on techniques or methods that do work knowing that compensation and recognition are attached to these goals. Six Sigma defines organization beliefs through a robust system of measures. Thus, it pulls quality into the organization rather than the classic TQM approach that tends to push quality into the organization.

2.3.13   Twelve Criteria for Performance Metrics

1. *Aligned.* Performance metrics must always align with corporate strategies and objectives.
2. *Owned.* Performance metrics must be "owned" by those who are accountable for their outcome.
3. *Predictive.* Performance metrics must be a leading indicator of business value.

4. *Actionable.* Performance metrics must reflect timely, actionable data so users can meaningfully and effectively intervene.

5. *Minimal/Few in Number.* Performance metrics must focus users on high-value tasks and not scatter their attention.

6. *Simple/Easy to Understand.* Performance metrics must be straightforward, not based on complex indices.

7. *Correlated/Balanced and Linked.* Performance metrics must be vertically correlated and reinforce each other, not compete and confuse.

8. *Transformative.* Performance metrics must trigger a chain reaction of positive changes in the organization.

9. *Standardized.* Performance metrics must be based on standard definitions, rules, and calculations.

10. *Contextual/Context driven.* Performance metrics must be contextually dependent so as to ensure relevancy.

11. *Reinforced.* Performance metrics must be tied to the reward and recognition system.

12. *Validated/Relevant.* Performance metrics must be periodically reviewed to ensure relevancy and validity.

2.4 APPLICATION PROJECTS

There is no substitute for on-the-job experience. This is also true in Six Sigma. In this context, an application project is the point where the proverbial rubber of training meets the road of reality. Through application projects, Six Sigma X-Belts demonstrate their skills while ferreting out added value for the sponsoring organization. A valid application project must return tangible, measurable, and verifiable benefits for the practitioners' employer or sponsoring organization. By convention, the sponsor assesses the viability of each potential project, and determines whether it will return measurable and verifiable benefits. In this context, cost avoidance activities do not qualify as valid X-Belt projects. Only those activities that can deliver "hard" value to the bottom line should be considered as viable Six Sigma projects. This means that only "accounting quantifiable" problems should be placed in the pool of assignable X-Belt projects. The proposed scope and depth of each potential project must be properly scaled or otherwise "sized" so that application of the full DMAIC process is required while being narrow enough to be completed within 2–3 months. When "jumbo-sized" projects are assigned to a single Black Belt, the outcome is usually failure. Worse yet is the case where large "training" projects are unwittingly assigned to a worker-level team led by a Green Belt trainee.

As expected, lower-ranked X-Belt projects produce smaller returns because they are usually smaller in scope and depth. For example, a Green Belt project should be scaled so that it can be completed in 4–8 weeks. While the returns of a single Green Belt project are normally less than those of a Black Belt, there are far more of them. In a typical large or midsized business, the total project impact of Green Belt projects frequently exceeds the total impact of Black Belt projects.

In general, Six Sigma application projects should focus on persistent or chronic problems that have previously defied solution. Six Sigma projects should avoid problems that can be solved through "common sense." They should, instead, focus on problems or opportunities that require "extraordinary effort" to resolve. This class of problems usually requires a data-centric, science-based approach. Thus, the typical Six Sigma X-Belt project is more substantive and involved than a "just do it" (JDI) project or *kaizen* event. It means that the outcome and process must be mapped in advance of execution. The plan should be flexible enough to take advantage of unanticipated opportunities, but without losing sight of the overall project goal.

2.5   DEPLOYMENT TIMELINE

What are the key milestones of a Six Sigma deployment? While moving from its current state to one with a new performance trajectory, an organization passes through a number of milestones. Assuming a "typical" business, our discussion begins at time T_1.

T_1—when the organization is functioning reasonably well, is a stable force in its industry, and is capable of maintaining itself reasonably well.

T_2—when a member of the senior leadership team, often the CEO or COO (chief executive or operations officer) begins to contemplate a level of performance beyond the existing knowledge and power base. They evaluate alternative methodologies to reach the desired improved state. During this phase, the leadership studies best-in-class organizations, often discovering that successful companies commonly use Six Sigma.

T_3—this marks a company's initial exposure to Six Sigma. Initial exposure may result from intentional actions of a conscientious executive, but more often occurs by a chance event such as information acquired encounters that energize the need for due diligence.

T_4—at this point, the leadership team become aware that Six Sigma could be a viable intervention alternative and initiates a formal period of due diligence. Frequently at this point, the leadership engages a consulting organization that will guide them "through the progressive layers of the onion." This is often accomplished by an "executive summit" where the senior executive team develops the collective will and personal commitment to pursue Six Sigma.

T_5—this occurs when the corporation "initializes" its Six Sigma initiative. At this juncture, the corporation selects its senior champion, trains executives, appoints project champions, forms needed deployment structures, installs support systems, and primes the various deployment platforms for Six Sigma implementation.

T_6—training begins for the first group of Black Belts.

T_7—occurs when results are documented from the first wave of Black Belt projects. This is the "moment of truth." It is when tangible proof of the Six Sigma principle is first demonstrated. Of course, this happens directly after the improve phase of DMAIC.

T_8—the point at which Six Sigma has reached critical mass. The Black Belt infrastructure has become firmly rooted and several waves of application projects have been completed. It is at this point that a corporation fully recognizes the benefits of Six Sigma and contemplates expanding the initiative into other aspects of the business.

T_9—when the performance trajectory for an organization shifts from its historical slope to its future slope. Here the scope of Six Sigma is materially extended across all of the corporate functions (sales, finance, legal, etc.).

T_{10}—*Six Sigma has reached a state where it is self-sustaining.* This is a critical period for a corporation. It is the time when Six Sigma will "be born again" for another round of breakthroughs or "die a natural death".

Training is key to achieving Six Sigma. Figure 2.4 displays a typical Six Sigma training deployment.

Master Black Belt. In addition to completing Black Belt training (5 weeks over 4 months, including a project), he/she attends an additional session of 2 weeks.

Black Belt. The mandatory Black Belt training takes 5 weeks: week 1 is a kickoff week (e.g., corporate Six Sigma vision, basic statistics review and MINITAB overview). Weeks 2–5 are the 1-week sessions on the *measure, analyze,* improve, and control phases. *Note:* For Black Belts who need help in mastering the PC, there is optional training on Windows, Excel, PowerPoint, and other critical programs.

Six Sigma training days—average

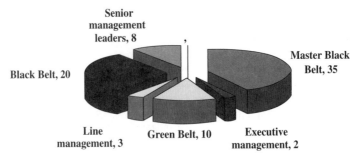

Figure 2.4 Six Sigma training to kick-start deployment: example of Six Sigma average training in days.

Green Belt. Green Belt training takes about 10 days (usually four $2\frac{1}{2}$-day sessions spread over 4–6 months after completing a project, and usually with a Black Belt).

Executive Management (Presidents, EVPs). This briefing takes a minimum of 3 days.

Senior Management Leader(s) of Six Sigma. The senior management leaders of Six Sigma attend the first week of master Black Belt training (they can also attend the second week of master Black Belt training if they are able). As senior managers, they would probably also attend the 3-day senior management briefing.

Line Manager. Line managers should receive a basic 1-day introduction to Six Sigma. Some line managers, who are managing projects, attend a more detailed 3-day session that gives them more background on Six Sigma tools, concepts, and deployment.

2.6 DESIGN FOR SIX SIGMA (DFSS) PRINCIPLES

Design for Six Sigma (DFSS) focuses on reducing risk attributable to the design of a product, service, system, process, transaction, activity, or event. DFSS is concerned with two equally important and often interrelated aims. The first is the reduction in the relative number of risk opportunities and consequential exposures inherent in the functional performance and physical attributes of a design (customer satisfaction issues). The risk consequence of a design subjected to marginal overstress, or the risk of a design feature not having been assigned an adequate performance specification, are examples of such risk. The second is the reduction of the relative number of risk opportunities and consequential exposures associated with the "processing viability" of a design (provider satisfaction issues). An example of such risks is the assignment of overly conservative tolerances that result in more expensive and lower yield production processes. It should be emphasized that any configuration or plan constitutes a design. Thus the idea of design can be applied to a wide array of "deliverables" such as proposals, books, products, services, and software. Examples include a strategic plan, the proposed office layout in a new building and the execution of an annual meeting or conference. All of these must be planned and designed before they are realized in time and space.

Inversely, use of DFSS increases confidence that a planned product, service, system, process, transaction, or activity will perform to its entitlement expectation level and be robust to uncertainties. For example, consider an automobile design that performs reliably in a wide range of climates. The goal of DFSS is to exercise the breakthrough strategy within the total design lifecycle such that each critical value opportunity associated with the functional and physical properties of a design, as well as its inherent "processability," is limited to no more than 3.4 risk exposures per million value opportunities (i.e., CTQs). Chapter 23 discusses in detail the design for Six Sigma methodology.

2.7 PROCESSING FOR SIX SIGMA (PFSS) PRINCIPLES

Processing for Six Sigma (PFSS) is concerned with reducing the value-related risk from the ongoing operation of systems, processes and supporting activities. Specifically, the objective of PFSS is to reduce the consequences of a risk exposure. In this context, every risk exposure generated by a design has some probability of occurring as a defect, loss, error, or quality-related problem during the value creation process. PFSS is frequently used to improve process capability. It is also used to improve other important metrics such as cycle time, labor cost, inventory, and material cost. Unfortunately, many corporations focus PFSS almost solely on customer satisfaction issues, which inevitably translate into defect reduction initiatives. This restricted application encourages many to conclude that Six Sigma is no different from TQM. In general, PFSS maximizes our confidence that a system, process, or activity will realize its entitlement capability and capacity in the short term and sustain that capability and capacity in the long term. The goal of PFSS is simple: exercise the breakthrough strategy to the extent that each value opportunity related to the operation of a system or process experiences no more than 3.4 risk consequences per million value opportunities. Note that this goal is often expressed in the language of classic quality management, no more than 3.4 defects per million opportunities (DPMO). Although DPMO is frequently used to measure risk, it is quite restrictive in that not all forms of risk will appear as defects. For example, certain economic risks are not related to the creation of defects.

2.8 MANAGING FOR SIX SIGMA (MFSS) PRINCIPLES

Managing for Six Sigma (MFSS) forms the foundation of the Six Sigma initiative. It is concerned with the creation, installation, initialization, and utilization of deployment plans, reporting systems, and implementation processes that support DFSS and PFSS. The ultimate goal of MFSS is to attain best-in-class business performance by improving the operational capability and capacity of an organization at an annualized rate of approximately 78% (Six Sigma learning curve).

We view MFSS as the unifying component of Six Sigma because it combines the goals of DFSS and PFSS to realize value for both customer and provider. MFSS implements planning, organizing, initializing, executing and sustaining breakthrough improvement for lost value, critical-to-value characteristics (value opportunities) and provides a basis for formulating business improvement plans. Six Sigma identifies points in the customer/provider relationship where value is at risk; these points include all aspects of the business cycle from problems in design, in manufacturing, in workmanship, in paperwork, and in transactions. To reduce business risk and increase business confidence, we must identify, characterize, optimize, and validate them using key aspects of Six Sigma (ideas, needs, value, design, determinism, measurement, leverage, opportunity, transformation, error, probability, and power), the breakthrough strategy and various Six Sigma technologies. This value/risk orientation made visible by DFSS and MFSS represents the next generation of Six Sigma thinking and practice. In the spirit of effective communication, we have diligently couched Six Sigma in the language of quality management and practice. With the passage of time, a critical mass of companies will truly achieve value entitlement between customer and provider in every key aspect of the business relationship. When this occurs, there will be a much stronger linkage between minimizing business risk and maximizing customer satisfaction, and Six Sigma will have played a key role in making this connection.

2.9 PROJECT REVIEW

We believe that the X-Belt's sponsoring organization should manage their Six Sigma projects through a formalized set of management tollgates. A tollgate is a set of success criteria that must be satisfied before a project can proceed to the next scheduled execution phase.

2.9.1   Tollgate Criteria

Tollgate reviews ensure timely and proper planning, approval, execution, review, verification, and closure of Six Sigma projects. The sponsoring organization should establish the success criteria associated with each tollgate before the fact. Prevailing business needs, management philosophy, operational requirements, and organizational policies often influence the identification and timely adoption of success criteria. These criteria will vary from organization to organization, depending on a host of business variables and constraining circumstances. Because of the business-centric nature of the criteria, only the sponsoring organization's management can definitively say what constitutes project success.

2.9.2   Project Closure

After an X-Belt has successfully completed a Six Sigma DMAIC project, the project must pass through a final management review before it can be officially closed. During this review, project activities and resulting benefits must be validated. The closure criteria will vary from organization to organization. However, it will usually include one or more of the following, sustainable benefits, accounting verification, management approval, and technical validity.

2.9.3   Project Documentation

On project completion, the sponsoring organization should provide the X-Belt with a document that confirms the successful completion and closure of a Six Sigma project. This document can take the form of a letter, an internal memorandum, or a certificate.

2.9.4   Personal Recognition

Generally speaking, these documents do not reveal specific details about the project (in the interests of business confidentiality), but do acknowledge the X-Belt's contributions and accomplishments. Needless to say, such "artifacts of success" are quite important to the individual X-Belt. These documents are also helpful to the sponsoring organization during future personnel reviews.

2.9.5   Authenticating Agent

Normally, a recognized member of the sponsoring organization's management team must formally certify a successful project. For example, this could be the X-Belt's immediate manager, the appropriate process owner, or one of the organization's senior Six Sigma leaders such as the champion or master Black Belt. After being certified, the X-Belt can declare a project successful in a verifiable way. Thus, an external consultancy or professional institution is not meaningfully positioned to authenticate, validate, or otherwise certify the successful closure of an X-Belt project. Such tasks should reside with the sponsoring organization.

2.10   SUMMARY

In summary, leadership is the leverage variable when it comes to the success of Six Sigma. Time and time again we have observed that the absence of a leader leads to failure. A group working toward a common goal can survive just about anything except not having a leader. This is especially true of application projects. Without leadership, all is for naught. Leaders are the glue that bond employees' dreams and aspirations together. Leaders create the synergy that ensures the whole is greater than the sum of parts. Simply stated, leaders adapt and improvise—they turn liabilities into assets. Always recognize that an answer must be preceded by a question. We now recognize that an answer is dependent on the question, not the other way around. Therefore, it can be said that $A = f(Q)$, where A is the answer and Q is the question. Based on this relationship, leaders should be consumed with the design, detail, and delivery of questions, not with the provision of answers. To lead is to question, and to follow is to answer. To achieve power, create the questions. If you seek praise, provide the answers.

GLOSSARY

Black Belt   Classification of individuals who embrace Six Sigma. Every completed Black Belt project, by virtue of solid leadership and sound planning, yields benefits back to the firm in the form of immediate "hard" cost savings validated by independent accounting.

champion leader   Deployment champion—an individual who is trained in Six Sigma and is deeply involved in establishing Six Sigma infrastructure.

champion project   An individual who already has a thorough functional knowledge of the business and has established a strong leadership track record. Mikel Harry and Rich Schroeder recommend having one project champion for every 10–20 Black Belts.

CT   "critical to," a term that can be expanded to express concepts such as "critical to satisfaction" (CTS).

CTC   "Critical to cost" (CTC)—a parameter for product, service, and/or transactional characteristics that significantly influence one or more CTSs in terms of cost.

CTD   "Critical to delivery"—a parameter for product, service, and/or transactional characteristics that significantly influence one or more CTSs in terms of delivery.

CT matrix   A tool that links the process or CT tree (columns of the matrix) and the product or CTY tree (rows). See **process tree (CTY)** for a discussion of CTY.

CTP   "Critical to process" CTP characteristics are process parameters that significantly influence a CTQ, CTD, and/or CTC. In the equation $Y = f(x_1, x_2, \ldots, x_n)$, the CTQ, CTD, and CTC represent the dependent variable (Y), and the CTPs represent the independent variables (X).

CTQ   "Critical to quality"—a parameter for product, service, and/or transactional characteristics that significantly influence one or more CTSs in terms of quality.

CTS   "Critical to satisfaction"—CTS characteristics are the expression of the customer's vital needs. These needs are translated into critical requirements for quality, delivery, and cost.

CT tree   A tool that enables us to represent the "need–do" interaction by translating needs considered vital by the customer into product or service characteristics, and to link these characteristics to our business processes.

DCCDI   *Define* the project goals; *customer* analysis is completed; *concept* ideas are developed, reviewed, and selected; *design* is performed to meet the customer and business specifications; and *implementation* is completed to develop and commercialize the product/service.

DIDOV   *Define, identify, design, optimize, verify,* and *monitor.* Slight modification of the DMADV methodology.

DMADOV   *Define, measure, analyze, design, optimize,* and *verify.* A slight modification of the DADV methodology.

DMADV   *Define, measure, analyze, design,* and *verify.* The phases that some companies use for DFSS methodology.

Green Belt   Part-time resources who assist Black Belts to complete projects and help to extend the reach of Black Belts.

IDOV   *Identify, design, optimize,* and *validate.* Well-known design methodology especially in the manufacturing world.

master Black Belt   These are the full-time trainers of our organization as well as the coaches for Black Belts.

primary metric   Also called *process metric*; a measure of a process and not a measure of a high-level business objective.

process tree (CTX)   A hierarchical breakdown of our product or service that allows us to visualize the CTQ, CTD, and CTC characteristics at each level of the hierarchy.

product tree (CTY)   A breakdown of our engineering, manufacturing, service and transaction processes. CTPs are identified at the lowest level of this tree.

secondary metric   Consequential metric that can be both business and process metric. Metric that measures anything that goes wrong as a result of improving the primary metric. There can be multiple consequential metrics in a Six Sigma project.

White Belt   Project-specific part-time resources that provide process and cross-functional knowledge, as well as help, to sustain the gains.

REFERENCES AND NOTES

1. Don Linseman, Vice President and Corporate Champion of DuPont, has global responsibility for DuPont's Corporate Strategic Accounts. He lives in West Chester, Pennsylvania.

2. Gary Reiner, SR Vice President and Chief Information Officer of GE, a position he has held since 1996. Reiner earned a B.A. in economics from Harvard and earned an M.B.A. from Harvard Business School. He is based in Fairfield, Connecticut.

3. B. W. Galvin, former chairman, CEO and COB of Motorola, in his book. *The Idea of Idea*, discusses the differences between managers, especially bureaucratic and leaders. *The Idea of Ideas*, Motorola University Press, April 1993, (ISBN 1569460027).

4. A. Maslow, *A Theory of Human Motivation*, 1943. Abraham Maslow's "the hierarchy of needs" is a theory in psychology that contends that as humans meet "basic needs," they seek to satisfy successively "higher needs" that occupy a set hierarchy. Maslow studied exemplary people such as Albert Einstein, Jane Addams, Eleanor Roosevelt, and Frederick Douglass rather than people with psychological or psychiatric disorders.

5. D. Leighton, former General Director of Operations and Administration of HSBC-Mexico published several papers on leaders and change management, 2007.

6. J. Germaine, Motorola, was hired and worked for B. W. Galvin former chairman, CEO and COB of Motorola.

7. M. J. Morin, former chairman and CEO of Drake Beam Morin, Inc., is author of numerous books and articles on business matters. In his book discusses the difference between managers and leaders and change management. Silent Sabotage, American Management Association, 1995 (ISBN 0-8144-0300-X).

8. Rich Schroeder is president of the Six Sigma Academy. A former vice president at Motorola, he joined Mikel Harry to create the Academy. Mikel Harry and Richard Schroeder implement Six Sigma programs at major corporations throughout the world. In 1998, corporate consulting contracts from their Six Sigma training topped $100,000,000.

9. isixsigma definitions for primary and secondary metric.

10. M. Harry and R. Schroeder, Six Sigma. The breakthrough management strategy, doubleday, 2000-ISBN 0-385-49437-8, 1st. edition.

3 Lean Sigma Projects

3.1 OVERVIEW

One of the most frequently asked questions when organizations try to implement Six Sigma is whether all programs or activities are Six Sigma projects. For this reason it is important to know the definition of a project. A project is a temporary group of related tasks undertaken to create a unique product, service, or result.[1] The following are unique characteristics to projects. A project is time-limited, is unique, consists of interrelated activities, and is undertaken for a purpose. A program is a larger effort than a project, because it is a group of related projects coordinated together. Many organizations often group projects into programs in order to realize some benefits that could not be achieved if those projects were not taken in concert. After reading this chapter, the practitioner will have enough information to know what a Six Sigma project is. Also, the S^4/IEE^2 roadmap assists in clarifying the decisionmaking of what constitutes a Six Sigma project.

3.2 INTRODUCTION

Lean Sigma projects are identified, planned, executed, reviewed, and verified to achieve a predetermined benefit for the sponsoring organization and its customers. To this end, a Lean Sigma project must be properly chartered and initialized. To realize a lasting solution, it must flow smoothly through the define, measure, analyze, improve, and control (DMAIC) improvement phases. In short, Lean Sigma projects are formal investigations guided by scientific knowledge, not a collection of loosely organized ad hoc activities with indeterminate outcomes or a "let's see what falls out" focus. For Lean Sigma projects, success results from thoughtfully designed outcome, not a chance event. This makes Lean Sigma projects both effective and efficient. To achieve these ends, Lean Sigma projects must be carefully planned, resourced, managed, and judiciously reviewed. On completion of a project, the accounting function must verify derived benefits, before management officially closes the project.

Today corporations seek both top-line growth and total cost reduction. Organization business leaders want to increase capability and capacity with minimal capital investment. They recognize that continually improved customer satisfaction is required to achieve these objectives. When you add enhanced investor relations, the demands on business executives seem endless. Some business professionals are unaware of the huge gains made by companies such as General Electric, Honeywell, Sony, Ford Motor Company, and DuPont, but many know that these fine organizations have found a "magic formula" in the chemistry of Lean Sigma. Each discovered the power of highly focused benefit-centric projects. Since Motorola introduced Six Sigma in the 1980s as a quality initiative, it has evolved into a world-class business management system. Over time, it has blended with many of the best practices of Lean manufacturing. Thus, Lean Sigma provides an effective means for industrial and commercial organizations to improve company performance. The combination of Lean Sigma

Practitioner's Guide for Statistics and Lean Six Sigma for Process Improvements. By Mikel J. Harry, Prem S. Mann, Ofelia C. de Hodgins, Christopher J. Lacke, and Richard Hulbert
Copyright © 2010 John Wiley & Sons, Inc.

thinking and the DMAIC problem-solving methodology offers the capability to reform the critical processes of a business.

Lean Sigma forces an organization to reexamine how it gets work done and not simply modify or augment the existing system to realize marginally acceptable improvements. While the Lean method targets process efficiency through waste reduction, Six Sigma targets process effectiveness through variation reduction. Taken together, the two deliver the proverbial one–two punch against waste and variation that translate into improved overall business efficacy. Lean Sigma is more than a simple quality improvement or cost reduction initiative. It is a management system that can produce quantum improvements wherever applied.

It offers a formal way to identify, execute, and manage many simultaneous improvement projects that contribute to a common set of unified business objectives. When such projects are deployed and implemented, they collectively impact the performance and image of the business. In this sense, Lean Sigma is both a strategic and tactical system for managing the total business enterprise. The increased popularity of Lean Sigma projects confirms its effectiveness.

3.3 PROJECT DESCRIPTION

Projects should be selected based on the company's strategic goals and direction. The larger Black Belt projects target improvements in customer satisfaction, cost, capacity, and top-end growth. At the operations level, Lean Sigma projects offer local impact that may not immediately impact corporate-level objectives. However, these projects should be consistent with the overall strategic goals and direction. When selecting Black Belt projects, frequently asked questions include

- What is the nature of a Six Sigma project?
- What is the scope of a Six Sigma project?
- How many Lean Sigma projects should be identified?
- What results should be expected?

Project selection can be done from either a "top–down" or "bottom–up" approach. The top–down approach considers a company's major business issues and objectives and then assigns a senior champion. This senior manager, most affected by these business issues, will broadly define the improvement objectives, establish performance measures, and propose strategic improvement projects with specific and measurable goals that can be met in a given time period. Then the teams (led by a Six Sigma champion) identify processes and critical-to-quality characteristics, conduct process base lining, and identify improvement opportunities. This is the favored approach and the best way to align local business needs with corporate goals.

Production managers facing budget reductions, specific quality problems, or a need to improve process flow usually choose projects by the bottom–up approach. These projects, called "targets of opportunity," rarely integrate well with the company's strategic business goals. For example, managers may be asked to identify specific areas of waste, part shortages, supplier quality issues, and unclear or impractical engineering requirements, and then assign Black Belts to solve specific problems. During deployment planning, an organization should consider how it allocates resources between these two approaches. While the top–down approach has structural and managerial appeal, the bottom–up approach can ensure that key opportunities at lower levels of the organization are promptly addressed. Usually, a rational combination of the two is the optimal approach.

At the process level, Lean Sigma projects should focus on those processes and critical-to-quality characteristics that offer the greatest financial and customer satisfaction leverage. Each project should address at least one key business objective. It is the champion's responsibility to gain executive support so that roadblocks are cleared and projects are successfully aligned and completed regardless of the selection process.

Six Sigma Value-Based Projects. Lean Sigma value-based projects proliferate in companies that have adopted Lean Sigma on a global scale. These projects are selected, prioritized, configured, and coordinated such that taken together; they can significantly impact higher-level operational and business performance metrics. This vertical correlation between projects and business performance is managed by the support structure, training initiative, and information technology system that enable a corporation to do business.

3.4   PROJECT GUIDELINES

What are some of the most common project selection guidelines? Project selection methods differ by corporation. While the overall goal of each Lean Sigma breakthrough project is to improve customer satisfaction and profitability, projects may focus on industrial or commercial processes.

> *Project Identification.* Lean Sigma application projects focus on business value, plain and simple. A valid project must yield some form of tangible value to the sponsoring organization such as cost savings, yield improvement, defect reduction, cycle time reduction, inventory reduction, headcount reduction, and increased customer satisfaction. Of course, all of these value-based improvements can be translated into the language of business—money and time.

> *Project Returns.* Regardless of magnitude, any true improvement must be viable, visible, and verifiable in terms of money. If a Lean Sigma practitioner cannot translate his/her efforts and outcomes into some form of economic benefit, then the true value of that practitioner's contribution is suspect. If this vital monetary translation cannot be made, it is doubtful that the proposed "improvement" was real. Because of this business-minded approach, Lean Sigma projects have traditionally been focused on high-impact value-centric problems and issues that are also financially traceable and verifiable. More specifically, Lean Sigma projects are concerned with creating measurable value for both the customer and provider in every aspect of the business relationship.

> *Project Responsibilities.* While, the X-Belt must take responsibility for learning the Lean Sigma tools and methods, the champion must identify, select, assign, review, approve, and close Lean Sigma projects. When these two conditions are realized and prudently blended, success usually follows.

> *Project Scale.* A Lean Sigma project must have the right scope and depth for a Black or Green Belt. Only those projects that are capable of delivering hard benefits should be considered. In other words, a Lean Sigma project should be declared, activated, and executed only if it meets certain predefined decision criteria.

> *Project Criteria.* The general criteria for selecting a Six Sigma project are as follows:
> 1. The project has the proper scope, depth, and timing.
> 2. The DMAIC improvement process can be applied to realize its projected benefits.
> 3. The targeted process and forecasted benefits can be clearly defined and are rationally measurable.

> A project will usually be considered "viable" and "feasible" when these three criteria can be reasonably satisfied in both spirit and intent. Only then should that project be assigned to a qualified Lean Sigma practitioner.

> *Project Activation.* By convention, a proficiency certified Black Belt (or Green Belt) must have completed at least one live Lean Sigma DMAIC project as part of his/her professional development. Such projects are intended primarily to reinforce the X-Belt's training program. In this context, the training project is very similar to posttraining projects. Regardless of the project's nature, the scope and depth of a Lean Sigma project should always be consistent with the X-Belt's skill and rank. Furthermore, it is recommended that the X-Belt prepare a

Lean Sigma project charter, receive management approval, and then execute the project in accordance with the defined plan. In this manner, the overall improvement system remains measurable, manageable, and accountable.

Project Charter. A project charter is the cornerstone of successful execution. Approval of a project charter activates a Lean Sigma project. Once a project charter has been created and sanctioned by the sponsoring organization, the key elements of execution must then be identified, prioritized, and committed to a timeline. The charter is the document that officially starts the project, and this is the process that creates it. The charter is one of the most important documents on a project because it is essential for creating a project. The project chart is necessary for documenting the business needs and the new product, service, or other result that is intended to satisfy those requirements.[1]

Project Execution. Each Lean Sigma project will follow the DMAIC improvement strategy to realize its aims. The planning and execution of this strategy can involve up to 15 independent milestones that are usually interrelated in a progressive and interactive manner.

3.5 PROJECT SELECTION

Project selection is one of the most critical and challenging activities faced by all organizations pursuing Lean Sigma projects. Most organizations identify a host of project opportunities, but have difficulty identifying opportunities/challenges or defects that are key for the creation of a meaningful business project.

3.5.1 Project Selection Guidelines

Among existing Six Sigma organizations, there is no single best method for project selection. Practitioners do not agree on which approach works best, nor is a standard being developed. Most Lean Sigma projects are selected because they will have a positive impact on the customer. As described in the overview in Section 1.1, the motto is simple. If you make an improvement, then by day's end, the gains should be verifiable." In this context, the elimination of a system defect, mistake, fault, or error must directly translate to a measurable benefit—reduction in headcount, less material, or lower overhead cost. In many firms beginning their Six Sigma process, projects are selected because they are quick-win opportunities with strong financial returns.

Once a company has exhausted this initial set of beneficial projects, the hard work of project selection begins. Practitioners must face a large mix of potentially marginal projects. Many of these projects have been delayed by questions about their feasibility or financial risk. Senior management must consider a new set of issues: What selection methods can help you cope with a large number of recommended projects? How do you compare dissimilar projects? Is it possible to forecast which projects will provide the best return given uncertainties in scheduling, investment, and success?

How can senior management select and prioritize projects that are consistent with your organization practices and preferences? Listed next, in no particular order, are some of the project selection methods. Each method or approach has been employed with varying degrees of success and each has a corresponding set of strengths and weaknesses.

Pareto Analysis. A strong benefit of Pareto analysis for Six Sigma project selection is that it is usually based on data: warranty costs, rework costs, production costs, defect measurements, and other related factors that have been tracked by historical data and accepted by management. A Pareto chart can be compiled by an analyst and does not utilize senior management resources. The downside of this problem-solving approach is that the project selection is based only upon one or a few measured criteria that may not improve customer satisfaction. Also, the Pareto priority index (PPI) offers no insight into which projects will provide the best opportunities.

Cost of Poor Quality. Similar to Pareto analysis, cost of poor quality (COPQ) measures costs of defective product from data gathered from prevention controls, testing controls, or warranty/scrap/rework activities. However, COPQ focuses solely on cost; again, it offers no insight into customer satisfaction levels. The primary disadvantage of COPQ is that the many of these costs are intangible and not measured by traditional accounting systems.

Project Prioritization Matrix. This is a top–down approach that defines the potential projects that line up directly with the firms' strategic initiatives. This method, based on the cause–effect (CE) matrix, requires senior management participation. The downside of this approach is that it requires senior management and Six Sigma leadership (champions and/or master Black Belts) to meet together. The inability to routinely schedule this activity may become an obstacle to success.

Project Clustering. This is also a top–down approach where projects are "clustered" around the firm's strategic initiatives. Like the project prioritization matrix, this technique requires senior management and Six Sigma leadership participation to ensure the top–down nature of the decisionmaking. This means that a concurrent block of time must be devoted for team activity.

Value-based Management. This technique relies on determining where value lies within an organization. Then management sets project priorities based on a flowdown from the top-level strategy where shareholder needs are satisfied by business units and value streams through to the individual projects. The advantage of this approach is that it requires a thorough assessment of the competition and ensures that the company never loses sight of where economic value resides. The disadvantage is that such a value-based process does not account for projects that cannot easily be applied to bottom-line shareholder value.

Quality Function Deployment (QFD) and Strategy Deployment Matrix. This method creates a matrix around customer needs that exposes relationships between the corporate goals and the individual projects. Like the project prioritization (CE) matrix, this methodology is a top–down approach that requires senior management and Six Sigma leadership participation. More advanced Six Sigma organizations use QFD for product development (design for Six Sigma). One disadvantage of this approach is that it requires experienced facilitators and more training than for CE approaches. It also requires multiple sessions and substantially more time for participants.

Theory of Constraints. The theory of constraints (TOC) approach uses resource constraints to prioritize improvement projects. It is superior to traditional total quality management (TQM) approaches in that TOC can identify opportunities below-the-macro-operations level. The primary disadvantages are that analysts must be trained and practiced in the technique and that the time required performing the analysis might be longer than management desires. It is not widely used in Six Sigma programs.

Political Choices. In this method, the structure of a company and Six Sigma initiative is such that certain individuals or groups of individuals strongly influence projects selections. Sometimes, these individuals are simply the ones who are loudest or present the glitziest presentations. The lack of an objective project selection process usually has negative consequences for morale and impedes the development of a team-based culture.

These methods are popular because Six Sigma projects can be aligned with organization strategic plans and customer concerns. However, none of these approaches directly addresses the key metrics, such as probability of project completion or potential financial return, which are deemed critical to the long-term success of any program. Although these metrics prove difficult to forecast, decisionmakers can use one relatively new selection technique—Monte Carlo simulation and optimization—to define project lists that include project cost and measure potential probability of success. The following section discusses the details of this methodology and where it can be applied.

Where to find projects?

1. A project could be any process that addresses wasted activities.
2. Any department outputs that are not meeting their CTQs.
3. Any tasks that you or your department must rework (execute more than one time) to get it right.
4. Assignments that normally take you more than a week to complete because *you not found a solution yet.*
5. GB/BB projects, in your department, where data are not available or where help is required to define the process.

(Note: If a task is simple and has an obvious solution, *Just do it!* However, this is not a Six Sigma or Lean project.)

3.6   PROJECT SCOPE

What does a Lean Sigma project look like? The following discussion of the breakthrough strategy applies at the process or Black Belt level and represents how "low-hanging fruit" can be gathered using the breakthrough strategy. Our story takes place within General Electric's Plastics Division. GE Plastics, by applying the breakthrough strategy to over 3000 projects, saved $137 million in 1997. One of these projects took place in Polymerland, a division within GE Plastics that distributes thermoplastic resins. Polymerland ships these plastic products to a variety of customers, including manufacturers of computers, compact disks, digital videodisks, and automobiles.

The employees had already pulled together a Black Belt team to focus on a suspected problem. The team was in the midst of the *define* phase of their "damage-free delivery project" when one of their largest customers came to their sales force and said, "Look, we are ready to take our business elsewhere. Too many of the products you are sending us have been contaminated." The customer was frustrated on continually receiving damaged products that necessitated use its own resources to return the defective product back to Polymerland for replacement. Between January 1, 1997, and June 30, 1997, Polymerland (nonexistent organization used to illustrate the concept) delivered 275,501,855 lb of the product to its customers. During this period, 176,381 lb of product were contaminated, meaning that the packaging had been damaged during shipping and a damaged product was delivered to the customer. This resulted in a damage-free reported sigma of 4.72 and a DPMO (defects per million opportunities) of 640. Realizing that one unhappy customer probably meant that there were many more who were not voicing their unhappiness, and that lost customers mean lost sales. Polymerland employees informed the customer that they were aware of the problem and that they were seeking a solution. They then applied the DMAIC procedure to the problem.

DMAIC Improvement Process. Lean Sigma derives its power from the DMAIC application strategy— define, measure, analyze, improve, and control. Figure 3.1 illustrates the DMAIC methodology and its phasewise milestones.[3] Through this strategy, it is possible to improve almost anything because we reduce the problem to its deterministic root. In other words, we apply the fundamental transfer equation: $Y = f(X)$.

In the transfer equation, Y is the outcome and X represents all of the essential inputs, and f is the ways and means (process) by which the critical inputs (x's) affect the output (Y). For any given problem-centric situation, we must define, measure, and analyze the critical Y (i.e., the CTQ) and then improve and control the critical x's (i.e., the CTP's). In the measure phase, Polymerland's (assumed names organization) a Black Belt team looked closely at each of their 10 warehouses from which the product was shipped. From the 10 warehouses, 2 warehouses (Maumee and Piedmont) produced the highest number of defects. On the other hand, one of the warehouses, the one in Brampton, Ontario, was found operating at Six Sigma levels. The team calculated that with 10

	Phase	Goal	Milestones		
			1	**2**	**3**
1	**Define**	Establish focus	Identify problematic output	Review output history	Describe output process
2	**Measure**	Create baseline	Evaluate process efficiency	Qualify measurement system	Establish output capability
3	**Analyze**	Discover causes	Diagnose output variation	Identify possible causes	Isolate critical factors
4	**Improve**	Devise solution	Examine factor effects	Optimize factor settings	Prescribe factor tolerances
5	**Control**	Sustain benefits	Verify solution repeatability	Ensure process stability	Monitor output capability

Figure 3.1 Tabular representation of the DMAIC goals and related milestone activities (*Source:* M. Harry and R. Schroeder, Six Sigma, Doubleday, 2000.)

warehouse locations, 50 different product lines, 14 different types of packaging, 100 transportation carriers, and nearly 700,000 combinations the organization could be studied if an exhaustive search approach was used.

In the analyze phase, the team focused on the Maumee and Piedmont warehouses (hypothetical examples). They found that 95% of the damaged boxes had two punctures at the base of one side of the box, causing the product to leak. The organization's Black Belts also found that 87% of the damage occurred when forklifts moved the boxes. Soon the team concluded that there was a strong correlation between the number of boxes damaged in-house and the number of damaged products received by customers. The team discovered a relationship between how the product was packaged and the type of damage. Bags were more likely to be damaged if they were scraped across the floor, whereas boxes were more likely to be damaged when being moved by forklift. Why, they asked, did forklifts damage the boxes? The answer turned out to be fairly simple. The forks on the forklift protruded 6–9 in. beyond the box, so that when the fork was inserted under the box being retrieved, the box behind it was punctured. But if this was the case, why was the Brampton warehouse able to operate at Six Sigma using the same forklift?

Several possibilities were offered, but the most tenable seemed to be the drivers' experience, forklift drivers not following forklift pressure standard operating procedures (SOPs), and the lack of control they wielded over the forklift operation. When a box was lifted, the forks were tilted back for load stabilization. If the driver did not lower the fork tilt, the protruding fork, instead of sliding under the first box, would puncture the second box as the first box was placed next to it. Because drivers could not see over the lift, they were forced to blindly place the boxes on the pallets.

In the *improve phase*, now that the problem had been analyzed, the team could focus on several possible solutions. More experienced forklift operators could be hired into the Maumee and Piedmont plants at a significant cost to the company, assuming it was possible to find forklift operators with nearly 20 years' experience. A second option required extensive training for the forklift operators in the Maumee and Piedmont warehouses. An expensive solution without a guaranteed outcome. In the end, the team decided to install shorter forks, eliminating the fork protrusion altogether. In December 1997, the Piedmont plant replaced the 42-in. forks with 36-in. forks at a relatively nominal cost of $350 per truck. Within a short time, the 42-in. forks had been replaced at all 10 warehouses and following SOPs was reinforced.

In the *control phase*, the Black Belt team was ready to move into the final phase, where steps are implemented to ensure that the processes stays fixed. GE has a mandate that before any Lean Sigma project is completed; it must pass through data points. A data point shows intervals and can reflect how often an event occurs within a given time period. A data point might be weight, inches, or the number of hospital deaths that fall within a specific time period. In Polymerland's case, a data point was a period of time, anywhere from 2 weeks to 2 months, during which Black Belts must show beyond a doubt that their projects have a workable and permanent solution.

3.7   PROJECT LEADERSHIP

What is a Lean Sigma project leader? In general eight leadership roles are essential to effectively install a Six Sigma initiative: advocate, coach, mentor, leader, enabler, facilitator, manager, and consultant. These skills are commonly associated with the role of steward.

Project leaders must practice effective stewardship to effectively guide work teams through the project lifecycle. To get the most from each team, the project leader must be flexible and adaptive to ever-changing circumstances. The appropriate leadership style is highly dependent on the leader's style, organization culture, prevailing competencies, resource constraints, and the goals to be achieved. Early in a Six Sigma project, the leader (steward) must serve as an advocate. Following this, the leader may revert to a coaching role to transfer critical knowledge. In some cases, the leader might find it necessary to mentor high-potential team players. After this, it is likely that a steward will personally lead the team. The project leader must frequently exploit unforeseen opportunities, therein driving meaningful and productive adaptations of Six Sigma during the DMAIC cycle. Usually, the project leader is expected to manage the activities of one or more work teams, and will often directly or indirectly provide consulting along the way. Therefore, Six Sigma project leaders must frequently adjust their leadership styles to meet team needs.

3.8   PROJECT TEAMS

How can employees be motivated to do projects? From our perspective, this is a management problem, not a Six Sigma problem. Motivation will improve with proactive leadership. The difficulty in relating local issues to the bottom line likely results from the nature of your accounting system (i.e., a mismatch between the key issues and the chart of accounts). Consequently, you may have to amend, modify, or otherwise augment your accounting system to track financial correlations, and this takes time. In the interim, management must "edict" these projects, rather than wait for volunteers to choose them. Remember, not everything in this world will happen with the word please. Some things require one to *just do it*.

3.9   PROJECT FINANCIALS

How should Lean Sigma projects be measured? Identifying and improving those performance metrics that will best boost a company's financial success and positively impacting the customer base is the key challenge. Projects are often measured through the following key metrics:

1. *Defects per million opportunities*—the total number of defects per unit divided by the total number of opportunities for defects per unit multiplied by 1,000,000. This number can be directly converted into a sigma value.
2. *Net cost savings*—verifiable reductions in fixed or variable costs.

3. *Cost of poor quality*—the cost of failing to produce 100% quality the first pass through the production process.
4. *Capacity*—the number of good units a process is able to produce in a given period of time.
5. *Cycle time*—the length of process or elapsed time required to produce a product or service.

By analyzing the performance of these five key metrics and placing the data in a single database, companies can better understand their operations. This creates a baseline to

- Show how well a current process is working
- Determine theoretically how well a process should work
- Determine how much processes can be improved
- Determine magnitude and direction of each process improvement on customer satisfaction
- Establish how much of this impact will be realized in cost saving.

How well a process should work is also known as "entitlement." For some companies, entitlement is "best possible performance" and usually occurs intermittently and for short periods of time. The logic behind entitlement is that if something functions well for even a short period of time, by implementing simple process improvements, it should be able to function at the "best possible performance" level all the time. Entitlement does not involve creating new technologies or significantly redesigning current processes.

3.10   PROJECT MANAGEMENT

How can I drive successful Lean Sigma projects in sales and marketing? There is a wide array of reasons why people are not willing to "do" Lean Sigma projects. For example, incentives do not offset the added risk and responsibility. It is well known that a lack of dedicated resources can deter seasoned veterans from accepting the challenge of a Black Belt project. There are as many reasons as branches on a tree. You simply have to separate the "vital few" motives from the "trivial many." Perhaps the quickest way to solve a problem would be to "guess" the correct answer. Even though it consumes both time and resources, you should consider the design and execution of a Pareto study. This type of investigation constitutes a more assured path to the root of your problem. To execute such a study, you identify the potential reasons why people will not do a project, then design a survey to interrogate these reasons, deploy the questionnaire, tabulate the results, and analyze the data. Let the data lead you to the solution. The approach does not always work, but it is certainly far more effective than guessing.

Project Reviews. It is firmly suggested that the X-Belt's sponsoring organization manage their Lean Sigma projects through a formalized set of management tollgates. A *tollgate* is simply a set of success criteria that must be fully satisfied before a project is allowed to continue on the next scheduled phase of execution. Normally, a management tollgate review will coincide with the completion of each phase of the DMAIC improvement process. However, in some cases, it may be necessary to incorporate in-phase reviews (i.e., management and technical) so as to ensure adequate oversight and governance.

In this case, the presiding master Black Belt would assume responsibility for the planned technical reviews, while the champion (and perhaps process owner) would have responsibility for the planned management reviews. It is also recommended that one or more of the organization's executive management team participate in the management reviews. This should be done periodically to provide senior leadership visibility and encouragement.

Project Tollgates. Tollgates ensure the timely and proper planning, approval, execution, review, verification, and closure of Lean Sigma projects. The X-Belt's sponsoring organization should establish success criteria for each tollgate before the fact. Prevailing business needs, management philosophy, operational requirements, and organizational policies can often influence the identification and adoption of certain success criteria. As one might expect, such criteria will vary from organization to organization, depending on a host of business constraints. Because of the business-centric nature of such criteria, only the sponsoring organization's management can definitively say what constitutes project success as they are concerned with the *quality of business*, rather than the *business of quality*.

3.11   PROJECT PAYBACK

Ignoring Six Sigma can cost you money. A recent *iSixSigma Magazine* benchmarking study found that the median value of hard financial benefit of using Black Belts (business professionals who lead Six Sigma process improvement projects full time) is $188,000. That figure represents savings delivered per project, across all industries and company sizes. Companies with revenue of more than $1 billion see median results of $335,000. Companies with revenue of less than $100 million see median results of $117,000. Similar results were observed when segmented by number of employees in an organization (<1000, 1000–10,000, and >10,000). Fewer companies track the "soft" financial savings (no direct impact to a company's financial statement), but for those that do, the median value of soft savings was $107,500.

Michael Cyger, publisher and founder of *iSixSigma Magazine* and *i*SixSigma.com, said, "This study validates that companies of all sizes can reap significant financial benefits from Six Sigma. Enormous positive results from Six Sigma at companies like GE, Motorola, Allied Signal, and DuPont are well documented. Now that we have the data across all company sizes (by both revenue and employee count) and industry, it's significant enough that CEOs cannot afford to ignore Six Sigma. It would be a disservice to their shareholders and customers.

Given that most Black Belts complete at least three Lean Sigma projects per year, Cyger continued, the cumulative benefit of all projects is over $500,000 per Black Belt per year just measuring the 'hard' savings that can be directly tied to the company's top and/or bottom lines on the financial statement.

Erin Ducceschi, editor of *iSixSigma Magazine*, also noted that significant financial results can be achieved at any stage of a Six Sigma initiative. There is no statistical difference in financial benefits based on the number of years a company has been using Six Sigma, meaning that the return on a company's investment can be achieved within a short timeframe. The benchmarking data was gathered from the *i*SixSigma.com readership from October 7–15, 2004. Of the 819 individuals who responded, 48.7% were Black Belts. All data reported was for the 12-month period prior to the survey.

> SEATTLE, WA—(MARKET WIRE)—07/18/2005—Companies applying the innovation and design techniques of Six Sigma (called Design for Six Sigma, or DFSS) report median "hard" gains (direct tie to financial statement) of $200,000 per project with $100,000 in "soft" benefits, according to the latest research by Six Sigma Magazine
> (http://www.isixsigma.com/dfss_research).

The numbers for larger companies are even more impressive, added Michael Cyger, CEO and publisher of *iSixSigma Magazine* and *i*SixSigma.com, "Companies near or over $1 billion in annual revenue reported median 'hard' gains of $500,000 and 'soft' gains of $200,000. But even small companies benefited from DFSS (design for Six Sigma), Cyger continued, the smallest companies in our survey reported combined gains ranging from $10,000 to $300,000. That's a big impact when you have a small budget." Cyger said that these numbers should convince more companies to use DFSS techniques, all of the 1112 respondents in our survey are using Six Sigma, but only 41% are using DFSS methods. That is because companies that say they use Six Sigma users really mean they use the process improvement methodology of Six Sigma, or what is termed DMAIC, according to Cyger.

Cyger also said that DFSS is the innovation arm of Six Sigma, that it is a rigorous roadmap and toolset for understanding exactly what customers want and translating that knowledge into nearly perfect new products and services, with reduced delivery time and lower development costs.

A survey was designed to ask companies that did not use DFSS why they did not, Cyger explained that most of these companies said that they either didn't know much about it or didn't think there was a need for it. He thinks that these companies are missing out. Survey respondents used DFSS for both improving process capability and developing new products. These kinds of uses apply to any company, no matter what sector, size or geography. A person or company can learn design for Six Sigma without having to learn the traditional Six Sigma DMAIC methodology (Solectron used DFSS for the design of printed-circuit boards (PCBs) for years without using the traditional DMAIC methodology). Cyger concluded that in the wake of the survey showing the financial value of DFSS, perhaps more companies will consider learning and using DFSS.

3.12   PROJECT MILESTONES

Project milestones are usually deliverables to upper management. During the executive management briefing, expectations are set. Milestones record the progress toward those expectations (see Fig. 3.2 illustrating DMAIC milestone activities). On its way to disrupting the historical business continuum and changing its performance trajectory, an organization passes through a number of milestones. Assuming a "typical" business, our discussion begins at time T_1, when the organization is functioning reasonably well as a stable force in its industry, capable of maintaining itself reasonably well.

The next milestone is T_2, a time when a member of the senior leadership team, often the CEO or COO, begins to contemplate a level of performance beyond the existing knowledge and power base.

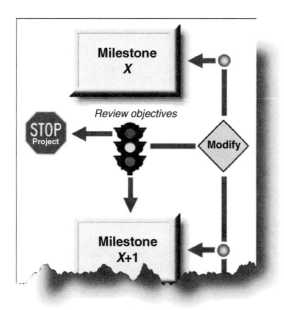

Figure 3.2   Illustrating DMAIC milestone activities (Mikel Harry).

This recognition is followed by a period of exploration during which various interventions are investigated to assess their capability to induce positive change. During this phase, the leadership team studies best-in-class organizations frequently discovering that are active users of Six Sigma.

Once the need for quantum change has been recognized, we progress to T_3. This point marks a company's initial exposure to Six Sigma. Such an exposure sometimes happens by the intentional design of a conscientious executive, but more often by a chance eruption (e.g., information is acquired and exchanged in a network of interdependent encounters that energize the need for due diligence).

The next milestone is T_4, at which point the leadership team become aware that Six Sigma could be a viable intervention alternative and initiates a formal period of due diligence. To facilitate this effort, the leadership team normally engages a consulting organization to guide them through the progressive layers of the onion, so to speak. This is often accomplished by way of an executive summit. Through such a summit, the senior executive team develops the collective will and personal commitment to pursue Six Sigma.

Projecting forward in the business continuum, we reach T_5. This is the time when the corporation initializes its Six Sigma initiative. At this juncture, the corporation selects its senior champion, trains executives, prepares project champions, forms needed deployment structures, installs support systems, and primes the various deployment platforms for Six Sigma implementation.

The next milestone is T_6 —training the first wave of black belts. Related to this, T_7 occurs when the results are documented from the first wave of Black Belt projects. This is the moment of truth. It is when tangible proof of the Six Sigma principle is first demonstrated. Of course, this happens directly after the improve phase of DMAIC.

After this we move to T_8, the point at which the Six Sigma intervention has reached critical mass. This is when the Black Belt infrastructure becomes firmly rooted and several waves of application projects have been completed. It is at this point that a corporation fully recognizes the benefits of Six Sigma and begins to contemplate expanding the initiative into other facets of the business.

Milestone T_9 is the hinge-point in time when the performance trajectory for an organization shifts from its historical slope to its future slope. In other words, this is when the scope of Six Sigma is materially enlarged, a period when it is taken across all the various functions of the corporation (sales, finance, legal, etc.). Finally, the last milestone is T_{10}, the point of renewal. This is a very critical period for a corporation. It is a time in which Six Sigma will die a natural death or be born again for another round of breakthroughs.

3.13 PROJECT ROADMAP

Project roadmaps are graphical illustrations of a project and its lifecycle. Examples of a roadmap are shown in Figure 3.3. Another example of a roadmap illustration is the case study outlines shown in Figures 3.4 and 3.5. These illustrations of case studies show the project name, the problem definition, the anticipated performance improvement, the tools used for the case, and the benefits realized.

3.14 PROJECT CHARTERS (GENERAL)

People are often skeptical of the benefits of Six Sigma; how can we counter this type of argument and show them the benefit? The project charter is often drafted by the project champion or master Black Belt and then refined by the Black Belt or Green Belt. Naturally, the leadership team must approve the charter. It is not uncommon for the charter to be revised several times over the life of the project is better understood and data become available. The typical elements of a project charter are illustrated in Figure 3.6.

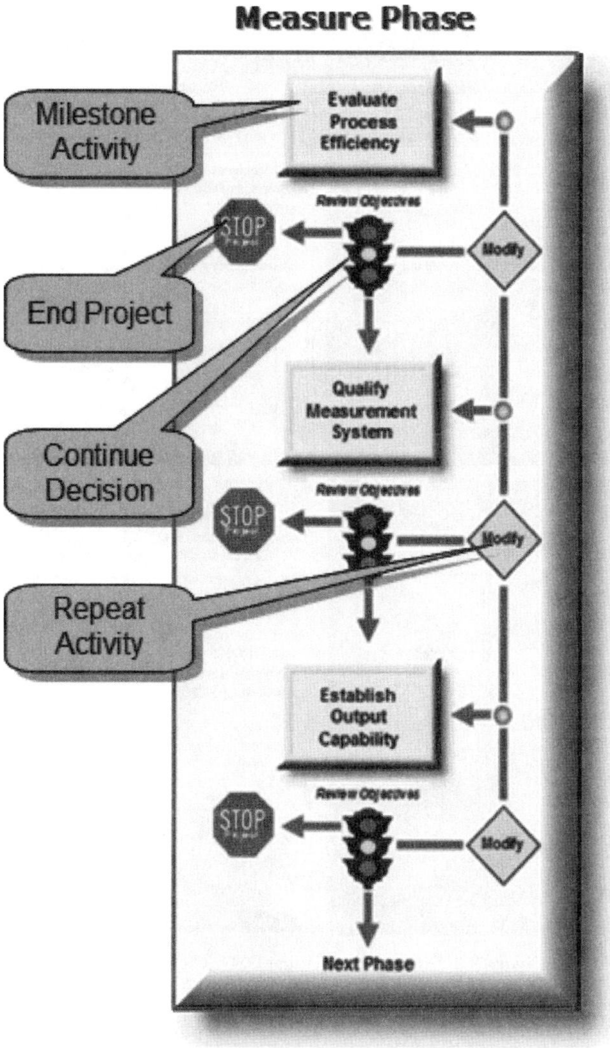

Measure Phase

Figure 3.3 Flowchart for a project roadmap.

Success factors, such as setting business goals, global deployment planning, Black Belt project selection criteria, and Six Sigma curriculum design, must be merged to form a common and uniform structure. If the guiding blueprint does not properly specify how the components fit together, it is likely that the final assembly will function in an asynchronous and dysfunctional manner, increasing the probability that the Six Sigma thrust will quickly stall and fade from corporate consciousness.

Having said this, the focus must be on the creation of a guiding curriculum, as it pertains to the key Six Sigma roles and responsibilities—senior champion, deployment champion, project champion, master Black Belt, process Black Belt, design Black Belt, and Green Belt.

A Six Sigma *deployment champion* is a hybrid of Henry Kissinger and Xena, the Warrior Princess, say Mike Carnell and Scott Shank. It may actually be easier to find the mutant offspring of these two leaders than it is to find the complete champion. Being the complete champion requires

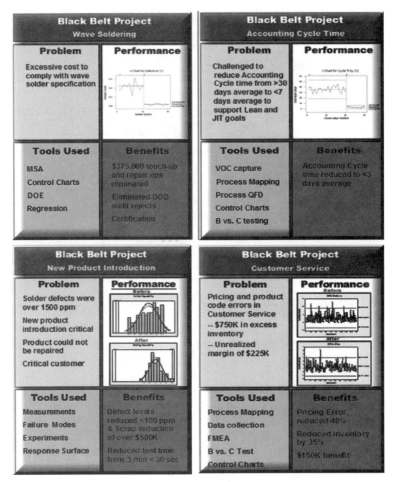

Figure 3.4   Summaries of some DMAIC case studies.

more than diplomatic and warrior skill sets. Deployment champions must be proficient on four more other areas:[4]

1. Business–operations interface
2. Project selection
3. Pace mediation
4. Results implementation

At a fundamental level, we can say that these roles constitute the backbone of what makes Six Sigma tick. Of course, these roles are specifically designed to support the appropriate identification, organization, rapid dissemination, and effectual execution of knowledge-intensive, application-focused improvement projects. Without careful planning and governed execution, it is easy for a Six Sigma project to experience mission creep. When this occurs, it is possible that the lines between success and failure will become blurred. In addition, mission creep can easily cause resources to become diluted, misguided, and ineffectually applied. If this occurs, the anticipated benefits must be reevaluated, adjusted downward, or pushed into the future to compensate for the escalating chaos. In turn, this drives the unnecessary modification of roles and responsibilities, thereby forcing various

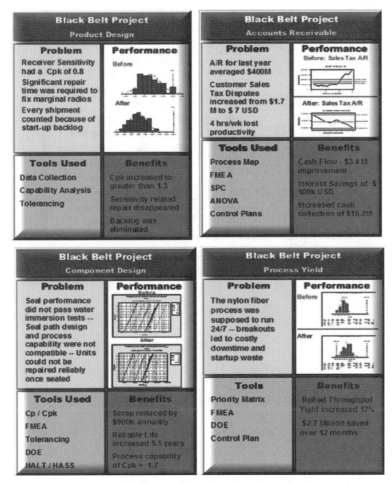

Figure 3.5   Summaries of some additional case studies. HALT = Highly accelerated life testing; HASS = Highly accelerated stress screening.

alterations to the curriculum. Obviously, this represents a never-ending cycle of reactionary change. To avoid the failure spiral, it is essential that the executive management pick the best-of-the-best to fill the Six Sigma roles. In addition, that management must fully empower each Six Sigma role with autonomous authority and dedicated resources. Finally, there must exist:

- Valid project selection criteria
- A time-tested process for governing project execution

It goes without saying that the interactive effect of these things is everything. Given the installation of high-quality people in all the Six Sigma roles, we must now recognize the 15 critical planning factors. These must be carefully considered prior to, during, and after the execution of a Six Sigma application project:

1. Scope
2. Depth
3. Duration

Figure 3.6   Typical elements of a Lean Sigma project charter (Mikel Harry).

4. Resources
5. Solidarity
6. Metrics
7. Benefits
8. Roles
9. Leadership
10. Objectives
11. Strategy
12. Reporting
13. Accounting
14. Knowledge
15. Closure

These factors must not only be considered individually but also be jointly deliberated and then made relative to the various people, products, processes, policies, and practices that are germane to the improvement situation. Only in this manner can their pragmatic implications and consequences be fully understood, appreciated, and acted upon. An example of a project charter is shown in Figure 3.7.

Another option for the Six Sigma project management is to create an Excel workbook partitioned into several separate pages labeled the same way as the sections in the project charter (see Fig. 3.8). The first sheet in the workbook (sheet 1) is labeled "dashboard" and is used to summarize the progress of the active projects. Sheet 2 is labeled "scope"; sheet 3 is labeled "depth," and so on until sheet 16 is labeled "closure." Each Black Belt responsible for a project populates each page with the appropriate data enabling the manager to view a project in play. Generally, a project dashboard could have four sections, subdividing the screen into adjustable sections:

Section 1: Project Activity of the Week.   This is where the project managers place the work activity scheduled for the upcoming week. Simply the name of the project and a color bar indicating green for "go", yellow for pending issues, and red for immediate action.

Section 2: Schedules.   This is where each manager places pending schedules for meeting and other interactive sessions and the objectives of the interaction for either data gathering, data mining, validation, scheduling or knowledge transfer.

Six Sigma		Project Charter	
General Project Information			
Project Mentor	O. C. de Hodgins - MBB	Type of Project	**Improve Process**
Process / Platform	Finance	Project Champion	James Dunn
Sub-Process	Reporting Finance Data	BB Candidate	Janet O'Hara
Alignment	Operation Share of Wallet	Project Owner	William Johnson
Project Start Date	02/23/2008	Project Completion Date	01/01/2009
Title of Project	Organization Monthly Financial Closing		
Statement of Problem	How to Improve Organization Monthly Financial Closing		
Objective of the Project	Mapping and Benchmarking Current Process, Standardize Financial Closing Across The Organization. Spread Best Practices.		

Project Metrics	Name	Equation	Baseline	Target
Primary Metric(s)	Financial Closing CycleTime Units = Number of Days.	Reporting Cycle Time = Entities Submission Time + Group Accounting Consolidation + Central Corrections	4-5 Days	3-5 Days Approx.
Secondary Metric(s)	Correctness of Financial Data According to Standards.	Group Accounting Scores (A_1-E)	Today We are in B	A_J

Type of Financial Benefit	Hard / Soft Savings	Annual Estimated Savings	$			
Project Scope	37 European Entities					
Defect Opportunity	Opportunity to Improve Cycle Time and Reporting Information Quality	Project Transferability	TBD			
Boundaries and Requirements	Scope: Mapping of Current European Financial Reporting Parameters : National Sales Organizations,Factories and HQs. Short Time Rationalization of Process Activities Best Practices Sharing Standardization of Processes.					
Locations	1	Austria	2	Belgium	3	Bulgaria	4	

Resource Requirements				
Resource	Project Role	Number FTE,s	Required Time	Other
Diego Sanchez	Champion	1	1%	
Rene Azul	Process Owner	1	5%	
Thomas Brown	Team Member	1	1%	
Luigi Santini	Team Member	1	2%	
Elizabeth Stephanie	Team Member	1	2%	
William Jones	Team Member	1	2%	
Donald Cody	Team Member	1	2%	
Adam Miller	Team Member	1	1%	
Martin Garry	Team Member	1	5%	
Peter Jordan	Project Leader	1	95%	

Required Signature Approvals			
Title	Signature		Date
Executive Champion			
Project Champion			
Process Leader			
Site / Process Manager			
Finance Representative			

Figure 3.7   Illustration showing example of a project charter.

Section 3: Project Meetings. This is where the managers prepare lists of Six Sigma–related meetings for the week.

Section 4: Financials. All projects are about money. This section could be a tracking tool showing the estimated saves or a balance against an approved budget.

Section 5: Hot Issues. Any roadblocks for other issues that hamper completion of a project task are addressed here.

3.15   SIX SIGMA PROJECTS

Six Sigma projects are key to any business generating return on investment and the organizational momentum for more return of investment. Six Sigma projects are the vehicles for reaching strategic goals. Major lessons learned for launching projects are as follows:

- Launching a few big projects ("save the world" projects) rather than many well-focused projects targeted at specific process elements

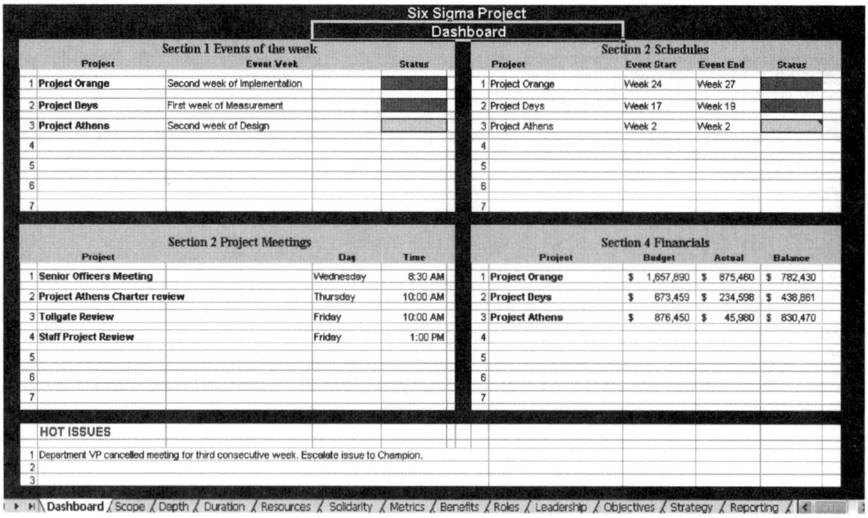

Figure 3.8   A Six Sigma project dashboard used in one organization.

- Defining projects without using best available data along with expert judgment
- Defining initial projects for training (Black Belt's first project) that are too large or complex

There is no single methodology that guarantees project success. However, starting a project correctly (e.g., good problem statement, objectives, planning and scope) provides great advantage to improve the likelihood of success.

Six Sigma projects focus on CTQs, CTCs, and CTDs, which offer the greatest financial and customer satisfaction leverage and address at least one element of the organizations key business objectives. Six Sigma projects in general, are properly scoped, planned, and have buy-in and commitment from senior management (champions) and all the organization (stakeholders).

Three main sources for identifying projects can be used are: (1) process/product baselining, (2) major business issues, and (3) operational needs. CTQ, cost, and delivery characteristics are clearly identified regardless of the original sources and the characteristics eventually flow down to relevant processes and their critical inputs (CTPs) as the project progresses. Regular reviews between the Champion, Master Black Belt and Black Belt are critical to ensuring barriers are identified and removed. Additionally, reviews will provide moral support for the Black Belt who must take on the role of "change Black Belt." Figure 3.9 illustrates this principle.

Change Black Belt.   Some organizations label their Black Belts as "change agents." Change agents inherit the responsibility for implementing projects with identified potential savings.

Often, projects are cross-functional, and will have a need to cut across different business areas. These cross-functional projects can sometimes be complex and therefore, difficult to navigate because of project manager, business champion, and business Black Belt lack of authority in other organization's business areas. The primary sources (also known as the "4Ps") of projects are as follows:

- *Process Quality Focus.* Overall, this is the best method to attack the root causes of defects and customer satisfaction issues. The key is identifying the processes that are critical to satisfaction and are operating at a low sigma level. This approach requires good cross-functional coordination since many of the processes will cross the traditional boundaries of departments.
- *Product Focus.* A different way to focus efforts is to identify the product family or system that is contributing most to poor customer satisfaction, but is also a strategically important product or

Six Sigma projects

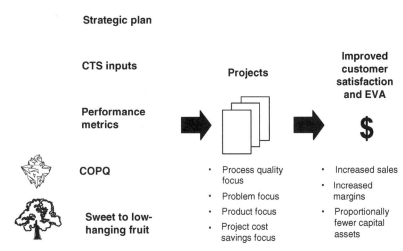

Figure 3.9 Schematic representation of the key concepts involved in Six Sigma projects.

service. Such an approach usually requires looking at a number of processes that feed into that product or service.

- *Project Cost Savings Focus.* While tracking the reductions in cost of poor quality of projects, recognize that, depending on the priorities of a business, cost savings might become the main focus. In this situation, projects are selected on how much management thinks that it will save in costs, not necessarily which projects will most affect the root drivers of process capability. One must recognize the limitations of this approach with respect to establishing new mindsets about quality.

- *Problem Focus.* Worst of all is a focus on the biggest problem or fire. It is short–sighted and seldom very focused, since it is not clear what process should be addressed.

Example 3.1   A Six Sigma champion and master Black Belt discuss the various sources of projects. "Project ideas can come from many sources," said the master Black Belt. They can come from high cost of poor-quality areas, from the CT tree, and from known problems. Other important sources include the strategic plan and areas identified by the Six Sigma performance metrics. "There are lots of problems to tackle," said the champion. "I agree," replies the master Black Belt, "However, we should not forget the four primary sources of projects (4Ps) and that the best projects are those that focus on process quality." The champion and master Black Belt conclude that regardless of which approach is adopted in project by project, the improvement of critical processes is key to reducing defects and improving economic value-added (EVA) performance. The approach to project selection can take one of the listed below, four forms. The focus of the project could use any of the forms listed below, combined or separate from each other (this depends on the project itself and also on the culture of the organization) (typically, projects will have some combination of these; a project might focus on a specific process, but also might focus initially on a certain key product line as shown in Fig. 3.10).

A Six Sigma project

1. Focuses on CTQs, CTCs, and CTDs that offer the greatest leverage
2. Addresses at least one element of the organizations key business objectives

Ways to focus

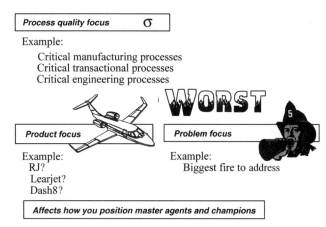

Figure 3.10 Illustration of two ways to focus: by process quality and by product.

3. Is properly scoped, planned, and has buy-in and commitment from Champions and stakeholders
4. Benefits from the rigorous four-phase breakthrough strategy (MAIC).

A *master agent* is someone who is results-oriented, a go-getter, a strong listener with the ability to facilitate change, particularly strong in people skills, has the ability to see downstream, and shows leadership and management skills.

Example 3.2 Projects should be selected wherever possible to maximize leverage. If we analyze manufacturing as an example, a manager responsible for the Dash 8 assembly line has 200 people on structural assembly. Most of these 200 people locate parts, drill holes, and install fasteners. In looking at statistics from quality, the manager realized that 70% of the defects found by inspection, were related to hole drilling. Therefore, by somehow improving and controlling the drilling process, this manager would be able to leverage the benefits to most of the employees and eliminate 70% of the defects. This would improve quality, cost (hours, scrap), and cycle time, not to mention leveraging to other areas, programs, and sites. Leverage is the subsidy of a specific project on multiple processes improving *quality*, *cost,* and *delivery*, delivering simultaneous benefits to internal and/or external customer's CTQs, CTCs, and CTDs. Figure 3.11 illustrates a simple approach to project selection.

Project Selection Approaches. A number of approaches that can be adopted when selecting projects. The preferred method is related to the CT tree and product/process baselining, which will naturally become the primary source as the organization learns and business metrics become more developed. Initially, customer focus (internal and external) and COPQ will drive the selection from top and bottom. CTQ, cost, and delivery characteristics are clearly identified, regardless of the original source, and these characteristics eventually flow down to relevant processes and their critical inputs as the project progresses.

In any case, it is always best to make good use of credible data during project selection. Do not rely on hearsay or folklore for this process. "Should priority be given to those projects that address CTCs, CTDs, and CTQs simultaneously to maximize leverage?" an organization's champion asks. Projects that address all three CTCs, CTDs, and CTQs should be given special consideration. However, common business sense should be used in evaluating other areas where major business objectives may

Project selection approaches

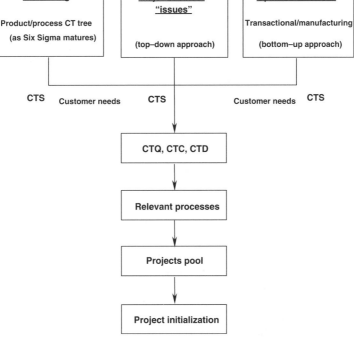

Figure 3.11 Flowchart of general approaches project selection.

be effectively addressed with a focus on one or two of the CTs. A general approach is illustrated in the Figure 3.11.

Example 3.3   A Six Sigma champion asks: "How do we ensure the selection of a good Six Sigma project?" "Projects must focus on our business priorities and commitments, and the project selection process must begin by identifying these," explained the master Black Belt. "It is important to also consider characteristics that are critical to customer satisfaction (CTs) at this point. Through this process, each champion can identify the priorities for their functional area(s). You may have already completed this task through the annual performance management process (PMP). This will enable you to define your deliverables—for example, assemblies, parts, drawings, documentation, etc.," continued the master Black Belt. "Then you can define the processes that produce these deliverables. This is important because by focusing on process improvement, we can maximize the leverage of projects across functions and sites, which is why it is important to identify. One can do this by grouping processes under the (for example), BMS/BES/Picasso/. . . processes. On completion of this activity, we can then prioritize processes for improvement on the basis of business benefits and identify the process owner who will be responsible for the improved process. Remember that core processes may be affected by these projects."

A process owner is someone who is responsible for the output of the process. It may be you if the process crosses functional boundaries, or one of your managers if the process resides within a single function. It is important that the process owners works with the champion and the master Black Belt to baseline the process and identify project opportunities. We must verify that someone else before has not undertaken the project. Figure 3.12 will assist in the project selection process. During the project

Project selection process

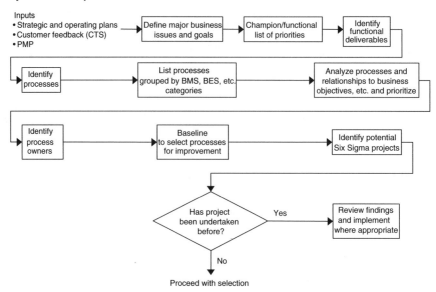

Figure 3.12   Flowchart showing simple approach for project selection process.

selection process, we must identify any resource limitations that could affect our ability to complete the project. For projects that require *information technology* (IT) resources, two aspects must be considered: (1) if the proposed project uses software that is not company-compliant, it should be rejected; and (2) if IT resources are not available, it should also be rejected. Where resources are available and the benefit of the project is likely to outweigh the required cost and effort, then the project should be given further consideration. One key step in project selection and definition of the project scope. Projects should be scoped such that they can be completed within the training period (for new Black Belts) or within 4–6 months (for existing Black Belts). Proposed projects that could take much longer than this (more strategic projects) should be divided into smaller projects to facilitate earlier completion and provide more immediate benefits.

For the Black Belt's first project, data must be available to facilitate initial measurement and analysis before the measure phase begins. The master Black Belt and champion must review the availability and adequacy of the data before and assigning the project to a new Black Belt. For experienced Black Belts, the Black Belt in conjunction with a master Black Belt can undertake the data collection effort before project launch. Indeed, for some transactional processes, data collection could be a miniproject for a Black Belt.

The final stage of project selection and definition is the completion of the *project authorization* form by the champion. The authorization form is a formal document that defines the purpose and goals of the project, as well as the key people who will be involved in the improvement activity. In particular, the nature of the problem or opportunity should be clearly defined, and project goals and business benefits should be specifically stated. The current metrics in use and the improvement targets for each metric must also be defined, as well as the estimated project costs and savings.

(*Note*: Improvement targets for new projects taking more than 6 months to achieve (e.g., improvement from 3σ to 6σ should be divided into separate project phases.)

On completion of the Authorization form, it must be signed by the champion and forwarded to the master Black Belt for review and approval. If the project is adequately defined and properly scoped, the master Black Belt will authorize the project and forward the form to the local IT director and site controller for review and approval. Finally, the business unit head/senior executive manager should review the project and authorize its launch. This final approval ensures the project is clearly linked to the business, and that its completion will bring clear benefits to the organization. The authorization form would be the basis for the project charter (Fig. 3.7).

Key Success Factors. A successful project is one that results in process performance reaching its targeted level, with the key stakeholders, including customers being satisfied with the results. If a project has been properly scoped with the required amount of data collected to launch the project, then there should be no scope changes. However when there is a need to rescope, these changes should be minimized. Any process improvement must be maintained for the project to be considered successful and the costs of the project should be contained within budgeted levels. Finally, recognizing that day-to-day work needs to continue as our projects proceed, disruption due to projects should be kept to a minimum bearing in mind the importance of completing projects on time.

Summary

- Projects are the vehicles for continuous incremental and radical quality, cost, and delivery improvements and consequently financial return.
- The nature of a Six Sigma project is that it offers the greatest financial or customer satisfaction leverage.
- Three main sources of projects can be used: baselining, major business issues, and operational needs.
- Generally, a selected project must address at least one element of the organization's key business objectives, make good business sense, and address key customer's needs.
- Properly scoped and successful Six Sigma projects achieve targeted process objectives with customer satisfaction and overall benefits to the business.
- The five project planning questions can help the Black Belt focus on the correct areas with the required tools for the most efficient resolution.
- A formal project authorization process allows us to verify that all elements of the project have been addressed allowing a good project start.

3.16   PROJECT SUMMARY

In essence, Six Sigma uses a divide-and-conquer strategy focused on effectiveness, whereas Lean practices pursues efficiency through continuous improvement. Thus Lean Sigma projects are selected by a well-defined process, not according to a vague notion of improving everything we do forever, followed up by a sporadic and disconnected set of actions. Rather, the practice of Lean Sigma divides the proverbial *business pie* into several understandable compartments. In turn, these compartments of improvement opportunity are portioned into application projects. By design, these projects collectively impact all levels of the business enterprise. It should now be clear why many noted industry (and governmental) leaders have taken up Lean Sigma; it allows an executive to reach the control function of an enterprise through highly focused, data-centric projects. The collective power of such projects unites management (and the workforce) in a pragmatic way. Lean Sigma indirectly forces everyone to synchronously raise their company above the limitations commonly associated with schizophrenic, disconnected, firefighting types of improvement activities. Through a circumspect pattern of interrogation and action, the application of Lean Sigma arranges the alphabet soup of improvement initiatives (i.e., the various acronyms associated with corporate programs) into coherent words, sentences, and paragraphs. In doing so, Lean Sigma takes the idea of quality to a more executable level that is more closely aligned with the aims of business. It elevates quality to the boardroom, demanding the kind of executive attention and action that cascades to the lowest reaches of an enterprise.

GLOSSARY

business management system (BMS)   A system that forms the basis for controlling, managing, and improving the business operation. In a company where mass production consists of sequences of short runs, management influences the quality of the end product. The importance of project management in quality is shown by the fact that ISO 9000 requires firms seeking ISO certification to have key project management processes in place.

BMS/BES/Picasso/...   Example of grouping processes under the business management system—for example, an organization could enter the projects as BSM/business economic system/Picasso as example of the assigned name of the project.

change Black Belt   Some organizations label their Black Belts as "change agents." Change agents inherit the responsibility for implementing projects with identified potential savings.

COPQ   "Cost of poor quality"—Dollar amount spent on non-value-added activities while producing a good service—costs involved in activities or situations such as finding and reworking defects, failure to meet customer expectations, missed opportunity for increasing efficiency, loss of market share, increase in process cycle time, labor for ordering replacing parts, cost associated with rework.

CT   "Critical to," a term that can be expanded to express concepts such as "critical to satisfaction" (CTS), "critical to quality" (CTQ), "critical to delivery" (CTD), "critical to cost" (CTC), and "critical to process" (CTP). The CT matrix is the tool that links the process or CT tree (columns of the matrix) and the product or CTY tree (rows).

CTC   "Critical to cost" a parameter for product, service, and/or transactional characteristics that significantly influence one or more CTSs in terms of cost.

CTD   "Critical to delivery"—a parameter for product, service, and/or transactional characteristics that significantly influence one or more CTSs in terms of delivery.

CTP   "Critical to process"—characteristics are process parameters that significantly influence a CTQ, CTD, and/or CTC. In the equation $Y = f(x_1, x_2, \ldots, x_n)$, the CTQ, CTD, and CTC represent the dependent variable (Y), and the CTPs represent the independent variables (Xs).

CTQ   "Critical to quality"—quality from a customer's perspective. CTQs have two characteristics: (1) important to customer and (2) measurable.

CTS   "Critical to Satisfaction"—characteristics that are the expression of the customer's vital needs. These needs are translated into critical requirements for quality, delivery, and cost.

customer   Person or organization that will use the project's product or service or result.

date   Term representing the day, month, and year of a calendar; in some instances, the time of the day.

deliverable   Output/input—any unique and verifiable product, result, or capability to perform a service that must be produced to complete a process, phase, or project. This term is often used more narrowly in reference to an external deliverable, a deliverable that is subject to approval by the project sponsor or customer.

deployment champion   A Six Sigma hybrid of Henry Kissinger and Xena, the Warrior Princess.

DMAIC cycle   The DMAIC cycle provides a systematic five-phase approach: (1) *define*—the target of this phase is a clear description of the problem as well as identification of the target and scope for the improvement area or improvement initiative; (2) *measure* –– the target of this phase is data collection according to customer requirements, (3) *analyze* –– the target of this phase is identification and verification of assumed root causes, (4) *improve*—the target of this phase is to find and implement solutions that address the root causes, and (5) *control*—the target of this phase is to track whether the new process has been sustainably implemented.

earned value (EV)   The value of work performed expressed in terms of the approved budget assigned to that work for a schedule activity or work breakdown structure (WBS) component. In project management referred to as the budgeted cost of work performed (BCWP).

earned value management (EVM)   Management methodology for integrating scope, schedule, and resources for objectively measuring project performance and progress.

effort   Number of labor units required to complete a schedule activity or work breakdown structure component. Usually expressed as staff hours, staff days, or staff weeks. Contrast with *duration*.

entitlement   The theoretical view of how well a process should work.

EVA performance   A metric that estimates a particular type of economic profit, which states that to earn genuine profits, it is necessary for a company to earn sufficient profit to cover both the firm's operating costs and the cost of capital.

Execute   Directing, managing, performing, and accomplishing the project work, providing the deliverables, and providing work performance information.

Lean Sigma projects   Lean Sigma projects focus on business value, plain and simple. A valid project must yield some form of tangible value for the sponsoring organization, for example, hard cost savings, yield improvement, defect reduction, cycle time reduction, inventory reduction, headcount reduction, and increased customer satisfaction. All such value-based improvements can and should be translated into saving money.

leverage   If both operating and financial leverage allow us to magnify our returns, then we will get maximum leverage through their combined leverage. Operating leverage affects primarily the asset and operating expense structure of the firm, while financial leverage affects the debt–equity mix. From an income statement viewpoint, operating leverage determines return from operations, while financial leverage determines how the "fruits of labor" will be divided between debtholders (in the form of payments of interest and principal on the debt) and stockholders (in the form of dividends and retained earnings).

management tollgate   A tollgate is simply a set of success criteria that must be fully satisfied before a project is allowed to continue on the next scheduled phase of execution.

master agent   Someone who is results-oriented, a go-getter, a strong listener with the ability to facilitate change, particularly strong in people skills, who has the ability to see downstream, and shows leadership and management skills.

PMP process   Performance management program—helps employees provide better service through the organization of work and the delegation of responsibility using a system requiring planning and assessment of work by supervisors and each respective subordinate who manually establishes work standards and performance measures.

Process   Combination of people, equipment, materials, methods, and environment that produce an output—in general a product, transaction, or service.

process owner   Someone who is responsible for the successful management of an improvement area; a person who coordinates the various work activities at all levels of a process. A process owner has the authority or ability to make changes in the process as required and manages the entire process cycle to ensure performance effectiveness.

process performance management (PMP)   The overseeing of process instances to ensure their quality and timeliness. Can also include proactive and reactive actions to ensure good results.

process quality   Value of percentage defective or of defects per hundred units in product from a given process.

process reengineering   Strategy directed toward major rethinking and restructuring of a process.

project charter   Summarizes the project and the official authorization to proceed. A management tool where project responsibility and accountability are formally declared.

project financials   Project success can often be measured using *defects per million opportunities*, *net cost savings*, *cost of poor quality*, and *capacity*.

project milestone   Deliverables are a form of product for upper management.

project scope   Boundaries of a project that define what is included and excluded from the work in a project charter.

project scoping   Defining a project to complete the proposed statement of work defined in the project authorization form within 6 months.

project team   Manages the work of a project.

project teams   Selected groups of "subject matter experts" that work together to solve business challenges.

Six Sigma value-based projects   These projects are selected, prioritized, configured, and coordinated such that taken together significantly impact higher-level operational and business performance metrics.

quality factor   Factor that is continuous. For example, the solder paste in the printed-circuit board is with a temperature factor ranging from 90°C to 130°C.

quality function deployment (QFD)   Technique used for voice of customer in the design of a product.

scope   Sum of products, services, and results to be provided as a project.

service   Useful work performed that does not produce a tangible product or result, such as performing any of the business functions supporting production or distribution. Contrast with *product* and *result*.

targets of opportunity   Projects selected by production managers under pressure to make budget reductions, resolve specific quality problems, or improve process flow.

user   A person or organization that will use the project's product or service.

validation   Technique of evaluating a component or product during or at the end of a phase or project to ensure that it complies with the specified requirements.

verification   Technique of evaluating a component or product at the end of a phase or project to ensure or confirm that it satisfies the conditions imposed. Contrast with **validation**.

VOB   Voice of business—requirements of the business as detailed in the operating agenda or financial goals of the business,

VOC   Voice of customer—requirements of the customer as determined through various tools.

REFERENCES

1. PMPBOK® Guide: A Guide to the Project Management Body of Knowledge, third edition, Project Management Institute (PMI), Global Standard, ANSI/PMI 99-001-2004; ISBN 10; 1-930699-45-X (Paperback).

2. S⁴/IEE; "Implementing Six Sigma: Smater solution – Statistical Methods," 2nd edition; hardcover.

3. M.J. Harry and R. Schroeder,"*Six Sigma*", - The Breakthrough Management Strategy Revolutionizing the World's Top Corporations – M. Harry and R. Schroeder – Doubleday, 2000; ISBN 0-385-49437-8.

4. M. Carnell, and S. Shank,"*The Champion's Role in Successful Six Sigma Deployments*" isixsigma article, website publication.

4 Lean Practices

4.1 OVERVIEW

Work is a never-ending journey. Lean systems make businesses more competitive and therefore more likely to survive. There is a potential for tremendous improvement in manufacturing and in office settings. Lean seeks to eliminate waste; it is not an initiative but a commitment to continuous effort to eliminate non-value-added activities. In this chapter the practitioner will learn how world-class organizations operate by cost reduction principles to meet quality, cost, and delivery requirements. Eliminating waste (non-value-added activities) from the customer's value stream is controlled by the workers—not the reverse.

4.2 INTRODUCTION

Today, many corporations are adopting Lean Sigma as a strategic initiative to realize a breakthrough in one or more aspects of their business performance. The leaders of such progressive corporations are looking beyond common sense and historical convention to understand and control the underlying determinants of business success. The key stakeholders of contemporary corporations are no longer satisfied with incremental improvements of a symptomatic nature, but now seek quantum changes to significantly alter their historical trends. They recognize that their business operational and process problems are interrelated in a domino-like progression of cause–effect relationships. Thus, many now understand that such problems cannot be solved by compartmentalized incremental improvements that do not demonstrate immediate and visible economic benefits. Such noted corporations as General Electric, Ford, Sony, and DuPont have deeply probed their organizations for intellectual leverage and then exploited these ideas to encourage and activate the full power of Six Sigma quality, in everything they do, in every corner of their businesses.

As a first step, these corporations significantly raised their performance aims and optimized their use of resources by reexamining and reconfiguring their knowledge, capital, business structures, operating systems, and key processes. Today, *Lean Sigma* and the *Breakthrough Strategy* have become the proverbial rubber that met the road. Each of these corporations has learned that tweaking operations and employing classical quality improvement methods led only to transient improvements.

Many of today's top executives have questioned the beneficial effect and unifying capability of continuous improvement programs, total quality management, initiatives and statistical process control mandates. They concluded that TQM and similar programs resulted in shortlived incremental extensions that seldom improved actual business performance or customer perception of their business. While such efforts have been successfully applied to sporadic problems and have generally supported the image of a quality-conscious organization, they have proved of little value in the pragmatic improvement of business quality.

Practitioner's Guide for Statistics and Lean Six Sigma for Process Improvements. By Mikel J. Harry, Prem S. Mann,
Ofelia C. de Hodgins, Christopher J. Lacke, and Richard Hulbert
Copyright © 2010 John Wiley & Sons, Inc.

While this does not mean that classic quality approaches lack merit, it does mean that their use often leads to disconnected improvements that rarely improve corporate profitability. If an improvement is "real," an immediate economic benefit should be visible in the labor, material, and overhead cost structure of the company. In this sense, Six Sigma is a distinct break from the past as it demands the direct involvement of top management, focuses organization resources on business performance breakthroughs, and contributes to business quality. Simply stated, a relatively sudden and quantum improvement in business performance requires high-quality outcomes at all levels of the corporation, not the reverse as quality professionals have been trained to believe and advocate.

4.3   THE IDEA OF LEAN THINKING

Corporate business leaders drive change. They spend much of their time improving the change process after it has been set in motion. If support functions (like many of the quality organizations in existence today) insist on retaining archaic knowledge, intellectual antiquities, outdated ideas, counter-productive beliefs, and competing aims, they will be excluded from the overall process of change.

In this context many senior business leaders have asked what it really means to "do" Six Sigma. From the executive perspective, Six Sigma entails the institutionalization of those practices that ultimately ensure realization of quantifiable improvements to business operations and processes of the corporation. At the business level, Six Sigma requires definition and realization of certain stretch goals that drive key measures of business success such as customer satisfaction, margin growth, market share, and after-tax profit. At the operations level, Six Sigma realizes improvements at an annualized rate of 78% per year for such metrics as defects per unit, and delivery delinquencies. At the process level, Six Sigma corresponds to no more that 3.4 errors per million opportunities for error over many process cycles. This particular process-level goal may be expressed in other equivalent measures of quality. Interconnections and relationships between the various business operations and processes must be analytically validated to assure full realization of the global aims of Six Sigma. From this holistic perspective, it is the interplay between the key components of Six Sigma that produce quantum change. Six Sigma evolution and our knowledge of its methods have led to a revolution in practice as it has passed through several distinct but interdependent phases: exploration, approbation, initialization, installation, realization, and continuation. The practitioner will understand that as a corporation progresses toward institutionalization of Six Sigma, it passes through the aforementioned phases. In this sense, the path of Six Sigma in a large-scale corporation resembles the natural human life path. The corporation discovers, learns, grows, adapts, and develops as it encounters events and experiences within its operating environment. Over time the corporation learns how to leverage its knowledge and beneficially alter its environment. In short, it is the force of knowledge that helps shape the life path of a Six Sigma initiative and ultimately the destiny of a corporation. Just as science analyzes knowledge derived from observation and experiment to understand nature, quality systematizes the way an organization progresses toward what is "good." Quality is no longer grounded in the subjective skills of the artisan. The notion of quality has entered a new age by integrating, practicing and embodying the tenets of science.

Together with engineering and production, quality has evolved into a science that must be organized and optimized. As an organization evolved, the initial concern about the quality of products expanded to include the quality of cost, time, actions, and tasks. Fredrick Taylor,[1] well known for his innovative contributions to the field of time and motion measurements, focused principally on efficient use of time,[2] money, and resources. For an industrial business to be successful, it must deliver products at a lower cost and somewhat higher quality to keep pace with increasing customer and shareholder demand.

Lean thinking is directed toward eliminating waste in the production of a product or delivery of a service. However, the idea of systematization and measurement shows that value-added (VA) activities extend beyond manufacturing production. Thus, the fundamental concepts can be applied to financial transactions and delivery of services. Lean thinking is driven by the precise definition of value-added activities and non-value-added activities. In this chapter, we will describe key lean principles, the five laws of Lean Sigma, and the methods used to realize lean thinking.

Key Lean Principles. Key Lean principles ensure that the correct products reach the correct place at the appointed time in the requested quantity while minimizing waste and maintaining flexibility to adapt to change. It is a holistic, comprehensive, enterprisewide program designed to be integrated into the organization's core strategy. Experts in this field believe that the philosophy-based lean manufacturing strategy is the most effective way to launch and sustain lean activities. The "Toyota Way," popularized by Dr. Jeffrey Liker's[3] book of the same name, emphasizes the need for the right kind of environment in which to grow and support lean thinking. These lean thinking principles are as follows:

- Perfect first-time quality—quest for zero defects, revealing and solving problems at the source.
- Waste minimization—eliminate all activities that do not add value and maximize the use of scarce resources such as capital, people, and land.
- Continuous improvement—reduce costs, improve quality, and increase productivity and information sharing.
- Pull processing—pull products from consumer demand, not push them from the production side.
- Flexibility—produce different mixes or greater diversity of products quickly at lower production volumes without sacrificing efficiency.
- Building—maintain long-term relationships with suppliers by collaboratively sharing risk, cost, and information.

Five Laws of Lean Sigma. The five laws of Lean Sigma have been formulated to guide improvement efforts. The laws combine the key ideas of both Six Sigma and Lean Sigma:

Law 0. *The law of the market*—customer critical to quality defines quality and is the highest priority for improvement, followed by return on invested capital (ROIC) and net present value. It is called the *zeroth law* as it is the basis for the other laws.

Law 1. *The law of flexibility*—the velocity of any process is proportional to the flexibility of the process.

Law 2. *The law of focus*—20% of the activities in a process make up 80% of the delay (related to the Pareto principle).

Law 3. *The law of velocity*—the velocity of any process is inversely proportional to the amount of work-in-process (WIP). This is also called "Little's law".[2] More work in progress (unfinished tasks) slows the speed of progress, due to various work-related issues.

Law 4. *The fourth law*—the complexity of the service or product adds more non-value-added costs and WIP than either poor quality (low sigma) or slow speed (nonlean process problems).

These laws complement each other and concentrate our efforts to reduce human labor, inventory, and product development time.

Figure 4.1 illustrates simple summary of lean thinking:

Defining Value. Value is determined when a customer buys the right product at the right price. Lean companies continually work to create products that are "right." The value of any product or service is thus defined by the customer and not by the supply chain or manufacturer. Thinking lean may redefine how improvements are made. The first step in defining improvements is the thought process.

Lean process

Pull	Flow/kanban	Total productive maintenance
Mistakeproofing	Score (kaizen)	Quick changeover
Standardized work	Batch reduction	Waste reduction
5S	Visual management	Cellular layout

Value stream mapping

Figure 4.1   A schematic summary of lean thinking.

4.4   THEORY OF CONSTRAINTS (TOC)

The *theory of constraints* (TOC) was developed by Dr. Eli Goldratt[4] and described in the book *The Goal*. TOC is an overall management philosophy that aims for excellence and continued system improvement. In a for-profit business, the TOC goal becomes one of improving profits now and in the future. According to TOC, every profit-seeking organization must have at least one constraint that prevents the system from achieving higher performance relative to its goals. The essence of this theory is that all processes (P) are composed of many activities or process steps. Each of these individual processes has its "own capacity" for transforming individual inputs into outputs. Therefore, the total process (made up of all individual processes) can produce only as fast as its slowest process step.

Figure 4.2 is a graphical representation of inputs–process–output for a model to reduce the costs and improve operational efficiencies in the process of acquisition and distribution of digital images to the news media.

To increase the output of the process, the slowest individual process step must be identified and fixed. This slowest individual process or activity is known as a "bottleneck" or in physics as a *constraint*. From this fundamental concept came the name *theory of constraints*. Dr. Goldratt refers to fixing the individual slow process as elevating the constraint or lifting the workload of the constraint through alternative methods. Fixing means finding alternative paths to perform an activity by

Figure 4.2   Process of acquisition and distribution of digital images.

providing greater capacity to the slowest process step to increase its productivity. Goldratt emphasizes the following principles to reach the business goal:

- Increase throughput (total output of the process)
- Decrease expenses
- Decrease inventories

By understanding these concepts, you can improve your organizational effectiveness and productivity, free up capital driven by excess inventory, and reduce customer frustration and organizational anxiety. Both Lean and regular Sigma drive process transformation, and identifying root causes of process problems. To help identify the root causes of these process problems, Taiichi Ohno[8] of Toyota identified seven types of waste that can cripple any process. (see Section 4.15):

4.5   LEAN CONCEPT

The Lean concept concentrates on rapid process flow (RPF). Process waste is identified and removed at each stage of the process. In creating a map of the process flow, team members physically walk through the process sequence and record the distances, work times, idle times, and processing times of non-value-added activities. Non-value-added distances are reduced by physically changing the plant layout. Manufacturing floorplans should be designed in cells rather than functional groupings so that distance traveled during the process is minimized. The objective is continuous flow. This requires producing and moving one item or a small and consistent batch of items through a series of processing steps as continuously as possible, with each process step making only what is requested by the next step. This is also called "one-piece flow," "single-piece flow," or "make one, move one."

4.6   VALUE-ADDED VERSUS NON-VALUE-ADDED ACTIVITIES

Value-added activity transforms the product, service, or transaction in a way meaningful to the customer. In other words, value-added activities are those actions that are valuable from the external customer's viewpoint.

Non-value-added activity is "we do this because the boss wants it." This does not mean that this task is adding value to the customer.

And... always remember: The critical place to begin any improvement effort is with a clear specification of the value (see definition of **value** in the Glossary for this chapter) of a product as perceived by the end customer. Customer specifications are occasionally illustrated in process diagrams. Typical process data may include cycle time, changeover time, and uptime, and production batch sizes, number of operators, number of product variations, pack size, working time, and scrap rate. The process diagram box stops wherever processes are disconnected and material flow stops. (See Chapter 6 for process mapping.)

4.7   WHY COMPANIES THINK LEAN

Lean thinking applies to any enterprise from meat packing to complex equipment assembly. Lean provides us with a structured work environment with defined processes and tools enabling businesses to concentrate on profits. The enterprise must change to achieve a Lean operating environment. Lean is a *way of life;* to achieve the expected breakthroughs, we must think about our processes in an entirely new way. Lean philosophy emphasizes operational efficiency, elimination of non-value-added activity, continuous improvement, and a rebirth of employee value. Lean principles focus on streamlining the flow of production material throughout the enterprise. Lean practices support lean principles by reducing production variation. The best measure for tracking lean progress is total cycle time (the time materials spend in the production system). Shorter cycle times reduce production costs. The same factors that reduce cycle times also decrease production costs.

Lean is based on a customer-focused-view containing six steps. We have added an additional step, thought process mapping, to make projects easier to sell internally and to gain executive management support. These "six plus one" steps ease the introduction of lean concepts into an operating environment. The producer usually evaluates the lean thinking steps shown below as each sequential process step is reviewed and verified:

- Value
- Thought
- Value stream
- Flow
- Pull
- Perfection
- Replication

The value of any product, service or transaction is defined by *customer needs* and not by the non-value-added (NVA) activities of the supplier, producer, or provider. The customer is willing to pay for operations by producers or their suppliers that *transform the* product in a way that is *meaningful to the customer*. Customers do not wish to pay for waste. Companies and organizations that fail to follow this fundamental philosophy are not lean.

4.8   VISUAL CONTROLS—VISUAL FACTORY

Visual controls are signs, information displays, layouts, material storage and handling tools, color-coding, and mistakeproofing devices. These controls ensure that there is a place for everything and everything is in its place. The visual control system makes product flow, operations standards, schedules, and problems instantly identifiable to even the casual observer.

Every step in a manufacturing process is labeled with custom labels and signage to create a workplace free of errors. A factory is visually managed when systems are in place that enables anyone to assess the current status of an operation or process at a glance. Both the American National Standards Institute (ANSI) and the Occupational Safety Hazard Administration (OSHA) have recommended standards for warning signs. Their regulations should be consulted before making your facility's signs.

4.9   THE IDEA OF PULL (KANBAN)

The idea of "pull" is directed toward removing excess capacity (inventory) or increasing the demand rate. The essence of a pull system is that work enters a process at the same rate at which it exits. Since the "exit rate" is controlled by customer demand, the process is paced to meet customers' needs.

Kanban is frequently known as a "pull" system, as everything is pulled in response to demand. From the production perspective, a pull system makes products at the same rate as they are being sold. From the supply perspective, a pull system ensures that resources enter the production process at the same rate they are being used—a just-in-time (JIT) production system.

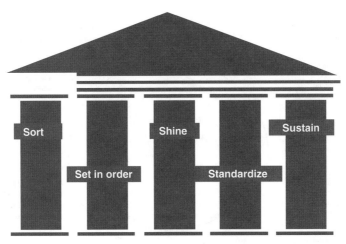

Figure 4.3 Illustration showing where human labor is separated from machines.

4.10 5S–6S APPROACH

The 5S approach is a systematic method to organize, order, clean, and standardize a workplace—and to keep it that way. The objective of 5S philosophy is to eliminate NVA activities, focus on root causes, not just symptoms, 5S needs to use total participation to check processes. 5S provides structure to sustain system. 5S is the basis for lean and the foundation for a disciplined method to achieve continuous improvement, not only in the workplace but lifestyle as well. This process includes everyone—no exceptions. 5S is part of lean and is not a standalone method; 5S denotes the Japanese words, seiri, seiton, seison, seiketsu, and shitsuke (sort, straighten, shine, standardize, and sustain). If diligently pursued, their use will change every aspect of one's life. Figure 4.3 illustrates the main columns of the 5S method.

The 5S method contains visual displays that convey important information, but do not control what people or machines do; they are passive in nature. Visual controls convey information so that activities are performed in accordance with standards.

Seiri (*sort*)—sorting the contents of the workplace and removing unnecessary items. This is a positive action to identify and eliminate unnecessary items (tasks) from the workplace.

Seiton (*straighten*)—placing or arranging the necessary items in their place and providing easy access and clear identification.

Seiso (*shine*)—cleaning everything; keeping it clean and using the cleaning process to inspect the workplace and equipment for defects.

Seiketsu (*Standardize*)—creating visual controls and guidelines to ensure that the workplace and space are organized, orderly, and clean. In other words, maintaining the seiso or shine.

Shitsuke (*Sustain*)—instituting training and discipline to ensure that everyone follows 5S standards.

Many practitioners believe that 5S is a good tool to achieve a lean work environment. Trust us; it is much more than a tool. 5S is not just another methodology for reducing waste; it is a state of mind, it is a lifestyle that when practiced faithfully can change one's life. 5S is a logical process that uses visual management techniques to create a work environment that is clean, ordered, and disciplined—And because the 5S method includes visual management, the abnormalities become obvious, and therefore, their resolution becomes straightforward.

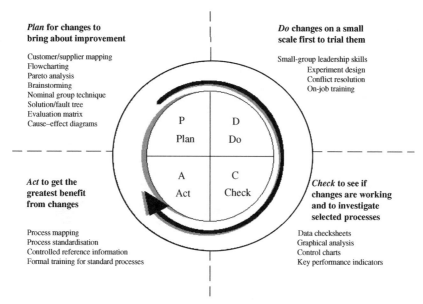

Plan **for changes to bring about improvement**

Customer/supplier mapping
Flowcharting
Pareto analysis
Brainstorming
Nominal group technique
Solution/fault tree
Evaluation matrix
Cause–effect diagrams

Do **changes on a small scale first to trial them**

Small-group leadership skills
Experiment design
Conflict resolution
On-job training

Act **to get the greatest benefit from changes**

Process mapping
Process standardisation
Controlled reference information
Formal training for standard processes

Check **to see if changes are working and to investigate selected processes**

Data checksheets
Graphical analysis
Control charts
Key performance indicators

P Plan   D Do   A Act   C Check

Figure 4.4   The Deming cycle. (*Source*: W. E. Deming, *Out of the Crisis*, MIT press, 1989).

Visual management concepts and tools make possible to tell how a process is operating by simply observing the process operation. Through the use of visual (and audible) feedback mechanisms such as cards, color-coding schemes, speakers, and layout design. The ideal situation is for the processes to be "talking" to us so that we can easily discern between "good" conditions and "bad" conditions. Nonnormal situations—abnormalities—are made obvious, and are eliminated through the use of countermeasures.

The *Deming cycle*, or *plan do check act (PDCA) cycle*, is a continuous quality mprovement model consisting of a logical sequence of four repetitive steps for continuous improvement and learning: plan, do, (check), and act. It is an effective model for excellence and one way to initiate 5S thinking. Figure 4.4 describes the Deming cycle. The Deming cycle phases can be outlined as follows:

Phase 1: Seiri (*sort*)—sort through the contents of a workplace and remove any unnecessary items (tasks).

1. Search your workplace with your colleagues and identify those items that are not necessary to deliver quality products to your customers.
2. During the identification process, pose the following questions:
    - Is the item needed?
    - How is it used and how often?
    - Who is responsible for the item?
3. Find an area to hold these items and segregate any items found to be unnecessary or have suspect value.
4. A departmental manager has to evaluate your choices.
5. Follow your firm's policy for disposal (sell it, move it, or discard it).
6. Celebrate success.

Phase 2: Seiton (straighten or systematize)—put all necessary remaining items in their place and focus on efficiency. Place items so the user has easy access.

1. Ensure that unnecessary items do not return.
2. With your colleague's participation, decide which things go where for efficient operation.
3. Ask these questions:

4. Do I need that item for my job?

5. How often is it needed?

6. Where should I locate this item?

7. Who is responsible for the item?

8. Are there other unnecessary items that muddle my space?

9. Use the five "why"s[5] to decide where to place the items.

10. Explain to the workplace the positioning of the item.

11. Label each locker and drawer as to what is inside.

12. Identify all items with labels.

Phase 3: Seison (shine/sweep or clean everything daily and use this time to inspect the workplace).

1. Make cleaning a daily activity as part of an inspection. Clean before you start work and before you close the workstation.

2. Put aside 15 minute daily for this task.

3. Develop methods to prevent contamination

4. Use five "why"s[5] to find contamination and develop preventive action.

5. 5S owner checksheets should be maintained on a daily basis (see Fig. 4.5).

Phase 4: Seiketsu (standardize)—create visual controls[6] and guidelines to keep your workstation in order. This phase is where cleanliness is maintained. This phase is where routine becomes natural behavior.

1. Establish routines and practices to regularly repeat the first three phases.

2. Standardize procedures for visual controls.

3. Establish a maintenance system for housekeeping.

4. Celebrate good practices and good performers

Phase 5: Shitsuke (sustain, self-discipline)—maintain training and discipline to make certain that everyone follow 5S standards.

1. Ensure everyone treats their workplace like their home.

2. Encourage management participation in checking the first four Ss.

			Mon	Tues	Wed	Thur	Fri
Workstation No:			**Mon**	**Tues**	**Wed**	**Thur**	**Fri**
Network ID:			[Date]	[Date]	[Date]	[Date]	[Date]
Location:							
No.	**Checks**	**Frequency**	**Initials**				
		Daily					
		Daily					
		Daily					
		Tuesdays					
		Fridays					
Check	**Supervised by Finman**	**Daily**					

Figure 4.5   Illustration showing example of a 5S checksheet—daily maintenance of a sample log for a cleaning workstation.

3. Senior management should celebrate success.

4. Commitment is required to implement the first four phases daily.

On implementation, the 5S process creates positive impressions on customers and visitors, and improves efficiency and pride in the organization. Using 5S can reduce lead times, decrease waste, and improve quality. It also increases production, reduces cost, and improves quality. 5S is not only for cleanliness, but is an integrated approach to improve the overall business. When one more step, safety, is added to this 5S approach it becomes the *6S approach*. The steps then become

- *Sort*
- *Stabilize* (also *straighten* or *systematize*)
- *Shine* (also *sweep*)
- *Standardize*
- *Sustain* (also *self-discipline*)
- *Safety*

4.11   THE IDEA OF PERFECTION (KAIZEN)

Kaizen is a Japanese term that means continuous improvement, taken from the words *Kai,* which means continuous, and *zen,* which means improvement. Kaizen is the key to Japan's competitive success; it was introduced to the Western corporate world in 1986. Masaaki Imai defined it as "a means of continuing improvement in personal life, home life, social life, and working life." The kaizen business strategy involves everyone in an organization working together to make improvements without large capital investments. Instead of allocating more money for additional machinery or running them longer, kaizen steers an organization's attention toward small but significant details. Managers are encouraged to improve the efficiency of existing infrastructure instead of investing in more of the same. Mikel Harry, the coauthor of this book, says:

- We don't know what we don't know.
- We can't act on what we don't know.
- We won't know until we search.
- We won't search for what we don't *question*.
- We don't question what we don't measure. Hence, we just don't know.

Mikel adds, "Questions lead and answers follow. The same questions most often lead to the same answers which invariably, produce the same result. To change the result, means to change the questions. The most effective way leadership conveys expectations is through the questions they ask. . ."

The philosophy of continual improvement is that every process can and should be continually evaluated and improved in terms of time required, resources used, resultant quality, and other aspects relevant to the process.

4.12   REPLICATION—TRANSLATE

Replication is the extension of achieved improvements. This lean step extends improved system processes, tools, and methods across the organization. For replication to occur, the practitioner must adapt to global issues such as culture, laws, business regulations, and other local characteristics. The main benefit of replication is that benefit analysis need only be done once. Translate in this context means, move the process to other business sites and configurate/taylor for business needs accordingly.

4.13   POKA-YOKE SYSTEM—MISTAKEPROOFING

Mistakeproofing was developed by Shigeo Shingo[7] to achieve zero defects. Shigeo Shingo is translated into "mistakeproofing" and is used to avoid inadvertent errors. Poka-yoke is the Japanese term for mistakeproofing, derived from *yokera* (to-avoid) and *poka* (inadvertent errors). Shigeo Shingo, one of the industrial engineers at Toyota, formalized the approach and the uses of poka-yoke devices. A poka-yoke device is any mechanism that either prevents a mistake from being made or makes the mistake obvious at a glance. Some examples of prevention are some cameras that will not function when there isn't enough light to take a picture, some laundry dryers equipped with a device that shuts them down when overheating is detected, gas stations that still offer leaded gasoline in addition to unleaded gasoline, the nozzle on the unleaded pump and the hole for the gas tank were smaller than those for leaded gasoline, and fruit orchards that take great pride in oversized apples and ensures that only the biggest ones get to customers by passing all apples through a sizer. Those apples that don't make it are sent to the discount outlet, cars that have warning systems to alert the driver that not all seatbelts have been fastened, smoke detectors provide a warning that smoke has been detected and that there's a possible fire, and so on.

A *Poka-yoke* mechanism or system must have the following characteristics:

- Be inexpensive
- Be easy to use
- Be able to effectively prevent or significantly reduce the possibility of mistakes being made

The purpose of the poka-yoke system is to take over repetitive tasks or actions that depend on vigilance or memory, thus freeing workers' time and minds to pursue more creative and value-added activities. A second purpose is to provide capability for performance of source inspection. A third purpose is to prevent mistakes from occurring, and to provide information for immediate feedback and action, should the mistakes occur, or prevent them from continuing in the process and therefore, moving on to the customers. The general idea is to design processes that are robust and tolerant to mistakes and thus avoid inadvertent errors. Shigeo Shingo developed it as a method to achieve zero defects (mistakeproofing). Again, poka-yoke means to avoid (yokeru) and (poka) inadvertent errors. Figure 4.6 shows an example of a mistakeproofing sheet.

Components in Figure 4.6

1. Details of operation/process step, product/service, or part identification and date.
2. Description of the defect that is the subject of the mistakeproofing exercise.
3. Separate descriptions of where in the process, the defect is first made and detected.
4. Graph of the rate of defect production as a function of time.
5. Itemized list of process steps where defects are occurring. One step is entered in each box.
6. For each defect-producing process step entered in the previous box, list the defects which occur at the specific step.
7. For each defect-producing process step, list the possible causes for the production of the defect.
8. Red-flag conditions—provision for special accounting for particularly dangerous or costly defect/error situations.
9. Root cause of the error or defect.

The mistakeproofing planning sheet helps to identify and implement mistakeproofing devices to prevent errors or defects. This chart assumes a reactive effort, that is, that a problem already exists. We should, however, use mistakeproofing (poka-yoke) in a proactive mode, preventing defects from ever occurring by mistakeproofing products, services, and processes during the design stage. *Mistake prevention* is the ability to catch and prevent mistakes before they occur, and *mistakeproofing* is trying to make impossible errors to be passed from the former to the next step in the process.

Mistakeproofing planning sheet 1 of 2

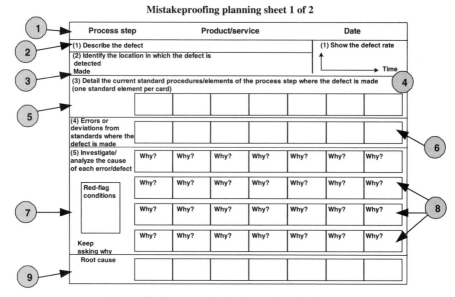

Figure 4.6 Example of a mistakeproofing planning sheet.

In general, mistake prevention and mistakeproofing are used when rework to correct errors or process delays are observed to be the cause of process cycle efficiency. The elements of a poka-yoke systems are (1) inspection method—where and how do you check for mistake, (2) setting approach—what to check for, and (3) regulatory function—how to provide feedback to the operator/associate.

Mistake prevention and mistakeproofing are systematically used in the manufacture of semi-conductor devices for cost-effectiveness purposes. Two common mistakeproofing approaches are (1) control or warning strategies such as shutting down the machine or system, turning on emergency lights, emitting warning sounds, message prevention, and (2) providing other preventive measures such as the examples shown in the mistakeproofing sheets, in Figures 4.6 and 4.7.

Components in Figure 4.7

1. Root cause of the error or defect.
2. For each defect-producing process step, list possible ways to eliminate the errors that have occurred at the specific step.
3. Indicate where in the process the defect or error should be detected or prevented, either before it is made, as it is being made, or after it is made.
4. Describe improvement(s) required to eliminate the defect or error.
5. Current state of the process.
6. Device or method for eliminating the defect or error.
7. Cost of implementing the improvement.
8. Length of time needed to install the improvement.
9. Team members involved in the mistakeproofing exercise.

4.14 SMED SYSTEM

Taichi Ohno[8] from Toyota in Japan worked with Shigeo Shingo to develop the *single minute exchange of dies* (SMED) system. This system reduces turnover time in production lines and enables small-lot

Mistakeproofing planning sheet 2 of 2

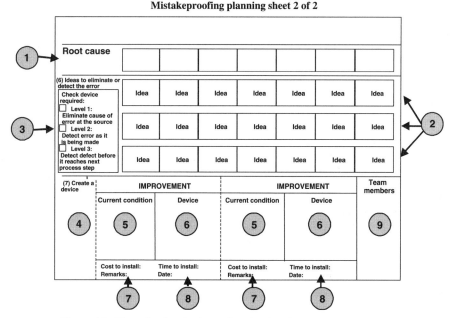

Figure 4.7   Example of a mistakeproofing planning sheet, page 2 (continued).

production. The principle evolved into the *just-in-time* (JIT) supply practice. There is a set of procedures to be followed for successful implementation. During one of Ohno's trips to the United States, he visited a Detroit supermarket where he observed how American customers chose from an array of merchandise; he also observed how supermarket workers immediately restocked the product space left by the just purchased merchandise. Ohno applied this "replacing principle" to manufacturing lines where items used to build the product would be replenished at each station or section of the production line. He called this replacement the "pull system" of production. Today the pull system has become a principal component of lean. Advantages of SMED include fast and predictable setups that enable lean manufacturing. Lower setup times reduce setup cost, allow small-lot production, smooth the flow, and therefore improve kanban.

4.15   7W + 1 APPROACH—SEVEN PLUS ONE DEADLY WASTE(S)

"One of the things I haven't ever itemized is waste," said Taiichi Ohno. Ohno identified seven types of waste now known as "7W." 7W is one of the most important aspects of lean—waste (or *muda*, in japanese). An eighth form of waste came from Womack and Jones.[9] Throughout this section we will refer to this method as the "seven plus one" approach. The elimination of the sources of waste throughout product, transaction, development, and service deliveries, targets unnecessary documentation and prototypes such as reuse of processes, concepts or outdated designs. It also reduces cost and time spent waiting for input from others in the process line, results, and data. For example, in the banking system, a major source of waste (muda) is the reentry of data, or even worse, the entry of conflicting data by different functional specialists. Lean identifies the "seven plus one deadly wastes" or defects as the source of waste. Producing a product that is not sold immediately is waste. In an office environment, a transaction that is not delivered immediately but must wait for delivery or movement is considered waste. In the production process there are three types of "work"—value-added (VA), non-value-added (NVA), and waste. The seven wastes that are the root of all nonprofitable activities within any organization are

D—waste from product defects

O—waste from overproduction

T—transportation waste

W—waste in waiting time

I—inventory waste

M—waste of motion

P—processing waste

The eighth waste is a *waste of skills.*

Many practitioners use the acronym DOTWIMP to remember the "*seven wastes of lean.*" A slightly different way to remember the seven deadly wastes is by remembering them as the "seven + one wastes" and by a different acronym "TIM WOODS," as follows:

- Transport—moving people, products, and information
- Inventory—storing parts, pieces, and documentation ahead of requirements
- Motion—bending, turning, reaching, lifting
- Waiting—for parts, information, instructions, equipment
- Overproduction—making more than is immediately required
- Overprocessing—tighter tolerances or higher-grade materials than are necessary
- DDefects—rework, scrap, incorrect documentation
- Skills—underutilizing capabilities, delegating tasks with inadequate training

The most serious of these "seven plus one wastes" is overproduction, because it contributes to each of the remaining six. It is the main driving force for the Toyota just-in-time (JIT) system. A smart way in lean to approach overproduction is to address the waste contributions of the other six (see Fig. 4.8). Remember that overproduction is producing more than the customer demands. The way to fix this is to manufacture/produce on the basis of a Pull system, or producing products/services just as the customer orders them. Anything produced beyond this ties up valuable labor and material resources.

The TIM WOODS acronym is broken down as follows:

T: Waste of Conveyance (Travel).   Supplies needed to do a job are not at hand so you have to either get them or take them to where they're needed.

I: Waste of Inventory.   Inventory is not an asset. While inventory has value, this value declines as it ages. If funds are used for excess inventory, they are not available to meet other needs. Someone mentioned that older PC equipment were assets. Why keep old equipment? If you sell off your tenured equipment to the gray market, you rarely get what you paid for it. It's not an asset, it's waste.

M: Waste of Motion.   There are three kinds of motion: actual work, ancillary work, and muda, (waste):

1. Actual work adds value to a product.

2. Ancillary work is work that supports actual work. Shipping and receiving are examples.

3. Muda is motion that creates no value. For example, time spent verifying inventory because of discrepancies in system counts has no value.

[Note: If an operation creates a repetitive stress injury (e.g., typing with an incorrect posture), Lean categorizes these types of injuries as waste.]

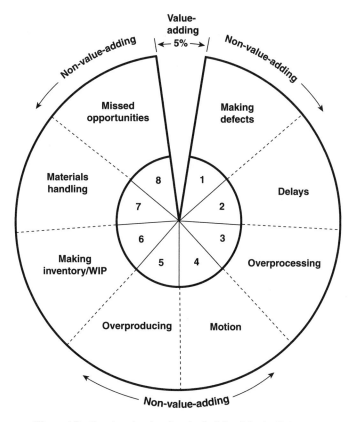

Figure 4.8   Drawing showing the wheel of the eight deadly wastes.

W: Waste of Delay (Waiting). Computer installers are waiting for the promised arrival of equipment. IT department technicians install a new PC only to find that the operating system was not shipped with the system. These are examples of increased operational expense or waste that directly decrease profitability.

O: Waste of Overproduction. Overproduction is a severe problem because you need space to store work-in-process (WIP). You'll need people to move your products around. Ideally, you should produce only those products required to satisfy orders received. Some industries naturally incur more overproduction waste than others. Clothing and agriculture are examples where overproduction is unavoidable due to batch processing.

O: Waste of Overprocessing. This is a particularly menacing form of waste because it causes the practitioner to examine which of the processes or products are necessary. An example is when a document needs five levels of signature to approve one customer order. On occasions, the waste of overprocessing result from including extra undocumented features in a product that the customer did not request and is not willing to pay for.

D: Waste of Correction (defects). "We don't have the time to do the job right the first time, but we have plenty of time to fix it afterward," said a product manager for air compressors. In his logic, it was less expensive to get the product to market (working or not) than take valuable line time to repair each problem in manufacturing. "Let warranty worry about it." (*Note:* This gentleman is no longer the product manager for air compressors.)

S: Waste of Knowledge Disconnection/Underutilization of Resources (Skills). This refers to the waste of administrative issues within a company, and its suppliers and customers. Such "issues" create excess costs, missed opportunities, and frustration. A customer's purchase order system requires customized input on your part. Your system doesn't have the capability to automate the customer request, so you send your input in the same legacy format that you have always supplied, ignoring the customer request. Your customer must now retype your quote into the purchase order manually. The customer's purchasing department places the order on the stack of work to be input and maybe they'll get to it in a week. You are upset because the customer didn't provide the purchase order as agreed. The customer is upset because you didn't send correctly formatted data and accounts payable will not process any invoices until the order data are correct. To solve the problem of waste, lean has several "tools" at its disposal. These include constant process analysis (Kaizen), Pull production (by means of kanban), and mistakeproofing (poka-yoke).

Poka-yoke can be applied in the following areas:

- Information/Instruction—public or confidential
- Materials—manufacturing, services or transactional process and uses
- Machines/tools—operations, safety, performance
- Manpower/Workers—follow defined methods
- Methods—defined by workers and managers
- Environment/mother nature

4.16   6M APPROACH

This graphical method identifies the different causes that could produce the errors/defects under study. Cause-and-effect and failure-mode analysis diagrams guide creation of checklists. The cause–effect (CE) diagram is also known as the fishbone or Ishikawa diagram. This diagram examines the process and lists all possible areas where the problem under investigation could occur. The purpose of this method of graphing is to simplify the material flow. This method analyses causes that could affect the process under study. This method is used for root cause analysis. The centerline of the diagram identifies the problem under investigation; the branches are the major areas that could cause the problem defined in the main branch or centerline. The next step is to label these branches under

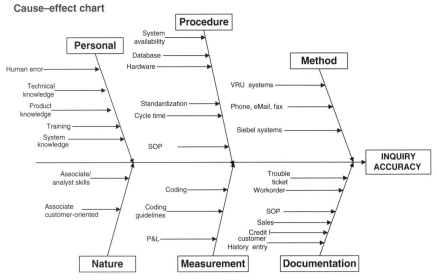

Figure 4.9   Example of a cause effect (CE) diagram.

workers/*m*anpower, *m*aterials, environment/*m*other Nature, *m*ethods, Tools/*m*achines, and *m*easurements or 6M—personnel, procedure, method, nature, measurement, and documentation.

Once a defect or error has been identified and isolated for further study, we must analyze the causes of this problem. In situations where causes are not clear, the CE diagram is the most frequently used to identify potential causes. The CE diagram can have from four to six main branches. Each branch can have one or more branches. The complexity of the CE diagram depends on the complexity of the problem under study.

The next step is to evaluate each of the branches for possible causes; the last step is to evaluate the possible causes under each branch. It must be remembered that charting is not the end result; many process designers spend a lot of time in the charting process and less in evaluating the end result. The objective is not to produce a beautiful chart but to evaluate the problems affecting service deliveries, transactions or/and products. Figure 4.9 is an example of a CE diagram showing the reasons for calls placed to a call center.

The CE or Ishikawa diagram examines a process and lists possible areas where a problem under investigation can occur. The purpose of this graphing method is to simplify the flow of the process information when searching for a root cause. This diagram is frequently used in root cause analysis (RCA).

4.17 SUMMARY

Lean Sigma is a body of flexible ideas that empower and sustain the quest for quantum change in business performance. These concepts, when fully implemented, slice through the ambiguity of complexity. In this sense, the journey is just as important as the destination, and can best be characterized by a repetitious cycle of carefully orchestrated events and actions that rely heavily on the empowerment of certain unconventional ideas, beliefs, aims, and goals. Each cycle progressively validates previous best-in-class practices and spawns new approaches through the continued implementation of application projects. Those practices, methods, and tools that prove beneficial are incorporated into the organization by knowledge acquisition and distribution. Naturally, the organization's culture must be capable of absorbing, integrating, and framing this knowledge, or the culture itself must be changed. This thought process is consistent with the principle that we cannot examine and change what we do if we do not know what we know. In the final analysis, it is always knowledge that changes the organization's historical continuum and places it on a new performance trajectory.

GLOSSARY

causative   Effective as a cause.

cause   That which produces an effect or brings about a change.

cycle time   The total time from beginning to end of a process as defined by you and your customer.

effect   That which is produced by a cause.

just in time (JIT)   Producing or conveying only the items that are needed by the next process when they are needed and in the quantity needed.

kaizen   Japanese term for continuous improvement.

kaizen event   Team event dedicated to quick implementation of a lean method in a particular area over a short time period.

kaizen plans   Lean improvement proposals presented to management by the team following analysis of their current-state map and mapping their future state.

kanban   A Japanese term that signals a cycle of replenishment for production and materials. A just-in-time delivery tool.

lead time   Time required for one piece to move all the way through a process or value stream from start to finish.

Lean nterprise   Organization that fully understands, communicates, implements, and sustains lean concepts seamlessly throughout all operational and functional areas.

Lean production   Manufacturing strategy that uses less of everything compared to traditional manufacturing.

Lean Sigma   A strategy aimed at the elimination of waste in every area of production, including customer relations, product design, supplier networks, and factory management. Its goal is to incorporate less human effort, less inventory, less time to develop products, and less space to become highly responsive to customer demand while producing top-quality products in the most efficient and economical manner possible.

line balancing   Process in which work elements are evenly distributed within a value stream to meet Takt time.

location indicator   Visual workplace element, such as a line, arrow, or label that shows where an item belongs.

milestones   Activities that have or should have a scheduled completion date or time, and that have a major effect on the timing or completion of a project.

muda   See **waste (muda)**.

non-value-added activities   Steps in a process that add no value as perceived by the customer and that they are not willing to pay for.

pitch   Multiple of Takt time that will allow one to create, maintain, and sustain a consistent and practical workflow throughout the value stream.

poka-yoke   Colloquial term used in reference to a mistakeproofing process.

preventive action   Action taken to prevent problem from occurring, based on an understanding of the product or process. Preventive action will address inadequate "conditions" that may produce nonconformance.

problem solving   Goal of a team working together to accomplish the following steps: (1) *define* the problem, (2) *analyze* possible causes, (3) *identify* possible solutions, (4) *develop* action plans, (5) *Evaluate* and *renew* action plans, and (6) *standardize* effective ideas.

process   Sequence of operations consisting of people, materials, methods, environment, and measurements required to create and deliver a product or service.

pull   System of creation and delivery in which nothing is produced by the upstream supplier until the downstream supplier signals a need.

pull system   The flow of resources in a production process by replacing only what has been consumed.

push   Conventional work in which products are pushed along in accordance with sales projections and availability of materials.

queue time   Amount of time that a work unit waits before a downstream process is ready to work on it.

red tag   Label used in a 5S implementation to identify items that are not needed or that are in the wrong place.

root cause analysis (RCA)   The process of identifying causal factors in a formal and structured approach to search for the "basic reason" for a problem's occurrence. Adhering to the definition of root cause, care must be taken to clearly distinguish symptoms from causes, as well as "perceived" root causes.

safety resources   Means of meeting customer demand when internal constraints or inefficiencies disrupt process flow.

shine   Third activity in the 5S system.

sort   First activity in the 5S system.

standardize Standardize for 5S— the fourth activity in the 5S system. Involves creating the rules for maintaining and controlling the conditions established after implementing the first three Ss.

standardize work Agreement on a set of work procedures that establishes the best method and sequence for each process. Standardize work is implemented to maximize efficiency while simultaneously ensuring safe conditions.

stream A sequential flow of activities needed to create product units and deliver them on time to the customer.

sustain Fifth activity of the 5S system where a team or person ensures adherence to 5S standards through communication, training, and self-discipline.

Takt time Pace of customer demand. Takt time determines how fast a process needs to run to meet customer demand.

theory of constraints A set of tools that examine the entire system for continuous improvement.

thought Process map Visual representation of thoughts that act as a roadmap to progress through DMAIC.

total cycle time Total of the cycle times for each individual operation or process or work area in the value stream. Total cycle time ideally equals total value-added time.

total lead time Total of all cycle times from all individual processes within the lean office value stream, plus the queue times that exists between each process.

value The exchange in which the customer pays. Value can also equate to quality over cost. It is useful to model a firm as a chain of value-creating activities. Michael Porter identified a set of interrelated activities common to most firms. The resulting model is known as the *value chain*. The primary components include inbound logistics, operations, outbound logistics, marketing and sales, and service.

value stream All the steps in a process that the customer is willing to pay for in order to bring the product or service to market.

visual controls Vehicles that convey information to ensure that activities are performed in accordance with standards

visual displays Vehicles that convey important information, but do not control what people or machines do; they are passive in nature.

visual workplace A work environment that is self-explaining, self-ordering, self-regulating, and self-improving.

waste Waste in a process is any activity that does not result in moving the process closed to the final output or adding value to the final output.

waste – (Muda) Anything within a value stream that adds cost or time without adding value.

REFERENCES AND NOTES

1. F. Taylor, Frederick Taylor formalized the principles of scientific management, and the fact-finding approach put forward and largely adopted was a replacement of what was known as the "old rule of thumb." Taylor was, by far, the most influential person of the time and someone who had had an impact on management service practice and well as management thought up to the present day. *The Taylor System of Scientific Management*, A. W. Shaw, Chicago, 1917.

2. *Little's law*—relating process, speed, capacity, and WIP—law to calculate (lead time) – (total lead time) amount of work in process/Average completion rate. Little's law, the fundamental long-term relationship between work-in-process, throughput, and flow time of a production system in steady state is expressed as: Inventory = throughput × flow time. For example, if we observe a milling machine that cuts 100 parts per hour with a queue of 800 parts in front if it, we say that it has "8 hours of WIP."

3. Dr. J. Liker, *The Toyota Way: 14 Management Principles from the World's Greatest Manufacturer*, McGraw Hill, Dec. 2003 (ISBN 0071292319).

4. E. Goldbratt and J. Cox, *The Goal*, 3rd ed. July 2004 (ISBN 0884271781).

5. "Five whys" method—used by root cause analysis (RCA) to find the root cause (ask "Why?" 5 times), O. de Hodgins, Root Cause Analysis Guideline, Solectron Corp. 2007, course 4125. The "Five whys" typically refers to the practice of asking, 5 times, why the failure has occurred in order to get to the root causes of the problem.

6. W. E. Deming, *Visual Controls*, Visual controls are a system of signs, information displays, layouts, material storage and handling tools, color-coding, and poka-yoke or mistakeproofing devices (See also W. E. Deming, out of the Crisis, MIT Press, 1989).

7. Shigeo Shingo, Toyota's quality control guru.

8. T. Ohno,*Toyota Production System*, Feb. 1998 (ISBN 0915299143).

9. J. P. Womack and D. T. Jones, Lean Thinking; Banish Waste and Create Wealth in Your Corporation, 2nd ed., June 2003 (ISBN 0743530489).

10. M. Harry and R. Schroeder, *The Vision of Six Sigma*, 5th ed., March 1997 (ISBN 0385494378).

5 Value Stream Mapping

5.1 OVERVIEW

This chapter presents the fundamental concepts of the Toyota production system. Toyota based much of its concepts on the US "supermarket" system, with just-in-time ordering, and based the construction of its auto lines based on a job-shop approach (build to order), which is similar to our "print and comp" model. Lean is a manufacturing philosophy that shortens the time between customer order and the product build/shipment by *eliminating sources of waste.*

The five principles of Lean are as follows:

1. The definition of value from the customer perspective
2. Value stream mapping—mapping information flow and product/service transformation
3. Continuous flow—having everything ready internally as needed for a symphony of service or production
4. Pull production—providing only when needed
5. Strive for perfection—never giving up the chase for perfection

Value stream mapping is one of the most powerful tools in Lean; it is frequently compared to a flowing river with smooth running water. The practitioner will learn how to map the current state by following the product's production path from customer to supplier and carefully draw the visual representation of each process with the material and information of the flow. Value stream mapping helps to visualize more than only the single level. The practitioner actually will be able to "see" the flow.

5.2 INTRODUCTION

According to the National Institute of Standards and Technology Manufacturing Extension Partnership's Lean Network, "Lean is a systematic approach to identifying and eliminating waste through continuous improvement, flowing the product at the pull of the customer in pursuit of perfection."

Sven lives in a small Norwegian village where he owns a bike shop. In his shop he sells fancy red bikes, each with a bright chrome basket on the front and a shiny silver bell on the handlebars; he also sells a basic blue model. Naturally he charges more for the red bike, and his profit margin is higher than that for the blue bike.

When people visit Sven's shop, he gives a marketing spiel that he devised to convince them they need the red bike. He tells them how they can ride the red bike to town on market day, " with the basket for food and things," and emphasizes that if they ride the bike instead of driving, they will be doing something good for the environment. They will also get in shape and stay healthy—and with all the money saved on gas, the bike would quickly pay for itself. Sven sells a lot of red bikes but eventually the

Practitioner's Guide for Statistics and Lean Six Sigma for Process Improvements. By Mikel J. Harry, Prem S. Mann,
Ofelia C. de Hodgins, Christopher J. Lacke, and Richard Hulbert
Copyright © 2010 John Wiley & Sons, Inc.

shoppers realize that they don't ride the red bike when they come to the village to shop; instead, that they drive. Around the neighborhood, fancy red bikes are gathering dust in garages, but you seldom see one in the village.

When new people move to the village and go to Sven's shop looking to buy a bike, Sven gives the spiel about the fancy red bike and tells them it's his best seller. Some people buy it, but others look outside and do not see any red bikes and decide the basic blue model will be fine, since it doesn't seem like anyone else is actually rides a red bike to the village to shop. Now Sven, who ordered his inventory months ago on the basis of earlier sales, suddenly realizes that he is stuck with too many fancy red bikes. He thinks about his problem, considers having a sale and lowering the price of the red bike to reduce inventory, but hates losing money on his best seller. It seems unreasonable to him that he should. He decides that the problem lies in the basic blue bike. If he did not stock that, everyone would have to buy the red bikes, and he would be back in business. Sven cancels his order for more blue bikes, revises his marketing spiel and implements a hard-sell campaign to sell the red bikes.

About this time, Hans moves to town, sees that the only bike store offers just one model and decides to open his own store with a wider variety of bikes that would appeal to a broader market. He stocks both red and blue bikes, but also decides to offer yellow and green models as well. Before long, Hans is doing a booming business while Sven isn't doing so well. He's alone in his shop with a lot of red bikes, and he can't figure out what went wrong. They were his best seller; so he decides to come up with a different marketing spiel for his red bikes and give his business another try.

Anyone with common sense knows how this story ends. Most of us have an emotional investment in our competence, but step back and think about the fable and relate it to this chapter. You can sell anything at the right price; just look on eBay. We must either lower the price of full equity or add value to the offering in the form of enhanced services and facilities to compete in the current market. There are many new competitors that offer everything we have and more. We must adjust to the conditions in the real world if we are to survive.

Why Lean? Lean systems help business become more competitive and therefore, more likely to survive in today's competitive world. We begin our journey in Lean by dividing Lean into different categories: operational, administrative, and strategic *improvements*. Most organizations that fail to implement Lean have the perception that Lean applies only to operations; however, based on experience, we have observed that Lean also applies to services and transactions. An example of Lean in transactions is Lean billing, which is the reduction of processing errors, reduction of paperwork, streamlining customer service functions, reducing rework of the billing process or materials involved in the rework, overtime, and other related issues. Lean in administrative areas brings solid strategic benefits and therefore significant cost reductions. This is consistent with the cost reduction principle, which states that to remain profitable, a business should not pass price increases on to the customer but should to save cost internally through the elimination of waste. It is for this reason that this chapter starts with a brief review of waste previously covered in Chapter 4.

5.3 VALUE STREAM MAPPING

Value stream mapping eases the Lean transition. Done right, Lean can make real improvements to manufacturing and administrative operations. Done wrong, it can be what the Germans call *Schlimmbesserung*: "a so-called improvement that just makes things worse."

One good way to obtain a new perspective on the plant's operations is by using a tool called *value stream mapping*. In essence, mapping the value stream means following the product or product family from beginning to end. Each value-added and non-value-added step is noted and timed. It may look simple, but this kind of attention to what adds value can be eye-opening, says Dave MacDuffee, an industrial extension specialist. "Managers may feel they know their plant inside out, but doing the mapping can help them to better see that value-added time." Value added (VA) activities are those that customers are willing to pay for. *Value-added* is also known as *customer value-added* (CVA). In contrast, *non-value-added* (NVA) are all those activities required by a business to execute work that adds no value from a customer point of view and that customers are not willing to pay for. Non-value-added activities are also known as *business non-value-added* (BNVA). *Non-value-added* is all those activities that can be categorized as "waste." Non-value added activities are those that add no value to customer such as waiting time and movement between processes, or not required for legal, financial, safety, environmental, or other reasons.

People, who are embedded in a process day by day and have a desire to become Lean, find it difficult to see the big picture. What becomes difficult is learning to see the flow. Value stream mapping is very simple to do once you see the flow." Celine Yamada Nomoto,[2] Lean expert, agrees when she said: "We mapped a product family during the training, the next day, after the training was over, we went back and did another product family ourselves." She credits consulting instructors for that rapid progress. Consulting instructors won't just give the participants the one answer to a problem; they'll describe other companies and how they approached the problem. They describe available choices and options. And if you have a question, they call back with immediate feedback.

To construct the value stream,[3] imagine a flowing river where water always runs smoothly, a river in which downstream is consistently receiving what flows from upstream with little or no effort. Now, imagine each process in the organization as a component of this river as it passes its *output* to the next downstream activity or internal customer. We define the process that is the farthest downstream as the customer who buys the product/service, transaction, and service produced by the organization. Just as many rivers flow across a continent into the ocean, there are many value streams that flow within an organization before reaching the customer.

Customer relationship management (CRM) is a term applied to processes implemented by a company to handle their contact with their customers. CRM is used to support these processes, storing information on customers and prospective customers. Information in the system can be accessed and entered by employees in different departments such as sales, marketing, customer service training, professional development, performance management, human resource development, and compensation. Details on any customer contacts can also be stored on the system. The rationale behind this approach is to improve services provided directly to customers and to use the information in the system for targeted marketing and sales purposes.

Process Steps that Create Value. A value stream encompasses all the actions both value-added [VA] and non-value-added (NVA) that are necessary to bring a service or product from concept to development, to "manufacturing" of the service, to the receipt of payment or completion of the transaction. Once value is specified by thinking as the customer, the right process (processes that only add value to the product, meaning a process that is "waste-free") the value stream consists in the following activities:

- Process steps that create value and those process steps that have no value but are necessary, due to the current state of the environment (documentation) or system (organization culture, organization tradition, etc.) needs are considered value-added steps.
- All those process steps that create NVA activities and therefore should be eliminated.

5.3.1 Waste Review

Knowing that Lean applies to the entire organization, not only in manufacturing, but also in the administrative areas we will first review the Toyota production system. Waste normally includes NVA activities and is known to Lean practitioners as to the "seven wastes". Taiichi Ohno[1] suggested that these NVA accounted for approximately the 95% of all costs in non-Lean manufacturing environments. Remember from Chapter 4 that these wastes (TIM WOODS) are as follows:

1. *Transportation* — Lean demands that the material be shipped directly from vendor to location. The Lean term for this is known as "point-of-use storage."

2. *Inventory* — excess inventory, related to overproduction, is the inventory beyond that needed to meet customer demands. Remember that overproduction negatively impacts cash flow and uses valuable floor space.

3. *Motion* — unnecessary motion caused by poor "workflow" (movement of people, material, or machines during process activities).

4. *Waiting* — waiting for material, information, equipment, parts, tools, diagrams, instructions, or signatures. Lean demands all components and resources on a just-in-time (JIT) basis.

5. *Overproduction* — the Lean principle is to produce on the basis of a pull system (Chapter 4) producing only what customer demands.

6. *Overprocessing* — overprocessing is related to inventory; it is the inventory beyond that needed to meet customer demands. The objective is to process products or services just as the customer orders them. Anything beyond this "standard" uses valuable labor and material.

7. *Defects* — production defects and service errors wastes resources, materials, overhead, and so on; materials consumed, labor used to overproduce, rework in redoing what went wrong the first time, and labor and cycle time required to address customer complaints. In other words, defects include everything related to poor quality, scrap, rework, and related activities.

8. *Skills* — underutilized skills include mental, creative, and physical skills and abilities wasted because of poor workflow, project management, organization culture, personal agendas, large amount of temporary workforce (constant retraining), lack of training, lack of skills, lack of experience, and other related factors.

The elimination of waste is critical to Lean. This chapter encourages the extension of the fundamental Lean concepts of demand, flow, and leveling of manufacturing processes to encompass all organization areas by creating a common understanding, daily use and application of these Lean principles. *Value stream management* (VSM) is a process for planning and linking the Lean initiatives through a systematic data collection and subsequent data analysis. Eliminating wastes and promoting a smooth flow of information and work reduce organization costs.

5.3.2   Value-Added and Non-Value-Added Activities

A **value-added activity** transforms the product, service or transaction in a way that is meaningful to the customer. In other words, value-added activities are those activities that are valuable from the external point of view of the customer. **Non-value added activity** is "we do this because the boss wants it." Frequently these tasks have no value to the customer.

The customer who wants to buy the right product with the right capabilities at the right price all the time determines "value." This product must be "right" every time from design to manufacture and from manufacture to delivery. Each of the operations must be "error-free." Lean organizations strive to make their processes right the first time by eliminating waste. By linking value to customer requirements, the following questions can be asked to review the value for the customer as it relates to the product.

The customer might ask the following questions:

- What is the problem that impacts me (the customer)?
- Do the stakeholders understand the problem?
- Stakeholders understand and agree to the problem and the impact to business?
- Do the stakeholders agree that fixing the problem is critical to the business?
- Are my needs (the customer's) identified?
- Are the key parameters to be fixed identified?

Once value is determined, the next step is to identify the correct process. The correct process is one that only adds value to the product, a waste-free process. The value stream for a product should contain

1. Steps that create value
2. Steps that create no value but are necessary (as specified above)
3. Steps that create no value and therefore can be eliminated.

Steps that create no value and are not required for the process are labeled for elimination; then any necessary NVA activities are labeled for improvement. At this point, detailed *process maps* should be generated for each product and product category. Brainstorming sessions are used to confirm the team's conclusions as to which steps contain value, which steps contain no value but can't be eliminated, and which steps should be eliminated.

To eliminate waste, the type of waste must be identified and where and under what conditions it is occurs. In this example the organization doesn't have an automated procedure for documenting processes. Thus, most of the processes are in managers' desks where others in the organization have no access to the documentation. In manufacturing, one of the most frequent waste types is over-production. This waste occurs when more products are produced than are required. It is important in manufacturing because when demand slackens, overproduction creates unsold inventory and byproducts associated with it such as: extra handling, extra space, and extra paperwork. Unlike overproduction, the waste from waiting is usually identifiable since idle workers are visible to all.

Organizations can profit from learning to think in terms of Lean. The main goal of Lean is to eliminate waste. Lean attacks waste by shipment. Lean enables organizations to eliminate waste and produce more when value is added at every step in the process. The value of any product is defined by customer needs and not by any of the NVA activities of the supplier of producer. Lean is a manufacturing philosophy that decreases the time between customer order, product build, and shipment by eliminating the sources of waste. The examples in this chapter demonstrate how the Lean effort can be directed into removing excess of inventory and increasing "pull." On one hand, Lean identifies the seven deadly wastes, "defects," while the "pull" system ensures that the product is made at the same rate at which it is being sold. From the supply chain viewpoint, a pull system flows resources into a production process by replacing only what has been consumed. On the other hand, value stream ensures that all product units "flow" to customers as smoothly as possible. In the real world, this ideal situation rarely exists as there are shortcomings such as poor documentation or restrictions within the process that impede this desired smooth flow. The objective of Lean is not to make the workforce work faster, but rather to streamline the workflow so that it moves more quickly through the value stream. Lean concepts utilize the proper thought maps and Lean tools to make this downstream work flow as smoothly as possible. The principle of Lean thinking is illustrated in Figure 5.1.

This step focuses on rapid product flow (RPF). Waste is identified with specific steps of the process and is eliminated. Non-value-added activities are identified and eliminated by physical changes to physical layout. These involve human and other resources such as fax (facsimile), personal computers, and telephones (landline and mobile). Transactions should be laid out in cells emulating factories to reduce distances between specific products such as loan applications, car loans, and mortgages. Value stream maps (VSMs) show the flow of the transaction or material, paperwork, and information through the process. It is at this point that Lean implements 5S (a tool that reduces the hidden slack produced in manufacturing a process). Value stream maps focus on identifying infrastructure inefficiencies and waste at the organization process level. Value stream reduces lead time, improves service quality, and reduces rework, scrap, and inventory levels and therefore reduces indirect labor costs.

Often value stream maps capture the current process and thus may not represent all the variations of the process or the evolution of the process over time. The value stream map does not typically capture the causal structure, but is not restricted from visually documenting causes of waste, bottlenecks and other relevant cycle time information delays. Transaction and service areas provide enormous and frequently overlooked opportunities for improvement.

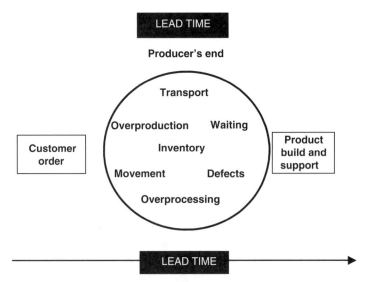

Figure 5.1   Illustration showing the principle of Lean thinking.

5.3.3   Elements of a Value Stream Map

Common failure modes of value stream maps are

- Failure to link the VSM to the question being studied
- Failure to effectively include the VSM in a sequence of improvement activities (rarely should the VSM be the first map used)
- Failure to understand the implications of the snapshot in time on the reported cycle time metrics
- Failure to consider multiple metrics

Typical process data for value stream mapping include cycle time, changeover time, uptime, production batch size, talk time, queuing time, lead time, number of operators, number of process variations, pack size, working time, rework time and scrap rate. It also addresses the concepts of waste, push, pull, perfection, replication, and flow. *Waste* could result from queuing time between activities or from any part of the process that takes time and resources but adds no value to the final product, transaction, or service. Waste includes something as small as extra footsteps to bring paper, documentation, and information to another part of the business for transaction completion. Value stream mapping can scrutinize business processes from beginning to end and a visually representation of every activity involved in the process including information flows. The most significant source of waste is overproduction, creating excess inventory, extra people, and extra equipment. Shortages result in additional processes that often collect the wrong data and therefore result in incorrect conclusions that lengthen the lead time, which impairs organization ability to quickly respond to customer demands and requirements.

The following definitions are necessary:

Push — production system that results in overproduction and shortages since the process flow does not reflect the actual needs of the downstream customer but builds to plan and "pushes" it ahead.

Pull — production system driven by actual consumption and controlled by synchronized replenishment. In Lean, this step is directed toward either removing excess capacity (inventory or increasing the rate of pull). Lean identifies the "seven deadly wastes" and inventory as sources of waste. Therefore, producing something that is not sold immediately and is waiting for any period of time before delivery is waste. A pull system, which is manufacturing/processing a product at a rate at which it is being sold, is also a waste-eliminating step. Pull systems only replace what has been consumed. Once value streams are identified, the next Lean step is *flow*. Flow produces a product, service, or transaction from customer request to completion, without unnecessary interruption or delay. The goal is to achieve single-piece flow in each activity of the process, ensuring that work "flows" smoothly from one activity to the other, one at the time, increasing flexibility, and as a consequence, reducing work in progress, parts movement, parts handling, quality errors, defects, and lead time. The goal of flow is to align processes to deliver a service, transaction (product) to fulfill customer requirements while reducing waste in the process. The key tools for implementing flow are

- Takt time
- Standardize work
- 5C or 5S—6S system (Chapter 4)
- Work balancing
- Leveled production

Producing "Takt Time" requires concentrated effort to provide rapid response to problems, eliminating changeover and unplanned downtime. Developing continuous flow using a combination of continuous flow and some pull is a good approach can be to begin with. Continuous flow gives the biggest bang for the dollar by eliminating waste and shortening lead time.

Takt Time — available production time divided by customer demand (see case study in Section 5.16). For example, assume that a manufacturing plant operates 8 hs per day, and customers demand is 15 complex circuit boards per day. The purpose of Takt time' is to match production with demand. "Takt time" was first used as a production management tool in the German aircraft industry in the 1930s. It was the precise time when an aircraft was moved ahead to the next production station. *Takt* is a German word that translates as conductor's baton. Toyota adopted this concept in the 1950s. Since then, Toyota typically reviews Takt time for a process each month. Takt time is the working time available (effective working time) divided by customer demand. Thus, Takt time is the "pace" of customer demand:

$$\text{Takt time} = \frac{Production\ Time\ Available^{*}}{Customer\ Demand}$$

* Minus lunch breaks, time briefing, other

Example: Production Time Available/day = 8 hours = 480 minutes
Prodcution Time Available/day = 480 − (lunch + breaks)
Lunch = 45 minutes/day
Breaks = 2 breaks of 15 minutes each/day = 30 minutes

$$\text{Production time availabel/day} = \frac{(480)-(75)}{Customer\ Demand}$$

Assume that daily customer demand is = 100 parts per day

$$\text{Takt Time} = \frac{405\ minutes}{100\ parts/day} = 4.05 \text{ parts per minute}$$

Production lead time — throughput time or total product cycle time is the time required for a product to move through the entire process from start to finish. In manufacturing language this is known as "door-to-door time."

Pull system — is a system driven by actual consumption and controlled by synchronized replenishment signals. The shorter the production lead time, the shorter the time between paying for raw information and getting paid for the transaction that used this information. For maps with multiple upstream flows, the longest time path is used to calculate the total lead time.

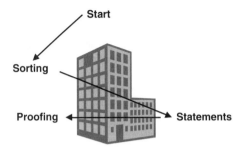

Figure 5.2   Drawing showing how physical distances interrupts continuous flow—this example graphically shows that waste is being produced (waste of motion and waste of transportation).

Supermarket — when obstacles are found in the continuous flow, moving one work unit at a time, we can use supermarket. Supermarket is a method used to store a level of completed work units, or partially completed work units (WIP) and replenish the units that have been pulled, in order to satisfy customer demands (internal and external customer demands). In Lean methodology, the supermarket approach is used when circumstances make it difficult to sustain the continuous flow. Toyota found that the supermarket method is the best approach to scheduling upstream processes when they don't or can't flow continuously. As flow is improved, the need for the supermarket approach will decrease. Supermarkets are therefore, a compromise to the ideal state.

For example, in some South Korean restaurants the supermarket approach is used. Customer tables are located around a table that continuously rotates containing all the items of the menu. The customer can select and "take out" from this table any item of the menu. The kitchen immediately replenishes that item (the empty space is replaced immediately and therefore, the rotating table is never empty). The signal in this case, is the empty space (kanban).

The supermarket pull system provides accurate production instruction to the upstream process without predicting downstream demand or scheduling the upstream process. It is used to control production where continuous flow is interrupted by physical distance or when continuous flow does not extend upstream (Fig. 5.2). Send the customer schedule to only one production process, the "pacemaker" process. There should be no supermarkets or "pulls" downstream of the "pacemaker" process.

When multiple process permutations take place, the value stream map may not be the first analysis tool used because the value stream considers everything including non-value-added (NVA) activities that make the transformation from raw information so Lean that the customer is willing to pay for it. Value stream communicates orders and forecasts throughout the supply chain. In other words, it is the network of process activities through which information flows in time and space as this information is being *transformed.*

While the ideal state may not be achieved overnight, organizations must work continually toward this ideal state. *Kanban* means sign, card, or billboard and in the context here refers to the inventory control card used in the pull system.

5.4   FOCUSED BRAINSTORMING

Because brainstorming sessions (or whiteboarding) are a valuable means to formulate new ideas, engaging in the creative process, and involving a group not necessarily assigned to the team, these sessions are used in several reengineering methods such as process redesign, process optimization, root cause analysis, Lean, Six Sigma, value stream mapping, and other improvement approaches. For example, a methods analyst uses a brainstorming session as first and easiest method to begin process mapping. People enjoy brainstorming because they can display their knowledge, yet not be accountable for the results. There are many ways to conduct and compile the information, gained from a brainstorming session. The approach can be formal or informal, but flexibility must be maintained as each team or group will have an agenda. These meetings can take on personalities all their own. The brainstorming process described below should be adjusted to fit specific job and business needs. To begin the process, ask a group of people to assemble and discuss a problem you are having, in either a round table or informal setting. Discuss the problem and ask each individual member for suggestions. Present one idea at a time. This process should continue until contributions cease. It is acceptable for someone on the team to ask "Why?" You justify your response until you're out of responses. Ask the five "Whys" until you can no longer answer a question.

- Rule out all critical judgments.
- Encourage wild ideas. Enjoy the show.
- Encourage laughter and informality.
- Target for quantity, not quality.
- Look for improvements and combination of ideas.
- Encourage participants to extend the ideas of others.
- Make these meetings fun. Participation will never be an issue if you entertain the conferees.

Five Why's—Determine the Root Cause. To the extent possible, all direct contributor conditions should be identified. The process starts by asking "Why?" and continues by asking "Why?" for each condition. In general, it is important to ask "Why?" 5 times to get from the *effect(s)* to the *root cause(s)*.

Although this approach is subjective, it works extremely well if used as a simple, logical, and organized "thought map"—all that it is needed is ask to ask "Why?" during the brainstorm session every time a "Why?" is asked is considered a level. When the "*Why* does this cause. . .?" is repeated is considered, the next level of detail. In general, after asking *Why* does this cause. . .? 5 times is enough so that the event runs out of causes. This systematic approach is known as the "five "Why's rule of thumb"

Write drown the problem—writing the issue helps to define what is the problem. Start brainstorming the why of the problem and categorize the why's with the assistance of the cause-and-effect (CE) diagram (Chapter 6), ask again why, until there are no more other why's to ask. This in general takes place after asking why 5 times. Each why is considered a level. Figure 5.3 illustrates the concept of the five why's.[4]

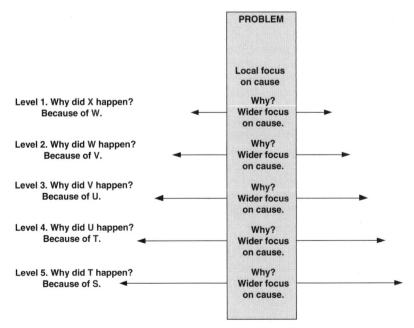

Figure 5.3   Figure showing the levels of the five why is: identifying the problem and then brainstorming "Why?" 5 times.

Steps for ensuring effective meetings are as follows:

- The problem needs to be simply stated.
- Document the ideas. Recognize every contribution.
- Document the name of the participant who suggested the idea.
- Try to produce around 100 ideas during one meeting.
- Schedule a follow-up session to "polish" the ideas of the previous meeting and create more ideas.
- Discuss advantages and disadvantages of these ideas.
- Sort ideas by categories and prioritize them (use the *XY* matrix method if possible).

Brainstorming is useful for completing a CE diagram, by reversing the process flow, performing a failure mode analysis, creating an *XY* matrix, and other important key problem-solving activities. The whiteboard is effective because it directs attention toward the board and away from the presenter. Inanimate objects are great at taking abuse. People will open up to a whiteboard before they grandstand to promote their idea. By promoting an idea, a person becomes the focus of the meeting. That is when the political scavengers appear and begin the process of stripping down the ideas to their basic elements and argue why this idea will fail. As all participants cannot take this type of cross-examination, it is best to propose ideas from a whiteboard.

5.5 GRAPHICAL REPRESENTATION OF A PROCESS IN A VALUE STREAM MAP

Value stream consists of everything, including non-value-adding activities that transform "raw information" and materials into what the customer is willing to pay for. Value stream consists of communication across the supply chain, regarding actual and order forecasted orders. Value stream mapping can be visualized as the network of processes and operations through which materials and information flow as they are being "transformed." Now imagine that each process in an organization is part of a river, as it passes its output(s) to the next downstream internal customer. Value stream can be applied to the manufacturing floor, to business processes and to service activities. It should be carefully constructed so that the future-state map could be drawn and implemented. Some organizations, when redesigning a process, draw the future-state value stream map first, to help ensure an effective handoff and continued implementation progress. The starting points for drawing the value stream are, (1) select a product family; (2) one person, personally, should lead the mapping activities; (3) start the door-to-door level—begin a quick walk along the entire door-to-door value stream; (4) include both the material and information flows—always collect current-state information while walking along the actual pathways of material and information and always, always start the first view of your current value stream map with the customer; (5) remember to bring your stopwatch and do not rely on standard times or information that you personally do not obtain—my electron microscopy professor in college used to say; "Don't trust anyone"; and... most of all (6) map the whole value stream yourself. Experience says always draw a draft of your value stream map by hand and in pencil, use the software latter. Figure 5.4 shows most of the commonly used symbols in the value stream map.

Exercise Questions

1. What are the five principles of Lean?
2. How efficiently does material and information flow through the process? How could the flow be streamlined?

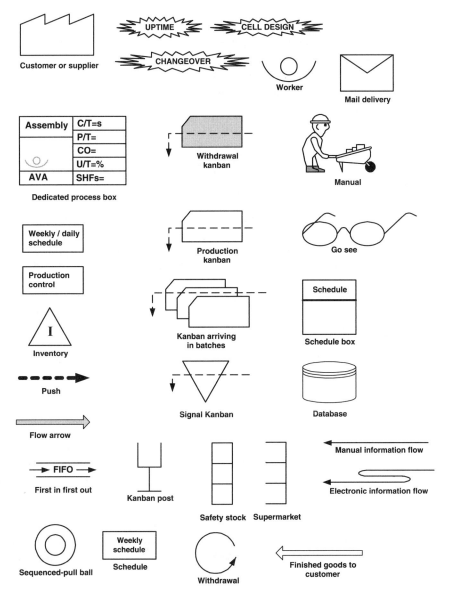

Figure 5.4 Some of the symbols used for the value stream map. Figure Acronyms: C/T = cycle time; P/T = production time; CO = changeover time; U/T = uptime (time that machine or software working); AVA = available working time per shift at that process step - (breaks + minutes + clean up or re-tooling) and machine or software uptime information; SHFs = shifts (number of shifts); LT = lead time; I = inventory [when inventory is observed in different locations between processe, a triangle with an I should be drawn for each location); W = waiting time between processes; FIFO = first in, first out; EPEx = every part/transaction, every...

3. Are there any obvious redundant or non-value-added (NVA) stages in the process that could be eliminated?
4. Which process stages should be given priority in process improvement activities?
5. What should the ideal process look like?

Answers

1. The five principles of Lean are as follows:
   - *Define value*—from customer perspective
   - *Value stream mapping*—map information flow and product/service transformation
   - *Continuous flow*—internally have everything ready as needed for a perfect symphony of service or production
   - *Pull production*—provide only when needed
   - *Perfection*—never give up continuous improvement

2. Because Lean improves the "flow" by eliminating waste (non-value-added activities), material and information flows like water downstream. Also, the flow is streamlined by eliminating variation. Remember that value stream are all actions, both value-added and non-value-added, required to bring a product through the main flows; therefore, the flow is streamlined by eliminating the NVA activities. Imagine that material and information are part of the same side of a coin; therefore, the practitioner should map both of them. The result is smooth and steady flow.

3. Yes, in any process there are obvious redundant NVA activities. However, when working the value stream the practitioner takes the big picture, not just the individual process steps. The practitioner should eliminate all those NVA activities.

4. The main activities of the process—the practitioner should start at the shipping end and work upstream, instead of starting at the receiving point and walking downstream. This will reveal the main processes linked most directly to the customers. These are main activities in the process that set the pace for other processes further upstream.

5. The ideal process should be a process that contains a minimum amount of NVA activities.

Some limitations of the value stream map (VSM) are as follows:

- Maps can become cumbersome to draw, if the process is complicated and has many converging and/or diverging paths.
- Frequently the VSM is a "snapshot in time" and may not represent all the variations of the process or the evolution of process knowledge over time.
- The VSM does not typically capture the causal structure, but it does document causes of waste and other relevant information.
- If there are multiple process permutations, VSM may not be the best initial tool.
- With value stream mapping, it is possible to graphically scrutinize business processes from beginning to end and draw a visual representation of every process involved in the material and information flows. Then you draw a future-state map to show how things should work for your best competitive advantage.

Value stream mapping is a powerful graphical tool that assists in analyzing information and material flow through an organization and identifies opportunities for improvements. Value stream mapping provides a visual approach to improve lead time, as well as planning and identifying kaizen events for effectiveness, and allows different areas of an organization to gain understanding of the overall material flow of the business. Value stream mapping is therefore a powerful tool for analyzing information and material flow throughout the organization and between organizations to identify and plan a Lean process.

Plan, do, check, act (PDCA), described in Chapter 4, Section 4.10, is part of value stream mapping; it is an improvement approach based on a scientific method of proposing a change in the process, implementing the change, measuring the results, and taking the appropriate action.

Plan the goals and changes for the process.

Do implement the changes.

Check and evaluate the results in terms of performance.

Act by standardizing and stabilizing the change.

5.6   EFFECTIVE WORKING TIME

The effective working time is the available time for work calculated after subtracting lunchtime, time for meetings, break times, and other work-related time(s).

Example 5.1   The operational definition is as follows:

- One workday $= 8\,\text{h} = 480\,\text{min}$
- Lunchtime $= 30\,\text{min}$
- Meetings $= 45\,\text{min}$

Therefore

$$\text{Effective working time} = 8\,\text{h} - \text{lunch} - \text{meetings}\ (1\,\text{h} = 60\,\text{min};\ 8\,\text{h} = 480\,\text{min})$$
$$= 480\,\text{min} - 30\,\text{min} - 45\,\text{min} = 405\,\text{min}$$

Example 5.2   The effective working time in an assembly line for two shifts is the available time for work after subtracting two breaks per shift, two lunchtimes, and two shift-change meetings:

- One workday $= 480\,\text{min}$ per shift $= 900\,\text{min}$ for two shifts
- Lunchtime $30\,\text{min} \times (2\,\text{shifts}) = 60\,\text{min}$
- Meetings $= 45\,\text{min} \times (2\ \text{shifts}) = 90\,\text{min}$
- Break $= 15\,\text{min} \times (2\ \text{breaks per shift}) = 30\,\text{min}$ per shift $\times 2 = 60\,\text{min}$ for the two shifts

$$\text{Effective working time} = 900 - (60 + 90 + 60) = 900 - 210 = 690\,\text{min}$$

5.7   CUSTOMER DEMAND

Customers demand is the number of transactions, services, or parts per week (per day, per month, or other established metric). An example of customer demand in a bank call center is telephone customer inquiries per time period. The operational definition is one business week $= 5$ days.

Example 5.3   Assume that in a branch bank customers make 300 deposits per week. Therefore, customer demand is calculated as follows:

$$\text{Customer demand} = \frac{\text{number of units}}{\text{week business week}} = \frac{300}{5}$$
$$= 60 \text{ deposits per day}$$

5.8   TAKT TIME

Takt time is the "available production time" divided by customer demand; in other words, Takt time is the pace of customer demand.

$$\text{Takt time} = \frac{\text{effective working time}}{\text{customer demand}}$$

Example 5.4   A second bank needs to improve deposit cycle time. To apply Lean methodology to minimize cycle time, the analysts must calculate Takt time. Customer demand for this second bank is 4000 deposits per day.

- Takt time = (available working time)/(customer demand)
- Takt time = 405 min per 4000 deposits
- Takt time = 0.101 min per deposit − rounding up
- Takt time = 0.10 min per deposit

To apply Takt time, the process of producing deposits must be stable. To achieve this stability, we must know the what, the who, the how, the where and the when required to satisfy customer demand. In Lean methodology we call this "standardized work."

5.9   PITCH TIME

Pitch time is a multiple of Takt time that allows the practitioner to consistently create, maintain, and sustain practical workflow through the value stream. To calculate pitch time, we must multiply the Takt time by the number of work units to be grouped to flow through the system. In general, pitch time is used do determine the optimal workflow quantity in those cases when manufacturing cannot to move work units through the value stream at the true Takt time pace. Pitch establishes workflow for a work unit to be moved from the beginning to the end of the value stream. Pitch creates a sense of urgency for the Lean office. Without it everyone will produce at his/her own schedule, or what he/she feels is reasonable. As soon as customer demand is determined, the organization must make the commitment to meet it. If demand can't be met with current resources and processes, Lean tools such as buffer or safety resources should be used. A *buffer resource* ensure meeting customer demand when customer trends vary. *Safety resources* ensure that customer demand is met when internal constraints disrupt the process flow. It is safe to say that pitch helps to create a sense of urgency. In other words

$$\text{Pitch} = (\text{Takt time}) \times (\text{number of workunits})$$

Example 5.5   Assume Takt time = 7 min; then

$$\text{Production time available} = 8\,\text{hs} \times 60\,\text{min/day}$$
$$= 480\,\text{min} - 30\,\text{min lunch} - 30\,\text{min meetings}$$
$$= 420\,\text{min}$$
$$\text{Customer demand} = 300\,\text{deposits per week (5 days peer week)}$$
$$= \frac{300}{5}$$
$$= 60\,\text{deposits/day Takt time}$$
$$= \frac{420}{60} = 7\,\text{min}$$
$$\text{Pitch} = \text{Takt time} \times \text{number of work units}$$
$$\text{Pitch} = 7\,\text{min} \times 60\,\text{deposits/day}$$
$$\text{One order} = 60\,\text{deposits/day}$$
$$\text{Pitch} = 420\,\text{min/deposit order}$$

The bank can meet the customer's demand. (*Note:* Pitch is recommended to be no more than 8 h and no less than 2 h. Pitch is used to set the pace for the completion of a unit of work in order to satisfy customer demand.)

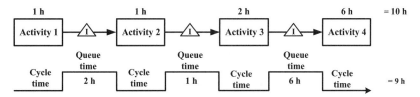

Total cycle time (10 h) + total queue time (9 h) = total lead time of 19 h

Figure 5.5   Value stream map showing the total lead-time in a high-level current-value-stream map. The total lead time for this graphical representation is $= 19\,h = 1140\,min$.

When drawing the future value stream map, we suggest forming an outer shell and then drawing the icon for the customer, then the first and last processes, including notes of communication between these processes and the customer. Then customer requirements can be added, including Takt time and pitch. To implement the demand phase successfully, the practitioner must

- Understand customer's demand
- Not attempt to microdesign the future state at this phase
- Be flexible
- Create a plan with the agreement of the team (getting buy-in from everyone)

5.10   QUEUING TIME

Queuing time is the time that a "work unit" waits for a downstream process to begin work on it. In the case of our banking example, one deposit waits for a downstream process to be ready to start work on it. See Figures 5.5–5.7.

5.11   CYCLE TIME

Cycle time is the elapsed time for an individual activity from start to completion. Cycle times are measured by time and motion studies.

5.12   TOTAL CYCLE TIME

Total cycle time is the sum of the cycle times for all of the individual activities in the process in the value stream. This is also referred to as value added time (VAT) because it is the time during which value is added as the work unit flows through the process.

5.13   CALCULATION OF TOTAL LEAD TIME(S)

Lead time is the total of all cycle times from all individual activities within the process for the Lean value stream, plus the queuing times that exist between activities in the process.

$$\text{Total lead time} = \text{total cycle time} + \text{total queuing time}$$

Example 5.6   The sales department of a steel company wants to apply Lean principles to reduce the time of their sales order process.

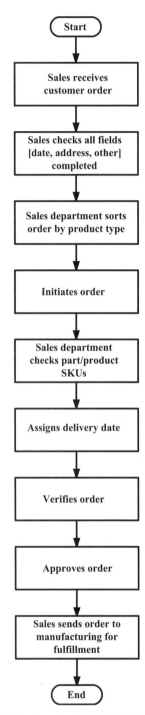

Figure 5.6   Example of a high-level flowchart.

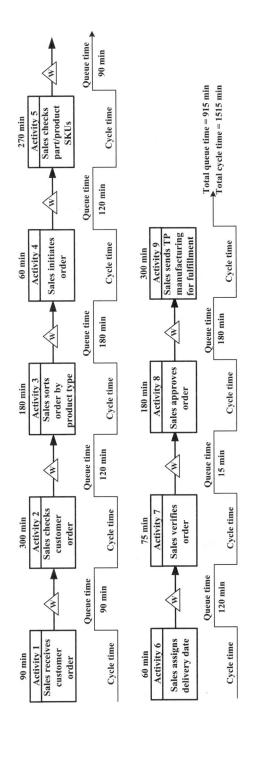

Total cycle time (1515 min) + total queue time (915 min) = total lead time of 2430 min = 40.5 h

Figure 5.7 Example of current-value-stream map illustrating waste and non-value-added Activities in the process flow.

103

Solution   List process steps:

- Sales receives customer order.
- Sales checks all fields (date, time, customer's address, phone number, name of company, etc.).
- Sales sorts order by product type.
- Initiates order.
- Checks product SKU (stock keeping unit) for availability.
- Assigns delivery date.
- Verifies order one more time.
- Approves order.
- Sends order to manufacturing for fulfillment.

Construct a high-level flowchart

Perform time and motion studies. From time to time, perform time and motion studies and collect data such as the following:

$$\text{Total cycle time} = 90 + 300 + 180 + 60 + 270 + 60 + 75 + 180 + 300$$
$$= 1515 \, \text{min} = 25 \, \text{h} + 15 \, \text{min}$$

where

$$\text{Total cycle time} = 25 \, \text{h} + 15 \, \text{min} \text{Total queuing time}$$
$$= 200 + 90 + 120 + 460 + 30 + 30 + 20 + 10$$
$$= 915 \, \text{min}$$
$$= 15 \, \text{h} + 15 \, \text{min}$$

From these observations, the current-value-stream map (CVSM) is constructed and total lead time is calculated.

Therefore, total lead time = total cycle time + total queuing time = 40 h + 30 min   ■

Example 5.7   A bank branch is having difficulty in improving deposit cycle time. After calculating Takt time, the bank branch manager needs to determine how many deposits could be done in one hour.

Solution   *Takt time*-available work time (available production time)/(customer demand) rate per shift or (available production time) ÷ (total daily quantity required)

- *Available work time* = 8 h − (1 h for lunch + 30 min for breaks + 30 min for meetings) = 6 h = 360 min
- *Customer demand for this branch* = 180 work units

$$\text{Takt time} = \frac{360 \, \text{min}}{180 \, \text{work units}} = 2 \, \text{min per unit}$$

- *Pitch* = Takt time × number of work units = 2 min per unit × work units = 360 min = 6 h

The bank has to move 1 unit every 2 min. *Pitch* is the time element that determines how often or when the bank deposits—work units—are moved throughout the value stream.   ■

5.14 VALUE-ADDED PERCENTAGE AND SIX SIGMA LEVEL

From Figure 5.7, lead time results, the following calculations were performed to obtain the value added percentage.

- Total lead time: 2970 min = 49 h 30 min
- Total cycle time (Process time): 1515 min. = 25 h 15 min = 4 days 1 h 15 min
- Total queuing time: 915 min = 15 h = 1.9 days
- Available working time = 8 h − lunch − meetings
- Units: 1 h = 60 min
- 8 h = 480 min
- Available working time = 480 min − 30 min − 45 min = 405 min per day
- Therefore, available working time = 405 min per day = 6 h 45 min
- Takt time = (available working time)/(customer demand)
- Takt time = 405 min/customer demand
- Value-added percentage = (total cycle time)/(total lead time)
- Value-added percentage = 2240 min/2970 min = 0.754 = 75.4%

Value-added percentage indicates how efficiently the process is converting workflow into exits (outputs)/completions. A low *process cycle efficiency* will have large NVA costs and great opportunities for cost reduction. From the calculation above we concluded that NVA activities make up 24.6% of the total lead-time. If we consider the percentage of NVA activities as defects per 100, we can calculate the Six Sigma level that the Six Sigma methodology uses and find that 24.6 is 0.7–0.8 long-term sigma and 2.2–2.3 short-term sigma. The reader should be reminded that this example illustrates the "as is" current-value-stream map. The improvement phase of the project removed in the downstream waste and resulted in the final future-value stream with an increase of $\sim 1\sigma$, meaning that after some waste was eliminated, the value-added percentage increased to 95.53%. Long-term sigma improved by 1.7–1.8 while short-term sigma improved by 3.2–3.3. The project was closed after the practitioner created the project control documentation. See Table VIII in Appendix A.

5.15 DRAWING THE CURRENT-VALUE-STREAM MAP

Imagine a SIPOC diagram as a total value stream. Now imagine the part played of the process and the part played by the output of the process; it is these two roles that will to be mapped to show the flow between processes and activities. Value stream mapping is an essential tool that helps visualize more than one single activity of the process and allows us to follow the flow. It also identifies the waste defined in Chapter 4 and how to eliminate it. The value stream map ties together the Lean concept and helps avoid "non-value-added" (NVA) activities. It is a communication tool since it links the information flow and cycle and queuing times.

5.15.1 Drawing Tips
- Draw by hand and pencil.
- Try to collect current information.
- Check for integrity of the data.
- Start at the end and work upstream.
- Bring your stopwatch to perform needed or missing time and motion studies.
- Do the value stream yourself.

5.15.2   Common Failure Modes

- Failure to link the VSM used with the type of question being asked on the thought map
- Failure to effectively locate the VSM in a sequence of improvement activities (*rarely* should the VSM be the first map used)
- Failure to understand the implications of the snapshot in time on the cycle time metrics reported
- Failure to consider multiple metrics

5.15.3   Common Definitions

The term *value* assumes creating something of value that a customer is willing to pay for.

The term *stream* refers to a sequential flow of activities needed to create product units and deliver them on time to the customer.

In transactions, each activity follows a separate value stream. A transaction is typically a family of activities/components that share a common processing sequence. For example, month-end closings, sales order lead-time process, on-transit process, billing process, and availability process are known as order-to-cash processes.

Order-to-cash-Processes Example.   This area of the value stream begins with the incoming customer order, which is usually received through the marketing or sales department (depending on the organization), customer service, or the Internet and ends up with the receipt of payment. Remember that transactions fall into the category of costs associated with product manufacturing that are nonproduction costs. These administrative processes ensure that the product is ordered, available, released, shipped, and paid for. These transactional processes play a crucial role in maintaining the business of any organization.

Let us not forget that in *Lean management the purpose is not to make someone work faster, but rather to streamline the flow to have the work move faster through the value stream.* Depending on the flow of the transactional process, there may be some overlap with the raw material-to-finished product/transaction value stream. In drawing the CVSM, the following assumptions and operation definitions should be noted:

1. 1 day $=$ 1 business day $=$ 8 h.
2. Process time $=$ time required for each of the activities within a process.
3. Queuing time $=$ time delayed between one activity and another.
4. Cycle time $=$ time required for one activity to perform one unit of work.
5. Queuing time for this example refers only to those times delayed due to waiting that happens during the working hours of newspaper team.
6. Waiting time of newspaper team.
7. Process time and waiting time of other business departments and editing company.
8. The process time and queuing time of other business departments in the editing company that happen after the working hours of newspaper team until the staff of the next working hours are excluded from the queuing time of newspaper team.
9. The time required by the editing company to print out newspapers after all the processes of newspaper team are excluded from the queuing time.

5.16   DRAWING THE VALUE STREAM MAP

An organization has to deliver the corporate newspaper twice a week on time. However, Mr. Hank, [5] the senior corporate executive of communications, finds that his team has to work overtime for the on-time-delivery of the *Corporate Communications Newspaper*. After taking a course in Lean Sigma that included value stream mapping he decides to apply what he learned in class. He starts by drawing the current-state map.

- *Background of System Development* — by performing activities related to issuing newspapers such as identifying news sources and writing articles in a single system, reduce contact time and process lead time.
- *System overview* — the system consists of three phases: register sources of articles, writing articles, and approving articles. When each of these phases end, a message is automatically sent to the email addresses and mobile phones of the responsible person.

The staff of the business department registers sources of articles when they feel the need to post an article and the newspaper team determines and informs the staff whether they will accept it as a news article. When responsible person of either the newspaper or business teams writes and/or revises news articles, a message is sent automatically to counterparties. Therefore, it is expected that the waste of time or queuing time arising from making contact can be reduced with the establishment of the system. In terms of article approval, the newspaper team leader can approve news articles in the system once revisions are done by responsible staff, eliminating the activities of printing out and reviewing the news articles. Instead of starting at the newspaper meetings with his team and editor, he started at the shipping end of the newspaper and "walked upstream" to "see" the processes that were most linked to his customers, which set the pace for other activities further upstream. He carried with him his stopwatch and didn't rely on the standard times given to him by his people. He decided that he was going to rely only on information he gathered himself because reported data may reflect times when everything is running well. He learned that his ability to envision the future state of the value stream depends upon personally going where the action is and understanding and timing what is happening. He also decided to draw the current-value stream himself even if several of his people remained involved. He learned during his training that if different people map different segments, no one would understand the "whole" picture.

His instructor advised him to always draw by hand and pencil and then use the software to transfer the map. He resisted the temptation to use the software first. By drawing by hand and pencil, he could sketch the process and clean it up latter, then transferred it to value stream map software.

Drawing the Current-State Map. The boundaries that the corporate newspaper first map are the door-to-door flow of the product (completion, release, and shipping) of the corporate newspaper.

Mr. Hank's *operational definitions* are as follows:

One day $= 8\,$h

Process time $=$ time required for each activity in the Process

Cycle time $=$ time required for one activity to perform one unit of work

Queuing time $=$ time delayed (waste $=$ delay) from one process activity to the next one. *Queuing time* refers only to time delays due to

- Waiting that takes place during the working hours as defined above for the newspaper team
- Waiting time of newspaper team within their own shift
- Process time and waiting time of other corporate departments and editing of the newspaper
- Process time and queuing time of other business departments and editing that took place after working hours and waiting for the newspaper team until staff of the next shift arrives and are excluded from the queuing time

Figures 5.8–5.11 show Mr. Hank's current-value stream for the release and shipping of the *Corporate Communications Newspaper. Note:* Time required for the outsourced editing company

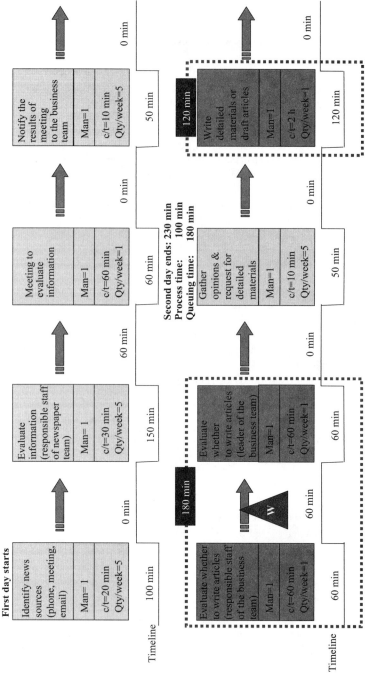

Figure 5.8 Value stream map showing the first view of the current-state map (part 1 of 4).

108

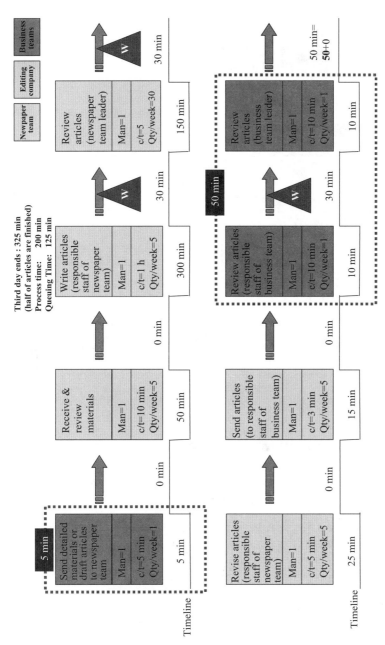

Figure 5.9 Value stream map showing the first view of the current-state map (part 2 of 4).

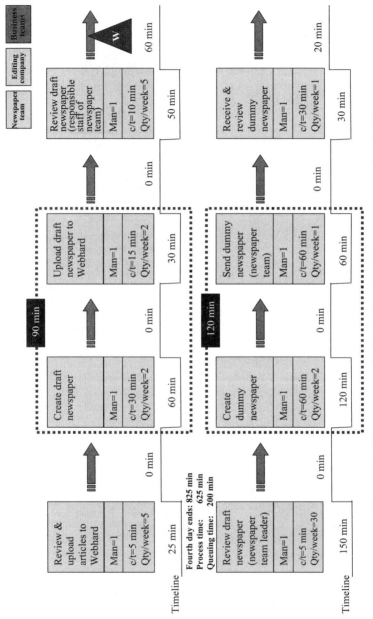

Figure 5.10 Value stream map showing the first view of the current-state map (part 3 of 4).

110

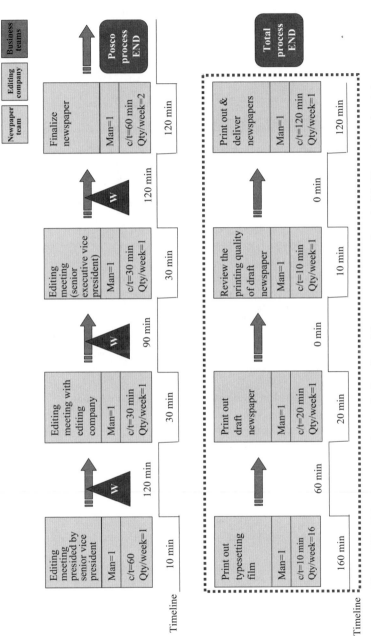

Figure 5.11 Value stream map showing the first view of the current-state map (part 4 of 4). The customer is not shown.

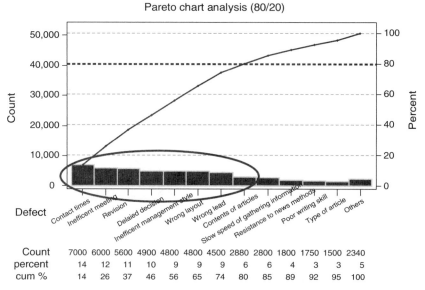

Figure 5.12   Results of the Pareto analysis for the current-value stream map.

and printing of the newspaper after the completion of the Corporate team activities (research and writing of the articles, approvals, etc.) was excluded from the operational definition of queuing time for this case study value stream map.

Results of the Current-Value-Stream Map.   It was observed that the activities to contact business departments or outsourced editing company in identifying news sources and writing articles had a large amount of waste. The results were:

- Queuing time due to inefficient meetings and delays in decisionmaking.
- Analysis of time required to perform the processes

Results of the CVSM, times are as follows:

- Total lead time = 2280 min = 40 h = 5.0 days
- Process time = 1515 min = 25 h = 3.1 days
- Queuing time = 915 min = 15 h = 1.9 days

Following the analysis of the CVSM, a Pareto analysis was performed (Fig. 5.12).

Results of Pareto Analysis.   The Pareto analysis showed that his longest cycle times were for contact times, time spent in inefficient meetings, revision of the drafts, delayed decisions, inefficient management styles, and incorrect decisions resulting in an incorrect newspaper layout. As a result rework of the layout was often necessary. Root cause analysis was performed, and the following recommendations were issued: "Reduce contact time and process lead time by performing activities related to issuing newspapers such as identifying news sources and writing articles by unifying in a single system."

In the CVSM example, it was observed that the activities to contact business departments or editing company in identifying news sources and writing articles generated a large amount of waste. The *queuing time* due to inefficient meetings and delays in the process of decisionmaking is also wasteful.

5.17   WHAT MAKES A VALUE STREAM LEAN

Value stream mapping requires a number of future-state maps, each a little leaner and closer to the ideal. We are seeking a process that makes only what the next process needs, when it needs it. We are trying to link all processes, from the final customer back to the raw material, in a smooth, seamless flow without detours that generates the shortest lead time, highest quality, and lowest cost. Future-state maps need to eliminate the sources, or "root causes" of waste (the "eight wastes of Lean") or TIM WOODS in the value stream.

Remember that one of the most significant sources of waste is overproduction:

- Watch for the waste of excess inventory
- Watch for waste of materials handling and storage.

Waste is everything that requires extra people and equipment, waiting for systems to become available and other related issues. Wastes are shortages, such as processes that are busy making the wrong parts as in the case of banking working on the wrong adjustments or working on the wrong transactions or delivering the wrong services. Waste is "everything that lengthens the lead time," which impairs the ability of business to respond to changing customer requirements.

5.18   THE FUTURE VALUE STREAM MAP

The purpose of value stream mapping is to highlight sources of waste and eliminate them by implementation of a future-state value stream that can become a reality within a short period of time. The first iteration of your future-state map should take product designs, process technologies, services as provided, transactions "as is," and in the case of manufacturing, plant locations as given. Another question pertinent to future-state VSM would be: "What can we do to improve what we have?" The purpose of value stream mapping is to highlight the sources of waste (muda) as Mr. Hank did in the case study presented above. Eliminate waste. The goal is to construct a chain of production in which an individual process or activity closely approaches the goal of producing only what its customer need only when they need it. Pertinent questions might be as follows:

- What is the Takt time, based on the available working time of the downstream activities that are the closest to customer?
- When can continuous flow processing be used?
- At which single point in the production chain ("pacemaker process") production should be scheduled?
- What increment of work should be consistently released and taken away at the pacemaker process?
- What process improvements are necessary for the value stream to flow as the future-state design specifies?
- Calculate available working time on the basis of corporate operational definitions.
- Calculate Takt time:

$$\text{Takt time} = \frac{\text{available work time}}{\text{customer demand}}$$

- The plan for achieving the future-state value stream should be a compact document that includes
  The future-state map
  Detailed process-level maps or layouts
  Yearly value-stream plans

5.19 SUMMARY

1. The critical place to begin any improvement effort is a clear specification of the product value as perceived by the end customer.

2. The process box stops wherever processes are disconnected and material flow stops.

3. Typical process data include cycle time, changeover time, uptime, production batch sizes, number of operators, number of product variations, pack size, working time, and scrap rate.

4. "Push" means that a process produces something regardless of the actual needs of the downstream customer and "pushes" it ahead. To qualify as "pull," parts must not be produced or conveyed when there is no kanban or signal. The quantity of parts produced must be the same as specified on the kanban.

5. The shorter your production lead time, the shorter the time between the time when services were requested and the time they were paid for at delivery.

6. For maps with multiple upstream flows, use the longest time path to complete total lead time.

GLOSSARY

benchmarking Structured approach to identifying a world-class process.

current value stream map A graphical representation that includes waste of the service process or transaction process or product process to visually determine current conditions.

continuous flow Producing or moving one item, information, activity, or other at a time through the process steps, with as much continuity as possible.

cycle time Time in which an activity is completed from beginning to end. Lapsed time from start to end.

demand—customer demand Quantity of work units required by the customer.

excess inventory Related to overproduction. Inventory beyond that needed to meet customer demand.

excess motion Unnecessary motion caused by poor workflow.

extended team members Individuals who provide special skills or expertise to the core implementation team but who are not responsible for the implementation

FIFO *First in–First out* is a work control method used to ensure that the oldest work (first in) is the first to be processed (first out).

flow Movement of material or information.

future value stream map Graphical representation that shows opportunities of improvement (waste removal), identified in the current value stream map.

heijunka Heijunka is also known as *load leveling*—balancing the amount of work to be done (load of work) during one day with the capacity to do the work.

Just in Time (JIT) production systems that produce just what it is needed and just when it is needed, in the just needed amount.

kaizen Continuous improvement to create more value by taking all those non-value-added added activities out of the process, to produce a transaction or deliver a service with less waste.

kanban Device that "signals" authorization and instructions for the production or conveyance of activities in a pull system.

Lean Paradigm based on the fundamental goal of eliminating waste and maximizing flow.

Lean production System in business for organizing and managing product development information, operations, suppliers, and customer relations that will require minimum human effort, less space, at a lower cost, and at a minimum cycle time with the minimum amount of mistakes to provide exactly what the customer desires.

Lean thinking.   Five steps proposed by James Womack and Dan Jones in 1996.

line balancing   Process in which work elements are evenly distributed within a Value Stream to meet Takt time

muda   See **waste (muda).**

non-value-added-processing   Reworking — the product or service should have been done correctly the first time. An example would be deburring—parts should have been produced without burrs.

overproduction   Producing more than customer demands.

pitch   Multiple of Takt time that will allow to create, maintain, and sustain a consistent and practical workflow throughout the entire value stream.

plan, do, check, act (PDCA)   Improvement cycle based on scientific methods of proposing changes in the process. Implement the change, *measure the results*, and take appropriate action.

problem solving   Team working together to follow the defining of the problem, analyzing possible causes, identifying possible solutions, developing action plans, and evaluating and renewing action plans. Finding the *root cause*.

process   Sequence of operations consisting of people, materials, methods, environment, and measurements for the creation and delivery of a product or service.

production lead time   Throughput time and total product cycle time—time required for information to provide a service or perform a transaction moving all the way through the process, from start to finish. In manufacturing, this term is known as "door-to-door time."

pull   System of creation and delivery in which nothing is produced bv the upstream supplier until the downstream supplier signals a need.

pull system   The flow of resources in a production process by replacing only what has been consumed.

push   Conventional work in which schedules are pushed along according to on sales projections and availability of materials.

queue time   Amount of time that a work unit waits before a downstream process is ready to work on it.

red tag   Label used in a 5S implementation to identify items that are not needed or that are in the wrong place.

safety resources   Means of meeting customer demand when internal constraints or inefficiencies disrupt process flow.

set in order   Second activity in the 5S method.

shine   Third activity in the 5S system.

sort   First activity in the 5S system.

standardize   Standardize for 5S—The fourth activity in the 5S system. Involves creating the rules for maintaining and controlling the conditions established after implementing the first three Ss. Visual control.

standardize work   Agreement upon a set of work procedures that establishes the best method and sequence for each process. Standardize work is implemented to maximize efficiency while simultaneously ensuring safe conditions.

stream   A sequential flow of activities needed to create product units and deliver them on time to the customer.

supermarket   System used to store a set level of finished-goods inventory or WIP and replenish what is pulled to fulfill customer orders (internal and external)

sustain   Fifth activity of the 5S system where a team or person ensures adherence to 5S standards through communication, training, and self-discipline.

takt time   Pace of customer demand. Determines how fast a process needs to run to meet customer demand. Available production time divided by customer demand.

team leader   Person who facilitates the value stream management process from beginning to end until a complete kaizen plan is created).

theory of constraints   A set of tools that examine the entire system for continuous improvement.

thought process map   Visual representation of thoughts that act as a roadmap to progress through DMAIC.

total cycle time   Total of the cycle times for each individual operation or process or work area in the value stream. Total cycle time ideally equals total value-added time.

total lead time   Total of all cycle times from all individual processes within the Lean office value stream, plus the queue times that exist between each process.

toyota production system   Toyota's system to provide the best quality at the lower cost. This model grew, thanks to the publication entitled *The Machine That Changed the World*, 1990.

underutilized people   Term referring to underutilization of mental, creative, innovative, and physical skills and abilities of human resources.

U-shaped cells   U-shaped work area layout that allows one or more workers to transfer and process work units or one small group of pieces at a time.

value   Value is the exchange in which the customer pays. Value can also equate to quality over cost.

value-added percentage   The percentage of the total lead time that is spent actually adding value to a work unit. To calculate value-added percentage, divide the total cycle time by the total lead time.

value stream   All value-added activities within a process, required to deliver a service, complete a transaction, or complete a product from launch to delivery. Collection of all steps (value added and non-value-added) involved in making the transformation from raw material to what the customer is willing to pay for.

value stream champion   Person with the authority and responsibility to allocate the organization's resources during the life of the project.

value stream management   Sequential eight-step process used to implement Lean concepts and tools derived from the Toyota production system.

value stream mapping   A graphical representation of every step within the information flow needed to bring a service from order to delivery. See also **value stream process mapping**.

value stream process mapping   Visual representation of the material and information flow of a specific product family.

visible pitch board   Visual control method that will control the flow of work throughout the day. Shows how each person in the value stream receives his/her portion of work

waste (MUDA)   All those activities that use resources (human, stems, or other) with no value to customer; also any process step activity that uses or consumes resources and creates no value to the customer.

work balance chart   Visual display or the work elements, times, and workers (associates) at each location.

work unit   Specific measurable amount of work that is customized and treated as a whole (unit).

work unit family   Group of parts that share common equipment and processing.

REFERENCES AND NOTES

1. Taiichi Ohno—considered to be the father of the Toyota production system, also known as *Lean manufacturing*, wrote several books, the most popular of which is T. Ohno, *Toyota Production System: Beyond Large-Scale Production*. Feb.–May 1990.

2. Celise Yamada Nomoto—Lean practitioner; Gerente Distrital de Vendas Pleno; Eli Lilly do Brasil LTDA.

3. M. Rother and J. Shook, *Learning to See*, Lean Enterprise Institute, June 2003.

4. O. C. Hodgins, *flextronics root cause analysis training course*, 2007.

5. Mr. Hank, corporate communications senior executive, collects information (time and motion studies) by walking through the actual pathways and information flow himself. He starts with a quick walk along the entire door-to-door value stream to get a "feeling" of the flow of the process. He knows the sequence well. After his quick walkthrough, he returns and gathers information at each activity. Mr. Hank's customers are the executives of the corporation and the employees of the corporation who maintain communication with the vision and mission of their organization through the corporate newspaper.

6 Introductory Statistics and Data

6.1 OVERVIEW

In this chapter, the practitioner will learn the value of organized information gained through experience, observation, and experiment. We will introduce techniques that best present these data. Once facts are presented clearly and accurately, they can be critically examined to guide the design of practical, economical, and effective products, processes, and events.

Each craft and "craft-person" uses a set of tools to improve performance. Just as the machinist has micrometers and calipers and the patternmaker has chisels and planes, the methods analyst uses tools to quickly achieve superior results. Good and timely information includes the identification of the service to be produced, the transaction to be done, or the product to be manufactured; information is also required for delivery schedules, operational times, resources, machines, capacity, special materials, and tools. Facts must be presented clearly and accurately, and then critically examined to define and implement practical and cost-effective methods.

6.2 INTRODUCTION

Data may consist of numbers, words, images, graphical representations, measurements, or information on sets of variables. Analysis of procedures and processes is a critical aspect of Six Sigma and Lean Sigma. Much of this analysis involves large quantities of data necessitating knowledge and use of statistics. Large amounts of information can be overwhelming unless the data are carefully organized and interpreted. Statistics is used in many professions. Knowledge of statistics is critical to effectively apply Six Sigma and Lean methodology. The word *statistics* has two meanings. Many think of *statistics* as being numerical facts. Numbers that represent the value of a person's 401K account, the age of an employee, and the percentage of defective items in a product shipment are examples of statistics applied in this manner. Following are a few more examples of numeric information that we call *statistics* in everyday life:

1. According to *ESPN Magazine*, 13% of fans believe that God affects the outcome of sporting events [*ESPN Magazine*, p. 80 (Dec. 2006), sample size = 275,000].

2. According to *ESPN Magazine*, 54% of fans believe that Lance Armstrong did not use performance-enhancing drugs.

3. According to ESPN, a poll of sports fans resulted in the overrating of Terrell Owens by 30%, Kobe Bryant by 28%, Tiger Woods by 13%, Barry Bonds by 29%, Dale Earnhardt Jr. by 19%, Jeff Gordon by 17%, Alex Rodriguez by 18% and Dwyane Wade by 5% (sports fans sample size = 50,000).

Practitioner's Guide for Statistics and Lean Six Sigma for Process Improvements. By Mikel J. Harry, Prem S. Mann, Ofelia C. de Hodgins, Christopher J. Lacke, and Richard Hulbert
Copyright © 2010 John Wiley & Sons, Inc.

4. Before the Democratic National Convention in late August 2008, Barack, Obama held an 8% point lead among white women voters, 50% to 42%, but after the Republican Convention in early September, a poll found that John McCain was ahead by 12 points among white women, 53% to 41% (*USA Today*/Gallup Poll).

Statistics is also the science that helps us collect, analyze, and interpret data and then make decisions on the basis of the data. To use data, you must learn how to compress, summarize, and describe data. A few numbers computed from data include averages and percents. Numbers computed from data are the raw material for statistical inference, the science of drawing conclusions from data with the aid of probability. Summary statistics, such as an average, can be calculated from data, but the science of statistics is much broader than the common notion that statistics and summary statistics are the same thing. For individuals and corporations, decisionmaking is a daily occurrence. Many of these decisions carry a degree of uncertainty. Whether the decision is a simple one, such as choosing a new coffee flavor at the convenience store, or a more complicated one as in choosing a set of investments for a retirement fund, uncertainty is there. Obtaining and analyzing data can aid the decision process. Knowing that 85% of those who tried a specific new flavor, liked it, would probably lead you to try that flavor. Analyzing the performance of mutual funds could steer you away from poorly performing investments. Statistical methods help us make intelligent decisions in such uncertain situations. Without statistical methods, decisionmaking would be equivalent to basing a decision on the roll of dice. This text is concerned with *applied statistics* as used in Six Sigma and Lean methodology; it deals with applications of statistical theory to solve real-world problems. Here, you will learn how to think statistically, how to make educated guesses, and how to apply statistics to solve Six sigma and Lean problems.

6.3 GENETIC CODE OF STATISTICS

Assume that we have information on the credit scores of all people who applied for a specific credit card during the past year. The set of numbers that represents the applicants' credit scores is called a *dataset*; the name of each applicant is an *element*, and each applicant's score is an *observation*. Datasets can be very small or quite large. A local bakery might track the number of doughnuts of each flavor, sold each day. Assuming there are six flavors, they will record six observations per day. On the other hand, the federal government's Medicare database is enormous. Regardless of size, it is easier to draw conclusions from summary tables and graphs than from the original data set. Thus, we usually summarize data using *descriptive statistics*, which includes constructing frequency tables and graphs, and/or calculating summary measures such as averages. Therefore, a well-constructed graphical representation should not need much verbal explanation. We will learn about these concepts and techniques in Chapter 8.

In statistics, the set of all elements of interest is called a *population*, and a group of elements selected from a population is called a *sample*. A major part of statistics deals with making decisions, inferences, predictions, and forecasts about populations on the basis of results obtained from samples. For example, to estimate the medical costs of all employees, we collect information on medical claims for a sample of 100 employees. The area of statistics that deals with such decisionmaking procedures is known by a variety of terms, such as *inferential statistics*, *inductive reasoning,* or *inductive statistics.*

6.4 POPULATIONS AND SAMPLES

In statistics, we frequently use data from a *sample* to make one or more decisions about the corresponding *population*. Consequently, understanding the meaning of each of these two terms and the difference between them is crucial. Suppose that a statistician is interested in knowing:

1. The percentage of all adults who have life insurance policies
2. The average length of all long-distance phone calls made by customers of a telephone company
3. The average daily balance of all credit cards issued by a bank

In these examples, the statistician is interested in the population of *all* adults, *all* long-distance telephone calls, and the balances of *all* credit cards. Note that in statistics, a population is not limited to a collection of living things; instead, it can be a collection of any entity such as houses, cars, or banks. The population, whose characteristics are being studied, is usually called the *target population*. In most cases, obtaining data on every member of a population is not practical. Therefore, decisions are based on a portion of the population. For example, when a food company develops a new product, it is initially sold in a few selected geographical regions, called *test markets*, to estimate sales and decide whether the company should market the product on a national basis. In this case, the population consists of marketing regions across the entire country. The sample is made up of a few representative regions, selected as test markets, and sales data are collected from only these test markets. In other words, the set of elements selected from a population is called a *sample*. Collecting data from every element of a target population is called a *census* and is usually impractical, if not impossible. Typically, we perform a *sample survey*, in which we select a sample population, collect information from this subgroup, and then make decisions on the basis of this sample information. The method chosen to select the sample population is critically important to both statisticians and decisionmakers. For example, if we collect information on a variable from every customer that enters a bank on a given day, it would be referred to as a *census* for that day. However, if we collect information on 30 customers selected from all customers for that day (assuming that there are more than 30 customers), it would be called a *sample survey*.

The purpose of a sample survey is to make decisions about the entire population. It is important that results obtained from a sample survey closely match the results that we would have obtained by conducting a census; otherwise, any decision based on a sample survey will not be valid for the entire population. Suppose that a marketing firm wants to predict demand for a new product according to the information obtained from a sample of 1000 households. For the results obtained from this sample to be valid, the sample must contain families who belong to different demographic groups in almost the same proportion as exist in the population. A sample that closely represents the characteristics of the corresponding population is called a *representative sample*. If the sample is not representative, it is likely that actual product demand will differ from projected demand, causing the company to make an inappropriate decision regarding production level or even in the decision to offer the product.

A sample can be random or nonrandom. In a *random sample*, each element of the population has some chance of being included in the sample. However, in a *nonrandom sample*, this may not be the case. For example, if we handpick a set of elements or individuals from a population, it will be a nonrandom sample. Thus, a sample survey based on a few coworkers or friends may not be typical for all employees of the company. It is extremely unlikely that a nonrandom sample would truly represent the entire population. Using a random process, like a lottery, is one way to ensure that a random sample will be selected. For example, to select 12 employees from a company with 103 employees, we would write each of the 103 names on a separate piece of paper, put all 103 slips in a box and mix them thoroughly, and finally, randomly select 12 slips from this box. The 12 names drawn are a random sample. On the other hand, if we arrange all 103 names alphabetically and then select the first 12 names from the list, it is a nonrandom sample because the employees, listed 13th to 103rd, had no chance of being included in the sample. In practice, we use statistical software (such as MINITAB) and spreadsheet packages to select random samples. A sample may be selected with or without replacement.

In sampling *with replacement*, each time we select an element from the population, we return it to the population before we select the next element. Thus, in sampling with replacement, the population contains the same number of items each time a selection is made. As a result, we may select the same item more than once. Consider a production run of a medicine that is in pill form. Suppose that we

TABLE 6.1   Market Values of Five Companies

Company	Market Value (Billion Dollars)
A	239
B	176
C	384
D	46
E	11

select a pill at random, record whether it is broken, and put it back in the pile before selecting the next pill. Every time we select a pill from this pile, the pile contains the same number of pills. This is an example of sampling with replacement.

Sampling *without replacement* occurs when the selected element is not returned to the population. Thus, the size of the population is reduced by one element after each selection. Therefore, we cannot select the same item more than once in this type of sampling. Usually, samples selected in real life are without replacement. Consider the previous pill example. You would not want to select the same pill over and over again because you could misrepresent the proportion of broken pills in the production run. In this case, no pill is selected more than once. If we sample voters to seek their opinions on an issue, we would not return the selected voter to the population to ensure that the same voter would not be counted more than once. In statistics, we often use the terms *element* (or member), *variable*, *observation*, and *dataset*. We explain these four terms below.

Consider Table 6.1, which gives the total market value (in billions of dollars) for five companies. We have denoted the five companies by A, B, C, D, and E.

The information in this table as a whole is a *dataset*. Each company listed in the table is a *member* or *element*. Thus, company B is a member of this sample. We have collected information on the market value of each company. The market value is called the *variable* for this dataset. In other words, we expect each company to have a different market value. When we collect information on the market values of these companies, we observe the numbers listed in the second column of the table. Each of these numbers is referred to as an *observation* or a *measurement*. Thus, $384 billion, the market value of company C, is an observation or a measurement.

6.5   THE IDEA OF DATA

"All data is not numerical."

—Unknown

From the moment we are born, we become the data of the universe. We are an integral part of the statistical world, in infinitesimal and global dimensions. The data are the foundation on which all statistical inquiry must rely. We will provide you with an in-depth look at data and statistics and offer practical suggestions for implementation. In statistics, we extract information from data, such as facts, concepts, or instructions in a formalized manner suitable for communication, interpretation, or processing by humans or computers. Any representation, such as characters or analog quantities to which meaning is or might be assigned, is information. Data are raw information and thus have little meaning. However, manipulation of these raw data into management information is essential for decisionmaking. *Data* can be regarded as a sequence of quantified or quantifiable symbols. For example, text is a piece of data in which the letters and characters contained in the text are quantifiable symbols because there can be a finite number of them. Therefore, any letter of the alphabet or language character or symbol may be quantifiable data. Pictures, figures, recorded sounds, and animation are also examples of (quantifiable) data, because they may be quantified (using digital scanners, cameras, or recording devices) to the point that it is eventually difficult to distinguish them from their originals, since their reproduction is made from the quantified representation.

Therefore, data are observations or measurements, information (characters, symbols, etc.) and results of images, sounds, pictures, animations, or simple results of surveys in which opinions on a subject or subject area have been collected. For example, a dataset or a dashboard full of data remains lifeless information without some effective use of statistics. Data, without the application of statistical methods, will continue to be a list of gathered information. Two definitions, parameter and statistic, are required to proceed further:

- *Parameter* is a numerical measurement describing some characteristic of a population.
- *Statistic* is a numerical measurement describing some characteristic of a sample.

What is the difference between data and information? Merriam Webster defines *data* as factual information (as measurements or statistics), especially the factual information used for analysis based on reasoning or calculation. Merriam Webster defines *information* as the communication or reception of knowledge or intelligence obtained from investigation, study, or instruction. Data alone have no utility, but become information when they are interpreted; information is a collection of facts or data communicated through some "channel(s)." However, in many contexts, data and information are considered to be synonyms. Data, is the plural of *datum*. Information comes from Latin *informationem,* meaning concept, idea, or outline. In simple terms, data is the information that specifies *what, when, where* and *who*. By understanding the relations between different data items, information is obtained on the *how*. By recognizing patterns (through the use of statistics) within data, knowledge is acquired. For example, the knowledge of *why* some event occurs.

Available facts can be defined as data; processed data can be defined as information. The most common way of representing and transmitting information is through data. For example, the statement "Determining sample size is important" is a piece of data that can be received by someone, and is incorporated as information if he/she understands the concept. Statisticians and Six Sigma practitioners know what sample size means. Finally, it can be said, one can pray for wisdom, work for knowledge, and learn from information, but only data can be managed. When data are collected, the information obtained from each member of the population or sample is recorded in the sequence in which it becomes available. This data-recording sequence is random and unranked. Such data, before being grouped or ranked, are therefore called "raw data." To manage data, they must first be organized. The first step in this organizing process is to describe some important characteristics of data:

1. *Center*—value that represents the location of the middle of a dataset
2. *Variation*—measure of the differences between individual data values within a dataset
3. *Distribution*—shape of the "distribution" of the data
4. *Outliers*—data values that are located far away from most data values of the dataset

6.6   NATURE OF DATA

Data do not provide information. To analyze numeric data, we must understand the various types of data, their important characteristics, and their relationship to problems that we need to resolve and the statistical tools available to us. We must always ask ourselves what do we want to know. Once this is clearly defined, we can collect the appropriate type of data, select a scale of measurement (nominal, ordinal, interval, or ratio), and then analyze the data using the appropriate tools. We then have information. A plan is necessary, if we are to avoid collecting useless data. The statistical tools available to us are dependent on the available data. For instance, if we wish to perform an analysis of variance (ANOVA), we must collect continuous data, not discrete data. If the data collection plan is correctly followed and we properly analyze these data, we will be able to support decisions with facts rather than opinions. In subsequent chapters, we observe that the structure and nature of data influences the choice of the method used for the data analysis. This is why data collection is such an important step in the Six Sigma approach. Data may be categorized in many ways; dividing data into qualitative and quantitative categories is one approach:

- *Qualitative, categorical,* or *attribute data* include such attributes as make of a computer, hair color, and gender.
- *Quantitative data* consist of numbers (number of houses, cars, or accidents) or measurements.

Data can also be classified using scales or levels of measurement. The four scales of measurement are nominal, ordinal, interval, and ratio. The nominal-scale data are the lowest level, while ratio-scale data are of the highest level:

Nominal-Scale Data. The nominal scale applies to data that can be divided into different categories, used only for identification purposes. An example of a nominal scale would be different makes of cars. The names given to different makes, such as Ford, Cadillac, GM, Toyota, Jaguar, Hummer, and Mercedes, are for identification purpose only. We can neither rank these categories nor do any arithmetic operations such as addition, subtraction, multiplication, or division. However, the categories for such a variable can be listed in any order. Other examples of nominal data are names of companies, gender, marital status, or computer model. In addition, we can assign numeric codes to different categories such as male $= 1$, female $= 2$. However, in this case, the coding does not mean that one code is superior to the other.

Ordinal-Scale Data. Ordinal scales apply to data whose values or categories can be ranked. For example, in a survey people are asked to evaluate a product as excellent, good, or poor. These categories possess the characteristic that they can be ranked. We know that excellent has the highest rank and poor has the lowest rank. However, as in the case of a nominal scale, arithmetic operations are meaningless for ordinal-scale data. However, the categories can be coded in this scale of measurement, and these codes can be assigned superior rankings. For example, to code "excellent," we could choose a 1, good $= 2$, and poor $= 3$. With this coding, 1 is the most superior and 3 indicates the most inferior.

Ordinal scales are used to rank data into ordered lists or sequences, which yield information about relative importance. The Six Sigma "critical to" (CT) characteristics for size and color are variables measured on an ordinal scale.

Another example application of the ordinal scale is the manufacturing of a product, the *dash 8 rudder*; the team, after collecting data on the painting process of this *dash 8 rudder*, prepared a Pareto chart that ranked in order of importance, the possible causes or reasons for poor paint adhesion.

The ordinal scale places nominal data into ordered categories without implying any information about the distance between categories, and, like the nominal scale, it is applied only to discrete (or integer) data. Examples include (1) Mary took second place in a cross-country race or (2) Wiley & Sons rates some manuscripts excellent and others good. Note that there is no specific quantitative difference between excellent and good. Ordinal scales provide information about relative comparisons but contain no quantitative information; therefore, ordinal-scale data are useless for calculations.

Interval-scale data can be ranked, and the difference between two values can be calculated and interpreted. There is no inherent natural zero starting point where none of the quantity is present. For example, temperature possesses an interval scale. Assume on a day in December that the temperature is 40°F in Charlotte, NC and 20°F in upstate New York. We can find the difference between two values; we can say that Charlotte, NC is 20°F warmer than upstate New York.

On an interval scale, a zero temperature in itself may have an arbitrary or carry no meaning; that is, it does not mean complete absence of warmness. For the Fahrenheit scale, zero represents an arbitrary temperature below the freezing temperature of water; for the Celsius scale, zero corresponds to the freezing temperature of water. Similarly, a zero score does not mean that a person who scored zero has no intelligence, because the zero point is chosen arbitrarily. However, we could say that the student who scored 90 is probably a better student compared to that student who scored 45. We can say that the difference between the scores of these two students is 50. Another characteristic of interval data is that a ratio does not make sense for such data.

We use the interval scale to accurately measure small variations in a large object. For example, we may use a finely graduated scale sitting on a reference surface or a dial gage mounted on a rigid reference. If we consider the velocity of a car on a highway relative to the rotational velocity of the earth, we would be using an interval scale of measure. In this case, even when the car is at rest near the equator, the velocity would be approximately 1050 mph.

Ratio-scale data can be ranked and for which all arithmetic operations (addition, subtraction, multiplication, and division) give meaningful results. Suppose that the 1993 gross sales of organization A are $120 million and those of organization B are $60 million. We can say that 1993 gross sales of organization A sales are larger than those sales of B. A few other examples of ratio scales are family incomes, expenditures, and product prices. (*Note*: A zero point has meaning in ratio scale data. In the preceding example, we know that an organization that reports zero sales did not sell any products. A person with zero income has no income. The zero value on the Kelvin temperature scale means absolute zero where no heat remains. Thus, Kelvin values for temperature are examples of data measured on a ratio scale.)

We use the ratio scale of measure when we measure the length of an object using a ruler or tape measure. Other ratio-scale examples include position as a function of time at a constant speed, or weight as a function of the number of objects. In these cases, zero is considered to lie within the range of measurement. Interval and ratio scales are very important to Six Sigma practitioners because most data fall into these categories and these data types are rich with information. Table 6.2 summarizes this information.

To understand more about our processes, we must measure those parameters that characterize critical features. Without this understanding, we cannot improve our processes. Many quantities cannot be measured with the precision of a micrometer, and yet it is important that we measure them. The study of scales of measure helps us understand how these quantities could be evaluated to yield useful information. Some statistical tools require a certain type of data, and it is important to recognize this because the wrong type of data will be of no use.

The four scales of measure are summarized as follows:

1. *Nominal Scale.* This is a relatively crude way of classifying information such as yes/no, pass/ fail, grouping observations, or defect types where no precise scale is required.

2. *Ordinal Scale.* This method is somewhat more precise than nominal and is used to rank or order features (or items) such as causes for defects.

3. *Interval Scale.* This method offers more precision, but the scale may use a displaced or artificial zero. Many scientific measurements use interval scales; thus, one must be careful when applying mathematical operators to values from this scale type.

4. *Ratio Scale.* A ratio is a quotient of two quantities (a/b) with the same units. Ratios provide a way of comparing two *numbers or quantities.* The ratio of the number a to the number b may be written as a to b, $a : b$ or (a/b). *Ratio* is used when values are proportional throughout its range and includes an absolute zero.

Nominal- and ordinal-scale data are generally discrete, whereas interval- and ratio-scale data are usually continuous.

Discrete vesus Continuous Data. Two kinds of data are used for measuring purposes: *discrete* (attribute) and *continuous* (variable). One way to examine data is to merely count the number of

TABLE 6.2 Scales of Measurement

Field of Practice				General		Technical	
Layperson	Quality Profession	Six Sigma	General Statistics	Example	Focus	Nature	Clarification
Count	Attribute	Discrete	Nominal	Player number 8 was given 14 awards	The number of times something is counted	Categorical order is not inherent to the scale, but the frequency of occurrence can be established	The number 8 is simply a label that serves to categorize or otherwise classify the performance data (i.e., 14 awards)
Count	Attribute	Ranked	Ordinal	Mary took 2nd place in a foot race	The rank of something	Categorical order is inherent to the scale, but subdivision of the scale is not conceptually meaningful	The number 2 is the rank (i.e., performance placing). It's not possible to have place 2.5 - only 2nd or 3rd place
Measurement	Variables	Continuous	Interval	Bills IQ is 126	The measured quanity of something	Categorical order is inherent to the scale and subdivision of the scale is conceptually meaningful, but there is no absolute zero point	An IQ of 126.34 is possible, but an IQ of Zero is not possible.
Measurement	Variables	Continuous	Ratio	Sue's pet weighs 26.4 lb	The measured quanity of something	Categorical order is inherent to the scale, subdivision of the scale is conceptually meaningful, and there is an absolute zero point	A weight of 26.4 lb is possible and it's also possible to have Zero weight

occurrences, such as the number of times that an event happens or fails to happen. Note that data of this type cannot be meaningfully subdivided into more precise increments and are discrete by nature. Go/no-go and pass/fail data are discrete data. Data that characterize a product or process feature in terms of its weight, voltage, and size are said to be continuous by nature. In other words, the measurement scale can be meaningfully divided into finer and finer increments of precision. Time is a continuous scale; however, meeting or not meeting a "due date" is discrete data. To apply the normal distribution, one must have continuous data. For example, when driving a car, the driver uses both discrete and continuous data. The speedometer provides continuous data about the speed; a warning light provides discrete data when it informs the driver that a door is not fully closed or that the coolant fluid is too hot. Since a temperature gage indicates the actual temperature, the driver may observe the temperature trend and thus be able to avoid a serious problem. Thus, we see that continuous data provide much more information than do discrete data. Because of this, it usually costs less to collect continuous data than to collect discrete data. The reason is that a small amount of continuous data is rich in information, making it more effective in gathering meaningful observations and therefore, more information.

Example of the Advantage of Continuous Data. A Six Sigma champion asks why it is necessary to seek continuous data when several forms of discrete data are already available. The organization's Six Sigma master Black Belt explains that the validity of inferences made using discrete data is strongly dependent on the number of observations. It may take many discrete data values to have the same degree of confidence as we can obtain from just a few continuous measurements. In other words, the sample size required for discrete measurements is much larger than the sample size required for continuous data. After noting that, in many cases, it is very difficult to gather an adequate number of data points, the champion thought for a moment and then realized that discrete data tend to be

- Defect-related, available only after the defect has occurred and been recognized
- Reflective of long-term trends with no guidance as to what went wrong
- Expensive to collect because so much is needed
- Deficient in the amount of information obtained

whereas continuous data is often

- Available earlier in the process (one can even design measurement strategies to monitor the "ingredients" of the process)
- Reflective of short-term trends, allowing immediate corrective action
- Less expensive than discrete data, because less is required for a given level of understanding
- Rich with useful information

Figure 6.1 illustrates the advantage of continuous data compared to discrete data.

In summary, measurement leads to data that, in turn, lead to information and knowledge. Real knowledge of a process is achieved only when it is possible to express this knowledge in numerical form. Thus, once this knowledge is formulated with numbers, it can be analyzed using statistical tools. We tend to think of data as pages of numbers; however, data may be nonnumeric, and numerical data may apply to different categories with different characteristics. The nature and structure of data can affect the result of data analysis, or even preclude analysis. The correct combination of three factors: (1) the problem, (2) types of data, and (3) types of tools available to give the desired result.

6.6.1 Quantitative Variables and Data

A *quantitative variable* is a variable that can be measured numerically, such as the length of a checkout line at a grocery store, credit scores, or asset : liability ratio. For instance, the income of a family may be

The advantage of continuous data

To obtain the same level of understanding

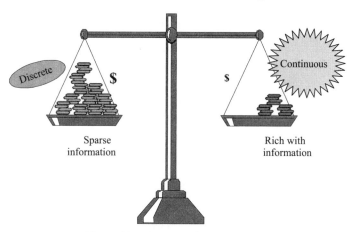

Sparse
information

Rich with
information

Figure 6.1   Advantage of continuous data.

$78,326.34 per year, or the asset : liability ratio for a company may be 1.84. Such quantitative variables may be classified as either *discrete* or *continuous variables*. The information collected on a quantitative variable constitutes a quantitative dataset.

Discrete Variables and Data. Quantitative variables are classified by the set of values that they can assume when counted. For example, we can count the number of people in a grocery store checkout line or the number of cars assembled by a manufacturer or the number of cars sold during a month by a car dealer. On the other hand, we cannot count the asset–liability ratio of a company or the width of a room. A variable that assumes a countable set of values is called a *discrete variable*. Note that there are no intermediate values between consecutive values of a sorted discrete-variable list. For example, the number of customers in the checkout line at a grocery store at any given time is a discrete variable because the number of customers must be 0,1,2,3, ..., n, and we can count these values. Other examples of discrete variables are the number of cars passing through an intersection during a green light, the number of homes in a city block, the number of newspaper subscriptions ordered during a one-week period, and the number of shares of a particular stock traded on a given day. The set of information collected on a discrete variable constitutes discrete data.

Continuous Variables and Data. Some variables cannot be counted because they can assume any numerical value between two numbers. Such variables are called *continuous variables*. Time taken by a customer to pass through the checkout line at a grocery store is an example of a continuous variable because it can assume any value, let us say, between 2 and 20 mins. The time taken may be 3.3 mins, 3.29 mins, or 3.292 mins. (Practically, we can measure continuous variables, such as time, as precisely as we want. However, we are limited theoretically by the limitations imposed by quantum mechanics.) Similarly, the asset : liability ratio of a company, which is the total value of a company's assets divided by the total value of that company's liabilities, can be measured to as many decimal places as one chooses (Here, a practical limit is imposed by the significant digits of the numerator and denominator that, in turn, limit the number of meaningful digits in the ratio.) However, neither time nor the asset: liability ratio can be counted in a discrete fashion. The entire set of information collected on a continuous variable constitutes continuous data. Other examples of continuous variables are the amount of gasoline dispensed when the meter reads one gallon (note that gas pumps aren't perfectly calibrated), time taken to commute from home to work, price of a house, height of a person, and the yield of cotton in pounds per acre. Note that any variable that involves money is considered a continuous variable.

Finally mathematical notation concisely expresses a relationship. This paragraph describes rounding decimals. Suppose that you are asked to round 15.067:

Step 1. Find the rounding place. For financial data, the rounding place is usually the hundredths place; in this case the digit is 6.

Step 2. Look at the digit to the right of the rounding place. If it is less that 5, leave the digit in the rounding place unchanged. If it is ≥ (greater than or equal to) to 5, increase the digit in the rounding place by 1.

Step 3. Write the rounded number. In this example, the original number was 15.067; rounded to the nearest hundredth, it becomes 15.07.

6.6.2 Qualitative/Categorical Variables and Data

A variable that cannot assume a numerical value but can be classified into two or more nonnumeric categories is called a *qualitative* or *categorical* variable. The data collected on such a variable are called *qualitative data*. These variables cannot be measured numerically but can be divided into different *qualitative* or *categorical* categories (e.g., make of a computer, hair color, gender). The quantitative variables are classified into *discrete* (e.g., number of houses, number of cars, number of accidents) and *continuous* (e.g., income, age, sales, profits, time). For example, an employee's wireless provider is a qualitative variable since it is drawn from the set of companies that provide cellphone services. Other examples of qualitative variables are the brand of a computer, the risk level of a loan applicant, and the status of an undergraduate college student (freshman, sophomore, junior, or senior).

Although nonnumeric variables are qualitative variables, variables represented by numbers are not always quantitative. For example, telephone area codes and postal zipcodes are numeric codes that identify different sections of a country or the world. Similarly, some software packages use numbers to encode qualitative variables. For example, to enter the gender variable into a spreadsheet may require assigning codes for female and male. Here, we could assign 1 to female and 2 to male and then enter these codes in the spreadsheet. Note that these numbers represent categories as opposed to numeric values. The set of information collected on a qualitative variable constitutes a qualitative dataset.

6.7 DATA COLLECTION

Proper data collection determines our ability to determine where we are and where we want to go. The precision, consistency, and strategy employed during data collection strongly influence our ability to accurately analyze, improve, and control; in other words, garbage in–garbage out (GIGO). This section deals with the tools and concepts needed to ensure the integrity of the data collected. Data collected using data collection sheets and control cards are entered into a database for further analysis. A control card is a tailor-made form for a specific component or product type. It uses graphics to tell how and where to measure critical-to (CT) characteristics. Control cards are vital to achieve measurement consistency across employees, shifts, and departments; they provide graphical details on how measurements are made and recorded. A data collection sheet is used to record the measurement when graphics are not required. A database must be logical, expandable, flexible, and traceable. It is a collection of alphanumeric information on critical-to-quality (CTQ), critical-to-defect (CTD), and critical-to-cost (CTC) characteristics. The database provides a structure; when populated and analyzed, we can define required performance levels and diagnose problems in products, processes, services, and transactions.

6.8   THE IMPORTANCE OF DATA COLLECTION

The data collection strategy is frequently established during brainstorm sessions. This encourages buy-in from participants and lessens the pain by using existing data sources where possible. For example, the information collected for an education study should include student, family, community, age, background, and other demographic information. Six Sigma practitioners assigned to a project should ensure that adequate data collection occurs during the measurement phase.[1] Practitioners of design for Six Sigma (DFSS) should develop a data collection plan during the design phase (D-IDOV-M): define—identify–*design*–optimize–verify–monitor. Certain crucial steps must be followed to ensure that data integrity and that measurement systems are stable and reliable.

- Clearly define the objective of data collection (define the problem).
- Search for existing or create operational definitions related to the objective.
- Ensure data integrity (accuracy).
- Develop a data collection plan that addresses required data types, sources, and frequency of collection; the plan should ensure that information on the data collector and the time of collection are included with the measured values.
- Ensure measurement repeatability and reproducibility.

Example 6.1   A Black Belt is assigned to assist the tool designers, toolmakers, production supervisor, and tooling supervisor as they analyze a problem with an assembly fixture. The team identified the problem, and is eager to fix it. They have already met and have several ideas on how to fix the problem.

Solution   The Black Belt redirects the energy of the team, as fixing the problem is the objective of the project, not the initial step. Without data, they do not know where they are, or what they need to accomplish. After the data were collected, the Black Belt explained to the team that an accurate analysis of the problem, based on collection of data with integrity, directly influences our ability to improve and control a process. We must understand what must be collected and its required precision before populating the database. If the data collected has integrity, then our ability to make well-founded process improvements will be enhanced.   ■

6.8.1   Control Cards

By using control cards and data collection sheets, employing the correct sampling strategies, and implementing a well-developed database, we can convert raw data into meaningful information from which valid inferences can be drawn.

Example 6.2   A manager concerned about measurement error voiced her concerns to a Black Belt. Each operator is measuring 'spar height' CTQs. The manager knew that each individual had a unique way of measuring the CTQ and that there were variations in the location of the measurements; some operators even used different measuring instruments. In cases such as this, how do you ensure consistency?

Solution The Black Belt explained the various ways to take samples, and then let the team determine the best approach consistent with the production mode, the batch production schedule, and the size and timing of production lots. A data collection sheet was created specifying parameters and measurement methods. The first step in ensuring data integrity is to create a control card with clear specifications on the product ID, the component number, the operator, and the locations and times of the measurements. An example is given in Figure 6.2. ■

A Black Belt explains that since measurement is critical, inconsistent measurements make it difficult, if not impossible, to make statistical inferences from the data. To ensure measurement consistency, we use a set of instructions. These instructions are included in either a data control sheet or a control card. A control card lists the characteristics to be inspected, the method of inspection, the instruments to be used and the location of the measurement in a graphical format.

Developing the Control Card. Group participation improves decisions on where and how to measure CTQs One of the best ways to capture the expertise of the team and obtain their buy-in is through brainstorming sessions on the different aspects needed to measure the CTQs. After we develop the control card, it is used for data collection when visual instructions are required. Data entry should be made in the same format required for analysis. Each control card is unique to a specific component or product type. Does the control card work as a measurement procedure? The answer is "No"; the control card defines where to measure. The procedure defines when and how the measurements should be taken.

6.8.2 Data Collection Sheet

Imagine an environment where engineering changes are occurring at very high rates for many of the company's programs. The Black Belt decides that the best way to assist his/her organization to identify

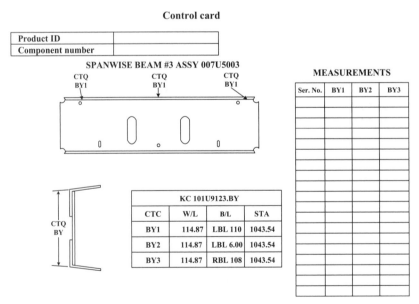

Figure 6.2 Illustration showing the relationship between control card and measurements.

Data collection sheet

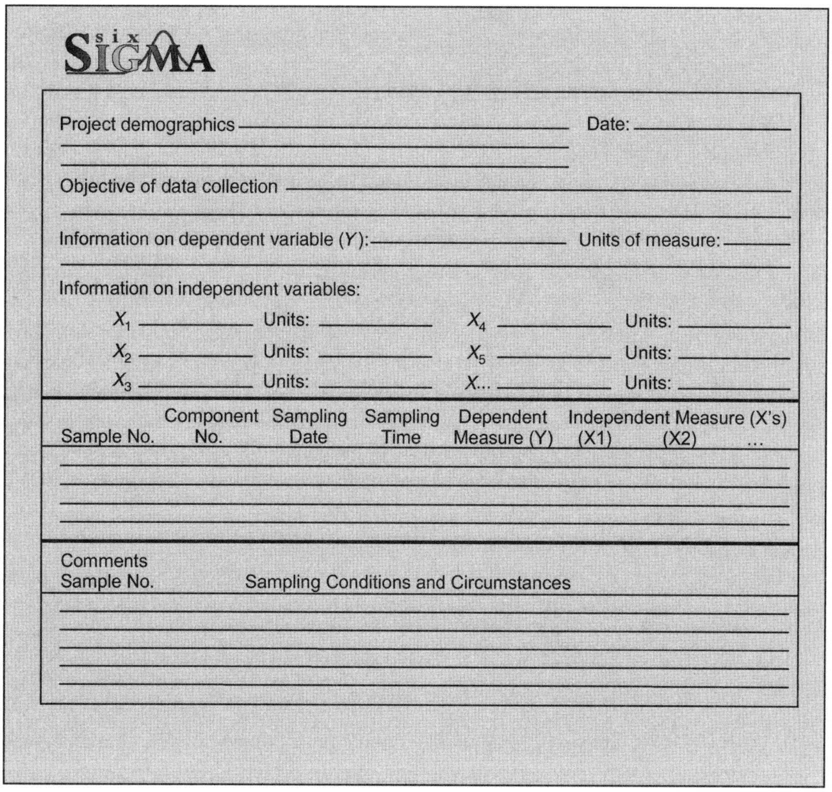

six SIGMA

Project demographics ______ Date: ______

Objective of data collection ______

Information on dependent variable (Y): ______ Units of measure: ______

Information on independent variables:

X_1 ______ Units: ______ X_4 ______ Units: ______

X_2 ______ Units: ______ X_5 ______ Units: ______

X_3 ______ Units: ______ $X...$ ______ Units: ______

Sample No.	Component No.	Sampling Date	Sampling Time	Dependent Measure (Y)	Independent Measure (X's) (X1)	(X2)	...

Comments

Sample No.	Sampling Conditions and Circumstances

Figure 6.3 Example of a data collection sheet.

improvements in the design process is to periodically collect data. The organization makes the Black Belt responsible for the project. In order to collect data, the team led by the Black Belt must create a data collection sheet. An example of a data collection sheet is shown in Figure 6.3.

Next, the Black Belt assembles a cross-functional team and ensures representation from all business areas (including engineering). The team maps the process and enters the engineering changes into the process maps. Once the team agrees on the process steps, brainstorming sessions identify possible causes for defects and the likely reasons for each occurrence. Then, the Black Belt integrates the team's ideas and assembles them into a draft data collection sheet similar to the one shown in Figure 6.3. After the team reviews the proposed data collection sheet, the Black Belt reminds them of the important role of the data collection sheet to ensure that data are entered in the same format as will be used for analysis. The current version was modified to address team concerns and to include additional team suggestions. The draft was presented again at the next team meeting. However, one engineer was concerned about how the data, once collected, would be used and said, "The data collected on engineering changes is an indicator of the different problem areas in engineering. With this in mind, do you think that people will voluntarily participate in the data collection process?" "The Black Belt replied: No; however, data with integrity must be collected following the principles of operation analysis."[1] Operation analysis is beyond the scope of this book. Once data are collected; the Black Belt should identify opportunities for improvement; then the team can move on to the next step, which is to improve the processes that prevent the organization from delivering high-quality products/services.

6.9 SAMPLING IN SIX SIGMA

Sampling results are greatly influenced by the selection method used to gather the sample. For example, in the semiconductor industry, wafer lots must be considered when developing a sampling strategy. A lot is group of wafers (in general 25 wafers per lot] that have undergone a specific process sequence, made by the same equipment and production staff, and completed the process during a given time period. Samples drawn from a specific lot should not be confused with a population. For example, population could be all the lots producing in the manufacturing floor during a week. (*Note*: The definition of *population* may vary from organization to organization.)

6.9.1 Random Sampling

Random sampling ensures that units selected are the result of pure chance. *Sequential sampling* is the selection of sample units in the order of production or receipt. Random sampling increases the likelihood that the measurements of the samples are representative of the process population. Its use improves the validity of population estimates and makes possible application of statistical tests. A random sampling is one where the conditions are such that each unit of the population has an equal chance of being chosen. Random numbers can be generated using calculators, computers, MINITAB, Excel, a bowl of numbered chips, or random throws of dice.

Stratified sampling is used when a production lot is the result of differences such as multiple employees, machines, and shifts; in other words, the product is assumed to be the combination of several smaller lots.

For example, an assembly department has encountered problems assembling an engine check valve because of the variation in the diameter of the check balls. A Six Sigma Black Belt investigated the problem. The manufacturing process data collected on ball diameter indicated that the process is capable of producing good parts. Therefore, there should not be an assembly problem. She reviewed the sampling plan and found that six parts were measured. The Black Belt observed that the operator chose the first six parts produced and asked, "Is this how you select your samples each time?" Yes, replied the operator; I always select the first six parts produced. Samples need to be selected at random to increase the likelihood that the measurements are representative of the process population. The validity of population estimates and statistical tests improve if samples are randomly chosen. Sampling the first few parts may not capture changes in process performance over time.

Changes in temperature (heat) and equipment wear may impact process performance. So, how do we ensure that the sample is random? Together, the team and the Black Belt generated random numbers and collected the data from parts chosen randomly. MINITAB and Excel assist in generating random numbers from tables, In MINITAB, the code is `Go to Calc > Random Data > Uniform`. Figure 6.4 illustrates this random sampling concept.

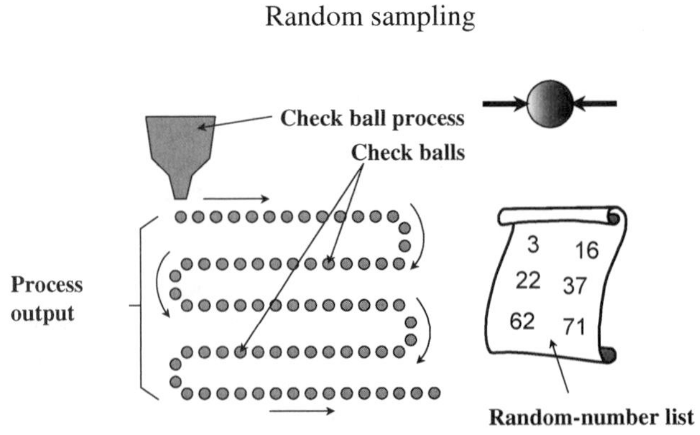

Random sampling

Figure 6.4 Schematic representation of the concept of random sampling.

6.9.2   Sequential Sampling

Sequential sampling differs from simple, double, or multiple sampling covered in Chapter 8. In sequential sampling, a sequence of samples are selected from a *lot*. If the length of the sequence is one, then the sample process is usually called *item-by-item* sequential sampling. For a sample size greater than one, this sample process is usually called *group-sequential sampling*. Item-by-item sampling is the more widely used form of sequential sampling.

Example 6.3   A ball-bearing manufacturer has identified the spherical diameter of the bearings as one of the CTQs for this product. The manufacturer has three manufacturing lines that simultaneously produce the bearings and needs to choose the proper sampling technique.

Solution   To determine baseline performance of the CTQ process, a cross-functional team prepares a sampling plan. The Six Sigma Black Belt assigned to the project is familiar with the process. He suggests that collecting data from samples be done in sequence (we will learn later that by comparing the variance between samples, we can gain insight into short-term process capability). The process was then sampled throughout the day to ensure that the samples represented the true output of the process. The Six Sigma Black Belt explained that samples would be collected using *sequential sampling*. To ensure that process variations over time would be detected, the Black Belt suggested that the operator test a sequential sample of five bearings each hour. The team used a data collection sheet to document its measurements; detailed instructions were included on the back of the sheet. Space was provided on the sheet to record traceability information on the selected samples as it was important to know when each sample was selected. Figure 6.5 shows the sequential sampling concept for the ball-bearing process. Sequential sampling can also be applied to transactional processes.   ■

Sequential sampling minimizes the number of items tested and clearly shows when a batch meets, or fails to meet, required standards.

6.9.3   Stratified Sampling

Suppose that we need to sample the population of a city such that households with different income levels are proportionately represented in the sample. Instead of selecting the simple and systematic

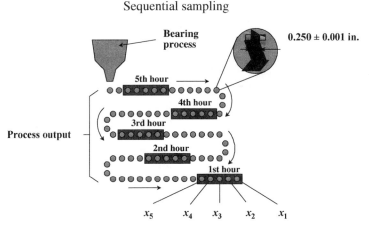

Figure 6.5   Illustration showing the sequential sampling for the bearing process.

method of random sample, we choose a slightly more complex random sample technique. First, we divide the population into different groups according to income levels, such as low, medium, and high. We now have three subpopulations called *strata*. We then select one sample from each subpopulation or stratum. The collection of samples from each of the three strata creates a stratified random sample. Usually the sizes of the samples selected from different strata are proportional to the number of elements in the subpopulations of these strata. Note that the elements of each stratum share a common characteristic (in this case, household income). Therefore, in a *stratified random sample*, we first divide the population into subpopulations called *strata*. Then, samples are selected from each of these strata. The samples collected from the three strata comprise the *stratified random sample*.

Example 6.4   After learning that a lot is the result of multiple operators and machines, a Six Sigma Black Belt explained her proposed sampling plan to the team; the product actually consists of several smaller lots. Therefore, I propose using a stratified sampling plan. Figure 6.6 illustrates a stratified sampling plan.

Stratified sampling improves the likelihood that measurements taken on the samples are representative of the combined process population. The team developed a data collection sheet to document their findings. Detailed measurement instructions were included on the data collection sheet. Space was provided on the sheet to record traceability information on the samples selected. The completion of a *production order form* is another example that requires a stratified sampling plan. The team has determined that one of the entries on the form is a CTQ. Forms originate at three different workstations. Thus, the lot for production order forms consists of three smaller lots. The sampling strategy for the CTQ requires that 18 sample forms be collected daily. To ensure that the samples are representative of the combined lots' process population, six forms are selected at random from the daily output of each of the three stations. The combination of the individually selected documents forms the stratified sample. Other examples of transactional processing requiring stratified sampling plans include a survey of different departments, the number of errors on forms generated by three different shifts, and the number of customer billing errors occurring at different company sites.

Stratified sampling

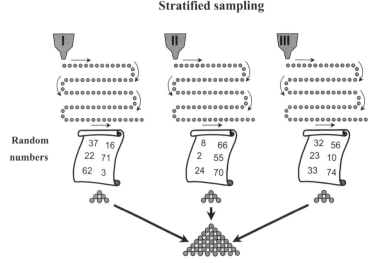

Figure 6.6   Illustration showing an example of stratified sampling.

6.10   SOURCES OF DATA

Sources of data include the following:

- *Internal sources* created within the company. Retailers collect data regarding sales and returns, while hospitals collect data on survival rates, waiting times, and costs resulting from waste such as expired medicines. Since internal sources are under company control, errors in the data can be easily identified and corrected.

- *External sources* are created outside the company's control. Sources could be other companies, government agencies, or various publications. A common external source is a credit bureau. Banks and credit card companies obtain credit reports from various credit reporting agencies to support lending decisions. The quality of the data obtained from an outside source is not within end-user control. This is not to imply that external sources always have lower-quality data than do internal sources, but since the end user is the customer for these data, any issues with an external data source must be dealt with through standard customer assistance procedures.

- *Surveys and experiments* are two primary methods used to obtain data. In a survey, a sample of individuals or objects is selected (hopefully at random) and responses to a set of questions are obtained. In some cases, individuals are asked a series of questions, as one might find in a poll or a questionnaire regarding consumer behavior. In other cases, the *questions* might require a series of measurements on the selected objects or individuals.

- *Census surveys* include every member of the population. The technique of collecting information from a portion of the population is called a *sample survey*.

- *Design of experiments* is a key Six Sigma methodology. An experiment is any test in which inputs (x's) are either controlled or directly manipulated according to a plan. Design of experiments is an approach to effectively and efficiently search for the cause and effect relationships between process variables (x's) and the process output (Y). The purposes for experimentation include (1) determining which factors (x's) are most influential to the output, (2) identifying the variables that are most influential to variability in order to reduce variability to the response, (3) obtaining the maximum amount of information using a minimum amount of resources, (4) finding factor (x's) settings that optimize the response and therefore minimize the cost, and (5) validating the proposed process improvement.

Experiments need not be *scientific*. A well-known example of a nonscientific experiment was the "Pepsi challenge." Individuals were given unmarked samples of Pepsi-Cola and Coca-Cola to taste and asked to state their preferences regarding the two sodas. The factors in this scenario were as follows: x_1 = sample of Pepsi and x_2 = sample of Coke. Two levels were selected for this pilot experiment: (1) different age groups (low = 18, high = 40 years old) and (2) geographic locations (low = California, high = New York).

6.11   DATABASE

A database provides organization and supports computational tasks for data collected from control cards and data collection sheets. A good database design is one that is logical, provides a meaningful input format to the user, identifies the originating sources (time, date, product, equipment, etc.), and is expandable and flexible.

Example 6.5   A new Black Belt recruit approached the master Black Belt for his first project review. The Black Belt in training spread an armful of papers on the table and began commenting on the problem at hand. The Black Belt stated confidently, that nothing seems to stand out as being a problem but it is obvious that the assembly jig needs some serious attention. "OK," replied the master Black Belt, "let's look at your data and see what they say." The Black Belt leafed through the pile of papers, struggling to locate the information he sought to present to the master Black Belt.

The master Black Belt commented that "there is great benefit to organizing the data; why haven't you done that?" There were several pages of data—not always consistent—and not always recorded in the same format. A database must be thoroughly planned, structured, and implemented so that it can be quickly accessed and manipulated. CTDs, CTCs, and CTQs must be identified, and you need to determine the format that will be used to record the data. The master Black Belt suggested that thought should be given to the possible long-term uses of the data. After defining and organizing the CTDs, CTCs, and CTQs, the master Black Belt accurately determined process variations that the Black Belt had not identified in his preparatory review. These discoveries led to additional questions; further analysis revealed other significant patterns and relationships.

6.12   SUMMARY

The following points were discussed in this chapter:

1. A well-defined control card is required to ensure measurement consistency between operators and shifts.
2. Data collection sheets and control cards are used to consistently collect data on CTQ, CTD, and CTC characteristics.
3. Data collection sheets and control cards can be used to gather data on both dependent and independent variables.
4. Sampling results are strongly influenced by the method of selecting the sample.
5. Samples should be selected so that they are representative of the process population.
6. Stratified sampling techniques are used when it is known that the lot results from different sources such as shift, employees, and machines.
7. In taking many stratified samples from a given lot, the average number of defectives will be the same as if the samples were drawn at random.
8. In random selection, each unit in a lot is assumed to have an equal chance of being included in the sample.
9. A database is used to capture, store, and analyze the information gathered from the control cards and data collection sheets.
10. Sequential sampling is simply the selection of a sample in sequence (i.e. as produced or received).

GLOSSARY

abscissa   Horizontal axis of a graph.

analysis of variance (ANOVA)   Statistical procedure that uses variance to determine difference of means. ANOVA is used when we have discrete x's with continuous y's.

applied statistics   Applications of statistical theory to solve real-world problems.

attribute data   Numerical information at the nominal level.

census   Collection of data for every element of a targeted population.

champion   Member of a senior management who is responsible for the logistical and business aspects of the program, including project selection and tollgates.

continuous variables   Some variables cannot be counted, but they can assume any numerical value between two numbers.

control card   Used when data collection instructions need to be visual. Lists the characteristics to be inspected, the method of inspection, the instruments to be used, and the location of the measurement in a graphical format

CT   "Critical to" characteristic.

CTC   "Critical to cost."

CTD   "Critical to defect."

CTQ   "Critical to quality" (defect-free order).

data   Factual information used as a basis for reasoning, discussion, or calculation. Often refers to quantitative information.

database   Basic organization and computational tasks on data collected by different methods such as control cards and data collection sheets. A good database design is logical; the format is meaningful, expandable, flexible, and provides traceability to originating sources.

data collection sheet   Log used to record data in the same format that is required for analysis or for entry into the database.

dataset   Group of numbers or information describing some characteristic.

dependent variable   Response variable. Y depends on the independent variable X; thus, $Y = f(X)$.

descriptive statistics   Methods and techniques used to reduce the quantity of data to a manageable size.

design of experiments   Commonly known as DOE or DoE—purposeful changes to the inputs (x's) or factors to a process or activity in order to observe corresponding changes in the outputs (y's) or responses. A means of extracting theoretical knowledge of the process or activity by the use of mathematical methods.

designated experiments   The primary method used to obtain data involving the use of a treatment and at least one measurement.

distribution   Tendency of large numbers of observations to group themselves around some central value with a certain amount of variation or scatter on either side.

element (or member)   the name of each applicant.

experiment   One of the primary methods to obtain data.

external data source   Data created outside the company's control.

grouped data   Dataset presented in the form of a frequency distribution.

histogram   Vertical display of a population distribution in terms of frequencies.

inferential statistics   The area of statistics that deals with decisionmaking procedures. This branch of statistics is also called *inductive reasoning* or *inductive statistics*.

internal data source   Data created within the company that is being analyzed.

interval scale   Data that can be ranked and for which we can find the difference between two values are said to have an interval scale.

nominal scale   Data that are divided into different categories that are used for identification pupose4s only are said to have a nominal scale.

nonrandom sampling   To handpick a set of elements from a population.

observation or measurement   The value of a variable for an element is called an *observation* or *measurement*.

ordinal scale   Data that can be divided into different categories that can be ranked are said to have an ordinal scale.

percentage   Percentage for a class or category is obtained by multiplying the relative frequency of that class or category by 100.

population   Group of similar items from which the sample is drawn. Often referred to as the *universe*.

qualitative or categorical data   Data generated by a qualitative variable.

qualitative variables (or categorical variables)   Variables that cannot be measured numerically but can be divided into different categories.

quantitative variable   A variable that can be measured numerically.

random sample   Each element of the population has some chance of being included in the sample.

raw data   Data recorded in the sequence in which they are collected and before they are processed.

representative sample   A sample that represents the characteristics of the corresponding population as closely as possible.

sample   A group of elements selected from a population.

sample survey   Collection of information for a sample selected for a population; a survey that includes elements of a sample.

sampling with replacement   Each time we select an element from the population, we return the selected element to the population before we select the next element. Thus, the population contains the same number of items each time a selection is made.

sampling without replacement   Each time we select an element from the population, the selected element is not returned the population. Thus, the size of the population is reduced by one element after each selection.

sequential sampling   Collecting samples in sequence.

stratified sampling   Samples are selected proportionately from each process.

stratified random sampling   Random selections of samples are selected proportionately from each process.

stratified sequential sampling   A sequential selection of samples proportionately selected from each subgroup, lot, or process.

survey   Collecting data on the elements of a population or sample.

target population   Population whose characteristics are being studied.

time-series data   Data that give the values for the same variable for the same element at different points in time or for different periods of time.

variable   Characteristic under study or investigation that assumes different values for different elements.

REFERENCES

1. E. S. Buffa, *Operations Management*, 6th ed., Wiley, New York, 1980.

2. H. Mikel and R. Schroeder, *Six Sigma*, Currency and Doubleday, Feb. 2000.

7 Quality Tools

7.1 OVERVIEW

This chapter will provide the practitioner with an array of classical quality improvement methods and diagnostic tools commonly associated with such initiatives as Six Sigma, Lean Sigma, total quality management (TQM), eight disciplines (8D), and other fine process improvement programs. Specifically, the practitioner will learn how to establish basic cause–effect relationships, solve ongoing operational problems, and discover how to enhance or otherwise improve daily operations. Furthermore, the practitioner will learn how data can be collected and graphically analyzed to track down unwanted sources of product and process variation, thereby reducing product defects, process cycle time, and operational costs.

These time-proven methods and tools can serve as the backbone for virtually any type of quality improvement effort. Naturally, the tools and methods presented in this course can serve as an excellent entry point for a Six Sigma X-Belt program or be linked to other initiatives such as Lean manufacturing to form a Lean Sigma initiative. Such an integrated approach can often represent a viable alternative for a commercial enterprise or small business that is dedicated to increasing customer satisfaction and business performance. Reinforcement of major concepts, techniques, and application is realized through exercises, Graphical representation of data, such as bar graphs, histograms, run charts, time charts, pie charts, Pareto charts, and Multi-Vari charts, which are frequently used to help us understand the true importance of data. By way of this training, the practitioner will gain insight into the logic and reasoning that underlies Six Sigma and the process of breakthrough improvement.

While this may be a review to some practitioners, the objective of this chapter is to ensure a common understanding of basic graphical representations used to analyze data. A brief discussion on the nature of Six Sigma variables is also included.

7.2 INTRODUCTION

Newcomers to the world of Six Sigma often ask: What are the best tools to use when practicing Six Sigma? To best answer this question, let us reshape its form: If you are allowed to keep only a few organs in the human body, how would you decide which ones to keep? An analogy would be: Which organs are among the "vital few" and which ones constitute the "trivial many"? Would the practitioner randomly decide which organs surgically to remove from those that reside at the bottom of the list? In the following list, some critical if not humorous strings of thought on the subject are found:

- Tool selection is much like picking a spouse; both the tool designer and the user make some assumptions.

Practitioner's Guide for Statistics and Lean Six Sigma for Process Improvements. By Mikel J. Harry, Prem S. Mann, Ofelia C. de Hodgins, Christopher J. Lacke, and Richard Hulbert
Copyright © 2010 John Wiley & Sons, Inc.

- There is no single toolbox that represents a one-size-fits-all solution. Every project and every problem to be solved is different in one or more aspects. Therefore, the set of tools will be slightly or largely different, depending on the specific application circumstances.
- In Six Sigma, Six Sigma Black Belts, in general, focus on ideas, while quality focuses on the use of tools.
- There are times in which a good idea combined with a good approach is better than the use of several Six Sigma tools.
- There are key analytical ideas that every Six Sigma Black Belt should ponder and explore.
- Many new Six Sigma practitioners are naive when it comes to Six Sigma training and tool application. When we start, we often want to use most of the tools that are at our disposal (not fewer), regardless of whether they will ever be needed.

What is needed is more creative problem solving and leadership in the Six Sigma world. We must clearly focus on the quality of business. However, we suggest being less concerned with tool mastery and more focused with the business of quality. Tools are essential, but ideas are more important. We can say that after spending more than two decades in the practice of Six Sigma, is only natural to develop and have favorite tools and methods. In fact, we emphasize some tools and methods over others, but never to the exclusion or expense of another. The array of intellectual tools evolved for a variety of valid reasons. In other words, every tool has its place. Of course, the ideal tool list should vary, depending on the application environment (engineering, manufacturing, services, healthcare, financial, transactional, etc.).

Statistical methods have traditionally been used to solve process variation problems in both manufacturing and finance, which raises a fundamental question. Does process variation affect an organization's financial performance, and if does, how? Six Sigma recommends that unnecessary process variation should be eliminated. In transactions and services, it has been observed that the elimination of process variation is necessary but not sufficient to improve performance. Studies have shown that after eliminating process variation, high levels of inherent variation still remain, due to factors such as commodity costs, customer diversity (preferences and expectations), and political events. It is well known that variation is present in all processes. Therefore, knowledge about the nature of process variables is invaluable.

7.3   NATURE OF SIX SIGMA VARIABLES

One of the primary objectives of any business is to maximize profit. However, complex organizational structures containing multiple frameworks make us wonder how businesses are going to provide largest profits for the lowest investments. At the same time, organizations having a need to optimize the investment that process improvements demand that they strive for common methods that would permit to empower business process owners to deliver the breakthrough strategy. Many studies have found that by satisfying customers, businesses ensure a large percentage of their future survival and prosperity.[1] The transfer equation $Y = f(x)$ or the simple expression Y is a function of X (where $X = x_1, x_2, x_3, \ldots, x_n$) recognizes that a causal relationship exists in any process or action performed. This equation is the one that contains the nature of the Six Sigma variables. Since it is rare to find a process that uses a single input, we extend this concept to say that Y is a function of one or many x's where Y is the dependent variable and the x's are assumed to be independent variables. The nature of Six Sigma variables is illustrated in Figure 7.1.

Thus, when we say that customer satisfaction is a function of the quality, delivery time, and cost of the product and/or service that is provided, we are saying that for customer satisfaction the dependent variable is $= (Y)$, and that the independent variables are $x_1 =$ quality, $x_2 =$ delivery time, and $x_3 =$ cost of the product. Therefore, the term *critical to satisfaction* refers to any variable that has significant influence on one or more of the three determinants of customer satisfaction, critical to quality,

Nature of Six Sigma variables

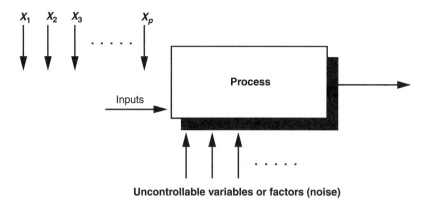

Figure 7.1 Graphical representation of the nature of Six Sigma variables.

critical to delivery, and critical to cost. Suppliers provide products and/or services to fulfill customer needs for defect-free products, on-time delivery, low cost, and quick response to changing requirements. Examples of critical-to-satisfaction variables, such as quality, cost, and cycle time, are controlled by process capability, which is, in turn, limited by process variation. Therefore, the expression Y is a function of X allows us to explain and ultimately control variation.

Some practices have a direct influence over cycle time and cost without directly impacting the processes (e.g., reduction in overhead cost/hour). Six Sigma focuses on the quality of processes; improved process quality will, in turn, improve cycle time and cost. In the equation $Y = f(X)$, the dependent variables include process accuracy, precision, time to fulfill requests and orders, introduce designs, and perform tests. Independent variables are the inputs to processes; they are the variables that influence the outcome. We also recognize that independent variables (x's) have unequal affects on the dependent variable (Y). The Pareto principle suggests that 80% of the total error or variance will be caused by 20% of the variables. The variables that account for 80% of the outcome are called the "vital few," while the remaining variables are called the "trivial many." Design of experiments (DOE) is an effective approach to identify and control these vital few variables. The purpose of process optimization in the breakthrough strategy is to identify and control the vital few variables. When successful, the variability of Y improves while the probability for defects decreases, yield increases, rolled throughput yield increases, mean time between failures (MTBF) increases, product reliability increases, work-in-process inventory decreases, quality improves, cycle time decreases, cost decreases, customer satisfaction improves, and, ultimately the business grows and prospers.

Exercise Questions and Answers

1. What are variables? *Answer*: Variables are the elements subject to variation described by a cause–effect (CE) system.
2. What is a dependent variable? *Answer*: A dependent variable (Y) is one whose value depends on the state or values of the independent variables of the system.
3. What is an "independent variable?" *Answer*: An independent variable (X) is one whose value may result from a prior process or be chosen by the process owners.
4. What other labels are synonymous with "dependent" and "independent" variables? *Answer*: A CE system may have one or more independent variables ($x_1, \ldots, x_n$) whose values do not depend on the state or value of the dependent variable (Y). Dependent variables are also known as *response variables* and can be the output of a process, the effect in a CE system, or the

symptom of a problem. Independent variables are also known as the inputs to a process, the causes of the system, or underlying problems.

5. What are leverage variables? *Answer*: Leverage variables exert a disproportionately large influence on the dependent variable (*Y*).

6. What strategies can be used to isolate leverage variables? *Answer:* According to the Pareto principle, 20% of the variables contribute 80% to the variation in *Y*. Leverage variables reside within this 20%. The objective of process optimization in the breakthrough strategy is to identify and control these vital few variables. When this is achieved, the variability of *Y* is minimized, probability of a defect decreases, and yield improves.

7.3.1   CT Concept

The abbreviation CT stands for "critical to," which can be expanded to express concepts such as critical to satisfaction (CTS), critical to quality (CTQ), critical to delivery (CTD), critical to cost (CTC), and critical to process (CTP). The CT characteristic levels resemble the branches of a tree with customer satisfaction at the top of the tree; customer needs are defined in terms of quality, delivery, and cost, and these variables constitute the CTS characteristics (see Fig. 7.2).

Example 7.1   Following a review meeting, one of the managers said: "I wonder how the CT concept relates to our activities and to customer satisfaction." "I don't exactly know," said one of his colleagues, "but I know a master Black Belt who could meet with us to clarify this concept." When asked, the master Black Belt replied that suppliers provide products and/or services to satisfy specific needs through defect-free deliverables, shortest possible cycle time, and lowest possible cost. To supply these products or services, organizations use processes whose characteristics directly affect each of the customer's needs. Some characteristics are then identified as critical-to-quality, whereas others are identified as critical-to-delivery or critical-to-cost. In Six Sigma, we focus on improving the quality of our processes which, in turn, improves cycle time and cost. The manager replied that the strength of the coffee is a function of the number of spoonfuls of coffee and the amount and brewing temperature of the water (see Fig. 7.2).

Figure 7.3 Components:

1. *Critical-to-satisfaction (CTS) characteristics*—expression of the customers' vital needs

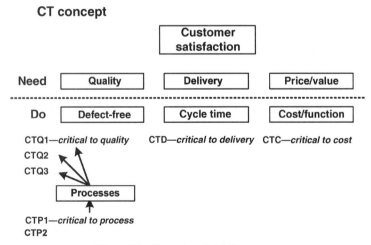

Figure 7.2   Illustration of the CT concept.

CT definitions

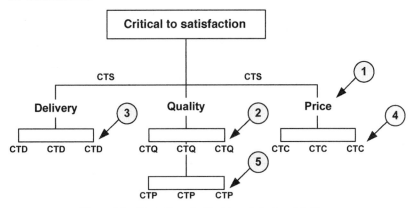

Figure 7.3 Illustration of critical-to-satisfaction definition.

2. *Critical-to-quality (CTQ) characteristics*—the product, service, and/or transactional characteristics that significantly influence one or more CTS in terms of quality

3. *Critical-to-delivery (CTD) characteristics*—the product, service and/or transactional characteristics that significantly influence one or more CTS in terms of delivery

4. *Critical-to-cost (CTC) characteristics*—the product, service, and/or transactional characteristics that significantly influence one or more CTS in terms of cost

5. *Critical-to-process (CTP) characteristics*—process parameters that significantly influence a CTQ, CTD, and/or CTC.

7.3.2 CTQ and CTP Characteristics

How do critical-to-quality (CTQ) characteristics and critical-to-process (CTP) characteristics apply to business? We refer to CTQ and CTP characteristics as *variables* and use them to describe the CE system that exists in any process. CTP characteristics are the inputs or independent variables (x's), and a CTQ characteristic is the dependent variable (Y), according to the master Agent. Moreover, process capability is limited by variation. Therefore, to improve process capability, we must first control the variations in the process inputs (see Figs. 7.3 and 7.4). The transfer function is a fundamental notion in Six Sigma; it implies that Y is a function of X. However, nothing in this world is simple enough to be determined by only one input. In reality, there are many independent

CTQ and CTP characteristics

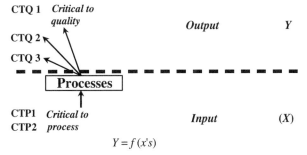

Figure 7.4 CTQ and CTP characteristics.

CTX tree (process tree)

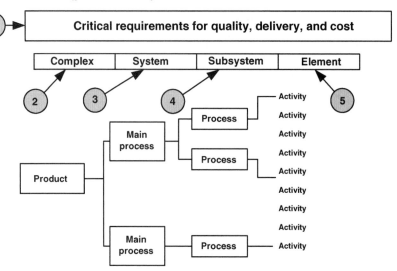

Figure 7.5   Flowchart of CTX process tree.

variables that affect an output. The Six Sigma journey begins with the transfer function when we define the equation $Y = f(x_1, x_2, \ldots, x_n)$ and use it to reduce variation and improve process quality.

7.3.3   CTX Tree (Process Tree)

Using the CTX or process tree (Fig. 7.5), a team breaks down the processes and identifies potential influence factors on the critical product/service delivery characteristics (see Figs. 7.3 and 7.4). Two major considerations are

- The definition of CTS characteristics must be understood across the organization.
- The main processes in the organization should be mapped.

Here, critical requirements for quality, delivery, and cost are the needs translated from the CTS characteristic. Process tree levels are as follows:

1. *Complex level*—the first level of the process tree. It is the last process performed to delivesssr the product/service to the customer.
2. *System level*—consists of the main processes of business.
3. *Subsystem level*—represents a breakdown of the processes listed at the system level.
4. *Element level*—represents a breakdown of the processes listed at the subsystem level. Critical-to-process characteristics (CTP) are identified directly below this level.

Example 7.2   Construct a process tree.

Solution

1. Identify the customer.
2. Identify the CTs.

3. Translate all CTs into product requirements in terms of CTQ, CTD, and CTC.

4. Define the complex level.

5. Define all system-level main processes that might influence product requirements.

6. Define all subsystem-level processes that might influence product requirements. Depending on the complexity of the process, the process tree can have more than one subsystem level.

7. Define all elements activity or activities that might have influence on product or service delivery requirements.

7.3.4 CTY Tree (Product Tree)

The purpose of CTY is to define all terms used in the CTX and CTY tree, as well as in the CT matrix concept. The term product tree (CTY tree) refers to a hierarchical breakdown of product or service deliveries that allows us to visualize the CTQ, CTD, and CTC characteristics at each level of the hierarchy (see Fig. 7.6).

The number of levels in the tree depends on the complexity of the product/service. The same is true for processes. In a multilevel tree, these levels have generic labels such as complex, system, subsystem or element. CTQs, CTDs, and CTCs can be found at any level of the product tree (CTY tree) and can be expressed as a function of the immediate lower-level characteristics. In other words, a CTQ complex is a function of (CTQ system$_1$,...,CTQ system), CTQ system $=f$ (CTQ subsystem$_1$,...,CTQ subsystem). At the element level, the CTQ element is a function of (CTP$_1$,...,CTP$_q$). In practice, CTPs are identified directly below the element level of the CT tree in which they directly affect the element CTQ, CTD, and/or CTC. However, they can influence these characteristics at other levels of the hierarchy. Further steps of the Six Sigma breakthrough strategy allow us to discover and subsequently control these CTPs. The CTQ, CTD, and CTC are *opportunities for nonconformance* that must be measured and reported (i.e., active opportunities), while CTP represents "control opportunities." If possible, the opportunities for nonconformance, which do not directly influence customer satisfaction, should remain in the passive mode (stable).

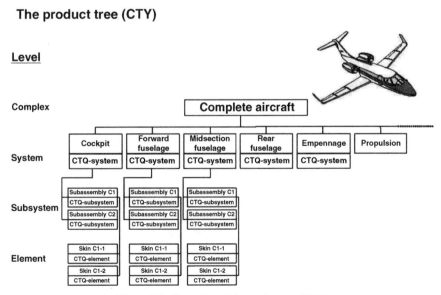

Figure 7.6 Illustration of the product tree (CTY).

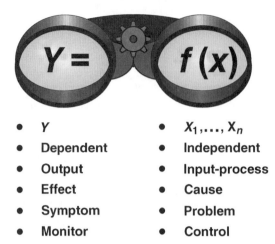

Figure 7.7   The focus of Six Sigma.

The CTX and CTY trees and the CT matrix are applicable to all areas of a business. The CT tree is a simplified quality function deployment (QFD) approach. As we have seen, these diagrams expose, describe, and communicate logical relationships between important events and goals. Using the insights gained from using these trees can guide the partition of a large problem into smaller components. By defining the terms used in CTX and CTY tree and the CT matrix concepts, a Six Sigma team can organize and summarize the critical-to-customer satisfaction characteristics.

7.3.5   The Focus of Six Sigma

Figure 7.7 shows the focus of Six Sigma.
A few examples of causal systems are included in Table 7.1.

7.3.6   The Leverage Principle

Do all independent variables affect the outcome equally? No, some variables have greater impact on the outcome of a process than others, as shown in Figure 7.8. In Six Sigma, variables that are likely to impact system performance are first identified; then the variables that exert strong influence on the outcome of a process are identified and controlled. According to the Pareto principle, 20% of the variables contribute 80% to the output or variation in Y. Those variables that exert undue influence on the outcome of a process are called the "vital few," whereas the ones that have lesser impact are called

TABLE 7.1   *X* and *Y* Examples

Y (Dependent Variable)	X (Independent Variable)
Etch rate in a chemical milling process	Bath temperature, metal alloy, dissolved metal in the bath, etc.
Time to fulfill a job requisition	Accurate job profile, interviewing availability, supply of qualified candidates, internal posting, external advertising, interviewing process, etc.
Machining precision	Quality of cutting tool, metal alloy, feed rate, quality of the cutting program, environmental conditions of the plant, etc.
Time to fulfill an order for an aircraft on ground (AOG)	Type of part, supplier, value of the part, geographic location, time of day when order is received, etc.
Number of errors in an engineering drawing	Designer's experience, complexity of the drawing, training, availability of manuals, etc.

The leverage principle

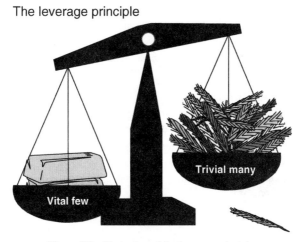

Figure 7.8   Illustration of the leverage principle.

the "trivial many." This is known as the *leverage principle*. Three variation reduction strategies for an example process are shown in Figure 7.9.

To illustrate the concept of leverage variables, consider the following. We know that the total variance of an output is the sum of the individual variances of the independent components. Imagine that we have a process with six independent variables (x's) and one CTQ characteristic or dependent variable (Y) expressed in the fundamental equation $Y = f(x_1, x_2, \ldots, x_6)$. Moreover, we make some measurements and determine the baseline condition of total variance presented on the first line of Figure 7.8. Since the objective is to reduce variability, we would ask the following question: What combination of individual variation minimizes total output variability? To illustrate how each variable influences the result, we could try three strategies: (1) reduce the variation of all x's by one unit, (2) reduce the variation of all x's to zero except for x_2, and (3) reduce the variation of x_2 by three units as shown in Figure 7.8.

Three variation reduction strategies

Strategy	Xs						Total Reduction Effort	Y σ_{tot} **	Changeover Baseline (%)
	σ_1	σ_2	σ_3	σ_4	σ_5	σ_6			
Baseline	4	11	3	2	1	1	0	12.33	N/A
Reduce each by 1	3	10	2	1	0	0	6	10.68	13.4
Reduce all to 0 except σ_2	0	11	0	0	0	0	11	11.00	10.8
Reduce σ_2 by 3	4	8	3	2	1	1	3	9.75	20.9

$$**\sigma_{tot} = \sqrt{\sigma_1^2 + \sigma_2^2 + \cdots + \sigma_6^2}$$

$$\text{Ex. } \sigma_{tot} \text{ for baseline} = \sqrt{4^2 + 11^2 + 3^2 + 2^2 + 1^2 + 1^2} = \sqrt{152} = 12.33$$

(*Note*: **Table relates to tolerance stackups.**)

Figure 7.9   Three variation reduction strategies.

From Figure 7.9, we observe that the fourth-row strategy (reduce σ_2 by 3) is the best choice of the illustrated strategies because a change of only three units improves total variation by 20.9% relative to the baseline. In addition, we observe that this option requires the least amount of change while yielding the most improvement, thus demonstrating a successful application of the leverage concept.

Summarizing:

1. A system of causation reflects the function or relationship between causes and their effects.

2. A dependent variable (Y) is influenced by the value or state of one or more independent variables (x's) included in the system of causation.

3. Independent variables (x's) take on values that do not depend on the state or value of any other variable included in the system of causation.

4. Customer needs are expressed in terms of quality, delivery time, and cost of the products and/or services that we provide.

5. Suppliers meet customers' needs through quality, cycle time, and cost. These depend on the level of performance of our processes.

6. Process performance is a function of probability for defects, and this, in turn, is a function of variability.

7. The Pareto principle states that 20% of the variables in a system of causation contribute 80% to the outcome or result. These are known as "leverage" or "vital few" variables.

8. The remaining 80% of the variables that contribute 20% to the outcome are known as the "trivial many" variables.

9. The purpose of process optimization in the breakthrough strategy is to identify and contain the vital few variables.

10. Six Sigma focuses on improving processes by identifying and controlling the "vital few" variables. Design of experiments (DOE) is used to achieve this objective.

7.4   QUALITY FUNCTION DEPLOYMENT (QFD)

Japan's Yoki Akao[2] conceived *quality function deployment* (QFD), a concept that serves as management tool for new product development. QFD converts customer demands into a quality design of the finished product or service because it addresses design targets and the major quality items. Today, quality deployment is used at all levels during process design. Therefore, the purpose of quality deployment is to ensure quality from the beginning to reduce future production costs. Benefits of QFD include lower product development and manufacturing costs, and as a consequence, increased market share. Many of the organizations that employ QFD have gained the following advantages:

1. Reduction of development cycle time by as much as 50%

2. Reduction in the number of engineering changes by as much as 50%

3. Reduction of startup and engineering costs by as much as 30%

4. Reduction in warranty claims by as much as 50%

5. Systematic retention of product development knowledge, so that it can be later applied to similar product or product family designs

6. Improved customer satisfaction

Value engineering, originally used to define product functions and to deploy efficient business process functions, has been extended by the QFD[2] concept. Therefore, the practitioner could apply this principle to the process of product development, rather than to the functions of the product itself. QFD uses the job functions that create product quality as quality functions. QFD covers all process activities, from product and technology development planning, system-level design, prototype design, and detail design to production preparation and mass production. In summary, the practitioner should view QFD as the most complete and systematic approach for designing products with the quality level that fulfills expressed and latent customer requirements. QFD is the fundamental concept of design for Six Sigma (DFSS).

7.5   SCALES OF MEASUREMENT

The concept of measurement is simple; we seek to compare or otherwise contrast a physical attribute of something to a rational and invariant standard. This provides us with a performance gap. Almost innately, we seek to quantify such gaps for purposes of communication, verification, and analysis. Without this ability, it would be difficult to improve or control an object, situation, or phenomenon.

By using numbers, we further refine the concept of measurement; otherwise a measurement (i.e., contrast) would reside only in the minds of those making the comparison. The relative magnitude of such a contrast cannot be reliably communicated without the aid of numbers. Without standards, numbers, and *scale*, the concept of measurement could not exist and therefore, improvement and control would not be possible. Perhaps our emphasis on measurement can best be understood by the Lord Kelvin comment extended in 1891. He stated: "When you can measure what you are speaking about and express it in numbers, you know something about it, but when you cannot express it in numbers, your knowledge is of a meager and unsatisfactory kind."

7.5.1   Likert Scale

The Likert scale, shown in Figure 7.10, evaluates customer satisfaction using an ordinal scale with a range of ratings or degrees of satisfaction arranged in order to determine what is being done right, as well as wrong. The Likert scale is frequently used to count survey responses.

Figure 7.10 Components

1. Ordinal scale arranged in ascending or descending order
2. A series of questions to be rated by the customer
3. Simple set of instructions
4. Collect names/addresses so that future follow-ups can be conducted

Measurement scale—Likert scale

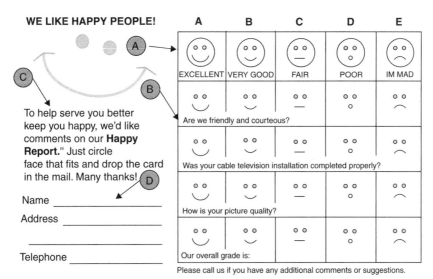

Figure 7.10   Graphical representation of the application of Likert scale.

When the Likert scale is used to count survey responses, the following should be considered:

- Keep the survey simple.
- Make it easy for customers to tell you when they are not happy.
- The goal is to understand how to better satisfy the customer.

Example 7.3   Construct the measurement scale for a survey questionnaire.

Solution

1. Survey design considerations:

   Length (not too long)

   Appearance (simple, not busy)

   Types of questions (statements of fact or measures of performance and importance)

   Open-ended questions/probes (enable respondents to easily suggest issues and provide explanations)

2. Type of question formats:

   Closed-ended (yes/no)

   Rating scales (even or odd numbers?—both can be useful)

   Open-ended (free response)

3. Other considerations:

   Focus on one theme; it is usually best to include a midpoint in rating scales (e.g., 3/5/7 categories).

   Try to solicit your customer's feelings toward your competitors.

4. Identify specific target control groups:

   At least 10% of total customer base.

   Stratify various customer segments.

   Give prior notice, before delivering survey.

   Personalize the survey and cover letter.

   Address confidentiality.

   Offer an incentive or token of appreciation for completion.

   Follow up with a friendly collection strategy.

   Develop action plans that are based on results.

   Communicate results to customers.

   Follow up with repeat surveys to monitor changes over time.

■

7.5.2   Logarithm Scale

A measurement scale displays nonlinear data in a format that spans several orders of magnitude and provides a presentation technique that can be read with a degree of precision for a wide range of values. If the data are nonlinear (i.e., with a very wide range of values), then a logarithmic scale is appropriate. The base 10 logarithmic scale is a favored choice, but other bases can be selected. Base 2 logarithmic scales are frequently used in the computer industry where binary technology prevails. Figure 7.11 is a graphical representation of logarithm scale.

Measurement scale—logarithm

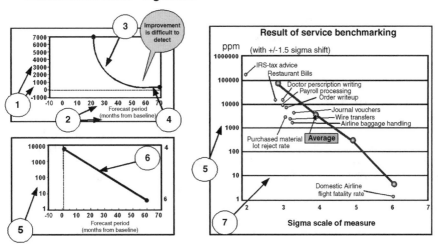

Figure 7.11   Graphical representation showing example of the logarithm scale.

Figure 7.11 Components

1. Linear vertical scale $\langle\langle Y \rangle\rangle$
2. Linear horizontal scale $\langle\langle X \rangle\rangle$
3. Curve representing the function: $Y = f(X)$
4. Difficult to detect any variation of Y for different Xs
5. Logarithm vertical scale
6. New line representation of the function: $\log Y = b + mX$ (if linear)
7. Parts per million (ppm) vs. sigma (values ranging from <1 to $>100,000$ can be read from this scale)

Example 7.4   Construct a scale of measurement (see Fig. 7.12).

Example of logarithm scale

data	number	log 10	log 6	ln
1	1	1	1.2850	2.3025
2	10	2	2.5701	4.605
3	100	3	3.8552	6.9077
4	1000	4	5.1403	9.210

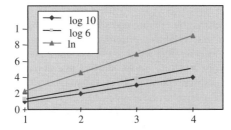

Figure 7.12   Example of a logarithm scale.

Solution

1. Enter data in Excel spreadsheet.
2. Select log 10, log *X* line, as required.
3. Plot data.

■

7.6 DIAGNOSTIC TOOLS

Newcomers to the world of Six Sigma often ask: "What are the best tools to use when practicing Six Sigma?" To best answer this question, let us reshape its form: "If you are allowed to keep only a few organs in the human body, which ones would you say are best to keep?" In other words, which organs are among the vital few and which ones constitute the trivial many? Would you surgically remove those that reside at the bottom of the list? On a more serious note, here is some critical (if not humorous) thinking on the subject:

- Tool selection is much like choosing a spouse; both the designer and the user must make assumptions.
- Six Sigma Black Belts focus on ideas; quality focuses on tools.
- There are key analytical ideas that every Six Sigma Black Belt should ponder and explore.
- If tools were the ticket to success, then statisticians would be CEOs.
- A simple idea can often negate the need for a tool. (A new concept can replace extensive analysis.)
- The majority of a physician's curriculum is about how the human body functions, not scalpels.

7.6.1 Elements for Problem Solving—Diagnostic Tools and Methods

There is no single toolbox that represents a one-size-fits-all. For example, the practitioner will find that there might be times in which one tool, for example, *failure mode–effects analysis* (FMEA), which is used to identify and assess risks associated with potential or process (service) failure modes, could be used as a diagnosis or analysis tool or both.

Every tool or method has its place, and the ideal list of tools and methods will depend on the application and familiarity with certain tools by engineering, manufacturing, service, and transaction processor groups. However, there are two basic types of diagnostic tools: graphical and analytical. Graphical tools include Pareto charts scatterplots, time-series plots, boxplots, probability distribution plots, main-effects plots, Multi-Vari charts, time series plots, and run charts that help the practitioner identify data distributions and statistical properties. Graphical methods include the cause–effect (CE) diagram, *XY* matrix, failure mode–effects analysis (FMEA), affinity diagrams, fault tree analysis (FTA), and other, diagrams. This chapter focuses on the most widely used methods.

For example, Pareto charts, by ranking and sorting problems by count and category, identify the most frequent problems, establish priorities, and focus attention on the most pressing issues. ScatterPlots help us determine whether two variables are correlated or act independently. Multi-Vari charts, boxplots, and time-series plots enable classification of the output variation into three categories: positional, cyclical, and temporal. This allows the practitioner to identify and classify the independent variables $(x_1, x_2, \ldots, x_n)$ and provides a starting point for further analyses, such as design of experiments, response surface design, and Monte Carlo simulations.

In the final analysis, Mikel Harry says that it is frequently better to "blow and go" than it is to "roast and toast." What he means, I do not know, but it sure sounds good (like some of the tools). For the practitioner, the advice is that for Lean Sigma, and Lean Sigma success, there might be the need to combine tools. This is, just like saying that for success, we might need on our side Roy Rogers, Dale Evans, Clint Eastwood, Humphrey Bogart, John Wayne, James Bond, and probably the Dixie Chicks. All we can say is that the truth lies somewhere out there. Finally, it is important to remark that graphical and analytical tools are both diagnostic tools.

7.6.2   Problem Definition—Defining Project Objective

The question "Why?" immediately triggers other questions, including "What?", "How?", "Who?", "Where?", and "When?". Therefore, when using improvement methods, we should ask

- What is the purpose of the process?
- How can the process be improved or performed better?
- Who can best perform the process?
- Where could the process be performed at a lower cost or be improved by removing non-value-added activities while maintaining process quality?
- When should the process be performed to be Lean or in plain English, require the least effort?

How we define a problem makes the difference between success and failure. Problem definition influences our approach to the problem, project goals, and even solutions. Just remember: Any time that you have to solve a problem, the problem definition that you adopt will influence the solution that you will find.

- *Example of poor problem definition*—reduce restorable time for severity 1 telephone banking problems, focusing on human errors as root cause.
- *Example of good problem definition*—reduce restorable time from 51% per month for severity 1 telephone banking problems to 30% or less per month by the third quarter of this year, resulting in annual savings of $4.5M.

Poor problem definition may cause a problem to go unsolved. Initially a problem will likely be vaguely defined. A good problem definition will include quantified information about the problem, the magnitude of the problem, the baseline, and the gap remaining to reach a benchmark or desired state. A good problem definition does not speculate about future actions to solve the problem." To achieve this, simply follow these steps:

Step 1. Be specific: State where and when the problem occurs. Locate the "pain" or dissonance points.

Step 2. Use declarative Format: State the problem clearly and definitely, not as a question or incomplete sentence.

Step 3. Quantify: State the difference between "what is" and "what should be" in measurable terms.

Step 4. Be factual: Avoid assumptions. State only the facts.

In summary, after clearly defining the problem, the practitioner must decide which methods and tools are needed to solve the problem. With time, and after spending some decades in the practice of Six Sigma, he/she will have favorite tools and methods. In fact, please notice that we emphasize some tools and methods over others, but never to the exclusion of another. A wide array of intellectual tools exists for a variety of good valid reasons. In other words, every tool has its place—and one more time, we emphasize that, of course, it is understood that the ideal tool list will vary, depending on the application environment, particular project, and circumstances.

7.7   ANALYTICAL METHODS

Data must be available to establish which analytical methods should be used for accuracy and reliability. The objective of using an analytical method or tool is to demonstrate that the procedure, when correctly applied, produces results that are fit for the purpose. Cause–effect diagrams, XY matrix, and failure mode analysis are examples of some widely used analytical tools in quality and Six Sigma. All analytical methods are of equal importance from a validation perspective. In general, validated analytical methods should be used irrespective of whether they are for in-process, release, acceptance, or process stability testing. Each analytical method should be designed to minimize variation.

Methods validation includes an assessment of the adequacy of the analytical procedure. Methods validation means establishing through documented evidence, a high degree of assurance that an analytical method will consistently yield results that accurately reflect the quality characteristics of the product (service) tested. Statistical analysis (e.g., linear regression analysis, relative standard deviation) of methods validation data is often used to demonstrate the validity of the method. The method followed, including the amount of data to collect and the criteria used in determining the acceptability of the analytical method, should be specified.[3]

7.7.1   Cause–Effect (CE) Analysis

The cause–effect (CE) diagram is an analysis tool used in Six Sigma, Lean manufacturing, and kaizen. The (CE) diagram is also known as the "fishbone diagram." Kaoru Ishikawa, who pioneered the quality management process in the Kawasaki shipyards in Japan, developed it. The CE diagram is a graphical representation that offers a clear visualization of causes, the effect and the relationship between causes. In addition, it highlights problem "root causes" and reasons for variation. The CE diagram is a structured framework that uses collective knowledge to identify the main causes of the effect under study. It is an example of the classic "divide and conquer" approach to solve complex problems. There are two types of CE diagrams: (1) manufacturing diagrams that usually include the labels measurement, manpower, machines, materials, methods, and environment/mother nature (six Ms) and (2) transactional diagrams usually include four labels for policies, procedures, personnel, and environment (4Ps). This diagram examines a process and lists all possible areas where a problem under investigation could occur. The purpose of this graphing method is to simplify the flow of information through complicated processes and find the cause of a problem. Figure 7.13 shows an example of the use of the CE analytical tool for order fulfillment.

Cause–effect diagrams are simple graphical ways to show the relationships between inputs and outputs of the process. The head of the diagram is the output/response (Y) of the problem for which source variations are to be generated; causes are also known as sources of variation or variables. CE diagrams include branches or "bones" that contain the major categories of failure causes (six Ms). Within each of the major categories, inputs, or factors of the CE diagram we find the independent variables $(x_1, x_2, x_3, \ldots, x_n)$. These inputs are potential causes of the effects, such as service failure, product failure, rework, product scrap, or stop production. For a full understanding of a process, it is important for the practitioner to create a CE diagram, by gathering, displaying, classifying, and analyzing data produced by the process to appreciate its power. When gathering data for a specific problem, we should ask how well the data align with the problem statement. For example, if the problem statement suggests method or equipment failures (machines) as potential root causes of the problem, then data collection should focus on failures occurring within these categories.

7.7.1.1   How to Construct the CE Diagram

Cause–effect diagrams are flexible and may contain less than the conventional six Ms. Sometimes the causes of the CE diagram are labeled as "Five Ps" (policies, plans, procedures, people and plant/manufacturing plant) instead of the conventional six Ms. The number of branches of the CE diagram is determined by the problem definition. To create a CE diagram, clearly define the problem

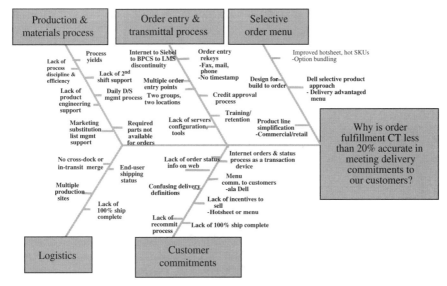

Figure 7.13   Example of a cause–effect diagram. BPCS = business planning and control systems; CT = cycle time; LMS = learning management systems.

(see Section 7.6.2) and record each brainstormed idea individually. During the "thought map" brainstorm session, "sticky notes" are suggested to record the ideas. Focusing on the problem through the evaluation of the cause is extremely important as other problems may frequently surface that are not directly related to the cause of the particular problem. A simple way to define the problem is by viewing the problem as a deviation from a requirement or when "actual" is different from "should" or when an undesirable situation or event occurs. Poor problem definition may result in failure to reach the proper resolution. Additional brainstorming sessions might be needed to solve the problem. For example, start with the six major categories and proceed as follows:

Step 1. Have the team perform a brainstorm root cause analysis.

Step 2. *Clearly define the problem.* The problem definition should contain both *what* the problem is and what is not, *where* the problem is, *when* did the problem occur, *how* it occurred, and the extent of the problem.

Step 3. Review the focus and problem statement

Step 4. Draw the CE diagram problem centerline (Y = problem).

Step 5. Write the problem in the centerline of the CE diagram. Place the problem statement in a box on the right side of the centerline or "trunk" of the CE diagram.

Step 6. Determine the major categories (six Ms) of the branches.

Step 7. Draw the branches following examples shown in Figure 7.14.

Step 8. Remember that the branches (six Ms) major categories are

- *M*anpower/people
- *M*ethods
- *M*achines/tools
- *M*easurements
- *M*other nature/environment
- *M*aterials

Step 9. Perform the branch breakdown within each one of these major categories.

Order fulfillment less than 20% to customer commitment

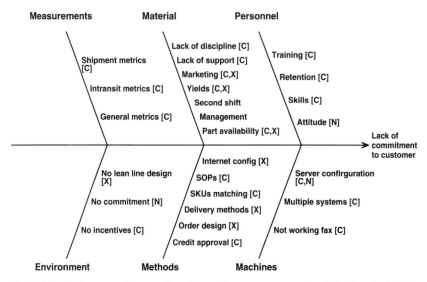

Figure 7.14   MINITAB-generated cause–effect diagram in response to question: "Why is order fulfillment CT less than 20% accurate in meeting delivery commitments to our customers?"

Step 10. Add title, date, and contact team

Step 11. Evaluate/find through brainstorming possible causes for each categorized branch and check for data integrity.

Step 12. It is important to complete the entire CE diagram for every category before moving on.

Step 13. Brainstorm the Five why's—Begin with the major category that the team has identified as the most likely to produce the actionable root cause and begin to ask "Why?"

Step 14. Label each cause with a "C," "N," or "X" – see examples in Figure 7.14.

Step 15. Identify the most actionable root cause(s) and circle it (them).

Step 16. Do a "might cause" check.

Step 17. Verify with data.

Note: MINITAB can help construct CE diagrams. The MINITAB procedure for constructing CE diagrams is outlined in Section 7.7.1.2.

7.7.1.2   How to Construct the CE Diagram Using MINITAB

1. Open a MINITAB worksheet.
2. Label each MINITAB column (C1, C2, C3, . . . , C6) with the six major CE categories:
   - *M*anpower/people
   - *M*ethods
   - *M*achines/tools
   - *M*easurements
   - *M*other nature/environment
   - *M*aterials
3. Enter the contributing causes by category in each column.

4. Select ``Stat'' > Quality Tools > Cause and Effect.

5. Dialog box appears.

6. Go to "causes" column.

7. Place cursor in the "causes" column and click once.

8. Data column appears on the left of the dialog box.

9. Place cursor again in the "causes"—first column, for example, "personnel."

10. Double-click on the data corresponding to "personnel"—column C1 Personnel.

11. Continue this with the remaining columns—repeat procedure.

12. Enter problem title.

13. Enter title of CE diagram.

14. Click OK.

The same CE diagram of Figure 7.13 drawn in PowerPoint is shown in Figure 7.14, generated with MINITAB following the instructions of Section 7.7.1.2.

In Figure 7.14, $C =$ fixed variable, controllable factors (process inputs are held fixed), $N =$ noise (uncontrolled independent variable), and $X =$ experimental independent variables (x's).

[*Note*: For each noise N, design an approach to convert it into a C (this step is just "good engineering").]

After filling in the CE diagram, each independent variable x is labeled with a C, N, or X as defined previously.

7.7.2   Failure Mode–Effects Analysis (FMEA)

Failure mode–effects analysis (FMEA), as we said before, is widely used to identify and assess risks from potential product, service, transaction, or process failure modes. FMEA is generally applied early in the design of a process or when process redesigns occur. This graphical method is widely used in the *analyze* and *improve* phases; however, it can also be used in the *control* phase. FMEA can identify the failure causes of a process and is capable of quickly testing weak points or links already present in a product or process design. Undertaking a FMEA after completion of the CE diagram will deepen our understanding of the cause analysis. Topics for further analysis can be more easily identified and prioritized. In the analyze phase, FMEA provides valuable insights into the causes and effects of variation. The FMEA assists organizations in

- Improving or designing more robust products, services, and processes
- Designing safer products and processes
- Designing safer delivery systems
- Receiving fewer complaints and reducing the organization's guarantee costs
- Creating fewer problems or minimizing them in everyday business processes
- Provide improvement teams with prioritized causes and identifying which causes need to be eliminated urgently

FMEA helps assess the failure severity and risk of certain failures, while allowing us to take advantage of collective knowledge and teamwork.

Example 7.5   A Six Sigma team works on a project to eliminate rework on engineering drawings. To prioritize its activities, the team conducts a failure mode–effects analysis (FMEA) exercise. The Black Belt reminds the team that the output of this multidisciplinary tool concentrates efforts by focusing on problems that represent higher risk to the customer. A team member looks at a general FMEA form and states that the preset criteria are difficult to apply to their particular project,

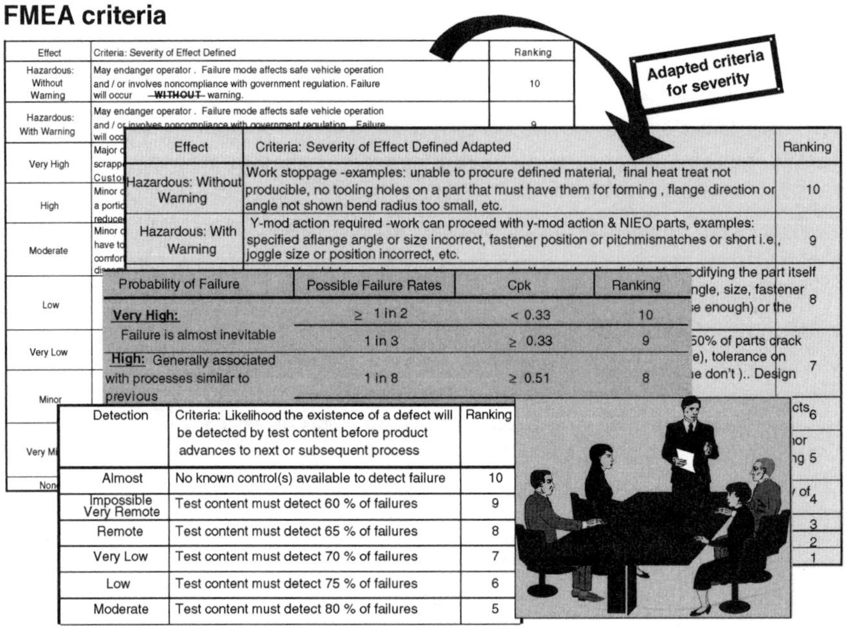
FMEA criteria
Effect
Criteria: Severity of Effect Defined
Ranking
Hazardous: Without Warning
May endanger operator. Failure mode affects safe vehicle operation and / or involves noncompliance with government regulation. Failure will occur WITHOUT warning.
10
Hazardous: With Warning
May endanger operator. Failure mode affects safe vehicle operation
Very High
High
Moderate
Low
Very Low
Minor
Very Minor
None
Adapted criteria for severity
Effect
Criteria: Severity of Effect Defined Adapted
Ranking
Hazardous: Without Warning
Work stoppage -examples: unable to procure defined material, final heat treat not producible, no tooling holes on a part that must have them for forming , flange direction or angle not shown bend radius too small, etc.
10
Hazardous: With Warning
Y-mod action required -work can proceed with y-mod action & NIEO parts, examples: specified aflange angle or size incorrect, fastener position or pitchmismatches or short i.e, joggle size or position incorrect, etc.
9
Probability of Failure
Possible Failure Rates
Cpk
Ranking
Very High:
Failure is almost inevitable
≥ 1 in 2
< 0.33
10
1 in 3
≥ 0.33
9
High: Generally associated with processes similar to previous
1 in 8
≥ 0.51
8
Detection
Criteria: Likelihood the existence of a defect will be detected by test content before product advances to next or subsequent process
Ranking
Almost Impossible
No known control(s) available to detect failure
10
Very Remote
Test content must detect 60 % of failures
9
Remote
Test content must detect 65 % of failures
8
Very Low
Test content must detect 70 % of failures
7
Low
Test content must detect 75 % of failures
6
Moderate
Test content must detect 80 % of failures
5

Figure 7.15 Criteria for creating the needed failure mode–effects analysis. NIEO = New international economic order (to order parts with the same part number in all countries of the world).

and asks whether they can be reformulated to better match the project characteristics; the practitioner created the needed "adjustments" in the general template of the FMEA, and the redesigned FMEA is shown in Figure 7.15.

The FMEA method is flexible and can be tailored to fit all process or product needs as long as the fundamental concepts of the method apply. Once FMEA has been tailored, the project team proceeded to rank the severity, occurrence, and detectability; they then calculated the *risk priority number* (RPN), which is the product of the severity, occurrence, and detectability rankings. Adapting the form to your own project/needs makes it easier and meaningful for any team to conduct an FMEA. After all, tools are meant to be helpful, not torturous.

7.7.2.1 FMEA Diagram

This method helps us address those characteristics whose variation has the greatest impact on our customer. In this particular case, it helps provide customers with better processes. Figure 7.16 shows an example template widely used in transactions, services, and manufacturing to redesign existing products and design new products.

To manually enter data in the FMEA template, populate the appropriate columns of the FMEA table shown in Figure 7.16.

Product or Process. Enter a simple, but precise, description of each process step.

Failure Mode. Enter every potential failure that could emerge in the process step or process. Several failures can be indicated per process step. In doing so, it is assumed that the failure may not necessarily emerge.

Failure Effects. Enter the actual or potential effects of a failure. Once again, there can be several potential failure effects.

Causes. Enter potential causes. Here, too, several potential causes can be given for every failure. The list of causes should be as detailed as possible, while symptoms should be deemphasized.

Product on Process	Failure Mode	Failure Effects	S E V	Causes	O C C	Controls	D E T	R P N	Actions	Plans	P S	P O	P D	P R P N

Figure 7.16   Example of general FMEA diagram template.

Occurrence (Occ). Enter how frequently the potential failure can occur on a scale from 1 to 10. It is important to estimate the frequency of occurrence for each cause.

Controls. Enter all opportunities to detect and eliminate potential causes in the early stages of the *CURRENT* process (before the process step in concluded).

Detection (Det). Enter the probability that the potential cause is detected within the process step on a scale from 1 to10.

Risk Priority Number (RPN). Enter the RPN; it is the product of *severity, occurrence,* and *detection* (probability of detection).

The rating position determined by the RPN prioritizes the areas where *action needs to be taken.* If the RPN *is high*, then an in-depth analysis must be undertaken:

$$\text{Risk } R = P \times S \quad \text{or} \quad R = P \times S$$

Component failures can be ranked from highest to lowest or from lowest to highest. In the example shown at the beginning of this section, brake failure and engine damage showed the highest RPN. The empty columns show the PRPN (*probability* of RPN) resulting from CA (corrective actions).

Organizations may select their FMEA ratings following their organization's culture. The scales can be changed from 1 to 5 or from 5 to 5 ticks (HSBC Bank—Mexico). The particular ranking scale chosen is less important than applying a common scale across the organization. Some example ratings are provided in Figures 7.15 and 7.17–7.19. Figure 7.20 shows a FMEA template for the failure causes of a lawnmower.

An example of a FMEA form for the failure mode of a lawnmower is presented in Figure 7.20.

Rating scale—*severity*

Rating	Criteria
1	Is not noticed by anybody, has no effect
2	Is not noticed, has only an insignificant effect
3	Causes only small irritations
4	Modest loss of performance
5	Fall in performance level; consequence is customer's complaints
6	Fall in performance level that disturbs the capacity to function
7	Disturbed capacity function that leads to an increase in customer dissatisfaction
8	Product or service becomes useless
9	Product or service is illegal
10	Customer or employee is injured or killed

Figure 7.17   Example of a rating scale for severity selected by an organization.

Rating scale—*occurrence*

Rating	Criteria	Probability
1	Once every 6 – 100 years	<2/1,000,000,000
2	Once every 3 – 6 years	<3/10,000,000
3	Once every 1 – 3 years	<6/1,000,000
4	Once a year	<6/100,000
5	Once every 6 months	<1/10,000
6	Once very 3 months	<0.03%
7	Once a month	<1%
8	Once a week	<5%
9	Every 3 to 4 days	<30%
10	More than once a day	>30%

Figure 7.18   Example of criteria used by a service organization for rating their FMEAs throughout their business areas.

Rating scale—*probability of detection*

Rating	Criteria
1	The cause of the failure is obvious and can be hindered simply
2	All units are to be inspected automatically
3	Statistical process controls (SPCs) with a systematic failure cause test and corresponding ***Actions***
4	Statistical process controls are to be conducted with a systematic failure cause test
5	Statistical process controls are to be conducted
6	All units are to be inspected manually and avoidance actions installed
7	All units are to be inspected manually
8	Manual inspections are to be conducted frequently
9	Manual inspections are to be conducted occasionally
10	The defect caused by the failure is not detectable

Figure 7.19   Illustration showing the probability of detection criteria of a service organization for all their business areas.

Product or Process	Failure Mode	Failure Effects	SEV	Causes	OCC	Controls	DET	RPN	Actions	Plans	PS	PO	PD	PRPPN
Engine	Overheat	Engine damage	8	Radiator hose worn	5	Check frequently	1	40	Replace radiatior hose					
Battery	Doesn't retain charge	Car doesn't start	8	Dry/dead battery	4	Service battery	1	32	Change battery					
Brakes	Brake failure	Can't STOP car	10	Worn brakepads	2	Replace worn pads	8	160	Revise brakes frequently					

Figure 7.20   Illustration showing example of the general FMEA template of a lawnmower.

And now, what? When using a CE diagram or FMEA it is important to keep the following points in mind:

1. If the results were generated in a brainstorming session (CE diagram) and an objective evaluation (FMEA), then verification based on actual data is urgently needed!
2. If the verification through data is only empirical, then quantitative and qualitative verification of statements is required.
3. It is recommended that a *tick* for "proven" and a *cross* for "disproved" *be entered* behind every piece of information in the *C&E diagram* and the *FMEA diagram.*

7.7.3 *XY* Matrix

The primary focus of the analysis of the strategy is to investigate the output or Y (level and variation) for insight into the nature of the x's (factors of variation) that control Y. Therefore, the practitioner must assume the role of a detective in solving a crime—meaning looking at the nature of the crime, collecting evidence, and analyzing the patterns in that evidence to identify the suspect. Good detectives know how to collect and use evidence. Researchers must learn what kind of data they need and how to collect them. An *XY* matrix assesses the effect of each input factor $(x_1, x_2, x_3, \ldots, x_n)$ on each output (y) of the process.

Benefits of the XY Matrix. The *XY* matrix allows everyone involved with a process to agree on outputs $(y$'s) critical to the survey, transaction, and/or customer. Through numerical rating, the matrix allows the project team to assign the level of importance of each variable to each output. In addition, through association, the *XY* matrix allows the project team to numerically assess the effect of each input (x) on each output (y).

The *XY* matrix is frequently used to guide a subsequent FMEA. An example application of the *XY* matrix method explores input variables and their contributions to the formula for brewing an outstanding cup of coffee using the transfer equation $Y = f(x$'s) for the association formulas. Figure 7.21 shows the *XY* matrix of input variables $(x_1, x_2, x_3, x_4, x_5, x_6, x_7)$ for the outputs (y_1, y_2, y_3, y_4) that make an outstanding cup of coffee (Y).

The input variables $(x$'s) are

$x_1 = $ coffee type
$x_2 = $ amount of coffee
$x_3 = $ grind time
$x_4 = $ water temperature
$x_5 = $ cup type
$x_6 = $ cup size
$x_7 = $ brew time

and the outputs $(y$'s) are

$y_1 = $ taste
$y_2 = $ aroma
$y_3 = $ cost
$y_4 = $ acidity

7.8 GRAPHICAL TOOLS

Most of us are familiar with the phrase "a picture is worth a thousand words." Often, a graphical display can reveal at a glance the main characteristics of data. During brainstorming sessions,

XY matrix

Project: **Date:**

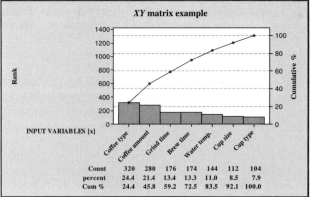

	Output Variables (*Y*s)	Flavor 1	Aroma 2	Price 3	Acidity 4	5	6	7	8	Rank 9	% Rank 10
	Output Ranking	10	10	10	2						
Input variables (*X*s)					Association Table						
Coffee type		10	10	10	10			1		320	24%
Coffee amount		9	7	10	10			2		280	21%
Grind time		9	6	2	3			3		176	13%
Water temp.		9	3	2	2			4		144	11%
Cup type		2	4	4	2			5		104	8%
Cup size		2	4	5	1			6		112	9%
Brew time		9	6	2	2			7		174	13%
										1310	

Figure 7.21 Figure showing the *XY* matrix method with the corresponding Pareto cumulative % analysis.

it is important to explore all phases of the project and processes the practitioner to enable to apply graphical representations (tools) that will provide information required to answer process questions. Critical questions include "What?", "How?", "Why?", "Who?", "Where?", and "When?" To achieve process improvements, it is important to ask:

- What is the purpose of the process?
- How can the process be improved or be performed better?
- Who can best perform the process?

(*Note*: Please note that the same principle of the what, how, who, where, why, and when was also used for the section of problem definition.)

Timely and quality information is required to successfully complete a product or transaction. Facts must be presented clearly and accurately and examined critically so that the most practical, effective, and cost-effective methods can be instituted. A well-constructed graphical representation should not require verbal explanation. This section describes frequently used graphical representations:

Bar Graph. To construct a bar chart or graph, we mark the various categories on the horizontal axis (abscissa). The frequencies are marked on the vertical axis (ordinate). A bar is drawn for each category such that the height of the bar represents the frequency of the corresponding category. A small gap usually separates adjacent bars that exist.

Pareto Charts. These are used to identify problems encountered in a process. To construct this graph, draw the horizontal and vertical scales. The horizontal scale shows reasons for defects in descending order; the vertical scale shows the percentage of the total number of occurrences for each group of occurrences. This method of grouping data helps identify where effort should be applied to reduce or eliminate problems. One variation of the Pareto analysis is a scale added to the right of the graph that shows cumulative sums. This scale starts at "0" on the bottom and ends with 100% on the top. These scales help identify the point of diminishing return for the problem groups. Diminishing return occurs when the expense to fix the problem equals the projected savings resulting from the fix.

Pie Chart. A pie chart is a circle divided into portions that represent the relative frequencies, or percentages of a population or sample belonging to different categories. Since a circle contains 360°, we multiply 360 by the relative frequency for each category to obtain the angle in degrees for the corresponding category.

Probability Plots. These plots identify the probability distribution underlying a population. If the data displayed closely match a known distribution, we can conclude that the sample data are drawn from a population described by that probability distribution.

Multi-Vari Charts, BoxPlots, and Time Series Plots. These enable us to classify variation into three categories: positional, cyclical, and temporal. This allows us to classify the independent variables (X_1, X_2, X_n) and provide a starting point for further analysis such as design of experiments (DOE).

Scatterplots. Scatterplots help us determine whether two variables are correlated. If the data points are randomly distributed, then we can conclude that there is no relationship between the variables tested.

[*Note*: MINITAB's graphs introduces the practitioner with the *p* value (probability value) or *probability value approach*. The *p* value is the smallest significance level at which the null hypothesis is rejected. The probability value (*p* value) is discussed in detail in this book in Chapters 16 and 17.]

7.8.1 Graphical Summary

The graphical summary combines useful basic statistics and graphs. This summary includes (1) histogram with normal curve; (2) boxplot with the first quartile, the median, and the third quartile; (3) normality test (dataset normality); (4) mean of the dataset; (5) standard deviation of the dataset, and (6) confidence intervals for the population mean (μ). A graphical summary provides several graphs that summarize data, as well as a statistical summary.

The graph in Figure 7.22 is a MINITAB example of this graphical summary. The graphical summary can be obtained with other statistical software such as JMP.

Example 7.6 A call center has collected data for one hour on the average speed of answer and would like to see the summary of this dataset. To obtain a graphical summary in MINITAB, choose Stat > Basic Statistics > Display Descriptive Statistics. Then

- Complete the dialog box by selecting the desired variable.
- Click on graphs.
- Check graphical summary.
- Click OK in each dialog box.

Figure 7.22 shows a graphical summary of the call center telephone answer time.

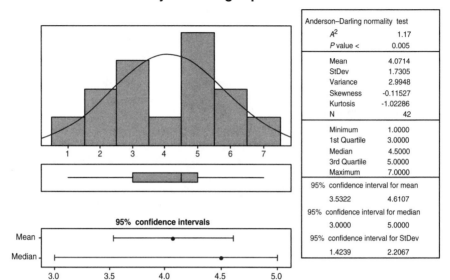

Figure 7.22 Graphical summary for a one-hour dataset of the variable average speed of answer.

Solution   The mean average speed of answer is 4.07 s. The median speed of answer is 4.5 s. The median is often reported rather than the mean for a sample with extreme values. The standard deviation is 1.73 s. The five-number summary is as follows:

1. Minimum value $= 1$ s
2. Median or $Q_2 = 4.5$ s
3. Maximum value $= 7$ s
4. $Q_1 = 3$ s
5. $Q_3 = 5$ s

Because the standard deviation is strongly influenced by extreme values, the interquartile range (IQR), which in this case is the difference between Q_3 and $Q_1 = 5 - 3 = 2$ s, may be a better measure of data spread. As can be easily seen, the majority of calls are being answered in 4 s.

Finally, if the p value is less than or equal to α, H_0 is rejected. On the other hand, if the p value is greater than α, H_0 is not rejected (fail to reject H_0). The typical p value is 0.05.

The p value obtained in this summary is 0.005, which is less than α, which means that H_0 is rejected and therefore, the data population is not normal.

7.8.2   Boxplot or Box-and-Whisker Plot

When exploring datasets, it is useful to know the key data characteristics such as the mean or median of the dataset, the standard deviation, the shape of the data distribution, and the minimum and maximum values of the dataset. A boxplot or box-and-whisker plot presents these data in a compact graphical format; in addition to the above mentioned characteristics, it also shows the center, the spread, and the skew of a dataset. MINITAB even identifies outliers by the asterisk symbol *. Boxplots are a convenient way to compare different distributions.

Box/whisker plot

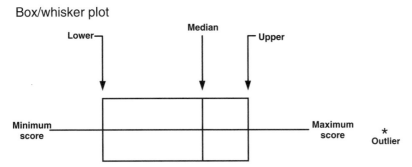

Figure 7.23   Drawing showing the five-number summary and example of an outlier.

A *boxplot* is constructed by first obtaining the minimum value, the maximum value, and quartiles of the dataset:

1. Minimum value of the dataset
2. Three values represented by a box and a line that shows the minimum value of the box or first quartile Q_1, a line inside the box or median or second quartile Q_2, the third quartile Q_3, and the maximum value of the box
3. The maximum value of the dataset

The practitioner should note that these add to five values. Therefore, some authors call these the "five-number summary" as stated in the following definition.

Five-Number Summary. This summary consists of the minimum value of the dataset, the first quartile Q_1 (left boundary of the boxplot); the median, also known as the second quartile Q_2; the third quartile Q_3 (right boundary of the boxplot); and the maximum value of the dataset.[4] This information can be shown in one "box-and-whisker" plot (Fig. 7.23).

A box-and-whisker plot gives a graphic representation of data using five measurements:

1. Median (second quartile)
2. First quartile
3. Third quartile
4. Minimum value
5. Maximum value

Example 7.7   A bank wants to improve its customer service; it first documents current performance by collecting monthly data for customer complaints at three different branches in the same city. The bank found significant variation in customer service between branches.

Solution   Figure 7.24 shows customer complaint data for the same month from three different bank branches. The box on the left is the branch that reported the minimum number of complaints, while the branch located in the middle reported the most complaints. The horizontal line in the middle box is the median value for that branch bank. The *median* of a dataset is the value of the middle term when the dataset has been ranked in increasing order. The boxplot shows that customers are served faster in upper-scale neighborhoods. The data of the first box in this example are skewed to the right, in the second box data are skewed to the left, and in the third box data are slightly skewed to the left.   ■

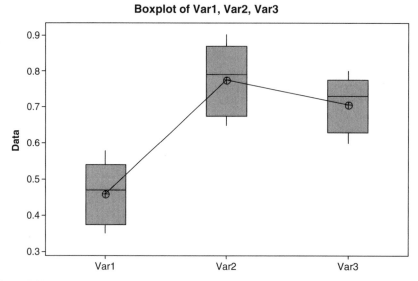

Figure 7.24 MINITAB graphical representation of a boxplot showing comparison of three variations.

7.8.3 Normal Probability Plot

Probability plots are used to test the validity of normality in data distributions. Statistical software such as JMP and MINITAB perform validity tests to determine whether the data fit a normal distribution. When the probability plot is a straight line, it is assumed that the data came from a normal distribution. This is the most important continuous probability distribution in statistics. Its familiar bell shape when plotted with linear vertical and horizontal axes frequently indicates that the sample values are normally distributed.

Measures of Normality. The Anderson–Darling test is the default test for normality (Fig. 7.25) in MINITAB. It compares the distribution of the actual data with a theoretical normal distribution created by using the dataset estimates for the population mean and the standard deviation. The p value for the Anderson–Darling test is a probability ranging from 0 to 1.0, where $0 =$ no chance and $1 =$ certain. A detailed discussion can be found in later chapters of this book.

To create the normal probability plot with MINITAB, go to `Calc > Probability Distributions > Normal`.

Interpretation. If the data are fairly normal, the points in the plot will fall close to the diagonal reference line. The normal probability plot for the data given above (see Fig. 7.25) indicates that the distribution is fairly normal; all the points are observed that fall fairly close to the center line. The p value from the Anderson–Darling test is used to determine whether the data deviate significantly from a normal distribution. The p value for the data (0.328) is larger than the 0.05 α level. Therefore, there is not enough evidence to suggest that the data are not normal. The MINITAB graph in Figure 7.26 shows the Anderson–Darling results for the same data.

Anderson–Darling Normality Test. The results of this test indicate that neither distributions (Fig. 7.25 or 7.26) departs significantly from normality; both of p values are greater than the 0.05 α level.

If the data are perfectly normal, then the data points on the probability plot will form a straight line. The centerline forms an estimate of the cumulative distribution function for the population from

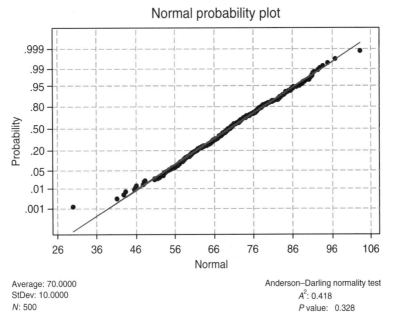

Figure 7.25 MINITAB normal probability plot showing that the dataset is close to a normal distribution.

which the data are drawn. The graph pictured in Figure 7.26 was generated with an Anderson–Darling normality test; however, the same graph could be generated regardless of which test the practitioner chooses. The plot shows that the points fall reasonably close to the reference line, indicating that the data follow a normal distribution.

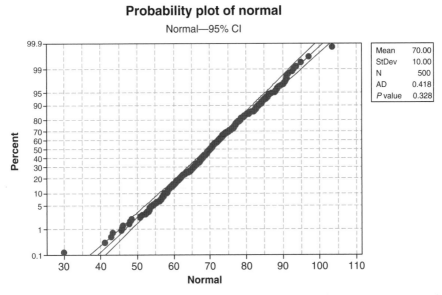

Figure 7.26 MINITAB normal probability plot showing that the data distribution is fairly normal; 90% of the points fall fairly close to the central line. In this graph AD stands for the Anderson–Darling normality test.

The probability plot in Figure 7.26 was generated using MINITAB by going to `Graph >` `Probability Plot > Single > OK`. One can also

- Select `Stat > Basic Statistics > Normality Test`
- In `variable` enter the column of data
- Select `test for normality: Anderson-Darling`
- Click `OK`.

Example 7.8 A Six Sigma Black Belt collected data on the process cycle to determine the nature of the data distribution.

Solution Raw data were first plotted in a histogram. The results indicated that the data closely followed a lognormal distribution. In general, data are tested against the normal distribution. If data fail normality, which in this case did, then they are tested using other distributions. The probability plot results shows four graphs (Fig. 7.27), where we see that the data are not normal but are in a lognormal distribution, thereby confirming the earlier assumptions that these data didn't follow a normal distribution. A detailed discussion on normal probability distributions is found in later chapters of this book. ∎

(*Note*: Because we are looking for patterns that are randomly distributed around the perfect line, any pattern that appears to be non-random suggests that the data are not normal, and then the plot should re-created using MINITAB's "individual distribution identification.")

To create a MINITAB probability plot

1. Go to `Stat > Quality Tools > Individual Distribution Identification`.
2. Note that MINITAB calculates the best fit.

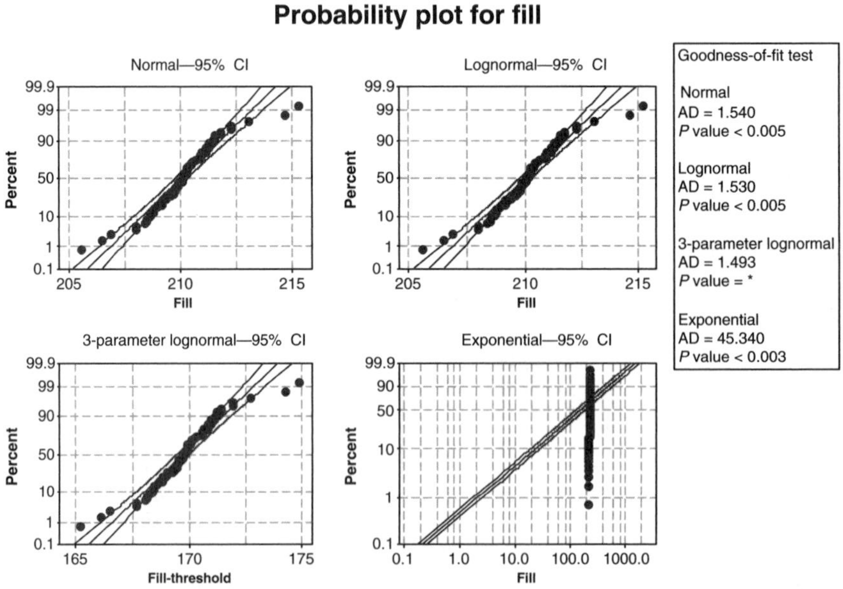

Probability plot for fill

Figure 7.27 MINITAB probability plot showing a lognormal distribution.

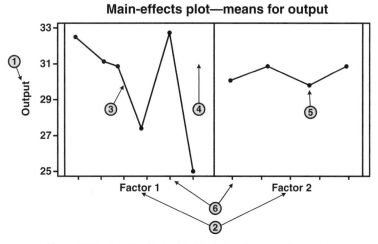

Figure 7.28   A main-effects plot, illustrating the means for output.

7.8.4   Main-Effects Plot

The main-effects plot (Fig. 7.28) graphically compares the level of a process output variable at various states of process factors, to gain an understanding of the main effect of a change in the factor on the output. Main-effects plots for standard deviation show variation changes with factor levels. For the main-effects plots of the mean, slope interpretation is relative. For example, lines with steeper slopes have larger impact on the output compared to those lines with little or no slope. In general, the main-effects plot is used to present results from analysis of variance (ANOVA). This plot is used to

- Examine the level means for each factor
- Compare the level means for several factors
- Compare the relative strength of the effects across factors

Figure 7.28 Components

1. Output variable being studied.
2. Factors whose effect on the output is being studied
3. The mean of the output variable at different levels (or values) of factor 1 (more than one data point may have been collected for each level of the factor, so the mean is used)
4. The mean of the output variable at different levels (or values) of factor 1 (more than one data point may have been collected for each level of the factor, so the mean is used); the mean of the output variable at different levels (or values) of factor 2
5. The grand mean of all the output values
6. Factor levels

A second main-effects-plot example is shown in Figure 7.29.

Main-effects plots also assist in comparing several factors at one time. A relatively flat line indicates that the factor has little effect on the output (drying profile and switch off criteria), while a line with heavy up/down movement (sample weight) indicates that the factor has a greater "main effect." Main-effects and interaction plots are standard methods of ANOVA or when analyzing designed experiments (DOE), which use ANOVA for analysis.

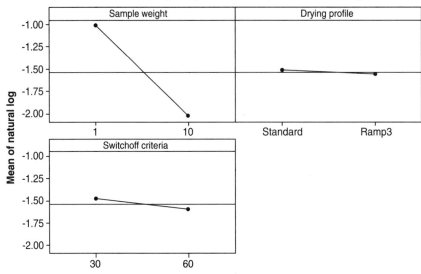

Figure 7.29   A main-effects plot.

Constructing Main Effects Plots is MINITAB

1. Gather data and present in tabular form with output values being matched with corresponding levels of the chosen factors.

2. When more than one output value is presented for each factor, calculate the mean of the output value for each factor.

3. Calculate the grand average of all the output values, and plot it as a straight line on all main-effects plots.

4. For each factor, plot the mean of the output values at each level of the specific factor.

5. If more than one factor is being analyzed, the plots for the different factors should be presented together for comparison.

6. Go to `Stat > ANOVA > Main Effects Plot` function.

Example 7.9   A Black Belt is assigned to a project to reduce the variation in the time required to collect account receivables. The company's policy states 30 days net, but some clients take up to 45 days, causing the company to incur additional collection costs. The company's clients are located in two regions (A and B); company employees work in two shifts to prepare the invoices. The team collects and displays data to see if there is a relationship between delay in payment and geographic region and/or shift when the invoices are prepared. The Black Belt and a master Black Belt interpret the main-effects plot. The objective of a diagnostic chart is to "listen" to the process without interference to analyze the relationships between independent and dependent variables, claims the master Black Belt.

Solution   On the left side of Figure 7.30, we observe a flat line indicating that performance is not affected by the shift, while the sloped line on the right indicates performance differs between regions. The lack of sensitivity to shift was expected since employees of each shift received the same training and follow the same procedures. Also, customers in region B take on average about 5 days longer to pay their invoices. However, the most revealing issue is that, on average, customers take almost 10 days longer to pay than stipulated on our invoices. Therefore, the Black Belt say, "although region A seems to include customers who pay more quickly than those in region B, further investigation is required to understand the reasons for the delayed payments." ■

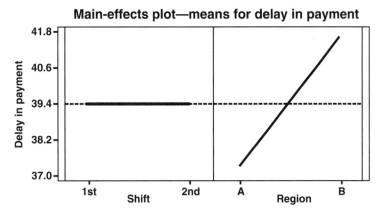

Figure 7.30   Main-effects plot showing delay in payment.

7.8.5 Pareto Chart

A *Pareto chart* is a bar graph for quantitative data, with the bars arranged in order by frequency of occurrence. As in histograms, vertical scales in Pareto charts represent frequencies or relative frequencies. The tallest bar is at the left, and the smaller bars are farther to the right. A Pareto chart organizes data to show the major factor(s) that make up the subject being analyzed. It is a search for significance. The bars in a Pareto chart are arranged side-by-side (touching) in descending order from the left. The basis for Pareto analysis is the "80/20 rule," meaning that 80% of the problems result from 20% of the causes. Arranging data on a Pareto chart highlights the vital few in contrast to the trivial many. Selecting categories, tabulating data, ordering data, and constructing the Pareto chart can enhance communication among team members and with management. Before and after improvement, Pareto diagrams provide a good basis for measuring the success of an improvement effort. Selecting categories, tabulating data, ordering data, and constructing the Pareto chart can enhance communication among team members and management. Pareto charts can be of first, second, third levels, and more. Figure 7.31 shows a Pareto chart for customer requests to a bank call center.

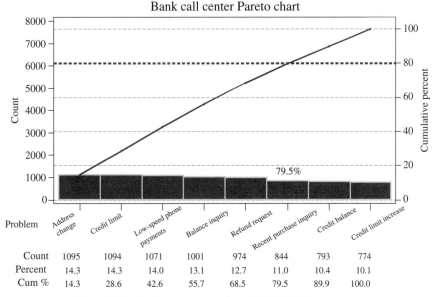

Problem	Address change	Credit limit	Low-speed phone payments	Balance inquiry	Refund request	Recent purchase inquiry	Credit balance	Credit limit increase
Count	1095	1094	1071	1001	974	844	793	774
Percent	14.3	14.3	14.0	13.1	12.7	11.0	10.4	10.1
Cum %	14.3	28.6	42.6	55.7	68.5	79.5	89.9	100.0

Figure 7.31   Graphical representation of a MINITAB Pareto chart.

Example 7.10   A Manager attends a Six Sigma project review for a specific area in the company. "We need immediate action," exclaims the manager. "We have so many component shortages that we even schedule shortage meetings to firefight daily problems." "Precisely, replies another attendee, but where could we start?"

Solution   The Six Sigma Black Belt displays a Pareto chart that the team created and says, "as you can see, we collected data and counted the frequency of occurrence for each problem source (shortages/defects). We then sorted the relative frequencies in descending order and created a Pareto chart to depict the frequencies."

Supplier lateness accounted for 68.5% of shortages, thus giving us a clear vision of where immediate action was required. Moreover, we concluded that supplier and receiving inspection accounted for approximately 80% of the shortages. The production manager found the results quite interesting and said: "I am relieved to see that production is not always the source of all shortages. For example, in this case, production accounts (observe "Production" in Fig. 7.32) accounts for only 5.2% of the shortages."

Steps in Manual Construction of a Pareto Chart

1. Clearly define the problem.
2. Define boundaries of study for example, by week, by month, by day, by banking branch, by bank geographic location, by shift per branch bank, by quarter, or by branch bank.
3. On a piece of graph paper draw a vertical scale and a horizontal scale. The vertical scale (ordinate) shows the frequencies of each characteristic. The horizontal scale (abscissa) shows the data (problem) characteristics.
4. Gather data on frequency characteristic, i.e., address change.
5. Rank characteristics by number of occurrences.
6. Plot with tick marks the number of frequencies of each characteristic.
7. The resultant pattern shows which groups make the "vital few."
8. Calculate cumulative percent.

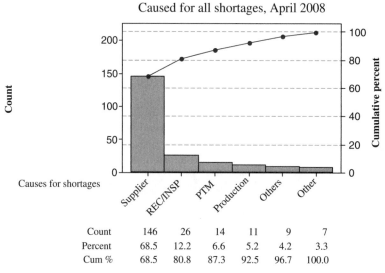

	Supplier	REC/INSP	PTM	Production	Others	Other
Count	146	26	14	11	9	7
Percent	68.5	12.2	6.6	5.2	4.2	3.3
Cum %	68.5	80.8	87.3	92.5	96.7	100.0

Figure 7.32   MINITAB Pareto chart showing Pareto chart used for solution of the exercise.

In MINITAB, proceed as follows:

- Go to Stat > Quality Tools > Pareto Chart.
- Select Chart defect table.
- Labels in—enter labels.
- Frequencies in—enter count.
- Select Options >—enter chart title.
- Click OK.

7.8.6 Run Chart

There is currently a strong trend toward improving the quality of American goods and services. The methods presented in this section are used by a growing number of businesses. Evidence of the increasing importance of quality is found in its greater role in the financial services, delivery of services, manufacturing, advertising, transactions, and banking industries. A *run chart* is a simple graphical representation of process characteristics showing values gathered from the process. Often these are individual values are averages for short time periods (e.g., daily averages for monthly chart). A central line, often the median of the values, can be analyzed for patterns. By monitoring the characteristics of a process, we are able to obtain a pattern of the data and identify changes in process characteristic values over time. Maintaining a stable process is critical to the production of goods and delivery of services at consistent levels of quality.

Figure 7.33 Shows a run chart of 22 data events (22 points) of a banking call center for *average speed of answer* showing a mean $= 7.045$ s, a standard deviation $(\sigma) = 1.914$, a median $= 6.5$ s, and a minimum speed of answer of 5 s and maximum speed of answer of 12 s.

Figure 7.34 illustrates the number of abandoned calls per day for a call center. Run charts monitor process changes associated with a particular characteristic over time. The practitioner uses run charts to identify patterns in the process data and perform tests for nonrandom behavior.

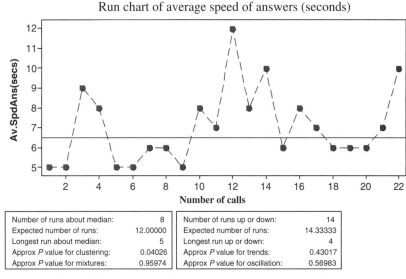

Figure 7.33 Run chart showing a call center average speed of answer.

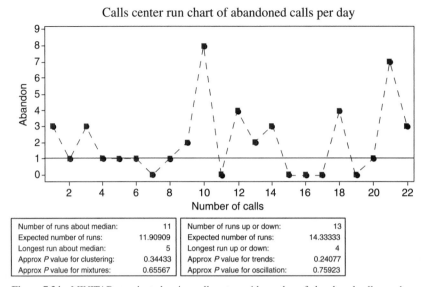

Figure 7.34   MINITAB run chart showing call center, with number of abandoned calls per day.

7.8.7 Time-Series Plot

Time-series data are collected on the same element or variable at different points in time or for different periods of time. Time-series data contain information on the same element for different periods of time. The time-series plot is a graphical representation of observations taken at specified, usually at equal intervals, to measure the performance of a process over time, and the existence of any trends. Time-series plots are graphical representations of the relationship $Y = f(t)$. The data collected for this graph are called *time-series data*. In a time-series plot, the data are entered in the y axis and the time data in x axis.

Figure 7.35 is an example of a time-series-plot. This graph shows the total value of bank-error fee reversals per month from data collected at monthly intervals over one year beginning in July.

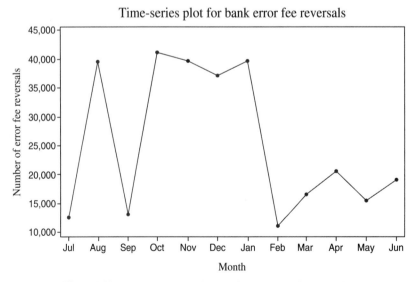

Figure 7.35   MINITAB time-series plot for bank-error fee reversals.

From Figure 7.35 we observe how the number of error fee reversals fluctuated during the one-year period. Generally, a time-series plot reveals the trend of a variable over a period of time.

How to Manually Create a Time-Series Plot

1. Gather the time-based data in tabular format, ensuring that the independent (x) variable is evenly spaced.
2. Plot the data manually using MS Excel, or MINITAB's `Graph > Time Series Plot` function.
3. Points should be connected with straight lines.

How to Create a Time-Series Plot Using MINITAB

- Go to `Stat > Time Series > Simple`.
- Click `OK`.
- Dialog box appears.
- In series, enter column data frequencies.
- Click `OK`.

7.8.8 Multi-Vari Charts

Multi-Vari charts are powerful graphical representations for analysis of variance. In general, these graphical representations are used in the preliminary stages of data analysis to see the variation pattern of the data. The chart displays the means of every factor (x) at each level (see Fig. 7.36). Multi-Vari studies are extremely important because they allow simultaneous analysis of different types of variation. In this section, we will concentrate on the graphical analysis part of the Multi-Vari. This section covers the study of multiple x's on the Y characteristic. Multi-Vari charts can be constructed in a variety of ways to offer the "best view" of the data. This graphical representation allows the practitioner to identify sources with different types of variability. Multi-Vari charts allow us to obtain a first look at the process stability over time and to provide direction for the design of new processes or

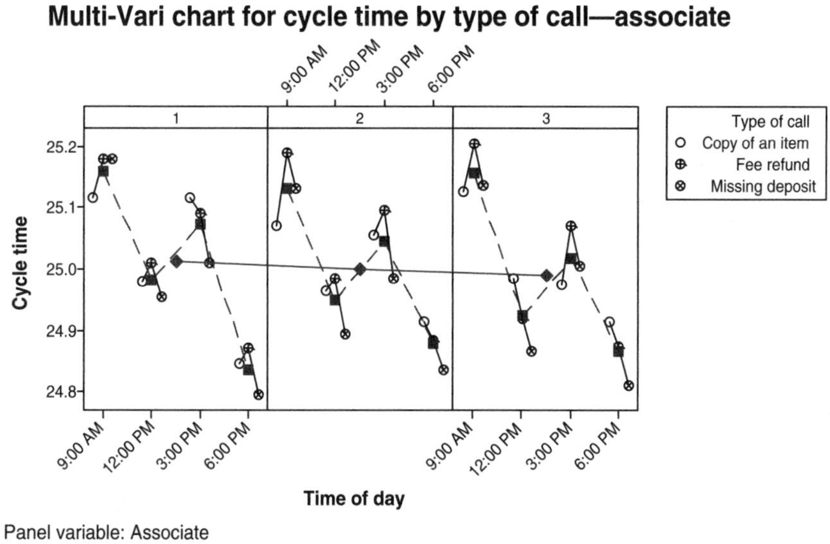

Figure 7.36   Multi-Vari chart for cycle time by type of transaction, associate, and time of day.

Multi-Vari chart for cycle time by type of call—time of day

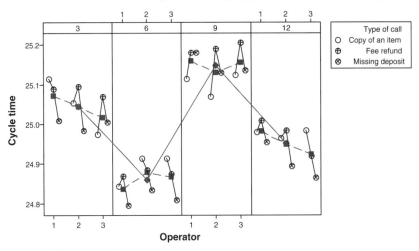

Panel variable: Time of day

Figure 7.37   Multi-Vari chart showing different operators and cycle time by type of call.

redesign of existing processes. In general, Multi-Vari charts graphically show the following types of variation:

Positional—variation within a part or process. For example, you could look at different types of transactions or different associates within a bank branch.

Cyclical—variation between consecutive parts or process steps, for example, vault dollar volume.

Temporal—time variability, for example, the number of transactions within one selected branch or the entire bank.

The Multi-Vari chart in Figure 7.36 shows the cycle times for various types of transactions received by the bank branch.

It was found that variability increased because each associate handles multiple transaction types. This is explored in Figure 7.37.

To create a Multi-Vari chart using MINITAB

- Go to `Stat > Quality Tools > Multi Vari Chart.`
- Dialog box appears.
- Enter (Y) = response (for the preceeding example Y = response = cycle time).
- Enter factor(s).
- Click `OK`.

7.8.9 Scatterplot

Scatterplots show the relationship between two variables; this relationship can take such forms as linear and quadratic. In the following chapters, we will discuss, in detail, the concepts of linear correlation and regression. In this section, we use scatterplots to determine whether potential process factors (variables) are related to a key outcome, and then identify the underlying probability distribution. Frequently, by using this graphical approach, specific courses of action become clear. When necessary, more advanced statistical tools may be used to quantify the relationship between the

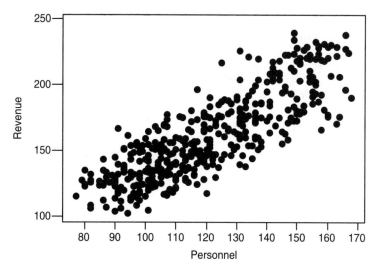

Figure 7.38 MINITAB's scatterplot showing a weak positive linear correlation; r is positive but close to 0.

two variables (factors). Scatterplot fundamental concepts are discussed in more detail in Chapter 8 of this book.

Example 7.11 Following data collection, the raw data should be plotted as a simple Y vs. X, or scatter, plot. This provides a quick way to qualitatively determine whether there is a relationship between the two variables, and the nature of this relationship. Consider the results of the revenue–vs personnel plot shown in Figure 7.38. We wish to determine whether a relationship exists between revenue and personnel of an organization. We also want to identify this relationship with an equation so that we can predict the number of personnel as the organization's revenue changes. After plotting the data, it was found that there was no obvious relationship between the two variables as the data points were, for all intents and purposes, randomly placed.

Regression Analysis: Revenue versus Personnel. The regression equation is `Revenue = 16.1 +` `1.20 Personnel`:

Predictor	Coef	SE Coef	T	P
Constant	16.108	4.904	3.28	0.001
Personnel	1.19593	0.03995	29.94	0.000

S=18.0373 R-Sq=68.2% R-Sq(adj)=68.1%

Although we can explain the linear correlation using the population parameter ρ, we will use the sample correlation r. The linear correlation coefficient r measures the strength of the linear relationship between the paired x and y values in the sample. The linear correlation coefficient is referred as the Pearson correlation r. If r is close to 0, we say that there is no significant linear correlation between the pair data x and y. When all points in the scatterplot lie on a straight line that slopes *upward* from left to right, we have a positive correlation between x and y. In the case that all points in the scatterplot fall on a straight line that slopes *downward* right to left, we say that we have a perfect negative linear correlation between x and y.

In addition, we say that there is a strong positive linear correlation when r is close to $+1$. But when r is closer to -1, we say that there is a strong negative linear correlation. Finally, it is important to know that the value of r is always between -1 and $+1$. This is: $-1 \leq r \leq +1$

Graphically, a strong correlation indicates that the points in the scatterplot are very close to the regression line and a weak correlation indicates that the points in the scatter plot are widely spread around the regression line.

In Example 7.11, data were collected for revenue and personnel and a scatterplot was drawn to find any relationship between revenue and personnel. R^2 is the multiple coefficient of determination. (*Note*: Here the multiple coefficient is applied to a single-variable case.) R^2 is a measure of how well the multiple regression equation fits the sample data. A perfect fit would be when $R^2 = 1$. A poor fit is considered when R^2 is $= 0$ or close to 0.

The regression equation is

```
Revenue = 16.1 + 1.20 Personnel
S = 18.0373 R-Sq = 68.2% R-Sq(adj) = 68.1%
```

The MINITAB results include the regression equation, which is expressed as $\hat{y} = 16.1 + 1.20x$. The MINITAB result includes `R-Sq = 68.2%`, which is the coefficient of determination; this value indicates that 68.2% of the variation in revenue is explained by the number of personnel, and 31.8% of revenue remains unexplained. The result also shows other terms that are covered in detail in later chapters of this book. In addition, latter chapters of this book contain detailed sections of correlation and regression and multiple regressions. We conclude this brief introduction into how to interpret the results of the scatterplots by saying that the multiple coefficient of determination R^2 is a measure of how well the regression equation fits the sample data. As more variables are included, R^2 increases; in some cases R^2 could remain the same, but in general it increases. Consequently, it is better to use the adjusted coefficient of determination when comparing multiple regression equations because the R^2 value is being adjusted on the number of variables and the sample size. Figure 7.40 shows a plot performed using MINITAB. Observe that MINITAB's results above the scatterplot in Figure 7.39

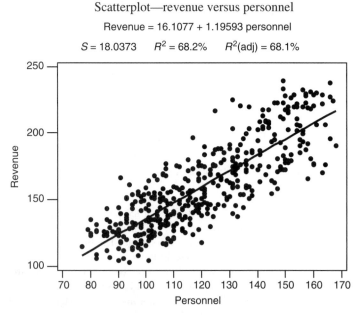

Scatterplot—revenue versus personnel

Revenue = 16.1077 + 1.19593 personnel

$S = 18.0373 \quad R^2 = 68.2\% \quad R^2(\text{adj}) = 68.1\%$

Figure 7.39   MINITAB scatterplot—with line—showing weak positive linear correlation.

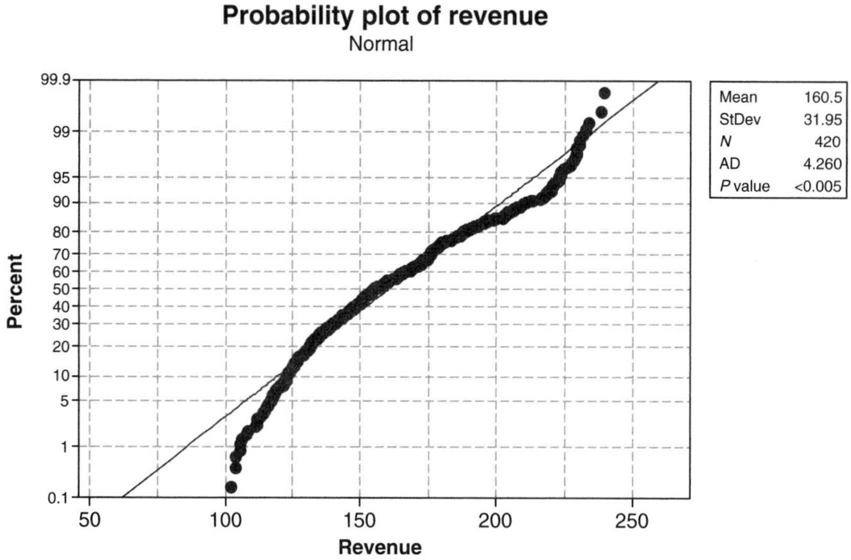

Probability plot of revenue
Normal

Figure 7.40   MINITABs normal probability plot showing that the data points do not correspond to a normal distribution (data do not form a straight line). The p value is < 0.05 α-level. The departure from the normality straight line is significant.

show the coefficient of determination as `r - sq (adj)` $= 68.1\%$. [*Note*: When comparing this multiple regression equation to others, it is better to use the adjusted R^2 of 68.1% (0.681).] The figures above show positive correlation between revenue (X) and personnel (Y). The objective was to try to find if the revenue results were important for the personnel results. An important assumption that is difficult to check is that the distribution of the paired data (X, Y) follows a bivariate normal distribution. The key feature of such distribution is that for any fixed value of X, the values of Y have a normal distribution. We have seen in a previous section how to test for normality.

To create the normality test in MINITAB

- Select `Stat > Basic Statistics > Normality test`.
- Select the variable.
- Click `OK`.

If the data came from a normal distribution, the data points should follow a straight line. However, in this case the data do not follow a normal distribution. This is consistent with the Anderson–Darling test, value suggesting that the data points are not normally distributed.

7.9   GRAPHICAL REPRESENTATION OF A PROCESS

Independent of which analysis method is used on a process or phase, it is important to present factual data and conclusions in a clear logical form. In methods work, the first step is to gather all facts necessary to perform the operation or process. Well-researched information such as the quantity to be produced, delivery schedules, operational times, facilities, type of services, machine capacities, special materials, special tools, and standard operating procedures (SOPs) may be important to solve a problem or complete a Lean process redesign. Once facts are presented clearly and accurately, they must be examined critically so that the most economical and effective method can be designed (Lean design).

Among the most important tools of the methods engineer are the process map and the value stream map. A *process map* is a graphical representation of a manufacturing, service, transaction, or business process. Methods engineers frequently use different process maps for different applications, including

- The flowchart
- The process map
- The cross-functional process map
- The value stream mapping chart
- The operations process map
- The worker–machine process map
- The gang process map
- The PERT (program evaluation–review technique) chart

In this section, we will limit our discussion to flowcharts, process maps, and cross-functional maps. We have addressed value stream maps in a previous chapter. The operation process map, worker machine process map, gang process map, and PERT charts are beyond the scope of this book.

7.9.1 Process Flowcharts

Flowcharting has matured to the point that standardized construction symbols are supported by various standards organizations (ANSI, OSI). Most of these symbols are available in software tools such as Clarify, Microsoft Excel, PowerPoint, VISIO, iGrafx, and Abode. Frequently used symbols are shown in Figure 7.41.

A *flowchart* (see Fig. 7.42) is a visual representation of the major process steps. It improves our understanding of a process by identifying critical problem areas and improvement opportunities. When building a flowchart, one should use the four standard symbols: start/stop, decision point, activity, and connector (in general, used to continue to a different page or section of the page, part of the diagram, or the continuation of the flowchart). Flowcharts facilitate process investigation, but rarely provide insight into the mechanisms driving process levels and variability. On the other hand, the potential causal structure captured in the cause–effect diagram is not tied to location in the process, nor is the current state of process knowledge depicted. Seldom do we realize the full potential of the combination of process flowcharts and CE diagrams. However, neither of these two diagrams

This Symbol	Represents	Some examples are ...
 (rounded rectangle)	Start/stop	Receive trouble report Machine-operable
 (diamond)	Decision point	Approve/disapprove Accept/reject Yes/no Pass/fail
 (rectangle)	Activity	Drop off travel voucher open access panel
 (circle)	Connector (to another page or part of the diagram)	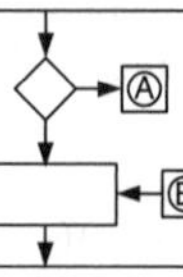 (diagram)

Figure 7.41 Symbols commonly used to illustrate the activities of the flowchart.

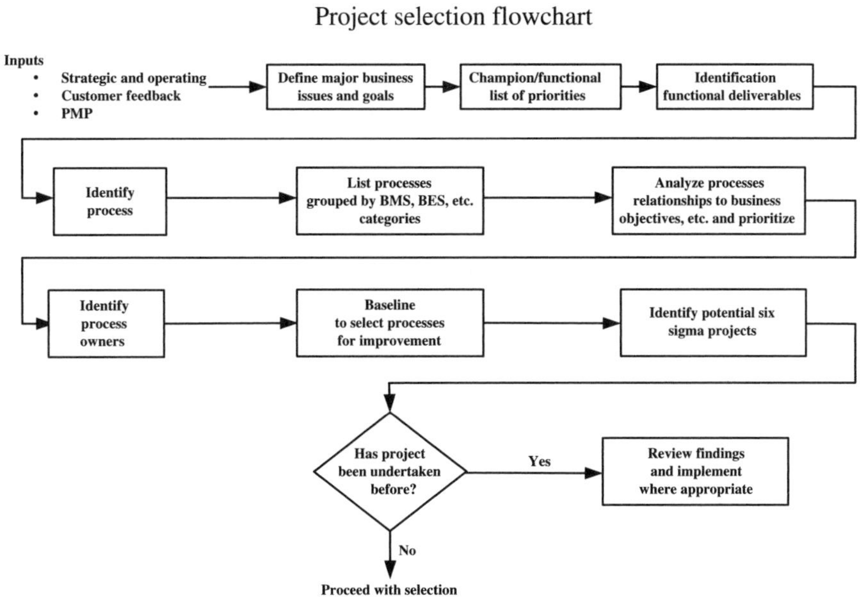

Project selection flowchart

Figure 7.42 Example of a flowchart.

indicates whether the potential sources of variation are currently being managed or if they can be managed at all. A further drawback is that flowcharts are one-time documents that are seldom updated as new knowledge is acquired.

The purpose of the process chart is to compare the "as is" process with the "should be" process. A process chart assesses the complexity of the process and identifies non-value-added operations, possible simplification and standardization opportunities, areas where defects may occur, and data collection points. The graphical symbols shown in Figure 7.41 are used in Figure 7.42. When constructing a flowchart, it is important to keep the following definitions in mind:

1. *Terminal symbol* (rounded rectangle) is used to designate the beginning or end of a process flow. Usually identified as start, end, beginning, or stop.
2. *Activity symbol* (rectangle) is used to designate an activity. A brief description of the activity is also included within this symbol.
3. *Decision symbol* (diamond) is used to illustrate the point where a decision must be taken. Subsequent activities are dependent on the decision taken at this point.
4. *Flowline* (arrow) indicates the direction of the process and connects its elements.
5. *Connector symbol* (circle) is used to illustrate a break and its continuation elsewhere on the same page or another page.
6. *Document symbol* indicates a printed document pertinent to the process.
7. *Delay symbol* is used to indicate that there is a delay or waiting period in the process.

The process flow diagram should be accurate and reflect the true process, not the "ideal" process. This is an important tool for continuous process improvement and should not be neglected. The same level of detail should be used for all the steps of the process.

How to Successfully Construct a Flow Diagram

1. Determine the limits of the process. Clearly define where the process begins and ends and the level of detail to be shown in the process map.

2. Determine the steps in the process. In a brainstorming session, list major inputs, outputs, activities, and decisions.

3. Identify the sequence of the steps. Draw the steps in the same order as they are carried out without any arrows. Define what is and not what should be flow diagrams. By comparing these two flow diagrams, probable causes of problems may be identified and solved.

4. Draw the process flow diagram using the standard symbols. Label each process step so that the labels are understandable by all concerned. Add arrows to show process flow direction. Identify the process map with its name, date, and names of the team members.

5. Verify that the process flow diagram is complete. Are the symbols used correctly? Are the process steps clearly identified? Is every feedback loop complete?

6. Does every continuation point have a corresponding entry point in the process map? Do all activity boxes have a single output? If not, a decision diamond may be required. Review the process map with colleagues who are not part of the team and who perform the activities depicted in the process map. Highlight and incorporate their comments.

7. Finalize the process flow diagram. Is this process performing as it should? Are people/departments following the process as mapped? Are there obvious redundancies or complexities that can be reduced or eliminated?

8. How does the current process map differ from an ideal one? Draw an ideal process map. Compare the two to identify discrepancies and opportunities for improvement.

7.9.2   Process Mapping

One of the most important tools for the process engineer or designer is mapping. Mapping requires creating teams whose members are selected from every department involved in the new service from marketing through production to customer. The next phase involves mapping each step in the proposed process from start to finish. For a successful redesign of a process, the five characteristics of a good process must be understood:

1. *Design Clarity.* The end-to-end process is well defined; activities/steps are named and interrelated; inputs and outputs are clearly identified. The value of the end-to-end process is well understood.

2. *Roles/Teams.* Each person must understand how his/her work and the work of teammates fits together to implement the end-to-end process. The decisionmaking process is well understood.

3. *Measurement.* The required measurements must be defined and understood before implementation. Measurements are used to evaluate performance of the entire process, not to measure individual performance. On the basis of outcomes that matter most to customers, components of a process must be understood and improved.

4. *Improvement.* There is a systematic ongoing effort to enhance end-to-end process performance by problem resolution or redesign.

5. *Ownership/Leadership.* All employees who have management responsibility relating to the end-to-end process must work together to plan, build, and operate the process and exhibit process thinking, process relationship, process leadership, and process improvement in their work.

To design a new call center or to improve one already in operation, it is helpful to present in clear, logical form the factual information related to the business. Process maps offer a visual description of how process activities flow within an organization. In general, a process map contains considerably more detail than a operation process chart. A process map is a graphical representation that shows a sequence of activities/steps represented by flowchart symbols. A process map forms a base for team breakthrough activities; it can identify and eliminate non-value-added activities, combine operations, and assist in root cause analysis. In addition, a process map identifies areas where data collection exists and ascertains appropriateness of these locations. The objective of a process map is to identify the input and output variables that define the Six Sigma transfer equation:

$$Y = f(X)$$

The process map becomes a living document used to monitor process changes. It is the baseline for other Six Sigma tools such as the *XY* matrix and the failure mode–effects analysis (FMEA). Everyone with roles in defining, executing, designing, or changing a process should participate in the creation of a process map. This includes operators, workstation owners, associates, engineers, supervisors, managers, design engineers, customers, and suppliers. Therefore, the process map is a graphical representation of the process showing the chronological sequence of all activities or steps in the process. A detailed process map contains information to improve the process, for example, cycle times, quality costs, inputs (*x*'s) and outputs (*y*'s). Any product item or feature deemed to be customer-critical is denoted as $y_1, y_2, y_3, \ldots, y_n$. Any item that has an impact on *Y* is denoted as $x_1, x_2, x_3, \ldots, x_n$. Each independent variable (*X*) is also known as a knob variable or *input* to the process that can be controlled [i.e., through design of experiments (DOE)]. An input can be changed to measure its effect on *Y*. The *C*, *N*, and *X* variables are defined in Section 7.7.1.

Before we can *Lean-design, Lean-redesign* or *improve an existing process*, we must graphically represent the product or process "as is." The process map is especially valuable at recording hidden cost activities, such as temporary delays and bottlenecks. Once these nonproductive activities (non-value-added) are highlighted, analysts can propose steps to improve the process. Process maps capture current process knowledge and provide valuable information about the different mechanisms that drive process variability. They capture information from the cause–effect analysis and relate it to the different process activities (see Fig. 7.43). A process map is a graphical representation of the process flow. In general, the process map contains more detail than the corresponding flowchart. Therefore, the flowchart is not used for complicated or complex process assemblies. A process map allows everyone involved with improving or designing a process to agree on the steps necessary to produce not only a good product, transaction, or service but also a Lean product, transaction, or service.

Example of process map

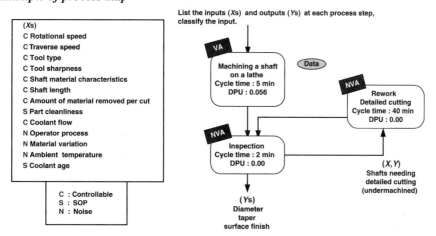

Figure 7.43 Process map activities providing steps in obtaining and classifying the input (*X*), the inputs *x*'s and the output (*Y*).

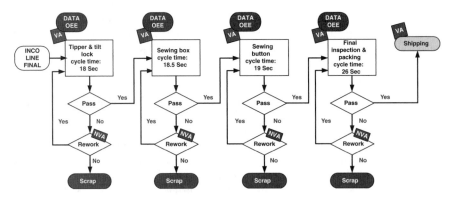

Figure 7.44   High-level cross-functional process map for inspection of seatbelts manufacturing, backend of the line.

7.9.3   Cross-Functional Mapping

It is widely known that a process map gives a visual process description of a process. Within a process map everyone can see whether there is a clear path for a process or whether obstacles exist in the form of non-value-added activities/rework, which is identified as waste (waiting time, queuing, overproduction).

Creating and working with process maps makes obvious how other business areas interact with or are part of the same process. The cross-functional process map directly addresses these situations. Figure 7.44 is an example of a high-level cross-functional map for the process inspection of seatbelt manufacturing.

7.9.4   Process Mapping—Deployment Diagram

The purpose of a process map is to graphically represent the process linkages, responsibilities, and functional interfaces to the stakeholders.

Section Definitions

- *People or departments*—responsible for the key activities.
- *Key process*—list steps or subprocesses included in the process.
- *Connections*—indicate the flow and sequence of the key activities with arrows.
- *Symbols*—indicate the nature of each step (see Section 7.9.1 for more details).
- *Cycle time*—measure each process step or subprocess.

The process mapping—deployment diagram shown in Figure 7.45 is a tool used for process definition, project scoping, and identification of the CTQs (critical-to-quality) key process inputs/factors (x's) or opportunities to eliminate steps, and data collection points.

In Figure 7.45, the first activity of the process (1) takes place in department A and then the process flows to department C (4). From department C the process flows to department D, in which a decision is made for "No," to continue the process in department B (3), or if "Yes," to go to department E. After a process activity is completed in department B, it flows to department F, where a second decision takes place. Note that when the activity of department E is completed, the process flows to department F for decisionmaking. From department F, if "No," the activity goes to department E, if "Yes," the activity flows to department G. All these process activities are time-tracked (5) in x days.

Process mapping—deployment diagram

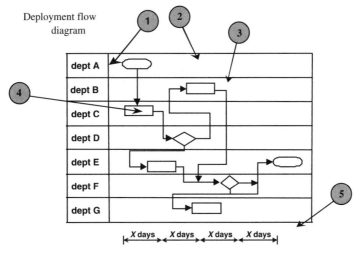

Figure 7.45 Deployment diagram showing days duration (*x* days) of process activities in different departments.

The deployment process map is a key tool for transactional processes and should reflect the process as is. There are various levels at which the process can be mapped: macro, process, and micro levels. It is important to maintain a consistent level of detail for the process being mapped. The following procedure is recommended:

1. Identify the customers of the process and their deliverables and requirements.
2. Determine the boundaries of the process to map.
3. Clearly define where the process begins and ends.
4. Clearly define what is considered outside the scope of the process.
5. Agree on the level of detail to show in the process map.
6. Determine the functions, departments, or roles involved throughout the process.
7. Brainstorm to determine the subprocesses or steps in the process and assign by function, department, or role.
8. Identify the sequence of activities.
9. Draw the deployment process diagram (see example shown in Fig. 7.46).
10. Label each subprocess or process step using verbs.
11. Add arrows to indicate the direction of the flow.
12. Validate the deployment process diagram to ensure its accuracy.
13. Assign cycle times where appropriate.
14. When mapping at a subprocess or process step level, and using process flow diagram symbols:
    - Identify value-added and non-value-added steps in the process.
    - Identify "critical to" characteristics at each step (where appropriate).
    - List and classify key process inputs.
    - Identify times for each step.
    - Validate the current situation with the team.

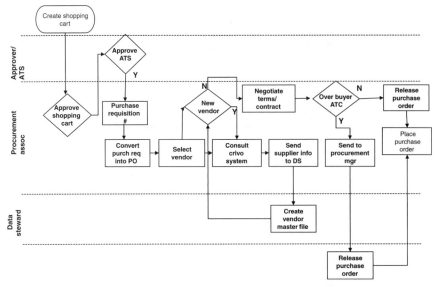

Figure 7.46 Example of a deployment process map for e-shopping. ATC=average total cost; ATS=authorized tracking system; DS=data systems.

7.10 SIPOC DIAGRAM

Another method for developing a process map is the *supplier–input–process–output–customer* (SIPOC) diagram. It is a simplified customer view of a process in an organization. It contains the following columns: supplier, input, process, output, and customer. The SIPOC diagram can "discover" pain points for the customer and pass this valuable information to the team doing a value stream map (VSM). Figure 7.47 shows an example SIPOC diagram. This visualization promotes a common understanding of the project scope (beginning to end). Simply stated, a SIPOC diagram is the simple overview of how a company satisfies customer requirements:

- *S*uppliers are the roles within the organization that support customer requirements.
- *I*nputs are the information used to execute customer requests.
- *P*rocess shows the value added activities that perform the work required for a process that will satisfy customer requirements.
- *O*utput is the service, transaction, or product sent to the customer.
- *C*ustomer is the end user.

S **Suppliers**	I **Inputs**	P **Process**	O **Output**	C **Customers**
Bank	Credit Report	Customer credit review	Credit approval	Bank customers
	Customer application	Customer credit application	Contract	New customers
	Data to process application	Check customer data for integrity	Terms and conditions	Other

Figure 7.47 Tabulated example of a SIPOC diagram.

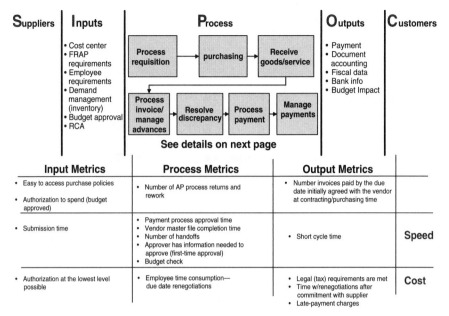

Figure 7.48   Example of a purchase SIPOC diagram. FRAP = facilitated risk analysis process; RCA = root cause analysis.

A second example for a simple SIPOC diagram is shown in Figure 7.48.

We use SIPOC and matrix analysis to find answers to the "who," "what," "when," and "why" in questions. When constructing a SIPOC diagram, we go from inductive to deductive reasoning (from specific to general). The main purpose of a SIPOC diagram is to develop a simple "thought map" of the different factors in the process expanded to include supplier and customer. Simple steps used to create the SIPOC diagram are

- Identify y's (key outputs) by brainstorming meetings with project team.
- Identify x's (independent variables or factors) by brainstorming during meetings with project team.
- Identify suppliers.
- Identify critical-to-quality (CTQ) requirements for the inputs, process activities, and outputs.
- Verify conclusions and ensure that supporting data is valid.

7.11 IPO DIAGRAM—GENERAL MODEL OF A PROCESS SYSTEM

Understanding an organizational process begins with creating a visual representation of a process activity or input–process–output (IPO) diagram. When creating this IPO diagram, the project team need not start at the beginning of the process. The project team should brainstorm the inputs (x's) in this diagram that affect the process (please note that Fig. 7.49 also appears in other sections and chapters to help to illustrate other different concepts).

From the IPO diagram, a project team can easily determine the outputs or quality characteristics of the process and finally what is called the "big Y" or title of the process being mapped. In Figure 7.49, we show a general model of a process diagram (IPO), the process to acquire and distribute digital

General model of a process

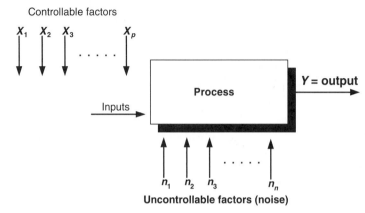

Figure 7.49 General model of a process showing the Six Sigma transfer equation $Y = f(X)$.

images to the news media (P), some of the inputs (I) to this process, and finally outputs I (O) of this process.

Then, how does the practitioner determine the Xs in the analyze phase of DMAIC? The brainstorming process remains one of the best ways to identify candidate variables. However, there are several other alternatives:

- Query process experts.
- Benchmark similar processes.
- Review the literature.
- Perform an Internet search.

[*Note:* In Fig. 7.49, there are two types of input factors that affect the outcome (n's are noise variables and the x's are fixed factors). In addition, Fig. 7.49 shows the general process diagram using the Six Sigma fundamental equation $Y = f(X)$. This diagram is also known as the *I*nput–*P*rocess–*O*utput diagram (IPO) diagram (see also Fig. 7.50).]

The choice of outputs to be measured is based on the inputs or factors of the process and the behavior of these factors during processing. Satisfying these customer requirements leads to

- Yield improvement
- Reduced variability and closer conformance to nominal
- Cycle time reduction
- Overall resource reduction

7.12 FORCE-FIELD ANALYSIS

Force-Field analysis identifies driving and restraining forces (factors) that affect the solution of a problem so that the impact from positive factors can be reinforced and/or those from negative factors can be reduced or eliminated. This technique forces brainstorming session participants to identify positive and negative aspects involved in the process and changes required to solve the problem.

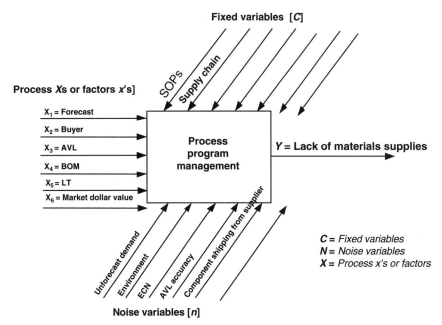

Figure 7.50 IPO diagram for the supply chain project—lack of material supplies. IPO Key factors area: AVL = approved vendor list; BOM = bill of materials; ECN = engineering change number; LT = lead time.

Example 7.12   How do you determine the x's in the analyze phase of DMAIC?

Establishing the initial list of potential causes is often more of an art than a science. The brainstorming process is still one of the best ways to surface candidate variables. However, there are several other alternatives:

1. Query process experts.
2. Benchmark similar processes.
3. Review the literature.
4. Perform internal search.
5. Conduct correlation studies.
6. Observe process.
7. Analyze physical models.
8. Conduct FMEA (failure mode–effects analysis).
9. Execute fault tree Analysis.
10. Develop *XY* matrix.

   Another effective approach is Force-field analysis, a simple technique for looking at all the forces for and against a decision. It is a specialized method for building an understanding of the forces that drive and restrain a project. It helps to weigh the importance of all the factors involved and decide whether the project is worth implementing. Force field analysis helps decide how to improve the probabilities of a project to be successful by analyzing choices such as (1) reducing the strength of the forces opposing the project or (2) increasing the forces pushing the project.

TABLE 7.2   Processing Mapping Using Six Sigma Methods

Work occurs in systems of interconnected processes	Process map, cross functional process map, value stream map, IPO (input–process–output), SIPOC, cause–effect (CE) diagram, failure mode–effects analysis (FMEA), quality function deployment (QFD), etc.
Variation	Descriptive statistics, measurement system analysis (MSA) Multi-Vari charts, CE diagrams, boxplots, process capability analysis, etc.
Reducing Variation	Pareto charts, Pareto diagrams with Lorenz curve, analytic statistics, statistical process control (SPC), control charts, design of experiments (DOE), hypothesis testing, ANOVA, chi-square, correlation, regression analysis, etc.

Force-field analysis is an analysis tool that uses a creative process for building an understanding of the forces that will drive and resist a proposed change. It consist of a two column form, with driving forces listed in the first column, and restraining forces on the second, thus forcing the agreement about all facets of a desired change. A force field diagram can be used to compare any kind of opposites, actions and consequences, different points of view, and so on.

7.13   MATRIX ANALYSIS—THE IMPORTANCE OF STATISTICAL THINKING

Statistical thinking provides a means to view processes from a holistic viewpoint. In Table 7.2 we map popular Six Sigma tools according to these principles.

Good results are achieved when statistical thinking principles are used as the discovery guideline. Guidelines are then used to create a roadmap. A *roadmap* traces the thought process for teams and leaders and distinguishes methodology from a great number of available tools.

7.14   CHECKSHEETS

Checksheets (see example in Table 7.3) are simple documents that systematically record and compile information about historical or current events. Checksheets are easy to prepare and use. Careful observation of the data contained in these sheets can enable us to recognize patterns and trends. The information is gathered by listing the problems in a table that is being tracked over time. It is important to insert a column to show where defects/errors occur; this will improve our ability to count how often each defect type occurs. The checksheet is similar to a frequency table, where problem classifications are listed in the first column and the following columns could be the collection of occurrences over time. For the checksheet example, this period was assumed to be 4 weeks.

TABLE 7.3   Example of a Checksheet

Problem	Problem Description	Week 1 (# of Occurrences)	Week 2 (# of Occurrences)	Week 3 (# of Occurrences)	Week 4 (# of Occurrences)	Total (# of Occurrences)
A	Credit balance	300	0	500	293	1093
B	Credit limit	190	384	120	400	1094
C	Credit limit increase	250	300	276	198	1024
D	Refund request	500	150	220	104	974
E	Low-speed phone payments	300	450	223	98	1071
F	Balance inquiry	220	278	403	100	1001
G	Address change	450	250	200	195	1095
H	Recent purchase inquiry	200	250	236	158	844

7.15 SCORECARDS

Scorecards provide performance summaries of the critical to quality (CTQ) measures. Balanced scorecards are constructed around four key performance metrics:

1. Financial
2. Customer
3. Internal process
4. Learning and growth

Examples of the cost-of-poor quality (COPQ) categories include the cost of not accepting change, doing nothing and the hidden factory. Examples in the customer category include on time delivery, service quality delivery [key process output variables (KPOVs)], and good communications. Examples in the internal process category include defects in data inspection, DPMO sigma quality level, rolled throughput yield, cycle time, key process input variables (KPIVs), and baseline measurements. Examples of the learning-and-growth category are Six Sigma tool utilization, number of projects successfully completed, cost-effective projects, and total savings from successful projects during the measurement time period.

In general, balanced scorecard metrics are reported as a single number within a dashboard. There are different levels of scores within this dashboard. These levels are ranked from low to high. Some organizations assign colors to each one of the levels, i.e., red for nonconformance, yellow for warning, and green for conformance. If scorecards are not balanced across the organization, the business could be driving its activities toward a single area that will affect the metric of another. For example, a manager appears successful since he/she is meeting his/her goal at the expense of others in the organization who suffer negative impacts or ramifications with the final result that the entire organization suffers.

7.16 AFFINITY DIAGRAM

The affinity diagram or KJ method was created in the 1960s and named after its author, anthropologist Kawakita Jiro;[5] this diagram was developed to discover meaningful groups of ideas and is widely used today for brainstorming purposes. In the seven QC Tools, Ishikawa recommends the use of the affinity diagram when thoughts are not clear and need to be organized.

The affinity diagram assists in providing structure to the brainstorm process by organizing ideas and grouping them by themes. To create an affinity diagram

- Conduct a brainstorming session on the problem or topic under investigation
- Create a list of ideas
- Sort ideas into groups by themes
- Insure that the ideas are described with phrases or sentences
- Minimize discussions while sorting – discuss while developing the theme classification
- Discuss the grouping
- Sorting should not start until all meeting members agree with the themes
- Try to understand how one group relates to another

The final list or diagram should show a clear relationship between issues and themes. Categories should be ranked and duplicate issues combined to create a simpler overview. If one issue belongs in two different themes, place the issue in both groups. The affinity diagram is useful because it encourages a diverse range of ideas[5].

7.17 CONCEPT INTEGRATION

- Diagnostic tools display and analyze data collected from processes to enable us to recognize the signs and symptoms emitted by a process. For example, if a process is not stable over time, then the use of time-series plots might help identify trends or cycles that help us to understand the process' behavior.
- Diagnostic tools use data collected when processes operate without interruption. In this sense, we say that a diagnostic tool helps us "listen to the processes."
- There are two main types of diagnostic tools: graphic and analytic.
- Graphic tools include Pareto Charts, cause–effect diagrams, scatterplots, time-series plots, boxplots, probability distribution plots, and Multi-Vari charts.
- Analytic tools include failure mode–effects analysis (FMEA).
- Multi-Vari charts, boxplots, and time-series plots help us classify the variation of an output into three categories: positional, cyclical, and temporal.
- Pareto charts are powerful tools that identify the problems most frequently encountered in a process. They help establish priorities and focus on the most pressing issues.
- Cause–effect analysis represent a structured framework that takes advantage of collective knowledge to identify the main causes of the effect under study.
- Failure mode–effects analysis (FMEA) assesses the severity and risk if certain failures should they occur.
- Probability plots help identify the type of probability distribution that a dataset follows.
- Scatterplots help determine whether two variables are correlated.
- Processes provide a link between the diagnostic tools and the CT characteristics. CT characteristics represent the Y in the Six Sigma transfer equation $Y = f(x_1, x_2, \ldots, x_n)$, and diagnostic tools helps us understand and classify the x's.
- Classifying the independent variables (x's) provides an excellent first use of tools such as design of experiments (DOE).
- The tools presented in this chapter represent a general framework. Each project might have certain particularities; therefore, the tools might be applied in specific ways for each project.

GLOSSARY

activity symbol   Rectangle used to designate an activity. A brief description of the activity is also included within this symbol.

affinity diagram   Tool used to organize and present large amounts of data (ideas, issues, solutions) into logical categories on the basis of perception.

alpha risk   Probability of accepting the alternative hypothesis (H_1), when in reality the null hypothesis (H_0) is true.

alternative hypothesis   Tentative explanation indicating that an event does not follow a chance distribution, in contrast to the null hypothesis.

boxplot   Graph that is particularly useful for showing the distribution characteristics of data. This graph shows the center, spreads, and skewness of a dataset. It is constructed by drawing a box and two "whiskers" that use the median, the first quartile, the third quartile, and the smallest and the largest values in the dataset between the lower and upper inner fences.

cause   That which produces an effect or brings about a change.

complex level   The first level of the process tree. It is the last process performed to deliver the product/service to the customer.

confidence level   The probability that a random variable x lies within a defined interval.

confidence limits   The two values that define the confidence level.

connections   Arrows indicating the flow and sequence of the key activities.

connector symbol   Circle used to illustrate a break and its continuation elsewhere on the same page or another page.

critical requirements for quality, delivery, and cost   The needs translated from the CTS characteristics.

critical-to-cost (CTC) characteristics   The product, service, and/or transactional characteristics that significantly influence one or more CTS in terms of cost.

critical-to-delivery (CTD) characteristics   The product, service, and/or transactional characteristics that significantly influence one or more CTS in terms of delivery.

critical-to-process (CTP) characteristics   The process parameters that significantly influence a CTQ, CTD, and/or CTC.

critical-to-quality (CTQ) characteristics   The product, service, and/or transactional characteristics that significantly influence one or more CTS in terms of quality.

critical-to-satisfaction (CTS) characteristics   Parameters for expression of the customers' vital needs.

cycle time   Time taken to measure each process step or subprocess.

data   Factual information used as a basis for reasoning, discussion, or calculation. Often refers to quantitative information.

decision symbol   Diamond used to illustrate the point where a decision must be taken. Subsequent activities are dependent on the decision taken at this point.

delay symbol   Used to indicate that there is a delay or waiting period in the process.

document symbol   Indicates a printed document pertinent to the process.

element level   This level represents a breakdown of the processes listed at the subsystem level. Critical-to-process (CTP) characteristics are identified directly below this level.

factors   Independent variables.

flow line   Arrow indicating the direction of the process and connecting its elements.

independent variable   A controlled variable, a variable whose value is independent of the value of another variable

interval scales   Measure used to express numerical information on a scale with equal distance between categories, but no absolute zero. Examples of interval scales include temperature (°F and °C), a dial gage sitting on top of a gage block, and comparison of differences.

key process   Series of steps or subprocesses included in the process.

lower inner fence   Value in a dataset that is $1.5 \times$ IQR (interquartile range) below the first quartile.

lower outer fence   Value in a dataset that is $3.0 \times$ IQR below the first quartile.

mean   Measure of central tendency calculated by dividing the sum of all values by the number of values in the dataset.

measures of central tendency   Measures that describe the center of a distribution; the mean, median, and mode are three such measurements.

measures of dispersion   Measures that give the spread of a distribution.

nominal scales   Measure used to classify elements into categories without considering any specific property. Examples of nominal scales include "causes" on fishbone diagrams, yes/no, and pass/fail.

null hypothesis   Tentative explanation indicating that a chance distribution is operating; contrast to the **null hypothesis**.

ordinal scales   Measure used to order or rank nominal data on the basis of a specific property. Examples of ordinal scales include relative height, Pareto charts, and customer satisfaction surveys.

outliers or extreme values   Values that are very small or very large relative to the majority of the values in a dataset.

Parameter   Summary measure calculated for population data.

pareto diagram   Chart used to plot percent defectives in a sample.

percentiles   Ninety-nine values that divide a ranked dataset into 100 equal parts.

process spread   Range of values that gives process characteristic displays; this particular term most often applies to the range but may also encompass the variance. The spread may be based on a set of data collected at a specific point in time or may reflect the variability across a given amount of time.

project   A problem, usually calling for planned action.

quartiles   Three summary measures that divide a ranked dataset into four equal parts.

random   Characteristic of a sample that is selected such each item in the population has an equal chance of being selected; lack of predictability; without pattern.

random variations   Variations in data that result from causes that cannot be pinpointed or controlled.

range   Difference between the highest and the lowest values in a dataset or values of a subgroup.

ratio scales   Measures used to express numerical information on a scale with equal distance between categories, but with an absolute zero in the range of measurement.

standard deviation   Statistical index of variability that describes the spread.

statistic   Summary measure calculated for sample data.

symbols   Terms indicating the nature of each step (see Section 7.9.1 for more details).

terminal symbol   Rounded rectangle used to designate the beginning or end of a process flow. Usually identified as "start," "end," "beginning," or "stop."

variance   Measure of spread.

The eight disciplines (**8D**) process steps are:

D1   Form the team. Team members are to be have experience and know the process, service or product,

D2   Clearly describe the (project) problem

D3   Develop temporary containment process

D4   Search and verify the root cause

D5   Clearly define and validate permanent corrective action(s) or PCAs

D6   Train and implement validated PCAs

D7   Design control plan to prevent problem recurrence

D8   Always recognize your team and individual project contributions

REFERENCES

1. Y. Akao and G. H. Mazur, "The leading edge in QFD, present and future," *International Journal of Quality & Reliability Management*, **20**(1): 20−35 (2003).

2. D. Clausing and S. Pugh, "Enhanced quality function deployment," *Proceedings of the Design Productivity International Conference*, 1991, pp. 15−25.

3. P. de Bievre et al., *EURACHEM Guidance Document—the Fitness for Purpose of Analytical Methods—A Guide to Method Validation and Related Topics*, (2000).

4. M. Prem, *Introductory Statistics*, 2nd ed., Wiley & Sons (1994).

5. J. Kawakita,The KJ Method or Affinity Diagram: A Technique For Analyzing Data Derived from Japanese Ethnology, ISSN 0018-7259 CODEN Huoray−Source 1997, vol 56 no 2, pp. 233−237 (10 ref.)

8 Making Sense of Data in Six Sigma and Lean

8.1 OVERVIEW

Night after night, around the world, children say to their parents, "Mommy/Daddy, will you please tell me a story?" Your boss sends you a dataset and asks, "Can you please explain this to me?" Reading a story to a child is easier, because the author has already chosen the words and arranged them to create the story. However, for the dataset example, your boss has given you the words, but you have to be the author.

The statistical methods you will learn in this chapter fall into the category of *descriptive statistics*. As the name suggests, these methods will help you describe a dataset and its characteristics. There are many such methods for quantitative and categorical data, including graphical and numerical methods. We will discuss the importance of recognizing patterns and numerical relationships among datasets. The ability to explain these patterns and relationships is paramount for any business.

8.2 SUMMARIZING QUANTITATIVE DATA: GRAPHICAL METHODS

This chapter explains how to organize and display data using tables and graphs. We will learn how to prepare frequency distribution tables for qualitative and quantitative data and how to construct dotplots, boxplots, bar graphs, pie charts, histograms, polygons, Pareto charts, and Multivariate charts, Usually the datasets in their original form are very large and it is not only difficult but also often impossible to base inferences on such large datasets. We can use descriptive measures (tables, graphs, and summary measures) to organize and summarize such datasets. It is much easier to base decisions on these organized datasets. Here we will discuss procedures to organize and summarize datasets.

8.2.1 Analytical Charts

When data are collected, the information obtained from each member of the population or sample is stored in a dataset, database, or other data recipient. Sometimes data are collected following a sequence of time or ranks. Most of the time data are labeled by geography, entry date, time entry date, type of data, group of data, time of day, day of the week, month, and so on. Data can enter a database with as many tags as the organization might need to perform for the business; otherwise data lose their validity. They do enter a database, but will never be distinguished one from the other and data extraction becomes simply impossible.

After data are processed, and/or ranked, the business organization can summarize the data or graphically present the data in analytical charts. For example, when collecting data on ages of employees of a bank, the data values are recorded in the order they are collected as shown in Table 8.1.

Practitioner's Guide for Statistics and Lean Six Sigma for Process Improvements. By Mikel J. Harry, Prem S. Mann,
Ofelia C. de Hodgins, Christopher J. Lacke, and Richard Hulbert

TABLE 8.1   **Ages of Bank Employees**

25	29	24	33	40	55	23	35
38	43	45	29	34	26	37	40
48	51	32	22	28	58	37	39
42	38	50	23	30	30	27	58
23	32	25	38	20	21	23	29
21	25	27	29	46	55	54	33
40	29	38	36	25	22	27	40
25	37	22	42	34	27	29	22
58	46	25	32	27	29	58	39
60	40	42	23	34	23	27	57

TABLE 8.2   **Departments in which the Bank Employees Work**

Economics	MIS	IT	Other	IT	MIS	Economics	Strategy
MIS	IT	Strategy	Sales	MIS	Sales	Marketing	Marketing
MIS	Sales	MIS	IT	Sales	Finance	Strategy	MIS
Marketing	MIS	Sales	MIS	Auditing	Finance	MIS	IT
Finance	Sales	Auditing	Other	Strategy	IT	IT	Auditing
Strategy	MIS	IT	IT	Sales	MIS	Securities	Sales
Securities	IT	Sales	Sales	MIS	Auditing	MIS	Marketing
Finance	IT	MIS	MIS	IT	Sales	Finance	Securities
Strategy	MIS	Finance	Other	Finance	Finance	Marketing	IT
IT	MIS	Sales	Sales	Securities	MIS	Sales	Securities

Assume that the first employee is 25 years old, the second employee is 29 years old, and so forth. This is an example of quantitative data.

Assume that we ask the same 80 employees about the department in the bank they work in. The responses of these employees to this question are recorded in Table 8.2. Assume that for this example we decide to classify the departments as economics, business, management information systems (MISs), finance, information technology [IT], strategy, sales, marketing, securities, and so on. This data classification is an example of qualitative or categorical data.

The data shown in these two tables are also called ungrouped datasets. An ungrouped dataset contains information on each member of the sample or population individually.

8.2.2   Dotplots

One of the simplest graphs to create for a dataset on a single quantitative variable is called a *dotplot*. A dotplot consists of a numbers line and a set of dots, and each of these dots represents one observation. These dots are placed above the numbers line at their respective values. If two or more observations have the same value, the dots representing them are stacked on top of each other. The following example shows how to make a dotplot.

Example 8.1   Thirty-five customers were selected from two banks that were situated in the same block. The following dataset gives the times (in minutes) spent by these 35 customers waiting for service at these banks:

8	12	13	9	14	11	7	6	14	9
7	11	10	8	12	12	5	8	9	5
14	12	8	12	7	8	9	10	7	10
12	10	8	16	6					

Construct a dotplot for these data values.

Figure 8.1   Dotplot of waiting times.

Solution   Figure 8.1 shows the dotplot for the data on waiting times. This figure is made by using the MINITAB commands `Graph > Dotplot > Simple`.

As we can notice, 5, which is the lowest value in the dataset, occurs twice, and these two values are plotted by marking two dots stacked above 5. Similarly, there are two dots stacked above 6 representing the two values of 6, and so on.

What main characteristics of this dataset do we observe from this graph? Here are a few of them:

1. The values 8 and 12 occur the maximum number of times in the dataset; each value occurs 6 times.
2. There is only one customer who waited for more than 14 min.
3. No customer waited for less than 5 min.   ■

If one or a few observations in a dataset are either much smaller or much larger than the remaining data values, we refer to these as *outliers*. Later in this chapter, we will describe a numeric criterion for classifying outliers. We can use dotplots to identify outliers. Visually, the existence of a large gap at either end of a dotplot (or any other graph) would imply the possible existence of an outlier. In the dotplot of Figure 8.1, we might consider 16 to be a mild outlier (which is not too far from the remaining data values), but not an extreme outlier (which will be very far from the rest of the values).

Graphs can also be useful when comparing the same variable for two or more groups. For example, suppose that we separate the data on waiting times for two banks given in Example 8.1 and create what is called a *stacked dotplot*. Such a dotplot allows us to see the data for the two groups on separate axes within the same graph. Example 8.2 shows how to make a stacked dotplot.

Example 8.2   The waiting times given in Example 8.1 were for two banks: Commercial Bank and Consumer Bank. Below are the same 35 data values separated by the two banks:

				COMMERCIAL BANK				
12	13	9	14	11	14	11	10	8
12	8	9	14	10	10	8	16	

				CONSUMER BANK				
8	7	6	9	7	12	5	5	12
8	12	7	8	9	7	10	12	6

Make a stacked dotplot for these data.

Solution   Figure 8.2 shows the stacked dotplot for the data on waiting times for two banks separately. This figure is constructed using the MINITAB commands `Graph > Dotplot > With Groups`.

What type of information on the characteristics of the data from these two banks can we obtain from this graph? Four such observations are mentioned here.

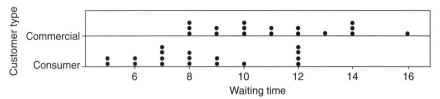

Figure 8.2   Stacked dotplot of waiting times by bank type.

1. The Consumer Bank customers tend to have shorter waiting times than do the Commercial Bank customers. This does not imply that all customers of the Consumer Bank have shorter waiting times than all customers of the Commercial Bank.
2. The majority of the Consumer Bank customers waited for 10 min or less.
3. No customer of Commercial Bank waited for less than 8 min.
4. Four customers of Consumer Bank waited for 12 min.

Observation 4 might result in a question from the branch manager. Why did all of these customers have this waiting time? Did a new teller serve these four customers? At what time of the day did they come to the bank? Is there something we could have done to shorten their waiting times?   ■

8.2.3   Stem-and-Leaf Plots

A "stem-and-leaf plot" is another simple, but useful, plot for displaying quantitative data. As in the case of dotplots, each observation is individually represented in a stem-and-leaf plot. However, a stem-and-leaf plot does not require a ruler or computer software, so it is easy to create and use while taking measurements in the field. The data in their original form, when each value is listed individually, are called *ungrouped data*. If these original ungrouped data have information on many hundreds or thousands of members, it will be spread over many pages. It will be very difficult to draw conclusions from these data. It is easier to make inferences and draw conclusions from what are called *grouped data*. A stem-and-leaf plot is one way to group data.

To create a stem-and-leaf plot, each observation is divided into two parts—a stem and a leaf. A common method of specifying the leaf is to use the digit in the lowest placeholder that you wish to report. The stem is the set of digits that precede the leaf. For example, for 57, the stem is 5 and the leaf is 7. For 243, either 2 can be used as the stem and 43 as the leaf, 24 can be used as the stem and 3 as the leaf, or we could truncate (cut off) the one digit and use 2 as the stem and 4 as the leaf. This depends on the spread of the data. If all values in a dataset are between 200 and 299 or so, we can use 20, 21, 22, 23, 24, 25, 26, 27, 28, and 29 as stems. However, if data are spread from, let us say, 155 to 1260, the stems will be, 1, 2, 3, 4, 5, 6, 7, 8, 9, 10, 11, and 12. Example 8.3 illustrates how to make a stem-and-leaf plot.

Example 8.3   The following dataset gives rents paid per month by 30 tenants selected from a small city:

625	750	920	605	840	680	930	975	755	830
850	1070	1160	940	830	1300	1070	690	1240	1150
745	1185	635	1205	960	1040	1080	915	1240	720

Make a stem-and-leaf display for these data.

Solution   To make a stem-and-leaf plot, we divide each value in the data set into two parts—a stem and a leaf. Here, we will use the last two digits as the leaf. The stems will contain one or two digits,

6	25 05 80 90 35
7	50 55 45 20
8	40 30 50 30
9	20 30 75 40 60 15
10	70 70 40 80
11	60 50 85
12	40 05 40
13	00

Figure 8.3   Stem-and-leaf display for rents.

depending on whether the observation has three or four digits. In Figure 8.3, we write all the stems on the left side of the vertical line. As we observe, here stems are 6, 7, 8, 9, 10, 11, 12, and 13. Then we write the leaves on the right side of this vertical line next to their stems. For example, for the first number of 625 in the data, 6 is the stem and 25 is the leaf. We write 25 next to 6 on the right side of the vertical line. We continue this procedure until all the data values have been entered in Figure 8.3.   ■

Note that in Figure 8.3, the leaves are unranked. Typically, we rearrange the leaves and list them in increasing order from left to right next to their corresponding stems. For example, the rearranged leaves for the stem of 6 will be 05 25 35 80 90. Figure 8.4 contains the ordered stem-and-leaf plot for the rent data.

6	05 25 35 80 90
7	20 45 50 55
8	30 30 40 50
9	15 20 30 40 60 75
10	40 40 70 80
11	50 60 85
12	05 40 40
13	00

Figure 8.4   Ordered stem-and-leaf plot for rents.

From Figures 8.3 and 8.4, by combining the stems and leaves, we can rewrite the actual values of all observations of the original data set.

8.2.4   Frequency Tables

Another way to group data is by using frequency tables. In a frequency table, to group data, we use what are called *classes* or *class intervals*. Each class has a *lower limit* and an *upper limit*. An example of a class is 0–9. For this class, 0 is the lower limit and 9 is the upper limit. Then we will count all values in the given dataset that fall in the interval 0–9. This number will represent the *frequency* for this class, which is usually denoted by *f*. These classes could be of equal widths or unequal widths. In this book, we will use classes of the same width such as 0–9, 10–19, 20–29, and so on. Each of these classes contains 10 numbers and, hence, each has a width of 10. To make a frequency table, we have to decide how many classes we should use, the width of each class, and the lower limit of the first class. There are no rules for making these decisions. These are all discretionary decisions. For a dataset of 30 or 40 observations, five to eight classes are enough. The lower limit of the first class should be equal to or less than the smallest value in the dataset. Taking the difference between the largest and the smallest values in the dataset and dividing this difference by the number of intended classes, we obtain the approximate width of each class. However, we can round up this width to a nice number. For most datasets, the number of classes can be anywhere between 5 and 25. Example 8.4 shows how to make a frequency table.

Example 8.4   The following dataset, reproduced from Example 8.3, gives rents paid by 30 tenants selected from a small town:

625	750	920	605	840	680	930	975	755	830
850	1070	1160	940	830	1300	1070	690	1240	1150
745	1185	635	1205	960	1040	1080	915	1240	720

Construct a frequency table for these data.

Solution   Suppose we decide that about five classes are sufficient for grouping these data. To find the width, we find the difference between the largest and the smallest values, which is $1300 - 605 = 695$. Now we divide 695 by 5, which is the intended number of classes. Thus, the approximate width is $695/5 = 139$. We can round 139 to a nice number such as 150. Suppose that we decide to use 150 as the width of each class. Next, the smallest number in the dataset is 605. The lower limit of the first class can be taken as 605 or a number less than 605. If we decide to take 600 as the lower limit of the first class, the classes for this example will be as shown in the first column of Table 8.3. Next, we count how many of the values in our dataset belong to each class and we write these numbers in the frequency column denoted by f. For example, eight of the values in our dataset fall in the interval 600–749. So, we write 8 in the frequency column, which is the frequency of the class 600–749. Then we observe that 6 of the values in the dataset belong to the class interval 750–899. This is the frequency for the second class. We continue this procedure until Table 8.3 is completed.

TABLE 8.3   Frequency Table for Rents Paid by 30 Tenants

Rent Paid ($)	f
600 to 749	7
750 to 899	6
900 to 1049	7
1050 to 1199	6
1200 to 1349	4

It is easier to make decisions and statements by looking at the grouped data of Table 8.3 than from the given ungrouped data. For example, from Table 8.3, we can state that seven tenants pay rents from $600 to $749, while 10 tenants pay rents of $1050 or more.   ■

Once we have made a frequency table like Table 8.3, we can convert the frequencies to *relative frequencies* and *percentages*. The relative frequency for a class is obtained by dividing the frequency for that class by the sum of all frequencies, which is the sample or population size. Then, we obtain the percentages by multiplying the relative frequencies by 100. Note that the relative frequencies are also called *proportions*. Using Table 8.3, we make Table 8.4, which contains relative frequencies and percentages.

Sometimes we may have a data set that contains decimals such as $567.87, $875.25, etc. In this case we can use the *less-than method* to write classes. For example, we can write the classes as $0 to less than $100, $100 to less than $200, $200 to less than $300, and so on. Here, all observations from $0 to $99.99 will belong to the first class; all observations from $100 to $199.99 will fall in the second class,

TABLE 8.4   Relative Frequency and Percentage Distributions

Rent Paid ($)	Relative Frequency	Percentage
600 to 749	7/30 = .233	23.3
750 to 899	6/30 = .200	20.0
900 to 1049	7/30 = .233	23.3
1050 to 1199	6/30 = .200	20.0
1200 to 1349	4/30 = .133	13.3

and so on. Other than that, the procedure to make a frequency table in this situation is similar to the one described in Example 8.4.

Often we may have a dataset where the possible values of the variable are countable and there are only a few possible values that the variable can assume. For example, the data on the number of children per family will contain the values such as 0, 1, 2, 3, 4, etc. There will be very few families with five or more children. Here, we can use what are called *single-valued classes*, such as 0, 1, 2, 3, and 4. Then we can combine all families with five or more children and write this class as *5 or more*. Thus, here the classes will be 0, 1, 2, 3, 4, and 5 or more. Next, we count the number of families that belong to each of these classes and use these as frequencies in the frequency table.

8.2.5   Histograms and Performance Histograms

In dotplots and stem-and-leaf plots, each data point has an individual representation. As datasets get larger, dotplots and stem-and-leaf plots become more cumbersome. Imagine making a stem-and-leaf plot for the salaries of 5000 employees of a company. Even with statistical software, it would be awkward to work with a stem-and-leaf plot for such a large dataset. Thus, we need to consider graphs that are good for aggregating information. The first type of graph that we are going to discuss is called a *histogram*, which is a visual representation of a frequency table that was discussed in the previous subsection.

In a histogram, the variable of interest is measured on the horizontal axis, while the frequency (or relative frequency or percentage) is marked on the vertical axis. Bar heights represent frequencies of classes, but no space is left between bars unless there is an interval with a zero frequency. Figure 8.5 shows a histogram for the frequency distribution given in Table 8.3. This histogram is made using the MINITAB command `Graph > Histogram > Simple`.

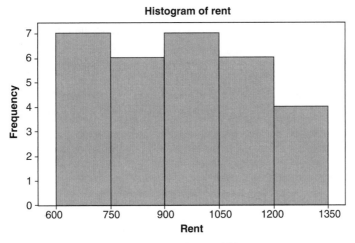

Figure 8.5   Histogram of rents paid by tenants.

Performance Histograms. It is often convenient and meaningful to modify the frequency histogram by plotting the class relative frequency rather than the class frequency. The relative frequency could also be readily plotted in terms of percentage of the total frequency. A performance histogram is a histogram that is a modified histogram and shows the performance of a process or a service or other such item as the one shown in Figure 8.6.

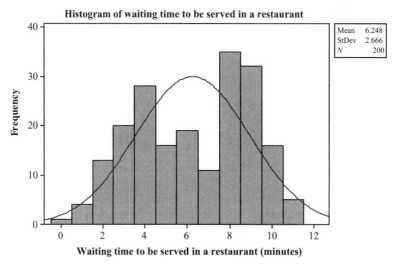

· **Figure 8.6**   A performance histogram of waiting time in minutes to be served in a restaurant.

A data table and its corresponding histogram are useful in presenting a general description of collected data. A histogram graphically shows the distribution of the collected data. In this example the data vary from 0.77 s to less than 12 min.

8.2.6 Run Charts

A *run chart* displays data collected at different times. Typically the data correspond to a process output or some other measure regarding the performance of a business or manufacturing activity over time. Here, time is plotted on the horizontal axis, and the variable of interest is placed on the vertical axis. Typically, statistical software packages will connect adjacent points with line segments in order to improve the visualization of any patterns in a run chart. Example 8.5 shows how to make a run chart.

Example 8.5   Fast-food restaurants are required to dispose off prepared food if it has been sitting around for too long. Whenever food is discarded, the items and quantities are noted. The following data represent the cost (in dollars) of disposed food for the 10 AM to 11 AM hour on 30 consecutive days of April at a restaurant. Note that the data are in chronological order reading from left to right:

2.19	0.77	4.24	2.40	5.27	4.95	5.81	3.23	2.54	4.51
4.27	3.63	1.87	1.18	1.80	5.61	5.45	3.27	2.57	1.49
3.30	5.69	1.12	5.49	11.87	1.27	3.69	2.10	5.65	4.85

Create a run chart for these data.

Solution   Figure 8.7 shows the run chart for the discarded food data. This figure is made using the MINITAB commands Graph > Time Series Plot > Simple.

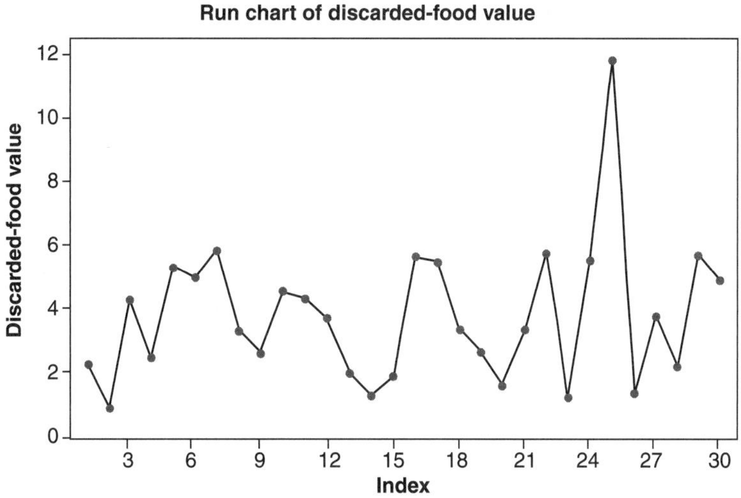

Figure 8.7   Run chart of discarded food value.

In analyzing this plot, we note the following characteristics of the data:

1. The value of food discarded during the one-hour period tends to be between $1.00 and $6.00.
2. On day 25 of the month, approximately $12.00 worth of food was discarded. Management would probably want to know the reason for this outlier.
3. During a 3-day period in the middle of the month, the value of the discarded food was below $2.00 each day. Can something be learned from these 3 days that will help reduce the amount of wasted food?   ■

8.2.7   Time-Series Plots

A run chart, as described in the previous section, is a specific type of *time-series plot*. Time-series data are data that are collected over a period of time. Unlike run charts, these data do not have to correspond to process outputs or performance measures. Data on many economic variables are collected as time-series data, such as the consumer price index, the unemployment rate, the level of inflation, and the overnight funds rate. Graphically, a time-series plot resembles a run chart with time on the horizontal axis and the variable of interest on the vertical axis. Table 8.5 gives an example of time-series data. It shows the total investment in hedge funds (in billions of dollars) for the years 2000, 2003, and 2006. Note that the investment for 2006 is an estimate. A time-series plot of total investment is given in Figure 8.8.

TABLE 8.5   Total Investment in Hedge Funds

Year	Total Investment in Hedge Funds (in billions of dollars)
2000	324
2003	600
2006	1200

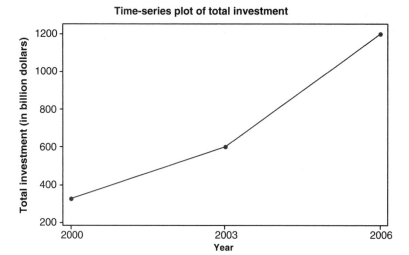

Figure 8.8   Time-series plot of total investment in hedge funds.

8.3   SUMMARIZING QUANTITATIVE DATA: NUMERICAL METHODS

In the previous sections we learned how to summarize data using stem-and-leaf displays and frequency tables, and how to display those data using graphs. We often summarize a dataset using one or a few summary measures such as the mean, median, range, standard deviation, and quartiles. In this section we will learn about such measures.

8.3.1   Measures of Center

You are asked to analyze the data on rents paid by 30 tenants given in Example 8.3. Then someone asks you, "How much rent is paid by a *typical* tenant?" What would your answer be?

Measures of center are useful for providing answers to such questions. There are many measures of center, but not all of these measures should be used in every situation. As we will discuss these measures, we will examine their usefulness under different situations.

Before discussing these measures, we need to establish notation to represent a dataset. Suppose that you have a dataset with a certain number of observations. For a sample, this number of observations is denoted by n. Usually we use letters x, y, z, etc. to represent the variable of interest. The notation x_i represents the ith observation in the dataset, with i representing any value from 1 to n. Thus, the observations are $x_1, x_2, x_3, \ldots, x_{n-1}, x_n$. The subscripts 1 through n are merely identifiers. They do not imply any numeric relationships, such as the relative sizes of the observations. Referring to the data on rents given in Example 8.3, the variable is the rent paid by a tenant, and the observations are $x_1 =$ \$625, $x_2 =$ \$750, $x_3 =$ \$920, $\ldots$, and $x_{30} =$ \$720. The sum of these x-values is denoted by Σx, where Σ is the Greek letter capital sigma and is called *summation notation*. We read Σx as *sigma x* or *sum of the x values*. This sum is obtained by adding all the x-values. In other words

$$\Sigma x = x_1 + x_2 + x_3 + \cdots + x_n$$

For example, if we add rents paid by all 30 tenants listed in Example 8.3, the sum is \$27,935. We will denote this sum by Σx; that is, we can write it as

$$\Sigma x = \$27,935$$

The two most common measures of center are the *mean* and the *median*. The term *mean* is simply another word for the *arithmetic average*. When talking about the mean of a dataset, we are typically talking about a *sample* mean, implying that the dataset comes from a sample as opposed to a census. The sample mean is usually denoted by $\bar{x}$, $\bar{y}$, etc., and pronounced as "*x* bar" or "*y* bar." The sample mean is obtained by dividing the sum of all values by the number of values in the dataset, which can be written as follows:

$$\bar{x} = \frac{\Sigma x}{n}$$

Example 8.6 shows the calculation of the mean for data on rents given in Example 8.3.

Example 8.6   The following dataset, reproduced from Example 8.3, gives rents paid per month by 30 tenants selected from a small town:

625	750	920	605	840	680	930	975	755	830
850	1070	1160	940	830	1300	1070	690	1240	1150
745	1185	635	1205	960	1040	1080	915	1240	720

Calculate the mean for these data.

Solution   To find the mean for these data, first we add all 30 values to find the sum of these values and then divide this sum by 30 as follows:

$$\bar{x} = \frac{\Sigma x}{n} = \frac{625 + 750 + 920 + 605 + \cdots + 720}{30} = \frac{27,935}{30} = \$931.17$$

Thus, we can state that these 30 tenants pay, on average, a rent of \$931.17 per month. This amount can also be referred to as the rent paid by a typical tenant.   ∎

If the given dataset contains information on a population instead of a sample, the mean calculated for that data will be denoted by μ. The number of observations in a population is denoted by N. To find the mean for population data, we find the sum of all values in the data set and then divide this sum by N. Thus

$$\mu = \frac{\Sigma x}{N}$$

Note that the notations are different but the procedure to find the population mean is the same as the one to find the sample mean.

The *median* is another important measure of center. The median of a dataset is analogous to the median on a highway—both medians divide their respective items in half. In order to find the median of a dataset (population or sample), first the values of the given dataset are arranged in increasing order. Then we find the value of the middle term, which gives the median. Note that whether we find the median of a population dataset or that of a sample dataset, the procedure is the same. Example 8.7 shows how to calculate the median of a dataset.

Example 8.7   The following dataset, reproduced from Example 8.3, gives rents paid per month by 30 tenants selected from a small town:

625	750	920	605	840	680	930	975	755	830
850	1070	1160	940	830	1300	1070	690	1240	1150
745	1185	635	1205	960	1040	1080	915	1240	720

Find the median for these data.

Solution   To find the median for these data, first we arrange these 30 values in increasing order as follows.

605	625	635	680	690	720	745	750	755	830
830	840	850	915	920	930	940	960	975	1040
1070	1070	1080	1150	1160	1185	1205	1240	1240	1300

If the number of total values in a data set is even, then the median is the average of the two middle terms. If the number of values is odd, then the median is the value of the middle term. To find the middle term, we add 1 to the number of values and divide by 2; For example, we find $(n + 1)/2$. In this example, $n = 30$. Hence, the middle term is $(30 + 1)/2 = 15.5$th term. Thus, the median is the value of the 15.5th term, which is in the middle of the 15th and 16th terms. Here, to find the median, we will take the average of the values of the 15th and 16th terms. In these arranged data, the 15th term is $920 and the 16th term is $930. Hence

$$\text{Median} = \frac{920 + 930}{2} = \$925$$

Therefore, the median rent paid per month by these 30 tenants is $925. Thus, half of these 30 tenants (i.e., 15 tenants) pay rents less than $925 per month and the other half (i.e., 15 tenants) pay rents that are higher than $925 per month.   ∎

One problem with using the mean as a measure of center is that if a dataset contains outliers, then these values affect the mean. For example, suppose that the balances of seven checking accounts (rounded to the nearest dollar) at a bank are as follows.

$763	890	344	1055	1406	578	783,640

Here, $783,640 is an outlier as this balance is much larger compared to the other six balances. The sum of these seven values is $788,676. The mean for these seven balances is

$$\text{Mean of balances with the outlier} = \frac{\Sigma x}{n} = \frac{788,676}{7} = \$112,668$$

This mean value of the seven balances is a very misleading number. Now let us drop the outlier and recalculate the mean for the remaining six balances. The sum of the six values (without the outlier) is $5306. Then

$$\text{Mean of balances without the outlier} = \frac{5036}{6} = \$839.33$$

Thus, we can observe that the value of the mean decreases from $112,668 to $839.33 when we drop the outlier.

However, the median as a measure of center does not suffer from this problem. To demonstrate this, the arranged data on the balances given above are

344	578	763	890	1055	1406	783,640

The value of median with the outlier included is $890, and without the outlier is $(763 + 890)/2 = 826.50$. The difference in these values of the two medians is very small, relative to the difference between the two averages.

One other measure of center that bears mention is called the *mode*. The mode is given by the value of the observation that has the highest frequency. The mode can be useful when working with categorical data, such as race or gender, or a quantitative variable that assumes a small number of unique values, such as the number of children per family or the number of defective items during one hour of production. For example, suppose that a branch of a bank has 15 employees and 9 of them are females. Because there are more females than males, female is the mode here. As another example, suppose in a sample of 200 families, 23 have no children, 49 have one child each, 68 have two children each, 38 have three children each, and the remaining 22 have four or more children. Since the largest numbers of families (68) have two children each, the mode is two children.

However, when the variable is continuous, such as the prices of 50 stocks, the mode is seldom well defined. In such a dataset, usually all 50 values will be different, so there will be no unique mode.

If you plan to use mode as a measure of center, you should recognize that its value is in identifying the most common value, so it can be useful in large datasets. However, just because you can calculate a summary measure does not mean that you should use it.

8.3.2   Measures of Variation

Assume that you have $5000 that you want to invest in a mutual fund. Your financial advisor tells you that there are two mutual funds and both these funds have a 15% expected annual rate of return. Can you conclude that you could simply flip a coin to determine which fund you should invest your money in? A wise investor knows that you should not judge any potential investment solely based on its expected rate of return. For example, if you are offered a CD (certificate of deposit) with a 15% annual yield, then you know that you will earn $750 interest on your investment of $5000 during the first year. However, if you invest the same money in a high-risk stock with an expected annual rate of return of 15%, then at the end of the year your actual return could be much higher than 15%, or it could be negative resulting in a loss of a part of your $5000 investment.

The difference between these two types of investments is not the center (mean or median) of the distribution of returns (the expected return); it is the variability of the returns that causes the difference. In the case of the abovementioned CD, there is no variability because the return is guaranteed to be exactly 15%. On the other hand, stock prices can have a wide range of variability levels. Low-variability stocks, also known as *low-risk stocks*, provide a fairly consistent level of returns. While it is possible to earn a large return or lose a part of your investment but these possibilities are not very likely with low-risk stocks. High-variability or high-risk stocks can have substantial swings in their values. As a result, the likelihood of making or losing a lot of money is much greater with such stocks than it is in case of low-risk stocks.

As with measures of center, there are a number of *measures of variation*. We will discuss only the following measures in this chapter:

1. Range
2. Variance and standard deviation
3. Coefficient of variation
4. Interquartile range (IQR)

8.3.2.1   *Range*

The *range* is obtained by taking the difference between the largest and the smallest values in a dataset. The following example demonstrates the calculation of the range.

Example 8.8   The following data give the lengths (in minutes) of six long-distance calls selected from all such calls made by a person during a month:

| 26 | 38 | 9 | 53 | 41 | 13 |

Find the range for these data.

Solution   As we can observe, the longest call in these six calls is for 53 min and the shortest call is for 9 min. Hence, the range is obtained by taking the difference between these two numbers. Thus

$$\text{Range} = 53 - 9 = 44 \text{ min}$$

Thus, we can state that the lengths of these six calls are spread over a range of 44 min.   ■

As we can notice, it is very easy to calculate the range for a dataset. But for practical purposes, the range is not of much use. Note that whether we are calculating the range for a sample dataset or for a population dataset, the procedure is the same.

Just like the mean, the range also has the disadvantage of being influenced by outliers. Reconsider the data on balances of seven checking accounts at a bank that were mentioned in the previous subsection of this chapter. Those data are reproduced here:

$763	890	344	1055	1406	578	783,640

Here, $783,640 is an outlier as this number is much larger compared to the other six balances. Then

$$\text{Range for these values with the outlier} = \$783,640 - \$344 = \$783,296$$

Now let us drop the outlier and recalculate the range for the remaining six balances. Then

$$\text{Range for these values without the outlier} = \$1406 - \$344 = \$1062$$

As we can observe, the value of the range drops from $783,296 to $1062 when we drop the outlier.

8.3.2.2 *Variance and Standard Deviation*

The variance and standard deviation are the measures that represent how far the various values are from the mean of a dataset. Statistically, a deviation is simply the difference between two values. In this situation, we would write the deviation between the ith data value and the sample mean as

$$x_i - \bar{x}$$

For example, if three values are 25, 36, and 44, their mean is $(25 + 36 + 44)/3 = 35$. The deviations of the three values from the mean are $25 - 35 = -10$, $36 - 35 = 1$, and $44 - 35 = 9$. Note that the sum of the deviations of all data points from their mean is always zero, as it is in this case. Similarly, the difference $(x_i - \mu)$ is the deviation of the ith value in a population data from the population mean.

The *sample variance* for a dataset is denoted by s^2 and is calculated using the following formula:

$$s^2 = \frac{\Sigma(x - \bar{x})^2}{n - 1}$$

Thus, to compute the sample variance, we square all of the deviations, add these squared deviations, and finally divide this sum by the sample size minus one. From this standpoint, you can think of the sample variance as being close to the average of the squared deviations of various data values from the sample mean.

The *sample standard deviation* is obtained by taking the (positive) square root of the variance. Thus the sample standard deviation, which is denoted by s, is obtained as follows:

$$s = \sqrt{\frac{\Sigma(x - \bar{x})^2}{n - 1}}$$

Example 8.9 demonstrates the computation of the variance and standard deviation for a sample dataset.

Example 8.9   The following data, reproduced from Example 8.8, give the lengths (in minutes) of six long-distance calls selected from all such calls made by a person during a month:

| 26 | 38 | 9 | 53 | 41 | 13 |

Find the variance and standard deviation for these data.

Solution   As we can notice, these data belong to a sample because these six calls are selected from a long list of calls. For these data values, the mean is

$$\bar{x} = \frac{\Sigma x}{n} = \frac{26 + 38 + 9 + 53 + 41 + 13}{6} = \frac{180}{6} = 30 \text{ min}$$

Now we compute the deviations and squares of deviations in Table 8.6.

TABLE 8.6   Deviations and Squared Deviations

x	$x - \bar{x}$	$(x - \bar{x})^2$
26	$26 - 30 = -4$	$(-4)^2 = 16$
38	$38 - 30 = 8$	$(8)^2 = 64$
9	$9 - 30 = -21$	$(-21)^2 = 441$
53	$53 - 30 = 23$	$(23)^2 = 529$
41	$41 - 30 = 11$	$(11)^2 = 121$
13	$13 - 30 = -17$	$(-17)^2 = 289$
$\Sigma x = 180$		$\Sigma(x - \bar{x})^2 = 1460$

Using the calculations made in the table, we obtain the values of the sample variance and sample standard deviation as follows:

$$s^2 = \frac{\Sigma(x - \bar{x})^2}{n - 1} = \frac{1460}{6 - 1} = 292$$

$$s = \sqrt{292} = 17.09 \text{ min}$$

Note that the value of the variance is not in minutes but squared minutes. When the deviations of the various data values from the mean are squared, the units of measurement are also squared. For the purpose of applications, we need to have a measure of spread that is in the same units as the original data. The standard deviation, which is obtained by taking the (positive) square root of the variance, is in the same units as the original data. Thus, in Example 8.9, the standard deviation is in minutes.

Just like the mean, the existence of outliers in a dataset affects the value of the standard deviation. For this reason, the mean and standard deviation are called *nonresistant* measures of center and spread, respectively, as outliers can have a large impact on their values.

If we are dealing with a population dataset, the variance and standard deviation calculated for this population dataset will be denoted by σ^2 and σ, respectively. The formulas used to calculate these measures will be as follows:

$$\sigma^2 = \frac{\Sigma(x - \mu)^2}{N} \quad \text{and} \quad \sigma = \sqrt{\frac{\Sigma(x - \mu)^2}{N}}$$

As we can notice from the preceding formulas, the population variance is the mean of the squared deviations of various data values from the population mean, and the population standard deviation is the positive square root of this mean of the squared deviations.

8.3.2.3   *Coefficient of Variation (CV)*

The *coefficient of variation* (CV) is a ratio measure of the spread of a dataset relative to its mean. Specifically

$$CV = \frac{s}{\bar{x}}$$

We can convert CV to percent by multiplying it by 100. One application of the coefficient of variation is to assess the level of variability in a production process relative to the mean. For example, suppose that a tire company produces tires for racing bicycles and for farm tractors. Racing bicycles have tires that are much thinner than farm tractor tires. If both production lines have the same standard deviation of tire thickness, then the coefficient of variation will be much larger for the bicycle tires, given that the average thickness is much smaller.

There is no single upper limit for the coefficient of variation that one can specify as acceptable in all production processes. Instead, what is acceptable will vary from one product to the next.

Refer to Examples 8.8 and 8.9 containing the data on the lengths of six long-distance calls. From those examples, the mean and standard deviation are

$$\bar{x} = 30 \, \text{min}$$

$$s = 17.09 \, \text{min}$$

Using these numbers, we can calculate the coefficient of variation for the data of Example 8.8 as follows:

$$CV = \frac{s}{\bar{x}} = \frac{17.09}{30} = .57 \, \text{or} \, 57\%$$

Thus, we can state that for these six values, the standard deviation is 57% of the mean value. In other words, if the mean is 100, then standard deviation will be 57.

8.3.2.4   *The Interquartile Range*

We learned earlier that the outliers in a data set do not impact the median. If we change the minimum and maximum values to anything we want, provided that they remain the minimum and maximum, the median will remain the same. Now we will introduce a measure of spread, called the *interquartile range*, which is *resistant* to outliers just like the median.

To calculate the interquartile range, we first calculate what are called the *quartiles*. Looking at the root of this word, one might suspect that the quartiles divide a dataset into four quarters, just as the median divides a dataset into two halves. There are three quartiles that divide an arranged dataset in four equal parts. In fact, the median is one of the three quartiles, because it identifies the upper boundary of the second quarter of the data.

The *three quartiles* are called the *first* (also called the *lower*) *quartile*, the *second quartile* (which is also the median), and the *third* (also called the *upper*) *quartile*. The first quartile divides the arranged data in such a way that 25% of the values are smaller than this quartile and 75% are larger than it. Similarly, 50% of the values are smaller than the second quartile and 50% are larger; and 75% of the values are smaller than the third quartile and 25% are larger. To calculate the three quartiles, first we arrange the given data in increasing order. The value of the middle term in this arranged dataset is the second quartile, which is the same as the median. The value of the middle term in the observations smaller than the median gives the first quartile. The value of the middle term in the observations larger than the median gives the third quartile. The three quartiles are denoted by Q_1, Q_2, and Q_3, respectively.

The *interquartile range* (IQR), is obtained by taking the difference between the third quartile and the first quartile:

$$IQR = Q_3 - Q_1$$

Example 8.10 shows how to calculate the three quartiles and the IQR.

Example 8.10   The following dataset, reproduced from Example 8.3, gives rents paid per month by 30 tenants selected from a small city:

625	750	920	605	840	680	930	975	755	830
850	1070	1160	940	830	1300	1070	690	1240	1150
745	1185	635	1205	960	1040	1080	915	1240	720

Find the three quartiles and the interquartile range for these data.

Solution   To find the three quartiles for these data, first we arrange these 30 values in increasing order as follows.

605	625	635	680	690	720	745	750	755	830
830	840	850	915	920	930	940	960	975	1040
1070	1070	1080	1150	1160	1185	1205	1240	1240	1300

The value of the middle term in these 30 observations is given by the average of the 15th and the 16th values as was discussed in Example 8.7. Thus

$$Q_2 = \frac{920 + 930}{2} = \$925$$

There are 15 values in this dataset that are less than $Q_2 = \$925$, and the middle term of these values is the 8th term. Hence, the value of the 8th term is the value of the first quartile, which is \$750. Similarly, there are 15 values in this dataset that are larger than $Q_2 = \$925$, and the middle term of these values is the 23rd term. Hence, the value of the 23rd term gives the value of the third quartile, which is \$1080. Thus

$$Q_1 = \$750 \quad \text{and} \quad Q_3 = \$1080$$

The value of the interquartile range is given by the difference between the values of the third and the first quartiles. Thus

$$\text{IQR} = Q_3 - Q_1 = 1080 - 750 = \$330 \qquad \blacksquare$$

8.3.2.5 *Boxplots*

Now we introduce a new type of graph called a *boxplot*. A boxplot provides a visual summary of the quartiles of a dataset. To create a boxplot, we use five summary measures: the minimum value, the first quartile, the second quartile or median, the third quartile, and the maximum value in the dataset under consideration. Taken together, these values comprise the *five-number summary*. For the data on rents paid by 30 tenants given in Example 8.10, these five summary values are

Minimum Value	Q_1	Q_2	Q_3	Maximum Value
\$605	\$750	\$925	\$1080	\$1300

To draw a boxplot, we use these five values. It is easier to draw a boxplot using any statistical software. A boxplot made for the data on rents paid by 30 tenants using the MINITAB command `Graph > Boxplot > Simple` is shown in Figure 8.9.

In the boxplot of Figure 8.9, the two lines on both sides of the box are called "whiskers." Because of this, many times this boxplot is also called the "box-and-whisker plot." The whisker on the left side of the box starts at \$605, which is the lowest value in the dataset. On the left side, the box starts at \$750, which is the first quartile. The vertical line that falls within the box

Boxplot of rent

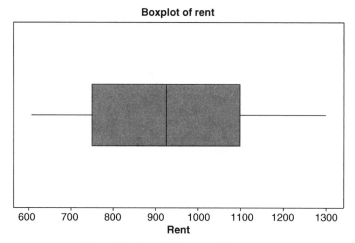

Figure 8.9   Boxplot of rents paid by tenants.

represents the second quartile or median, which is \$925. The right boundary of the box is at the third quartile, which is \$1080. The whisker on the right side of the box ends at \$1300, which is the largest value in the dataset. Thus, this boxplot has four parts, the whisker on the left side (this contains 25% of the lowest values in the dataset that are smaller than the first quartile), the part of the box that is to the left of the median (this contains another 25% of the values that are between the first and second quartiles), the part of the box that is to the right of the median (this contains 25% of the values that are between the second and third quartiles), and the largest 25% of the values (these are the values that are larger than the third quartile and are covered by the whisker on the right side of the box).

8.3.3   Identifying Potential Outliers

Outliers can have a substantial impact on any statistical analysis of data. In any such analysis, it is important not only to identify the outliers but also to determine why they exist. Sometime outliers may be a result of a transcription or typographical error, in which case one can return to the source of the data and correct the values of the outliers. In some other cases, outliers may exhibit malfunctions in a production process or the existence of a greater level of variability in the observations. How are these latter issues handled will be discussed in a later chapter.

There exist some numeric criteria that can be used to determine if a dataset contains potential outliers. These outliers can be categorized as mild or extreme outliers. To identify such outliers, we need to calculate the *inner* and *outer fences* for our dataset as follows:

$$\text{Lower inner fence (LIF)} = Q_1 - (1.5 \times \text{IQR})$$

$$\text{Upper inner fence (UIF)} = Q_3 + (1.5 \times \text{IQR})$$

$$\text{Lower outer fence (LOF)} = Q_1 - (3.0 \times \text{IQR})$$

$$\text{Upper outer fence (UOF)} = Q_3 + (3.0 \times \text{IQR})$$

All those observations that fall between the two lower fences and between the two upper fences are called *mild outliers*. The observations that fall either below the lower outer fence or above the upper outer fence are called *extreme outliers*. Example 8.11 shows how to identify the outliers in a dataset.

Example 8.11   The following data give incomes (in thousand of dollars) for the year 2007 for a sample of 25 households selected from a city:

81	77	91	59	94	21	49	61	107	126
53	85	74	178	460	12	82	69	55	45
284	90	116	39	88					

Find the mild and extreme outliers, if any, for these data.

Solution   To find the mild and extreme outliers for these data, first we arrange these 25 values in increasing order and find the three quartiles:

12	21	39	45	49	53	55	59	61	69
74	77	81	82	85	88	90	91	94	107
116	126	178	284	460					

The value of Q_2 in these 25 observations is given by the value of the 13th observation. The value of Q_1 is given by the value of the middle observation among the values that come before the 13th observation (the location of Q_2 in the ordered list). Similarly, the value of Q_3 is given by the value of the middle observation among the values that come after the 13th observation in the ordered list. (*Note*: we cannot make the statement that Q_1 is the value of the middle observation among the values that are less than Q_2 because it is possible that the data set could contain multiple observations that have the same value as the median.) Hence

$$Q_2 = 81$$

$$Q_1 = \frac{53 + 55}{2} = 54$$

$$Q_3 = \frac{94 + 107}{2} = 100.50$$

$$IQR = Q_3 - Q_1 = 100.50 - 54 = 46.50$$

Now we calculate the lower and upper, inner and outer fences as follows:

$$LIF = 54 - 1.5(46.50) = -15.75$$

$$UIF = 100.50 + 1.5(46.50) = 170.25$$

$$LOF = 54 - 3.0(46.50) = -85.50$$

$$UOF = 100.50 + 3.0(46.50) = 240$$

By looking at the 25 values of our dataset, we observe that there is no value between the lower inner and lower outer fences (i.e., between -15.75 and -85.50). But there is one value, 178, that is between the upper inner and upper outer fences (i.e., between 170.25 and 240). Thus, the income of \$178,000 is a mild outlier. For extreme outliers, we observe that there is no value in our dataset that is smaller than the lower outer fence of -85.50, but there are two values (284 and 460) that are larger than the value of the upper outer fence of 240. Thus, the incomes of \$284,000 and \$460,000 are extreme outliers in this dataset. ∎

8.3.4 Measures of Position and the Idea of z Scores in Six Sigma

At some point during your schooling, you probably took a standardized test such as Scholastic Aptitude Test (SAT) or American College Test (ACT). When you received your grade, you also received information regarding how your score compared with the scores of the other examinees. A measure that is used to provide this kind of information, generally called a *measure of position*, provides some knowledge about the location of any single observation in a dataset. Measures of position can involve an observation's place relative to the other observations, or relative to a summary measure such as the mean, median, or quartiles. Below we describe two types of measures of position—percentiles and z scores.

8.3.4.1 *Percentiles*

The kth *percentile*, denoted by P_k, is the value such that k% of the observations in the given dataset fall below this value. As a passing note, we can state that the first and third quartiles are the 25th and 75th percentiles, respectively, and the median is the 50th percentile. Note that there are a total of 99 percentiles that divide the whole (arranged) dataset in 100 equal parts.

To find the value of the kth percentile, where k is an integer from 1 to 99, we first calculate $(k \times n)/100$. For example, if a dataset contains 48 observations and we wish to determine the 64th percentile, we first calculate $64(48)/100 = 30.72$. This means that the value of the 30.72th observation in arranged data gives the 64th percentile. Since 30.72 is not an integer, there are numerous methods to obtain the value of the 30.72th observation. The simplest method would be to approximate the 64th percentile by using the value of the 31st observation in the arranged data set. Another method, used by statistical software, would be to take a weighted average of 30th and 31st observations, going 72% (.72) of the way from the 30th observation to the 31st. Example 8.12 shows how to find the value of a percentile.

Example 8.12   The following dataset, reproduced from Example 8.11, gives the incomes (in thousands of dollars) for the year 2007 for 25 households selected from a city:

81	77	91	59	94	21	49	61	107	126
53	85	74	178	460	12	82	69	55	45
284	90	116	39	88					

Find the 43rd percentile.

Solution   To find the required percentile for these data, first we arrange these 25 values in increasing order as follows:

12	21	39	45	49	53	55	59	61	69
74	77	81	82	85	88	90	91	94	107
116	126	178	284	460					

To find the 43rd percentile, first we find $kn/100$ as follows:

$$\frac{k \times n}{100} = \frac{43(25)}{100} = 10.75$$

Thus, the 43rd percentile is given by the value of the 10.75th observation in the arranged dataset. The simple estimate is to use the 11th observation, which is 74. We can also estimate the 43rd percentile by finding the value that is 75% of the way between the 10th and the 11th observations. Thus

$$P_{43} = 69 + (.75 \times (74-69)) = 72.75$$

Using this calculation, the value of the 43rd percentile is 72.75. In other words, incomes of (approximately) 43% households in this dataset are less than $72,750, and those of 57% are higher than $72,750.   ■

In many practical situations, people are just as concerned with the observations that fall in the extremes of a dataset as they are with the observations that fall in the middle. Typically, the 1st, 5th, 10th, 90th, 95th, and 99th percentiles are of great interest to us.

8.3.4.2 *The Use of z Scores*

A *z score*, which is also known as a *standardized score*, is used to identify the distance between a data value and the mean of a dataset in terms of the standard deviation of that dataset. Specifically, to calculate the *z* score for an *x* value, we use one of the following two formulas depending on whether the data belong to a population or a sample:

$$z = \frac{x-\mu}{\sigma} \qquad \text{or} \qquad z = \frac{x-\bar{x}}{s}$$

Here, *x* is the data value of interest, μ is the population mean, σ is the population standard deviation, $\bar{x}$ is the sample mean, and *s* is the sample standard deviation. Since σ and *s* are always positive, a data value below the mean will have a negative *z* score, a data value above the mean will have a positive *z* score, and the data value equal to the mean will have a *z* score of zero. Furthermore, as a data value moves farther away from the mean, the *z* score will move farther away from zero.

Example 8.13   A large dataset on commuting times for a sample of workers has a mean of 24 min and a standard deviation of 4 min. Find the *z* scores for commuting times of 7, 18, 31, and 45 min.

Solution   We find the *z* scores for these four values of *x* as follows:

$$\text{For } x = 7, \qquad z = \frac{x-\bar{x}}{s} = \frac{7-24}{4} = -4.25$$

$$\text{For } x = 18, \qquad z = \frac{x-\bar{x}}{s} = \frac{18-24}{4} = -1.50$$

$$\text{For } x = 31, \qquad z = \frac{x-\bar{x}}{s} = \frac{31-24}{4} = 1.75$$

$$\text{For } x = 45, \qquad z = \frac{x-\bar{x}}{s} = \frac{45-24}{4} = 5.25$$

Thus, the point representing $x=7$ is 4.25 standard deviations below the mean, $x=18$ is 1.50 standard deviations below the mean, $x=31$ is 1.75 standard deviations above the mean, and $x=45$ is 5.25 standard deviations above the mean.   ■

The concept of *z* scores is very useful for comparing datasets measured on different scales, as *z* scores are unitless.

8.3.5 Measures of Spread and Lean Sigma

In the Six Sigma environment, data often form the basis for actions and decisions. Once the raw data are collected, they are analyzed and converted into information through the use of statistical methods, some of which we have seen in this chapter. For example, in the histogram we plotted the frequency of occurrences of each value; the resulting graphical display is called the *frequency histogram*.

Graphical methods such as the histogram and other graphical methods presented here are useful in showing a general view of the collected data such as variation or variability of the data distribution. The *variation or variability* of a distribution can be expressed using two measures as shown before: the range and the standard deviation.

The most useful measure of dispersion is the *standard deviation*, a way of defining how the numbers in a distribution vary from the center or mean of the distribution.

The Normal Distribution. As we said before, the natural behavior and variation of many industrial processes, services, and transactions are truly random. Although the distributions of many processes may assume a variety of shapes, many random variables observed in nature possess a frequency distribution that is approximately a normal probability distribution.

The normal distribution is described by two characteristics: (1) average (or mean) and (2) standard deviation. Since the normal curve is the distribution of a continuous random variable, prediction of the probability of any particular given values can be based on the area under the curve, which is known as the *probability density function*.

To find the area under the normal curve, we must know two numerical values, the mean μ, and the standard deviation σ.

As shown above, the area under the standard normal distribution can simply be determined for a specified number, z (Section 8.3.4.2), of standard deviations from the center, where:

$$z = \frac{x - \mu}{\sigma}$$

Example 8.14   Find the area below two standard deviations to the left of the mean.

Solution   From Table II of Appendix A, for $z = -2$, the area under the curve is shown to be .0228 or 2.28%, with $a = 2.28\%$ (in any actual situation, this proportion is only approximate). Note that we have not learned yet how to read Table II of Appendix A. We will do so in Chapter 12.   ■

Let a = the proportion of process output beyond a particular value of interest such as a specification limit. With a data normal distribution it is important to note (Fig. 8.10) that 99.7% of the population should fall within ± 3 standard deviations (3σ).

This method enables practitioners to predict the percent of the population contained within any interval of a normal distribution so that when the ability to meet specification tolerances is being assessed, the z value is computed by subtracting the average from the upper and/or lower specification limits and dividing by σ: We will see this in more detail in Chapter 9 "Process Capability."

$$z_{\text{USL}} = \frac{\text{USL} - \bar{x}}{\sigma} \quad \text{and} \quad z_{\text{LSL}} = \frac{\bar{x} - \text{LSL}}{\sigma}$$

As can be easily seen, the sample standard deviation s can also be substituted for σ (or process sizing purposes).

The standard normal distribution is a special case of the normal distribution. For the standard normal distribution, the value of the mean is equal to zero and the value of the standard deviation is equal to 1. The normal distribution with $\mu = 0$ and $\sigma = 1$ is called the *standard normal distribution*.

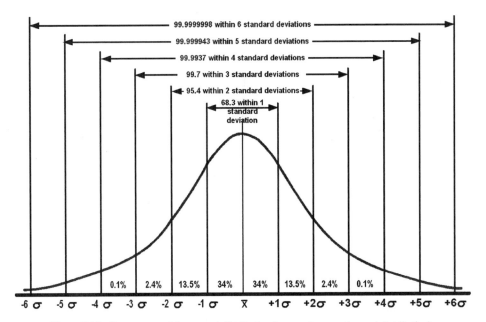

Figure 8.10   Percentages of the normal distribution (area under curve/normal distribution).

Figure 8.10 displays the Standard Normal Distribution curve. Note that z denotes the random variable that possesses the standard normal distribution. In other words, the units for the standard normal distribution curve are denoted by z and are called the z values or z scores. They are also called *standard units* or *standard scores.*

Distributions can differ in (a) location, (b) spread, and (c) shape as shown in Figure 8.11 on page 218.

8.4   ORGANIZING AND GRAPHING QUALITATIVE DATA

We know from the discussion in an earlier chapter that *qualitative data* are the data collected on a variable that does not assume numeric values but can be divided into categories or for which the numeric values are effectively labels, such as a zipcode. Examples of qualitative data include makes of cars, brands of computer, types of accounts a bank offers, and opinions of adults on an issue. Usually the raw dataset is quite large, so we need to organize them using frequency tables and display them using graphs. It is much easier to make decisions using the organized data and graphs.

8.4.1   Organizing Qualitative Data

To create a frequency table for qualitative data, we write the categories in the first column of the table and frequencies in the second column. To find the frequencies, we count the number of observations that belong to each category. The following example shows how to construct a frequency table for qualitative data.

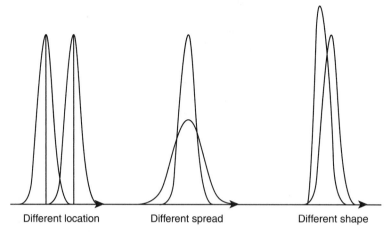

Different location   Different spread   Different shape

Figure 8.11   Tracings of different normal distributions varying in location, shape, and spread.

Example 8.15   A sample of 30 adults was selected and these adults were asked for their opinions on privatizing the US Social Security System. The following dataset lists the opinions of these 30 adults, where *F* represents those in favor, *A* indicates those against, and *N* stands for no opinion.

A	A	F	N	A	F	A	A	A	A
F	N	F	N	A	A	F	A	F	N
F	A	A	A	F	N	A	A	F	A

Create a frequency table for these data. Calculate the relative frequencies and percentages for various categories.

Solution   To construct the required frequency table, we list the three categories: F, A, and N, in the first column of Table 8.7. Then we count the number of observations that belong to each of these three categories and enter these frequencies in the second column of the table. In the third column of this table, we calculate the relative frequencies by dividing the frequencies listed in the second column by the sample size. The fourth column lists the percentages, which are obtained by multiplying the relative frequencies by 100.

TABLE 8.7   Frequency Table for Opinions of 30 Adults

Category	Frequency	Relative Frequency	Percentage
F	9	9/30 = .300	30.0
A	16	16/30 = .533	53.3
N	5	5/30 = .167	16.7
	Total = 30	Sum = 1.0	Sum = 100

Using Table 8.7, we can state that 30% of adults in this sample are in favor of privatizing Social Security, 53.3% are against it, and 16.7% have no opinion. ■

8.4.2   Graphing Qualitative Data

Qualitative data can be displayed using pie charts and bar graphs. Below we describe how to create these graphs.

8.4.2.1   Pie Chart

One of the most basic charts for analyzing qualitative data is called a *pie chart*. A circle denotes the pie, representing all of the observations, with wedges representing the pieces of the pie that correspond to different categories. The size of a wedge indicates the percentage of all observations that this category contains.

The pie chart shown in Figure 8.12 is created for the percentage distribution of opinions of 30 adults shown in Table 8.7. This pie chart is made using MINITAB commands `Graph > Pie Chart > dialogue box (enter data column) > OK`.

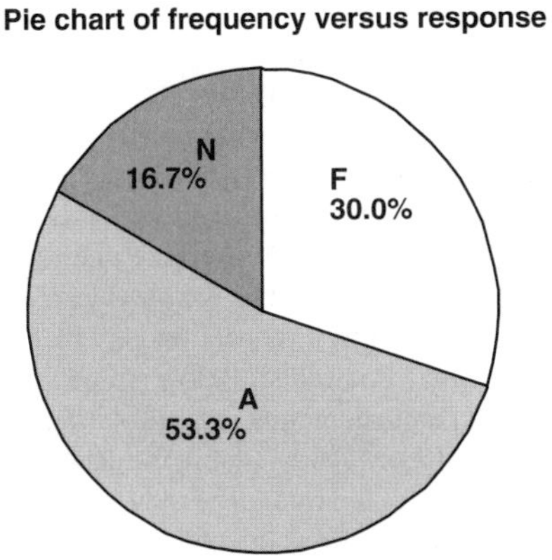

Pie chart of frequency versus response

Figure 8.12   Pie chart for opinions of adults.

8.4.2.2   Bar Graph

A bar graph is very similar to a histogram with the only difference that in a bar graph we leave a space between bars because data are not continuous. Figure 8.13 shows a bar graph for the frequency

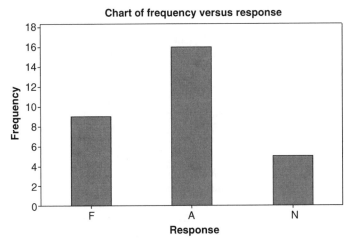

Figure 8.13   Bar graph for opinions of adults.

distribution of opinions given in Table 8.7. This bar graph is made using the Excel command Chart > Chart Type > Column > Chart sub-type (simple).

8.4.3 Pareto Analysis with Lorenz Curve

A well-constructed graphical diagram should not need much verbal explanation. If any of the elements is missing, the graph does not represent a complete picture/analysis. Selecting categories, tabulating data, ordering data, and constructing a Pareto chart can enhance communication among team members and with management.

A Pareto chart is a bar graph for qualitative data, with the bars arranged according to frequencies. As in histograms, vertical scales in Pareto charts represent frequencies or relative frequencies. The tallest bar is at the left, and the smaller bars are farther to the right. A Pareto chart organizes data to show the major factors that make up the subject being analyzed. It is a search for significance. The bars in a Pareto chart are arranged side by side (touching) each other like the ones in a histogram in descending order from left to right.

"Vital Few" and "Trivial Many".   When we apply the Pareto principle, this means that all potential problems do not occur at the same rate. Some of them will occur more often than others. In fact, studies have been conducted on this principle, and it has been found that about 20% of all potential problems occur about 80% of the time. This is known as the 80/20 rule or the "vital few" and "trivial many." The basis for Pareto analysis is the "80/20 rule," meaning that 80% of the problems result from 20% of the causes. Arranging data on a Pareto chart helps highlight the "vital few" in contrast to the "trivial many."

Pareto charts can be of first, second, third levels and more, meaning that more than one Pareto diagram covering the same process may be needed. Below we will learn how to construct a Pareto chart. The before/after improvement Pareto diagrams show a good basis for determining the success of an improvement effort. To construct a Pareto chart, identify the data to be analyzed, such as errors, and defects (i.e., consumer complaints). Select the categories to be used such as product(s), and sort data into these categories. Draw two vertical axes and a horizontal axis to make the graph, arranging the bars in decreasing or increasing order. When utilizing Pareto charts, remember that there is a second-stage and a third-stage Pareto, if applicable.

Example 8.16   A company manufacturing appliances received 486 executive escalation complaint letters during a 4-week period and needed to know which products were the source of most of the executive escalation customer complaints. The information conveyed concerning the relationship among data would be more effectively conveyed if we were to use a Pareto chart. (See Table 8.8 and Fig. 8.14.)

To make a Pareto Chart using MINITAB, in the main menu select Statistics > Quality Tools > Pareto Chart > Charts defects table > Labels in C1 > Frequencies in C2 > OK.

Pareto charts are used in problem solving to categorize defects. In this case the defects were the executive escalation customer complaints, as was used in the case of the customer executive escalation complaints.

In order for the Pareto to be accurate, all categories in the data should be inspected against other categories before performing the Pareto analysis. For example, the category "Dishwasher" should be compared against a category appearing in the dashboard entitled DW.

Interpretation - In the case of the Executive Escalation Complaint Letters, the major complaint was about refrigerators followed by about dishwashers, washing machines, dryers, built-in refrigerators, and electric ovens. The Lorenz curve (cumulative sum line) shows the percentage of total data, which assists in finding the vital-few groups. The Pareto chart shows that the first six groups are the vital-few groups.

Second- and third-level Pareto charts are simply additional Pareto charts used to further and in more depth analyze the "vital few."

TABLE 8.8   Summarized Data of Executive Customer Complaints

Product Name	Number of Customer Escalations
Refrigerator	137
Dishwasher	119
Washing machine	38
Dryer	36
Electric oven	25
Built-in refrigerator	28
Microwave oven	14
Cooktop	21
Bottle wine cooler	12
Icemaker	10
Self-cleaning stainless steel gas range	9
Stainless-steel—electric range	8
Freezer	6
Double oven	5
Built-in double oven	4
Refrigerator with icemaker	4
Combo microwave	3
Efficiency washer	3
Countertop microwave	2
Range—stainless-steel microwave	2
Total	486

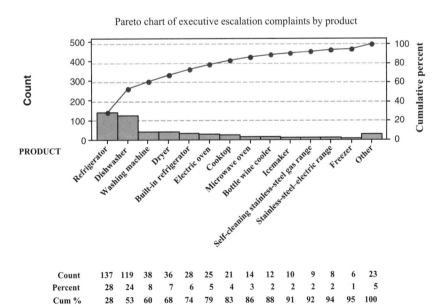

Count	137	119	38	36	28	25	21	14	12	10	9	8	6	23
Percent	28	24	8	7	6	5	4	3	2	2	2	2	1	5
Cum %	28	53	60	68	74	79	83	86	88	91	92	94	95	100

Figure 8.14   Pareto chart for executive escalation complaints by product.

8.5 SUMMARIZING BIVARIATE DATA

Analyses of data that we have discussed so far in this chapter are examples of one-variable (also called *univariate*) analyses. Many times an analysis of data may involve the relationship between two variables. These types of analyses are called *bivariate analyses*. In the following sections, we will examine *descriptive bivariate* analysis, which means that we will be describing the relationship between two variables.

8.5.1 Scatterplot

A *scatterplot* is a two-dimensional plot that is used to examine the relationship between two quantitative variables. Time-series plots and run charts (studied earlier in this chapter) are similar to scatterplots; however, the points on a scatterplot (to be studied in this section) are not connected. Each point on a scatterplot represents one pair of values, such as the dividend and asset liability ratio of a company. The primary purpose of a scatterplot is to identify the relationship between two variables. The following example shows a scatterplot of a bivariate dataset.

Example 8.17   Table 8.9 shows the total advertisement expenditure (in millions of dollars) and the total sales (in millions of dollars) for 2009 for seven companies of the same type. Construct a scatterplot for these data.

TABLE 8.9   Advertisement Expenditure and Sales of Seven Companies

Advertisement Expenditure (Millions of Dollars)	Total Sales (Millions of Dollars)
1.2	230
.8	193
2.3	400
.5	160
1.9	220
2.8	670
3.3	845

Solution   Figure 8.15 shows the scatterplot made for the data of Table 8.9. This scatterplot is made using MINITAB commands `Graph > Scatterplot > Simple`. As we can observe in this plot,

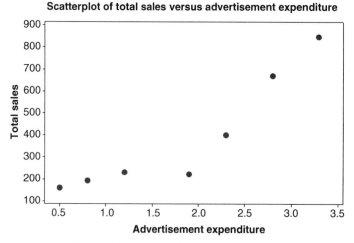

Figure 8.15   Scatterplot for the data of Table 8.9.

the advertisement expenditure is measured on the horizontal axis and the total sales are marked on the vertical axis. Each point in the graph represents one company's advertisement expenditure and sales.   ■

A scatterplot shows the pattern of relationship between two variables. This relationship can be described as *increasing (positive), decreasing (negative),* or *neutral (no relationship).* An increasing or positive relationship exists if an increase in one variable corresponds with an increase in the other variable and, hence, they move in the same direction—increase or decrease. The scatterplot of Figure 8.15 is an example of a positive relationship between the two variables. Here, an increase in the advertisement expenditure corresponds with higher sales. A decreasing or negative relationship exists when an increase in one variable corresponds with a decrease in the other variable and, hence, they move in opposite directions. An example of this is that the value of a car decreases as it gets older (as long as it does not become an antique). In other words, as the age of a car increases, its value depreciates. A scatterplot of such a dataset will show a downward pattern (from left to right) among the dots. A relationship between two variables is said to be neutral or have no relationship if a change in one variable does not correspond with any predictable change in the other variable. In the scatterplot of such a dataset, the points will appear to be randomly scattered throughout the plot. An example of such a relationship is the ages of houses and the values of cars owned by people. How old the house of a family is will not be related to the value of the car owned by this family. If we collect data on these two variables for a few randomly selected families, the dots in the scatterplot of that dataset will probably not show any pattern. Another insight into the relationship between two variables that a scatterplot can exhibit is the type of relationship. For example, we can determine whether the relationship shown by a scatterplot is of a linear or nonlinear nature. In other words, do the dots in a scatterplot show a straight-line relationship or that of a curve? Figures 8.16–8.19 show these aforementioned situations.

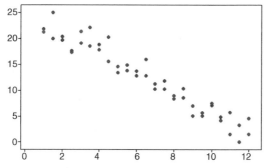

Figure 8.16   A negative and linear relationship.

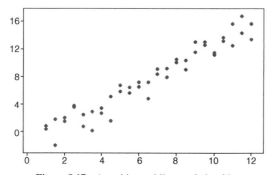

Figure 8.17   A positive and linear relationship.

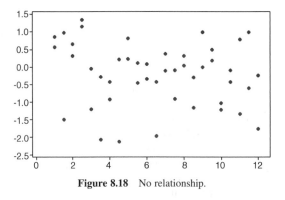

Figure 8.18   No relationship.

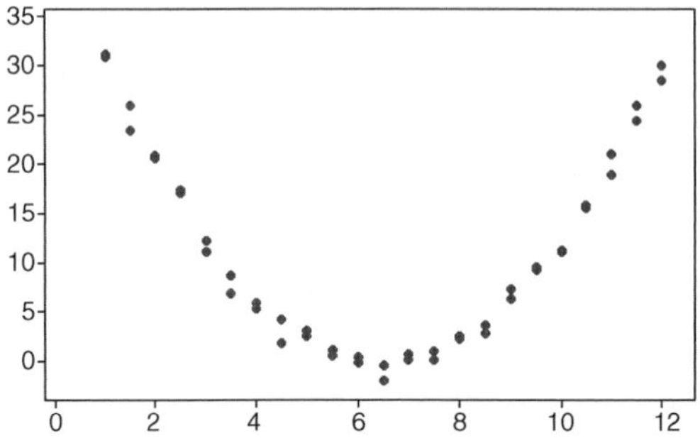

Figure 8.19   A nonlinear relationship.

Regardless of the type of relationship exhibited in a scatterplot, one has to be beware of assuming that the association is *causal* in nature. For example, if you were to look at a scatterplot for a correlation between a person's shoe size and that person's income, you might notice that larger shoe sizes are associated with higher income levels. Having bigger feet does not *cause* one to have a higher salary. Aging causes one to have bigger feet, and older people have more work experience and, hence, higher salaries. Although this example sounds somewhat absurd, when one discovers the existence of a relationship, it is always important to establish whether the relationship is causal.

8.5.2   Correlation Coefficient

One of the most widely used statistical terms is *correlation*. In fact, this term is also widely misused because many people use it when they actually mean *association*. In fact, the term *association* implies any type of relationship between two variables, while *correlation* means the strength of the linear relationship between two variables. In other words, correlation measures the proximity of the points to a line.

Correlation is most commonly measured between two quantitative variables. It can be measured between categorical variables that can be ordered or ranked. The correlation coefficient is usually denoted by ρ when calculated for a population dataset and by r when calculated for a sample dataset. The following are the properties of the correlation coefficient:

1. The value of a correlation coefficient r always lies in the interval -1 to 1.
2. The sign of the correlation coefficient identifies the direction of the linear association. A negative value of the correlation coefficient implies a negative relationship in a scatterplot, while a positive value implies a positive relationship in a scatterplot. The value of correlation coefficient being very close to zero usually means no linear correlation.

3. Correlation is unitless. In other words, correlation does not have units such as minutes, feet, or dollars.

4. Correlation is unit invariant as long as the change in units requires simply multiplying by, dividing by, adding, or subtracting a constant from each value of the data. For example, you calculate the correlation coefficient between advertising expenditures in millions of dollars and sales in billions of dollars, and your European office calculates it with advertising expenditures measured in millions of euros and sales in billions of euros for the same companies and the same dataset. Although the units of money are different, the value of the correlation coefficient in both cases will be the same.

Below we discuss three common measures of correlation, and explain under which situations each measure should be used.

8.5.2.1 *Pearson Correlation Coefficient*

The *Pearson correlation coefficient*, which is also known as the *Pearson product–moment correlation coefficient*, is used to measure the strength of the linear association between two quantitative variables. Following is the formula to calculate the Pearson correlation coefficient. Note that we have denoted this correlation coefficient with *r*, assuming that it is calculated for the sample data. In case we are calculating this correlation coefficient for the population data, we will denote it by ρ (Greek letter rho) and we will change *n* to *N* in this formula.

$$r = \frac{\Sigma xy - \frac{(\Sigma x)(\Sigma y)}{n}}{\sqrt{\left(\Sigma x^2 - \frac{(\Sigma x)^2}{n}\right) \times \left(\Sigma y^2 - \frac{(\Sigma y)^2}{n}\right)}}$$

If the value of *r* is -1 or 1, then all data points in a scatterplot fall on the same line. As the value of *r* gets closer to zero, the corresponding data points move farther and farther from a line, which means that the relationship is less linear. Also, if *r* equals zero, there is no linear association. Note that this does not imply a lack of an association. There are many examples of strong associations between variables having linear correlation coefficient values near zero. In many cases the variables may have nonlinear relationships. Example 8.18 shows the calculation of the correlation coefficient.

Example 8.18   Table 8.10, reproduced from Example 8.17, shows the total advertisement expenditure (in millions of dollars) and the total sales (in millions of dollars) for 2009 for seven companies of the same type. Calculate the Pearson correlation coefficient.

TABLE 8.10   Advertisement Expenditure and Sales of Seven Companies

Advertisement Expenditure (Millions of Dollars)	Total Sales (Millions of Dollars)
1.2	230
.8	193
2.3	400
.5	160
1.9	220
2.8	670
3.3	845

Solution   Let x denote the advertisement expenditure and y the total sales. Table 8.11 shows all the calculations for the data of Table 8.10 that are needed to compute the Pearson correlation coefficient.

TABLE 8.11

x	y	x^2	y^2	xy
1.2	230	1.44	52,900	276.00
.8	193	.64	37,249	154.40
2.3	400	5.29	160,000	920.00
.5	160	.25	25,600	80.00
1.9	220	3.61	48,400	418.00
2.8	670	7.84	448,900	1876.00
3.3	845	10.89	714,025	2788.50
$\Sigma x = 12.8$	$\Sigma y = 2718$	$\Sigma x^2 = 29.96$	$\Sigma y^2 = 1{,}487{,}074$	$\Sigma xy = 6512.90$

Substituting all the values in the formula for correlation coefficient, we obtain

$$r = \frac{\Sigma xy - \frac{(\Sigma x)(\Sigma y)}{n}}{\sqrt{\left(\Sigma x^2 - \frac{(\Sigma x)^2}{n}\right) \times \left(\Sigma y^2 - \frac{(\Sigma y)^2}{n}\right)}}$$

$$= \frac{6512.90 - \frac{(12.8)(2718)}{7}}{\sqrt{\left(29.96 - \frac{(12.8)^2}{7}\right) \times \left(1,487,074 - \frac{(2718)^2}{7}\right)}}$$

$$= \frac{1542.842857}{\sqrt{(6.554286) \times (431713.4286)}} = \frac{1542.842758}{1682.13355} = .9172$$

Thus, the correlation coefficient between these two variables is .9172, which is close to 1.0. This indicates that the correlation between these two variables is positive and very strong. However, it is always important to look at the scatterplot in addition to the value of the correlation coefficient, as some nonlinear relationships can still have a correlation that is close to 1 or −1.   ∎

8.5.2.2   *Spearman's Rho (ρ)*

Spearman's rho is a measure of correlation that can be used on a variety of data that may not necessarily have a linear relationship. Specifically, Spearman's rho is calculated using the ranks of the data on two variables. If you rank the data, from smallest to largest, separately for both variables, and plot the corresponding ranks on a scatterplot, Spearman's rho would identify how close the ranks are to being linear.

Note that if the dataset belongs to a sample, Spearman's rho is denoted by r_s. But if the dataset belongs to a population, we denote it by ρ_s, where ρ is the Greek letter rho, and, hence, the name rho. Example 8.19 demonstrates the calculation of Spearman's rho.

Example 8.19   The following data represent the grade-point averages (GPAs) and starting salaries (in thousands of dollars) of the top eight graduates from the same business school:

GPA	3.99	3.97	3.93	3.92	3.91	3.85	3.84	3.77
Salary	57.7	61.2	57.3	54.6	64.7	55.3	52.2	54.1

Calculate the value of Spearman's rho for these data.

Solution   First we need to rank data on each variable individually. To rank these data, we always assign the rank of 1 to the lowest value, 2 to the second lowest value, and so on. Table 8.12 replaces the data values with their ranks for each variable.

TABLE 8.12

GPA rank	8	7	6	5	4	3	2	1
Salary rank	6	7	5	3	8	4	1	2

Once the data have been ranked, and assuming that there are no ties, there are two options for calculating Spearman's rho. The first option is to use the ranks of Table 8.12 to calculate the Pearson correlation coefficient as was done in Example 8.17. That value of the Pearson correlation coefficient will be the same as that of the Spearman's rho. The second option is to use the following formula:

$$r_s = 1 - \frac{6\Sigma D^2}{n(n^2-1)}$$

In this formula, D is the difference in the ranks for each pair of values, and n is the number of pairs. Thus, to use this formula, first we find the difference D for each pair of ranks listed in Table 8.12. These differences are calculated and listed in the third row of Table 8.13. Next we square each of these differences and find the sum of all squared differences to obtain ΣD^2.

TABLE 8.13

GPA rank	8	7	6	5	4	3	2	1	
Salary rank	6	7	5	3	8	4	1	2	
D	2	0	1	2	−4	−1	1	−1	
D^2	4	0	1	4	16	1	1	1	$\Sigma D^2 = 28$

Therefore, the value of Spearman's rho is

$$r_s = 1 - \frac{6\Sigma D^2}{n(n^2-1)} = 1 - \frac{6 \times 28}{8(8^2-1)} = 1 - \frac{168}{504} = .667$$

■

Suppose that you have calculated Spearman's rho and found the value to be .96. We have already stated that this implies that the scatterplot of the ranks is almost linear. However, this does not necessarily imply the same thing about the scatterplot of the original data. Instead, a value of Spearman's rho that is close to 1 or −1 implies that the data can be well represented by a *monotonic* function. In order for a function to be monotonic, it must always go in the same direction. In other words, if it is increasing in one location, it must be increasing in all locations. Similarly, if it is decreasing in one location, it must be decreasing in all locations. However, the rate at which the function is increasing or decreasing does not have to remain constant, as it does in case of a line.

8.5.2.3 Kendall's Tau (τ) Rank Correlation

Kendall's tau is another example of a rank correlation. However, unlike Spearman's rho, Kendall's tau can be used if there are ties in the data. Since the calculations are based on ranks, we must rank the data first, separately for each variable. For each pair, we need to determine the number of pairs that have higher ranks in both variables. For example, if a pair of values has ranks 3 and 5 for the x and y values, respectively, then we need to count the number of pairs having an x rank above 3 and a y rank above 5. This quantity is labeled P_i. The formula for calculating Kendall's tau rank correlation is as follows, where r_τ is the notation used to denote Kendall's tau rank correlation:

$$r_\tau = \frac{4\Sigma P}{n(n-1)} - 1$$

Example 8.20   The data in Example 8.19 representing the GPAs and starting salaries (in thousands of dollars) of the top eight graduates from the same business school are reproduced here:

GPA	3.99	3.97	3.93	3.92	3.91	3.85	3.84	3.77
Salary	47.7	51.2	47.3	44.6	54.7	45.3	42.2	44.1

Calculate the value of Kendall's tau for these data.

Solution   We need to rank each variable individually, as we did with Spearman's rho. Table 8.12, which listed ranks, is reproduced here as Table 8.14.

TABLE 8.14

GPA rank	8	7	6	5	4	3	2	1
Salary rank	6	7	5	3	8	4	1	2

We need to calculate the values of P for each of the eight pairs of ranks. The first person in the list has a GPA rank of 8 and a salary rank of 6. Is there any other person who has a GPA rank above 8 and a salary rank above 6? The answer is "No." Since P_1 is the number of people who have a GPA rank above 8 and a salary rank above 6, $P_1 = 0$.

The second person has both ranks equal to 7. There is no person who has both a GPA rank of more than 7 and a salary rank of more than 7. Hence, $P_2 = 0$.

The third person in the list has a GPA rank of 6 and a salary rank of 5. Is there any other person who has a GPA rank above 6 and a salary rank above 5? The answer is "Yes." Person 1 has a GPA rank of 8, which is larger than 6, and a salary rank of 6, which is larger than 5. Person 2 has a GPA rank of 7, which is larger than 6, and a salary rank of 7, which is larger than 5. No other person has a pair of ranks that exceed 6 and 5, respectively, so $P_3 = 2$.

The fourth graduate has a GPA rank of 5 and a salary rank of 3. The first graduate has a GPA rank of 8, which is greater than 5, and a salary rank of 6, which is greater than 3. The second and third graduates also outrank the fourth graduate. The final four graduates in the list do not outrank the fourth graduate in both GPA and salary, so $P_4 = 3$. Using the same process, we obtain Table 8.15.

TABLE 8.15

GPA rank	8	7	6	5	4	3	2	1	
Salary rank	6	7	5	3	8	4	1	2	
P_i	0	0	2	3	0	4	6	6	$\Sigma P = 21$

$$r_\tau = \frac{4\Sigma P}{n(n-1)} - 1 = \frac{4(21)}{8(8-1)} - 1 = \frac{84}{56} - 1 = .50$$

8.6 MULTI-VARI CHARTS

Multi-Vari Charts were developed by Leonard Sedar of Gillette Safety Razor Co., in the 1950s. The main purpose of a Multi-Vari study is to aid in reducing the number of Xs to a vital few. Some of the advantages of these charts are the following:

- These charts graphically assist in studying the system in its natural state without interruption to the process.
- These charts allow for simultaneous analysis of different types of variability.
- Such a chart provides a graphical analysis of multiple Xs on the Y characteristic of concern.
- The Multi-Vari chart can be reconstructed in a variety of ways that allows for the "best view" of the data.

In order to understand a Multi-Vari chart, we should understand that there are many different types of variations of these charts such as

- *Positional*—a variation within a part or process. For example, you could look at different types of transactions or different operators within a call center.
- *Cyclical*—a variation between consecutive parts or process steps, for example, dollar volume in a vault.
- *Temporal*—a time variability, for example, call volume.

Example 8.21 A Six Sigma Green Belt (GB) is examining the cycle times of various types of calls coming into a call center. The assumption by the GB is that the types of calls create variability. To fully examine the calls, the GB collects cycle time data on types of calls, operator, and time of day. Data are collected over 2 days, and the day variable is also captured.

It is easy to construct Multi-Vari charts. First, data should be collected in an ordered manner, allowing for analysis of variability. Then enter data in MINITAB (see Fig. 8.20).

The Multi-Vari chart (see Fig. 8.21 on page 230) shows that

- Type of call shows variability by type of call and operator handling the call.
- Time of day shows most variability.
- Cycle time is a function of operator, type of call and time of day.

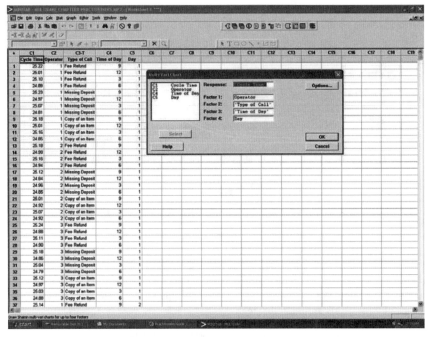

Figure 8.20 Multi-Vari data in MINITAB.

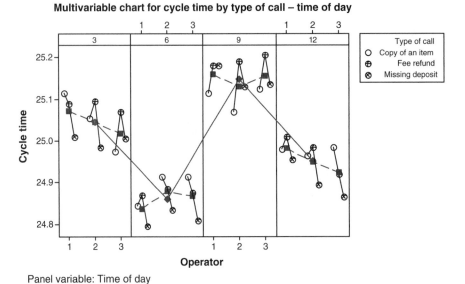

Figure 8.21 Multi-Vari chart for cycle time by type of call and operator in a call center.

GLOSSARY

bar graph A graph made of bars whose heights represent the frequencies of respective categories.

boxplot or box-and-whisker plot A plot that shows the center, spread, and skewness of a dataset by drawing a box and two whiskers using the median, the first quartile, the third quartile, and the smallest and the largest values in the dataset.

bivariate analysis   A method of data analysis in which two variables are examined simultaneously for the purpose of discovering whether they are related to each other, or independent of one another.

class   An interval that includes all the values in a (quantitative) dataset that fall within two numbers, the lower and upper limits of the class.

coefficient of variation   A measure of relative variability that expresses standard deviation as a percentage of the mean.

correlation   The strength of the linear relationship between two variables.

descriptive bivariate analysis   Relationship between two variables.

descriptive statistics   Collection of methods that are used for organizing, displaying, and describing data using tables, graphs, and summary measures.

dotplot   plot consisting of a numbers line and a set of dots, where each dot represents one observation.

extreme outliers   Observations that fall either below the lower outer fence or above the upper outer fence.

fences   Inner and outer fences are used to determine whether a set of data contains potential outliers.

first quartile   The value in a ranked dataset such that about 25% of the measurements are smaller than this value and about 75% are larger. It is the median of the values that are smaller than the median of the whole dataset.

grouped data   A dataset presented in the form of a frequency distribution.

histogram   Graph in which classes are marked on the horizontal axis and either frequencies, relative frequencies, or percentages are marked on the vertical axis. The frequencies, relative frequencies, or percentages of various classes are represented by bars that are drawn adjacent to each other.

interquartile range (IQR)   The difference between the third and the first quartiles; obtained by taking the difference between the third and first quartiles.

lower limit   Frequency table groups data into what is called classes or class intervals. Each class has a lower limit. An example of a class is "0 to 9" or 0–9. In this class, 0 is the lower limit.

mean   A measure of central tendency calculated by dividing the sum of all values by the number of values in the dataset.

measures of position   Measures that determine the position of a single value in relation to other values in a dataset. Quartiles and percentiles are examples of measures of position.

median   The value of the middle term in a ranked dataset; it divides a ranked dataset into two equal parts.

mode   A value (or values) that occurs with highest frequency in a dataset.

outliers or extreme values   Values that are very small or very large relative to virtually all of the values in a dataset.

percentage   The percentage for a class or category is obtained by multiplying the relative frequency of that class or category by 100.

percentiles   Ninety-nine values that divide a ranked data set into 100 equal parts.

pie chart   A circle divided into portions that represent the relative frequencies or percentages of different categories or classes.

qualitative or categorical data   Data generated by a qualitative variable.

quartiles   Three summary measures that divide a ranked dataset into four equal parts.

range   A measure of spread obtained by taking the difference between the largest and the smallest values in a dataset.

relative frequencies   The relative frequency for a class is obtained by dividing the frequency for that class by the sum of all frequencies, which is the sample or population size. Then, obtain the percentages by multiplying the relative frequencies by 100.

second quartile   Middle or second of the three quartiles that divide a ranked dataset into four equal parts. About 50% of the values in the dataset are smaller and about 50% are larger than the second quartile. The second quartile is the same as the median.

single-valued class   Often you may have a dataset where the possible values of the variable are countable and there are only a few possible values that the variable can assume. For example, the data on the number of bedrooms per home will contain the values such as 1, 2, 3, or 4. Very few homes will have 5 or more. A class with only 1 bedroom will be called a single-valued class.

standard deviation   A measure of spread that is given by the positive square root of the variance.

stem-and-leaf display   A display of data in which each value is divided into two portions, a stem and a leaf.

summation notation   The Greek letter capital sigma Σ is used for summation notation. Sum is obtained by adding all the x values.

third quartile   Third of the three quartiles that divide a ranked dataset into four equal parts. About 75% of the values in a dataset are smaller than the value of the third quartile, and about 25% are larger. It is the median of the values that are greater than the median of the whole dataset.

time-series data   Data that give the values for the same variable for the same element at different points in time or for different periods of time.

upper limit   Frequency table groups data into *classes* or *class intervals.* Each class has an upper limit. An example of a class is "0 to 9" or 0–9. In this class, 9 is the upper limit of the class.

variance   A measure of spread.

EXERCISES

8.1 The following data give the results of a sample survey where letters Y, N, and D represent the three categories—in favor (yes), against (no), and do not know.

D	N	N	Y	Y	Y	N	Y	D	N
Y	Y	Y	Y	N	Y	Y	N	N	Y
N	Y	Y	N	D	N	Y	N	Y	Y
Y	Y	N	N	Y	Y	N	N	D	Y

(a) Prepare a frequency distribution table.

(b) Calculate the relative frequencies and percentages for all categories.

(c) What is the percentage of the elements in this sample that belong to category Y?

(d) What is the percentage of the elements in this sample that belong to categories N or D?

(e) Draw a pie chart for the percentage distribution.

8.2 Twenty workers of a company were asked if they thought the salaries of CEOs of US companies were too high. The responses of the workers are listed below (H, N, and D indicate that a worker considers the salaries of CEOs to be too high, not too high, or has no opinion/does not know, respectively).

H	H	D	N	H	H	N	H	H	H
D	H	N	H	H	N	D	D	H	H

(a) Prepare a frequency distribution table.

(b) Calculate the relative frequencies and percentages for all categories.

(c) What is the percentage of the elements in this sample that belong to category H?

(d) What is the percentage of the elements in this sample that belong to categories N or D?

(e) Draw a pie chart for the percentage distribution.

8.3 The following data give the number of computer keyboards assembled at the Twentieth Century Electronics Company for a sample of 25 days:

45	52	48	51	56	46	44	42	
48	53	51	53	51	48	46	43	
52	50	54	47	44	47	50	49	52

(a) Construct a frequency distribution table using the classes 41–44, 45–48, 49–52, and 53–56.

(b) Calculate the relative frequencies and percentages for all classes.

(c) Construct a histogram for the percentage distribution.

(d) For what percentage of the days does the number of computer keyboards assembled fall in the interval 53–56?

(e) Construct a dotplot for these data.

(f) Create a stem-and-leaf display.

8.4 The following data give the number of new cars sold at a dealership during a 20-day period:

8	5	12	3	9	10	6	3	8	8
4	6	10	11	7	7	3	5	9	11

(a) Prepare a frequency distribution table with five classes.

(b) Calculate the relative frequencies and percentages for all classes.

(c) Construct a histogram for the percentage distribution.

(d) Construct a dotplot.

8.5 The following data give the charitable contributions (in dollars) made during the last year by randomly selected 30 households:

505	2250	100	200	400	1550	260	390	120	800
116	1750	364	400	173	60	200	1900	736	1000
50	1200	90	1300	90	150	290	700	75	2400

(a) Construct a frequency distribution table. Take $1 as the lower limit of the first class and 400 as the width of each class.

(b) Calculate the relative frequencies and percentages for all classes.

(c) What percentage of the households in this sample made charitable contributions of more than $1200?

(d) Construct a dotplot.

8.6 The following data give the number of orders received for a sample of 30 hours at the Timesaver Mail Order Company:

34	44	31	52	41	47	38	35	32	39
28	24	46	41	49	53	57	33	27	37
30	27	45	38	34	46	36	30	47	50

(a) Construct a frequency distribution table. Take 23 as the lower limit of the first class and 7 as the width of each class.

(b) Calculate the relative frequencies and percentages for all classes.

(c) For what percentage of the hours in this sample was the number of orders more than 36?

(d) Construct a histogram for relative frequencies.

(e) Create a stem-and-leaf display.

8.7 The following data give the annual incomes (in thousands of dollars) for 40 production managers randomly selected from large companies:

97.6	103.3	87.3	112.5	81.2	106.1	99.6	108.5	113.3	79.4
84.5	124.9	93.7	77.7	103.3	117.4	100.2	95.9	83.1	75.6
89.3	107.4	119.2	111.9	88.8	113.2	116.0	104.3	91.8	113.5
88.8	103.5	121.5	112.7	109.4	91.5	117.5	107.9	86.1	105.1

(a) Construct a frequency distribution table with the classes as $70,000–$79,999, $80,000–$89,999, . . . and $120,000–$129,999.

(b) Calculate the relative frequencies and percentages for all classes.

(c) What is the width of each class?

8.8 The following data give the number of bedrooms in homes owned or rented by 30 families randomly selected from a town:

3	5	2	3	2	3	1	2	1	3
4	1	4	3	1	3	3	2	2	3
3	4	3	1	2	4	2	2	5	3

(a) Prepare a frequency distribution table for these data using single-valued classes.

(b) Calculate the relative frequencies and percentages for all classes.

(c) How many homes in this sample have two or three bedrooms?

(d) Draw a bar graph for the frequency distribution.

8.9 The following data give the time (in minutes) that each of 20 workers took to complete an assembly job:

55	49	53	59	38	56	39	58	47	53
58	42	67	43	47	44	55	51	46	45

(a) Construct a stem-and-leaf display for these data. Arrange the leaves for each stem in increasing order.

(b) Calculate the mean, median, mode, range, variance, standard deviation, coefficient of variation, the quartiles, and the IQR.

8.10 The following data give the time (in minutes) taken to commute from home to work for 20 workers:

10	50	65	33	48	5	11	23	37	26
26	32	17	7	13	19	29	43	21	22

(a) Construct a stem-and-leaf display for these data. Arrange the leaves for each stem in increasing order.

(b) Calculate the mean, median, mode, range, variance, standard deviation, coefficient of variation, the quartiles, and the IQR.

8.11 The following data give the monthly grocery expenditures (in dollars) for 35 households selected from a town:

735	691	461	730	409	561	409	365	774	452	626	1017
430	387	430	604	583	583	804	691	648	561	761	952
648	926	583	787	452	474	474	517	874	376	496	

(a) Prepare a stem-and-leaf display for these data using the last two digits as leaves.

(b) Condense the stem-and-leaf display by grouping the stems as 3–4, 5–6, 7–8, and 9–10.

(c) Calculate the mean, median, mode, range, variance standard deviation, coefficient of variation, the quartiles, and the IQR.

8.12 The following data give the amount (in dollars) of personal taxes paid last year by seven households selected randomly from a town:

1038	12,604	895	37,411	4400	9428	5447

(a) Find the mean and median for these data.

(b) Do these data contain an outlier? If they do, drop this value and recalculate the mean and median. Which of the two summary measures changes by a larger amount when you drop the outlier?

(c) Is the mean or the median a better summary measure for these data?

8.13 The following data give the savings account balances (in dollars) at the end of last year for randomly selected nine accounts at a bank:

12,500	2450	70,000	1200	150
2200	7005	2000	3200	

(a) Find the mean and median for these data.

(b) Do these data contain an extreme outlier? If they do, drop this value and recalculate the mean and median. Which of the two summary measures changes by a larger amount when you drop the outlier?

(c) Is the mean or the median a better summary measure for these data?

8.14 The following data give the estimated market value (in dollars) of all stocks, bonds, mutual funds, and other such securities held at the end of last year by eight households selected at random from a town:

3250	11,028	18,900	90,000
2250	18,000	6000	12,000

(a) Find the mean and median for these data.

(b) Do these data contain an outlier? If they do, drop this value and recalculate the mean and median. Which of the two summary measures changes by a larger amount when you drop the outlier?

(c) Is the mean or the median a better summary measure for these data?

8.15 The following data give the gross sales (in billions of dollars) for a sample of 20 companies:

50	62	55	10	13	65	11	30	14	38
20	8	22	7	8	5	7	25	17	13

(a) Determine the values of the mean, median, mode, range, variance, standard deviation, coefficient of variation, the three quartiles, and the IQR.

(b) Find the (approximate) value of the 80th percentile. Give a brief interpretation of this percentile.

8.16 The following data give the number of times 10 persons used their credit cards during the past three months:

9	6	22	14	2	18	7	3	11	6

Calculate the mean, median, mode, range, variance, standard deviation, coefficient of variation, the three quartiles, and the IQR.

8.17 The following data give the number of books purchased by 16 adults during the past one year:

8	12	20	16	0	11	18	4
10	6	17	24	15	9	2	6

(a) Calculate the mean, median, mode, range, variance, standard deviation, coefficient of variation, the three quartiles, and the IQR.

(b) Find the (approximate) value of the 68th percentile. Give a brief interpretation of this value.

8.18 The following are the prices (in thousands of dollars) of 10 homes sold recently in a city:

$$179 \quad 366 \quad 458 \quad 207 \quad 487 \quad 649 \quad 793 \quad 2534 \quad 363 \quad 238$$

Calculate the mean, median, mode, range, variance, standard deviation, coefficient of variation, the three quartiles, and the IQR.

8.19 The following data give information on the ages (in years) and the number of breakdowns during the past month for a sample of seven machines at a large company:

Age	12	7	2	8	13	9	4
Number of breakdowns	9	5	1	4	11	7	2

(a) Construct a scatterplot.

(b) Calculate the Pearson's correlation coefficient, Spearman's rho correlation, and Kendall's tau rank correlation coefficient.

8.20 The following table gives information on the number of hours that eight bank loan officers slept the previous night and the number of loan applications that they processed the next day:

Number of hours slept	8	5	7	6	4	8	6	5
Number of applications processed	14	10	16	11	8	15	10	8

(a) Create a scatterplot.

(b) Calculate the Pearson's correlation coefficient, Spearman's rho correlation, and Kendall's tau rank correlation coefficient.

8.21 The management of a supermarket wants to know whether there is a correlation between the number of times a specific product is promoted on the intercom system in the store and the number of units of that product sold. To experiment, the management selected a product and promoted it on the intercom system for 7 days. The following table gives the number of times this product was promoted each day and the number of units sold:

Number of Promotions per Day	Number of Units Sold per Day (Hundreds)
15	11
22	18
42	26
30	24
18	17
12	15
38	21

(a) Construct a scatterplot.

(b) Calculate the Pearson's correlation coefficient, Spearman's rho correlation, and Kendall's tau rank correlation coefficient.

9 Fundamentals of Capability and Rolled Throughput Yield

9.1 OVERVIEW

This chapter introduces the *process capability* concept. We will provide guidelines, describe fundamental capability statistics, demonstrate short-and long-term capability analysis methods, and calculate process capability for a variety of processes.

9.2 INTRODUCTION

Few organizations have good estimates of process capability. Their activities are focused on continuously adjusting marginally capable processes to produce acceptable yields. Well-designed capability studies can help identify the causes of process variation; thus they can guide the elimination or reduction of this variation. These studies provide realistic expectations for the processes. There are multiple causes, called *components*, of process variation. In manufacturing, variation could be caused by machines, materials, measurements, methods, employees, or even changes in the weather. After we understand these causes, we can improve our process by control variation, our processes. Six Sigma practitioners appreciate the value gained from using proper metrics, operating robust processes and having defect-free units of output—the first time. During the implementation of Six Sigma, company management and employees cannot immediately identify or offer information to estimate the existing process performance level. Considerable effort would be required to measure the existing situation and to identify trends, strengths, and areas for improvement. However, most companies measure, but do not widely distribute, certain activities and record data that provide valuable starting points for capability assessment. Locating this information is a critical first step toward improved process performance.

Capability analysis[1] provides the practitioner with a measure of how well a process satisfies customer requirements. In much the same way that an archaeologist searches for artifacts of past civilizations to discover how ancient societies lived, Six Sigma practitioners can dig through existing records to assess the performance of existing processes and products. Each of these sources provides a ballpark figure of the number of defects per unit. From this number, we can calculate *throughput* or *rolled throughput yields*, which are capability analysis metrics and, therefore, can lead to Sigma values. *Yield*, which is the traditional process metric has a flaw; this flaw is the hidden factory. In a typical organization, the hidden factory consumes approximately 30% of the organization's resources—making and repairing defects.

Practitioner's Guide for Statistics and Lean Six Sigma for Process Improvements. By Mikel J. Harry, Prem S. Mann,
Ofelia C. de Hodgins, Christopher J. Lacke, and Richard Hulbert
Copyright © 2010 John Wiley & Sons, Inc.

9.3 WHY CAPABILITY

Measuring capability is important because it provides data to guide the allocation of resources, quantifies the defect rate, and prioritizes improvement opportunities. Measuring capability allows organizations to understand the true quality levels of their products and services. Capability statistics report the relationship between the range of values acceptable to customer and the range of values that the process can naturally deliver. In contrast, capability metrics report how well the process output satisfies customer specifications.

Process capability is defined as the spread within which almost all of the values of a data distribution will fall, generally described as $\pm 3\sigma$ or 6σ. This baseline definition compares the process capability with the process or specification tolerances. In general, manufacturing processes follow the normal probability distribution; when this is true, a high percentage of process measurements will fall within $\pm 3\sigma$ of the process mean or center of the distribution. Thus, some $\cong 0.27\%$ of the process measurements would naturally fall outside the $\pm 3\sigma$ limits and the remaining $\cong 99.73\%$ would be within the $\pm 3\sigma$ limits. Figure 9.1 shows percentages for various $\pm \sigma$ limits for a normal distribution.

The main objective of 6σ (Six Sigma) is to reduce process output variation such that, on a long-term basis, no more than 3.4 defect part per million opportunities or 3.4 defects per million opportunities (DPMO) occur. *Parts per Million* (ppm) reports the number of defective parts expected in a group. As a consequence, ppm reports the defective and compliance rate. For example, if a bank reported 100,000 defective invoices (parts) in a group of 1,000,000, then the invoice process reflects 90% compliance to specifications. Consider a process with a single specification limit, either the upper or the lower specification limit; this 3.4 DPMO results in Six Sigma standard deviations between the mean of the process and the customer's specification limit (6σ). For a process with two specification limits (upper and lower), this translates into slightly more that six standard deviations between the mean and each specification limit such that the total defect rate corresponds to the equivalent of six process standard deviations as can be easily seen in Figure 9.2.

It is known that processes are easily influenced by special or assignable causes that impact the overall performance. Therefore, as long as the process remains in statistical control, it will continue to produce the same proportion of specification parts. In a few words, the process must first be brought into statistical control by detecting and eliminating special causes of variation so that its performance

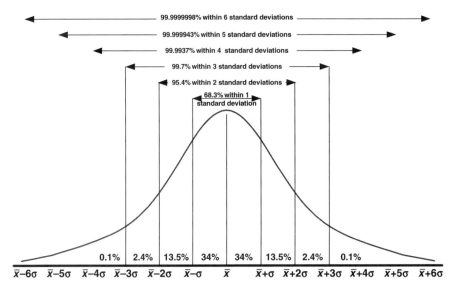

Figure 9.1 Graph showing percentages of the normal distribution (area under the curve).

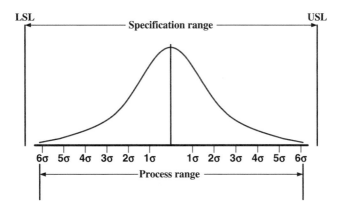

Figure 9.2 Graph showing specification range versus process range.

becomes predictable and capable of consistently meeting customer expectations. The overall performance of a process as the customer views it is 3.4 DPMO; this corresponds to long-term performance or 4.5σ (see 1.5 shift discussed later in this chapter). The difference between the best process measured by short-term process capability and the customer's expectation, long-term capability is known as *shift*, symbolized as z_{shift} or σ_{shift}. In general, when authors refer to DPMO, they refer to long-term capability. Since the process limits extend from -3σ to $+3\sigma$, the total spread amounts to about 6σ total variation. Process capability is determined by the total variation that comes from common causes—the minimum variation that can be achieved after all special causes have been eliminated. Therefore capability represents the performance of the process itself when the process is operating in a state of statistical control. Capability is frequently described as the proportion of output that will be within product specification tolerances.[1]

9.3.1 Performance Specifications

Assuming a normal distribution, process capability can be defined as the spread within which almost all of the parts or values will fall, generally defined as within $\pm 3\sigma$ deviations or 6 standard deviations or 6σ.

The process capability index is frequently used to make comparisons between process capability and process or specification tolerance. For bilateral tolerances, it is typically defined by the following relationship:

$$\text{Capability index} = \frac{\text{specification tolerance}}{\text{process capability or } 6\sigma}$$

Several organizations have established minimum required capability index objectives of 1.33 or better (the greater the better). Also, some corporations use the reciprocal of the capability index (6σ/tolerance). In this case the minimum capability ratio is 0.75 (1/1.33), and a lower ratio is better. A rules of thumb is

$$C_p \geq 1 \qquad \text{process capable}$$
$$C_p < 1 \qquad \text{process not capable}$$
$$C_p = 1.33 \qquad \text{minimum value for process to be considered capable}$$

(*Note:* C_p is usually evaluated over a short period of time (20–30 subgroups), which might miss some sources of variability that could affect a manufacturing line over a long period of time.)

However, by formulating the capability index to include the corresponding long-term standard deviation, the resulting performance measure estimates the sustainable reproducibility, or longitudinal capability, of the process. All too frequently, the Six Sigma practitioner has observed that performance metrics (indices of capability) are often improperly formulated, incorrectly applied, or reported in such a way that the measurement is counterproductive to management aims. This occurs because the analytical aspects of a capability metric are often found to be inappropriate or somehow deficient in their construction or compilation. The net effect is a performance metric that does not have the ability to report what it purports to measure. For example, this Six Sigma practitioner has been told (on numerous occasions) that the C_p of a particular process has a certain magnitude, only to later discover that the data used to compute the underpinning short-term standard deviation were gathered over a relatively lengthy period of time without regard to the fundamental principles of rational subgrouping. On such occasions, it is usually discovered that the standard deviation should have been classified as long-term in nature. When this occurs, management is presented with a misleading indicator of instantaneous reproducibility. Without elaboration, it should be obvious how such an understatement of entitlement capability (owing to improper or insufficient partitioning) could easily mislead those using the metric to support decisions. Numerous times, Mikel Harry has witnessed precious resources being squandered on new capital-intensive technology because the true capability of the existing technology had not been correctly estimated. Virtually every day across the world, many key quality and financial decisions are based on biased capability indices. The most common error in the use of C_p is the inclusion of a standard deviation that, unknown to the analyst, was confounded with or otherwise contaminated by sources of nonrandom error (black noise), thereby decreasing the short-term capability estimate. Only when these experimental and theoretical principles are understood and linked to an intimate knowledge of the process, can the practitioner properly calculate and subsequently interpret a given measure of capability, such as z_{ST}, z_{LT}, C_p, C_{pk}, P_p, and P_{pk}.

Example 9.1   A process has a mean $= 0.738$, $\sigma = 0.725$, LSL $= 0.500$ and USL $= 0.900$. Since the specification tolerance is 0.900–0.500, calculate the potential process capability index.

Solution

$$C_p = \frac{\text{tolerance}}{6\sigma} = \frac{\text{USL}-\text{LCL}}{6\sigma} = \frac{0.900-0.500}{6 \times 0.725} = \frac{0.400}{0.4350} = 0.92$$

A $C_p = 0.92$ means that if the process is centered, it barely complies with customer specifications. Any additional variation causing a shift in the mean will increase the number defects.

9.3.2   Fundamental Concepts of Defect-Based Measurement

To understand the following sections on defect measurement, we need some fundamental definitions:

- *Unit*—item (car, house, fruit, vegetable, or screw), service (customer request or inquiry), product (computer, telephone or TV set), transaction (bank deposit, mortgage, or car loan application) and final product or service delivered to customer.
- *Defect*—failure to meet customer requirements or a performance standard (delay in closing a car loan or mortgage, lost hotel reservation, or wrong airline seat assignment).
- *Defective*—unit that contains a defect (a deposit slip with six entries, where each entry is an opportunity for creating a defect). A car, with any type of defect, is a defective unit.
- *Defect opportunity*—most products, services, and transactions have hundreds of customer requirements or engineering specifications that provide a great number of chances for incurring a defect. For example, the number of opportunities for a car, software, an IT system, computer, a space shuttle printed-circuit board (PCB), or a telecommunications satellite, might each contain more than 100 defect opportunities.

Organizations seeking good Six Sigma levels must ensure that data collected are adequate to measure performance against customer requirements. For example, if a service agreement specifies on-time delivery while collected data collections measures cost per order, then key Six Sigma measurement information has been lost.

9.4   SIX SIGMA CAPABILITY METRIC

In the language of Six Sigma, it is a well-known axiom that you can't improve what you don't measure. What isn't so well known is that business metrics are more than a scale and a standard. They reflect what companies truly value and, therefore, elicit expected behavior and form the foundation of organizational culture. While this may seem like common sense, let's take a look at business metrics. By doing so, we will appreciate how Six Sigma metrics pull rather than push the business through change toward improvement. We will see that, to truly initiate and sustain change, performance metrics at every level of the organization must meet the following six criteria:

1. The metric must have a scale, such as frequency of occurrence, rate of occurrence, units produced correctly over time, number of defects, and dollars. To be effective, the measurement scale must be meaningful, valid, and reliable.

2. The metric must have a goal, or standard. If we are measuring injuries, the goal could be no more than 2 missed days of work per person per year from injury on the job. Alternative, the goal is to have an injury rate of no more than 0.05% each month for the entire workforce. In other words, 99.95% of the workforce in any given month has not experienced an injury that caused an absence of more than one day from work.

3. Compensation and other forms of recognition must be attached to actual performance compared to goal. If we measure safety, we must also monetarily reward and recognize people for improving safety. This is where many companies fail to support their measurement systems. While they may claim to value customer satisfaction, they do not compensate employees for improving customer satisfaction. While they have a scale of measure and a goal, they do not reward or recognize actions supporting the goal.

4. The metric is reported and reviewed vertically and horizontally throughout the organization on a regular basis. There is a system in place to distribute performance data to all executives, managers, and employees who impact the metric. High-level metrics are cascaded down through the business, operations, and process levels of the enterprise as a family of supporting measures. In turn, they are monitored and reported up, down, and across the organization as needed.

5. The metric must be able to be pooled horizontally and vertically within the enterprise. This means that the metric must have meaning and impact across various functions and at many levels of the organization.

6. The metric must be strongly correlated with one or more of the 12 dimensions of quality at the business, operations, and/or process level of the organization. These are discussed in Section 9.4.1.

For a metric to initiate and sustain change, it must possess all of above six components. In summary, Six Sigma is a quest for the quality of business, rather than a preoccupation with the business of quality. It defines, through a robust system of measures, the true values of an organization.

9.4.1   Criteria for Performance Metrics

Why measure capability? The answer is because measurement quantifies defect rate, identifies opportunities, assesses capability, and leads to improvement. This enables organizations to predict the true quality levels for their product, services, or transactions and identifies the nature of the process problem (nature of variation). Performance metrics should have the following characteristics:

1. Aligned—performance metrics must always be aligned with corporate strategies and objectives.
2. Owned—performance metrics must be "owned" by those held accountable for their outcome.
3. Predictive—performance metrics must be a leading indicator of business value.
4. Actionable—performance metrics must reflect timely, actionable data so that users can meaningfully and effectively intervene.
5. Minimal—performance metrics must focus on high-value tasks and not scatter user attention.
6. Simple—performance metrics must be straightforward, not based on complex indices.
7. Correlated—performance metrics must be vertically correlated and reinforce each other, not compete and confuse.
8. Transformative—performance metrics must trigger a chain reaction of positive changes in the organization.
9. Standardized—performance metrics must be based on standard definitions, rules, and calculations.
10. Contextual—performance metrics must be contextually dependent to ensure relevancy.
11. Reinforced—performance metrics must be consistent with the reward–recognition system.
12. Validated—performance metrics must be periodically reviewed to ensure relevancy and validity.

9.4.2 Computing the Sigma Level from Discrete Data

Capability metrics report how well a process output meets customer specifications by

1. Determining the process defective rate as a proportion
2. Calculating the compliance rate $= 1 -$ proportion of defectives
3. Calculating the z score from a normal distribution that corresponds to the compliance rate.

Six Sigma practitioners use a metric known as the *sigma level*; it is the capability goal for any organization, corporation, or other business searching to become a Six Sigma organization. The Six Sigma level for a process is a way to report ppm or DPMO. Once the sigma level is known, from tables the DPU is known, and vice versa. (*Note:* The z score will be discussed in later chapters of this book; therefore, we will not use it to calculate the Six Sigma level in this chapter.)

The sigma level of a process shows the proportion of the process output that complies with specifications. The practitioner should understand that the normal distribution is used to provide this report. In addition, the practitioner should understand that the area under the normal curve is statistically constituted by 100% of the area under the curve of the normal distribution. Given this, we naturally recognize that the "tails" of a normal distribution bilaterally extend to infinity. However, conventional quality practice often "trims the tails" of this distribution and declares that unity exists between the three sigma limits. This is done for analytical convenience. Of course, this convention logically assumes that the area extending beyond the three-sigma limits is trivial and, therefore, inconsequential. Perhaps, this assumption is reasonable when balancing statistical precision against the demands of quality reporting. The following example shows the steps for computing defects/unit or DPU and calculating from this, the sigma level.

Example 9.2   This is an example of DPU and DPMO. A bank completed an inspection of 6780 mortgage applications (units). These mortgage applications revealed a total of 3000 errors. Calculate defects parts per million opportunities DPMO or ppm and from this the sigma level for this mortgage application process.

Solution

$$DPU = \frac{N \, defects}{N \, units}$$

$$= \frac{3000}{6780} = 0.4425 = 44.25\% \, DPU$$

$$DPMO = ppm = 1{,}000{,}000 \frac{total \, N \, mistakes}{total \, N \, opportunities} = 1{,}000{,}000 \times 0.4425 = 442{,}477.876 \cong 442{,}478$$

where N = "number of."

From the ppm and the Sigma Conversion Table (Table VIII in Appendix A), we obtain

- Long-term sigma $= 0.2-0.1\sigma$
- Short-term sigma $= 1.6-1.7\sigma$
- If the estimate of the defect rate (DPU) is available, the sigma level can be determined. The defective rate usually represents the long-term process behavior, as opposed to a quick look, which might measure short-term behavior.

(*Note:* The normal distribution is used to provide a standard for reporting. Thus, anyone can take a reported sigma level and determine the DPMO, and vice versa.)

Example 9.3   A bank document has 7 key sections. A group of 150 documents are inspected. 18 sections are incorrect. The bank needs to know the DPU.

Solution

$$DPU = \frac{N \, defects \, found \, in \, group \, of \, units}{N \, units \, in \, group}$$

$$= \frac{18}{100} = 0.18 = 18\% \, DPU$$

Example 9.4   *Defects per opportunity* (DPO) is the proportion of defects over the total number of opportunities in a group. For example, a DPO of 0.150 would signify a 15% chance of a defect in the group.

A bank branch receives 300 mortgage applications per business day. At the end of the day, the report showed that these mortgage applications contained 50 defects. Each mortgage application contains 14 key fields (defect opportunities) per application. Calculate the number of defects per opportunity.

Solution   Summary: 50 defects, 300 mortgage applications, 15 defect opportunities per mortgage application:

$$\text{DPU} = \frac{N \text{ defects in total } N \text{ units}}{\text{total } N \text{ defects in group}} = \text{average number of defects per unit}$$

$$= \frac{50}{300} = 0.1667 \cong 0.167 \text{ DPU}$$

$$\text{DPO} = \frac{50 \text{ defects in all mortgage applications}}{(300 \text{ mortgage applications}) \times (14 \text{ defect opportunities/application})}$$

$$= \frac{50}{300 \times 14} = \frac{50}{4200} = 0.012 = 1.2\% \text{ DPO}$$

To obtain the sigma level, determine DPMO by multiplying DPU by 1,000,000:

- DPMO = DPU ×1,000,000
- DPMO = $0.012 \times 10^6 = 11,904.76$
- Now look up the short-term sigma and long-term sigma in the Sigma Conversion Table
- Short-term sigma = $3.8\sigma - 3.7\sigma$
- Long-term sigma = $2.2\sigma - 2.3\sigma$

9.4.3   Defective Proportions

Example 9.5   This is an example of proportion defectives. A bank receives 250 mortgage applications per day in one branch. The loan department reported 57 defective applications for this branch. The bank would like to determine the percent defective of these mortgage applications.

Solution

$$\frac{N \text{ defective applications}}{N \text{ units}} = \frac{57}{250} = 0.228 = 22.8\% \text{ defective}$$

A *defective* is a unit that contains one or more defects. Therefore, a mortgage application with one defect is technically as defective as a mortgage application with six defects. Each mortgage application is defective. The *proportion defective* refers to the fraction or percentage of items that had one or more defects.

9.4.4   Six-Sigma-Level Calculations (DPU, DPO, DPMO, PPM)—Examples

A well-known capability statistic in Six Sigma is the *sigma level*. This statistic can be used for continuous data and measurements with one or two specifications and with attribute measurements, defects per unit, or defective rate. For example, Six Sigma is a capability goal for a Six Sigma organization. If the process sigma level increases from 0 to 6, the variation around the mean value of the process decreases. When the process sigma level increases, the variation decreases and the process is identified as a process with a decreasing number of defects.

When considering the process performance tolerance of a critical feature, 25% of the cushion is not sufficient to absorb a sudden shift in the process centering. The manufacturing manager believes that the typical long-term shift is about 1.5σ (relative to the target value). While a 4σ level of capability is normally considered to be sufficient, should the center of the process move somehow in time from its central location (on the order of 1.5σ), the initial process capability of 4σ would decrease from 4σ to

2.5σ $(4\sigma - 1.5\sigma = 2.5\sigma)$. This shift will increase the number of defects, and in turn, this sudden increase in defects will have an adverse effect on reliability.

Defects per Million Opportunities (DPMO). Most defect opportunity measures are translated into DPMO format. This format indicates how many defects would arise if there were one million opportunities. DPMO is also called *parts per million* (ppm; the term ppm is used mostly in manufacturing):

$$\text{DPMO} = 1{,}000{,}000 \, \frac{\text{total } N \text{ defects}}{\text{total } N \text{ opportunities}} \text{ or}$$

$$= 1{,}000{,}000 \frac{\text{total } N \text{ mistakes}}{\text{total } N \text{ opportunities}}$$

$$= \frac{\text{DPU}}{\text{opportunities/unit}} \times 1{,}000{,}000$$

where opportunity = a value-added feature of a part, product, or service that must met specifications or be done correctly

unit = any part, component, subassembly, or complete product, service, or transaction for which a quality per performance measure is desired

$$\text{DPU} = \frac{N \text{ errors}}{\text{total } N \text{ units produced}}$$

DPMO is DPU normalized for product complexity. To use DPMO as a *trend* metric, applying the current process inspection scorecard might be the most practical approach, although it does not evaluate all opportunities. A different approach for calculating ppm uses descriptive statistics to derive a data distribution that can be used to estimate proportions in the population. In the following example, the data distribution appears normal. Therefore, for a given estimated mean and standard deviation, the normal probability tables (see Table II in Appendix A) can be used to determine ppm. Capability measurements can also be determined from descriptive statistics providing estimates for data location, spread, and shape. Also, defects per opportunity (DPO) multiplied by 1,000,000 is equal to DPMO. (*Note*: Defects per opportunity (DPO) expresses the proportion of defects over the total number of opportunities in a group.)

Example 9.6   In a microdevice manufacturing environment a DPO of 0.00037 was found, and the manufacturing manager wants to convert the DPO into DPMO.

Solution   DPMO = DPO $\times 10^6$, where $10^6 = 1{,}000{,}000$. Thus

$$\text{DPMO} = 0.00037 \times 10^6 = 370$$

Example 9.7   A bank needs to know the sigma level for deposits at one of its branches. A deposit contains a total of 12 critical fields (each critical field is an opportunity for a defect, mistake, or error). A final inspection of 4500 deposits revealed a total of 300 errors.

Solution

$$\text{DPMO} = 1{,}000{,}000 \frac{\text{total } N \text{ defects}}{\text{total } N \text{ opportunities}}$$

Total N opportunities $= 12 \times 4500 = 54{,}000$ opportunities

$$\text{DPMO} = \frac{300}{54{,}000} = 0.0056 \times 1{,}000{,}000 = 5{,}555{,}555$$

- Short-term sigma $= 4\sigma{-}4.1\sigma$
- Long-term sigma $= 2.5\sigma{-}2.6\sigma$

9.5   DISCRETE CAPABILITY

The objective of *discrete capability* is to assess the likelihood that a defect will occur. The only way to predict how a defect will occur is to determine the probability of defects. Various statistical distributions can be used for both discrete and continuous data. Figure 9.3 illustrates the probability of a defect.

When working with continuous data, the expected probability of a defect is usually established by applying the normal distribution. However, when working with discrete data, we are merely counting (go/no go, good bad, yes no, pass fail, etc.), the frequency in which something takes place. In this case ratio should be used to establish the observed probability of a defect. The expected probability of defect can be obtained by discrete distribution such as the Poisson or binomial distribution. In Figure 9.3 $p(d)$ is the probability of a defect occurring below the lower specification limit or above the upper specification limit or for the case of discrete data, the probability of defect $p(d)$ occurring above the specification limit:

- Probability of defect $p(d)$ occurring below the lower specification limit
- Probability of defect $p(d)$ occurring above the upper specification limit
- Probability of defect $p(d)$ occurring above the specification limit

Example 9.8 shows a calculation using discrete data.

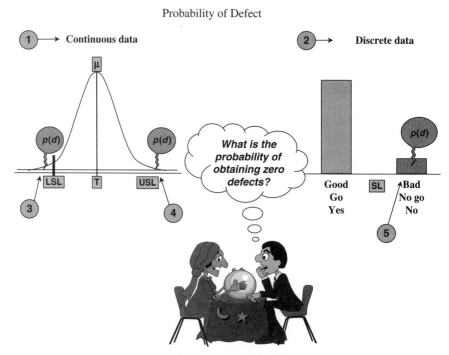

Figure 9.3   Cartoon illustrating the probability of a defect for continuous and discrete data.

TABLE 9.1   Data Table

Date	Pass	Defect Number	Total
Jan. 1, 2207	15	3	18
Feb. 1, 2207	17	1	18
March 1, 2007	16	2	18
April 1, 2007	16	2	18
May 1, 2007	15	3	18
June 1, 2007	15	3	18
July 1, 2007	18	0	18
Aug. 1, 2007	16	2	18
Sept. 1, 2007	17	1	18
Oct. 1, 2007	15	3	18
Nov. 1, 2007	14	4	18
Dec. 1, 2007	15	3	18
$\Sigma = \text{Sum}$	192	24	216

Example 9.8   A financial organization has selected a pilot-test site to calculate the sigma level of their services. Data collected over a 12-month period are shown in Table 9.1.

From these data:

Number of units processed	$N = 216$
Total number of defects	$D = 24$
Number of defect opportunities/unit	$O = 1$
Defect per million opportunities	DPMO
Process sigma level	?

Solution   By convention, DPMO represents long-term variation:

$$\text{DPU} = \frac{\text{total defects}}{\text{total units}}$$

$$= \frac{24}{192} = 0.125$$

$$\text{DPMO} = \frac{0.125}{1} \times 1{,}000{,}000 = 0.125 \times 1{,}000{,}000 = 125{,}000$$

From the conversion table:

- Short-term sigma level for 115,000 DPMO $= 2.7\sigma$
- Long-term sigma level for 115,000 DPMO $= 1.2\sigma$
- Short-term sigma level for 135,000 DPMO $= 2.6\sigma$
- Long-term sigma level for 135,000 DPMO $= 1.1\sigma$

Therefore process sigma level $=$ sigma $\cong 1.1\sigma - 1.2\sigma$

9.6   CONTINUOUS CAPABILITY—EXAMPLE

Example 9.9   Calculate parts per million for continuous data using MINITAB. A bottle company decided that the volume specifications for orange juice should be between 1.4 and 1.7 oz.

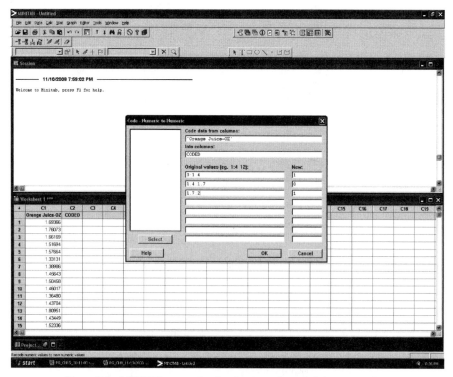

Figure 9.4 MINITAB screen showing data entry for coded data conversion.

Solution Using MINITAB to count the number of orange juice containers not meeting specifications, go to `Data > Code > Numeric to Numeric`.

MINITAB codes data as 0 or 1 depending on whether the measured value meets specifications.

MINITAB calculates the mean of the coded data (column C2 in the screen captured from MINITAB in Fig. 9.4). The nonconforming proportion is used to calculate parts per million (ppm) and then is used in the Sigma Conversion Table (Appendix A, Table VIII) to determine the sigma level.

Descriptive Statistics: CODED.

```
Variable  N  N*  Mean  SE Mean  StDev  Minimum   Q1    Median     Q3
CODED    117  0  0.2222  0.0386  0.4175 0.0000  0.0000 0.0000  0.0000

Variable  Maximum
CODED     1.0000
```

Mean = sum/count = proportion of defects = 0.22:

- PPM = 220,000
- Compliance rate = 78%

We obtain the sigma level from the Sigma Conversion Table:

- Short-term sigma = $2.2\sigma - 2.3\sigma$
- Long-term sigma = $0.7\sigma - 0.9\sigma$

TABLE 9.2 Process Capability Data Collection

Continuous Data (Short-Term Variation)	Discrete Data (Usually Long-Term Variation)
Collect Data	Collect Data
Calculate capability metrics	Calculate capability metrics
Long-term	Long-term
Sigma level z_{LT}	DPMO
P_{pk}	Sigma level z_{LT}
Short term	P_{pk}
C_p	Short-term
C_{pk}	C_p
Sigma level z_{ST}	C_{pk}
	Sigma level z_{ST}

For discrete counts, we obtain

$$\text{PPM} = \frac{\text{defective units}}{\text{total units produced}} \times 1,000,000$$

9.6.1 Data Collection for Capability Studies

Six Sigma capability studies can be performed with both discrete and continuous data. For each data type, traditional statistics are being utilized for process capability studies. A higher-level of statistics is now used for process capability studies of nonnormal data. When data follow a nonnormal distribution, such as a Weibull or lognormal distribution, defect rates are calculated by using the properties of the distribution and the specification limits. MINITAB (versions 14 and 15) transforms the raw data while calculating process capability. It transforms the specifications limits as it calculates the DPMO from the transformed data. MINITAB provides the process capability calculation using the Box-Cox transformation with $\lambda = 0$. When there are no identifiable subsets of transactions, or no method for transforming the data into an approximately normal distribution, the best approach is to collect data on the defects themselves and summarize the results using discrete tallies. Table 9.2 provides a simple summary of discrete and continuous data collection for capability studies.

9.7 FUNDAMENTALS OF CAPABILITY

Process capability can be assessed after all special causes have been corrected and the process is running in statistical control. If the variation from common causes is excessive, the process will not produce output that consistently meets customer expectations and needs.

The process capability concept is not new. It has been well documented over the years and appears as a major topic in most technical sources on quality process improvements. Essentially, process capability is the relative ability of a process to realize one or more CTQs in accordance with their respective standards. The extent to which a process can do this is most often reported by conventional metrics, such as C_p and C_{pk}. As a convenience to the reader, we will recall these metrics and generally describe them as

$$C_p = \frac{\text{USL} - \text{LSL}}{\pm 3\sigma} = \frac{|T - \text{SL}|}{3\sigma}$$
$$= \frac{\text{USL} - \text{LSL}}{\text{NT}} = \frac{\text{USL} - \text{LSL}}{6\sigma}$$

where   NT = natural tolerance

C_p = capability index

USL = upper specification limit (upper boundary condition of specification)

LSL = lower specification limit (lower boundary condition)

T = nominal specification (i.e., target)

SL = specification limit

The major weakness in C_p is that few processes remain centered on the process mean. C_p measures process potential, the spread of the data points, not centering. Therefore, to get better measurements of the current performance of a process, one must consider where the process mean is located relative to the specification limits. The index C_{pk} was created to address this issue. With C_{pk}, the location of the process center relative to the USL and the LSL is included in the computations, and a worst-case scenario is computed in which C_p uses the closest specification limit to the process mean. In general, C_{pk} is used for continuous data and assumes that the process is statistically stable and that its data is approximately normally distributed. If the data distribution is skewed, we should transform the data.

Although C_{pk} considers process centering and the short-term variation in the process, it should not be used alone to describe process capability. It should be used in conjunction with C_p, which is the short-term process capability index

$$C_{pk} = \min\left(\frac{\text{USL} - \mu}{3\sigma} \quad \text{and} \quad \frac{\mu - \text{LSL}}{3\sigma}\right)$$

where C_{pk} is the centered capability ratio and

$$C_{pk} = C_p(1 - k)$$

where k = a factor

μ = process center

σ = short-term standard deviation

3 = constant used to define distributional unity (i.e., z = 3)

$\text{SL}_{\min}$ = minimum specification limit, i.e., boundary condition (upper or lower) that minimizes the difference $T -$ SL in the event of an asymmetric specification

m = midpoint of specification

and

$$m = \frac{\text{USL} + \text{LSL}}{2} = \text{midpoint of specification.}$$

$$K = \frac{|m - \mu|}{(\text{USL} - \text{LSL})/2}$$

where $0 \leq k \leq 1$. (*Note:* When $k = 0$, the process is centered at the midpoint of the specification.)

The capability index for a characteristic compares the process bandwidth to that of the specification and modifies the former for any offset from the process center. Thus, as the replication error diminishes, C_p increases in value and, as the centering error improves, k decreases in value; this results in a strengthened value of C_{pk}. C_{pk} considers process centering.

In summary, C_p measures the inherent capability of the process. It describes the process capability under the constraint of pure error (white noise only). Thus, the standard deviation must be short-term by nature and, as a consequence, free of assignable or systematic causes. In this sense it is a reflection of entitlement capability and as such, reports on the adequacy of the process technology independent of how well that technology is managed over time.

9.8   SHORT- VERSUS LONG-TERM CAPABILITY

The output of every process has variation. We can divide the causes of this variation into two classes, those that drive short-term and those that drive long-term variation. Short-term variation is analyzed by holding all inputs constant—same associate, batch, and shift. Long-term variation is analyzed by selectively changing inputs—different associate, different batch, or different shift. Both short-term and long term capability metrics help identify areas of improvement. When the process is not short-term-capable, a defective rate exists even when the input levels are optimized; this only stabilizes the inherent defective rate. On the other hand, when the process is short-term-capable, even without optimizing long term input levels, a stable capable process will retain short-term-capability.

9.8.1   Short-Term Capability

Let us consider a random performance variable (Y) with a symmetric bilateral specification (two-sided tolerance with a centered nominal specification). From a design engineering perspective, a Six Sigma level of capability can theoretically be prescribed by the a priori assignment of 50% design margins. Of course, such guard banding of the specification limits is imposed to counterbalance the influence of uncertainties that induce process repeatability errors. These uncertainties will naturally occur during production. In light of such variation, we establish a bilateral design margin of $m = M = 0.50$ to provide a measure of resilience. Given this, the magnitude of necessary guard banding can be realized by hypothesizing a *Six Sigma model of reproducibility* during design. Naturally, such a postulated performance distribution would have a normal distribution with an infinite *degrees of freedom* (df). The 3σ limits of this distribution are pragmatic boundaries that prescribe unity. Given these factors, we are able to establish an operating margin with respect to the performance specification that is theoretically equivalent to 50%.

Short-term standard deviation (root mean square) is a statistical measure of random error that reports on the relative extent to which random background variation (extraneous noise) influences the expected mean deviation at any given moment in time. In this sense, it measures the magnitude of instantaneous error that emanates from the system of causation and is, therefore, an inherent capability measure of entitlement capability. As quality engineers, we understand that the instantaneous reproducibility of a design feature can be described by several different but related indices of *short-term capability*. For example, the *short-term (instantaneous) capability* of a process can be generally described by the relation $z_{ST} = (T - SL)/\sigma_{ST}$, where z_{ST} is the short-term standard normal deviate, T is the specified target value, SL is a specification limit (upper or lower), and σ_{ST} is the short-term standard deviation. This particular performance metric assumes that $\mu = T$ (centered process). Under the assumption, that $\mu = T$ and σ_{ST} is a measure of instantaneous reproducibility, z_{ST} represents the best-case estimate of inherent performance capability of the corresponding process.

Another common and closely related index of capability is given by the relation $C_p = |T - SL|/3\sigma_{ST}$. It can be algebraically demonstrated that $C_p = z_{ST}/3$, where 3 is a statistical constant that defines the corresponding limit of unity. We understand that the process capability ratio C_p is merely one-third of the quantity z_{ST}. Thus, a Six Sigma level of short-term capability (instantaneous repeatability) is given as $z_{ST} = 6.0$, or $C_p = 2.0$ if preferred. Consequently, a Six Sigma level of instantaneous reproducibility is described by a $\pm 6\sigma_{ST}$ random normal distribution that is centered between the limits of a symmetric, bilateral performance specification, thus realizing the design expectation $M = 0.50$. In this context, the $\pm 6\sigma_{ST}$ limits exactly coincide with the corresponding design limits—the upper specification limit (USL) and the lower specification limit (LSL), respectively. The practitioner must recognize that the given figure only provides the right side of a symmetric bilateral performance specification. Since the left side is a mirror image of the right side, there is little need to discuss both.

To better visualize the Six Sigma model of instantaneous reproducibility, the practitioner's attention is directed to Figure 9.5.

Although stated earlier, it should again be recognized that Y is an independent random normal performance variable in the context of a symmetric bilateral specification. Thus, we naturally

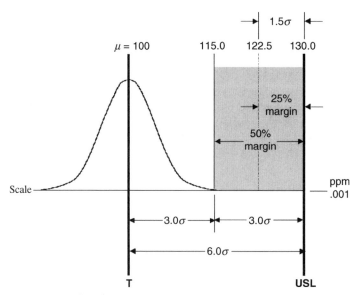

Figure 9.5   Illustration showing Six Sigma model of instantaneous reproducibility. In this figure the $\pm 6\sigma_{ST}$ limits exactly coincide with the corresponding design limits.

understand that $Y \sim$ Normally and Independently Distributed (Statistics) (NID) ($\mu\sigma_{ST}$) such that $\mu = T$, where T is the specified target value (nominal specification). On the basis of this model, the short-term quality goal of a Six Sigma characteristic is statistically translated to reflect one random error per 500 million chances for such an error, or simply two defects per billion opportunities for the centered, symmetric, bilateral case. The unilateral case (no target specified) would reflect only one defect per billion opportunities.

9.8.2   Long-Term Capability

Long-term capability is based on total process variation. In this case we include both the within-subgroup variation along with the characteristics of any shifting between subgroups, where C_p denotes short-term and P_p denotes long-term. Frequently, there is confusion in the use of short-term and long-term variation for continuous and discrete data. Discrete data are always considered long-term shift while short-term shift is always determined with continuous data. Therefore, it is necessary to differentiate between *within-* and *between-variation* with continuous data. The purpose of this section is to graphically illustrate long-term process performance and to statistically set the basis of the preceding estimations of the ppm, DPMO. In Figure 9.6 we can see the following fundamental concepts: (1) the general equation of $z = SL\sigma - \lambda$, (2) long-term z for upper specification limit, (3) long-term z for lower specification limit, and (4) the specification limits, which are an expression of the CTs, the central tendency to be used in the calculation of λ as follows:

- $\lambda = \mu$ for long-term naturally centered process
- $\lambda = T$ for long-term artificially centered process
- Long-term standard deviation, σ_{LT}

The long-term capability, based on total process variation, reflects future process performance that is known as the P_{pk} index. In this case, the total sigma should be estimated such that it includes both the within-subgroup variation and the effects of any shifting between subgroups. On the other hand, it is important to use s, the average sample standard deviation, by using all of the individual readings obtained from control charts or process studies.

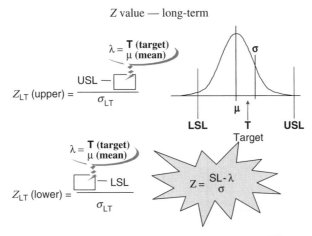

Figure 9.6   Illustration showing z value—long-term capability.

9.8.3   Introduction to Calibrating the Shift

To simplify the inflationary effect of transient and random temporal errors, an *equivalent mean offset* is often applied to the model distribution of Y. In Six Sigma, the offset is expressed in the form of $\delta = 1.5\sigma_{ST}$. Although the relative direction of such a linear correction to σ can be positive or negative, it is most often applied in the "worst case" direction when analyzing design performance. Applying this compensatory correction to the short-term distribution (illustrated in Fig. 3.4.1) reveals a long-term performance expectation of $6\sigma_{ST} - 1.5\sigma_{ST} = 4.5\sigma_{LT}$. Alternatively, the resulting long-term capability is given as an "equivalent" figure of merit and expressed in the form $z_{LT} = 4.5$. Under this condition, the design margin is reduced to $M = 0.25$. The remaining safety margin of 25% should still be large enough to absorb a fairly substantial shock to process centering. Such a shock may or may not be manifested as a momentary disturbance to the process center. Statistically translating the $4.5\sigma_{LT}$ level of capability into defects per million opportunities shows that DPMO = 3.4. For the reader's convenience, the long-term "shifted" model of Six Sigma capability is shown in Figure 9.7 and discussed in the following section.

9.9   CAPABILITY AND PERFORMANCE

Confusion about capability and performance is common; however, these are two different concepts. Many quality practitioners treat product and process capability as if they were interchangeable concepts. Presumably, this convention has evolved over time because of the naturally strong correlation between the two. Although related, these two concepts should not be viewed in the same light. It is important to recognize two things: (1) we must distinguish the concept of statistical process control (SPC) from that of statistical process monitoring (SPM)—in doing so, we will understand that SPC is focused on charting process variables, while SPM is focused on charting product variables and (2) both forms of process governance use the same statistical methods and graphing conventions; the two are often discussed within the same context even though the philosophical differences are many. To better illustrate this difference, consider a product characteristic denoted Y and its corresponding process, X, linked by the familiar equation, $Y = f(X)$. The practitioner strives to control X while concurrently monitoring Y. With this in mind, we concede that the capability of a product characteristic (Y) is dependent on the capability of its corresponding process capability (X). When reporting the capability of a product or process characteristic, we normally use one or more of the four commonly accepted indices of quality performance, C_p, C_{pk}, P_p, and P_{pk}, each its own statistical character and application nuances.

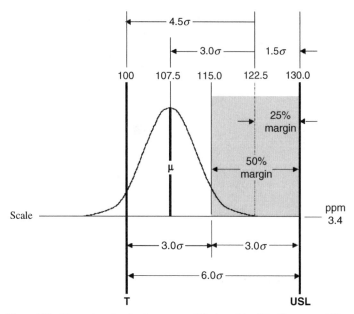

Figure 9.7   Illustration showing long-term shifted model of Six Sigma capability.

The capability index measures continuous improvement over time by prioritizing the order in which processes will be improved and determining whether a process is capable of meeting customer requirements. Remember that a standard score, or z score, is the number of standard deviations that a given value x is above or below the process mean. Therefore, the units marked on the horizontal axis of the standard normal curve are denoted by z. A specific value of z gives the distance between the mean and the point represented by z in terms of the standard deviation. On this basis, the process capability can be described in terms of the distance of the process average from the specification limits in standard deviation units: $z_{min} = minimum$ of z_{USL} or z_{LSL}. The z index assists yield prediction when the normal distribution describes the data. On the other hand, capability can also be measured with the C_{pk} index, which accounts for process centering and is defined as the minimum of C_{PU} or C_{PL} because it relates the scaled distance between the process mean and the closest specification limit to one-half of the process spread. This half of the process spread is the Six Sigma range of a process variation. The two indices are related by $C_{pk} = Z_{min}/3$.

Process capability may be calculated from short-term measurements, perhaps from a single run, or the process may be calculated from data collected over a long period of time. Thus, two kinds of variation must be distinguished (see discussion in Chapter 15):

- *Inherent process variation*—process variation due to common causes
- *Total process variation*—variation due to both common and special causes

The inherent process variation, estimated from control charts and sigma s, should be used to calculate the capability index C_{pk}, the long-term capability, on the basis of total process variation reflects future process performance, known as the P_{pk} index. In this case, the total standard deviation should be estimated such that it includes both the within-subgroup variation and the effects of any shifting between subgroups. On the other hand, it is important to calculate sigma s, the average sample standard deviation, by using all individual readings obtained from control charts or process studies. Therefore, to evaluate *performance*, it is important to allow the sample size n and to consider $\sigma = s/c_4$, where c_4 is found in statistical tables and has a value of $c_4 = 0.94$ for a sample size $n = 5$. Therefore, after obtaining the DPMO it may be converted into a decimal value by dividing in 1,000,000. To obtain the yield (%), the decimal is multiplied by 100. Once converted, the decimal is obtained from a normal curve (z table). The z value obtained in this manner is corresponds to the long-term z value. To convert to short-term z, sigma level, the following formula is applied:

$$z_{ST} = z_{LT} + 1.5$$

Therefore, C_{pk} can then be determined by

$$C_{pk} = \frac{z(\text{short-term})}{3}$$

Shift and Drift. Most processes show a natural behavior and vary with time with a recognized approximate value of 1.5σ. To better visualize the concept of shift and, the reader's attention is directed to Figure 9.8.

Figure 9.8 Components

1. Long-term central tendency of the process (i.e., overall mean)
2. Sample distribution (typical)
3. Sample mean (typical)
4. Dynamic shift between consecutive means (typical)
5. Sample number
6. Equivalent short-term distribution
7. Short-term standard deviation σ_{ST}, result of white-noise effect
8. Short-term mean centered (i.e., the best you may expect)
9. Shift and drift $\pm 1.5\sigma_{ST}$ ($\pm 1.5\sigma$ short-term) band, in which the short-term means are expected to lie, resulting from the dynamic variation in the process mean over time due to black noise
10. Long-term distribution illustrating the total variation due to white noise plus black noise

Shift and drift - 1.5σ

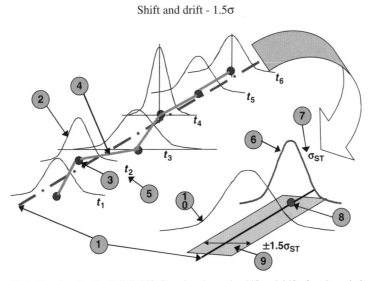

Figure 9.8   Illustration showing the [1.5σ] shift. Drawing shows the shift and drift after six periods of time (t_6).

9.10   INDICES OF CAPABILITY

A process capability index is a numerical summary that compares the behavior of a product or service to specifications. Operating with the philosophy of never ending quality improvement, the Japanese formalized the comparison of 6σ (specification range) with a process in statistical control. This is a comparison of the natural tolerances of the process with the specification ranges—upper specification limit − lower specification limit (USL−LSL). These measures are known as *capability* or *performance indices* or *ratios. Capability index* is a generic term related to voice of customer (VOC) or specifications limits (in the semiconductor industry, these indices are the engineering specifications) to the voice of the process. A large value of the index indicates that the current process is capable producing parts or services that will meet or exceed customer requirements. Thus, a capability index is convenient because it summarizes important information about the process to a single number.

9.10.1 C_p Index

Regardless of the widespread use of capability indices, not only in the manufacturing environment but also in financial organizations, there is much confusion and misunderstanding regarding their interpretation an appropriate use. Process capability index is frequently used to compare process capability with process or specification tolerances. Bilateral tolerance, shown in Figure 9.9 is typically defined as

$$\text{Capability index} = \frac{\text{specification tolerance}}{\text{process capability or } 6\sigma}$$

$$C_p = \frac{U - L}{6\sigma} = \frac{\text{tolerance}}{6\sigma} \quad \text{or} \quad C_p = \frac{USL - LSL}{6\sigma}$$

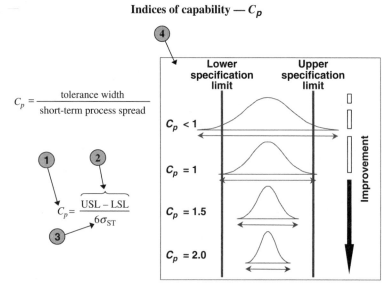

Figure 9.9   Illustration showing fundamental principles of index of capability C_p.

Figure 9.9 Components

1. Process capability index C_p indicates the short-term performance level that a process can potentially achieve.
2. Tolerance width is the upper specification limit (USL) minus the lower specification limit (LSL).
3. Short-term process spread equals 6 times the short-term standard deviation ($\pm 3\sigma_{ST}$).
   - $C_p < 1$: Process output exceeds specifications. The process is incapable.
   - $C_p = 1$: Process barely meets specifications. There is a probability that at least 0.3% defects will be produced and potentially more if the process is not centered.
   - $C_p > 1$: Process output falls within specifications, but defects might be produced if the process is not centered on the target value.
   - $C_p = 2$: This represents the short-term objective for process capability. Since $z_{ST} = 3 \times C_p$, we achieve 6σ when $C_p = 2$.

Capability index is used for continuous data and is based on several assumptions; the process is statistically stable and its data are approximately normally distributed. If the data distribution is skewed, the data should be transformed. Since C_p does not account for process centering, it should be used in conjunction with C_{pk}, which does account for process centering.

To calculate C_p indices using MINITAB, the following information is needed:

- Subgroup size, lower and upper specification limits, and the method for estimating standard deviation.
- In MINITAB, go to `Stat > Quality tools > Capability Analysis` (see Fig. 9.10).

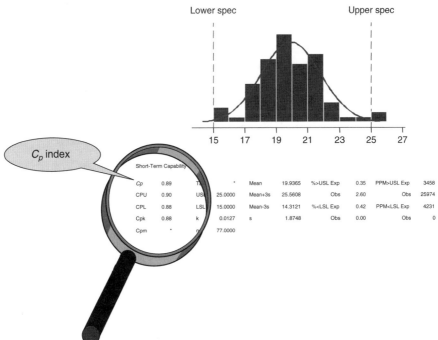

Figure 9.10   MINITAB capability analysis process capability printout.

Another common and closely related capability index is given by the equation $C_p = |T - \text{SL}|/3\sigma_{\text{ST}}$. For comparison purposes, it can be algebraically demonstrated that $C_p = z_{\text{ST}}/3$, where 3 is a statistical constant that defines the corresponding limit of unity. In this context, we naturally understand that the process capability ratio C_p is merely one-third of the quantity z_{ST}. Thus, a Six Sigma level of short-term capability (instantaneous repeatability) is given as $z_{\text{ST}} = 6.0$, or $C_p = 2.0$ if preferred.

Thus, the Six Sigma level of instantaneous reproducibility is described by a $\pm 6\sigma_{\text{ST}}$ normal random distribution that is centered between the limits of a symmetric bilateral performance specification realizing the design expectation $M = 0.50$. In this context, the $\pm 6\sigma_{\text{ST}}$ limits exactly coincide with the corresponding design limits—the upper specification limit (USL) and lower specification limit (LSL), respectively. The practitioner must recognize that the given figure provides only the right side of a symmetric bilateral performance specification. Since the left side is a mirror image of the right side, there is little need to discuss both. Consequently, the ensuing discussion is simplified without loss of specificity. On the basis of these model circumstances, the short-term quality goal of a Six Sigma characteristic is statistically translated to reflect one random error per 500 million chances for such an error, or simply two defects per billion opportunities, but *only for the centered, symmetric, bilateral case*. The unilateral case (no target specified) would reflect only one defect per billion opportunities.

9.10.2 C_{pk} Index

To determine whether a process, given its short-term variation, meets established customer requirements or specifications, we must calculate C_{pk}. C_{pk} considers process centering; it is the ratio of the difference measured between the process mean and the closest specification limit to one-half of the total process spread (see Fig. 9.11).

Figure 9.11 Components

1. *Process capability index*: Indicates the level of performance that a process can achieve taking into account the location of the process mean. It is equal to the smaller of either C_{pl} or C_{pu}. When the process is centered, $C_{pu} = C_{pl} = C_{pk} = C_p$.

Indices of capability — C_{pk}

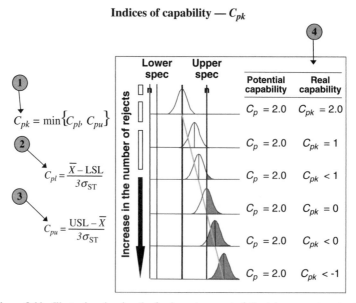

Figure 9.11 Illustration showing the fundament concept of C_{pk} (observe, not centered).

2. *Process capability index* (*lower*): Represents the difference between the lower tolerance and the mean divided by $3\sigma_{ST}$.

3. *Process capability index* (*upper*): Represents the difference between the upper tolerance and the mean divided by $3\sigma_{ST}$.

4. $C_{pk} = C_p$: The process mean is equal to the target.

5. $C_{pk} = 0$: The process mean falls on one of the specification limits; therefore, 50% of the process output falls outside the specification limits.

6. $C_{pk} < -1$: The process mean is completely out of the specification limits; therefore, 100% of the process output falls outside the specification limits.

Major Considerations. C_{pk} is generally used for continuous data and is based on several assumptions. C_{pk} assumes that the process to be statistically stable and that its data are approximately normally distributed. C_{pk} assumes that the data are approximately normally distributed. If the distribution of the data are skewed, we should transform the data. C_{pk} considers process centering and the short-term variation in the process. However, it should not be used alone to describe process capability. It should be used in conjunction with C_p, which is the short-term process capability index.

Calculation of C_{pk}

1. To calculate C_{pk} indices using MINITAB, the following information is needed: subgroup size, lower and upper specification limits, and the method for estimating standard deviation.

2. In MINITAB, go to `Stat > Quality tools > Capability Analysis`.

Figure 9.12 shows the MINITAB screen for process capability analysis. Within the same MINITAB screen, the practitioner can find the results for both C_p and C_{pk} (short- and long-term) calculations.

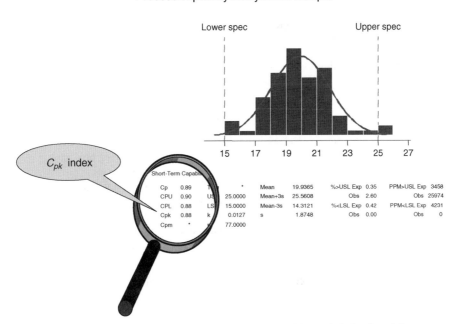

Figure 9.12 MINITAB screen showing the process capability analysis for C_p and C_{pk}.

Indices of capability — P_p

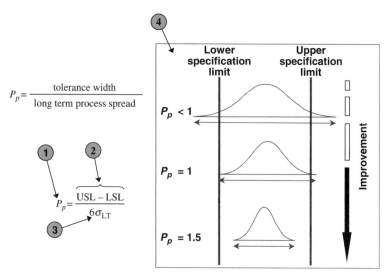

Figure 9.13   Illustration showing long-term variation.

9.10.3 P_p Index

We must calculate P_p to determine whether a process, given its long-term variation, has the capability to meet established customer requirements or specifications. P_p is a ratio of the tolerance width to the long-term spread of the process. P_p does account for shifts in the center of the process. Figure 9.13 illustrates the fundamental concepts of long-term variation. In figure 9.13, $P_p = 1$ Indicates that the process barely meets specifications—there is a probability that at least 0.3% defects will be produced (the number of defects increases if the process is not centered), while $P_p = 1.5$ represents the long-term objective for process capability.

Figure 9.13 Components

1. *Process performance index*: Measures the long-term level of performance that a process can potentially achieve.
2. *Tolerance width*: upper specification limit (USL) minus lower specification limit (LSL).
3. *Long-term process spread*: represents 6 times the long-term standard deviation ($\pm 3\sigma_{LT}$).
4. $P_p < 1$: The process variation exceeds specifications. Defects are being produced.

Please observe that P_p is based on the same equation as C_p, with one exception: P_p employs the long-term standard deviation while C_p employs the short-term standard deviation.

$$P_p = \frac{\text{USL} - \text{LSL}}{\pm 3\sigma} \quad \text{where} \quad P_p = \frac{\text{maximum allowable range of characteristic}}{\text{normal variation of process—long-term}}$$

To calculate P_p indices using MINITAB, the following information is required:

- Subgroup size, lower and upper specification limits.
- In MINITAB, select Stat > Quality tools > Capability Analysis (see Fig. 9.14).

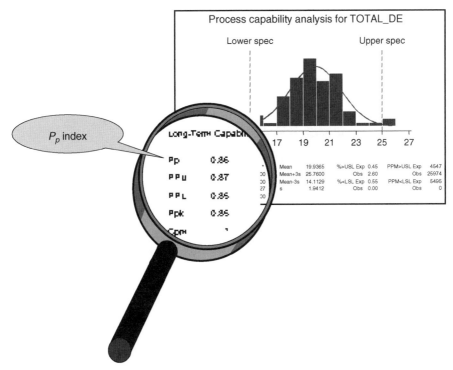

Figure 9.14 Illustration showing P_P calculation in MINITAB window.

9.10.4 P_{pk} Index

To determine whether a process, given its long-term variation, meets established customer requirements or specifications, P_{pk} accounts for the centering of the process. P_{pk} is a ratio of the measured distance between the overall mean of the process and the closest specification limit to half of the total process spread (see Fig. 9.15).

Figure 9.15 Components

1. *Process performance index*: Indicates the level of long-term performance that a process can achieve taking into account the location of the process mean. It is equal to the smaller of either $P_{p,l}$ or $P_{p,u.}$
2. $P_{p,u} = P_{pl} = P_{pk} = P_p$: Indicates that the process is centered.
3. *Process performance index (lower)*: Represents the distance between the lower tolerance and the mean long-term divided by $3\sigma_{LT}$.
4. *Process performance index (upper)*: Represents the distance between the upper tolerance and the mean long-term divided by $3\sigma_{LT}$.
5. $P_{pk} = P_p$: The process mean is on target. Since $z_{LT} = 3 \times P_{pk}$, we achieve $4.5\sigma_{LT}$ when $P_{pk} = 1.5$.

Also in Figure 9.15:

- $P_{pk} = 0$: The process mean falls on one of the specification limits. Therefore, 50% of the process output is out of the specification limits.
- $P_{pk} < -1$: The process mean falls completely out of the specification limits. Therefore 100% of the process output is out of the specification limits.

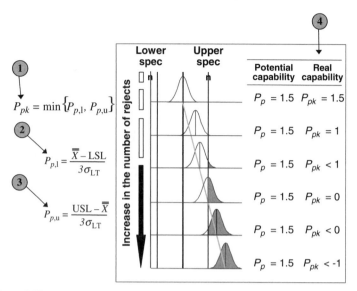

Figure 9.15   Illustration of long-term centering process and long-term process variation.

If the distribution of the data is not normal, the data should be transformed. To calculate P_{pk} indices using MINITAB, the following information is needed:

- Subgroup size and lower and upper specification limits.
- In MINITAB, go to `Stat > Quality tools > Capability Analysis`.

Figure 9.16 shows the MINITAB process capability Analysis screen and results for P_p, P_{pk}, C_p, C_{pk}, and other important capability analysis results and graphically illustrates an example of a MINITAB process capability analysis.

9.11   CALIBRATING THE SHIFT

So as to conceptually simplify the inflationary effect of transient and temporal errors (of a random nature), and to enjoy a more convenient form of application, an *equivalent mean offset* is often applied to the model distribution of Y. In the spirit of Six Sigma, such a quantity is expressed in the form of $1.5\sigma_{ST}$. Of course, the relative direction of such a linear correction to σ can be positive or negative, but not both. However, it is most often applied in the "worst case" direction—when testing or otherwise analyzing the performance of a design. Applying this compensatory correction to the short-term distribution illustrated in Figure 9.17 reveals a long-term performance expectation of $6\sigma_{ST} - 1.5\sigma_{ST} = 4.5\sigma_{LT}$.

Expressed differently, the resulting long-term capability is given as an equivalent figure of merit and expressed in the form $z_{LT} = 4.5$. Under this condition, the design margin is consequentially reduced to $m = M = 0.25$. Of course, the remaining safety margin of 25% is still large enough to absorb a fairly substantial shock to process centering, owing to some type or form of transient or temporal perturbation of a nonrandom nature. Of course, such a shock may or may not be manifested as a momentary disturbance to the process center. Statistically translating the $4.5\sigma_{LT}$ level of capability into defects per million opportunities reveals that DPMO (ppm) $= 3.4$. Thus, whenever we refer to a system, product, process, service, event, or activity as being Six Sigma, what we are really saying is

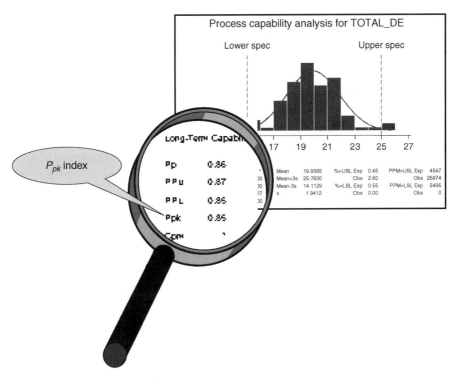

Figure 9.16   Illustration of P_{pk} index.

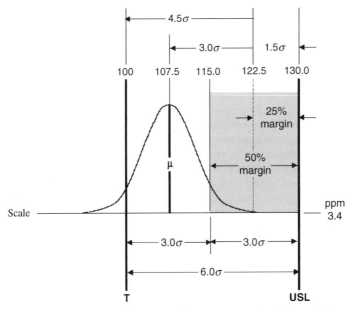

Figure 9.17   Illustration revealing a long-term performance expectation of $6\sigma_{ST} - 1.5\sigma_{ST} = 4.5\sigma_{LT}$.

that any given CTQ (critical to quality) characteristic related to that deliverable will maintain a short term capability (instantaneous reproducibility) of $\pm 6\sigma_{ST}$ and will exhibit no more than 3.4 DPMO over the long term (after many cycles or iterations of the corresponding process).

9.12   APPLYING THE 1.5σ SHIFT CONCEPT

To better understand the implications of the $1.5\sigma_{ST}$ shift, let us consider a simple example. Suppose that a particular system is characterized by $K = 2500$ defect opportunities. For this system, we assume only one opportunity per CTQ. We also assume that each opportunity is independent and the system sigma level capability is $4.5\sigma_{LT}$ (in the long term). With this level of sustained capability, we expect (on average) one nonconformance out of every $1/0.0000034 = 294{,}048$ opportunities.

According to these assumptions, the total defects per unit would be DPU $= p(d) * K = .0000034 * 2500 = 0.0085$. Given this, we recognize $p(d)$ as the statistical probability of nonconformance (per opportunity, over many cycles of process operation). We also understand that K is the total number of independent opportunities contained within the system (unit). With these facts in mind, we note that about one out of every 118 systems (units) could be expected to contain at least one CTQ that is defective (verified to be in a state of nonconformance). Owing to this level of long-term quality, it is often desirable to approximate the probability of zero defects, called "throughput yield." We use the Poisson function to estimate this yield value. Thus, $Y = (np)^r e^{-np}/r!$, where n is the number of trials, p is the event probability, and r is the number of such events. By 1 substitution, we observe that $Y = (dpu)^r e^{-dpu}/r!$, where DPU (dpu) is the defects per unit. Thus, for the special case of $r = 0$ (zero defects), we are able to ascertain the throughput yield (probability of zero defects) by the simple relation $Y_{TP} = e^{-dpu}$.

For this example, throughput yield is as follows:

$$Y_{TP} = e^{-dpu} = e^{-.0085} = .9915, \text{ or } \sim 99\%$$

There is 99% confidence that all ($K = 2500$ CTQ) opportunities will "yield" during the course of production, assuming that each characteristic (opportunity) is fully normal, is independent, and exhibits a long-term capability of $4.5\sigma_{LT}$. Thus, the long-term, first-time yield expectation is $Y_{FT} = .9999966$. Since there exist $K = 2{,}500$ independent yield opportunities per unit, the throughput yield should be given as $Y_{TP} = Y_{FT}^{2,500} = .9915,\backslash$ or $\sim 99\%$. In turn, the defects per unit expectation would be

$$DPU = -\ln(Y_{TP}) = -\ln(.9915) = .0085.$$

Here again, the Poisson distribution is used for this approximation. To provide a *first-order approximation* of the short-term capability, we add the "standard shift correction" of 1.5σ to the long-term capability. Thus, the short-term capability becomes $z_{ST} = z_{LT} + 1.5 = 4.5 + 1.5 = 6.0$. At this point, the reader is admonished to recognize that z_{ST} is merely a general figure of merit (performance index) that constitutes merely a high-level approximation of the prevailing instantaneous reproducibility (per CTQ opportunity). In this example, we would naturally recognize that the approximated value of $z_{ST} = 6.0$ would statistically translate to an *equivalent bilateral short-term throughput yield expectation* of 99.9999998% (per opportunity). Considering this level of short-term yield, we could expect about one out of every $1/(1 - 0.999999998) = 1/0.000000002 = 201{,}996$ defect opportunities to be validated in a state of nonconformance. Thus, the total defects per unit expectation becomes

$$DPU = p(d) \times K = .000000002 \times 2500 = .00000495.$$

Therefore, we could expect that about one out every 201,996 units contain al least one defective.

In addition, and generally speaking, the shift factor is *added* to an estimate of long-term capability to remove the long-term influences, to provide an approximation to the short-term capability. Conversely, the shift factor is *subtracted* from an estimate of the short-term capability to approximate the long-term capability.

9.13 YIELD

In previous sections the practitioner had the opportunity to observe that metrics are essential to establish product and processes baselines, track process improvements, track the quality of a business, and report business improvements. Some of the Six Sigma metrics already seen are DPU (defects per unit), DPMO (defects per million opportunities, and the sigma level (and z score). The following sections discuss rolled throughput yield (RTY).

The Nature of "Yield" and Why It Is Important. Two philosophies may be used to interpret and manage yields—active and passive (opportunity of counting). *Active* organizations count every measurement, test, or inspection as one opportunity. Thus, a single defect can cause several measurements to fail to meet specifications. For an organization pursuing the *passive* philosophy, parts and connections are counted. The philosophy selected is not critical, but it becomes critical when an organization changes from one to the other. It is important that organizations consistently implement the selected approach across the board so that comparisons between processes are possible. Still another way to scrutinize yield would be inductive reasoning vis-à-vis the individual considera- tion of each unit rather than a holistic consideration of the collective set. In this case, yield is dependent on how success is defined rather than the quantity of units processed. In this context, the realization of a "successful" unit would mean that it is free of any given type of nonconformity. In this case, a unit of product or service can be declared a success only if it was realized without a single quality defect. This type of yield is often referred to as "final-time yield," or simply *final time Y*. In this case, a given unit will fail in terms of either quality (number of defects, $d > 0$) or yield ($d = 0$).

Classical Yield Calculation. In general terms, the idea of "yield" can be viewed as the fruit of a process where yield is merely the ratio of output to input. In some processes, the only measurement is at final test. Other authors define yield as the area under the probability density curve between tolerances. From the first definition, this relationship is $Y = O/I$, where $Y =$ yield, $O =$ output, and $I =$ input.

Example 9.10   Suppose that we process 100 units ($I = 100$) and 90 units ($O = 90$) are successfully completed, where *unit* is defined as final product or service delivered to customer.

Solution   From these findings, the final yield of our example process is

$$\text{Yield} = \frac{90}{100} = .90 \text{ or } 90\%$$

This result is also known as *final test yield* (FTY), and this equation can be elaborated as

$$\text{FTY} = \frac{\text{units passing final test or inspection (units successfully completed)}}{\text{units tested (input)}}$$

Example 9.11   Assume that a bank processes 300 mortgage applications, 100 of which contain defects. Unit $= 1$ mortgage application.

Solution

$$\text{Yield} = \frac{200}{300} = .6667 \text{ or } 66.67\%$$

This result is the final test yield (FTY). Final test yield is a function of the number of defective units reaching final test and the ability to detect the units with defects. A high value for FTY indicates a process in control or a poor-quality inspection process. Since virtually no process is 100% efficient, it is reasonable to expect that what one gets out is most often less than what one puts in. From Examples 9.10 and 9.11, we recognize that yield does not measure the defects created while producing the number of units. It only counts those units that successfully passed inspection. Therefore, yield does not offer "management visibility" into quality. Yield is a metric that measures what goes into the process and what comes out. Observe that this metric does not consider what occurs within the process. Such an odd way of counting efficiency is attributable to the hidden process within the mainstream process, i.e., inspection/test points, established to detect, analyze, and fix defects inline and/or offline. When such defects are eliminated, it is possible that all units will ultimately be labeled as successful. Since it takes effort to detect, analyze, and fix a defect, it is intuitively obvious that yield does not have a consistent statistical or practical relationship (correlation) with key business factors such as cycle time, labor cost, material cost, and work in process (WIP). From this perspective, the process owner may have no visibility into the true efficiency of the operation. Thus we could argue that yield is based on units produced, and not on defects created. After all, it is reasonable and rational to assert that an individual unit should not yield if it was created with one or more defects during the course of its processing. Perhaps such a unit could yield at the backend of the process (after some amount of inline rework), but it would certainly not be considered as inline yield.

From the reasoning described above, we believe that yield should be dependent on defects. Thus, we define yield by the relation $Y = f(d)$, where Y is the yield and d is the defect rate. Given this relation, we reason that if $d > 0$, then $Y < 1.0$ and if $d = 0$, then $Y = 1.0$. Hence, this modified yield concept is superior because it is dependent on quality and not on the proportion of units successfully realized at the end of the process line. Quality focuses on the ability to meet customer expectations within specifications. By recognizing this quality dependency, we can correctly state that yield is a symptomatic effect and quality is the causal agent. To proceed further, we must be more specific on the idea of defects. By convention, we define a defect as being any type or form of deviation from expectation. In this sense, a defect is simply a nonconformance to standard. Interestingly, defects are most often randomly distributed throughout a unit as contrasted to being uniformly distributed. The random model holds that the various types of defects and their consequential location on a unit cannot be predicted. We cannot predict the creation of defects; nor can we know when or where a particular type of defect occurs and could be observed. This is in contrast to the uniform model that relies on the idea that any given defect type, and its consequential location, can be known in advance of its creation—and that the same type of defect always appears in the same spot on every unit realized by the process. However, the world of quality contains both types without an existing disproportional relationship.

We also assume that the various classes of defects are independent. This means that one kind of defect cannot be used to predict another defect type. When dependence exists, the defects are correlated. Fortunately, the independence assumption is quite reasonable. When a process owner considers various categories of defects, very few categories are correlated, and those correlations that do exist have little statistical or practical impact. Simply stated, the first-order influences of k defect categories tend to swamp out any second-order effects that may be present. Therefore, the assumption of independence is considered reasonable and practical when aggregating defects for yield analysis. Empirically, quality practitioners know that the overwhelming number of defects produced during processing are of the random variety. In fact, this researcher's experience is that the quantity of random defects inherent to a particular type of unit is much greater than that of the uniform variety. Most often, there is no detectable difference that is practically significant when contrasting the pooled sources of defects with that of a random model. For this reason, we will assume that defects are randomly distributed within and across the units of the product, service, or transaction. In short, a *random, model rather than a uniform model, should be used to study defects.*

9.13.1   Final Test Yield (FTY)

In summary, final test yield is the number of accepted units to the number of units tested. Final test yield has traditionally been used to assess process performance. It may sound redundant, but this concept is flawed since the rework and replacement of scrapped units is not measured, blinding management of their failure to produce first-pass quality products and services. It is rare to find products or services that are the direct result of a single process step. Final yield is the calculation of final test yield (FTY) at the last process step and is not an accurate measure of process performance. This is why FTY is not a Six Sigma metric. It is like saying that there is a 50% chance that any given unit will contain zero defects. Such a yield value can be directly obtained by

$$Y = \frac{O}{I} = \frac{50}{100} = .50, \text{ or } 50\%$$

where Y = yield, O = output, and I = input.

Example 9.12   If 100 units were started and 80 units passed through the process with zero defects d at each of the various "verification points," then final test yield would be $\text{FTY} = \frac{80}{100} = .80$ or 80%. Hence the *quality-centric yield* was computed as: $Y = O/I = \frac{80}{100} = .80$ or 80%.

From these results, we could say that 20% of the units has one or more randomly distributed defects; and that the defects were randomly distributed; however, we can only say that 80% of the units had no defects. Despite this limitation and under the correct circumstances, FTY could be an operationally attractive performance metric. Earlier, we said that FTY is a function of the number of defective units reaching final test because FTY counts defects at inspection. Also it was concluded that yield was $Y = O/I$, where O = output = number of units that passed inspection and I = input = number of units started. On the other hand, for *process yield*, Y = number of cycles completed and I = number of cycles initiated.

   This method of calculation shows that the output is created in a *quality way*. Because of this, *quality-centric yield* is often referred to as *rolled throughput yield* or final yield. The *distinction between final yield* and *first-time yield is very important* from the business perspective because any unit of product (or service) that was created with defects ultimately costs more to produce than a unit created without defects. Consequently, *rolled throughput yield* correlates more closely with total cost. From this perspective, we conclude that yield is *quantity-centric, but not quality-centric.*

9.13.2   Yield Related to Defects

The quality-centric perspective is related to defects. For example, consider a hospital that counts only those patients who received poor care, instead of the specific ways in which inadequate care was delivered. The Six Sigma practitioner must track defects to improve patient treatment quality and satisfaction. Preparation is required before the process capability can be improved. A detailed map of the process is required, and data on defects must be collected for each process step. In this way, we can track and identify every opportunity for the creation of a defect and monitor the number of defects produced at each opportunity. Yield should be dependent on producing a unit, service, or transaction that is "defect free." Only those units created without defects should be classified as yield.

   At this point, we need to emphasize the following definitions:

Defective—a unit that contains at least one defect

Defect—the output of a process that does not meet defined specifications

There are two ways to count opportunities for defects: (1) *product*-related, called *active* opportunities, and (2) *process*-related, called *passive* opportunities. Passive opportunities indicate process complexity and are related to process design. Active opportunities are those opportunities that can be measured.

Opportunity-Based Measurements

Defects per Opportunity (DPO)—the proportion of defects over the total number of opportunities in a unit. For example, DPO = .05 would mean that there is a 5% chance of a defect occurring in that unit category.

$$\text{DPO} = \frac{N \text{ defects}}{N \text{ units} \times N \text{ opportunities}}$$

Defects per Million Opportunities (DPMO)—a measure that is often translated into DPMO format, which indicates how many defects would arise if there were one million opportunities. [DPMO is also called *parts per million* (ppm)]:

$$\text{DPMO} = \text{DPO} \times 1{,}000{,}000 = \text{DPO} \times 10^6$$

Defective parts per million (Dppm)

Sigma Measure—obtaining the sigma performance equivalent is easy to accomplish. Just obtain DPMO and then use the Sigma Conversion Table (Table VIII in Appendix A) to translate it into a business sigma level.

The following examples illustrate the above definitions.

Example 9.13   A document contains a total of eight entries (number of opportunities). In this unit, each entry is an opportunity for the creation of a defect. The inspection of 5000 documents (total number of units in the group) revealed 175 defects; thus, 96.5% of the documents were good (defect-free). Calculate the defects per unit (DPU), defects per million opportunities, and (DPMO) or parts per million (ppm).

Solution

$$\text{DPU} = \frac{N \text{ defects in group of units}}{N \text{ units in group}}$$

$$\text{DPMO} = \text{DP0} \times 1{,}000{,}000$$

$$\text{DPPO} = \text{ppm}$$

$$\text{DPU} = \frac{175}{5000} = 0.035$$

$$\text{DPO} = \frac{N \text{ defects}}{N \text{ units} \times N \text{ opportunities}}$$

$$= \frac{175}{5000 \times 8} = \frac{175}{40{,}000} = 0.0044 \cong 0.4\% \text{ chance of defect in documents}$$

$$\text{DPMO} = 0.0044 \times 1{,}000{,}000 = 4375$$

Example 9.14   Difference between defects and defectives—quality measurements are about deliverable units and their compliance to specifications. A document contains the following fields:

1. Customer name: Cirk de Soleil
2. Address: 201 Jetton Strit
3. City: Begas
4. State: Nebada

Solution   For this example we conclude that we have one defective document with four defects in the document.

First-Time Yield (FTY).   This is the ratio of the number of units accepted to the total number of units inspected. This calculation of process performance represents the classical view of Yield. The calculation of Process Yield using the first time-yield (FTY) formula does not reflect the true process effectiveness because FTY is a function of the number of defective units reaching final tests and the effectiveness of the final tests in detecting those units. It often leads to an optimistic calculation of the true yield, and has no practical meaning other than the ratio of successes over units tested. Therefore, first-time yield should not be used in practice, and FTY is not recognized as a sigma metric.

FT Calculation

1. Count the number of units inspected or tested (U).
2. Count the number of units that pass inspection/test requirements (S).
3. Apply the FTY formula, expressing the result as a percentage.

First-time yield does not consider the effects of the "hidden factory" because these defects are not detected during the different steps of the process and therefore aren't repaired or replaced. Finally, each defect should be considered waste [time, money (rework), resources, equipment, and manpower].

9.13.3   Rolled Throughput Yield (RTY)

Final test yield [FTY] and rolled throughput yield [RTY] differ because of losses in the "hidden factory." RTY and its associated metrics are defects per unit (DPU), defects per million opportunities (DPMO), and the sigma level (z score). The purpose of rolled throughput yield is to measure each process step's effectiveness including the effects of the hidden factory. Because this approach includes the hidden factory, it is said to be a complete and true assessment of process effectiveness. Rolled throughput yield is the product of all subprocess yields. We calculate it by multiplying together all of the in-process yields (IPYs):

$$\text{RTY} = \text{IPY}_1 \times \text{IPY}_2 \times \text{IPY}_3 \times \cdots \times \text{IPY}_n$$

9.13.4   In-Process Yield (IPY)

In-process yields are often used to highlight throughput issues or missed-commitment issues. Sometimes yield losses are not addressed until it becomes a throughput or missed-commitment issue. The best way to control IPY is to track "first-time-good," or the proportion of units completed without rework to the total number of units started:

$$\text{IPY} = \frac{\text{unit shipped to the next process}}{\text{units started in this process}}$$

or

$$IPY = \frac{\text{units processed without defects}}{\text{units started in this process}}$$

To calculate IPY for process steps in which *only one* defect opportunity per unit exist, use the following simple equation:

$$IPY = 1 - DPU$$

From this equation, $DPU = 1 - IPY$.

To calculate IPY for process steps in which *multiple* defect opportunities per unit exist, use the following approach:

$$IPY = e^{-DPU}$$

For this second case $DPU = -\ln(IPY)$. In other words, IPY is the number of units processed without any defects and shipped to the next process step divided by the number of units in the process. The best way to track defect reduction is to track first-time good (defect-free) or the proportion of units completed without rework to the number of units in the group.

9.13.5 In-Process Yield (IPY) and Rolled Throughput Yield (RTY)

In-process yield is calculated by dividing the number of defects in a process step by the number of opportunities. RTY is calculated by multiplying together all the IPYs. Rolled throughput yield is the probability of any unit having zero defects.

Example 9.15 The headquarters of a bank has noticed that one branch has received more complaints related to its deposit services than the other branches in the same city. In response to customer complaints, the bank assigns a Six Sigma team to investigate the deposit process and reduce the number of complaints at the branches. The bank provides the practitioners with 280 deposit slips (units) and informs them that each deposit slip should be considered as a unit. Every unit (deposit slip) provides six (6) defect opportunities.

Solution

Step 1. Construct a high-level flowchart of the process (Fig. 9.18).

Step 2. Collect data [number of defects in 280 deposits (units)] (Fig. 9.19).

Step 3. Calculations:

$$\begin{aligned} DPU &= N \text{ defects found in group of unit} / N \text{ units in group} \\ &= \frac{N \text{ defects}}{\text{total } N \text{ units}} \end{aligned}$$

(*Note*: DPU denotes defects, not defectives.)

Then

$$IPY = \frac{N \text{ units passed to next process}}{N \text{ units in group}}$$

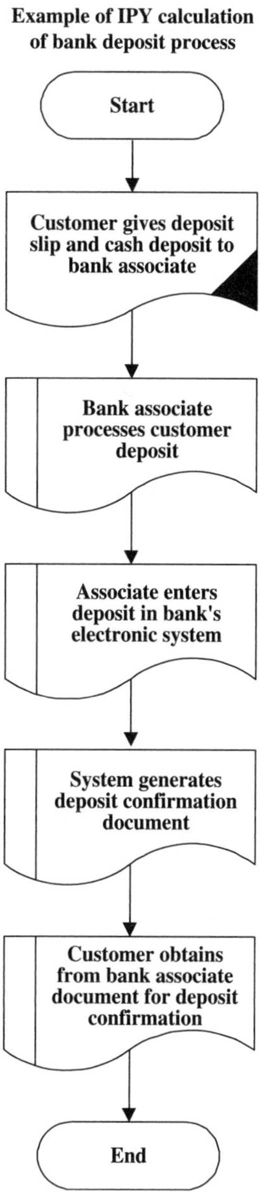

**Example of IPY calculation
of bank deposit process**

Figure 9.18   Example flowchart of high-level bank deposits.

IPY and DPU for one defect opportunity per unit:

$$IPY = 1 - DPU$$
$$DPU = 1 - IPY$$

IPY and DPU for more than one defect opportunities per unit:

$$IPY = e^{-DPU}$$
$$DPU = -\ln(IPY)$$

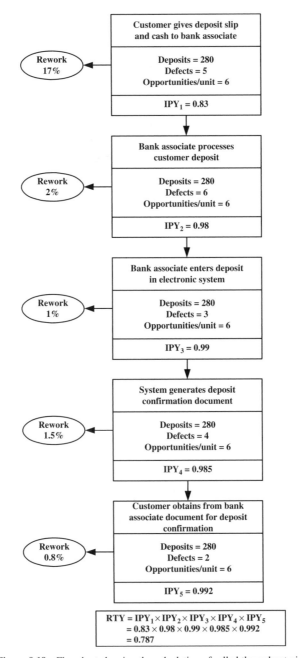

Figure 9.19 Flowchart showing the calculation of rolled throughput yield.

and

$$\text{RTY} = \text{IPY}_1 \times \text{IPY}_2 \times \text{IPY}_3 \times \cdots \times \text{IPY}_n$$
$$\text{DPU} = -\ln(\text{RTY})$$

From constructing the flowchart and collecting data for each process step we know the following:

1. Customer gives deposit slip, cash and checks to bank teller (5 defects)
2. Bank teller (associate) processes customer's deposit (6 defects)
3. Bank Associate enters deposit in Bank's electronic system (3 defects)
4. System generates deposit confirmation(s) document(s) (4 defects)
5. Customer receives deposit confirmation document from Bank Associate (2 defects)

Calculation of DPU for each process step proceeds as follows:

$$DPU_1 = \tfrac{5}{280} = 0.0179 \cong 0.18$$
$$DPU_2 = \tfrac{6}{280} = 0.021$$
$$DPU_3 = \tfrac{3}{280} = 0.010$$
$$DPU_4 = \tfrac{4}{280} = 0.014$$
$$DPU_5 = \tfrac{2}{280} = 0.007$$
$$DPU_{total} = 0.18 + 0.021 + 0.010 + 0.014 + 0.007 = 0.232$$

Because the deposit process involves multiple opportunities, the equation that we should select for these calculations is the one for *multiple defect opportunities per unit:*

$$IPY_{total} = e_{total}^{-DPU} = e^{-0.474} = 0.622 \cong 0.62$$

Calculation of IPY for each process step proceeds as follows:

$$IPY_1 = e^{-DPU1} = e^{-0.18} = 0.83 \qquad (rework = 17\%)$$
$$IPY_2 = e^{-DPU2} = e^{-0.021} = 0.86 \qquad (rework = 2\%)$$
$$IPY_3 = e^{-DPU3} = e^{-0.010} = 0.989 \quad (rework = 1.1\%)$$
$$IPY_4 = e^{-DPU4} = e^{-0.014} = 0.985 \quad (rework = 1.5\%)$$
$$IPY_5 = e^{-DPU5} = e^{-0.007} = 0.94 \quad (rework = 0.8\%)$$
$$RTY = 0.83 \times 0.98 \times 0.99 \times 0.985 \times 0.992 = 0.786 \cong 0.79$$

or

$$RTY = e^{-DPU} = e^{-0.232} = 0.79$$

The practitioner found the "hidden factory" and now the bank knows where and when the defects are created. Figure 9.20 graphically summarizes this concept.

9.14   HIDDEN FACTORY

Customer surveys, rejection reports, returned goods, customer problems referred to call centers but not resolved, and warranty claims provide companies with data on how well their processes are performing. They are able to count the number of product or service units that entered a process

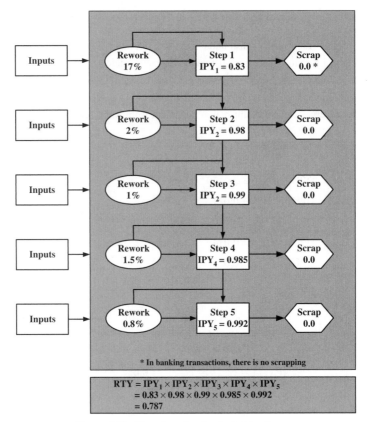

Figure 9.20   Flowchart showing the hidden factory.

and the number that successfully came out. Traditionally, companies use these data to calculate the "metric" called *yield*, the number of units that pass through that point in the process. Essentially, this metric is a simple concept, output divided by input, or the amount of good stuff that companies are able to produce, contrasted to the total amount put through the system. This measure of operational efficiency has very strong intuitive appeal and is quite easy to calculate, but it is a "major managerial trap." It has the ability to gravely mislead and deceive. It is embedded in many commonly used indices of performance, such as first-time yield, first-pass yield, and final yield. Comparing output to input lies at the heart of such metrics. To better understand how this can mislead and deceive, let us consider the following scenarios:

Scenario 1. Imagine two units of service delivered by a bank, say, transaction 2451 and 2452, respectively. These two transactions have just been completed. Transaction 2451 is error-free; however, transaction 2452 contains two errors (date and customer account number).

Scenario 2. Consider transaction 2453 and 2454. In this case, transaction 2453 is error-free, but transaction 2454 contains eight errors.

When considering the two transactions in scenario 1, the first-time yield would be computed as output (1) over input (2), or .5, which in percentage terms is equal to 50%. In scenario 2, the first-time yield would also be − 50%. From this perspective, both scenarios are rated at the same level of performance. However, from the quality perspective, they differ.

In scenario 1, the total defects per unit, or the TDPU (total defects per unit), would be computed as two defects divided by two transactions (units):

$$\text{TDPU} = \frac{\text{total } N \text{ defects observed}}{\text{total } N \text{ units processed}}$$

or $\frac{2}{2} = 1$. In scenario 2, the TPDU (total defects per unit) is computed as follows: eight defects divided by the two transactions (units), or $\frac{8}{2} = 4$.

Although the yield is the same for each scenario, the first scenario has superior process capability. When rework is feasible, it also incurs less cost to correct the defects. However, the production in the first scenario is less than that of the second scenario. Thus, the total cost is correlated to TDPU, and not to first-time yield. Using final throughput yield ignores the hidden factory. We repeat the summary of this section to emphasize the hidden factory concept.

Each defect must be identified, reworked, or replaced while in process. Each defect is waste because it takes time, resources, equipment, material, floor space, and employee efforts to correct or replace the defective units. Performing these activities, adds operational costs.

9.14.1 Hidden Factory Composition

The hidden factory is composed of all unnecessary activities or waste such as rework and scrap. FTY ignores the hidden factory. Since it counts *defectives* at inspection, while rolled throughput yield counts *defects* throughout the process. Observe how easily we can be seriously deceived when we gage our success on the basis of classic yield calculations. Stated more logically, yield is a function of errors, not output and input. Thus, the usual yield measurement method could easily mask the hidden factory. Therefore, the term "hidden factory" is used to designate all work performed above and beyond entitlement to produce a good unit of output and is not explicitly identified in business reports. Finally, we can say that the term "hidden factory" is closely related to the concepts of productivity, value, and money. Figure 9.21 is a flowchart of the hidden factory.

Today, there is no single definition of what constitutes a hidden factory. As a general guideline, its cost includes costs incurred for rework, scrap, excess floor space and equipment, a portion of the inspection and failure analysis, and a host of indirect and allocated costs. Constructing process maps is the most effect way to identify contributors to the hidden factory. The team must first discover where the actual process differs from the documented process. Frequently, the actual process includes undocumented inspections and rework steps that are regularly practiced on the manufacturing floor. However, these informal inspections, and workaround(s) frequently create non-value-added (NVA) activities. These hidden factories are seldom tracked or reported to management. The traditional yield calculation ignores the effect of the hidden factory; however, rolled throughput yield (RTY) captures rework and scrap and is thus an indicator of overall process quality.

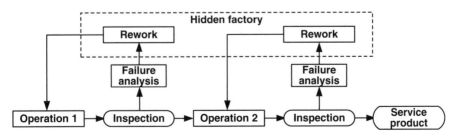

Figure 9.21 Illustration of a hidden factory.

GLOSSARY

abscissa   Horizontal axis of a graph.

activity symbol   Rectangle used to designate an activity. A brief description of the activity is also included within this symbol.

affinity diagram   Tool used to organize and present large amounts of data (ideas, issues, solutions) into logical categories.

alpha risk   Probability of accepting the alternative hypothesis (H_1), when in reality the null hypothesis (H_0) is true

alternative hypothesis   Tentative explanation indicating that an event does not follow a chance distribution, in contrast to the null hypothesis.

attribute data   Numeric information at the nominal level.

cause   That which produces an effect or brings about a change.

centerline   Line on statistical process control charts that represents the characteristics of central tendency.

characteristic   Measurable feature of process, product, or service.

classification   Differentiation of factors or variables into groups.

complex level   The first level of the process tree. It is the last process performed before delivery of the product/service to the customer.

confidence level   The probability that a random variable x lies within a defined interval.

confidence limits   The two values that define the confidence level.

Connections   Arrows indicating the flow and sequence of the key activities.

connector symbol   Circle used to illustrate a break and its continuation on the same or different page.

C_p **index**   Process capability index.

C_{pk} **index**   Measure of process dispersion and its centering about the average.

critical requirements for quality, delivery and cost   Needs translated from CTS requirements.

critical-to-cost (CTC) characteristics   The product, service, and/or transactional characteristics that significantly influence one or more CTS in terms of cost.

critical-to-delivery (CTD) characteristics   The product, service, and/or transactional characteristics that significantly influence one or more CTSs in terms of delivery.

critical-to-process (CTP) characteristics   Process parameter that significantly influences a CTQ, CTD, and/or CTC.

critical-to-quality (CTQ) characteristics   Product, service, and/or transactional characteristics that significantly influence one or more CTS in terms of quality.

critical-to-satisfaction (CTS) characteristic   Criterion for expression of the customers' vital needs.

cycle time   Measured time for each process step or subprocess.

data   Factual information used for reasoning, discussion, or calculation. Often refers to quantitative information.

defect   The output of a process that does not meet customer specifications.

defective   Unit with one or more defects.

defects per million opportunities (DPMO)   Number of defects multiplied by 1,000,000 divided by the total number of opportunities; DPU × 1,000,000.

defects per unit (DPU) $\dfrac{\text{Total } N \text{ Defects}}{N \text{ units in group}}$

dependent variable   Response variable Y: $Y =$ output $=$ response variable from transfer function $Y = f(x)$.

design for Six Sigma (DFSS)   The design process for development of new products or processes or redesign of existing processes.

detection   After-the-fact (ex post facto) strategy to identify unacceptable output and then separate it from the good output (*prevention*).

document symbol   Term that indicates a printed document pertinent to the process.

element level   This level represents a breakdown of the processes listed at the subsystem level. CTP characteristics are identified directly below this level.

factors   Independent variables.

flowline   Arrow indicates the direction of the process and connection between its elements.

independent variable   A variable whose value is independent of the value of another variable.

interval scales   Measure used to express numerical information on a scale with equal distance between categories, but with no absolute zero. Examples of interval scales include temperature (°F and °C), a dial gage sitting on top of a gage block, and comparison of differences.

key process   List of steps or subprocesses that make up the entire process.

lower inner fence   Value in a dataset that is $1.5 \times$ IQR (interquartile range) below the first quartile.

lower outer fence   Value in a dataset that is $3.0 \times$ IQR below the first quartile.

lower specification limit (LSL)   Minimum acceptable value for a CTQ, following voice of customer.

mean   Measure of central tendency calculated by dividing the sum of all values by the number of values in the dataset.

measures of central tendency   Measures that describe the center of a distribution. The mean, median, and mode are three such measures.

measures of dispersion   Measures of the spread of a distribution.

median   If the number of values is odd, the median value is the middle value when values are arranged from lowest to highest. If the number of values is even, it is the average of the middle two values.

never-ending improvement in quality and productivity   Operational philosophy that makes the best use of the talents within the organization to produce products or services of increasing quality for our customers in an increasingly efficient manner that protects the return on investment of stockholders.

nominal scales   Measure used to classify elements into categories without considering any specific property. Examples of nominal scales include "causes" on fishbone diagrams, yes/no, and pass/fail.

normal distribution   Continuous, symmetric density function characterized by a bell-shaped curve.

null hypothesis   Claim or statement about a population parameter that is assumed to be true until it is declared false.

operation definition   Means of clearly communicating quality expectations and performance.

opportunity   Count of potential failures, defects, or errors that impact ability to meet customer requirements.

ordinal scale   Measure used to rank nominal data on the basis of a specific property.

outcome   Response-dependent variable. The result of a trial with a given treatment is called *response*.

outliers or extreme values   Values that are very small or very large relative to the majority of the values in a dataset.

parameter   Summary measure calculated for population data.

pareto diagram   Chart used to plot percent defectives in a sample.

percentiles   Ninety-nine values that divide a ranked dataset into 100 equal parts.

process   Sequence of activities required to accomplish a task. Combination of people, equipment, materials, methods, and environment to produce a given product or service.

process average   Central tendency of a given process characteristic.

process control   Application of statistical methods and procedures relative to process and a given set of standards.

process spread   Range of values, which gives process characteristic displays; this particular term most often applies to the range but may also encompass the variance. The spread may be based on a set of data collected at a specific point in time or may reflect variability across a given amount of time.

project   A problem, usually calling for planned action.

quartiles   Three summary measures that divide a ranked dataset into four equal parts.

range   Difference between the highest and lowest values in a dataset or data subgroup.

random   Characteristic of a sample that is selected such that each item in the population has an equal chance of being selected; lack of predictability; without pattern.

random variations   Variations in data that result from causes that cannot be pinpointed or controlled.

range   Difference between the highest and the lowest values in a dataset or values of a subgroup.

ratio scales   Measure used to express numerical information on a scale with equal distance between categories, but with an absolute zero in the range of measurement.

sample   In process control applications, this term is a synonymous with *subgroup*. This use is totally different because it provides an estimate of a larger group of people or items.

sigma (σ)   Greek letter used to designate standard deviation.

spread   General concept for the extent by which values in a distribution differ from one another. Data dispersion.

stable process   Process that is in statistical control.

standard deviation   Statistical index of variability that describes the spread.

statistic   Summary measure calculated for sample data.

statistical control   The condition describing a process from which all special causes of variation have been eliminated and only common causes remain.

subgroup   Logical grouping of objects or events that display only random variations. One or more events or measurements used to analyze the performance of a process.

symbols   Terms indicating the nature of each step (see Section 7.9.1 for more details).

system   Something that is connected according to scheme.

upper specification limit (USL)   Maximum acceptable value for a CTQ following voice of customer.

variable   Characteristic that may take on different values.

variance   Measure of spread.

variation   Quantifiable difference between individual measurements such as differences that can be attributed to common causes or special causes.

yield   Amount of material that is processed by an operation less the scrap.

REFERENCES

1. J. Hillkirk, "Top quality is behind comeback," cover story, *USA Today* (March 28, 1989).

10 Probability

10.1 OVERVIEW

The workday is over, you are tired, and you do not feel like cooking dinner. You pull into the drive-through of a local fast-food restaurant, and, lo and behold, there is no one in line. "Wow!" you think. "What is the chance of that happening?" *Probability*, which is a scientific method for analyzing uncertain events, allows us to answer that question.

Probability has a major role in inferential statistics, which is the process of making decisions under conditions of uncertainty. Specifically, probability allows us to assess the level of uncertainty associated with such decisions. For example, the fast-food restaurant you went to has to maintain its reputation for fast service, so they need to project the demand for different types of food. They want to make enough food to have a high probability of satisfying customer demands, but they do not want to prepare so much food that they have to throw a lot of food away after it remains unsold. They use inferential models to help increase profits by wasting less food and creating happy customers who typically do not have to wait for a long time to get their food.

Inferential statistics combines probability and probability distributions (which are discussed in Chapters 11–13) with descriptive statistics. Using information gleaned from appropriately obtained samples, inferential statistics will help us make decisions about populations on the basis of information obtained from our samples. This chapter presents the basic concepts of probability and the rules for computing probability.

10.2 EXPERIMENTS, OUTCOMES, AND SAMPLE SPACE

Flip a coin and observe which side is faceup. Roll two dice and calculate the sum of the spots on two faces. Perform a heart transplant and identify whether the organ is rejected. When one refers to an *experiment* in a statistical setting, it is not the same as a mad scientist saying "Igor, bring me the brain." In fact, a statistical experiment implies that one is performing an activity that involves a *random phenomenon*. In a random phenomenon, we know the different outcomes that could occur, but we do not know which one will occur. Therefore, we perform an experiment and identify the *outcome*. In the case of the coin, the outcome will be heads or tails, while the outcome of the transplant will be "rejected" or "accepted." The collection of possible outcomes for any statistical experiment constitutes the *sample space*.

Example 10.1 List the sample spaces associated with the following random phenomena:

1. A customer enters a store and the customer's gender is noted.
2. A client is asked whether she is satisfied with the service provided.
3. A product with a rebate offer is purchased. Whether the rebate is claimed is noted.

Practitioner's Guide for Statistics and Lean Six Sigma for Process Improvements. By Mikel J. Harry, Prem S. Mann, Ofelia C. de Hodgins, Christopher J. Lacke, and Richard Hulbert
Copyright © 2010 John Wiley & Sons, Inc.

Solution   A sample space is denoted with an uppercase *S*, and the outcomes are listed inside a set of curly brackets { }. The sample spaces for the examples mentioned above are as follows:

1. $S = \{$female, male$\}$
2. $S = \{$satisfied, not satisfied, no opinion$\}$
3. $S = \{$claimed, not claimed$\}$ ■

Once we have established a sample space, we can calculate probabilities of collections of outcomes from the sample space. Any collection of outcomes from a sample space is called an *event*. Events involving only one outcome are called *simple events,* while events involving two or more outcomes are called *compound events*. For example, the event that you receive no phone call in the next 15 min is a simple event, because the only outcome included is "receiving zero phone calls." However, the event that you receive four or more phone calls in the next 15 min is a compound event, as it includes the outcomes "receiving four phone calls," "receiving five phone calls," "receiving six phone calls," and so on.

Events are typically represented by capital letters: *A, B, C,* and so on. A common notation for a simple event is the letter *E*.

Example 10.2   An employee of a mail-order company can pack a maximum of 10 orders in a day. Thus, the sample space for this example is $S = \{0, 1, 2, 3, 4, 5, 6, 7, 8, 9, 10\}$. Identify the outcomes in the following events.

1. $E =$ This employee packs exactly nine orders on a given day.
2. $A =$ This employee packs at least seven orders on a given day.
3. $B =$ This employee packs at most three orders on a given day.

Solution

1. $E = \{9\}$
2. $A = \{7, 8, 9, 10\}$
3. $B = \{0, 1, 2, 3\}$ ■

10.3   CALCULATING PROBABILITY

As mentioned at the beginning of this chapter, the probability of an event quantifies the likelihood that this event will occur. For an event *A*, the probability that *A* will occur is denoted by *P(A)*. There are some basic properties of probabilities, regardless of how simple or complex the events are.

1. *The probability of an event always lies in the range 0–1*:

$$0 \le P(E) \le 1 \qquad \text{and} \qquad 0 \le P(A) \le 1$$

Whether it is a simple or a compound event, the probability of an event is never less than 0 or greater than 1. If it is impossible for an event *A* to occur, then $P(A) = 0$. On the other hand, if an event *B* is guaranteed to occur, then $P(B) = 1$. If every event had either probability 0 or 1, then there would be no need to study probability. However, most events in the world have some degree of uncertainty; therefore the study of probability is quite important.

2. *The sum of the probabilities of all simple events (or outcomes) in a sample space, denoted by* $\Sigma P(E)$, *is always 1*. This rule, when combined with the first rule, identifies how the probabilities

of the outcomes in a sample space must be assigned. Each probability must be a number in the range 0–1, with all of the values adding up to 1.

3. *The probability of a compound event is equal to the sum of the probabilities of the outcomes that constitute the compound event.* This rule is very important because it implies that the probability of any compound event can be calculated if you list the probabilities of all of its outcomes and sum them. However, as we move along in the study of probability, under many circumstances it will be impractical to list all of the outcomes in a compound event, so we will have to find other methods to calculate probabilities. Nevertheless, this rule allows us to develop some other methods, and its importance cannot be understated.

10.3.1   Equally Likely Events

The simplest situation in probability occurs when all of the outcomes in a sample space have the same probability of occurrence. Such outcomes are called *equally likely outcomes.*

If a sample space S contains n outcomes, all of which are equally likely, then

$$P(E_i) = \frac{1}{n}$$

for every outcome E_i in S.

Whenever an experiment has equally likely events, finding the probability of a compound event can be performed using the third property of probability. If A is a compound event in an experiment with equally likely outcomes, then

$$P(A) = \frac{\text{number of outcomes in } A}{\text{total number of outcomes}}$$

Example 10.3   Thurow & Thurow employs 100 workers, of whom 60 are men and 40 are women. One of the 100 employees is Jenn Bard. Suppose that one of the employees is randomly selected to represent all employees on the management consultation committee.

1. What is the probability that Jenn Bard will be selected?
2. What is the probability that a woman will be selected?

Solution   Because the selection is to be made randomly, each of the 100 employees has the same probability of being selected. Consequently this experiment has a total of 100 equally likely outcomes:

1. P(Jenn Bard is selected) $= \frac{1}{100} = .01$
2. P(a woman is selected) $= \frac{40}{100} = .40$ ■

10.3.2   Probability as Relative Frequency

Companies in the cellphone industry often promote their reliability, which is measured by the probability that a cellphone call will not be dropped. Is this an experiment that can make use of equally likely outcomes? Does it seem reasonable to assume that a person making a cellphone call from a remote valley in the middle of the Rocky Mountains would have the same chance of losing the connection as a person calling from a flat area with a cellphone tower less than one mile away? Probably not. So, how can we develop probabilities for a situation like this?

Communication companies test cellphones and coverage by making calls from various locations across the United States. If one company makes 2000 calls from one area, and 10 of those calls are dropped, then one method for estimating the probability of having a cellphone call dropped in that area is to calculate the fraction of calls that are dropped, which is 10/2000, or .005. If another area was tested by making 800 calls and 14 of these were dropped, then an estimate of the probability of having a call dropped in that area would be 14/800, or .0175. From these estimates, one can conclude that the likelihood of losing a call in the first area is lower than the one in the second area.

Generalizing this idea, if an experiment is repeated n times and an event A is observed f times, then, according to the *relative frequency concept of probability*, we obtain

$$P(A) = \frac{\text{number of times } A \text{ occurs}}{n} = \frac{f}{n}$$

Thus, the relative frequency approach is used to calculate probabilities of events for experiments that

1. Do not have equally likely outcomes
2. Can be repeated again and again

Example 10.4   In a sample of 500 employees of a large company, 280 are in favor of a flexible working time schedule, 170 are against it, and 50 have no opinion. What is the probability that a randomly selected employee from these 500 employees is in favor of a flexible working time schedule?

Solution   Let n denote the total employees in the sample and f the number of employees in favor of the flexible working time schedule. Then

$$n = 500 \quad \text{and} \quad f = 280$$

Using the relative frequency approach to probability, we obtain

$$P(\text{randomly selected employee in favor}) = \frac{f}{n} = \frac{280}{500} = .56$$

Note that this probability is actually the relative frequency of this outcome. Table 10.1 lists the frequencies and relative frequencies of the three outcomes for this example.

TABLE 10.1   Frequency and Relative Frequency Distributions

Outcome	f	Relative Frequency
In favor	280	$\frac{280}{500} = .56$
Against	170	$\frac{170}{500} = .34$
No opinion	50	$\frac{50}{500} = .10$
	$n = 500$	Sum $= 1.00$

The column of relative frequencies in Table 10.1 is used as column of approximate probabilities. Thus, using the column of relative frequencies, we obtain

$P(\text{in favor}) = .56$

$P(\text{against}) = .34$

$P(\text{no opinion}) = .10$ ■

Note that relative frequencies are not probabilities but approximate probabilities. However, if the experiment is repeated again and again, this approximate probability of an outcome obtained from the relative frequency will approach the actual probability of that outcome. This is called the *Law of Large Numbers.*

10.3.3   Subjective Probability

There are many situations in which it is very hard, very expensive, or impossible to perform an experiment more than once. As an example, consider what would happen if crash-test dummies did not exist. If a car company wanted to test how safe their cars are, with respect to different types of injuries, they would have to ask people to be willing to be involved in a crash so that they could measure the extent of their injuries. It is safe to say that people would not be willing to participate in such a program, so measuring the relative frequency of broken legs, concussions, and other injuries would be infeasible.

This type of scenario relies on *subjective probability* to make an assessment of the probability of an event. Subjective probability is based on the knowledge and experience of an individual or a collection of individuals. Although anyone can provide probabilities in a *subjective* manner, common practice involves obtaining probabilities from someone who has extensive knowledge about the subject.

In some cases, the relative frequency method and subjective probability can be combined. For example, suppose that a grocery store has to decide how many turkeys they should order for the Thanksgiving season. They can use data from previous years to determine an average for those years. However, suppose that the store's customer base has expanded following two new housing developments. The store has to use subjective probability to determine how to adjust their order to account for a potential increase in demand.

10.4   COMBINATORIAL PROBABILITY

When we discussed equally likely outcomes, we noted that the probability of an event A can be determined by dividing the number of outcomes in A by the total number of outcomes in the sample space. If the number of outcomes in the sample space is small, we will usually list the outcomes, and count how many of them satisfy event A. However, when a sample space is large, listing all the outcomes will take too much time, so we have to use other methods to count the outcomes.

The traditional counting methods in probability are based on the *multiplication principle*, which works as follows. Suppose that a project has a series of k tasks, where n_1 is the number of ways that task 1 can be completed, n_2 is the number of ways that task 2 can be completed, regardless of how task 1 was completed, and so on through n_k denoting the number of ways that task k can be completed, regardless of how tasks 1 through $k - 1$ were completed.

Then, the number of ways that the series of tasks can be completed is $n_1 \times n_2 \times \cdots \times n_k$.

Example 10.5   It is a casual day at work. For your day, you have to choose a shirt, a pair of pants, a pair of shoes, and a pair of socks. You have 10 shirts, six pairs of pants, three pairs of shoes, and eight pairs of socks. Without worrying about whether the various outfits match, how many possible outfits are there?

Solution   There are $k = 4$ tasks in this problem, corresponding to the four components of your outfit. Specifically, $n_1 = 10$ shirts, $n_2 = 6$ pairs of pants, $n_3 = 3$ pairs of shoes, and $n_4 = 8$ pairs of socks. Thus

$$\text{Total outfits} = 10 \times 6 \times 3 \times 8 = 1440 \qquad ■$$

Example 10.6   Refer to Example 10.5. As fate would have it, you are running late for work, so you have asked your 6-year old son to pull out your clothes while you are finishing up in the bathroom. Three of your 10 shirts are red, and one of your six pairs of pants is green. You will be happy with your outfit as long as it does not involve a red shirt *and* a green pair of pants. What is the probability that you will be unhappy with your outfit?

Solution   Let the event A represent being unhappy with your outfit. Therefore

$$P(A) = \frac{\text{number of outfits containing a red shirt and green pants}}{\text{number of possible outfits}}$$

The number of tasks has not changed, but now $n_1 = 3$, $n_2 = 1$, $n_3 = 3$, $n_4 = 8$. We have to include the number of sock and shoe selections because all of them are part of a bad outfit, as long as it includes a red shirt and green pants. Therefore

$$P(A) = \frac{3 \times 1 \times 3 \times 8}{1440} = \frac{72}{1440} = \frac{1}{20} = .05$$

Our answer can be interpreted in a couple of ways. First, in a list of all 1440 outfits, 72 would include a red shirt and green pants. Next, if a person were to enter many closets, all having the same distribution of clothing as the aforementioned closet, and select an outfit at random from each closet, one out of every 20 outfits would include a red shirt and green pants.   ∎

10.5   MARGINAL AND CONDITIONAL PROBABILITIES

During the previous 3 months, Never-A-Lemon Motors sold 151 cars. Each buyer had the option to purchase an extended warranty. Table 10.2 provides a two-way classification of the buyers by their warranty decision and the type of car (new or used) that each person bought.

TABLE 10.2   Contingency Table

	Purchased Warranty	Did Not Purchase Warranty	Total
Bought a used car	26	17	43
Bought a new car	73	35	108
Total	99	52	151

This type of table is called a *contingency table*. It shows the distribution of 151 buyers with two variables or characteristics. The four boxes containing the numbers 26, 17, 73, and 35 (but not row or column totals) are called *cells*. Each cell identifies the number of items or buyers that have a pair of characteristics in common. For example, 17 buyers bought used cars and did not buy an extended warranty.

The row and column containing totals are referred to as the *margins* of the contingency table. If we choose a buyer at random and ask for the probability that the person bought a specific type of car, or her/his warranty purchase decision, we will need only the values given in the margins. For this reason, such probabilities are called *marginal probabilities*. Marginal probabilities are calculated by taking the appropriate row or column total and dividing it by the grand total for the contingency table.

Example 10.7   Refer to the information given in Table 10.2. What is the probability that a buyer selected at random from these 151 buyers purchased a new car? What is the probability that a buyer selected at random did not purchase an extended warranty?

Solution   From the information given in the table, we observe that 108 buyers out of 151 bought new cars. Also, 52 of the 151 buyers did not purchase an extended warranty. Hence

$$P(\text{selected buyer purchased a new car}) = \tfrac{108}{151} = .7152$$

$$P(\text{selected buyer did not purchase an extended warranty}) = \tfrac{52}{151} = .3444$$ ∎

Suppose that one of the buyers is selected at random. Furthermore, suppose that we know that this person bought a used car. What is the probability that she/he purchased an extended warranty? In this case, we have information that allows us to limit our analysis to the people who purchased used cars because we know that the selected person bought a used car. Specifically, we write this probability in the following form:

$$P(\text{purchased extended warranty}|\text{bought a used car})$$

The proper way to read this statement is "the probability of selecting a person who purchased an extended warranty, given that the person bought a used car." This is actually the conditional probability that the person purchased an extended warranty when the event that this person bought a used car has already happened, which is the condition. The vertical bar in the above probability is read as "given that."

Whenever information is provided in a contingency table, like the one given above, calculating a conditional probability is no more complicated than calculating a marginal probability. The only additional component in the process is recognizing which column(s) or row(s) you will use. In the case above, we have been told that the selected person bought a used car. This event has already occurred. Therefore, we need only the first row (which is the row for used car) in Table 10.2. There are 43 used car buyers, and 26 of them purchased extended warranties. Hence

$$P(\text{purchased extended warranty}|\text{bought a used car}) = \tfrac{26}{43} = .6047$$

It is also important to recognize that the statement that comes after the vertical line (which is read as "given that") represents the given information. In many cases, changing the order of the two events around the vertical line will change the result. Using the "purchased extended warranty" column of Table 10.2, we find

$$P(\text{bought a used car}|\text{purchased extended warranty}) = \tfrac{26}{99} = .2626$$

We will explore conditional probability further in Section 10.7.

10.6 UNION OF EVENTS

Whenever someone applies for a loan, the potential lender will obtain information about the applicant. Suppose that a specific company will offer a lower interest rate if the applicant has a credit rating of at least 700 or if the applicant has at least 50% equity in his/her house. Thus, to obtain a lower interest rate, at least one of the following two (compound) events must occur:

A: the applicant has a credit rating of at least 700
B: the applicant has at least 50% equity in his/her house

In this situation, the lender is interested in at least one of these two events occurring. This is called *union* of the two events and is written as $(A \text{ or } B)$ [or $(A \cup B)$]. This union consists of all applicants who have a credit ratings of at least 700, or those who have at least 50% equity in their houses, or those who satisfy both criteria. This section will examine how to calculate probabilities in such situations.

10.6.1 Addition Rule

Using the aforementioned events, suppose that 22% of all loan applicants have credit ratings of at least 700, and 14% of all loan applicants have at least 50% equity in their homes. If we select a loan applicant at random, we have $P(A) = .22$ and $P(B) = .14$. In addition, there are some loan applicants who satisfy both events; that is, they have credit scores of at least 700 and have at least 50% equity in their homes. Suppose that 8.7% of all loan applicants satisfy both of these criteria. This is written as $P(A \text{ and } B) = .087$. Next we will calculate $P(A \text{ or } B)$.

If one were to add $P(A)$ and $P(B)$, the result would be $.22 + .14 = .36$. If correct, this would imply that 36% of all loan applicants satisfy at least one of the two criteria, A and B. However, each of the two percentages, 22% and 14%, includes the 8.7% of all loan applicants who satisfy both criteria. To understand what 36% actually represents, consider the breakdown shown in Table 10.3.

Assume that there are 1000 loan applicants in total. Then, using the percentages given above, we can write Table 10.3.

TABLE 10.3   Categorization of Individuals by Credit Rating and Equity Level

	Have at Least 50% Equity	Have Less than 50% Equity	Total
Credit rating of at least 700	87	133	220
Credit rating of Less than 700	53	727	780
Total	140	860	1000

Thus, the number of applicants with a credit rating of at least 700 is $87 + 133 = 220$. The number of applicants who have at least 50% equity is $87 + 53 = 140$. Now if we add the totals for applicants with these two characteristics, we obtain $220 + 140 = 360$. But in this total 87 is counted twice. The actual number of applicants who possess these two characteristics is $87 + 133 + 53 = 273$. This will give us the probability of the union of events A and B as $273/1000 = .273$. Thus, to calculate $P(A \text{ or } B)$, we use the following *addition rule*:

$$P(A \text{ or } B) = P(A) + P(B) - P(A \text{ and } B)$$

Thus, the union rule will give us the probability of the union of A and B for Table 10.3 as follows:

$$P(A \text{ or } B) = P(A) + P(B) - P(A \text{ and } B) = \tfrac{220}{1000} + \tfrac{140}{1000} - \tfrac{87}{1000} = .22 + .14 - .087 = .273$$

$$\text{or } \frac{140 + 220 - 87}{1000} = .273 ?$$

Example 10.8   A mail-order sporting-goods company that specializes in camping equipment notices that the probability that an order includes hiking shoes is .092, that it includes a cooler is .136, and that it includes both hiking shoes and a cooler is .017. What is the probability that a randomly selected order includes hiking shoes or a cooler?

Solution   Let A represent the event that the selected order includes hiking shoes, and let B represent the event that this order includes a cooler. From the given information, we obtain

$$P(A) = .092, \qquad P(B) = .136, \qquad \text{and} \qquad P(A \text{ and } B) = .017$$

Hence, using the addition rule, we have

$$P(\text{hiking shoes or a cooler}) = P(A \text{ or } B) = P(A) + P(B) - P(A \text{ and } B) = .092 + .136 - .017 = .211$$

■

10.6.2   Mutually Exclusive Events

The addition rule discussed above is a general rule that can be used to calculate $P(A \text{ or } B)$. However, there are some situations in which the addition rule can be simplified. For example, let us reconsider our example on loan applicants, but let us add a new event C, which is the event that a selected applicant has a credit rating of less than 600. Assume that $P(C) = .315$. Suppose that we wish to calculate $P(A \text{ or } C)$. Using the addition rule, we obtain

$$P(A \text{ or } C) = P(A) + P(C) - P(A \text{ and } C)$$

Working with the information that we have so far, we can say that

$$P(A \text{ or } C) = .22 + .315 - P(A \text{ and } C)$$

What is the value of $P(A \text{ and } C)$? Consider what this event represents—the applicant has a credit rating of at least 700 *and* the applicant has a credit rating of less than 600. Unless the person is living a double life, this is impossible, so $P(A \text{ and } C) = 0$. If two events have no outcomes in common (cannot occur simultaneously), then we say that the events are *mutually exclusive*.

Thus, two events are termed *mutually exclusive* if they have no outcomes in common. If two events, A and B are mutually exclusive, then $P(A \text{ and } B) = 0$. Since A and C are mutually exclusive, $P(A \text{ and } C) = 0$. Hence

$$P(A \text{ or } C) = .22 + .315 = .535$$

Thus, to calculate $P(A \text{ or } B)$ when A and B are mutually exclusive events, the *addition rule* simplifies to

$$P(A \text{ or } B) = P(A) + P(B)$$

Furthermore, we can extend this rule to any number of events, as long as every pair of events is mutually exclusive. If we have n pairwise mutually exclusive events, $A_1, A_2, \ldots, A_n$, then

$$P(A_1 \text{ or } A_2 \text{ or} \cdots \text{or } A_n) = P(A_1) + P(A_2) + \cdots + P(A_n)$$

10.6.3 Complementary Events

A special case of mutually exclusive events occurs when the two (mutually exclusive) events contain all the outcomes in the sample space. These events are called *complementary events,* and each event is referred to as the *complement* of the other event. Returning to our example of loan applicants, we defined A to be the event that the applicant has a credit score of at least 700. The complement of this event would be that the applicant's score is less than 700. The notation for this event is $\bar{A}$, which is read "A-bar" or "A-complement."

Complementary events have a very special and useful probabilistic relationship, but it is not complicated. Using our example, consider the event $(A \text{ or } \bar{A})$. This is the event that an applicant has a credit score of at least 700 *or* an applicant has a credit score of less than 700. Since there are no other possible credit scores, this is a *certain* or *definite* event. Thus its probability is 1.0. Therefore

$$P(A \text{ or } \bar{A}) = 1.0$$

By definition, complementary events are mutually exclusive, so that

$$P(A \text{ or } \bar{A}) = P(A) + P(\bar{A})$$

Putting these two facts together, we get

$$P(A) + P(\bar{A}) = 1.0 \quad \text{which gives} \quad P(A) = 1 - P(\bar{A})$$

Simply stated, this implies that the probability that an event occurs is one minus the probability that it does not occur. Similarly, the probability that an event does not occur is one minus the probability that it does occur. In our example, the probability that the applicant's credit score is less than 700 is one minus .22, or .78. In mathematical notation, this is

$$P(\bar{A}) = 1 - P(A) = 1 - .22 = .78$$

This relationship is so useful because there are many situations in which finding the probability of a specific event is difficult, but finding the probability of its complement is quite easy.

In Table 10.2, the events that a customer buys a used car *and* buys a new car are two complementary events, assuming that each buyer purchased only one car.

Example 10.9   Refer to Example 10.8. A mail-order sporting-goods company that specializes in camping equipment notices that the probability that an order includes hiking shoes is .092, that it includes a cooler is .136, and that it includes both hiking shoes and a cooler is .017. What is the probability that a randomly selected order does not include hiking shoes? What is the probability that it does not include a cooler?

Solution   Let A represent the event that the order includes hiking shoes, and let B represent the event that the order includes a cooler. Therefore, $\bar{A}$ represents the event that the order does not include hiking shoes and $\bar{B}$ represents the event that the order does not include a cooler. We know that $P(A) = .092$ and $P(B) = .136$. Therefore

$$P(\bar{A}) = 1-P(A) = 1-.092 = .908$$

$$P(\bar{B}) = 1-P(B) = 1-.136 = .864$$ ∎

10.7   INTERSECTION OF EVENTS

When we discussed the union of two events, our interest was in determining the probability that at least one of the two events occurs. In some situations we are interested in the probability that both events occur at the same time. In the latter situation, we are interested in the *intersection* of the two events, which is written as (*A* and *B*) [or (*A* ∩ *B*)], and consists of all the outcomes that are common to both events, *A* and *B*. As with the median, we can use an analogy from driving to help explain the intersection of two events. Using the following events

A: the car is on Main Street

B: the car is on Elm Street

the event (*A* and *B*) is the event that the car is on Main Street and on Elm Street at the same time, which occurs when the car is in the intersection of Main and Elm.

The probability of the intersection of two events, *A* and *B*, is written as *P(A* and *B*) and this is also called the *joint probability* of events *A* and *B*.

10.7.1   Independent Versus Dependent Events

In Section 10.5, we explored marginal and conditional probabilities using the distribution of the type of car purchased and the decision regarding the purchase of an extended warranty at Never-A-Lemon Motors. From the values in Table 10.2, we determined the following probabilities:

$P(\text{buyer purchased an extended warranty}) = \frac{99}{151} = .6556$

$P(\text{buyer purchased extended warranty}|\text{buyer bought a used car}) = \frac{26}{43} = .6047$

In this situation, knowing the type of car purchased provides additional information regarding the person's decision to purchase an extended warranty. As is obvious from these two probabilities, knowing that a used car was purchased lowers the probability that an extended warranty was purchased.

Table 10.2 provides the distribution of car type and warranty decision for the Never-A-Lemon Motors Company. Now consider a second company, Kick'n Tires, which also sells cars. Table 10.4 gives the distribution of car type and warranty decision for 180 cars sold at this company.

Calculating the probabilities of the same events as above, we find

$P(\text{buyer purchased an extended warranty}) = \frac{120}{180} = .6667$

$P(\text{buyer purchased extended warranty}|\text{bought a used car}) = \frac{100}{150} = .6667$

TABLE 10.4   Contingency Table for Kick'n Tires Company

	Purchased Warranty	Did Not Purchase	Total
Bought a used car	100	50	150
Bought a new car	20	10	30
Total	120	60	180

In this case, knowing the type of car purchased has no impact on the probability that an extended warranty was purchased.

These two examples demonstrate dependent and independent events, respectively. The events in Table 10.2 are dependent, and those in Table 10.4 are independent.

Two events, A and B, are *probabilistically independent* if the fact that one of the events has occurred does not affect the probability that the other event would occur. In mathematical notation, this means that

$$P(A|B) = P(A) \quad \text{and} \quad P(B|A) = P(B)$$

These two statements are logically equivalent, which means that if one of the equalities holds, the other also holds; and if one of the equalities fails, the other also fails. When data are provided in a contingency table, checking for probabilistic independence of two events involves comparing a conditional probability with its corresponding marginal probability, as we just did. To determine whether the two events are independent, one must calculate and compare the conditional probability and marginal probability for each pair of events. All of the comparisons for Tables 10.2 and 10.4 are shown in Table 10.5.

TABLE 10.5   Comparisons of Conditional and Marginal Probabilities for Tables 10.2 and 10.4

Events		Never-A-Lemon		Kick'n Tires	
A	B	P(A|B)	P(A)	P(A|B)	P(A)
Used car	Warranty	.2626	.2848	.8333	.8333
Used car	No warranty	.3269	.2848	.8333	.8333
New car	Warranty	.7374	.7152	.1667	.1667
New car	No warranty	.6731	.7152	.1667	.1667

At Never-A-Lemon, "car type" and "extended warranty purchase" are dependent events, as $P(A|B) \neq P(A)$ in at least one case. But $P(A|B) = P(A)$ for all cases at Kick'n Tires, so the events are independent in the latter case.

10.7.2   Multiplication Rule

In some cases, calculating the probability of the intersection of two events requires taking the product of the marginal probability of one event and the conditional probability of the other event, given that the previously mentioned event has occurred. The *multiplication rule* that is used to find the joint probability of two events, states that

$$P(A \text{ and } B) = P(A) P(B|A) = P(B) P(A|B)$$

Note that the rule is given with each event taking the position of the "given" event. This implies that you are able to choose the given event according to the information that you have. You do not have to worry about choosing the "right" event to be event A.

When using the multiplication rule, a *tree diagram* can often help visualize what it is you are trying to calculate. In a typical tree diagram for two events, the first set of branches coming from the "trunk"

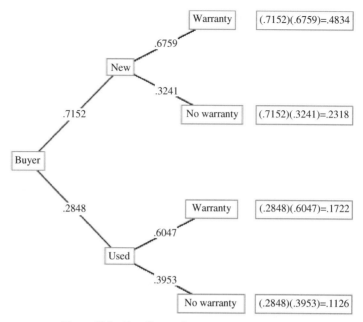

Figure 10.1   Tree diagram for Never-A-Lemon Motors.

of the tree will represent the possible events associated with the "given" variable. A set of branches will extend from each of the branches in the first set. The second set will contain the possible events for the conditional variable. Figure 10.1 is the tree diagram for the Never-A-Lemon Motors scenario described in Table 10.2.

Note that in this tree diagram, the first probability is the joint probability that a buyer buys new car and extended warranty:

$$P(\text{new and warranty}) = P(\text{new}) \cdot P(\text{warranty}|\text{new}) = (.7152)(.6759) = .4834$$

Similarly, we can interpret the remaining three joint probabilities given in this tree diagram.

An additional point regarding the multiplication rule is that we now have a formula for conditional probability. Specifically

$$P(A|B) = \frac{P(A \text{ and } B)}{P(B)} \qquad \text{given that} \qquad P(B) > 0$$

The restriction $P(B) > 0$ serves two purposes: (1) it prevents us from dividing by 0, and (2), it prevents us from specifying that the "given" event is actually impossible. As with the multiplication rule, choosing either A or B as the given event should depend on the information you have for solving the problem.

Now that we have a formula for calculating a conditional probability, we can also determine whether two events are independent. Using our loan application information given earlier in Section 10.6.1, we can determine whether having a credit score of at least 700 (event A) and having at least 50% equity in one's home (event B) are independent or dependent. Specifically

$$P(A|B) = \frac{P(A \text{ and } B)}{P(B)} = \frac{.087}{.14} = .6214$$

$$P(A) = .22$$

Since these two probabilities are not equal, the two events are dependent.

The addition rule had a very nice simplification when the two events were mutually exclusive. Similarly, the multiplication rule simplifies nicely when two events are independent. If A and B are independent, then $P(A|B) = P(A)$. Therefore

$$P(A \text{ and } B) = P(A)P(B)$$

This is the multiplication rule for independent events.

Example 10.10   A food supplier places bids for contracts. From past experience, the food supplier knows that the probability of winning a specific contract is .25, and this probability is the same for all contracts. Furthermore, the supplier believes that being awarded any contract is independent of winning any other contract. In other words, all contracts are independent. The supplier places bids on two contracts. Determine the probability that

1. The supplier wins both contracts.
2. The supplier loses both contracts.
3. The supplier wins at least one of the two contracts.

Solution   Let A_1 and A_2 represent the events that the supplier wins contracts 1 and 2, respectively. From the problem description, we have $P(A_1) = P(A_2) = .25$.

1. $P(A_1 \text{ and } A_2) = P(A_1)P(A_2) = (.25)(.25) = .0625$
2. When two events are independent, so are their complements. Thus

$$P(\bar{A}_1 \text{ and } \bar{A}_2) = P(\bar{A}_1)P(\bar{A}_2) = (1-.25)(1-.25) = .5625$$

Note that $\bar{A}_1$ is the complement of A_1, and $\bar{A}_2$ is the complement of A_2. In other words, $\bar{A}_1$ is the event that supplier loses contract 1, and $\bar{A}_2$ is the event that the supplier loses contract 2.

3. Winning at least one contract means that the supplier won either one contract or both contracts. In other words, the contractor did not lose both of the contracts. This is the complement of the event in part (b). Hence:

$$P(\bar{A}_1 \text{ or } \bar{A}_2) = 1-P(\bar{A}_1 \text{ and } \bar{A}_2) = 1-.5625 = .4375 \qquad \blacksquare$$

The multiplication rule can be extended to any number of events, provided that each event is independent of every combination of the other events. Thus, for three independent events, we obtain

$$P(A_1 \text{ and } A_2 \text{ and } A_3) = P(A_1)P(A_2)P(A_3)$$

Example 10.11   Another food supplier places bids for five contracts, assuming that the probability of winning each contract is .40. Again, the supplier believes that all these contracts are independent. Determine the probability that

1. The supplier wins all five contracts.
2. The supplier loses contracts 1, 2, and 4, and wins contracts 3 and 5.
3. The supplier wins at least one of the contracts.

Solution   Let A_i represent the event that the supplier wins contract i, where $i = 1, 2, 3, 4, 5$. From the problem description, we have $P(A_i) = .40$. and $P(\bar{A}_i) = .60$. Thus,

1. $P(A_1 \text{ and } A_2 \text{ and } A_3 \text{ and } A_4 \text{ and } A_5) = P(A_1)P(A_2)P(A_3)P(A_4)P(A_5)$

$$= (.40)(.40)(.40)(.40)(.40) = .01024$$

2. $P(\bar{A}_1 \text{ and } \bar{A}_2 \text{ and } A_3 \text{ and } \bar{A}_4 \text{ and } A_5) = (.60)(.60)(.40)(.60)(.40) = .03456$

3. $1 - P(\bar{A}_1 \text{ and } \bar{A}_2 \text{ and } \bar{A}_3 \text{ and } \bar{A}_4 \text{ and } \bar{A}_5) = 1 - P(\bar{A}_1)P(\bar{A}_2)P(\bar{A}_3)P(\bar{A}_4)P(\bar{A}_5)$

$$= 1 - (.60)(.60)(.60)(.60)(.60) = .92224 \qquad \blacksquare$$

GLOSSARY

complementary events Two events that taken together include all the outcomes for an experiment but do not contain any common outcome.

compound event An event that contains more than one outcome of an experiment; also called a *composite event.*

equally likely outcomes Two (or more) outcomes or events that have the same probability of occurrence.

event A collection of one or more outcomes of an experiment.

intersection of events Given by the outcomes that are common to two (or more) events.

law of large numbers If an experiment is repeated again and again, the probability of an event obtained from the relative frequency approaches the actual or theoretical probability.

marginal probability The probability of one event or characteristic without consideration of any other event.

mutually exclusive events Two or more events that do not contain any common outcome and, hence, cannot occur together.

outcome The result of the performance of an experiment.

probability A numerical measure of the likelihood that a specific event will occur.

relative frequency as an approximation of probability Probability assigned to an event on the basis of the results of an experiment or historical data.

sample space The collection of all sample points or outcomes of an experiment.

simple event An event that contains one and only one outcome of an experiment; also called an *elementary event.*

subjective probability The probability assigned to an event according to the information and judgment of a person.

EXERCISES

10.1 A box contains three items that are labeled A, B, and C. Two items are selected at random (without replacement) from this box. List all the possible outcomes for this experiment. Write the sample space S.

10.2 A test contains two multiple-choice questions. If a student makes a random guess to answer each question, how many outcomes are possible? List the outcomes. (*Hint*: Consider two outcomes for each question—the answer is either correct or wrong.)

10.3 Refer to Exercise 10.2. List all the outcomes included in each of the following events and mention which are simple and which are compound events:

(a) Both answers are correct.

(b) At most one answer is wrong.

(c) The first answer is correct and the second is wrong.

(d) Exactly one answer is wrong.

10.4 In a group of 50 executives, 27 have type A personalities. If one executive is selected at random from this group, what is the probability that this executive has a type A personality?

10.5 A large company has 840 employees, and 520 of them are union members. If one employee of this company is selected at random, what is the probability that this employee is not a union member? What is the probability that this employee is a union member? Do these two probabilities add up to 1.0? If "yes," why?

10.6 In a sample of 500 families, 95 have a yearly income of less than $40,000, 272 have a yearly income of $40,000–$80,000, and the remaining families have a yearly income of more than $80,000. Write the frequency distribution table for this problem. Calculate the relative frequencies for all classes. Suppose that one family is randomly selected from these 500 families. Find the probability that this family has a yearly income of

(a) Less than $40,000

(b) More than $80,000

10.7 A statistical experiment has 10 equally likely outcomes that are denoted by 1,2,3,4,5,6,7,8,9,10. Let event $A = \{3,4,6,9\}$ and event $B = \{1,2,5\}$.

(a) Are events A and B mutually exclusive events?

(b) Are events A and B independent events?

(c) What are the complements of events A and B and their probabilities?

10.8 The following table gives a two-way classification of all 1000 employees of a large company according to whether they are single or married and whether they own stocks and bonds.

	Own Stocks and Bonds	
	Yes	No
Single	120	280
Married	180	420

(a) If one employee is selected at random from these 1000 employees, find the probability that this employee

 (i) Is married

 (ii) Owns stocks and bonds

 (iii) Is single given he/she owns stocks and bonds

 (iv) Does not own stocks and bonds given he/she is married

(b) Are the events "single" and "yes" mutually exclusive? What about the events "yes" and "no"? Why or why not?

(c) Are the events "married" and "yes" independent? Why or why not?

10.9 Two thousand randomly selected adults were asked whether they considered themselves financially better off than their parents. The following table gives the two-way classification of the responses according to the education levels of the persons included in the survey and whether they are financially better off, the same, or worse off than their parents.

	Education Level		
	Less than High School	High School	More than High School
Better off	140	450	420
Same	60	250	110
Worse off	200	300	70

  (a) If one adult is selected at random from these 2000 adults, what is the probability that this adult is financially

  (i) Better off than his/her parents?

  (ii) Better off than his/her parents given he/she has less than high school education?

  (iii) Worse off than his/her parents given he/she has high school education?

  (iv) The same as his/her parents given he/she has more than high school education?

  (b) Are the events "better off" and "high school education" mutually exclusive? What about the events "less than high school" and "more than high school"? Why or why not?

  (c) Are the events "worse off" and "more than high school" independent? Why or why not?

10.10 In a city, a total of 8250 persons have more than one job and 5925 of these people are male. If one person is selected at random from these 8250 persons, what are the two complementary events and their probabilities?

10.11 Find the joint probability of A and B for the following.

  (a) $P(B) = .59$ and $P(A|B) = .77$

  (b) $P(A) = .28$ and $P(B|A) = .15$

10.12 Given that A and B are two independent events, find their joint probability for the following.

  (a) $P(A) = .20$ and $P(B) = .86$

  (b) $P(A) = .57$ and $P(B) = .32$

10.13 Refer to Exercise 10.8.

  (a) Suppose that one employee is selected at random from these 1000 employees. Find the following probabilities:

  (i) P(married and owns stocks and bonds)

  (ii) P(owns stocks and bonds and single)

  (b) Mention what other joint probabilities you can calculate for this table, and then find all those joint probabilities.

10.14 Refer to Exercise 10.9.

  (a) Suppose that one adult is selected at random from these 2000 adults. Find the following probabilities:

  (i) P(better off and high school)

  (ii) P(more than high school and worse off)

  (b) Find the joint probability of events "worse off" and "better off." Is this probability zero? Explain why or why not.

10.15 Refer to Exercise 10.8. Suppose that one employee is selected at random from these 1000 employees. Find the following probabilities:

  (a) P(single or owns stocks and bonds)

  (b) P(does not own stocks and bonds or single)

10.16 Refer to Exercise 10.9. Suppose that one adult is selected at random from these 2000 adults. Find the following probabilities:

  (a) P(better off or high school)

  (b) P(more than high school or worse off)

  (c) P(better off or worse off)

10.17   Of the customers who shop at a membership store, 25% are small-business owners and 75% of the customers shop for personal use. Of the small-business owners, 85% spend more than $300 per visit and 15% spend $300 or less. Of the customers who shop for personal use, 20% spend more than $300 per visit and 80% spend $300 or less. A randomly selected purchase showed a total sale of more than $300. What is the probability that this purchase was made by a small-business owner?

11 Discrete Random Variables and their Probability Distributions

11.1 OVERVIEW

Chapter 10 discussed the concepts and rules of probability. This chapter extends the concept of probability to explain probability distributions. As was seen in Chapter 10, any given statistical experiment has more than one outcome. It is impossible to predict which outcome will occur if an experiment is performed. Consequently, decisions are made under uncertain conditions. For example, a lottery player does not know in advance whether that play will result in a win. A potential player who knows that there is no chance of winning will definitely not play. This chapter shows that if the various outcomes and their probabilities for a statistical experiment are known, we can find out what will happen on average if that experiment is performed many times. For the lottery example, we can find out what a lottery player can expect to win (or lose) on average by playing this lottery again and again.

First, random variables and types of random variables are explained in this chapter. Then, the concept of a probability distribution and its mean and standard deviation are discussed. Finally, three special probability distributions for a discrete random variable—the binomial probability distribution, the Poisson probability distribution, the geometric distribution, and the hypergeometric probability distribution,—are developed.

11.2 SIX SIGMA PERFORMANCE VARIABLES

The objective of any business is to maximize profit. This is largely achieved by satisfying customers, and to the extent that we satisfy our customers we ensure our future survival and prosperity.

In the transfer function $Y = f(X)$; the expression "Y is a function of X" reflects the fact that a causal relationship exists in any process or action that the process performs. In reality, a result is seldom defined by a single input (X); therefore, we would extend this concept to say that Y is a function of one or many Xs, where Y is the dependent variable and the Xs are the independent variables or factors or inputs as can be seen in Figures 11.1 and 11.2.

The Transfer function can also be illustrated as the general model of a process. For example, when we think of customer satisfaction as a function of the quality delivery time and cost of the product and/or service that we provide, the term "critical to satisfaction" refers to any variable that bears significant influence on any of the three determinants of customer satisfaction.

The products and/or services that suppliers provide fulfill customer needs in terms of defect-free products, cycle time, and cost. The terms "critical to quality" (CTQ), "critical to delivery" (CTD), and "critical to cost" (CTC) refer to any variable(s) that bears significant influence on any of these characteristics. Quality, cost, and cycle time are controlled by process capability and, in turn, this is limited by variation. The expression "Y is a function of X" allows us to explain and ultimately control

Practitioner's Guide for Statistics and Lean Six Sigma for Process Improvements. By Mikel J. Harry, Prem S. Mann,
Ofelia C. de Hodgins, Christopher J. Lacke, and Richard Hulbert
Copyright © 2010 John Wiley & Sons, Inc.

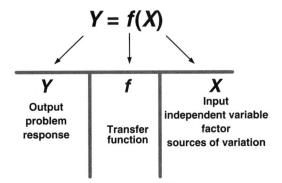

Figure 11.1 Graphical representation of the transfer function $Y = f(X)$.

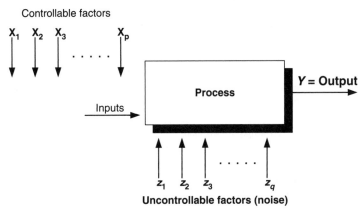

Figure 11.2 Graphical representation of the general model of a process or $Y = f(X)$.

process variation. Some practices have a direct influence over cycle time and cost without necessarily having an impact on our processes (e.g., reductions in costs per hour through a focus on overhead reduction). However, Six Sigma focuses on improving the quality of our processes that will, in turn, improve cycle time and cost.

In the equation $Y = f(X)$, the dependent variables are expressed in our processes as process accuracy, precision, time to fulfill requests, orders, designs, tests, and so on. Independent variables are expressed as causes, inputs to the processes, or any variable that bears influence on the outcome.

We must also recognize that not all independent variables (Xs) equally affect the dependent variable (Y) or the result. Some variables exert more influence over the outcome than others. This is known as the *Pareto principle* which suggests that 80% of the total error or variance can be accounted for by 20% of the variables. The variables that account for 80% of the outcome are called the "vital few," whereas the rest are called the "trivial many." The tool *design of experiments* constitutes a powerful tool to identify and control these vital few variables (also called the leverage variables).

The purpose of *process optimization* in the *breakthrough strategy* is to identify and contain the vital few variables. When this is done, the variability of Y decreases, probability for defects decreases, yield increases, rolled throughput yield increases, mean time between failure(s) or (MTBF) increases, product reliability increases, work-in-process (WIP) inventory decreases, quality improves, cycle time decreases, cost decreases, customer satisfaction improves, and, ultimately, we as a business grow and prosper.

EXERCISE QUESTIONS

1. What is meant by the term "variables?"
2. What is a "dependent variable?"
3. What is an "independent variable?"
4. What other labels are synonymous with "dependent" and "independent" variables?
5. What is meant by the phrase "leverage variable?"
6. What strategies can be used to isolate "leverage variables?"

Answers

1. *Variables* are the elements subject to variation that describe a cause and effect system.
2. A *dependent variable* is the response or output (Y) of the transfer function $Y = f(X)$. The value taken by a dependent variable (Y) depends on the state or condition of other variables within the system.
3. *Independent variables* are all the Xs, inputs, factors of the transfer function $Y = f(X)$. Independent variables are also known as the *input* of a process, or the *cause* of the system or the underlying problem.
4. In a cause–effect (CE) system, there might be one or more independent variables ($X_1, \ldots, X_n$). The value they take does not depend on the state or value of the dependent variable, output, or response (Y). In other words, a dependent variable is also known as a *response* variable and can be seen as the output of a process $Y = f(X)$. The effect is a CE system or the symptom of a problem.
5. *Leverage variables* exert a disproportionately large influence on the dependent variable (Y). According to the Pareto principle, these variables represent 20% of the variables and contribute 80% to the variation in Y.
6. The objective of *process optimization* in the *breakthrough strategy* is to identify and contain these "vital few" variables. When this is achieved, the variability of Y is minimized. In other words, as the probability of a defect decreases, the yield increases, and so on. An important method for finding the best performance of these independent variables Xs is design of experiments.

11.3 SIX SIGMA LEVERAGE VARIABLES

Leverage Principle. According to the Pareto principle, 20% of the variables contribute 80% to the output or variation in Y. We even have a name for these variables. The variables that exert undue influence over the outcome of a process are called "vital few," whereas the ones that have a lesser impact are called "trivial many." This is known as the *leverage principle*.

Variation Reduction Strategies. The trivial many, leverage principle allows us to explain how realistically this applies to processes. Let us illustrate the concept of leverage variables. We all know that the total variance of an output is given by the sum of the individual variances of the independent components. Assume that we have a process with six independent variables (Xs) and one critical-to-quality characteristic or dependent variable (Y). Expressing this in equation form, we would write $Y = f(X_1, X_2, \ldots X_6)$.

Moreover, let us say that we take some measurements and determine the baseline condition of total variance presented on the first line. Since the objective is to reduce variability, we would ask ourselves the following question: What combination of individual variation minimizes total output variability?

To illustrate how each variable influences the result, we could try one of three strategies: **(1)** reduce the variation of all Xs by one unit, **(2)** reduce the variation of all Xs to zero except for X_2, or **(3)** reduce the variation of X_2 by three units.

Example 11.1 Demonstrate how each variable influences the output $= (Y)$ or σ_{total} (see Table 11.1).

TABLE 11.1 Variation Reduction Strategies

	X						Y		
Strategy	σ_1	σ_2	σ_3	σ_4	σ_5	σ_6	Total Reduction Effort	σ_{total}[a]	Changeover Baseline (%)
Baseline	4	11	3	2	1	1	0	12.33	NA
Reduce each by 1	3	10	2	1	0	0	6	10.68	13.4
Reduce all to 0 except σ_2	0	11	0	0	0	0	11	11.00	10.8
Reduce σ_2 by 3	4	8	3	2	1	1	3	9.75	20.9

[a] $\sigma_{\text{total}} = \sqrt{\sigma_1^2 + \sigma_2^2 + \sigma_3^2 + \sigma_4^2 + \sigma_5^2 + \sigma_6^2}$.

Therefore, for baseline, we obtain

$$\sigma_{\text{total}} = \sqrt{4^2 + 11^2 + 3^2 + 2^2 + 1^2 + 1^2} = \sqrt{152} = 12.33$$

Similarly, we can find σ_{total} for other three rows in the table.

As we can see, the third strategy (given by the fourth row) is superior to the other two because with a change of only three units in total reduction it was obtained a 20.9% improvement over the baseline condition. Observe that X_2 exerts an undue influence in the total cause system; therefore, it is *leverage* in nature. ∎

11.4 RANDOM VARIABLES

Random variables are numeric representations of events. Each outcome in a sample space is assigned a numeric value, and events are expressed in terms of a set of values.

Example 11.2 Consider the next four customers who enter a bank. In each case, we identify whether the customer makes a deposit. Table 11.2 contains the possible outcomes, where D represents a customer who makes a deposit and N represents a customer who makes some other kind of transaction. We define our random variable x to be equal to the number of customers who make deposits. When $x = 0$, this means that none of the four customers made a deposit. Similarly, $x = 1$ means that one of the four customers made a deposit and three customers made other transactions, and so on. These values of x and the corresponding outcomes are listed in Table 11.2.

TABLE 11.2 Outcomes and Events for Customers

x	Outcomes
0	NNNN
1	DNNN, NDNN, NNDN, NNND,
2	DDNN, DNDN, DNND, NDDN, NDND, NNDD
3	DDDN, DDND, DNDD, NDDD
4	DDDD

The event "exactly one customer makes a deposit" can be represented by the event $x = 1$, while the event "three or fewer customers make deposits" can be represented by the event $x \leq 3$. Here, x is a random variable that can assume values from zero to 4 as shown in Table 11.2

11.4.1 Discrete Random Variables

A discrete random variable assumes values that can be counted. This implies that the possible values can be ordered and counted in some fashion. There will be a gap between consecutive values, and, while the gap does not have to be the same from one pair of values to the next, it is often the same.

In Table 11.2, the number of customers who make a deposit is an example of a discrete random variable, because the number of costumers can be counted as 0, 1, 2, 3, and 4. A few other examples of discrete random variables are

1. The number of insurance claims filed in one day
2. The number of cars sold by a specific dealership in one month
3. The number of employees who call in sick on a specific day
4. The number of expired packages of hamburger that a grocery store has to discard on a given day

11.4.2 Continuous Random Variables

A continuous random variable assumes values that cannot be counted. Typically, a continuous random variable takes on values over one or more intervals.

One of the most common continuous random variables is time. A few examples of time-based continuous random variables include the following:

1. The time taken to check out a customer at a grocery store
2. The number of years after which a computer monitor has to be replaced
3. The length of a business meeting
4. The time taken to commute from home to work

In each of these examples, we choose to measure time with a specific amount of precision. In the case of commuting time, one is probably interested in the time rounded to minutes, such as 27 min, as opposed to reporting a quantity such as 27.4315 min. Similarly, the checkout times are probably reported to the nearest second. Nevertheless, our choice of precision does not change these continuous variables to discrete variables.

A simple example of why these variables are still continuous involves one's own age. One who is 33 years old continues to state that one is 33 until the one's 34th birthday, although that person has continued to age over that year. One does not remain the same age for 365 days, instantaneously aging one year on one's birthday.

Many other variables are continuous, especially those that involve some measure of weight, volume, distance, or area. Often continuous random variables are used to approximate distributions that involve money, even though it is a discrete random variable. This occurs when there is a large set of possible values, such as the price of a house. Following are a few more examples of continuous random variables:

1. The weight of an outgoing shipment
2. The amount of gas dispensed when the fuel pump states one gallon
3. The distance traveled by a delivery truck in a single day
4. The price of a gallon of gas

This chapter is limited to discussion of discrete random variables. Chapter 12 contains discussion of continuous random variables.

11.5 PROBABILITY DISTRIBUTIONS OF A DISCRETE RANDOM VARIABLE

Let x be a discrete random variable. The *probability distribution* of x describes how the probability is distributed over the possible values of x. Example 11.3 demonstrates the concept of a probability distribution of a discrete random variable by extending Example 11.2.

Example 11.3   Recall from Example 11.2 the outcomes and values of x corresponding to the number of customers out of four who make a deposit. This information is reproduced in Table 11.3. In Example 11.2, we did not make any assumptions regarding how likely a customer is to make a deposit, but we will provide an example here. In addition, we will make a common assumption that customers arriving at a bank make deposits independently of each other.

Suppose that there is a probability of .50 that any given customer makes a deposit. On the basis of this information, Table 11.3 lists the various values of x, the corresponding outcomes, and the probabilities of various values of x. Note that this example contains 16 outcomes and all of them are equally likely (because of .50 probability of deposit), so that the probability of any specific value of x is equal to the number of outcomes corresponding to that value (or event) divided by 16.

TABLE 11.3   Outcomes and Probability Distribution

x	Outcomes	$P(x)$
0	NNNN	$\frac{1}{16} = .0625$
1	DNNN NDNN NNDN NNND	$\frac{4}{16} = .2500$
2	DDNN DNDN DNND NDDN NDND NNDD	$\frac{6}{16} = .3750$
3	DDDN DDND DNDD NDDD	$\frac{4}{16} = .2500$
4	DDDD	$\frac{1}{16} = .0625$

If we consider columns 1 and 3 of Table 11.3, we call it the probability distribution of x. Each probability distribution of a discrete random variable has two properties:

1. The probability assigned to each value of x, $P(x)$, lies in the range 0–1. In other words, $0 \leq P(x) \leq 1$ for each x.
2. The sum of the probabilities assigned to the various values of x must equal 1; that is, $\Sigma P(x) = 1$. (*Note*: Roundoff error can cause the sum to be different from 1, but it should be extremely close to 1.) ∎

Example 11.4   Using Table 11.3 of Example 11.3, calculate the following probabilities:

1. P(exactly one depositor in four customers)
2. P(two or more depositors)
3. P(fewer than four depositors)

Solution

1. P(exactly one depositor) $= P(x = 1) = P(1) = .2500$
2. P(two or more depositors) $= P(x \geq 2) = P(2 \text{ or } 3 \text{ or } 4)$
   In this example, the compound event connected by "or" requires us to determine whether the events are mutually exclusive. If they are not mutually exclusive, then we could have exactly two depositors at the same time that we have exactly three depositors, which is impossible.

This example demonstrates one of the beauties of using random variables. Since it is impossible for a variable to assume two different values simultaneously, different values of random variables always correspond to mutually exclusive events. This means that we can find this probability by simply adding the probabilities associated with the values of x that we are interested in. Thus

$$P(2 \text{ or } 3 \text{ or } 4) = P(x = 2) + P(x = 3) + P(x = 4)$$
$$= .3750 + .2500 + .0625 = .6875$$

3. $P(\text{fewer than 4 depositors}) = P(x < 4) = P(0) + P(1) + P(2) + P(3)$
$$= .0625 + .2500 + .3750 + .2500 = .9375 \qquad ■$$

Example 11.5   Table 11.4 lists the probability distribution of the number of breakdowns per week for a machine based on past data.

TABLE 11.4

Breakdowns per week	0	1	2	3
Probability	.15	.20	.35	.30

Find the probability that the number of breakdowns for this machine during a given week is

1. exactly 2   2. zero to 2
3. more than one   4. at most 1

Solution   Let x denote the number of breakdowns for this machine during a given week. Table 11.5 lists the probability distribution of x.

TABLE 11.5   Probability Distribution of Breakdowns

x	$P(x)$
0	.15
1	.20
2	.35
3	.30
	$\Sigma P(x) = 1.0$

Using Table 11.5, we can calculate the required probabilities as follows.

1. The probability of exactly two breakdowns is
   $P(\text{exactly 2 breakdowns}) = P(x = 2) = .35$
2. $P(0 \text{ to } 2) = P(0 \leq x \leq 2) = P(0) + P(1) + P(2) = .15 + .20 + .35 = .70$
3. $P(\text{more than } 1) = P(x > 1) = P(2) + P(3) = .35 + .30 = .65$
4. $P(\text{at most } 1) = P(x \leq 1) = P(0) + P(1) = .15 + .20 = .35 \qquad ■$

11.6   MEAN OF A DISCRETE RANDOM VARIABLE

The *mean of a discrete random variable*, denoted by μ and, is actually the mean of its probability distribution. The mean of a discrete random variable x is also called its *expected value* and is denoted by $E(x)$. The mean (or expected value) of a discrete random variable is the value that we expect to observe per repetition, on average, if we perform an experiment a large number of times. For example, we may expect a car salesperson to sell, on average, 2.4 cars per week. This does not mean that every week this salesperson will sell exactly 2.4 cars. (Actually she cannot sell exactly 2.4 cars.) This simply means that if we observe for many weeks, this salesperson will sell a different number of cars during different weeks. However, the average for all these weeks will be 2.4 cars.

To calculate the mean of a discrete random variable x, we multiply each value of x by the corresponding probability and sum the resulting products. This sum gives the mean (or expected value) of the discrete random variable x. The mean of a discrete random variable is denoted by μ and calculated as

$$\mu = \Sigma x P(x)$$

The mean of a discrete random variable x is also called its *expected* value and denoted by $E(x)$:

$$E(x) = \Sigma x P(x)$$

Example 11.6 illustrates the calculation of the mean of a discrete random variable.

Example 11.6   Recall Example 11.5. The probability distribution in Table 11.5 from that example is reproduced below. In this table, x represents the number of breakdowns for a machine during a given week and $P(x)$ is the probability of the corresponding value of x:

x	$P(x)$
0	.15
1	.20
2	.35
3	.30

Find the mean number of breakdowns per week for this machine.

Solution   To find the mean number of breakdowns per week for this machine, we multiply each value of x by its probability and add these products. This sum gives the mean of the probability distribution of x. The products $xP(x)$ are listed in the third column of Table 11.6.

The sum of these products gives $\Sigma x P(x)$, which is the mean of x. This mean is

$$\mu = \Sigma x P(x) = 1.80$$

Thus, on average this machine is expected to break down 1.80 times per week over a period of time. In other words, if this machine is used for many weeks, then for certain weeks we will observe no

TABLE 11.6

x	$P(x)$	$xP(x)$
0	.15	$0(.15) = .00$
1	.20	$1(.20) = .20$
2	.35	$2(.35) = .70$
3	.30	$3(.30) = .90$
		$\Sigma x P(x) = 1.80$

breakdowns, for some other weeks we will observe one breakdown per week, and for still other weeks we will observe two or three breakdowns per week. The mean number of breakdowns is expected to be 1.80 per week for the entire period.   ∎

11.7 STANDARD DEVIATION OF A DISCRETE RANDOM VARIABLE

The *standard deviation of a discrete random variable*, denoted by σ, measures the spread of its probability distribution. A higher value for the standard deviation of a discrete random variable indicates that x can assume values over a larger range about the mean. On the other hand, a smaller value for the standard deviation indicates that most of the values that x can assume are clustered closely about the mean. The basic formula to compute the standard deviation of a discrete random variable is

$$\sigma = \sqrt{\Sigma(x - \mu)^2 P(x)}$$

However, it is more convenient to use the following shortcut formula to compute the standard deviation of a discrete random variable:

$$\sigma = \sqrt{\Sigma x^2 P(x) - \mu^2}$$

Note that the variance σ^2 of a discrete random variable is obtained by squaring its standard deviation.

Example 11.7 illustrates how to use the shortcut formula to compute the standard deviation of a discrete random variable.

Example 11.7   Baier's Electronics manufactures computer parts that are supplied to many computer companies. Despite the fact that two quality control inspectors at Baier's Electronics check every part for defects before it is shipped to another company, a few defective parts do pass through these inspections undetected. Let x denote the number of defective computer parts in a shipment of 400 parts. The following table gives the probability distribution of x:

x	0	1	2	3	4	5
$P(x)$	.02	.20	.30	.30	.10	.08

Compute the standard deviation of x.

Solution   Table 11.7 shows all the calculations required for the computation of the standard deviation of x.

TABLE 11.7   Computations to Find the Standard Deviation

x	$P(x)$	$xP(x)$	x^2	$x^2 P(x)$
0	.02	.00	0	.00
1	.20	.20	1	.20
2	.30	.60	4	1.20
3	.30	.90	9	2.70
4	.10	.40	16	1.60
5	.08	.40	25	2.00
		$\Sigma xP(x) = 2.50$		$\Sigma x^2 P(x) = 7.70$

We perform the following steps to compute the standard deviation of x:

Step 1. Compute the mean of the discrete random variable.
The sum of the products $xP(x)$, recorded in the third column of Table 11.7, gives the mean of x:

$$\mu = \Sigma xP(x) = 2.50 \text{ defective computer parts in } 400$$

Step 2. Compute the value of $\Sigma x^2 P(x)$.
First we square each value of x and record it in the fourth column of Table 11.7. Then we multiply these values of x^2 by the corresponding values of $P(x)$. The resulting values of $x^2 P(x)$ are recorded in the fifth column of Table 11.7. The sum of this column gives

$$\Sigma x^2 P(x) = 7.70$$

Step 3. Substitute the values of μ and $\Sigma x^2 P(x)$ in the formula for the standard deviation of x and simplify.

By performing this step, we obtain:

$$\sigma = \sqrt{\Sigma x^2 P(x) - \mu^2} = \sqrt{7.70 - (2.50)^2} = \sqrt{1.45}$$
$$= 1.20 \text{ defective computer parts}$$

Thus, a given shipment of 400 computer parts is expected to contain an average of 2.50 defective parts with a standard deviation of 1.20. ∎

Remember: Because the standard deviation of a discrete random variable is obtained by taking the positive square root, its value is never negative.

Example 11.8   Loraine Corporation is planning to introduce a new makeup product. According to the analysis made by the financial department of the company, it will earn an annual profit of $4.5 million if this product has high sales and, an annual profit of $1.2 million if the sales are mediocre, and it will lose $2.3 million a year if the sales are low. The probabilities of these three scenarios are .32, .51, and .17, respectively.

1. Let x be the profits (in millions of dollars) earned per annum by the company from this product. Write the probability distribution of x.
2. Calculate the mean and standard deviation of x.

Solution

1. The following table lists the probability distribution of x—note that since x denotes profits earned by the company, the loss is written as negative profit in the table:

x	$P(x)$
4.5	.32
1.2	.51
−2.3	.17

2. Table 11.8 shows all the calculations needed for the computation of the mean and standard deviation of x.

TABLE 11.8   **Computations to Find the Mean and Standard Deviation**

x	$P(x)$	$xP(x)$	x^2	$x^2P(x)$
4.5	.32	1.440	20.25	6.4800
1.2	.51	.612	1.44	.7344
−2.3	.17	−.391	5.29	.8993
		$\Sigma xP(x) = 1.661$		$\Sigma x^2P(x) = 8.1137$

The mean of x is

$$\mu = \Sigma xP(x) = \$1.661 \text{ million}$$

The standard deviation of x is

$$\sigma = \sqrt{\Sigma x^2P(x) - \mu^2} = \sqrt{8.1137 - (1.661)^2} = \$2.314 \text{ million}$$

Thus, Loraine Corporation is expected to earn an average of $1.661 million profits a year with a standard deviation of $2.314 million from the new makeup product. ∎

11.8   THE BINOMIAL DISTRIBUTION

The *binomial probability distribution* is one of the most widely used discrete probability distributions. It is applied to find the probability that an outcome will occur x times in n performances of an experiment. For example, given that the probability is .05 that a digital video recorder (DVR) manufactured at a firm is defective, we may be interested in finding the probability that in a random sample of ten DVRs manufactured at this firm, exactly one will be defective. As a second sample, we may be interested in finding the probability that four of nine customers who visit a department store will make a purchase when 25% of all customers who visit this store make a purchase.

To apply the binomial probability distribution, the outcomes in the sample space must be divided into two complementary events. In other words, each repetition of the experiment must result in one of two possible mutually exclusive events. The binomial distribution is applied to experiments that satisfy the four conditions of a *binomial experiment*. (These conditions are described in Section 11.8.2.) Each repetition of a binomial experiment is called a *trial* or a *Bernoulli trial* (after Jacob Bernoulli). For example, if an experiment is defined as one toss of a coin and this experiment is repeated 10 times, then each repetition (toss) is called a *trial*. Consequently, there are 10 total trials for this experiment.

11.8.1   Factorials and Combinations

This section introduces factorials and combinations, which will be used in the binomial formula that will be discussed in Section 11.8.3.

Factorials.   The symbol "!" (read as *factorial*) is used to denote *factorials*. The value of the factorial of a number is obtained by multiplying all integers from that number to 1. For example, "7!" is read as *seven factorial* and it is evaluated by multiplying all integers from 7 to 1.

The symbol $n!$, read as "n factorial," represents the product of all integers from n to 1. In other words

$$n! = n(n-1)(n-2)(n-3)\cdots 3 \cdot 2 \cdot 1$$

By definition

$$0! = 1$$

Example 11.9   Evaluate 10!.

Solution   The value of 10! is given by the product of all integers from 10 to 1. Thus

$$10! = 10 \times 9 \times 8 \times 7 \times 6 \times 5 \times 4 \times 3 \times 2 \times 1 = 3{,}628{,}800$$

∎

Example 11.10   Evaluate $(12-4)!$.

Solution   The value of $(12-4)!$ is

$$(12-4)! = 8! = 8 \times 7 \times 6 \times 5 \times 4 \times 3 \times 2 \times 1 = 40{,}320$$

∎

Example 11.11   Evaluate $(5-5)!$.

Solution   As shown below, the value of $(5-5)!$ is 1.

$$(5-5)! = 0! = 1$$

∎

Note that 0! is always equal to 1.

Combinations.   Quite often we face the problem of selecting a few elements from a large number of distinct elements. As an example, a company may have to select two managers out of four to attend a conference. As another example, a lottery player may have to pick 6 numbers from 49. The question arises: In how many ways can we make the selections in each of these examples? For instance, how many possible selections exist for the company that is to select two managers out of four? Suppose that we name the four managers: A, B, C, and D. Then the possible selections are

(A,B)   (A,C)   (A,D)   (B,C)   (B,D)   (C,D)

The company can choose managers A and B, or A and C, or A and D, and so on. Thus, there are six possible ways to select two managers out of four.

Each of the six possible selections in this list is called a *combination*. All six combinations are distinct; that is, each combination contains a different set of managers. It is important to remember that the order in which the selections are made are not significant in the case of combinations. Thus, whether we write (A,B) or (B,A), both these arrangements represent only one combination.

Combinations give the number of ways x elements can be selected from n elements. The notation used to denote the number of combinations is $_nC_x$, which is read as "the combination of n elements selected x at a time."

Suppose that there are n elements from which we want to select x elements. Then, the number of combinations for selecting x from n distinct elements is given by the formula

$$_nC_x = \frac{n!}{x!(n-x)!}$$

where $n!$, $x!$, and $(n-x)!$ are read as "n factorial," "x factorial," and "n minus x factorial," respectively.

In the combinations formula,

$$
\begin{aligned}
n! &= n(n-1)(n-2)(n-3)\cdots 3 \cdot 2 \cdot 1 \\
x! &= x(x-1)(x-2)(x-3)\cdots 3 \cdot 2 \cdot 1 \\
(n-x)! &= !(n-x)(n-x-1)(n-x-2)(n-x-3)\cdots 3 \cdot 2 \cdot 1
\end{aligned}
$$

Note that in combinations, n is always greater than or equal to x. If n is smaller than x, then we cannot select x distinct elements from n.

Example 11.12   Reconsider the example of a company that is to select two managers from four to attend a conference. Using the combinations formula, find the number of ways this company can select two managers from four.

Solution   For this example,

$n = $ total number of managers $= 4$ and
$x = $ managers to be selected $= 2$.

Therefore, the number of ways this company can select two managers from four is

$$_4C_2 = \frac{4!}{2!(4-2)!} = \frac{4!}{2!2!} = \frac{4 \cdot 3 \cdot 2 \cdot 1}{2 \cdot 1 \cdot 2 \cdot 1} = \frac{24}{4} = 6$$

We listed these six combinations earlier in this section. ∎

Example 11.13   Three members of a jury will be randomly selected from five persons. How many different combinations are possible?

Solution   There are a total of five persons and we are to select three of them. Hence

$$n = 5 \quad \text{and} \quad x = 3$$

Applying the combinations formula, we obtain

$$_5C_3 = \frac{5!}{3!(5-3)!} = \frac{5!}{3!2!} = \frac{5 \cdot 4 \cdot 3 \cdot 2 \cdot 1}{3 \cdot 2 \cdot 1 \cdot 2 \cdot 1} = \frac{120}{6 \cdot 2} = 10$$

If we assume that the five persons are A, B, C, D, and E, then the 10 possible combinations for the selection of three members of the jury are

ABC, ABD, ABE, ACD, ACE, ADE, BCD, BCE, BDE, CDE ∎

11.8.2   The Binomial Experiment

An experiment that satisfies the following four conditions is called a *binomial experiment*:

1. There are n identical trials. In other words, the given experiment is repeated n times. All of the repetitions are performed under identical conditions.
2. Each trial has two and only two mutually exclusive events. These outcomes are usually called a *success* and a *failure*.
3. The probability of success is denoted by p and that of failure by q, and $p + q = 1$. The probabilities p and q remain constant for each trial.
4. The trials are independent. In other words, the outcome of one trial does not affect the outcome of another trial.

Note that one of the two events of a trial is called a *success* and the other a *failure*. Note that a success does not mean that the corresponding outcome is considered favorable or desirable.

Similarly, a failure does not necessarily refer to an unfavorable or undesirable outcome. Success and failure are simply the names used to denote the two possible outcomes of a trial. The outcome to which the question refers is usually called a success; the outcome to which it does not refer is called a failure.

Example 11.14   Consider the experiment consisting of 10 tosses of a coin. Determine whether it is a binomial experiment.

Solution   As described below, the experiment consisting of 10 tosses of a coin satisfies all four conditions of a binomial experiment:

1. There are a total of 10 trials (tosses), and they are all identical. All 10 tosses are performed under identical conditions.
2. Each trial (toss) has only two possible outcomes: a head and a tail. Let a head be called a success and a tail be called a failure.
3. The probability of obtaining a head (a success) is $\frac{1}{2}$ and that of a tail (a failure) is $\frac{1}{2}$ for any toss. That is,

$$p = P(H) = \tfrac{1}{2} \qquad \text{and} \qquad q = P(T) = \tfrac{1}{2}$$

The sum of these two probabilities is 1.0. Also, these probabilities remain the same for each toss.
4. The trials (tosses) are independent. The result of any preceding toss has no bearing on the result of any succeeding toss.

Consequently, the experiment consisting of 10 tosses is a binomial experiment.   ■

Example 11.15   Of all digital video recorders (DVRs) manufactured by a large electronics firm, 5% are defective. Ten DVRs are randomly selected from the production line of this firm. Each of the selected DVRs is inspected to determine whether it is defective or good. Is this experiment a binomial experiment?

Solution

1. This example consists of 10 identical trials. A trial represents the test result of a DVRs.
2. Each trial has two events: A DVRs is either defective or good. Let a defective DVRs be called a success and a good DVRs be called a failure.
3. It is found that 5% of all DVRs are defective. So, the probability p that a DVRs is defective is .05. As a result, the probability q that a DVRs is good is .95. These two probabilities add up to 1.
4. Each trial (test result) is independent. In other words, if one DVRs is defective, this does not affect the outcome of another DVRs being defective or good. This is so because the size of the population is very large as compared to the sample size.

Since all four conditions of a binomial experiment are satisfied, this is an example of a binomial experiment.   ■

11.8.3   The Binomial Probability Distribution And Binomial Formula

The random variable x that represents the number of successes in n trials for a binomial experiment is called a *binomial random variable*. The probability distribution of x in such experiments is called the *binomial probability distribution* or simply *binomial distribution.* Thus, the binomial probability distribution is applied to find the probability of x successes in n trials for a binomial experiment. The number of successes x in such an experiment is a discrete random variable. Consider Example 11.15. Let x be the number of defective DVR players in a sample of 10. Since we can obtain any number of defective DVR players ranging from 0 to 10 in a sample of 10, x can assume any of the values 0–10. Since the values of x are countable, it is a discrete random variable.

Binomial Formula.   For a binomial experiment, the probability of exactly x successes in n trials is given by the binomial formula

$$P(x) = {}_nC_x \, p^x q^{n-x}$$

where $n =$ total number of trials
    $p =$ probability of success
    $q = 1 - p =$ probability of failure
    $x =$ number of successes in n trials
  $n - x =$ number of failures in n trials

   In the binomial formula, n is the total number of trials and x is the number of successes. The difference between the total number of trials and the total number of successes, $n - x$, gives the total number of failures in n trials. The value of ${}_nC_x$ gives the number of ways to obtain x successes in n trials. As mentioned earlier, p and q are the probabilities of success and failure, respectively. Again, although it does not matter which of the two outcomes is called a success and which one a failure, usually the outcome to which the question refers is called a success.

   To solve a binomial problem, we determine the values of n, x, $n - x$, p, and q and then substitute these values in the binomial formula. To find the value of ${}_nC_x$, we can use either the combinations formula from Section 11.8.1 or a calculator.

   To find the probability of x successes in n trials for binomial experiment, the only values needed are those of n and p. These are called the *parameters of the binomial probability distribution* or simply the *binomial parameters.* The value of q is obtained by subtracting the value of p from 1.0. Thus, $q = 1 - p$.

Example 11.16   At the Express House Delivery Service, providing high-quality service to its customers is the top priority of the management. The company guarantees a refund of all charges if a package that it is delivering does not arrive at its destination by the specified time. It is known from past data that despite all efforts, 2% of the packages mailed through this company do not arrive at their destinations within the specified time. A corporation mailed 10 packages through Express House Delivery Service on Monday.

1. Find the probability that exactly 1 of these 10 packages will not arrive at its destination within the specified time.
2. Find the probability that at most 1 of these 10 packages will not arrive at its destination within the specified time.

Solution   Let us call it a success if a package does not arrive at its destination within the specified time and a failure if it does arrive within the specified time. Because the question refers to packages that do not arrive on time, that is why we have called it a success.

$$n = \text{total number of packages mailed} = 10$$
$$p = P(\text{success}) = .02$$
$$q = P(\text{failure}) = 1 - .02 = .98$$

1. For this first part

$$x = \text{number of successes} = 1$$
$$n-x = \text{number of failures} = 10 - 1 = 9$$

Substituting all values in the binomial formula, we obtain

$$P(x = 1) = {}_{10}C_1(.02)^1(.98)^9 = \frac{10!}{1!(10-1)!}(.02)^1(.98)^9$$

$$= (10)(.02)(.83374776) = .1667$$

Thus, there is a .1667 probability that exactly one of the 10 packages mailed will not arrive at its destination within the specified time.

2. The probability that at most 1 of the 10 packages will not arrive at its destination within the specified time is given by the sum of the probabilities of $x = 0$ and $x = 1$. Thus

$$P(x \leq 1) = P(x = 0) + P(x = 1)$$

$$= {}_{10}C_0(.02)^0(.98)^{10} + {}_{10}C_1(.02)^1(.98)^9$$

$$= (1)(1)(.81707281) + (10)(.02)(.83374776)$$

$$= .8171 + .1667 = .9838$$

Thus, the probability that at most 1 of the 10 packages will not arrive at its destination within the specified time is .9838. ■

Example 11.17   Companies that receive large shipments of products use acceptance sampling methods to check whether the products conform to a specified quality level. The quality control department selects a sample of n items from the shipment and observes the number of defective items in that sample. If the number of defective items is not more than a specified number x, then the shipment is accepted; otherwise, not. The sample of n items is called an acceptance sample.

Oaks Electronics Company makes food processors. The company buys motors for these food processors from Old Haven Motors Inc. Oaks Electronics receives motors in shipments of 500 each. The quality control department at Oaks Electronics randomly selects 20 motors from each shipment and inspects them for being good or defective. If this sample contains more than two defective motors, the entire shipment is rejected. Old Haven Motors promises that only 5% of its motors are defective.

1. Find the probability that a given shipment of 500 motors received by Oaks Electronics will be accepted.
2. Find the probability that a given shipment of 500 motors received by Oaks Electronics will not be accepted.

Solution   Let p be the probability that a randomly selected motor from the shipment is defective. From the given information

$$n = \text{number of motors in the sample} = 20$$
$$p = .05 \quad \text{and} \quad q = 1 - .05 = .95$$

1. The shipment will be accepted if the sample of 20 motors contains 0, or 1, or 2 defective motors. The required probability is calculated as follows:

$$P(\text{the shipment is accepted}) = P(x \leq 2) = P(x = 0) + P(x = 1) + P(x = 2)$$
$$= {}_{20}C_0(.05)^0(.95)^{20} + {}_{20}C_1(.05)^1(.95)^{19} + {}_{20}C_2(.05)^2(.95)^{18}$$

$$= .3585 + .3774 + .1887 = .9246$$

2. The shipment will not be accepted if the sample of 20 motors contains three or more defective motors. The two events "shipment is accepted" and "shipment is not accepted" are complementary events. Hence, using the complementary rule learned in Chapter 10, the required probability is

$$P(\text{the shipment is not accepted}) = 1 - P(\text{the shipment is accepted})$$
$$= 1 - .9246 = .0754 \quad\blacksquare$$

Using MINITAB to Calculate Binomial Probabilities. As mentioned earlier, probability calculations will typically be performed using software as opposed to hand calculations shown in the preceding examples. To use MINITAB, select `Calc > Probability Distributions > Binomial`. The dialog box of Figure 11.3 will appear. Note that the values from Example 11.17 part 1 are entered in the dialog box.

To calculate the probability of a specific number of successes, as in Example 11.17 part 1, click `Probability`. To calculate the cumulative probabilities, that is, the probability of having k or fewer successes, select `Cumulative probability`. Enter the value for the `Number of trials`, n, and the `Probability of success`, p, in their respective boxes. Click `Input constant` and enter the value of x for the specific probability you wish to compute. Click `OK`. The output will appear in MINITAB's session window and is shown in Figure 11.4.

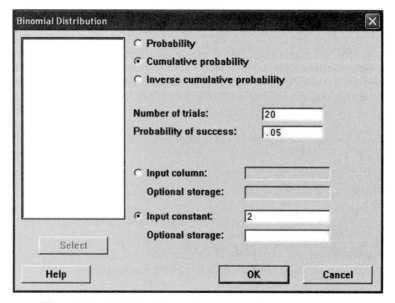

Figure 11.3   MINITAB dialog box for calculating binomial probabilities.

Cumulative Distribution Function

Binomial with n = 20 and p = 0.05

```
x    P( X <= x )
2      0.924516
```

Figure 11.4 MINITAB output for Example 11.17 part 1.

11.8.4 Probability of Success and Shape of the Binomial Distribution

For any number of trials n

1. The binomial probability distribution is symmetric if and only if $p = .50$.
2. The binomial probability distribution is skewed to the right if p is less than .50.
3. The binomial probability distribution is skewed to the left if p is greater than .50.

These three cases are illustrated next with examples.

1. Let $n = 4$ and $p = .50$. Using MINITAB, we have written the probability distribution of x in Table 11.9. As we can observe from Table 11.9, the probability distribution of x is symmetric.

TABLE 11.9   Probability Distribution of x for $n = 4$ and $p = .50$

x	$P(x)$
0	.0625
1	.2500
2	.3750
3	.2500
4	.0625

2. Let $n = 4$ and $p = .30$ (which is less than .50). Table 11.10, which is made using MINITAB, shows that the probability distribution of x for $n = 4$ and $p = .30$ is skewed to the right.

TABLE 11.10   Probability Distribution of x for $n = 4$ and $p = .30$

x	$P(x)$
0	.2401
1	.4116
2	.2646
3	.0756
4	.0081

3. Let $n = 4$ and $p = .80$ (which is greater than .50). Table 11.11, which is made using MINITAB, shows that the probability distribution of x for $n = 4$ and $p = .80$ is skewed to the left.

TABLE 11.11   Probability Distribution of x for $n = 4$ and $p = .80$

x	$P(x)$
0	.0016
1	.0256
2	.1536
3	.4096
4	.4096

11.8.5   Mean and Standard Deviation of the Binomial Distribution

Earlier in this chapter we learned how to compute the mean and the standard deviation for a probability distribution of a discrete random variable. When a discrete random variable has a binomial distribution, the formulas learned in Sections 11.6 and 11.7 could still be used to compute its mean and standard deviation. However, it is more convenient and simpler to use the following formulas to find the mean and standard deviation in such cases:

$$\mu = np \quad \text{and} \quad \sigma = \sqrt{npq}$$

where n is the total number of trials, p is the probability of success, and q is the probability of failure.

Example 11.18 describes the calculation of the mean and standard deviation for a binomial distribution.

Example 11.18   Refer to Example 11.16. It is given that 2% of all packages do not arrive on time. Ten packages are mailed. Find the mean and standard deviation of the probability distribution of x for the number of packages in 10 that do not arrive on time.

Solution

$$n = 10, \quad p = .02, \quad q = .98$$

Using the formulas for the mean and standard deviation of the binomial distribution, we obtain

$$\mu = np = 10(.02) = .20$$
$$\sigma = \sqrt{npq} = \sqrt{10(.02)(.98)} = .44$$

Thus, the mean of the probability distribution of x is .20 and the standard deviation is .44. The value of the mean is what we expect to obtain, on average, per repetition of the experiment. In this example, we expect that on average, .20 packages out of 10 will not arrive on time.   ∎

11.9   THE POISSON PROBABILITY DISTRIBUTION

The *Poisson probability distribution*, named after the French mathematician Simeon D. Poisson, is another important probability distribution of a discrete random variable that has a large number of applications. Suppose that a washing machine in a laundromat breaks down an average of 3 times a month. We may want to find the probability of exactly two breakdowns during the next month. This is an example of a Poisson probability distribution problem. Each breakdown is called an *occurrence* in Poisson probability distribution terminology. The Poisson probability distribution is applied to experiments with random and independent occurrences. The occurrences are random in the sense that they do not follow any pattern and, hence, are unpredictable. Independence of occurrences means that one occurrence (or nonoccurrence) of an event does not influence the successive occurrences or nonoccurrences of that event. The occurrences are often considered with respect to an interval. In the example of the washing machine, the interval represents one month. The interval may be a time interval, a space interval, or a volume interval. The actual number of occurrences within an interval is random and independent. If the average number of occurrences for a given interval is known, then, by using the Poisson probability distribution, we can compute the probability of a certain number of occurrences x in that interval. Note that the number of actual occurrences in an interval is denoted by x.

Conditions for Application of Poisson Probability Distribution. The following three conditions must be satisfied to apply the Poisson probability distribution:

1. x is a discrete random variable.
2. The occurrences are random.
3. The occurrences are independent.

The following are a few examples of discrete random variables for which the occurrences are random and independent; hence, these are examples to which the Poisson probability distribution can be applied:

1. Consider the number of customers arriving during a one-hour interval at the Eagleville branch of the United National Bank. In this example, an occurrence is the arrival of a customer at this branch, the interval is one hour (an interval of time), and the occurrences are random. The total number of customers who may arrive at this bank during a one-hour interval may be 0,1,2,3,4,.... The independence of occurrences in this example means that the customers arrive individually and the arrival of any two (or more) customers is not related.
2. Consider the number of bacteria in a sample of 10 milliliters (mL) of pasteurized milk. In this case, the interval is a volume interval (10 mL). The occurrences (number of bacteria) are random because there may be 0,1,2,3,..., bacteria in the 10 mL. We can assume the occurrences of bacteria to be independent of one another.
3. Consider the number of defects in a 5-ft-long iron rod. The interval, in this example, is a space interval (5 ft). The occurrences (defects) are random because there may be any number of defects in a 5-ft iron rod. We can assume that these defects are independent of one another.

The following examples also qualify for the application of the Poisson probability distribution.

1. The number of accidents that occur at a company during a one-month period.
2. The number of customers coming to a grocery store during a one-hour interval.
3. The number of television sets sold at a department store during a given week.

On the other hand, arrivals of patients at a physician's office will be nonrandom if the patients have to make appointments to see the physician. The arrival of commercial airplanes at an airport is nonrandom because all planes are scheduled to arrive at certain times, and airport authorities know the exact number of arrivals for any period (although this number may change slightly because of late or early arrivals and cancellations).

In the Poisson probability distribution terminology, the average number of occurrences in an interval is denoted by λ (Greek letter lambda). The actual number of occurrences in that interval is denoted by x. Then, using the Poisson probability distribution, we find the probability of x occurrences during an interval given the mean number of occurrences are λ during that interval.

According to the Poisson probability distribution, the probability of x occurrences in an interval is

$$P(x) = \frac{\lambda^x e^{-\lambda}}{x!}$$

where λ is the mean number of occurrences in that interval and the value of e is approximately 2.71828. This is the Poisson probability distribution formula.

Example 11.19   The automatic teller machine (ATM) installed outside Mansfield Savings and Loan is used on average by five customers per hour. One day the bank closed this ATM for one hour for repairs. What is the probability that during that hour eight customers came to use this ATM?

Solution   Let λ be the mean number of customers who use this ATM per hour. Then, $\lambda = 5$. Let x be the number of customers who came to use this ATM during the one-hour period when it was closed for repairs. We are to find the probability of $x = 8$. Substituting all the values in the Poisson probability distribution formula, we obtain

$$P(x = 8) = \frac{\lambda^x e^{-\lambda}}{x!} = \frac{(5)^8 e^{-5}}{8!} = \frac{(390625)(.06738)}{40320} = .0653$$

Thus, the probability is .0653 that eight customers came to use this ATM during the one-hour period when it was closed for repairs.

We can use a calculator to find the values of 8! and e^{-5}.   ■

Example 11.20: Using MINITAB to Calculate Poisson Probabilities   Calculate the probabilities in Example 11.19 using MINITAB.

Solution   Select `Calc -> Probability Distributions -> Poisson`. The dialog box in Figure 11.5 will appear. Note that the values from Example 11.19 are entered in the dialog box.

To calculate the probability of a specific number of occurrences, click `Probability` in the dialog box. To calculate the cumulative probabilities, that is, the probability of having x or fewer occurrences, select `Cumulative probability`. Enter the value for λ in the box next to `Mean`. Click `Input constant` and enter the value of x for the specific probability that you wish to compute. Click `OK`. The output will appear in MINITAB's session window, which is shown in Figure 11.6 that gives the required probability as .0652780 or .0653.

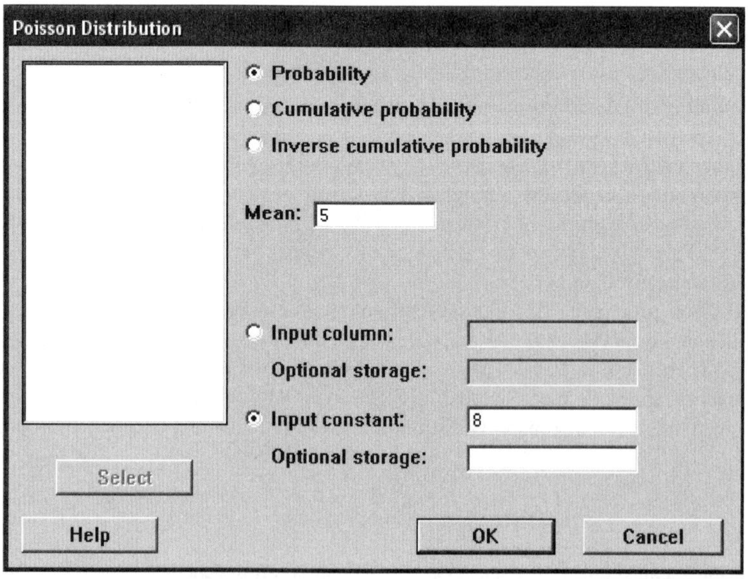

Figure 11.5   MINITAB dialog box for calculating Poisson probabilities.

```
Probability Density Function

Poisson with mean = 5


x    P (X = x)
8    0.0652780
```

Figure 11.6 MINITAB output for Example 11.18. ∎

Example 11.21 Cynthia's Mail Order Company provides free examination of its products for 7 days. If not completely satisfied, a buyer can return the product within that period and get a full refund. According to past records of the company, an average of 2 of every 10 products sold by this company are returned for a refund. Using the Poisson probability distribution formula, find the probability that exactly 5 of the 20 products sold by this company on a given day will be returned for a refund.

Solution Let x denote the number of products in 20 that will be returned for a refund. We are to find $P(x = 5)$. The given mean is defined per 10 products but x is defined for 20 products. As a result, we should first find the mean for 20 products. Because on average 2 out of 10 products are returned, the mean number of products returned out of 20 will be 4. Thus, $\lambda = 4$. Substituting $x = 5$ and $\lambda = 4$ in the Poisson probability distribution formula, we obtain

$$P(x = 5) = \frac{\lambda^x e^{-\lambda}}{x!} = \frac{(4)^5 e^{-4}}{5!} = \frac{(1024)(.018316)}{120} = .1563$$

Thus, the probability is .1563 that exactly 5 products out of 20 sold on a given day will be returned. ∎

Note that for the Poisson probability distribution, λ and x must be defined for the same interval. If they are not, then λ can be changed, as was done in this example, to make the intervals same.

11.9.1 Mean and Standard Deviation of the Poisson Probability Distribution

For the Poisson probability distribution, the mean and variance both are equal to λ and the standard deviation is equal to $\sqrt{\lambda}$; that is, for the Poisson probability distribution

$$\begin{aligned} \mu &= \lambda \\ \sigma^2 &= \lambda \\ \sigma &= \sqrt{\lambda} \end{aligned}$$

For Example 11.21, $\lambda = 4$. Therefore, for the probability distribution of x, the mean, variance, and standard deviation are

$$\begin{aligned} \mu &= \lambda = 4 \\ \sigma^2 &= \lambda = 4 \\ \sigma &= \sqrt{\lambda} = \sqrt{4} = 2 \end{aligned}$$

11.10 THE GEOMETRIC DISTRIBUTION

Many companies produce specialized equipment that has to meet very specific standards, such as the lasers used in vision correction surgery. As a result, each part is checked to make sure that it falls within the required standards. Moreover, if the object falls outside a certain set of specifications, the production line will be shut down and examined to find the source of the error. In such a situation, the producer is not interested in the number of defective items. Instead, we want to know how many items will be produced before the line has to be shut down and fixed. The geometric distribution is used to model the number of trials until the first of a specific type of outcome occurs.

The geometric and binomial distributions are similar in the sense that both involve Bernoulli trials. Outcomes are divided into two categories (success/failure), the probability of a success p is constant for each trial, and trials are independent. However, in the geometric distribution, the number of trials is not fixed. Instead, we keep repeating the process until a success occurs for the first time. This is equivalent to saying that you will toss a coin until a head appears for the first time. It could happen on the first toss, or you may keep tossing forever, although the latter is highly unlikely.

The geometric probability distribution is defined as follows:

$$P(x) = p(1-p)^{x-1} \quad \text{where} \quad x = 1, 2, 3, \ldots$$

Here, p is the probability of success and x is the number of trials needed to get the first success.

Example 11.22   A military aviation contractor produces nuts that must be within 0.04 mm of a specific diameter, otherwise they will not provide appropriate support for the landing gear. If a nut exceeds this limit, the production machine is shut down and adjusted. Past data indicate, the probability that the diameter of a nut exceeds the allowable error is .0014.

1. What is the probability that the machine will be shut down exactly after the 100th nut is produced?
2. What is the probability that the 200th nut produced is the first to shut down the machine?
3. The firm needs to produce 250 nondefective nuts per day. What is the probability that they can be produced without having to shut down the machine?

Solution   Based on the manager's information, $p = .0014$.

1. $P(100) = p(1-p)^{x-1} = .0014(1-.0014)^{100-1} = .0014(.9986)^{99} = .00122$
2. $P(200) = p(1-p)^{x-1} = .0014(1-.0014)^{200-1} = .0014(.9986)^{199} = .00106$
3. In order to produce 250 nondefective nuts without shutting down the machine, the first defective nut must occur after 250 nuts have been produced. In other words, we need $P(X > 250)$. In the case of the geometric distribution, $P(X > x)$ simplifies to a special form, namely

$$P(X > x) = (1-p)^x, x = 1, 2, 3, \ldots$$

therefore

$$P(X > 250) = (.9986)^{250} = .7045 \qquad \blacksquare$$

Probabilities based on the geometric distribution are easy to compute with a handheld calculator or basic functions within a spreadsheet. Some statistical software packages have built-in functions; while others have calculators. MINITAB has the latter, which can be accessed by selecting Calc > Calculator.

11.11   THE HYPERGEOMETRIC PROBABILITY DISTRIBUTION

In Section 11.8.3, we learned that one of the conditions required to apply the binomial probability distribution is that the trials are independent so that the probabilities of the two outcomes (success and failure) remain constant. If the trials are not independent, we cannot apply the binomial probability distribution to find the probability of x successes in n trials. In many such cases we replace the binomial by the *hypergeometric probability distribution*. Such a case occurs when a sample is drawn without replacement from a finite population.

As an example, suppose that 20% of all automobile parts manufactured at a company are defective. Four auto parts are selected at random. What is the probability that three of these four parts are good? Note that we are to find the probability that three of the four auto parts are good and one is defective. In this case, the population is very large and the probability of the first, second, third, and fourth auto parts being defective remains the same at .20. Similarly, the probability of any of the parts being good remains unchanged at .80. Consequently, we will apply the binomial probability distribution to find the probability of three good parts in four.

Now suppose that this company shipped 25 auto parts to a dealer. Later on, the company finds out that five of those parts were defective. By the time the company manager contacts the dealer, four auto parts from that shipment have already been sold. What is the probability that three of those four parts were good parts and one was defective? Here, because the four parts were selected without replacement from a small population (of 25 parts), the probability of a part being good changes from the first selection to the second selection, to the third selection, and to the fourth selection. In this case we cannot apply the binomial probability distribution. In such instances, we use the hypergeometric probability distribution to find the required probability.

Let

$N =$ total number of elements in the population

$r =$ number of successes in the population

$N - r =$ number of failures in the population

$n =$ number of trials (sample size)

$x =$ number of successes in n trials

$n - x =$ number of failures in n trials

The probability of x successes in n trials is given by

$$P(x) = \frac{{}_rC_x \times {}_{N-r}C_{n-x}}{{}_NC_n}$$

Example 11.23   A common use of the hypergeometric distribution involves *acceptance sampling*. This occurs when a buyer selects items at random from a production run that the buyer's company is considering to purchase. If the sample includes more than a specified number of defectives, then the company will not purchase the product. For example, suppose that a local ice cream parlor has an option to purchase 400 waffle cones for their store. They select 50 such cones and inspect them. The parlor will purchase the set of 400 cones if no more than 3 of the 50 cones are broken.

1. What is the probability that the parlor will buy the cones if 35 of the 400 cones are broken?
2. What is the probability if 50 of the 400 cones are broken?
3. What is the probability that they will buy the cones if the set of 400 contains 10 broken cones and they will only buy the cones if none of the 50 inspected cones are broken?

Solution

1. This is a hypergeometric problem with $N = 400$, $n = 50$, $r = 35$, $N - r = 365$, and $x \leq 3$. Therefore

$$P(x \leq 3) = P(x = 0) + P(x = 1) + P(x = 2) + P(x = 3)$$

$$= \frac{{}_{35}C_0 \times {}_{365}C_{50}}{{}_{400}C_{50}} + \frac{{}_{35}C_1 \times {}_{365}C_{49}}{{}_{400}C_{50}} + \frac{{}_{35}C_2 \times {}_{365}C_{48}}{{}_{400}C_{50}} + \frac{{}_{35}C_3 \times {}_{365}C_{47}}{{}_{400}C_{50}} = .3369$$

2. This is a hypergeometric problem with $N = 400$, $n = 50$, $r = 50$, $N-r = 350$, and $x \leq 3$. Therefore

$$P(x \leq 3) = P(x = 0) + P(x = 1) + P(x = 2) + P(x = 3)$$

$$= \frac{{}_{50}C_0 \times {}_{350}C_{50}}{{}_{400}C_{50}} + \frac{{}_{50}C_1 \times {}_{350}C_{49}}{{}_{400}C_{50}} + \frac{{}_{50}C_2 \times {}_{350}C_{48}}{{}_{400}C_{50}} + \frac{{}_{50}C_3 \times {}_{350}C_{47}}{{}_{400}C_{50}} = .0982$$

3. This is a hypergeometric problem with $N = 400$, $n = 50$, $r = 10$, $N-r = 390$, and $x = 0$. Therefore

$$P(x = 0) = \frac{{}_{10}C_0 \times {}_{390}C_{50}}{{}_{400}C_{50}} = .2588$$

As with the binomial and Poisson distributions, MINITAB has a built-in function for calculating probabilities on the basis of the hypergeometric probability distribution. The function is accessed by selecting `Calc → Probability Distributions → Hypergeometric`, and is operated in a fashion similar to the binomial and Poisson functions. ■

GLOSSARY

Bernoulli trial   One repetition of a binomial experiment; also called a *trial*.

binomial probability distribution   The probability distribution that gives the probability of x successes in n trials when the probability of success is p for each trial of a binomial experiment.

continuous random variable   A random variable that can assume any value in one or more intervals.

discrete random variable   A random variable whose values are countable.

hypergeometric probability distribution   The probability distribution that is applied to determine the probability of x successes in n trials when the trials are not independent.

mean of a discrete random variable   The mean of a discrete random variable x is the value that is expected to occur per repetition, on average, if an experiment is performed a large number of times. The mean of a discrete random variable is also called its *expected value*.

Poisson probability distribution   The probability distribution that gives the probability of x occurrences in an interval when the average occurrences in that interval are λ.

probability distribution of a discrete random variable   A list of all the possible values that a discrete random variable can assume and their corresponding probabilities.

random variable   A variable, denoted by x, whose value is determined by the outcome of a random experiment; also called a *chance variable*.

standard deviation of a discrete random variable   A measure of spread for the probability distribution of a discrete random variable.

EXERCISES

11.1   Indicate which of the following random variables are discrete and which are continuous:

(a) The number of new accounts opened at a bank during a certain month

(b) The time taken by a lawyer to write a real estate contract

(c) The price of a concert ticket

(d) The number of rotten eggs in a randomly selected box

(e) The number of workers employed at a randomly selected company

(f) The weight of a randomly selected package

11.2 One of the four gas stations located at an intersection of two major roads is an Exxon station. Suppose that the next six cars that stop at any of these four gas stations make the selections randomly and independently. Let x be the number of cars in these six that stop at the Exxon station. Is x a discrete or a continuous random variable? Explain.

11.3 Elmo's Sporting Goods sells exercise machines as well as other sporting goods. On different days, it sells different numbers of these machines. The following table, constructed using past data, lists the probability distribution of x where x is the number of exercise machines sold per day at Elmo's:

x	4	5	6	7	8	9	10
$P(x)$	.08	.11	.14	.19	.23	.16	.09

Determine the probability that the number of exercise machines sold at Elmo's on a given day is

(i) exactly 6   (ii) more than 8   (iii) 5−8   (iv) at most 6

11.4 Despite all safety measures, accidents do happen at Brown's Manufacturing Corporation. Let x denote the number of accidents that occur during a month at this company. The following table lists the probability distribution of x:

x	0	1	2	3	4
$P(x)$	.25	.30	.20	.15	.10

Determine the probability that the number of accidents that will occur during a given month at this company is

(i) exactly 4   (ii) at least 2   (iii) less than 3   (iv) 2 to 4

11.5 A consumer agency surveyed all 2500 families living in a small town to collect data on the number of television sets that they owned. The following table lists the frequency distribution of the data collected by this agency.

Number of TV sets owned	0	1	2	3	4
Number of families	120	970	730	410	270

(a) Construct a probability distribution table for the number of television sets owned by these families. (Note that probabilities will be given by relative frequencies.)

(b) Let x denote the number of television sets owned by a randomly selected family from this town. Find the following probabilities:

(i) $P(x = 1)$ (ii) $P(x > 2)$ (iii) $P(x \leq 1)$ (iv) $P(1 \leq x \leq 3)$

11.6 The following table, reproduced from Exercise 11.3, lists the probability distribution of the number of exercise machines sold per day at Elmo's Sporting Goods store:

x	4	5	6	7	8	9	10
$P(x)$	.08	.11	.14	.19	.23	.16	.09

Calculate the mean and standard deviation for this probability distribution. Give a brief interpretation of the value of the mean.

11.7 Despite all safety measures, accidents do happen at Brown's Manufacturing Corporation. Let x denote the number of accidents that occur during a month at this company. The following table, reproduced from Exercise 11.4, lists the probability distribution of x:

x	0	1	2	3	4
$P(x)$	.25	.30	.20	.15	.10

Calculate the mean and standard deviation for the number of accidents per month at this company. Give a brief interpretation of the value of the mean.

11.8 Refer to Exercise 11.5. Find the mean and standard deviation for the probability distribution that you developed for the number of television sets owned by all 2500 families living in a town. Give a brief interpretation of the values of the mean and standard deviation.

11.9 Which of the following are binomial experiments? Explain why.

(a) Drawing 3 balls (with replacement) from a box that contains 10 balls, 6 of which are red and 4 blue, and observing the colors of the drawn balls

(b) Drawing 3 balls (without replacement) from a box that contains 10 balls, 6 of which are red and 4 blue, and observing the colors of the drawn balls

(c) Selecting a few households from New York City and observing whether or not they own stocks when it is known that 28% of all households in New York City own stocks

11.10 It is known that 15% of people working in a large city moonlight (i.e., hold a second job). Using the binomial probabilities formula, find the probability that the number of workers who moonlight in a random sample of five workers is

(a) exactly 0   (b) exactly 2   (c) exactly 1

11.11 It is known that 30% of credit card holders pay off their balances in full each month. Using the binomial probability, find the probability that in a random sample of 8 credit card holders, the number who pay off their balances in full each month is

(a) exactly 3   (b) 0 to 1   (c) exactly 4

11.12 An Internal Revenue Service inspector is to select 3 corporations from a list of 15 for tax audit purposes. Of the 15 corporations, 6 earned profits and 9 incurred losses during the year for which the tax returns are to be audited. If the IRS inspector decides to select 3 out of 15 corporations randomly, find the probability that the number of corporations in these 3 that incurred losses during the year for which the tax returns are to be audited is

(a) exactly 2   (b) at most 1   (c) none

11.13 Bender Electronics buys keyboards for its computers from another company. The keyboards are received in shipments of 100 boxes, each box containing 20 keyboards. The quality control department at Bender Electronics first randomly selects one box from each shipment and then randomly selects five keyboards from that box. The shipment is accepted if not more than one of the five keyboards is defective. The quality control inspector at Bender Electronics selected a box from a recently received shipment of keyboards. Unknown to the inspector, this box contains six defective keyboards.

(a) What is the probability that this shipment will be accepted?

(b) What is the probability that this shipment will not be accepted?

11.14 A large proportion of small businesses in the United States fail during the first few years of operation. On average, 1.3 businesses file for bankruptcy per day in a large city.

  (a) Using the Poisson formula, find the probability that 3 businesses will file for bankruptcy on a given day in this city.

  (b) Using the Poisson formula, find the probability that the number of businesses that will file for bankruptcy on a given day in this city is

$$\text{(i) 2 to 3} \qquad \text{(ii) exactly 3} \qquad \text{(iii) exactly 5}$$

11.15 Despite all efforts by the quality control department, the fabric made at Benton Corporation always contains a few defects. A certain type of fabric made at this corporation contains an average of .4 defects per 500 yards.

  (a) Using the Poisson formula, find the probability that a given piece of 500 yards of this fabric will contain exactly one defect.

  (b) Using the Poisson formula, find the probability that the number of defects in a given 500-yard piece of this fabric will be

$$\text{(i) exactly zero} \qquad \text{(ii) exactly 2}$$

12 Continuous Random Variables and Their Probability Distributions

12.1 OVERVIEW

In any form of business, reliability is a major issue. Whether it involves delivering a pizza in 30 min or less, creating automobile batteries that will start when it is 20 degrees below zero ($-20°F$), or having a defibrillator that will send the appropriate charge when called on, reliability is crucial to the success of a business. In many reliability studies, time is an important variable. Delivery times, system lifetimes, and production times are commonly measured and evaluated.

Time is one very common example of a continuous random variable. As mentioned in Chapter 11, a continuous random variable is a variable that can take on any value in one or more intervals. Shipment weight and rate of growth are two other examples. One issue that can confuse the situation is that people measure time to the nearest second/minute/hour, or other time unit. However, it is impossible for a waiting time to go from 3 min to 4 min without passing through every amount of time between 3 and 4 min.

Some variables that are truly discrete variables are often represented by continuous random variables. For example, if you are looking at the price of a house, it can be measured to the nearest penny. There are so many possible different prices that a continuous model is more convenient than a discrete random variable. On the other hand, the price of a piece of bubble gum does not have many possible values, so a discrete random variable is more appropriate. In general, if the number of possible values is extremely large, a continuous random variable is often preferred to a discrete random variable.

12.2 CONTINUOUS PROBABILITY DISTRIUTIONS

Suppose Table 12.1 gives the work experience (in years) of all its 10,000 employees.

Figure 12.1 contains a histogram of the population distribution of Table 12.1. The relative frequencies given in Table 12.1 can be used as the probabilities of the respective classes. Note that these are exact probabilities because we are considering the population of all the employees of the company. Note that each class in Table 12.1 has a width equal to 5 years. When the width of classes is different than 1 unit, the bar heights are equal to the relative frequency divided by the class width. The relative frequency densities are calculated to make the sum of the areas of all rectangles in the histogram equal to 1.0. Note that we use relative frequency densities on the vertical axes of Figures 12.1 and 12.2 to draw histograms. The probability distribution curve of a continuous random variable is also called its probability density function.

Figure 12.2 shows the smoothed polygon for the data of Table 12.1. The smoothed polygon is an approximation of the probability distribution curve of the continuous random variable x.

Practitioner's Guide for Statistics and Lean Six Sigma for Process Improvements. By Mikel J. Harry, Prem S. Mann, Ofelia C. de Hodgins, Christopher J. Lacke, and Richard Hulbert
Copyright © 2010 John Wiley & Sons, Inc.

TABLE 12.1   Work Experience and Relative Frequencies

Work Experience	Relative Frequency
0 to < 5 years	.179
5 to <10 years	.170
10 to <15 years	.156
15 to <20 years	.136
20 to <25 years	.118
25 to <30 years	.095
30 to <35 years	.070
35 to <40 years	.047
40 to <45 years	.025
45 to <50 years	.004

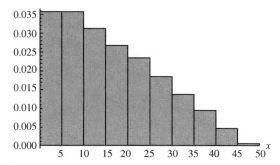

Figure 12.1   Histogram of work experiences shown in Table 12.1.

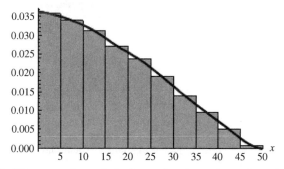

Figure 12.2   Histogram and smoothed polygon of work experiences shown in Table 12.1.

The probability distribution of a continuous random variable possesses the following two characteristics:

1. The probability that x assumes a value in any interval lies in the range 0 to 1.
2. The total probability of all the (mutually exclusive) intervals within which x can assume a value is 1.0.

The first characteristic states that the area under the probability distribution curve of a continuous random variable between any two points is between 0 and 1, as shown in Figure 12.3. The second characteristic indicates that the total area under the probability distribution curve of a continuous random variable is always 1.0 or 100%, as shown in Figure 12.4.

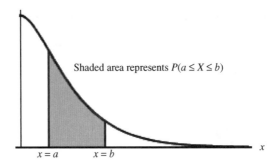

Figure 12.3   Area under density curve corresponds to probability.

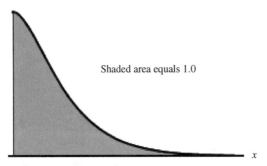

Figure 12.4   Total area under a density curve equals one.

Probability calculations for continuous random variables generally involve calculus; however, statistical software packages will perform the calculations in a similar fashion was shown for discrete random variables in Chapter 11.

12.3   THE NORMAL DISTRIBUTION

The normal distribution is one of the most common distributions for a continuous random variable. The normal distribution is widely used for modeling real-world phenomena, either exactly or approximately. Continuous random variables representing heights of people, the sizes of items coming off a production line such as the weights of food packages or amount of liquid (soda, milk) in a newly filled bottle, and the time taken to complete a certain type of job have all been observed to have a (approximate) normal distribution.

The normal probability distribution is represented by a bell-shaped curve, which is symmetric. The mean of a normal distribution is denoted by μ, and the standard deviation is denoted by σ. A continuous random variable x that has a normal distribution is called a *normal random variable*. Figure 12.5 shows an example of a normal curve.

Note that all bell-shaped curves are not normal curves. In addition to meeting the basic property that the area under the curve must equal one, normal curves have the following properties:

1. The curve is symmetric about the mean.
2. The mean, median, and mode are equal.
3. The tails of the curve extend indefinitely.

A normal distribution curve is symmetric, as shown in Figure 12.6. As a result, 50% of the area lies on the left side of the mean, and 50% lies on the right side of the mean.

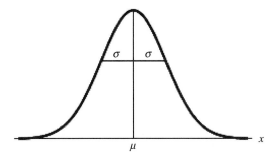

Figure 12.5   A normal distribution with mean μ and standard deviation σ.

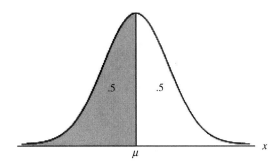

Figure 12.6   Symmetry of a normal distribution.

Since the mean divides the curve in half, the mean must be equal to the median. Moreover, the mean and the median occur at the value of x that corresponds to the highest point of the density curve, which is the mode.

The tails of a normal distribution extend infinitely in both directions without meeting the horizontal axis. This notion can be confusing at first, as we have mentioned that we can use the normal curve to represent the amount of soda in a 2-liter bottle. Certainly, we can not have an infinite or negative amount of soda, so one would wonder how we can use a normal curve.

All normal curves have the property that most of the values of x fall within three standard deviations of the mean; that is, the interval of x-values ranging from $\mu - 3\sigma$ to $\mu + 3\sigma$ contains close to 100% of the values of the variable. This is shown in Figure 12.7.

The mean and standard deviation of a normal distribution are the *parameters* of the distribution. Their values completely define the shape of the bell-shaped curve and where it is located along the horizontal axis. Figure 12.8 demonstrates three normal curves that have different means, but the same

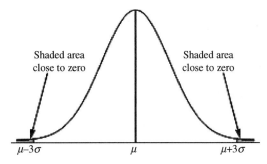

Figure 12.7   Regions outside three standard deviations in a normal distribution.

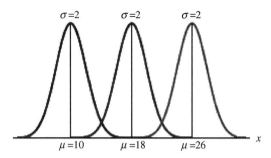

Figure 12.8   Normal distributions having different means and the same standard deviation.

standard deviations. As you can see, the shapes of the bells are the same, but the curves are centered at different locations.

Figure 12.9 demonstrates three normal curves having the same mean, but different standard deviations. In this case, all the bells are centered at the same location, but the bells have different heights and spreads. In general, normal curves with smaller standard deviations will be taller and narrower than curves with larger standard deviations.

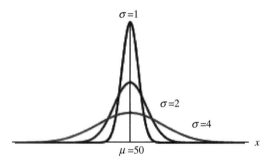

Figure 12.9   Normal distributions having the same mean and different standard deviations.

12.3.1   The Empirical Rule

A simple but powerful rule, called the *empirical rule*, provides some basic information about normal curves. Specifically, for a normal distribution with mean μ and standard deviation σ

1. The interval from $\mu - \sigma$ to $\mu + \sigma$ contains approximately the middle 68% of the observations.
2. The interval from $\mu - 2\sigma$ to $\mu + 2\sigma$ contains approximately the middle 95% of the observations.
3. The interval from $\mu - 3\sigma$ to $\mu + 3\sigma$ contains approximately the middle 99.7% of the observations.

Example 12.1   The diameter of a steel cable used in radial tires is supposed to be 5.8 mm; however, the diameter is not constant from one cable to another. The diameter is normally distributed with a mean of 5.8 mm and a standard deviation of .07 mm. Use the empirical rule to determine the intervals that contain the middle 68%, the middle 95%, and the middle 99.7% of the cable diameters.

Solution   We have $\mu = 5.8$ and $\sigma = .07$. Therefore, the interval containing the middle 68% of the cable diameters ranges from $5.8 - .07$ to $5.8 + .07$, or $(5.73, 5.87)$. The interval containing the middle 95% of the cable diameters ranges from $5.8 - 2(.07)$ to $5.8 + 2(.07)$, or $(5.66, 5.94)$. The interval

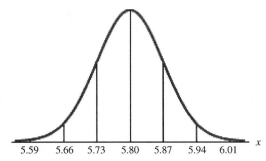

Figure 12.10   Regions corresponding to the empirical rule in Example 12.1.

containing the middle 99.7% of the cable diameters ranges from $5.8 - 3(.07)$ to $5.8 + 3(.07)$, or $(5.59, 6.01)$. The normal curve and the endpoints of these intervals are displayed in Figure 12.10. ■

Understanding the empirical rule can assist in determining whether something has gone wrong with a production process. Suppose that a company that produces the cables mentioned in Example 12.1 measures the diameters of the next 1000 cables produced. According to the empirical rule, approximately 95%, or 950, of the 1000 cables should have a diameter between 5.66 and 5.94 mm. The remaining 50 cables should be split evenly, with 2.5% having a diameter below 5.66 mm and the other 2.5% having a diameter above 5.94 mm. Suppose that 120 of the cables have a diameter of more than 5.94 mm. This amounts to 12%, when it should be close to 2.5%. This may indicate that something in the production process is causing the cables to be bigger than they should be.

Using the cable diameters again, suppose that 18 (1.8%) of the cables had a diameter below 5.59 mm and 21 (2.1%) had a diameter above 6.01 mm. According to the empirical rule, only 0.3% $(100\% - 99.7\%)$ of the diameters should fall in this range, yet almost 4% of our sample fell in this region. This may be a sign that the standard deviation of the process has increased, which would result in a larger range of diameters.

12.3.2   The Standard Normal Distribution

Our discussion regarding the empirical rule is a natural segue to the *standard normal distribution*. The empirical rule tells us that if we go to one standard deviation on each side of the mean, the percentage of observations that will fall in that interval is always approximately 68%. We are told that similar properties exist if we go exactly two or exactly three standard deviations to each side of the mean, regardless of the values of μ and σ. One might hope that this type of relationship holds for any number of standard deviations, regardless of the mean and standard deviation.

The standard normal distribution is a special form of the normal distribution, where the values on the horizontal axis are no longer in terms of the units of measurements, but in terms of z scores. Consider the cable diameters mentioned in Example 12.1. The z scores for the endpoints of the interval containing the middle 68% of the observations are

$$z = \frac{5.73 - 5.8}{.07} = -1.0 \quad \text{and} \quad z = \frac{5.87 - 5.8}{.07} = 1.0$$

Similar calculations with the other intervals will result in z score intervals of -2.0 to 2.0 and -3.0 to 3.0. Thus, the standard normal distribution identifies the probabilities associated with being in a specific range of z scores.

Since the standard normal distribution is a specific case of a normal distribution, it must have a mean and standard deviation, just like any other normal distribution. Since a normal distribution is always centered at its mean, the mean of the standard normal will be the z-score for the value of the mean. Using the cable diameters as an example, we see that the z score for $\mu = 5.8$ is

$$z = \frac{5.8 - 5.8}{.07} = 0$$

so the mean of the standard normal distribution is zero. The standard deviation of the standard normal distribution is the distance between the mean and the point that is one standard deviation above the mean. Since these points are being expressed in terms of their z scores, the values at these points are zero and one, implying that the standard deviation of the standard normal distribution is 1.0.

Let x be a normally distributed random variable with mean μ and standard deviation σ. Let $z = (x - \mu)/\sigma$. Here, z is also normally distributed with mean 0 and standard deviation 1.0, and z is called a *standard normal random variable*.

Suppose that we are interested in finding the probability that a normal random variable takes on a value that is more than three-quarters of a standard deviation below its mean; that is, $P(z < -.75)$. The empirical rule does not provide us with information about this, or most of the other intervals. How do we handle this?

Calculating probabilities associated with the standard normal distribution can be done using a normal probability table (see Table II in Appendix A) or by using either statistical software or spreadsheet programs.

12.3.2.1 *Using a Normal Probability Table to Calculate Normal Distribution Probabilities*

To use Table II from Appendix A to find a normal probability, you first need to locate the z value or values of interest. The values are given to two decimal places. The first digit and the first decimal place are located in the leftmost column. The second decimal place is located in the row at the top of the table. We need to look up the value $z = -.75$. To do this, move down the left column until you come to the row beginning with -0.7. Now, move to the column headed by .05, where .05 corresponds to the fact that the second decimal place in .75 is held by a 5. Move down the column and across the row that you located. At their intersection, you should find the value .2266.

What exactly does this value represent? It is the area under the standard normal curve and to the left of $z = -.75$. In other words, it is $P(z < -.75)$. The normal table provided in this text will always give the area to the left of the z-value of interest.

Example 12.2   Find the following probabilities for the standard normal distribution using Table II of Appendix A:

1. $P(z < 2.35)$
2. $P(z > -1.66)$
3. $P(z > .84)$
4. $P(-1.66 < z < .84)$
5. $P(-2.00 < z < 2.00)$

Solution

1. Because the problem requests us to find the probability to the left of $z = 2.35$, we need to locate the row corresponding to $z = 2.3$ and the column headed by .05. Finding their intersection, we see that $P(z < 2.35) = .9906$. This result is visualized in Figure 12.11.
2. Here we first locate the row corresponding to $z = -1.6$ and the column headed by .06. Finding their intersection, we see that $P(z < -1.66) = .0485$. However, we are interested in

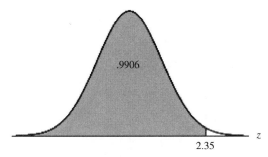

Figure 12.11   Region corresponding to Example 12.2, part 1: $P(z < 2.35)$.

$P(z > -1.66)$, which is the area to the right of -1.66. Since the areas to the left and right of -1.66 must sum to 1, we have $P(z > -1.66) = 1 - P(z < -1.66) = 1 - .0485 = .9515$. This result is visualized in Figure 12.12.

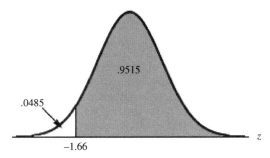

Figure 12.12   Region corresponding to Example 12.2, part 2: $P(z > -1.66)$.

3. We need to locate the row corresponding to $z = 0.8$ and the column headed by .04. Finding their intersection, we see that $P(z < .84) = .7995$. In this case, we are interested in $z > .84$, which is the area to the right of .84. Since the areas to the left and right of .84 must sum to 1, we have $P(z > .84) = 1 - P(z < .84) = 1 - .7995 = .2005$.

4. In this case we want to find the area between -1.66 and .84. The standard normal table provides us with the area to the left of .84, which includes the area to the left of -1.66. We can remove the area that we do not want (the area to the left of -1.66), by subtracting it from the area to the left of .84. Thus

$$P(-1.66 < z < .84) = P(z < .84) - P(z < -1.66) = .7995 - .0485 = .7510$$

This result is visualized in Figure 12.13 on next page.

5. In this case we want the area between -2.00 and 2.00. We need to locate the row corresponding to $z = 2.0$ and the column headed by .00. Finding their intersection, we see that $P(z < 2.00) = .9772$. We also need to locate the row corresponding to $z = -2.0$ and the column headed by .00. Finding their intersection, we see that $P(z < -2.00) = .0228$. Thus

$$P(-2.00 < z < 2.00) = P(z < 2.00) - P(z < -2.00) = .9772 - .0228 = .9544$$

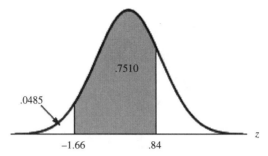

Figure 12.13 Region corresponding to Example 12.2, part 4: $P(-1.66 < z < .84)$. ■

The need for normal tables has dwindled substantially as a result of technology. Example 12.3 demonstrates how to use MINITAB to calculate probabilities associated with the standard normal distribution.

Example 12.3 Find the following probabilities using MINITAB:

1. $P(z < 2.35)$
2. $P(z > -1.66)$
3. $P(-2.00 < z < 2.00)$

Solution Begin each problem by selecting `Calc > Probability Distributions > Normal`. Figure 12.14 shows the information entered to find probability for part 1 of this example. Select `Cumulative Probability`. Enter 0 in the box next to `Mean` and 1 in the box next to `Standard deviation`. The remaining instructions cover the items that are different for each problem:

1. Select `Input constant` and enter 2.35. Click `OK`. The result will appear in the `Session` window, which is shown as the first probability in Figure 12.15. As with the normal table, the

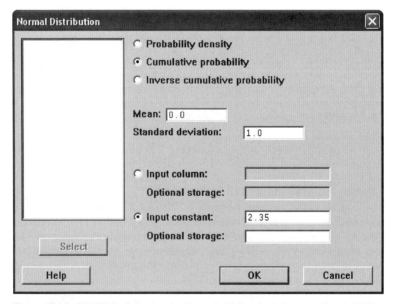

Figure 12.14 MINITAB dialog box for Example 12.3: calculating normal probabilities.

```
┌─────────────────────────────────────────────────────────┐
│  Cumulative Distribution Function                        │
│                                                          │
│                                                          │
│  Normal with mean = 0 and standard deviation = 1         │
│                                                          │
│                                                          │
│       x          P (X <= x)                              │
│     2.35           0.990613                              │
│    -1.66           0.048457                              │
│    -2.00           0.022750                              │
│     2.00           0.977250                              │
│                                                          │
└─────────────────────────────────────────────────────────┘
```

Figure 12.15 MINITAB session output box for Example 12.3: calculating normal probabilities.

cumulative probability represents the area to the left of the value in `Input constant`. Thus, the desired probability is .9906.

2. Select `Input constant` and enter -1.66. Click `OK`. The result will appear in the `Session` window, which is shown in Figure 12.15. As with the normal table, the cumulative probability represents the area to the left of the value in `Input constant`. Since we are interested in the probability to the right of -1.66, the desired probability is $1 - .0485 = .9515$.

3. Select `Input constant` and enter -2. Click `OK`. Repeat the process, changing the value of `Input constant` to 2.00. The results will appear in the `Session` window, which is shown in Figure 12.15. As with the normal table, the cumulative probability represents the area to the left of the value in `Input constant`. Since we are interested in the probability between $z = -2.00$ and $z = 2.00$, we need to calculate $P(z < 2.00) - P(z < -2.00) = .9772 - .0228 = .9544.$ ∎

Since the values produced by MINITAB (and many other software packages) correspond to the values in the normal table, the process of finding the probabilities in Example 12.2 mirrors the method that would be employed using MINITAB. For example, in order to find $P(-1.66 < z < .84)$, one would use the `Normal Distribution` function to calculate $P(z < .84)$ and $P(z < -1.66)$, and then calculate the difference between the two probabilities, just as was done in Example 12.2.

12.3.2.2 *Determining Percentiles of the Standard Normal Distribution*

Another issue that arises in Six Sigma is concern with items or issues that are far away from the average. In the case of a normal random variable, we will establish a probability that we consider to be extreme, such as the largest 10%, the smallest 3%, or the most extreme 1%. We can then use the normal distribution to determine which values fall in each region. In this Section, we will examine how to determine the z scores for these probabilities. In Section 12.3.3, we will examine how to convert the z scores to values of the variable of interest.

Example 12.4 Using Table II of Appendix A, find the z values corresponding to each of the following regions of the standard normal distribution; in each case, state what percentile the z value represents:

1. The smallest 30% of the standard normal distribution
2. The largest 8% of the standard normal distribution
3. The most extreme 2% of the standard normal distribution

Solution

1. In Example 12.2, we had to find the row and column in Table II of Appendix A that corresponded to the z value, and then we found the intersection of the row and column to find the area under the standard normal curve. In the current example, we work backward. First, we look in the inner

part of Table II to find the area that is closest to the area that we are interested in. After locating that area, we look at the z value at the beginning of the row and at the top of the column to determine the z score. In this part, we want to find the z value that has an area of 30% or .3000 to the left of it. In the standard normal table (Table II), we find that .3000 is not in the table; however, the closest values to .3000 are .2981 and .3015. Since .3015 is closer than .2981 to .3000, we will use the z value that corresponds to .3015. Moving to the left of the row containing .3015, we see that the first two digits of z are -0.5. Moving to the top of the column containing .3015 we see that the column header is .02. Therefore $z = -.52$. This is approximately the 30th percentile of the standard normal distribution. In other words, about 30% of the z values are less than $-.52$ and approximately 70% of the values are larger than $-.52$.

2. In this situation, we want to find the z value that has an area of .08 to the right of it. Note that we do not look for the area .08 in the table because the normal table contains the area to the left of a point. Since the total area must equal 1.0, the area to the left must be $1 - .08 = .92$, which is what we need to find. In the standard normal table, we find that .9200 is not in the table. The closest values to .9200 are .9192 and .9207. Since .9207 is closer than .9192 to .9200, we will use the z value that corresponds to .9207. Moving to the left of the row containing .9207, we see that the first two digits of z are 1.4. At the top of the column containing .9207, we see that the column header is .01. Therefore $z = 1.41$. This is approximately the 92nd percentile of the standard normal distribution; that is, about 92% of the z scores are less than 1.41 and about 8% are larger than 1.41.

3. In this situation, we want to find two z values on both sides of the mean that have a total area of .02 outside of them. Whenever we are concerned with the most extreme region, we split the probability in half. Therefore, we want to find the z value that has an area of .01 to the left of it, and the z value that has an area of .01 to the right of it. In the latter case, we need to look for the value corresponding to an area of $1 - .01 = .99$ in Table II.

In the standard normal table, we find that .0100 is not in the table; however, the closest values to .0100 are .0099 and .0102. Since .0099 is closer than .0102 to .01, we will use the z-value that corresponds to .0099. Moving to the left of the row containing .0099, we see the first two digits of z are -2.3. Moving to the top of the column containing .0099, we see that the column header is .03. Therefore $z = -2.33$. This is approximately the 1st percentile of the standard normal distribution; that is, about 1% of the z values are less than $z = -2.33$.

We also find that .9900 is not in the table; however, the closest values to .9898 and .9901. Since .9901 is closer than .9898 to .99, we will use the z value that corresponds to .9901. Moving to the left of the row containing .9901, we see that the first two digits of z are 2.3. Moving to the top of the column containing .9901, we see that the column heading is .03. Therefore $z = 2.33$. This is approximately the 99th percentile of the standard normal distribution; that is, about 99% of the z values are less than 2.33. ∎

The last example reveals a special property about the standard normal distribution. When determining percentiles, if the areas in the tails are the same, the corresponding z values are the same but with opposite signs. If we know the z score for the 78th percentile, the z score for the 22nd percentile is its opposite. Similarly, if we know the z score for the 36th percentile, the z score for the 64th percentile is its opposite, and so on.

As with normal probabilities, the process of determining percentiles of the standard normal distribution is simplified by using statistical software.

Example 12.5 Using MINITAB, find the z values that correspond to each of the following regions of the standard normal distribution:

1. The smallest 30% of the standard normal distribution
2. The largest 8% of the standard normal distribution
3. The most extreme 2% of the standard normal distribution

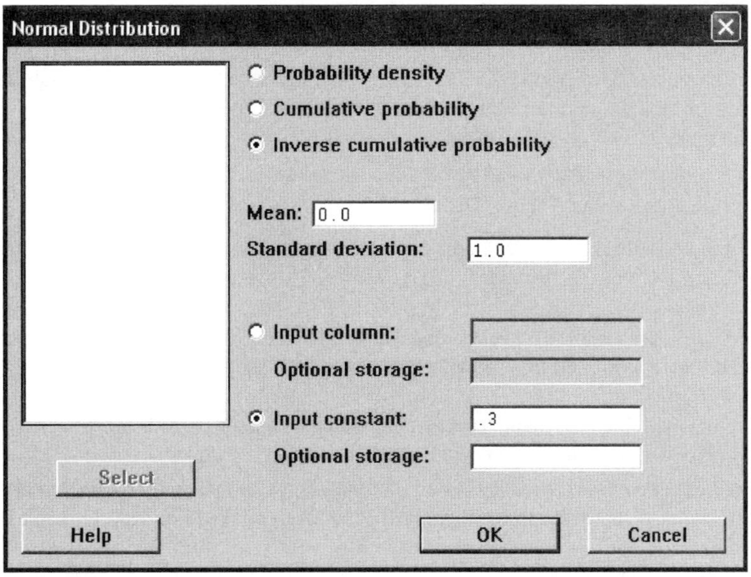

Figure 12.16 MINITAB dialog box for Example 12.5: calculating normal probabilities.

Solution We will use the same function here that we used in Example 12.3 to calculate normal probabilities, with a couple of changes in the entries. Begin each problem by selecting `Calc > Probability Distributions > Normal`. Figure 12.16, which has the correct entries for part 1 inserted, will appear. Select `Inverse Cumulative Probability`. Enter 0 in the `Mean` box and `1.0` in the `Standard deviation` box. The remaining instructions cover the items that are different for each problem

1. Select `Input constant` and enter .3 in the box next to it. Note that here you are to find the z value so that area to the left of it is .3000. Click `OK`. The result will appear in the `Session` window, which is shown in Figure 12.17. The z value for the 30th percentile is $z = -0.5244$. Note that Figure 12.17 contains answers to all three parts.

2. Select `Input constant` and enter .92 in the box next to it. (Note that here we are to z so that area in the upper toil is .08.) Click `OK`. The result will appear in the `Session` window, which is shown in Figure 12.17. The z value for the 92nd percentile is $z = 1.4051$.

3. Select `Input constant` and enter .01 (which is one-half of 2%) in the box next to it. Click `OK`. Repeat the procedure, entering .99 in the `Input constant`. The results will appear in the `Session` window, which is shown in Figure 12.17. The z value for the 1st and 99th percentiles are, respectively, $z = -2.3264$ and $z = 2.3264$.

Inverse Cumulative Distribution Function

Normal with mean = 0 and standard deviation = 1

P (X <= x)	x
0.30	-0.52440
0.92	1.40507
0.01	-2.32635
0.99	2.32635

Figure 12.17 MINITAB session output box for Example 12.5: calculating normal percentiles.

One advantage for calculating normal probabilities and finding normal percentiles using software instead of a normal probability table is that normal tables are limited to two decimal points for the values of z. The amount of roundoff error when using a normal probability table can be substantial in an application, so it is always preferred to use software whenever possible.

12.3.3   Applications of the Normal Distribution

In this section, we will take a few examples that show applications of the normal distribution.

Example 12.6   In Example 12.1, we stated that the diameter of steel cables used in radial tires is normally distributed with a mean of 5.8 mm and a standard deviation of 0.07 mm.

1. A cable is unusable if its diameter is smaller than 5.65 mm or larger than 5.98 mm. What is the probability that a randomly selected cable is unusable?

2. If a cable diameter is between 5.94 and 5.98 mm, it is usable, but there is an increased risk of the tire shredding. What is the probability that the diameter of a randomly selected cable falls in this range?

3. If a cable diameter falls in the most extreme 1% of the distribution of cable diameters, the machine is shut down and recalibrated. What diameters will cause the machine to be shut down and recalibrated?

Solution By Using the Normal Distribution Table   When using a normal table for calculating probabilities, one must convert the values of the variables to their corresponding z scores before using the table.

1. Letting x represent the diameter of a randomly selected cable, we need to find

$$P(x < 5.65 \text{ or } x > 5.98) = P(x < 5.65) + P(x > 5.98)$$

First, we will find $P(x < 5.65)$ as follows:

$$P(x < 5.65) = P\left(z < \frac{5.65 - 5.8}{.07}\right) = P(z < -2.14) = .0162$$

Now we will find $P(x > 5.98)$ as follows:

$$P(x < 5.98) = P\left(z < \frac{5.98 - 5.8}{.07}\right) = P(z < 2.57) = .9949$$

Thus, $$P(x > 5.98) = 1 - P(x < 5.98) = 1 - .9949 = .0051$$

Therefore, $$P(x < 5.65) + P(x > 5.98) = .0162 + .0051 = .0213$$

2. In this problem, we are interested in $P(5.94 < x < 5.98)$, which is calculated as follows:

$$
\begin{aligned}
P(5.94 < x < 5.98) &= P(x < 5.98) - P(x < 5.94) \\
&= P\left(z < \frac{5.98 - 5.80}{.07}\right) - P\left(z < \frac{5.94 - 5.80}{.07}\right) \\
&= P(z < 2.57) - P(z < 2.00) = .9949 - .9772 = .0177
\end{aligned}
$$

3. This situation is similar to Example 12.5 part 3. We need to find the diameters that identify the most extreme 1%. Once again, we need to divide 1% in half because we are equally concerned about cables that are too narrow and cables that are too thick. Therefore, we are looking for the diameter that has an area of .005 falling below it, and the diameter that has an area of .005 falling above it that corresponds to an area of .995 falling below it.

In the standard normal distribution table, we find that .0050 is not in the table; however, the closest values to .005 are .0049 and .0051. In this case, .0050 is halfway between .0049 and .0051. The z values that correspond to .0049 and .0051 are -2.58 and -2.57. Since .0050 is halfway between .0049 and .0051, we will use the value of z that is halfway between -2.58 and -2.57, namely, $z = -2.575$. As you may recall from Example 12.5, if the tail areas are the same; the corresponding z values are the same but with opposite signs. Therefore, the value of z that is associated with .9950 is $z = 2.575$.

Now we need to convert the values of z to values of x. Using the formula for z, we have

$$z = \frac{x - \mu}{\sigma} \Rightarrow z\sigma = x - \mu \Rightarrow \mu + z\sigma = x \Rightarrow x = \mu + z\sigma$$

For $z = -2.575$ we get $x = \mu + z\sigma \Rightarrow x = 5.8 + (-2.575 \times .07) = 5.8 - .18025 = 5.61975$ mm.
For $z = 2.575$ we get $x = \mu + z\sigma \Rightarrow x = 5.8 + (2.575 \times .07) = 5.8 + .18025 = 5.98025$ mm.
Therefore, if the diameter of a cable falls at or below 5.61975 mm or above 5.98025 mm, the machine is shut down and recalibrated.

Solution Using MINITAB   If you have become comfortable using the normal distribution calculator in MINITAB, as was demonstrated in Examples 12.3 and 12.5, then you should be ready to solve these problems. The choice between `Cumulative Probability` and `Inverse Cumulative Probability` is based on whether you are looking for a probability or a percentile of the normal distribution. The values that will go in the `Mean` and `Standard deviation` boxes are no longer 0 and 1, respectively, but the mean and standard deviation of the variable that is being measured. You do not have to convert the values to z scores before using this function.

Begin each problem by selecting `Calc > Probability Distributions > Normal`. Type 5.8 in the box next to `Mean`, and .07 in the box next to `Standard deviation`, as shown in Figure 12.18. The remaining instructions for each problem are given separately as follows:

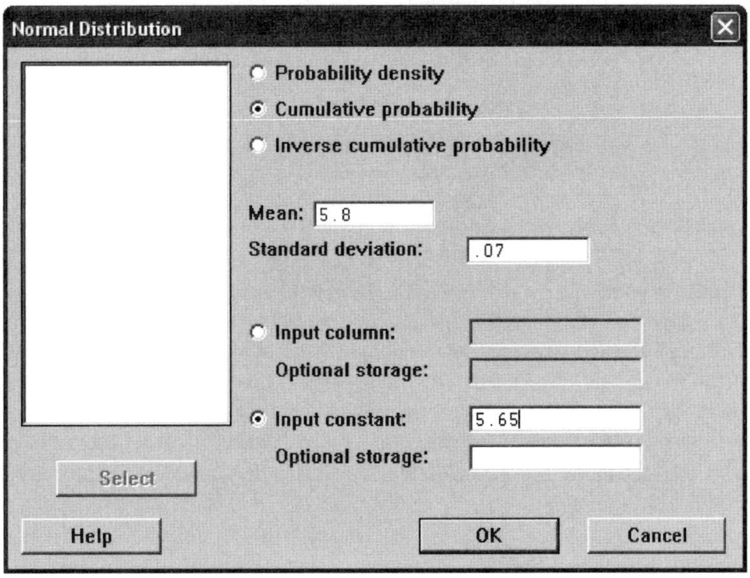

Figure 12.18   MINITAB dialog box for Example 12.6: calculating normal probabilities.

Cumulative Distribution Function

Normal with mean = 5.8 and standard deviation = 0.07

x	P(X <= x)
5.65	0.016062
5.98	0.994936
5.94	0.977250

Figure 12.19   MINITAB session output box for Example 12.6: calculating normal probabilities.

1. We are looking for a probability, namely, $P(x < 5.65 \text{ or } x > 5.98) = P(x < 5.65) + P(x > 5.98)$. To find this probability, select `Cumulative Probability`. Select `Input constant` and enter 5.65. Click OK. Repeat the procedure, changing the `Input constant` to 5.98. The results will appear in the `Session` window, which is shown in Figure 12.19. From the output we have $P(x < 5.65) = .0160$ and $P(x > 5.98) = 1 - P(x < 5.98) = 1 - .9949 = .0051$. The desired probability is $.0160 + .0051 = .0211$.

2. We are looking for $P(5.94 < x < 5.98)$. Select `Cumulative Probability`. Select `Input constant` and enter 5.94. Click OK. The result, $P(x < 5.94) = .9772$, is shown in Figure 12.19. The probability $P(x < 5.98)$ was calculated in part 1. The desired probability is

$$P(5.94 < x < 5.98) = P(x < 5.98) - P(x < 5.94) = .9949 - .9772 = .0177.$$

3. We are looking for percentiles of a normal distribution, specifically those with areas of .005 and .995 to the left of them, so select `Inverse cumulative probability`. Select `Input constant` and enter .005. Click OK. Repeat the procedure, entering .995 in `Input constant`. The results, 5.61969 and 5.98031 mm, will appear in the `Session` window and are shown in Figure 12.20. Therefore, if the diameter of a cable falls at or below 5.61969 mm or above 5.98031 mm, the machine is shut down and recalibrated.

Inverse Cumulative Distribution Function

Normal with mean = 5.8 and standard deviation = 0.07

P (X <= x)	x
0.005	5.61969
0.995	5.98031

Figure 12.20   MINITAB session output box for Example 12.6: calculating normal percentiles.   ■

Example 12.7   Double-fiber wheat breads are advertised as having 5 g of dietary fiber per slice. However, since large batches of bread dough are made and split into many loaves, the amount of fiber per slice varies from loaf to loaf according to a normal distribution with a mean of 5 g and a standard deviation of .14 g.

1. What is the probability that a randomly selected slice has more than 5.2 grams of dietary fiber?
2. What is the probability that a randomly selected slice has between 4.85 and 5.15 grams of dietary fiber?
3. If a slice has a fiber content that places it in the highest 2% of fiber content, how much fiber would be in this slice of bread?

Solution Using the Normal Table

1. Letting x represent the fiber content of a randomly selected slice, we find $P(x > 5.2)$ as follows:

$$P(x > 5.2) = P\left(z > \frac{5.2-5}{.14}\right) = P(z > 1.43) = 1-P(z < 1.43) = 1-.9236 = .0764$$

2. In this problem, we are interested in $P(4.85 < x < 5.15)$, which is obtained as follows:

$$P(4.85 < x < 5.15) = P(x < 5.15)-P(x < 4.85)$$
$$= P\left(z < \frac{5.15-5}{.14}\right)-P\left(z < \frac{4.85-5}{.14}\right)$$
$$= P(z < 1.07)-P(z < -1.07)$$
$$= .8577-.1423$$
$$= .7154$$

3. Since we are looking for the highest 2% of fiber content, we need the 98th percentile of the distribution of fiber content. In the standard normal table, we find that .9800 is not in the table; however, the closest values to .9800 are .9798 and .9803. In this case, .9800 is closer to .9798, so we will use the z value that corresponds to .9798, namely $z = 2.05$.

Now we need to convert the value of z to a value of x. Using the formula for z, we have

$$x = \mu + z\sigma \Rightarrow x = 5+ (2.05 \times .14) = 5+.287 = 5.287 \text{ g}$$

Therefore, if the dietary fiber in a slice of bread is at least 5.287 g, it falls in the highest 2% of dietary fiber levels for this type of bread.

Solution Using MINITAB   Selecting `Calc > Probability Distributions > Normal`. Type 5 in the box next to `Mean` and .14 in the `Standard deviation`, as is shown in Figure 12.21. The remaining instructions for each problem are given separately as follows:

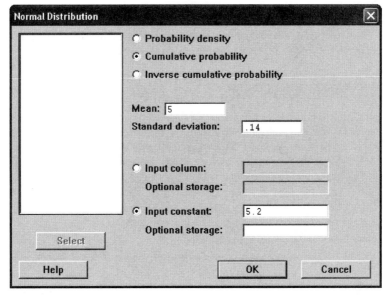

Figure 12.21   MINITAB dialog box for Example 12.7: calculating normal probabilities.

Cumulative Distribution Function

Normal with mean = 5 and standard deviation = 0.14

x	P(X <= x)
5.20	0.923436
4.85	0.141988
5.15	0.858012

Figure 12.22 MINITAB session output box for Example 12.7: calculating normal probabilities.

Inverse Cumulative Distribution Function

Normal with mean = 5 and standard deviation = 0.14

P(X <= x)   x
      0.98   5.28752

Figure 12.23 MINITAB session output box for Example 12.7: calculating normal percentiles.

1. We need to find $P(x > 5.2)$, so select Cumulative Probability. Select Input constant and enter 5.2. Click OK. The results will appear in the Session window, which is shown in Figure 12.22. From the output we have $P(x < 5.2) = .9234$. Thus $P(x > 5.2) = 1 - P(x < 5.2) = 1 - .9234 = .0766$.

2. We need to find $P(4.85 < x < 5.15)$. Select Cumulative Probability. Select Input constant and enter 4.85. Click OK. Repeat the procedure using 5.15 as the Input constant. The results, $P(x < 4.85) = .1420$ and $P(x < 5.15) = .8580$, are shown in Figure 12.22. The desired probability is $P(4.85 < x < 5.15) = P(x < 5.15) - P(x < 4.85) = .8580 - .1420 = .8160$.

3. Since we are looking for the highest 2% of fiber content, we need the 98th percentile of the distribution of fiber content. Select Inverse cumulative probability. Select Input constant and enter .98. Click OK. The result, 5.288 g, will appear in the Session window that is shown in Figure 12.23. Therefore, if the dietary fiber in a slice of bread is at least 5.288 g, it falls in the highest 2% of dietary fiber levels for this type of bread. ∎

12.4 THE EXPONENTIAL DISTRIBUTION

While the normal distribution is a very important continuous probability distribution, it is by no means the only important continuous probability distribution, as there are many random variables that do not have bell-shaped distributions. Variables that involve time are nonnegative and, quite often, skewed to the right. In many cases, these variables are represented by a distribution called the *exponential distribution*. Some examples of exponential random variables include

1. The lifetime of a part of a machine
2. The amount of time taken to check out at a grocery store
3. The concentration of a chemical in a solution
4. The amount of time between consecutive arrivals of vehicles at a gas station

Just as the normal distribution is always shaped like a bell (with different widths and heights), the exponential distribution has a common shape regardless of the parameter value. An exponential

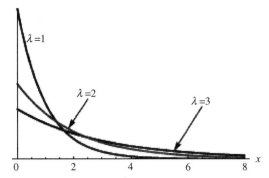

Figure 12.24 Exponential distributions with means equal to 1, 2, and 3.

distribution is completely defined by its mean, which we will denote by λ. The mathematical function that represents the exponential distribution is

$$f(x) = \frac{1}{\lambda}e^{-(x/\lambda)} \quad \text{where} \quad \lambda > 0, x \geq 0$$

Figure 12.24 Contains graphs of the exponential distributions for $\lambda = 1, 2,$ and 3.

As demonstrated in Figure 12.24, an exponential density curve is at its highest when $x = 0$, then it drops off rapidly as x increases. Exponential random variables with larger means (larger values of λ) have curves that are flatter and more spread out; however, intervals of values closer to 0 always have higher probabilities than similar width intervals that are farther away from 0.

Exponential probabilities can be calculated with a hand held calculator or statistical software. Unlike the normal distribution, there is no need for tables. The formula for the cumulative probability for an exponential distribution is

$$P(x < k) = 1 - e^{-(k/\lambda)}$$

where k is a specific value of x. This implies that

$$P(x > k) = 1 - (1 - e^{-(k/\lambda)}) = e^{-(k/\lambda)}$$

Example 12.8 demonstrates how to calculate probabilities that involve the exponential distribution. As with the normal distribution, MINITAB has a function for performing the calculations. We shall demonstrate how to use the function in Example 12.8.

Example 12.8   A human resources department for a large company wishes to study the need for hiring new secretaries. It is estimated that the amount of time that a secretary stays in the job can be described by an exponential distribution with a mean of 26 months. The company just hired a new secretary. Calculate the probabilities of the following events:

1. The secretary has to be replaced within the first year.
2. The secretary has to be replaced during the third year.
3. The secretary remains in the position for more than 5 years.

Solution   Since the mean is given in months, we shall express x, the length of time spent in the job, in months. The mean is 26 months, which implies that $\lambda = 26$.

Using the exponential probability distribution formula, we obtain the answers to our questions as follows:

1. Here, $k = 12$ months, $\lambda = 26$, and we are to find the probability $P(x \leq 12)$. Hence

$$P(x \leq 12) = 1 - e^{-(k/\lambda)} = 1 - e^{-(12/26)} = 1 - .6303 = .3697$$

2. Here, we are to find the probability $P(24 < x \leq 36)$. This probability is obtained as follows:

$$\begin{aligned}
P(24 < x \leq 36) &= P(x \leq 36) - P(x < 24)\\
&= (1 - e^{-(36/26)}) - (1 - e^{-(24/26)})\\
&= .7496 - .6027 = .1469
\end{aligned}$$

3. Here, we are to find the probability $P(x > 60)$, which is obtained as follows:

$$P(x > 60) = e^{-(60/26)} = .0995$$

Note that in case of a continuous probability distribution, the probability of x being equal to a specific single value is zero. For this reason, in the examples above, $P(x \leq 12)$ is the same as $P(x < 12)$, and $P(x \leq 36)$ is the same as $P(x < 36)$.

Using MINITAB, we obtain answers to this question as follows. Begin each problem by selecting Calc > Probability Distributions > Exponential. Figure 12.25, which has the correct entries for part 1 inserted, will appear. Select Cumulative Probability. Enter 26 (the mean) in the Scale box and 0 in the Threshold box. The remaining instructions cover the items that are different for each problem.

1. Select Input constant and enter 12 (as one year is equivalent to 12 months). Click OK. The result will appear in the Session window, which is shown in Figure 12.26. As with other probability functions, the cumulative probability represents the area to the left of the value in

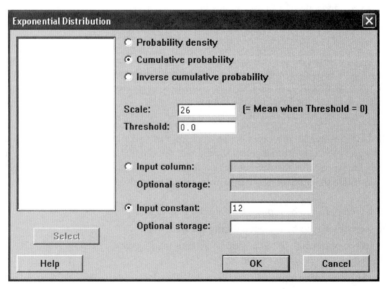

Figure 12.25 MINITAB dialog box for Example 12.8: calculating exponential probabilities.

Cumulative Distribution Function

```
Exponential with mean = 26

 x      P(X <= x)

12       0.369687

24       0.602705

36       0.749580

60       0.900509
```

Figure 12.26   MINITAB session output box for Example 12.8: calculating exponential probabilities.

Input constant. Thus, the desired probability $P(x \le 12)$ is .3697. This implies that approximately 37% of all secretaries have to be replaced during the first year.

2. Being replaced during the third year means that the person holds the position for 24–36 months. Select Input constant and enter 36. Click OK. Repeat the procedure using 24. The results will appear in the Session window, which is shown in Figure 12.26. We are interested in $P(24 < x \le 36) = P(x \le 36) - P(x < 24)$. Thus the desired probability is .7496 − .6027 = .1469.

3. Remaining in the position for at least 5 years means that $x \ge 60$. Select Input constant and enter 60. Click OK. The results will appear in the Session window, which is shown in Figure 12.26. $P(x > 60) = 1 - P(x \le 60) = 1 - .9005 = .0995$, which is the desired probability. This implies that approximately 10% of all secretaries remain in their positions for at least 5 years.

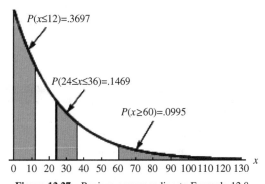

Figure 12.27   Regions corresponding to Example 12.8.

Figure 12.27 demonstrates the regions that correspond to the probabilities calculated in Example 12.8.   ■

Example 12.9   A Six Sigma consulting firm is analyzing waiting times for customers at a local bank. They have estimated that the waiting time before the service is provided by a customer service representative (not a teller) for an arriving customer is exponentially distributed with a mean of 1.8 min. A customer enters the bank and waits in line to be called. Calculate the probabilities of the following events:

1. The customer is served within 0.25 min.

2. The customer waits for at least 3 min before being served.

3. The customer waits for at least 8 min before being served.

Solution   Since the mean is given in minutes, we shall express x, the waiting time, in minutes. The mean is 1.8 min, which implies that $\lambda = 1.8$.

Using the exponential probability distribution formula, we solve this problem as follows:

1. Here, $\lambda = 1.8$, $x = .25$, and we are to find the probability $P(x \leq .25)$, which is obtained as follows:

$$P(x \leq .25) = 1 - e^{-(x/\lambda)} = 1 - e^{-(.25/1.8)} = 1 - .8703 = .1297$$

2. Here, we are to find the probability $P(x \geq 3)$, which is

$$P(x \geq 3) = e^{-(3/1.8)} = .1889$$

3. Here, we are to find $P(x \geq 8) = e^{-(8/1.8)} = .0117$

Now we find these same probabilities using MINITAB. Begin each problem by selecting `Calc >` `Probability Distributions > Exponential`. The screenshot in Figure 12.28, which has the correct entries for part 1 inserted, will appear. Select `Cumulative Probability`. Enter `1.8` (the mean) in the `Scale` box and `0` in the `Threshold` box. The remaining instructions cover the items that are different for each problem.

1. Select `Input constant` and enter `.25`. Click `OK`. The result will appear in the `Session` window, which is shown in Figure 12.29. Thus, the desired probability $P(x \leq .25)$ is .1297. This implies that approximately 13% of all customers do not have to wait for more than 0.25 min for service.

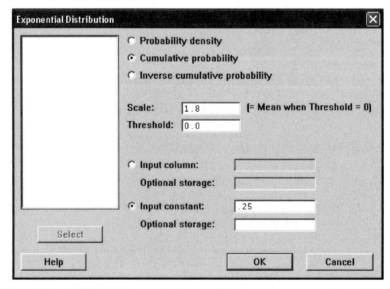

Figure 12.28   MINITAB dialog box for Example 12.9: calculating exponential probabilities.

```
Cumulative Distribution Function

Exponential with mean = 1.8

    x       P(X <= x)
 0.25        0.129675
 3.00        0.811124
 8.00        0.988256
```

Figure 12.29 MINITAB session output box for Example 12.9: calculating exponential probabilities.

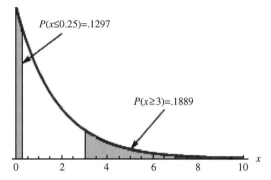

Figure 12.30 Regions corresponding to Example 12.9.

2. Select `Input constant` and enter 3. Click `OK`. The results will appear in the `Session` window, which is shown in Figure 12.29. We are interested in $P(x \geq 3) = 1 - P(x < 3)$. Thus the desired probability is $1 - .8111 = .1889$.

3. Select `Input constant` and enter 8. Click `OK`. The results will appear in the `Session` window, which is shown in Figure 12.29. We are interested in $P(x \geq 8) = 1 - P(x < 8)$. Thus the desired probability is $1 - .9883 = .0117$.

Figure 12.30 demonstrates the regions that correspond to the probabilities calculated in Example 12.9, parts 1 and 2. The region corresponding to $P(x \geq 8)$ is too small to display. ■

GLOSSARY

continuous random variable A random variable that can assume any value in one or more intervals.

exponential probability distribution The probability distribution of a continuous random variable that gives the probability of lapse time between two successive occurrences when the average number of occurrences per unit of time is known.

normal probability distribution The probability distribution of a continuous random variable that, when plotted, gives a specific bell-shaped curve. The parameters of the normal distribution are the mean μ and the standard deviation σ.

standard normal distribution The normal distribution with a mean of zero and a standard deviation of one. The units of the standard normal distribution are denoted by z.

z value or z score The unites of standard normal distribution that are denoted by z.

EXERCISES

12.1 Obtain the following probabilities for the standard normal distribution:

(a) $P(z > -.78)$

(b) $P(-2.47 \leq z \leq 1.09)$

(c) $P(0 \leq z \leq 2.25)$

(d) $P(-3.36 \leq z \leq 0)$

(e) $P(z > 3.07)$

(f) $P(z > -2.27)$

12.2 Obtain the following probabilities for the standard normal distribution:

(a) $P(z > -1.26)$

(b) $P(-.68 \leq z \leq 1.74)$

(c) $P(0 \leq z \leq 1.85)$

(d) $P(-2.34 \leq z \leq 0)$

(e) $P(z > 2.82)$

(f) $P(z > -1.27)$

12.3 Find the z value for each of the following x values for a normal distribution with $\mu = 30$ and $\sigma = 5$:

(a) $x = 37$

(b) $x = 19$

(c) $x = 23$

(d) $x = 44$

12.4 Find the following areas under a normal curve with $\mu = 20$ and $\sigma = 4$:

(a) Area between $x = 20$ and $x = 27$

(b) Area from $x = 23$ to $x = 25$

(c) Area between $x = 9.5$ and $x = 17$

12.5 Let x be a continuous random variable that is normally distributed with a mean of 80 and a standard deviation of 12. Find the probability that x assumes a value

(a) greater than 70

(b) less than 75

(c) greater than 100

(d) less than 89

12.6 The Bank of Connecticut issues Visa and Mastercard credit cards. It is estimated that the balances on all Visa credit cards issued by the Bank of Connecticut have a mean of $845 and a standard deviation of $270. Assume that the balances on all these Visa cards follow a normal distribution.

(a) What is the probability that a randomly selected Visa card issued by this bank has a balance between $1000 and $1400?

(b) What percentage of the Visa cards issued by this bank have a balance of $750 or more?

12.7 Fast Auto Service guarantees that the maximum waiting time for its customers is 20 min for oil- and-lube service on their cars. It also guarantees that any customer who has to wait for more than 20 min for this service will receive a 50% discount on the charges. It is estimated that the mean time taken for oil and lube service at this auto shop is 15 min per car and the standard deviation is 2.4 min. Suppose that the time taken for oil and lube service follows a normal distribution.

(a) What percentage of customers will receive the 50% discount on their charges?

(b) Is it possible that a car may take more than 25 min for oil and lube service? Explain.

12.8 Let x be a continuous random variable that follows a normal distribution with a mean of 200 and a standard deviation of 25.

(a) Find the value of x so that the area under the normal curve to the left of x is approximately .6330.

(b) Find the value of x so that the area under the normal curve to the right of x is approximately .05.

(c) Find the value of x so that the area under the normal curve to the right of x is .8051.

(d) Find the value of x so that the area under the normal curve to the left of x is .015.

(e) Find the value of x so that the area under the normal curve between μ and x is .4525 and the value of x is smaller than μ.

(f) Find the value of x so that the area under the normal curve between μ and x is .4800 and the value of x is greater than μ.

12.9 A machine at Keats Corporation fills 64-oz detergent jugs. The machine can be set to pour, on average, any amount of detergent into these jugs. However, the machine does not pour exactly the same amount of detergent in each jug; it varies from jug to jug. It is known that the net amount of detergent poured into each jug has a normal distribution with a standard deviation of 0.4 oz. The quality control inspector wants to set the machine such that at least 95% of the jugs have more than 64 oz of detergent. What should the mean amount of detergent poured by this machine into these jugs be? (Note: Here you are given the area and x but you are to find μ. You will use the formula $\mu = x - z\sigma$.)

12.10 The time taken by new employees at Shia Corporation to learn a packaging procedure is normally distributed with a mean of 24 hours and a standard deviation of 2.5 hours.

(a) What percentage of new employees can learn this procedure in less than 20 hours?

(b) Find the probability that a randomly selected new employee will take 26 to 30 hours to learn this procedure.

(c) Find the probability that a randomly selected new employee will take 21 to 28 hours to learn this procedure.

12.11 The management at Ohio National Bank does not want its customers to wait in line for service for a long time. The manager of a branch of this bank estimated that the customers currently have to wait for an average of 8 min for service. Assume that the waiting time for all customers at this branch has a normal distribution with a mean of 8 min and a standard deviation of 2 min.

(a) Find the probability that a randomly selected customer will have to wait for less than 3 min.

(b) What percentage of the customers will have to wait for 10 to 13 min?

(c) What percentage of the customers will have to wait for 6 to 12 min?

(d) Is it possible that a customer may have to wait for more than 16 min for service? Explain.

12.12 At Jen and Perry Ice Cream Company, the machine that fills one-pound cartons of Top Flavor ice cream is set to dispense 16 oz of ice cream in every carton. However, some cartons contain slightly less than and some contain slightly more than 16 oz of ice cream. The amounts of ice cream in all such cartons have a normal distribution with a mean of 16 oz and a standard deviation of 0.18 oz.

(a) Find the probability that a randomly selected such carton contains 16.25 to 16.5 oz of ice cream.

(b) What percentage of such cartons contain less than 15.75 oz of ice cream?

(c) Is it possible for a carton to contain less than 15.25 oz of ice cream? Explain.

12.13 A machine at Kasem Steel Corporation makes iron rods that are supposed to be 50 inches long. However, the machine does not produce all rods of exactly the same length. It is known that the probability distribution of the lengths of rods made on this machine is normal with a mean of 50 inches and a standard deviation of 0.06 inches. The rods that are either shorter than 49.85 inches or longer than 50.15 inches are discarded. What percentage of the rods made on this machine are discarded?

12.14 The time taken by new employees at Shia Corporation to learn a packaging procedure is normally distributed with a mean of 24 hours and a standard deviation of 2.5 hours.

(a) Find the time beyond which only 1% of the new employees take to learn this new procedure.

(b) What is the time below which only 5% of the new employees take to learn this procedure?

12.15 The shelf life of a cake is exponentially distributed with a mean of 3 days. What is the probability that a cake that is baked just now will still be good after 4 days?

12.16 At a department store, a customer has to wait an average of 4 min in line before being served. The time a customer has to wait is exponentially distributed.

(a) Find the probability that a customer will have to wait for more than 8 min in line.

(b) Find the probability that a customer will have to wait for 3 to 7 min in line.

12.17 On average, 20 telephone calls are received per hour at an office of the Better Business Bureau. The time between calls received at this office is exponentially distributed.

(a) What is the probability that no call will come in during the next 10 min?

(b) What is the probability that the next call will come in within 4 min?

12.18 Nicole Shooner works for a toy company and assembles five toys per hour on average. The assembly time for this toy follows an exponential distribution.

(a) Find the probability that the next toy will take more than 15 min to assemble.

(b) What is the probability that the next toy will take less than 8 min to assemble?

(c) Find the probability that the next toy will take 10 to 16 min to assemble.

12.19 On average, five cars are rented per hour at a particular office of National Rental Inc. Assume that the time between successive car rentals at this office follows an exponential distribution.

(a) Find the probability that the next car will be rented at this office after 30 min.

(b) What is the probability that the next car will be rented within 5 min?

(c) Find the probability that the next car will be rented in 15 to 25 min.

12.20 Ken Sisco sells life insurance policies. The past data show that he sells, on average, 10 life insurance policies per 4-week period. Assume that the time between successive sales of life insurance policies by Ken Sisco has an exponential distribution.

(a) What is the probability that the next life insurance policy will not be sold for 2 weeks?

(b) Find the probability that the next life insurance policy will be sold within 1 week.

(c) What is the probability that the next life insurance policy will be sold in 1 to 2 weeks?

13 Sampling Distributions

13.1 OVERVIEW

You go to get a soda at work, but it turns out that the dollar bill receiver is full, so you start emailing colleagues, asking if they have change for a dollar. Suppose you ask 10 people, and 4 of them tell you that they have it. If, for some reason, you were to ask another 10 people, would exactly 4 people in the second group have change for a dollar? It is possible, but it is not guaranteed. You could repeat this procedure many times, determining how many people in each group of 10 have change for a dollar.

In the aforementioned situation, we are developing the *sampling distribution* of the proportion of people in a sample of 10 who have change for a dollar. To obtain the sampling distribution of a statistic, one must take all possible samples of a given size and calculate the value of the statistic of interest for each sample. If you work for a company with 100 employees and try to form the sampling distribution in the aforementioned setting, you would have to take over 10 trillion different samples. Fortunately, we can develop an idea of what a sampling distribution will look like without going through that process! Note that a summary measure such as $\bar{x}$, s, median, or proportion that is calculated for sample data is called a *statistic*.

13.2 SAMPLING DISTRIBUTION OF A SAMPLE MEAN

The following example demonstrates the concept of the sampling distribution for a very small population.

Example 13.1 A company has six salespeople. The following are the annual bonuses for these six salespersons:

| $1800 | $2100 | $2400 | $1200 | $2100 | $2100 |

Let x denote the bonus of a salesperson. Table 13.1 lists the population frequency, relative frequency, and probability distributions for the population of salespeople in this company.

The population mean and population standard deviation, which are the mean and standard deviation of the probability distribution given in Table 13.1, are $\mu = \$1950$ and $\sigma = \$377.49$. These values are calculated using the formulas for the mean and standard deviation of a discrete random variable, as given in Sections 11.6 and 11.7.

Suppose that we were to select three salespeople at random and denote their bonuses. In this case, the number of possible samples is $_6C_3 = 6!/3!(6-3)! = 20$, which allows us to list all the samples and their corresponding values of $\bar{x}$, as is shown in Table 13.2

Practitioner's Guide for Statistics and Lean Six Sigma for Process Improvements. By Mikel J. Harry, Prem S. Mann, Ofelia C. de Hodgins, Christopher J. Lacke, and Richard Hulbert
Copyright © 2010 John Wiley & Sons, Inc.

349

TABLE 13.1   Population Frequency, Relative Frequency, and Probability Distributions of Salespersons' Bonus

x	f	Relative Frequency	$P(x)$
$1200	1	$\frac{1}{6}$	$\frac{1}{6}$
$1800	1	$\frac{1}{6}$	$\frac{1}{6}$
$2100	3	$\frac{1}{2}$	$\frac{1}{2}$
$2400	1	$\frac{1}{6}$	$\frac{1}{6}$
	$N = 6$	Sum $= 1$	$\Sigma P(x) = 1$

TABLE 13.2   Samples of Size 3 from Distribution of Salesperson's Bonuses

Sample	Bonuses in the Sample	$\bar{x}$	Sample	Bonuses in the Sample	$\bar{x}$
1, 2, 3	$1800, $2100, $2400	$2100	2, 3, 4	$2100, $2400, $1200	$1900
1, 2, 4	$1800, $2100, $1200	$1700	2, 3, 5	$2100, $2400, $2100	$2200
1, 2, 5	$1800, $2100, $2100	$2000	2, 3, 6	$2100, $2400, $2100	$2200
1, 2, 6	$1800, $2100, $2100	$2000	2, 4, 5	$2100, $1200, $2100	$1800
1, 3, 4	$1800, $2400, $1200	$1800	2, 4, 6	$2100, $1200, $2100	$1800
1, 3, 5	$1800, $2400, $2100	$2100	2, 5, 6	$2100, $2100, $2100	$2100
1, 3, 6	$1800, $2400, $2100	$2100	3, 4, 5	$2400, $1200, $2100	$1900
1, 4, 5	$1800, $1200, $2100	$1700	3, 4, 6	$2400, $1200, $2100	$1900
1, 4, 6	$1800, $1200, $2100	$1700	3, 5, 6	$2400, $2100, $2100	$2200
1, 5, 6	$1800, $2100, $2100	$2000	4, 5, 6	$1200, $2100, $2100	$1800

Now that we have the complete set of values of $\bar{x}$ for the 20 samples, we can form the sampling distribution of $\bar{x}$ when the sample size is 3. As you will notice in Table 13.3, the sampling distribution of $\bar{x}$ is simply the probability distribution of $\bar{x}$ for the given sample size.

Just like a single bonus x, the average bonus $\bar{x}$ is a random variable. It is very important to be specific when writing a probability statement. For example, $P(x = \$1800)$ is the probability that an individual's bonus is $1800, while $P(\bar{x} = \$1800)$ is the probability that the average of three bonuses is $1800. The two probabilities are .1667 and .20, respectively.

The probability distribution of $\bar{x}$ is called the *sampling distribution* of $\bar{x}$. This sampling distribution of $\bar{x}$ has a mean and a standard deviation and these are denoted by $\mu_{\bar{x}}$ and $\sigma_{\bar{x}}$, respectively. If we calculate the mean of all 20 sample means listed in Table 13.2, it will be called the *mean of the sampling distribution of the sample mean $\bar{x}$* when $n = 3$, and it will be denoted by $\mu_{\bar{x}}$. If we calculate the standard deviation of all 20 sample means listed in Table 13.2, it will be called

TABLE 13.3   Frequency, Relative Frequency, and Probability Distributions of the Sample Mean of Salespersons' Bonuses

$\bar{x}$	f	Relative Frequency	$P(\bar{x})$
$1700	3	$\frac{3}{20}$	.15
$1800	4	$\frac{4}{20}$	.20
$1900	3	$\frac{3}{20}$	.15
$2000	3	$\frac{3}{20}$	.15
$2100	4	$\frac{4}{20}$	.20
$2200	3	$\frac{3}{20}$	.15
	$N = 20$	Sum $= 1$	$\Sigma P(\bar{x}) = 1$

the *standard deviation* of the *sampling distribution of the sample mean* $\bar{x}$ when $n = 3$, and it will be denoted by $\sigma_{\bar{x}}$. ∎

13.2.1 Sampling and Nonsampling Errors

Usually, different samples selected from the same population will give different results because they contain different elements. This is obvious from Table 13.2, which shows that the mean of a sample of three bonuses depends on which three of the five bonuses are included in the sample. The result obtained from any one sample will generally be different from the one obtained from the corresponding population. The difference between the value of a sample statistic obtained from a sample and the value of the corresponding population parameter obtained from the population is called the *sampling error*. Note that this difference represents the sampling error only if the sample is random and no nonsampling error has been made. Otherwise only a part of this difference will be due to the sampling error.

Sampling error is the difference between the value of a sample statistic and the value of the corresponding population parameter. In the case of the mean

$$\text{Sampling error} = \bar{x} - \mu$$

assuming that the sample is random and no nonsampling error has been made.

It is important to remember that *a sampling error occurs because of chance*. The errors that occur for other reasons, such as errors made during collection, recording, and tabulation of data, are called *nonsampling errors*. Such errors occur because of human mistakes and not chance. Note that there is only one kind of sampling error—the error that occurs due to chance. However, there is not just one nonsampling error but many nonsampling errors that may occur due to different reasons.

The following paragraph, reproduced from the Current Population Reports of the US Bureau of the Census, explains how the nonsampling errors can occur:

Nonsampling errors can be attributed to many sources, e.g., inability to obtain information about all cases in the sample, definitional difficulties, differences in the interpretation of questions, inability or unwillingness on the part of the respondents to provide correct information, inability to recall information, errors made in collection such as in recording or coding the data, errors made in processing the data, errors made in estimating values for missing data, biases resulting from the differing recall periods caused by the interviewing pattern used, and failure of all units in the universe to have some probability of being selected for the sample (undercoverage).

The following are the main reasons for the occurrence of nonsampling errors:

1. If a sample is nonrandom (and, hence, nonrepresentative), the sample results may be too different from the census results.
2. The questions may be phrased in such a way that they are not fully understood by the members of the sample or population. As a result, the answers obtained are not accurate.
3. The respondents may intentionally give false information in response to some sensitive questions. For example, people may not tell the truth about drinking habits, incomes, or opinions about religious beliefs. Sometimes the respondents may give wrong answers because of ignorance. For example, a person may not remember the exact amount he spent on clothes during the last year and if asked in a survey, may give an inaccurate answer.
4. The polltaker may make a mistake and enter a wrong number in the records or make an error while entering the data on a computer.

Note that nonsampling errors can occur in either a sample survey or a census, whereas the sampling error occurs only when a sample survey is conducted. Nonsampling errors can be minimized by preparing the survey questionnaire carefully and handling the data cautiously. However, it is impossible to avoid sampling error.

Example 13.2 illustrates sampling and nonsampling errors using the mean.

Example 13.2   Reconsider the population of six bonuses given earlier in this chapter. The bonuses for the six salespersons are

$1800 $2100 $2400 $1200 $2100 $2100

The population mean is

$$\mu = \frac{1800 + 2100 + 2400 + 1200 + 2100 + 2100}{6} = \$1950$$

Now suppose that we take a random sample of three bonuses from this population. Assume that this sample includes the bonuses of $2100, $1200, and $2100. The mean for this sample is

$$\bar{x} = \frac{2100 + 1200 + 2100}{3} = \$1800$$

Consequently

$$\text{Sampling error} = \bar{x} - \mu = 1800 - 1950 = -\$150$$

In other words, the mean bonus estimated from the sample is $150 lower than the mean bonus of the population. Note that this difference occurred by chance, that is, because we used a sample instead of the population.

Now suppose, when we select the above mentioned sample, we mistakenly record the second bonus as $1500 instead of $1200. As a result, we calculate the sample mean as

$$\bar{x} = \frac{2100 + 1500 + 2100}{3} = \$1900$$

Consequently, the difference between this sample mean and the population mean is

$$\bar{x} - \mu = 1900 - 1950 = -\$50$$

However, this difference between the sample mean and the population mean does not represent the sampling error. As we calculated earlier, $-$150$ is the correct sampling error. The difference between the correct sampling error and incorrect sampling error, which is equal to $-\$150 - (-\$50) = -\$100$, represents the nonsampling error because it resulted from the error that we made in recording the second bonus in the sample. Thus, in this case

Sampling error $= -\$150$
Nonsampling error $= -\$100$ ∎

Now that we have an example of the sampling distribution of $\bar{x}$, let us see how it compares to the population distribution. Figure 13.1 includes graphs of both distributions on the same scale.

The mean and standard deviation of the population distribution of six bonuses are $1950 and $377.49, respectively. Looking at the sampling distribution of the average bonus, we can see that the center of the distribution is between $1800 and $2000. We can also see that the sampling distribution is

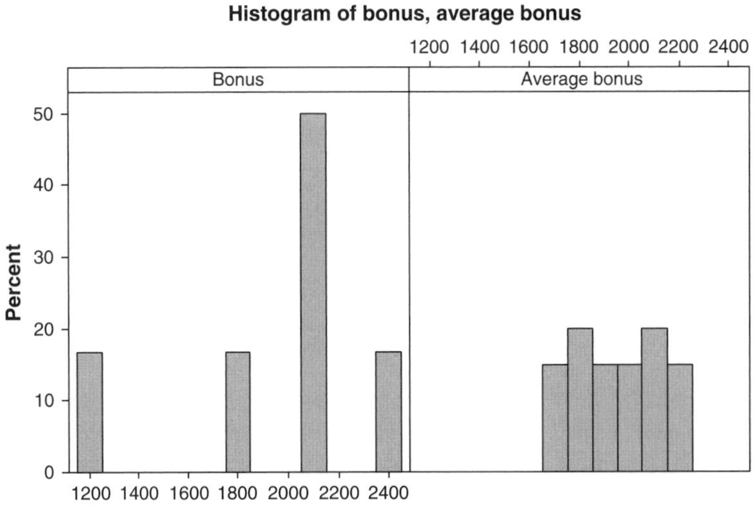

Histogram of bonus, average bonus

Figure 13.1 Histograms of the distributions of salespersons' bonuses and the sample mean of salespersons' bonuses.

much less variable than the population distribution. So, how do the mean and standard deviation of the sampling distribution, denoted by $\mu_{\bar{x}}$ and $\sigma_{\bar{x}}$, respectively, relate to the mean and standard deviation of the population distribution? Again using the formulas from Sections 11.6 and 11.7 by converting x to $\bar{x}$, we have

$$\mu_{\bar{x}} = \$1950 \qquad \text{and} \qquad \sigma_{\bar{x}} = \$168.82$$

In this case, the mean of the sampling distribution is the same as the mean of the population distribution. Fortunately, this example is not just a special case. Whenever we take every possible sample of a specific size, calculate $\bar{x}$ for each sample, then calculate the average of the $\bar{x}$ values ($\mu_{\bar{x}}$), we will get the population mean μ. Thus, for any population with mean μ, the mean of the sampling distribution of $\bar{x}$ is always equal to the mean of the population. In short,

$$\mu_{\bar{x}} = \mu$$

The sample mean, $\bar{x}$, is called an *estimator* of the population mean, μ. It is not the only estimator, but for many reasons, it is a very good estimator. Whenever the mean of the sampling distribution of an estimator is equal to the parameter that we are trying to estimate, the estimator is called an *unbiased estimator*. As we stated above, $\bar{x}$ is an unbiased estimator of μ. There are many other unbiased estimators of μ, but $\bar{x}$ is preferable for many reasons that are outside the scope of this book.

The relationship between the population standard deviation and the standard deviation of the sampling distribution is not as obvious as the relationship between the means. There is no obvious formulaic relationship between $\sigma_{\bar{x}} = \$168.82$ and $\sigma = \$377.49$. In this example, the standard deviation of the sampling distribution is smaller than the population standard deviation. As long as the sample size is greater than 1, this relationship will always hold. The reason for this is relatively simple. When we take an average of values, smaller and larger values will offset each other in the calculation. This reduces the chance of obtaining an extremely small or extremely large average, making $\bar{x}$ have less variability than x.

Specifically, the formula for $\sigma_{\bar{x}}$ depends on whether the sample size is small in relation to the population size. For any population with standard deviation σ, the standard deviation of the sampling distribution of $\bar{x}$ is

$$\sigma_{\bar{x}} = \frac{\sigma}{\sqrt{n}} \qquad \text{if} \qquad \frac{n}{N} \le .05$$

$$\sigma_{\bar{x}} = \frac{\sigma}{\sqrt{n}} \sqrt{\frac{N-n}{N-1}} \qquad \text{if} \qquad \frac{n}{N} > .05$$

The latter version includes $\sqrt{(N-n)/(N-1)}$, which is called the *finite population correction factor*. This will be necessary when the sample is selected without replacement and sample size is more than 5% of the population size.

Example 13.3 Determine the standard deviation of the sampling distribution of $\bar{x}$ for Example 13.1.

Solution The sample in Example 13.1 is taken without replacement, and the sample size $(n = 3)$ is larger than 5% of the population size $(N = 6)$ because $n/N = 3/6 = .50$, so we must use the finite population correction factor. We have $\sigma = \$377.49$; thus

$$\sigma_{\bar{x}} = \frac{\sigma}{\sqrt{n}} \sqrt{\frac{N-n}{N-1}} = \frac{\$377.49}{\sqrt{3}} \sqrt{\frac{6-3}{6-1}} = (\$217.9440)(0.7746) = \$168.82$$

which matches the result obtained from using the formula in Section 11.7. ∎

The majority of applications do not require the use of the finite population correction factor in the formula for $\sigma_{\bar{x}}$. Examples 13.4 and 13.5 are two such applications.

Example 13.4 The amount of gasoline that has been dispensed from a gasoline pump when it reads one gallon is a random variable with a mean of 1 and a standard deviation of 0.004 gal. Seven gasoline pumps are selected at random. What are the mean and standard deviation of the sampling distribution of $\bar{x}$, which is the average amount of gasoline distributed by the seven pumps?

Solution A sample of seven pumps is certainly much less that 5% of all of the gas pumps in existence. Since $\mu = 1$ and $\sigma = .004$, we have

$$\mu_{\bar{x}} = \mu = 1 \quad \text{and} \quad \sigma_{\bar{x}} = \frac{\sigma}{\sqrt{n}} = \frac{.004}{\sqrt{7}} = .0015 \qquad ∎$$

13.3   SAMPLING DISTRIBUTION OF A SAMPLE PROPORTION

We began this chapter by describing a situation in which you were asking people if they have change for a dollar bill so that you can purchase a soda from the machine. In that example, we stated that 4 out of the 10 people you asked had change for a dollar. This information allows us to calculate the sample proportion of people who have change for a dollar. Denoting the sample proportion using the symbol $\hat{p}$, we have

$$\hat{p} = \frac{\text{number of people with change for a dollar}}{\text{number of people asked}} = \frac{4}{10} = .40$$

Note that the population proportion is denoted by p and the sample proportion by $\hat{p}$. The *sample proportion* $\hat{p}$ is an estimator of the population proportion p which, in this case, represents the proportion of all employees who have change for a dollar. Just as $\bar{x}$ will vary from sample to sample, so will $\hat{p}$. Unlike $\bar{x}$, the sampling distribution of $\hat{p}$ requires that the underlying experiment follows the conditions of the binomial distribution. As was stated in Section 11.8.2, the conditions of the binomial distribution are

1. The number of identical trials n is a fixed quantity.
2. Each trial has two complementary events called a *success* and a *failure*.
3. $P(\text{success}) = p$ and $P(\text{failure}) = q$ for each trial.
4. Trials are independent of each other

Noting that $\hat{p} = (\text{number of successes})/(\text{number of trials}) = x/n$, we see that $\hat{p}$ is an estimate of the probability of success in a binomial experiment. Our example above about change for a dollar satisfies the binomial assumptions in the following fashion:

1. Here, $n = 10$ (we asked 10 people).
2. Each person either has change or does not have change for a dollar.
3. Probabilities of these two events are p and q, and $p + q = 1.0$.
4. The people were selected (independently) at random, which results in independent trials.

For any binomial experiment with n trials and probability p of success, the sampling distribution of $\hat{p}$, the sample proportion of successes in n trials, has the mean and standard deviation given by

$$\mu_{\hat{p}} = p$$

$$\sigma_{\hat{p}} = \sqrt{\frac{p(1-p)}{n}} \qquad \text{if} \qquad \frac{n}{N} \le .05$$

$$\sigma_{\hat{p}} = \sqrt{\frac{p(1-p)}{n}} \sqrt{\frac{N-n}{N-1}} \qquad \text{if} \qquad \frac{n}{N} > .05$$

Note that here $\mu_{\hat{p}}$ is the mean and $\sigma_{\hat{p}}$ is the standard deviation of the sampling distribution of $\hat{p}$. At this juncture, it is important to note three points: (1) as the statement $\mu_{\hat{p}} = p$ suggests, $\hat{p}$ is an unbiased estimator of p; (2) as the sample size increases, $\sigma_{\hat{p}}$ (given by the formula above) decreases (this relationship is similar to that between the sample size and the standard deviation in the case of $\bar{x}$); and (3) a finite population correction factor is needed if the sample size exceeds 5% of the population size. The correction factor has the same formula as in the case of $\bar{x}$.

Example 13.5   Assuming that 35% of all employees in your company actually have change for a dollar, what are the mean and standard deviation of the sampling distribution of $\hat{p}$, where $\hat{p}$ is the sample proportion of a random sample of 10 employees who have change for a dollar? The company you work for has 1000 employees.

Solution   Since $\frac{10}{1000} = .01 \le .05$, we do not need to use the finite population correction factor. Given that $n = 10$ and $p = .35$, we obtain

$$\mu_{\hat{p}} = p = .35 \qquad \text{and} \qquad \sigma_{\hat{p}} = \sqrt{\frac{p(1-p)}{n}} = \sqrt{\frac{.35(1-.35)}{10}} = \sqrt{.02275} = .1508 \qquad ■$$

Example 13.6   Refer to Example 13.5. How would the mean and standard deviation of $\hat{p}$ change if there were only 100 employees?

Solution   Since $\frac{10}{100} = .10 > .05$, we need to use the finite population correction factor for the standard deviation. The mean of the sampling distribution does not depend on the sample size. Hence

$$\mu_{\hat{p}} = p = .35$$

$$\sigma_{\hat{p}} = \sqrt{\frac{p(1-p)}{n}} \times \sqrt{\frac{N-n}{N-1}} = \sqrt{\frac{.35(1-.35)}{10}} \sqrt{\frac{100-10}{100-1}} = \sqrt{.02275} \times \sqrt{.9091} = .1438 \qquad \blacksquare$$

Example 13.7   A microchip manufacturer selects 100 chips at random from a very large production run for quality testing. Let $\hat{p}$ be the proportion of chips in this sample that are found to be defective. If 3% of the chips in the production run are defective, what are the mean and standard deviation of $\hat{p}$?

Solution   Since we are told that the production run is very large, we do need to use the finite population correction factor for the standard deviation. Given that $n = 100$ and $p = .03$, we obtain

$$\mu_{\hat{p}} = p = .03 \qquad \text{and} \qquad \sigma_{\hat{p}} = \sqrt{\frac{p(1-p)}{n}} = \sqrt{\frac{.03(1-.03)}{100}} = \sqrt{.000291} = .0176 \qquad \blacksquare$$

13.4   THE CENTRAL-LIMIT THEOREM (CLT)

So far in this chapter, we have explained what a sampling distribution is, and we have learned the formulas for the means and standard deviations for the sampling distributions of $\bar{x}$ and $\hat{p}$. But we have not yet discussed the shapes of the sampling distributions of $\bar{x}$ and $\hat{p}$. In this section we will discuss one of the most powerful theorems in statistics, the *Central Limit Theorem* (CLT), which states that the sampling distributions of $\bar{x}$ and $\hat{p}$ become approximately normal as the sample size increases. Since the rules regarding what constitutes a *large* sample differ for the two types of sampling distributions, we shall discuss these separately.

13.4.1   The CLT and Sampling Distribution of the Sample Mean

In some cases, the random variable x may be normally distributed; that is, the population from which the sample is drawn may be normally distributed. Whenever this is the case, the sampling distribution of $\bar{x}$ is also normally distributed, regardless of the sample size, with $\mu_{\bar{x}} = \mu$ and $\sigma_{\bar{x}} = \sigma/\sqrt{n}$. However, in many cases x is not normally distributed, that is, the population is not normally distributed. In any of these cases, the shape of the sampling distribution depends on a combination of the sample size and the shape of the population distribution, the latter of which is often unknown. Fortunately, the Central Limit Theorem allows us to develop an approximation of the sampling distribution with the only necessity being a *large* sample size.

   CENTRAL LIMIT THEOREM: Let x be a random variable representing a population with mean μ and standard deviation σ. The sampling distribution of $\bar{x}$ for a random sample of size n becomes approximately normally distributed, irrespective of the shape of the population distribution, as n increases. In other words, the sampling distribution of $\bar{x}$ can be approximated by a normal distribution with $\mu_{\bar{x}} = \mu$ and $\sigma_{\bar{x}} = \sigma/\sqrt{n}$. In many cases, this approximation works well as long as $n \geq 30$.

   It is crucial to remember that the sampling distribution does not become a normal distribution when n becomes 30. Instead, this means that the sampling distribution usually takes on a shape that is close to a normal curve at this point. Larger sample sizes will result in sampling distributions of $\bar{x}$ that are closer to a normal curve. In other words, the sampling distribution of $\bar{x}$ will be closer to a normal curve when $n = 75$ than when $n = 60$, which is closer to a normal curve than when $n = 40$, and so on. Figure 13.2

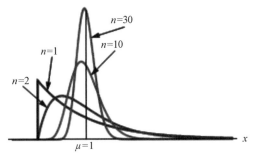

Figure 13.2   Sampling distributions of $\bar{x}$ for sample sizes of 2, 10, and 30 with an exponential population.

shows the population distribution for an exponential distribution, which was discussed in Section 12.4, and is nothing close to being a bell curve, as well as the sampling distributions of $\bar{x}$ for sample sizes of 2 and 10, respectively. However, the sampling distribution of $\bar{x}$ for $n = 30$ is close to being normal, although it is not exactly normal. Note that the distribution with $n = 1$ is actually the population distribution.

The graph of the sampling distribution when $n = 30$ appears to be a normal curve at first glance. However, if you look at the vertical line representing the mean, you will notice that it is slightly to the right of the peak of the curve. If the distribution were exactly normal, the mean would occur at the peak. Regardless, this approximation is still reasonable in many cases. If you are concerned about having a very close approximation, you can simply take a larger sample.

Example 13.8   The number of customers who enter a bank per 10-minute interval follows a distribution with a mean of 4.7 and a standard deviation of 2.168 customers. Forty-five nonoverlapping 10-minute intervals are selected at random.

1. What are the mean and standard deviation of $\bar{x}$, which is the average number of customers entering the bank during a 10-minute period? What is the shape of the sampling distribution of $\bar{x}$?

2. What is the approximate probability that the average number of customers will be below 4?

3. What is the approximate probability that the average number of customers will be 5 or more?

Solution   We are given that $\mu = 4.7$, $\sigma = 2.168$, and $n = 45$.

1. $\mu_{\bar{x}} = \mu = 4.7$ customers   and $\sigma_{\bar{x}} = \frac{\sigma}{\sqrt{n}} = \frac{2.168}{\sqrt{45}} = .323$ customers
2. Because $n = 45$, which is greater than 30, the central-limit theorem applies. Thus, the sampling distribution of $\bar{x}$ is approximately normal. Since $n > 30$, we can approximate $P(\bar{x} < 4)$ using the normal distribution:

$$P(\bar{x} < 4) = P\left(z < \frac{\bar{x} - \mu}{\sigma_{\bar{x}}}\right) = P\left(z < \frac{4 - 4.7}{.323}\right) = P(z < -2.17) = .0150.$$

Note that here, we have used the following formula to convert an $\bar{x}$–value to a z value:

$$z = \frac{\bar{x} - \mu}{\sigma_{\bar{x}}}$$

Figure 13.3 shows the probability of $\bar{x} < 4$.

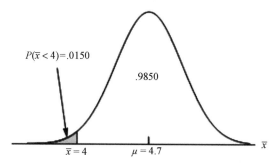

Figure 13.3   Region corresponding to Example 13.8 part 2.

3. Since $n > 30$, we can approximate $P(\bar{x} \geq 5)$ using the normal distribution:

$$P(\bar{x} \geq 5) = P\left(z \geq \frac{\bar{x} - \mu}{\sigma_{\bar{x}}}\right) = P\left(z \geq \frac{5 - 4.7}{.323}\right) = P(z \geq .93) = 1 - .8238 = .1762$$

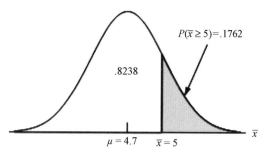

Figure 13.4   Region corresponding to Example 13.8 part 3.

Figure 13.4 shows the probability for part 3.   ■

Example 13.9   A company that produces a new synthetic motor oil for older cars states that it lasts longer than traditional motor oils. Suppose that the amount of time the oil lasts before it should be changed follows a distribution with a mean of 4800 miles and a standard deviation of 300 miles. A random sample of 35 older vehicles is selected and the new motor oil is tested on them. What is the approximate probability that the average distance traveled between oil changes will exceed 4900 miles?

Solution   We have been told that $\mu = 4800$, $\sigma = 300$, and $n = 35$. Thus

$$\mu_{\bar{x}} = \mu = 4800 \text{ miles} \qquad \text{and} \qquad \sigma_{\bar{x}} = \frac{\sigma}{\sqrt{n}} = \frac{300}{\sqrt{35}} = 50.71 \text{ miles}$$

Since $n = 35$, which is greater than 30, we can approximate $P(\bar{x} > 4900)$ using the normal distribution:

$$P(\bar{x} \geq 4900) = P\left(z \geq \frac{4900 - 4800}{50.71}\right) = P(z \geq 1.97) = 1 - .9756 = .0244$$

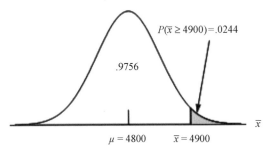

Figure 13.5 Region corresponding to Example 13.9.

Figure 13.5 shows this probability.   ■

13.4.2 The CLT and Sampling Distribution of the Sample Proportion

As was mentioned earlier in this chapter, the central-limit theorem can be applied to the sampling distribution of a proportion, but the rules regarding what constitutes a large sample differ dramatically from the rather simple $n \geq 30$ rule for the sampling distribution of $\bar{x}$.

Central-Limit Theorem as Applied to $\hat{p}$. Let $\hat{p}$ be the sample proportion of successes in a binomial experiment with n trials and $P(\text{success}) = p$. The sampling distribution of $\hat{p}$ for a random sample of size n becomes approximately normally distributed with

$$\mu_{\hat{p}} = p \quad \text{and} \quad \sigma_{\hat{p}} = \sqrt{\frac{p(1-p)}{n}} \quad \text{if} \quad np \geq 10 \quad \text{and} \quad n(1-p) \geq 10$$

Note that when $np \geq 10$ and $n(1-p) \geq 10$, we call it a large sample in the case of the sample proportion $\hat{p}$ to apply the CLT.

Example 13.10   Using the rules stated in the central-limit theorem as applied to a sample proportion, determine whether sampling distribution of $\hat{p}$ is approximately normal for the following situations:

1. $n = 50$, $p = .30$
2. $n = 80$, $p = .10$
3. $n = 500$, $p = .997$

Solution   For each situation, we have to determine whether both $np \geq 10$ and $n(1-p) \geq 10$ hold. If at least one of the two conditions fails to hold, the normal approximation cannot be used:

1. Here, $np = 50(.30) = 15 > 10$ and $n(1-p) = 50(.70) = 35 > 10$. Since both of the conditions are met, we can use the normal approximation.
2. Here, $np = 80(.10) = 8 < 10$ and $n(1-p) = 80(.90) = 72 > 10$. The second condition is met, but the first condition is not met, so the normal approximation cannot be used.
3. Here, $np = 500(.997) = 498.5 > 10$ and $n(1-p) = 500(.003) = 1.5 < 10$. Thus, the first condition is met, but the second condition is not met, so the normal approximation cannot be used.   ■

We may wonder why a product of np or $n(1-p)$ equaling 10 or more is so important in these conditions. To obtain a feel for this, consider part 3 of Example 13.10. The mean and standard deviation of the sampling distribution of $\hat{p}$ are

$$\mu_{\hat{p}} = p = .997 \quad \text{and} \quad \sigma_{\hat{p}} = \sqrt{\frac{p(1-p)}{n}} = \sqrt{\frac{.997(1-.997)}{500}} = .00245$$

If the sampling distribution were approximately normal, then the empirical rule would hold. Thus, approximately 99.7% of the $\hat{p}$ values would fall in the interval

$$\hat{p} \pm 3\sigma_{\hat{p}} = .997 \pm 3(.00245) = .990 \text{ to } 1.004$$

with approximately $(100-99.7)/2 = 1.5\%$ of the values falling outside each side of the interval. However, a proportion must always be between 0 and 1. The upper endpoint of the aforementioned interval is above 1, which is not possible. Thus, the conditions $np \geq 10$ and $n(1-p) \geq 10$ are set to make sure that the interval $\mu_{\hat{p}} \pm 3\sigma_{\hat{p}}$ always falls completely between 0 and 1.

Example 13.11   A microchip manufacturer selects 400 chips at random from a large production run for quality testing. Let $\hat{p}$ be the proportion of chips in this sample that are found to be defective. If 3% of all the chips in the production run are defective, what is the approximate probability that the sample proportion of defective chips will be between .02 and .04?

Solution   First, let us verify that the sample size is large enough for the normality approximation to hold. We have $n = 400$ and $p = .03$. Thus

$$np = 400(.03) = 12 \geq 10 \quad \text{and} \quad n(1-p) = 400(.97) = 388 \geq 10$$

so the normal approximation holds.

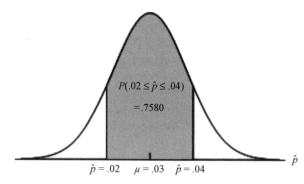

Figure 13.6   Region corresponding to Example 13.11.

The mean and standard deviation of the sampling distribution of $\hat{p}$ are

$$\mu_{\hat{p}} = p = .03 \quad \text{and} \quad \sigma_{\hat{p}} = \sqrt{\frac{p(1-p)}{n}} = \sqrt{\frac{.03(1-.03)}{400}} = .00853$$

We can approximate $P(.02 \leq \hat{p} \leq .04)$ using the normal distribution as follows:

$$P(.02 \leq \hat{p} \leq .04) = P\left(\frac{.02 - .03}{.00853} \leq z \leq \frac{.04 - .03}{.00853}\right) = P(-1.17 \leq z \leq 1.17)$$

$$= P(z \leq 1.17) - P(z \leq -1.17) = .8790 - .1210 = .7580$$

Note that here, we use the formula $(\hat{p} - p)/\sigma_{\hat{p}}$ to convert $p = .02$ and $p = .04$ to z values. Figure 13.6 shows this probability. ∎

Example 13.12   A credit card company selects 80 of its cardholders at random. Let $\hat{p}$ be the sample proportion of cardholders that make this month's payment on time. If 74% of all credit card–holders make their monthly payments on time, what is the probability that the sample proportion will be at least .80?

Solution   First, let us verify that the sample size is large enough for the normality approximation to hold. We have $n = 80$ and $p = .74$. Thus

$$np = 80(.74) = 59.2 \geq 10 \quad \text{and} \quad n(1 - p) = 80(.26) = 20.8 \geq 10$$

so the normal approximation holds.

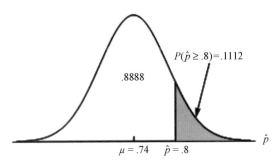

$P(\hat{p} \geq .8) = .1112$

.8888

$\mu = .74 \qquad \hat{p} = .8$

Figure 13.7   Region corresponding to Example 13.12.

The mean and standard deviation of the sampling distribution of $\hat{p}$ are

$$\mu_{\hat{p}} = p = .74 \quad \text{and} \quad \sigma_{\hat{p}} = \sqrt{\frac{p(1 - p)}{n}} = \sqrt{\frac{.74(1 - .74)}{80}} = .0490$$

We can approximate $P(\hat{p} \geq .80)$ using the normal distribution as follows:

$$P(\hat{p} \geq .80) = P\left(z \geq \frac{.8 - .74}{.0490}\right) = P(z \geq 1.22) = 1 - .8888 = .1112$$

Note that here, $\hat{p} = .80$ is converted to a z–value using the formula $(\hat{p} - p)/\sigma_{\hat{p}}$. This probability is shown in Figure 13.7. ∎

GLOSSARY

central-limit theorem   The theorem from which it is inferred that for a large sample size ($n \geq 30$), the shape of the sampling distribution of $\bar{x}$ is approximately normal. Also, by the same theorem, the shape of the sampling distribution of $\hat{p}$ is approximately normal for a sample for which $np \geq 10$ and $n(1 - p) \geq 10$.

estimator   A sample statistic that is used to estimate a population parameter.

finite population correction factor   Multiple denoted by $\sqrt{(N - n)/(N - 1)}$ that is used in calculating the standard deviation of the sampling distribution of $\hat{p}$ or $\bar{x}$ when the sample size is more than 5% of the population size.

mean of $\hat{p}$ The mean of the sampling distribution of $\hat{p}$, denoted by $\mu_{\hat{p}}$, is equal to the population proportion p.

mean of $\bar{x}$ The mean of the sampling distribution of $\bar{x}$, denoted by $\mu_{\bar{x}}$, is equal to the population mean μ.

population distribution   The probability distribution of the population data.

population proportion p   The ratio of the number of elements in a population with a specific characteristic to the total number of elements in the population.

sample proportion $\hat{p}$ The ratio of the number of elements in a sample with a specific characteristic to the total number of elements in that sample.

sampling distribution of $\hat{p}$ The probability distribution of all the values of $\hat{p}$ calculated from all possible samples of the same size selected from a population.

sampling distribution of $\bar{x}$ The probability distribution of all the values of $\bar{x}$ calculated from all possible samples of the same size selected from a population.

sampling error   The difference between the value of a sample statistic calculated from a random sample and the value of the corresponding population parameter. This type of error occurs as a result of chance.

standard deviation of $\hat{p}$ The standard deviation of the sampling distribution of $\hat{p}$, denoted by $\sigma_{\hat{p}}$, is equal to $\sqrt{p(1 - p)/n}$ when $n/N \leq .05$.

standard deviation of $\bar{x}$ The standard deviation of the sampling distribution of $\bar{x}$, denoted by $\sigma_{\bar{x}}$, is equal to $\sigma/\sqrt{n}$ when $n/N \leq .05$.

unbiased estimator   An estimator with an expected value (or mean) that is equal to the value of the corresponding population parameter.

EXERCISES

13.1   The following data give the years of experience for all five employees of a company.

$$20 \quad 8 \quad 25 \quad 6 \quad 12$$

(a) Let x denote the years of experience for an employee of this company. Write the population distribution of x.

(b) List all the possible samples of size four (without replacement) from this population. Calculate the mean for each of these samples. Write the sampling distribution of $\bar{x}$.

(c) Calculate the population mean. By how much does each sample mean differ from the population mean? In other words, calculate the sampling error for each sample mean.

13.2   A population of $N = 5000$ has a standard deviation of $\sigma = 20$. Using the appropriate formula, calculate $\sigma_{\bar{x}}$, the standard deviation of the sampling distribution of $\bar{x}$ for the following sample sizes.

(a) $n = 300$ (b) $n = 200$ (c) $n = 500$ (d) $n = 100$

13.3 The time taken to learn an assembly job for a new worker hired in the assembly department of a company is normally distributed with a mean of 80 hours and a standard deviation of 6 hours. Let $\bar{x}$ be the mean time taken to learn this job by a simple random sample of 16 new workers. Calculate the mean and standard deviation of $\bar{x}$ and describe the shape of its sampling distribution.

13.4 The amounts of electricity bills for all households in a city have an approximate normal distribution with a mean of $160 and a standard deviation of $20. Let $\bar{x}$ be the mean amount of electricity bills for a simple random sample of 25 households selected from this city. Calculate the mean and standard deviation of $\bar{x}$ and describe the shape of its sampling distribution.

13.5 The amounts of electricity bills for all households in another city have a skewed probability distribution with a mean of $145 and a standard deviation of $30. Find the probability that the mean amount of electricity bills for a simple random sample of 100 households selected from this city will be

(**a**) more than $138

(**b**) between $140–$152

(**c**) within $7 of the population mean

13.6 The balances of all savings accounts at a local bank have a distribution that is skewed to the right with a mean of $12,450 and a standard deviation of $4160. Find the probability that the mean of a simple random sample of 50 savings accounts selected from this bank will be

(**a**) more than $11,500

(**b**) between $12,000–$13,800

(**c**) within $1500 of the population mean

13.7 A survey of all medium-sized and large corporations showed that 65% of them offer retirement plans to their employees. Let $\hat{p}$ be the proportion in a simple random sample of 70 such corporations that offer retirement plans to their employees.

(**a**) Calculate the mean and standard deviation of $\hat{p}$ and describe the shape of its sampling distribution.

(**b**) Find the probability that the value of $\hat{p}$ will be between .54 and .61.

(**c**) Find the probability that the value of $\hat{p}$ will be more than .71.

13.8 Dartmouth Distribution Warehouse makes deliveries of a large number of products to its customers. It is known that 85% of all the orders that it receives from its customers are delivered on time. Let $\hat{p}$ be the proportion of orders in a simple random sample of 100 that are delivered on time.

(**a**) Calculate the mean and standard deviation of $\hat{p}$ and describe the shape of its sampling distribution.

(**b**) Find the probability that the value of $\hat{p}$ will be between .81 and .88.

(**c**) Find the probability that the value of $\hat{p}$ will be less than .87.

14 Single-Population Estimation

14.1 OVERVIEW

Why do we need confidence intervals? When you go shopping for a new car in the United States, the sticker on the car provides estimates of the car's gas mileage when used for highway driving and for city driving. So, when you see that a car gives an estimated 28 miles per gallon (mi/gal or mpg) on the highway, should you be certain that you will get the same mileage if you buy that car? As we discussed in Chapter 13, estimators such as the sample mean and sample proportion vary from one sample to another, so you cannot be certain that your car will match 28 mpg. If you are lucky, you might have purchased one that does better.

The estimated mpg comes from a sample of tested cars; specifically, it is the value of $\bar{x}$ for that sample. Statistics such as $\bar{x}$ and $\hat{p}$ are called *point estimators* of μ and p, respectively, because each value can be represented by a single point on a number line. We use point estimators to estimate *population parameters*, which are values that correspond to the entire population of interest. In Chapter 13 we discussed the fact that $\bar{x}$ and $\hat{p}$ are unbiased estimators of μ and p, respectively, but that does not imply that every sample will produce a value of $\bar{x}$ that is equal to μ. In fact, it is highly unlikely that a value of $\bar{x}$ will be the same as that of μ. Thus, while point estimators are very important, they do not tell the whole story about a parameter. We need to have something more than just a single value.

One other issue regarding sampling distributions is that they allow us to quantify the variability of a point estimator. In Example 13.9, we observed that the average lifetime of new synthetic motor oil in a random sample of 35 cars could be approximated by a normal distribution with a standard deviation $\sigma_{\bar{x}}$ of 50.71 miles. On the basis of the empirical rule, we would expect approximately 95% of the values of $\bar{x}$ to fall within two standard deviations of the population mean. Thus, we are fairly certain that a value of $\bar{x}$ will fall within 101.42 miles of either side of μ. If a random sample of 35 vehicles results in a sample mean lifetime of 4850 miles, then we could form an *interval estimate* of μ to be 4748.58 to 4951.42, which we could use to infer that the average lifetime of the new motor oil falls somewhere in this range. Moreover, we can state this inference with (approximately) 95% confidence.

The basic formation of a *confidence interval* for a population parameter involves a point estimator for the parameter and the standard deviation of the point estimator's sampling distribution. The standard deviation is multiplied by a constant, related to the desired *confidence level* and the type of distribution that the point estimator's sampling distribution takes on. This product is referred to as the *margin of error*. The confidence interval is formed by adding and subtracting the margin of error to and from the point estimator. Thus, an interval estimate for a population parameter that is constructed with regard to a given confidence level is called a *confidence interval*. Many types of confidence intervals can be expressed in the form

$$\text{Point estimator} \pm \text{margin of error}$$

Practitioner's Guide for Statistics and Lean Six Sigma for Process Improvements. By Mikel J. Harry, Prem S. Mann, Ofelia C. de Hodgins, Christopher J. Lacke, and Richard Hulbert
Copyright © 2010 John Wiley & Sons, Inc.

The confidence level associated with a confidence interval states how much confidence we have in the method producing an interval that contains the true population parameter. The confidence level is denoted by $(1 - \alpha) \times 100\%$ and $0 \leq \alpha \leq 1$. If the confidence level is expressed as a probability $(1 - \alpha)$, it is referred to as the *confidence coefficient*. We will revisit α, which is also known as the *significance level,* in future chapters on hypothesis tests.

Confidence levels can be any number between 0% and 100%, but the most commonly used confidence levels are 90%, 95%, 98%, and 99%. It should make sense that you would not want to use low confidence levels. For example, a 50% confidence interval would produce an interval that is just as likely to include the population parameter as it is to not include it. Similarly, while the notion of 100% confidence seems appealing, the only way to be completely confident that an interval contains the value of the population parameter is for the interval to go from negative infinity to positive infinity, which does not narrow things down at all.

In Section 14.2, we will examine what the confidence level means, as well as what it does not mean. Sections 14.3–14.5 discuss how to construct confidence intervals for the population mean, population proportion, and population variance, respectively. Section 14.6 examines how to determine appropriate sample sizes for forming confidence intervals for use in Six Sigma and Lean Sigma.

14.2 MEANING OF A CONFIDENCE LEVEL

We begin by explaining what a confidence level means by explaining one thing that it does not mean. In our example regarding the mean lifetime of the new synthetic motor oil, we determined an approximate 95% confidence interval to be 4748.58 to 4951.42. We cannot say that the probability that the population mean falls between 4748.58 and 4951.42 is .95. This is so because a population parameter is assumed to be a fixed value that is calculated using data from every member of a population. Although we typically do not know the value of a parameter, we do know that it is not subject to variability, at least at a point in time. Thus, the value of μ either falls inside the interval, or it does not fall inside the interval. In other words, the probability that it falls in the interval is either 0 or 1.

To explain how to interpret a confidence level, suppose that we have a population with a mean of 4850 and a standard deviation of 300. We can take repeated samples of 35 observations from this

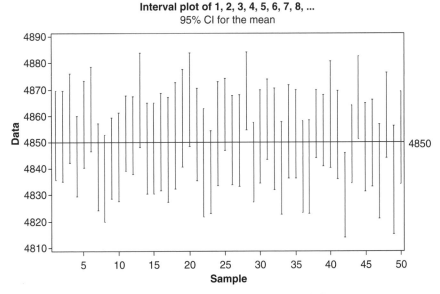

Figure 14.1 Plot of fifty 95% confidence Intervals for μ.

population, calculating a 95% confidence interval for the mean using each sample of 35 observations. Figure 14.1 shows the resulting confidence intervals for each of the 50 sets of 35 observations. A horizontal line has been added to the plot to identify the location of 4850.

The confidence interval formed from the first sample extends from a low of 4836 to a high of 4870. In this case, the population mean (4850) falls in this interval. The second interval, which has endpoints to those of the first, also includes 4850. The third interval also includes 4850, although the interval ranges from 4842 to 4876. As we continue to examine the intervals, note that the confidence interval formed from the 28th sample falls completely above 4850, as does the confidence interval formed from the 44th sample. The interval from the 42nd sample falls completely below 4850. Therefore, we have 47 samples that result in intervals that contain 4850 and three that do not. In other words, 94% of our 50 samples produce intervals that contain the population mean.

If we were to continue this process by taking a very large number of samples of 35 observations, we would find that the percentage of resulting confidence intervals that include 4850 would be very close to 95%. Thus, the confidence level is the percentage of all confidence intervals made using the same sample size that will contain the parameter of interest. Similarly, the confidence coefficient is the probability that the method of forming a confidence interval will result in an interval that contains the parameter of interest.

14.3   ESTIMATING A POPULATION MEAN

Most books will provide the use of two distributions (the normal distribution and the t distribution) for calculating a confidence interval for a population mean. The primary reason for demonstrating the z *interval for a population mean* is that it utilizes the normal distribution, which was covered in Chapter 12. This allows one to concentrate on the foundations of a confidence interval without having to worry about a distribution with which one is unfamiliar. The primary reason for demonstrating the t *interval for a population mean* is that it is the commonly used method in practice, as the z interval involves an assumption that is typically unrealistic in practice.

14.3.1   Confidence Interval for a Population Mean Using the Normal Distribution

As we mentioned earlier, the formation of a confidence interval involves a point estimate and the properties of its sampling distribution. In Chapter 13 we noted that if a random variable x is normally distributed with population mean μ and population standard deviation σ, the sampling distribution of $\bar{x}$ is normal with mean $\mu_{\bar{x}} = \mu$ and standard deviation $\sigma_{\bar{x}} = \sigma/\sqrt{n}$. In addition, according to the central-limit theorem, the sampling distribution of $\bar{x}$ is approximately normal with mean $\mu_{\bar{x}} = \mu$ and standard deviation $\sigma_{\bar{x}} = \sigma/\sqrt{n}$ when $n \geq 30$ even if the population from which the sample is selected is not normally distributed. If either of these conditions is met, then we can develop a $(1-\alpha) \times 100\%$ confidence interval for a population mean by using the formula

$$\bar{x} \pm z\frac{\sigma}{\sqrt{n}} \qquad \text{or} \qquad \bar{x} \pm z\sigma_{\bar{x}}$$

where z is the value obtained from the normal table (Table II in Appendix A) for a given confidence level. The process of determining the value z is similar to the process demonstrated in Examples 12.4 and 12.5. Example 14.1 demonstrates how to determine the value of z for a given confidence interval using the normal table, Table II of Appendix A.

Example 14.1   Determine the value of z for confidence levels of 90%, 95%, and 99%.

Solution   The region corresponding to the confidence coefficient is the area corresponding to the middle $(1-\alpha)$ under the normal curve. The area outside this region is α, which is the total area under

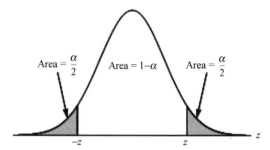

Figure 14.2 Critical values for a $(1-\alpha) \times 100\%$ z interval of a population mean.

the two tails. Since the normal curve is symmetric, the area in each tail is $\alpha/2$. This relationship is demonstrated in Figure 14.2.

Whether using the normal table (Table II of Appendix A) or a software package such as MINITAB, the area of interest is the area to the left of z (i.e., positive z). This area will always be $1-(\alpha/2)$. In Figure 14.2, total area to the left of z is $1-\alpha+(\alpha/2)=1-(\alpha/2)$. When we are creating a 90% confidence interval, we have $1-\alpha=.90$, so $\alpha=.10$, resulting in $1-(\alpha/2)=1-(.10/2)=.95$. The areas of interest for our three confidence levels are shown in Table 14.1.

TABLE 14.1 Determining $1-(\alpha/2)$ for a Confidence Level

Confidence Level (%)	$1-\alpha$	α	$\alpha/2$	$1-(\alpha/2)$
90	.90	.10	.05	.95
95	.95	.05	.025	.975
99	.99	.01	.005	.995

TABLE 14.2 Critical Values for Example 14.1

	90%	95%	99%
$1-(\alpha/2)$	.95	.975	.99
z using normal table	1.64 or 1.65	1.96	2.58
z to 3 decimals	1.645	1.960	2.576

To determine the appropriate z using the normal table, find the value closest to $1-(\alpha/2)$ in the region that contains the areas, then read to the left of the row and the top of the column to find the corresponding z value. For a 90% confidence interval, we want to find the value in the table that is closest to .95. Examining the normal table, we find .9495 and .9505. Reading back to the z column and the top row, we see that the corresponding z-values are 1.64 and 1.65, respectively. Since .95 falls in the middle of .9495 and .9505, we could use 1.64 or 1.65 as the value of z. However, we will note that the appropriate z-value to three decimal places is 1.645, which is noted in Table 14.2. ■

You might have noticed that the z value for a 95% confidence interval is 1.96. If you recall the empirical rule, we stated that any normal distribution contains approximately 95% of all observations within two standard deviations of the mean. We can now state a more precise version, namely, that 95% of all observations in a normal distribution fall within 1.96 standard deviations of the mean.

Critical values for any other distribution can be found using similar methods. However, statistical software packages contain functions that will calculate confidence intervals, so the process of finding a critical value is more important for those who wish to understand the relationship between the confidence level and the specific critical value than for those who wish to use confidence intervals.

14.3.1.1   Underlying Conditions for Using the z Interval for a Population Mean

To use the normal distribution to make a confidence interval for μ., the following conditions must hold true:

1. The data are obtained using a simple random sample.
2. The population standard deviation σ is known.
3. The population distribution is approximately normal or the sample size is large enough for the central-limit theorem to become valid. In case of the sample mean, $n \geq 30$ is considered to be a large sample.

Of these three assumptions, assumption 2 is exceedingly unlikely in practice. Some books will recommend using the z interval if the sample size is in the hundreds (or larger) even when the population standard deviation is unknown. This approach leads to intervals that are narrower than the corresponding t interval, which results in the actual confidence level being lower than the specified confidence level.

Example 14.2   The life (in hours) of a 75-watt lightbulb manufactured by a company has a standard deviation σ equal to 350 hours. A random sample of 40 such lightbulbs is selected, and the lives of these bulbs are measured. The average life of the 40 bulbs is found to be 840 hours. Let μ be the average lifetime of all such 75-watt lightbulbs.

1. Calculate a 90% confidence interval for μ.
2. Calculate a 95% confidence interval for μ.
3. Calculate a 99% confidence interval for μ.
4. What do you notice about the relationship between the confidence level and the width of the corresponding confidence interval?

Solution   Although we do not know the shape of the distribution of lives of such bulbs, the sample size is larger than 30 and σ is known. Hence, the central-limit-theorem allows us to approximate the confidence interval using the normal distribution. We have $\bar{x} = 840$, $\sigma = 350$, and $n = 40$. We can use the values of z for 90%, 95%, and 99% confidence levels from Table 14.2. The corresponding confidence intervals are as follows:

1. The 90% confidence interval for μ is

$$\bar{x} \pm z \frac{\sigma}{\sqrt{n}} = 840 \pm 1.645 \times \frac{350}{\sqrt{40}} = 840 \pm 91.03 = 748.97 \text{ to } 931.03$$

2. The 95% confidence interval for μ is

$$\bar{x} \pm z \frac{\sigma}{\sqrt{n}} = 840 \pm 1.960 \times \frac{350}{\sqrt{40}} = 840 \pm 108.47 = 731.97 \text{ to } 948.47$$

3. The 99% confidence interval for μ is

$$\bar{x} \pm z \frac{\sigma}{\sqrt{n}} = 840 \pm 2.576 \times \frac{350}{\sqrt{40}} = 840 \pm 142.56 = 697.44 \text{ to } 982.56$$

4. The width of a confidence interval is the distance between the two endpoints. The widths for the three confidence intervals are 182.06, 216.94, and 285.12, respectively. From this we observe

that whenever the confidence level is increased, and all other items remain the same, the width of the confidence interval increases. Note that the width of a confidence interval is obtained by taking the difference between the two values of the confidence interval. ∎

Example 14.3 A medical insurance company wishes to estimate the average claim amount for an angioplasty. A simple random sample of 50 claims had a mean claims of $\bar{x} = \$20,387$. Suppose that the standard deviation of all angioplasty claims amounts is $1845. Calculate a 95% confidence interval for the average claim amount of all angioplasty procedures.

Solution Although we do not know the shape of the distribution of angioplasty claim amounts, the sample size is larger than 30 and σ is known, so the central-limit theorem allows us to approximate the confidence interval using the normal distribution. We have $\bar{x} = \$20,387$, $\sigma = \$1845$, and $n = 50$. For a 95% confidence level, the value of z from the normal distribution table (Table II of Appendix A) is 1.96. The corresponding confidence interval is

$$\bar{x} \pm z \frac{\sigma}{\sqrt{n}} = 20,387 \pm 1.96 \times \frac{1845}{\sqrt{50}} = 20,387 \pm 511.41 = \$19,875.59 \text{ to } \$20,898.41$$

Thus, we are 95% confident that the average claim amount for an angioplasty is between $19,875.59 and $20,898.41. ∎

14.3.2 Confidence Interval for a Population Mean Using the *t* Distribution

The z interval discussed above requires that we know the population standard deviation σ. This is intriguing because the population standard deviation is a measure of distance from the population mean, yet we are creating a confidence interval to estimate the population mean because we do not know its value. Typically, the only information that we have about the standard deviation comes from the sample standard deviation s, which we use as an estimator of σ.

When we use s in place of σ, the sampling distribution of $\bar{x}$ is no longer normally distributed. Instead, it follows a distribution called the *Student's t distribution* (Student was the pseudonym of British statistician W. S. Gosset), which is often called the *t distribution*. The *t* distribution and the standard normal distribution have some properties in common. Both distributions are centered at zero; that is, both have the mean equal to zero. Both distributions are bell-shaped and symmetric. However, the standard normal distribution has a standard deviation of 1, while the standard deviation of the *t* distribution depends on the sample size through a quantity called the *degrees of freedom* (df). When calculating a confidence interval for a one population mean, the value of the degrees of freedom is Degrees of freedom $(df) = n - 1$. Figure 14.3 displays a graph of the standard normal distribution and graphs of the *t* distribution for sample sizes of 3, 6, 11, and 31, which give df values of 2, 5, 10, and 30, respectively.

Because a *t* distribution has a larger standard deviation than does the standard normal distribution, the critical value, *t*, for a specific confidence level, will always be larger than the corresponding z value. Table 14.3 lists values of *t* for 90%, 95%, and 99% confidence intervals for selected numbers of degrees of freedom. The *t* values are obtained from Table III of Appendix A for $\alpha/2$ area in the right tail of the *t* distribution and given df. As one can see, the *t* values approach the corresponding z values as the degrees of freedom increase.

In addition to the fact that z must be replaced by its corresponding *t* value, the quantity $\sigma/\sqrt{n}$ is now no longer relevant. The standard deviation of the sampling distribution of $\bar{x}$ is now replaced by $s/\sqrt{n}$, which is called the *standard error of $\bar{x}$*, and is an estimate of the standard deviation of $\bar{x}$ denoted by $\sigma/\sqrt{n}$. There are multiple notations for standard error; we shall use $s_{\bar{x}}$ to represent the standard error

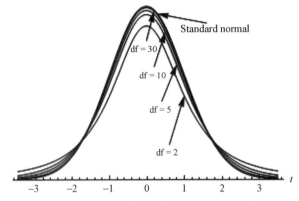

Figure 14.3   Comparison of the t distributions with 2, 5, 10, and 30 degrees of freedom versus standard normal distribution.

TABLE 14.3   Comparison of Critical Values between the t Distribution and the Standard Normal

df	90%	95%	99%
2	2.920	4.303	9.925
5	2.015	2.571	4.032
10	1.812	2.228	3.169
30	1.697	2.042	2.750
50	1.676	2.009	2.678
100	1.660	1.984	2.626
1000	1.646	1.962	2.581
z	1.645	1.960	2.576

of $\bar{x}$. With these substitutions, the formula for a *t interval for a population mean* is given by

$$\bar{x} \pm t\frac{s}{\sqrt{n}} \quad \text{or} \quad \bar{x} \pm ts_{\bar{x}}$$

Table III of Appendix A contains t values for a wide range of degrees of freedom and confidence levels. In addition, the bottom row of the t table gives the critical values for the standard normal distribution.

14.3.2.1   Underlying Conditions for Using the t Procedure

The underlying conditions for using the t procedure to make a confidence interval for a population mean are

1. The data were collected using a simple random sample (SRS).
2. The population standard deviation σ is *unknown*.
3. The population from which the sample is drawn is normally distributed.

Identifying that the first condition is satisfied simply involves knowing, or being able to find out, how the data were collected. If you are part of the team designing the experiment, then it is important to use a simple random sample for selecting the observational units/subjects that are involved in the study.

As mentioned numerous times, the population standard deviation is usually unknown, so the t procedure is the typical procedure for finding a confidence interval for a population mean.

To find out if the population is normally distributed is almost impossible without taking a census. Instead of verifying this condition, we can use a variety of methods to determine whether the assumption that the population is normally distributed is a reasonable one but these methods are not discussed in this text.

Examples 14.4 and 14.5 demonstrate using the t procedures under the assumption that the data come from a normal distribution.

Example 14.4   Corrugated cardboard boxes are rated according to the amount of weight they can hold. A random sample of 11 boxes having the same dimensions is selected. Weight is added to each box until it breaks. The sample mean and standard deviation of the breaking weights are 52.34 and 5.79 lbs, respectively. Assume that the breaking weights of these boxes are normally distributed.

1. Calculate a 95% confidence interval for μ, the average breaking weight of all such boxes.
2. What would the resulting confidence interval be if the data had been collected from a sample of 80 boxes? All other values remain the same.
3. What would the resulting confidence interval be if the population standard deviation was known to be $\sigma = 5.79$? Compute the interval using sample sizes of 11 and 80.

Solution   The t procedure works under the assumption that the data come from a normal distribution. We assume this to be the case in this example, but we will discuss the issue in greater detail after this example. We have $\bar{x} = 52.34$ and $s = 5.79$.

1. The sample size is 11, so there are $11 - 1 = 10$ degrees of freedom in this case. To use the t table (Table III of Appendix A) for finding the critical value, we need to determine the value of $\alpha/2$, which represents the area in each tail of the t distribution. The values in the top row of Table III of Appendix A corresponds to the value of $\alpha/2$. Once this value has been determined, we read down the column corresponding to $\alpha/2$ and across the row corresponding to the number of degrees of freedom. The value that appears at the intersection of the appropriate row and column is the needed t value. For a 95% confidence interval, $\alpha = .05$, so $\alpha/2 = .025$. Reading down the .025 column and reading across the row for df $= 10$, we observe that $t = 2.228$, resulting in the confidence interval for μ as

$$\bar{x} \pm t\frac{s}{\sqrt{n}} = 52.34 \pm 2.228\left(\frac{5.79}{\sqrt{11}}\right) = 52.34 \pm 3.890 = 48.45 \text{ to } 56.23$$

2. The only item that has changed in the problem is the number of degrees of freedom. Here, the df are $80-1 = 79$. Since the value of t is dependent on the number of degrees of freedom, we must return to Table III of Appendix A. Once again, we will use the column that corresponds to .025. However, Table III has no row for df $= 79$, although there are rows for df $= 75$ and df $= 80$. Typically we will use the row corresponding to the smaller number of degrees of freedom, so we will read across the row for df $= 75$, obtaining t $= 1.992$. The resulting confidence interval for μ is

$$\bar{x} \pm t\frac{s}{\sqrt{n}} = 52.34 \pm 1.992\left(\frac{5.79}{\sqrt{80}}\right) = 52.34 \pm 1.290 = 51.05 \text{ to } 53.63$$

   Thus, we are 95% confident that the average breaking weight of all boxes of this type is between 51.05 and 53.63 lbs.

3. Here, σ is known and other conditions that are required to use the normal distribution are met. Since we are calculating a 95% confidence interval, we use $z = 1.960$. The corresponding

confidence intervals for μ are

$$\text{For } n = 11: \quad \bar{x} \pm z\frac{\sigma}{\sqrt{n}} = 52.34 \pm 1.960\left(\frac{5.79}{\sqrt{11}}\right) = 52.34 \pm 3.422 = 48.918 \text{ to } 55.762$$

$$\text{For } n = 80: \quad \bar{x} \pm z\frac{\sigma}{\sqrt{n}} = 52.34 \pm 1.960\left(\frac{5.79}{\sqrt{80}}\right) = 52.34 \pm 1.269 = 51.071 \text{ to } 53.609$$

■

Let us compare the results obtained in Example 14.4 parts 1–3 above. These results are summarized in Table 14.4.

TABLE 14.4   **Comparison of Confidence Intervals using the t and z procedures**

	$n = 11$	$n = 80$
t interval	48.450 to 56.230	51.052 to 53.629
z interval	48.918 to 55.762	51.071 to 53.609

As we noted in Example 14.2, larger sample sizes result in narrower confidence intervals. Also, we note from Table 14.4 that t intervals are wider than the corresponding z intervals. If we compare the t-interval endpoints with the z-interval endpoints, we see that the endpoints for the t and z intervals get closer as the sample size increases.

Some of us may wonder why we should worry about using a t interval if the sample size is large and s is used as an estimate of σ. The t method accounts for the fact that we are only estimating the population standard deviation. In other words, it recognizes that we have more uncertainty, which implies that our interval should be wider in order to adjust for the additional uncertainty. While the difference may seem small, approximately 0.02 lb in the example with 80 observations, we do not know what will happen before the data are collected. Sound statistical procedures call for using the proper method for making a confidence interval, or performing any other type of analysis.

Example 14.5   A simple random sample of 27 full-time employees of a company was taken to estimate the average number of hours worked per week by all full-time employees of the company. The sample mean is found to be 40.74 hours and the sample standard deviation is 6.21 hours. Assuming that the number of hours worked by all full-time employees is normally distributed, determine a 99% confidence interval for the average time worked by all full-time employees of this company.

Solution   The t procedure works under the assumption that the data come from a normal distribution. We are told that this is the case in this example. We have $\bar{x} = 40.74$, $s = 6.21$, and $n = 27$. Here, the number of degrees of freedom is $n-1 = 26$, and $\alpha/2 = (1 - .99)/2 = .005$. The critical value appears at the intersection of the row for df $= 26$ and the .005 column in Table III of Appendix A. Thus $t = 2.779$. The resulting confidence interval for μ is

$$\bar{x} \pm t\frac{s}{\sqrt{n}} = 40.74 \pm 2.779\left(\frac{6.21}{\sqrt{27}}\right) = 40.74 \pm 3.32 = 37.42 \text{ to } 44.06$$

■

14.3.2.2   *Using the t Procedure When the Normality Assumption is Questionable*

Although the t procedure assumes that the data come from a normal distribution, it is *robust* to violation of that assumption. In other words, the procedure can still be used when the sample size is large enough to compensate for the reason that one may question the normality assumption. In general, as the sample size increases, the t procedure can be used with data that are more nonnormal. The following sample sizes are a good rule of thumb for using the t procedure:

1. $n < 15$: The t procedure can be used with data that are close to being normally distributed. Data that are heavy-tailed, which means that the tails of the distribution are thicker than a normal distribution, but symmetric will often work.
2. $15 \leq n < 40$: The t procedure can be used with data that have, at most, a moderate amount of skewness. The t procedure should not be used if the data are strongly skewed or have outliers.
3. $n \geq 40$: The t procedure can be used as long as the dataset does not contain outliers.

Example 14.6 will explain the process of examining a graph of the data and using the t procedure for making a confidence interval for the mean.

Example 14.6   A retailer that often hires retirees uses information on monthly Social Security payments to ensure that these retiree employees are paid a sufficient salary. The following data give the monthly Social Security payments (in dollars) for a simple random sample of 25 Social Security recipients. Determine if it is reasonable to use the t procedure, and if so, calculate a 98% confidence interval for the average monthly Social Security payment.

1276	1087	1150	1207	974
1040	942	1357	1004	1017
1287	1278	1318	1263	1262
831	1090	1188	864	1291
1150	935	1062	1237	899

Solution   We need to determine whether the normality assumption is reasonable for this dataset. Although a normal quantile plot is the best type of graph to use in this situation, we will use the histogram shown in Figure 14.4 to check the distribution of data.

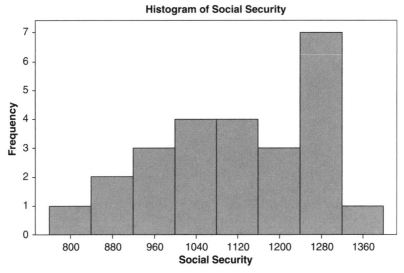

Figure 14.4   Histogram of Social Security payments for Example 14.6.

There are 25 data points in this dataset. Returning to the rules for using the t procedure, since 25 falls within the second group, we can use the t procedure as long as the data are not extremely skewed and/or do not have any outliers. The shape of the plot in Figure 14.4 implies that the distribution is somewhat skewed to the left, but the amount of skew is not extreme; therefore, the data meet the qualifications for using the t procedure.

The sample mean and sample standard deviation for the data are $\bar{x} = 1120.36$ and $s = 156.93$. Note that we find these values by using the formulas for x and s learned in Chapter 8. There are $n - 1 = 25 - 1 = 24$ degrees of freedom. Since we are interested in a 98% confidence interval, $\alpha/2 = (1 - .98)/2 = .01$. To determine the value of t for this interval, we look at the t value at the intersection of the row for df $= 24$ and the .01 column in Table III of Appendix A, which gives $t = 2.492$. The 98% confidence interval for μ is

$$\bar{x} \pm t\frac{s}{\sqrt{n}} = 1120.36 \pm 2.492\left(\frac{156.93}{\sqrt{25}}\right) = 1120.36 \pm 79.83 = 1040.53 \text{ to } 1200.19$$

We are 98% confident that the average monthly Social Security payment is between \$1040.53 and \$1200.19. ∎

14.4   ESTIMATING A POPULATION PROPORTION

There are two primary settings in which one wishes to develop a confidence interval for a population proportion p. One setting involves interest in the proportion/percentage of a population having a specific characteristic. Examples include the percentage of registered voters who plan to vote for a specific candidate, the percentage of people who have a specific disease, and the percentage of people who buy a foreign car. The other setting involves estimating the probability that a certain event will occur. Such examples include the probability that a specific baseball player will get a hit (Boston Red Sox fans will note Theo Epstein's use of statistics in developing the 2004 World Series Championship team), the probability that a person will be involved in an automobile accident during the next year, or the probability that a specific tropical storm will develop into a hurricane that will hit the continental United States.

When we discussed the sampling distribution of the sample proportion $\hat{p}$ in Chapter 13, we learned that the underlying experiment has to follow the principles of a binomial experiment. The same set of conditions must hold true here.

We shall discuss two types of confidence intervals for p. One type is the *traditional method*, based on a large sample and using the principle of the central-limit theorem, discussed in Chapter 13. The second method uses an estimator called the *Wilson estimator*. Although the Wilson estimator method can be used in any situation, it is quite helpful when p is suspected to be very close to 0 or 1 and very large samples are impractical or infeasible.

14.4.1   Traditional Large-Sample Method

The central-limit theorem states that the sampling distribution of $\hat{p}$ for a simple random sample of size n is approximately normally distributed with mean $\mu_{\hat{p}} = p$ and standard deviation $\sigma_{\hat{p}} = \sqrt{p(1-p)/n}$. The conditions required to use this formula for $\sigma_{\hat{p}}$ are that $np \geq 10$ and $n(1-p) \geq 10$. However, since we are trying to estimate p, we cannot use $\sigma_{\hat{p}}$, which is the standard deviation of $\hat{p}$. It turns out, however, that the normal distribution can be used as an approximation when the standard error $s_{\hat{p}} = \sqrt{\hat{p}(1-\hat{p})/n}$ is used as long as $n\hat{p} \geq 10$ and $n(1-\hat{p}) \geq 10$. Since the sampling distribution of $\hat{p}$ can be approximated using the normal distribution, the critical value will be z, just as it is in the z interval for a population mean, or any other confidence interval on the basis of the normal distribution. The formula to obtain a *large sample z interval for a population proportion* is

$$\hat{p} \pm z s_{\hat{p}} \quad \text{or} \quad \hat{p} \pm z\sqrt{\frac{\hat{p}(1-\hat{p})}{n}}$$

Example 14.7 demonstrates the procedure to make a confidence interval for p using the normal distribution.

Example 14.7   A food company is planning to market a new flavor of potato chips. Before starting the ad campaign, they wish to estimate the proportion of people who like the new flavor. A simple random sample of 350 people is selected, and 212 of them said that they like the new flavor. Calculate a 95% confidence interval for p, the proportion of all people who like the new flavor.

Solution   Let $\hat{p}$ be the proportion of people in the sample who like the new flavor. From the given information, we have

$$n = 350, \hat{p} = \tfrac{212}{350} = .6057, n\hat{p} = 350(.6057) = 212 \geq 10$$

and

$$n(1-\hat{p}) = 350(1-.6057) = 138 \geq 10$$

For the 95% confidence level, the z value from Table II of Appendix A is 1.96. The 95% confidence interval for p is

$$\hat{p} \pm z s_{\hat{p}} = .6057 \pm 1.96\sqrt{\frac{.6057(1-.6057)}{350}} = .6057 \pm .0512 = .5545 \text{ to } .6569$$

We can state with 95% confidence that the proportion of all people who will like the new flavor is between .5545 and .6569. We can also state that the percentage of all people who will like the new flavor is between 55.45% and 65.69%.   ■

14.4.2   Wilson Estimator

In many situations the value of p is close to zero. This causes a number of problems with the normal approximation, not the least of which is the need for a very large sample size. Suppose that a company wishes to estimate the proportion of their products that are defective. Furthermore, suppose that the company's actual defective rate is .005 (or .5%). According to the central-limit theorem, $n(.005)$ must be 10 or more, which implies that n must be at least 2000. Testing for defective products is often time-intensive, and often destructive, meaning that the product cannot be sold after it is tested. If the product is expensive, testing 2000 or more units is impractical, so the company will need to use a smaller sample size. If the company decides that they can afford to test only 25 units, then there is a very good chance that none of the 25 will be defective, resulting in $\hat{p} = 0/25 = 0$. If we were to use the large sample method, from previous subsection, the resulting 95% confidence interval would be

$$0 \pm 1.960\sqrt{\frac{0(1-0)}{25}} = 0 \pm 0 = 0 \text{ to } 0$$

which is not useful at all.

A confidence interval for p that resolves this issue is based on the *Wilson estimator* $\tilde{p}$, which is defined as

$$\tilde{p} = \frac{x+2}{n+4}$$

where x is the number of successes (or units with a specific characteristic) in the sample and n is the sample size. Then the confidence interval for p is based on the normal distribution, and is defined to be:

$$\tilde{p} \pm z\sqrt{\frac{\tilde{p}(1-\tilde{p})}{n+4}}$$

Here, the normal approximation works well for a sample size of 5 or more.

Example 14.8   Use the Wilson estimator procedure to calculate a 95% confidence interval for the proportion of all people who like the new flavor of potato chips using the data of Example 14.7.

Solution   From the information given in Example 14.7, we have

$$n = 350, \quad x = 212, \quad \tilde{p} = \frac{212+2}{350+4} = .6045, \quad z = 1.96$$

Then, the 95% confidence interval for p is

$$\tilde{p} \pm z\sqrt{\frac{\tilde{p}(1-\tilde{p})}{n+4}} = .6045 \pm 1.96\sqrt{\frac{.6045(1-.6045)}{350+4}} = .6045 \pm .0509 = .5536 \text{ to } .6554 \quad \blacksquare$$

Example 14.9   A company tests 50 units of their product, finding one of them to be defective. Use the Wilson estimator procedure to calculate a 99% confidence interval for the proportion of defectives in the population.

Solution   From the given information,

$$n = 50, \quad x = 1, \quad \tilde{p} = \frac{1+2}{50+4} = .0556, \quad z = 2.58$$

The 99% confidence interval for p is

$$\tilde{p} \pm z\sqrt{\frac{\tilde{p}(1-\tilde{p})}{n+4}} = .0556 \pm 2.58\sqrt{\frac{.0556(1-.0556)}{50+4}} = .0556 \pm .0804 = -.0248 \text{ to } .1360$$

We are 95% confident that the company's defective rate is between $-.0248$ and $.1360$. Since the defective rate cannot be negative, we can state that we are 99% confident that the company's defective rate is between 0 and .1360 or between 0% and 13.60%. $\blacksquare$

14.5   ESTIMATING A POPULATION VARIANCE

Just as we are interested in estimating measures of center, such as the mean and proportion, we are also interested in estimating measures of spread, such as the population variance σ^2 and the population standard deviation σ. The confidence intervals for σ^2 and σ are made using the sample variance s^2 and the sample standard deviation s, which were discussed in Chapter 8. Note that once we find the confidence interval for σ^2, we can take the positive square root of the two limits of that confidence interval to obtain the confidence interval for σ.

The confidence interval for σ^2 is based on the *chi-square distribution*. Like the t distribution, the chi-square distribution has a single value of degrees of freedom that determines the shape of the chi-square distribution. Furthermore, the value of the degrees of freedom for the chi-square distribution will depend on the estimation method being used, so degrees of freedom will be calculated differently for each specific test. The chi-square distribution is a nonnegative, right-skewed distribution. Figure 14.5 demonstrates the shapes of chi-square (χ^2) distributions with 2, 5, and 10 degrees of freedom, respectively.

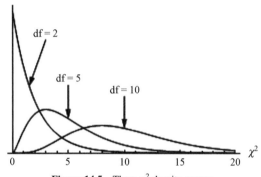

Figure 14.5   Three χ^2 density curves.

As we can see from Figure 14.5, the shape of a χ^2 distribution curve is skewed for very small degrees of freedom and changes drastically as the degrees of freedom increase. Eventually, for large degrees of freedom, the χ^2 distribution curve resembles a normal distribution curve. The peak (or mode) of a χ^2 distribution curve with df $= 1$ or 2 occurs at zero and for a curve with df ≥ 3 peak occurs at $\chi^2 =$ df -2. For instance, the peak of the χ^2 distribution curve with df $= 2$ in Figure 14.5 occurs at zero. The peak for the curve with df $= 5$ occurs at $\chi^2 = 5 - 2 = 3$. Finally, the peak for the curve with df $= 10$ occurs at $\chi^2 = 10 - 2 = 8$. Like all other continuous distribution curves, the total area under a χ^2 distribution curve is 1.0.

If we know the degrees of freedom and the area in the right tail of a χ^2 distribution, we can find the value of χ^2 from Table IV of Appendix A. Example 14.10 demonstrates how to read that table.

Example 14.10   Find the value of χ^2 for each of the following scenarios.

1. df $= 7$ and an area of .10 in the right tail of the χ^2 distribution curve
2. df $= 12$ and an area of .05 in the left tail of the χ^2 distribution curve

Solution

1. To find the required value of χ^2, we locate 7 in the column for df and .10 (which is written as .100 in the table) in the top row in Table IV of Appendix A. The required χ^2 value is given by the entry at the intersection of the row for 7 and the column for .10. This value is 12.017. The relevant portion of Table IV is presented as Table 14.5 here.

TABLE 14.5 χ^2 for 7 *df* and .10 Area in Right Tail

df	.995	...	**.100**	...	.005
			Area in Right Tail		
1	.00004	...	2.706	...	7.879
2	.010	...	4.605	...	10.597
.	...	...	...	...	...
.	...	...	...	...	...
.	...	...	...	...	...
7	.989	...	**12.017**	...	20.278
.	...	...	...	...	...
.	...	...	...	...	...
.	...	...	...	...	...
100	67.328	...	118.498	...	140.169

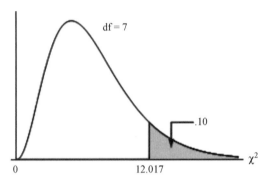

Figure 14.6   Graph for Example 14.10 part 1.

As shown in Figure 14.6, the χ^2 value for df = 7 and an area of .10 in the right tail of the χ^2 distribution curve is 12.017.

2. Table IV of Appendix A can be used only when an area in the right tail of the χ^2 distribution curve is known. When the given area is in the left tail, as in this example, the first step is to find the area in the right tail of the χ^2 distribution curve by subtracting the left-tail area from one. Therefore, for our example, the area in the right tail is $1 - .05 = .95$. Next, we locate 12 in the column for df and .95 in the top row in Table IV. The required value of χ^2, given by the entry at the intersection of the row for 12 and the column for .95, is 5.226. The relevant portion of Table IV is presented as Table 14.6 here.

TABLE 14.6 χ^2 for 12 *df* and .95 Area in the Right Tail

df	.995	...	**.950**	...	.005
			Area in Right Tail		
1	.00004	...	0.004	...	7.879
2	.010	...	0.103	...	10.597
.	...	...	...	...	...
.	...	...	...	...	...
.	...	...	...	...	...
12	.989	...	**5.226**	...	20.278
.	...	...	...	...	...
.	...	...	...	...	...
.	...	...	...	...	...
100	67.328	...	77.930	...	140.169

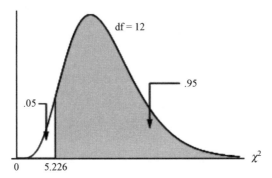

Figure 14.7   Graph for Example 14.10 part 2.

As shown in Figure 14.7, the χ^2 value for df $= 12$ and an area of .05 in the left tail of the χ^2 distribution curve is 5.226. ∎

The confidence interval for σ^2 is based on the χ^2 distribution because the sampling distribution of the quantity $(n-1)s^2/\sigma^2$ has a χ^2 distribution with $n-1$ degrees of freedom, provided that the data are obtained using a simple random sample from a population that is normally distributed. A normal quantile plot can be used to check to see if the normality assumption is reasonable; however, it is very important to recognize that this procedure is not robust to violation of the normality assumption, so the data have to be close to being normally distributed, regardless of the sample size, in order to use the method of this section.

The formula for the confidence interval for σ^2 is

$$\frac{(n-1)s^2}{\chi^2_{1-\alpha/2}} \quad \text{to} \quad \frac{(n-1)s^2}{\chi^2_{\alpha/2}}$$

where $\chi^2_{1-\alpha/2}$ and $\chi^2_{\alpha/2}$ are the critical values of the chi-square distribution having left-tail areas of $1-(\alpha/2)$ and $(\alpha/2)$. These values can be found in Table IV of Appendix A for a limited values of confidence levels and sample sizes. They can also be found using statistical software for any combination of confidence level and sample size. Moreover, statistical software packages contain functions that generate the confidence interval for σ^2.

The confidence interval for the population standard deviation σ can be obtained by taking the square root of the two limits of the above confidence interval for σ^2.

Example 14.11   An auto manufacturing company takes a simple random sample of 24 of its new model cars and measures the gas mileage per gallon. The sample variance of the gas mileage is obtained to be .47, and a histogram demonstrated that the normality assumption appears to be reasonable. Construct a 99% confidence interval for the variance of the miles per gallon for all cars of this model.

Solution   The χ^2 table (Table IV of Appendix A) works the same way as the t table, except for the fact that the chi-square distribution is not symmetric. As a result, you will need to obtain two critical values from the chi-square table, not just one. For the confidence interval procedure for σ^2, df are given by $(n-1)$. The confidence level is 99%. Thus

$$\alpha = .01, \quad \frac{\alpha}{2} = \frac{.01}{2} = .005, \quad 1 - \frac{\alpha}{2} = 1 - .005 = .995, \quad df = n - 1 = 24 - 1 = 23$$

Thus, we need to determine the values of $\chi^2_{.005}$ and $\chi^2_{.995}$ for df $= 23$, which we obtained using $n - 1 = 24 - 1$. Going to the intersections of the .005 and .995 columns with the row for df $= 23$ the resulting values of $\chi^2_{.005}$ and $\chi^2_{.995}$ are

$$\chi^2_{.005} = 9.260 \qquad \text{and} \qquad \chi^2_{.995} = 44.181$$

Hence, the 99% confidence interval for σ^2 is

$$\frac{(n-1)s^2}{\chi^2_{1-\alpha/2}} \text{ to } \frac{(n-1)s^2}{\chi^2_{\alpha/2}} = \frac{(23).47}{44.181} \text{ to } \frac{(23).47}{9.260} = .2447 \text{ to } 1.1674$$

Thus, we are 99% confident that the variance σ^2 of miles per gallon is in the interval .2447 to 1.1674. The 99% confidence interval for the standard deviation σ is obtained as

$$\sqrt{.2447} \text{ to } \sqrt{1.1674} \qquad \text{or} \qquad .4947 \text{ to } 1.0805 \qquad \blacksquare$$

GLOSSARY

chi-square (χ^2) distribution   A continuous distribution that is skewed to the right. One use of the chi-square distribution is to develop confidence intervals for the population variance and population standard deviation.

confidence interval   An interval constructed around the value of a sample statistic to estimate the corresponding population parameter.

confidence level   The confidence level, denoted by $(1-\alpha) \times 100\%$, states how much confidence we have that a confidence interval contains the true population parameter.

degrees of freedom (df)   For the estimation of μ using the t distribution and σ using the χ^2 distribution the degrees of freedom are $n-1$.

estimate   The value of a sample statistic that is used to find the corresponding population parameter.

estimation   A procedure by which numerical value or values are assigned to a population parameter on the basis of the information collected from a sample.

estimator   The sample statistic that is used to estimate a population parameter.

interval estimate   An interval constructed around the point estimate that is likely to contain the corresponding population parameter. Each interval estimate has a confidence level.

point estimate   The value of a sample statistic assigned to the corresponding population parameter.

t distribution   A continuous distribution with a specific type of bell-shaped curve with its mean equal to zero and standard deviation equal to $\sqrt{df/(df-2)}$.

EXERCISES

14.1   A consumer agency that proposes that lawyers' rates are too high wanted to estimate the mean hourly rate for all lawyers in New York City. A sample of 70 lawyers taken from New York City showed that the mean hourly rate charged by them is $550. Suppose that it is known that the standard deviation of hourly charges for all New York City lawyers is $100.

   (a) Explain why a z interval for a population mean can be used in this situation, despite the fact that we do not know whether the population distribution of hourly rates is normal.

(b) Construct a 99% confidence interval for the mean hourly charges for all lawyers in New York City.

14.2 A bank manager wants to know the mean amount of mortgage paid per month by homeowners in an area. A random sample of 40 homeowners selected from this area showed that they pay an average of $3100 per month for their mortgages. Suppose that the standard deviation of all mortgages is $582 per month. Construct a 98% confidence interval for the mean monthly mortgage payment for all homeowners in this area.

14.3 At Farmer's Dairy, a machine is set to fill 64-oz milk cartons. However, this machine does not put exactly 64 oz of milk in each carton; the amount varies slightly from carton to carton. It is known that when the machine is working properly, the mean net weight of these cartons is 64 oz. The quality control department takes a sample of 35 such cartons each week, calculates the mean net weight of these cartons, and makes a 99% confidence interval for the population mean. If either the upper limit of this confidence interval is greater than 64.3 oz or the lower limit of this confidence interval is less than 63.7 oz, the machine is stopped and adjusted. A recent sample of 35 such cartons produced a mean net weight of 63.88 oz, with a sample standard deviation of 0.3 oz. Assuming that the distribution of net weight is approximately normal, will you conclude that the machine needs an adjustment?

14.4 A tool manufacturing company wants to estimate the mean number of bolts produced per hour on a specific machine. The manager observed the machine for 20 randomly selected hours. The mean number of bolts produced per hour during these 20 hours was found to be 47.3 with a standard deviation of 2.65. Assume that the number of bolts produced per hour on this machine has a normal distribution. Construct a 90% confidence interval for the population mean μ.

14.5 The following data give the cost per room per day at 33 motels selected from a state.

$124	145	205	86	153	254	120	75	190	160	189
125	195	115	224	230	165	90	210	240	142	130
210	179	105	180	125	185	205	112	157	177	143

(a) What are the values of the sample mean and sample standard deviation of the daily costs per room?

(b) Create a histogram of the data. Use the histogram to justify the use of the t procedure for calculating a confidence interval for the mean daily cost per room for all motels in this state.

(c) Construct a 98% confidence interval for the mean daily cost per room for all motels in this state.

14.6 A company opened a new movie theater. Before setting the price of a movie ticket, the company wants to find the average price charged by other movie theaters. A random sample of 65 movie theaters selected by the company showed that the mean price of a movie ticket is $7.75 with a standard deviation of $0.80. The dataset does not contain any outliers. Make a 95% confidence interval for the population mean μ.

14.7 It is said that happy and healthy workers are more efficient and productive. A company that manufactures exercising machines wants to know the percentage of large companies that provide on-site health club facilities. A sample of 240 such companies shows that 80 of them provide such facilities on site.

(a) What is the value of the sample proportion of companies that provide on-site health club facilities?

(b) Construct a 99% confidence interval for the proportion of all such companies that provide such facilities on site using the large-sample procedure.

(c) What is the value of the Wilson estimator for this sample?

(d) Construct a 99% confidence interval based on the Wilson estimator for the proportion of all such companies that provide such facilities on site. How do the widths of the two intervals compare?

(e) Is the sample size large enough to use the large sample method for constructing a confidence interval?

14.8 An insurance company selected a sample of 50 auto claims filed with it and investigated those claims carefully. The company found that 18% of those claims were fraudulent.

(a) Is the sample size large enough to use the large sample method for constructing a confidence interval?

(b) Construct a 95% confidence interval for the proportion of all such companies that provide such facilities on site. Use the appropriate method based on your answer in part (a).

14.9 The management of a soft-drink company does not want the variance of the amounts of soda in 12-oz cans to be more than 0.01 square oz. (Recall from Chapter 3 that the variance is always expressed in square units.) The company manager takes a sample of certain cans and estimates the population variance quite often. A random sample of twenty 12-oz cans taken from the production line of this company showed that the variance for this sample was 0.014 square oz. Construct the 99% confidence intervals for the population variance and standard deviation. Assume that the amounts of soda in all 12-oz cans have a normal distribution.

14.10 The 2-inch long bolts manufactured by a company must have a variance of 0.003 square inches or less for acceptance by a buyer. A random sample of 29 such bolts gave a variance of 0.0028 square inches. Make the 98% confidence intervals for the population variance and standard deviation.

14.11 Harrods' Consultants, Inc. advises small businesses on energy-saving tips and procedures. A business manager wants to know the average money saved per month by all clients of the company who adopt its tips and procedures. The manager of the consulting firm took a sample of 35 clients and found that these clients saved an average of $85 per month by adopting the tips provided by Harrods. The standard deviation of the savings for these 35 clients was $12.

(a) Make a 99% confidence interval for the population mean.

(b) Suppose that the confidence interval obtained in part (a) is too wide. How can the width of this interval be reduced? Discuss all possible alternatives. Which of these alternatives is the best?

14.12 A company wants to estimate the mean net weight of its Top Taste cereal boxes. A sample of 16 such boxes produced the mean net weight of 31.98 oz with a standard deviation of 0.26 oz. Make a 95% confidence interval for the mean net weight of all Top Taste cereal boxes. Assume that the net weights of all such cereal boxes have a normal distribution.

14.13 The mean time taken to design a home plan by 20 architects was found to be 248 minutes with a standard deviation of 26 minutes.

(a) Assume that the time taken by all architects to design this home plan is normally distributed. Construct a 95% confidence interval for the population mean μ.

(b) Suppose that the confidence interval obtained in part a is too wide. How can the width of this interval be reduced? Describe all possible alternatives. Which alternative is the best and why?

14.14 According to a recent report by the U.S. Bureau of the Census, 26% of the households headed by single men own stocks, bonds, and mutual funds. Although Census Bureau estimates are based on very large samples, for convenience assume that this result is based on a random sample of 2000 households headed by single men.

(a) What is the point estimate of the proportion of all households headed by single men who own stocks, bonds, and mutual funds? What is the margin of error associated with this point estimate?

(b) Find a 95% confidence interval for the proportion of all households headed by single men who own socks, bonds, and mutual funds.

14.15 York Steel Corporation produces iron rings that are supplied to other companies. These rings are supposed to have a diameter of 24 inches. The machine that makes these rings does not produce each ring with a diameter of exactly 24 inches. The diameter of each of these rings varies slightly. It is known that when the machine is working properly, the rings made on this machine have a mean diameter of 24 inches. The standard deviation of the diameters of all rings produced on this machine is always equal to 0.02 inches. The quality control department takes a sample of 36 such rings every week, calculates the mean of diameters for these rings, and makes a 99% confidence interval for the population mean. If either the lower limit of this confidence interval is less than 23.975 inches or the upper limit of this confidence interval is greater than 24.025 inches, the machine is stopped and adjusted. A recent such sample of 36 rings produced a mean diameter of 24.015 inches. On the basis of this sample, can you conclude that the machine needs an adjustment? Explain.

15 Control Methods

> When you can measure what you are speaking about, and express it in numbers, you know something about it; but when you cannot measure it, when you cannot express it in numbers, your knowledge is of a meager and unsatisfactory kind
>
> —Lord Kelvin, 1839

15.1 OVERVIEW

Today, there is a strong effort to improve the quality of American goods and services, and increasing numbers of businesses are using the methods described here to achieve this objective. This chapter introduces the fundamentals of statistics and shows the short- and long-term process capability analysis results. Few organizations have good estimates of their process capabilities even though a great amount of activity is focused on efforts to produce acceptable yields. Evidence of the increasing importance of quality is found in its increased visibility of advertising and the growing number of books and articles on quality issues. In many cases, job applicants have a definite advantage when they have studied statistics and quality control methods. This chapter presents the basic tools commonly used to monitor process control. Statistical process control enables the practitioner to obtain and allocate resources on the basis of data that quantify defect rate, identify opportunities for improvement, predict true quality levels of products and services, and identify the nature and the location of the problem (once capability analysis is performed). MINITAB, Excel, and other software packages include programs that create charts of the type discussed in this chapter.

15.2 INTRODUCTION

President Reagan formally acknowledged that customers who form the enormous base of today's world market are sending a clear and undeniable message to corporate America. Customers are demanding, without compromise, higher levels of product quality, at a lower cost, improved responsiveness, and added value. The producer must struggle to satisfy technical, performance, schedule, and cost expectations of the customer. In turn, these objectives drive the need for control methods used in Six Sigma and total quality management (TQM). About the same time, IBM said that "product leadership is one of our goals for the 1980s." An essential element of product leadership is quality: the delivery of offerings, which are defect free at a minimum cycle time. This is how customers measure the quality of their competitors. Doing things right the first time is extremely important. Each stage in the process must produce defect-free output. Rising to the *TQM/Six Sigma* challenge is not an easy task. In fact, one of the biggest deterrents to meeting this challenge is the reduction and ultimate elimination of product variation. When variation acts on a product, the results are often catastrophic in terms of reliability, quality, performance, schedule, and cost. Although eradicating variation is a tough

Practitioner's Guide for Statistics and Lean Six Sigma for Process Improvements. By Mikel J. Harry, Prem S. Mann, Ofelia C. de Hodgins, Christopher J. Lacke, and Richard Hulbert
Copyright © 2010 John Wiley & Sons, Inc.

task, there are a number of things, which can be done. To this end, several key initiatives have been structured. These highly interactive initiatives are inclusive of, but not necessarily limited to

- Design
- Design to standard parts
- Design to standard materials
- Robust design
- Design for assembly
- Design for reliability
- Design for simplicity

Process

- Short-cycle manufacturing
- Process characterization
- Process standardization
- Statistical process control

Material and Components

- Part standardization
- Transaction(s) standardization
- Supplier statistical process control (SPC)
- Supplier certification
- Material requirements planning

Clearly, these initiatives have been devised to achieve *Six Sigma synergy*. Each of these Six Sigma initiatives and related tools are cross-functional by nature, enabling the whole to be greater than the sum of the parts. There are certain larger themes by which Six Sigma has progressed over the years as it has transitioned from a simple statistical target (3.4 DPMO) to a highly structured management system fully capable of deploying quality on a corporatewide, global scale. As one theme was embodied and perfected by a certain corporations, other companies built on that knowledge to expand the meaning and impact of Six Sigma. Motorola first proposed of Six Sigma as a statistical target for product quality. Around 1985, the company realized that its product quality was not well regarded by its customers. In fact, Motorola was a far cry from the benchmarks set by certain Asian competitors. In response to its newfound humility, then CEO Robert (Bob) Galvin, Motorola instituted a relentless drive for a virtually perfect level of production quality. In 1988, this pursuit won the company a Malcolm Baldrige Award, and Motorola established its reputation as the four-sigma producer that became a world-class quality benchmark (Six Sigma). Statistical process improvement methods were vigorously applied to achieve high-level stretch goals. In the past, the pursuit of quality was more a philosophy than an art or science. In essence, corporations such as IBM, Motorola, Sony Electronics, and General Electric have hardened the principle of quality to the point of giving torque and power to the idea of quality, thereby migrating that power from the realm of quality to the realm of business.

15.3   CONTROL LOGIC

Statistical process control is a method of analyzing data over time and using the results of the analysis to solve manufacturing and processing problems. The method can be applied to almost anything that can be expressed with numbers or data. The word *quality* means much more than the goodness or badness; it refers to the quality characteristics of the thing or process being analyzed. The word *control* means to keep something within boundaries. Remember, from previous chapters, that a process is any set of conditions or causes, which work together to produce an output or result.

In summary, this chapter defines a process as the organization of people, procedures, machines, and materials into the work activities needed to produce a specified end result (output). It is a sequence of activities characterized by

- Measurable inputs
- Value-added (VA) activities
- Measurable Outputs
- Repeatability

15.4   STATISTICAL CONTROL SYSTEMS

The term *process* refers to the operation of a single cause; in a wider sense, it may refer to the operation of a very complex system of causes. This broad definition permits process capability studies to be applied to many activities, including engineering, operating, management, clerical, services, accounting, financial, audit, information technology, systems, communications, sales, marketing, advertising, or other organizations that experience merchandise losses. Key techniques covered in this chapter include

- Mistakeproofing
- Process control and the use of control charts
- Process control charts that were originally used by manufacturing

15.4.1   Mistakeproofing

Mistakeproofing or poka-yoke, briefly discussed in Chapter 7, is derived from the Japanese *poka* (inadvertent error) and *yokeru* (avoidance). Dr. Shigeo Shingo,[1] an industrial engineer at Toyota, developed the poka-yoke concept. It is one of several control concepts where the solution is static as opposed to dynamic control obtained from a closed-loop feedback control system. It was originally named *foolproofing*, but Shigeo Shingeo was concerned that some workers might feel offended and the methodology was renamed. *Poka-yoke* is consistent with the fundamental aims and philosophy of Six Sigma and has wide applicability to manufacturing, engineering, and transactional processes. It requires design and implementation of actions to prevent errors, mistakes, or defects in our everyday activities and processes. The methodology requires complete understanding of the cause–effect relationship and identification of the simplest remedy to eliminate the occurrence of each particular error type. This frequently requires the creation of a checklist, a change in the sequence of operations, a highlighted field on a form, a software-generated reminder to the operator, or some other way to ensure that mistakes will be totally eliminated or substantially reduced. A major advantage of mistakeproofing is that it strives for zero defects. Often, mistakeproofing focuses on errors produced by humans, whether it is the machine operator, the person filling out a form, or someone packing materials. However, to say that most errors are human-induced is not to place blame. For example, failure of a human to consistently dunk a basketball is an error but should not be considered a human error. A change in the process of dunking a basketball will probably enable all humans to put the ball through the hoop most of the time. On the other hand, that most errors are human-induced is good news because humans can fix these errors. Most mistakes can be placed into one of the following generic types:

- Incorrect processing
- Workpieces placed incorrectly

- Missing parts
- Wrong parts
- Wrong process
- Wrong blue print or instructions
- Wrong piece processed
- Operation skipped or omitted
- Improper adjustment
- Equipment not set up properly
- Process improperly supervised
- Use of the wrong tool

Figure 15.1 lists some common mistakes.

Errors should not be accepted as inevitable. Experience has shown that by careful analysis of causes, any error type can be reduced considerably, if not eliminated altogether. Mistakeproofing reduces or removes error opportunities by replacing repetitive or error-prone tasks with tasks that can be done correctly only.

Mistakeproofing is a technique where we use wisdom and ingenuity to create procedures or devices that encourage us to do our job 100% defect-free most of the time. The practitioner observes that in the transactional and manufacturing worlds, mistakes occur mostly because there are many opportunities for defects. We know that 100% inspection will not catch all of these errors, which leads us to believe that little can be done. The following are examples of human induced errors; they include misunderstanding, lack of knowledge, lack of training, and lack of standards. Mistakes do not just happen. They are the result of failures to conform or to take the appropriate actions. Figure 15.2 graphically shows a symbolic example of taking action.

Examples of Poka-Yoke in Everyday Life. Examples include (1) spare tire mounted on a different type of rim by some carmakers; (2) car headlights that automatically turn off when the key is removed and driver side door is opened; (3) rain detector to automatically turn on wipers; (4) screen saver on

Types of Mistakes

1. Forgetfulness (not concentrating)	6. Inadvertent errors (distraction, fatigue)
2. Errors due to misunderstanding (jump to conclusions)	7. Errors due to delay in decision-making
3. Errors in identification (view incorrectly ... too far away)	8. Errors due to lack of standards (written & visual)
4. Errors made by untrained workers	9. Surprise errors (machine not capable, malfunctions)
5. Willful errors (ignor rules)	10. Intentional errors (sabotage - least common)

Figure 15.1 Humorous illustration of human-induced errors.

What is mistakeproofing?

Figure 15.2   Mistakeproofing spraying to avoid inadvertent mistakes is like spraying to avoid mosquito bites.

computers to prevent exposure of confidential material; (5) once the car engine is turned on, the gear can't be engaged unless the brake paddle and gearshift release button are depressed; (6) one key for all locks in a home; (7) musical chime at entrance of a shop to inform owner that a customer has arrived; (8) car key cannot be removed unless the transmission is in the park (P) position; and (9) road bumps at strategic speed locations to slow down traffic for safety reasons.

Instead of blaming humans for making mistakes, emphasis should be placed on modifying processes by training, standardization, skill building, experience, and discipline. In summary, we must improve all aspects of our business to be robust to mistakes or errors from any source. Some examples of the traditional application of mistakeproofing in a production environment to prevent errors include

- A stop added to a drill press
- A hydraulic ram added to align a component during assembly
- A lever designed into an assembly fixture to index the part
- A pin added so the part cannot be installed backwards

Mistakeproofing actions taken to improve transactional procedures include

- Critical fields on a data entry form are highlighted.
- An authorization procedure is introduced to control spending.
- A checklist is created to ensure that all items are addressed when planning a training session.
- A new policy is developed to ensure that expense claims are properly completed.

Example 15.1

1. What does the concept "mistakeproofing" mean?
2. What role does mistakeproofing play in Six Sigma?
3. Does mistakeproofing apply only to human error?

4. How can we put transactional processes in control?

5. How is mistakeproofing applied in engineering design?

6. Can mistakeproofing be applied with DOE (design of experiments)?

Solution

1. Mistakeproofing is a methodology for avoiding mistakes in any process.

2. Mistakeproofing can be used in Six Sigma to avoid potential problems, eliminate defects, or define improvement initiatives.

3. Mistakeproofing addresses potential causes of human error; it can also be applied to automated machines, computer software or to any process that contains variation.

4. For transactional processes, mistakeproofing actions include introduction of standard operating procedures, forms, and checklists.

5. Mistakeproofing can be applied in the engineering environment by using FMEA to document all possible failure modes of every component and system associated with the product.

6. Mistakeproofing can be applied in the improve and control phases of Six Sigma. In DOE, mistakeproofing can be used to control the CTPs (critical-to-process characteristics) identified through simulation. ■

A mistake is the same as an error, and errors cause defects, but an error may not necessarily result in a defect. Mistakeproofing (Fig. 15.3) ensures that errors do not occur, or that all potential errors are detected, removed, or corrected to produce a good final result. Once we determine the opportunities for defects in the process and the cause–effect relationships, it is usually a simple task to design a procedure or feature, which will prevent future errors. To show the simplicity of this concept, consider your last trip to the grocery store. As you probably noticed, the cashier used a barcode reader to correctly enter the product price into the cash register, reducing defects, and saving time. Previously, the cashier read the price and manually entered it, a repetitive task with several

Everyday examples of mistakeproofing

Figure 15.3 Cartoon examples of everyday mistakeproofing.

opportunities for error. Other examples of mistakeproofing include swimming pool fences and gates to prevent children from drowning, or the interlock switch on the clutch to prevent cars from being started in gear.

Example 15.2   A Master Black Belt assists a team in the methods department to reduce mistakes in the worksheets issued to the shop floor. Having already determined that the root causes of these errors can be traced back to the engineering, procurement, and vendor methods, the team is ready to determine ways to reduce the opportunities for errors. However, to reduce the number of mistakes, they must understand the different types of mistakes that can occur.

The Master Black Belt outlines 10 classes of mistakes that should be considered to improve the process. "There are many different types of mistakes; some are caused by (1) forgetfulness, (2) lack of understanding, (3) lack of training, (4) lack of standards, (5) delay in our decisionmaking, (6) lack of metrics, (7) lack of baseline, (8) data without integrity, (9) wrong data, and (10) other mistakes that may be complete surprises due to distraction and fatigue, by people who ignore the rules, or by people who do made them intentionally (sabotage)." However, this latter type is quite rare. Most mistakes result from the many opportunities for defects in our processes, which should really be the focus of our mistakeproofing efforts. We must identify these different mistake types to design controls that will address the underlying causes. The Master Black Belt also pointed out that understanding the various types of mistakes would help during the *analyze* phase of our project. We can trace the root cause of the mistake by using the *is/is not* technique, where you often arrive at the root cause by carefully examining when the mistake or problem does *not* occur. The results of this analysis will then be used to control the process.

Example 15.3   A Six Sigma champion asks, "I have heard about a concept called *mistakeproofing* that can be used to prevent mistakes from occurring. Apparently this is a very important tool for controlling transactional processes. Could it also be applied to manufacturing processes?" "Yes," replies the master Black Belt, "let me give you an example of how it is being applied. In our composite facility, there is a history of nonconformance from parts being trimmed undersize. We fail to meet the correct size specifications because we incorrectly cut the material." A Six Sigma project team analyzes the problem, and discovers that the inherent variability in the process is caused by the operator manually trimming parts with a knife to a scribe line marked on the tool. Using a *mistakeproof planning sheet*, the team evaluates various options that could eliminate this defect and decides that the most effective way to mistakeproof the process is to introduce a trim tool. They also realize that to fully mistakeproof the process, they must ensure that the trim tool is properly oriented. They decide that the best way to prevent this is to add a tooling hole and pin to the tool so that it can only be oriented one way. To fully mistakeproof the trim tool, they have it shaped such that it cannot be inverted. "I see," says the organization's champion after analyzing the source of the variation, the team mistakeproofs the process and improves the process by controlling the sources of errors. (See cartoon example of mistakeproofing a transactional process in Fig. 15.4.)

Use the highest principle possible.   In general, some references show only 7 types of waste. Some other references consider more than 8 (*t*ransportation, *i*ntellect, *i*nventory, *m*otion, waiting, *o*verproduction, *o*verprocessing, *d*elays, Rework, Motion) and some others as many as 12. For example some companies consider Safety/Injuries as a form of waste. However, the concept of the highest principle possible is always the same: identify and eliminate all efforts that do not add value.

The use of the highest principle possible is based on the concept of the "Lean seven deadly wastes" and therefore, the highest principle possible applies to (1) elimination, (2) replacement, (3) facilitation, (4) detection, and (5) mitigation. Figure 15.5 shows a simplified illustration of the highest principle.

Mistakeproofing transactional process

Figure 15.4   Example of how the difference in time zones could be the root cause of mistakes in a call center.

Example 15.4   A Six Sigma project team is focused on the dataplate of a molded coverplate. Some dataplate problems are related to its alignment (the alignment is crooked), its adhesive capacity (the adhesive fails and the parts fall off), or its orientation (the orientation is incorrect if it is upside-down). The team has considered various options and faced an overwhelming list of potential solutions. They contact a master Black Belt to help prioritize their mistakeproofing solutions. The master Black Belt introduces the team to the five principles used when choosing mistakeproofing solutions. Pointing to the list, he explains that there are some general rules of thumb for prioritizing mistakeproofing solutions. In fact, there is a descending scale of preference. The highest principle on this list would be to completely eliminate the possibility of error altogether by redesigning the process to eliminate the error-prone process step. This is the best solution because it directly achieves a better result. If this cannot be done, we consider the next level on the scale. We continue this thought process until we arrive at the lowest level on the scale, mitigation. In this case, we focus our efforts on minimizing the effect of the errors. This level of the scale would most likely incur some hidden factory expense.

One team member says, we will have to alter our thinking all together. Let's try to find a solution that will completely eliminate the possibility of errors. Reviewing their project findings from the Analyze phase, they discover that the coverplate is an injection molding made on an automatic machine, producing perfect quality every time. The team agrees that the best solution for the label is to simply

Use the highest principle possible

Elimination

Replacement

Facilitation

Detection

Mitigation

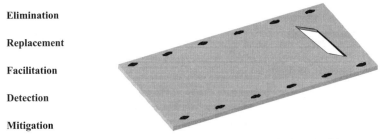

Figure 15.5   Drawing showing the use of the highest principle possible.

feed it into the automated molding process as one of the many layers of composite material. In this sense, a process step is removed and each potential problem from the old method is totally eliminated. This solution uses the highest principle possible criteria and optimizes the mistakeproofing solution.

Different Categories of Mistakeproofing. A Black Belt asks, "I understand the concept of mistakeproofing, but are there different types of mistake-proofing that we can apply?" The instructor replies, "there are many different types of mistakeproofing methods and devices. There is no inclusive list because this simple concept has wide applicability. However, we could list some of the main categories. Consider the ergonomic category as one type of mistakeproofing." Here, we look at the standard shape, position, and function of the controls and hopefully remove the opportunities for error. It is important that the principles of ergonomics be considered during the design to avoid mistakes. It is also important to evaluate potential failures that could dominate the design; in this case, the FMEA method should be used. *Failure mode–effects analysis* (see Chapter 7) can assist in identifying potential failures, errors, or defects that could occur during the design stage of a product or process. Another type of mistakeproofing is a checklist, which is essential in an area where complex sequences of events are infrequently performed. Checklists are a simple and inexpensive way to eliminate errors in any process. Other types of mistakeproofing are devices that grasp our attention; red flags so to speak, that we see everyday, such as buzzer alarms or devices that automatically shut off, such as sensors—and finally, computer software is another type of mistakeproofing especially useful for transactional processes. For example, a controlled data-entry field on a form would eliminate errors, and spellcheckers will *red-flag errors* so that we can correct them before they become defects. As we can see, examples of mistakeproofing are everywhere. Figure 15.6 shows an example of the different categories for mistakeproofing.

Defect Prevention Guidelines. Products, services, and processes should be designed with mistakeproofing concepts in mind. The key is to prevent defects from happening, instead of always reacting to them. In our company, there are already instances of this happening. In customer service, the shared vision of the technical publication team aims for error free publications. They created CD-ROM

Types of mistakeproofing

Figure 15.6   Cartoons illustrating different types of mistakeproofing.

Defect prevention guidelines

1. **Produce only what you need**
2. **Eliminate, simplify, or combine steps**
3. **Make some use of the deliverable ASAP**
4. **Involve everyone in defect prevention**
5. **Never stop your quest for perfection**

Figure 15.7   Illustration of some defect prevention guidelines.

(compact disk–read-only memory) versions of the manuals produced by the group. This, in itself, eliminates mistakes. Consider for a moment the task of printing the paper version. First, there is a tendency to print more copies than required. Then, one must consider the packaging, shipping, and postage costs. But this is only the beginning. Think about all of the work required to update all the paper copies of all the manuals by each of our operators. The support staff has a full-time job just keeping these manuals up-to-date. They must exchange one page at a time whenever there is an amendment. There is a high probability of making errors and these errors could have serious implications. Even with a low error rate, it is difficult to avoid mistakes because of the huge volume and repetitive nature of the task.

Now, let's think about the new process, cutting a CD. Think about the reduction in cycle time as we remove process steps. It is so much easier and less expensive to ship CDs. It is a just-in-time method and easier for everyone involved in the process, since substituting new material for the old is much easier. So, in fact, we have reduced potential mistakes in several different ways. All of these results from one relatively simple change. Figure 15.7 shows simple guidelines for defect prevention.

Section Summary

- Mistakeproofing methodology can help eliminate errors or defects when other, more analytical methods cannot be applied or do not produce results. The technique uses wisdom and ingenuity to create devices that will reduce or eliminate defects. Since mistakeproofing applies to such a wide range of topics, it is difficult to create hard-and-fast rules for applying the technique.

- An error (or mistake) does not always cause a defect unless it escapes detection and is delivered to the customer. However, even errors, which are caught and corrected, are undesirable because they contribute to the hidden factory activity required to produce the deliverable products.

- We must avoid blaming humans for defects. Most processes, even those prone to human error, can be redesigned to be robust to operator differences, boredom, and lack of experience.

- Mistakeproofing can reduce errors made by humans, but it can also be applied to many other types of mistakes or defects.

- There are many examples of mistakeproofing in our everyday lives. Few realize how much thought has gone into the design of products and services to ensure that they are user friendly.

- Mistakeproofing is the primary means to control transactional processes.

- There are various levels of control imposed by mistakeproofing, ranging from announcing that a mistake has been made (after the fact) to completely eliminating the possibility of making the mistake in the first place. Obviously, the latter situation is preferable.

15.5 STATISTICAL CONTROL

A process is within statistical control when the process contains only natural, chance variation; the process control charts (data plotted over time in a graph containing the mean, upper and lower control limits) show only typical patterns without cycles, trends, step changes, or extreme points. Thus, only when a process is *statistically stable* can its output be treated as population with constant mean, standard deviation, and distribution.

A process control system can be described as a feedback four element system:

1. *The Process.* The combination of people, equipment/machines, input materials, methods, and environment that work together to create the output. The total performance of the process and the quality of its efficiency depends on the way the process is designed and built and on the way it is operated. The rest of the process control system is useful only if it contributes to the improvement of the process.

2. *Information about Performance.* A great amount of information is learned by studying the process output (Y). In general the process output consists not only on the products are produced but also on the intermediate outputs ("small y's") that describe the operating state of the process. For example, when information is collected and correctly interpreted, it shows whether action is necessary to correct the process or the just produced output. Timely and appropriate actions should be taken; otherwise, the collected information effort should be considered waste.

3. *Action on the Process.* This is a future-oriented action because it is taken when necessary to prevent production to manufacture out-of-specification products. This action could consists in the operations (operation training, or, e.g., changes to the incoming materials) or the more basic elements of the process such as equipment that may be in need of maintenance or upgrading, or the design of the process as a whole that if not fully controlled could be vulnerable in some process steps to environmental changes such as temperature and humidity in the manufacturing floor (the semiconductor industry maintains a tight control of process variables such as humidity, type and amount of light, and temperature).

4. *Actions on the Output.* Action on the output should not be past-oriented (detecting out-of-specification output after being produced). Unfortunately when the final product does not satisfy specifications, this product is in the need of rework or is scrap. Root cause analysis should be carried out and corrective actions should be taken.

Finally, in order to use process control measurements data effectively, it is important to fully understand the concept of variation.

No two products or characteristics are exactly alike because any process is free of variation. The difference between products could be large or could be so small that it would be difficult to measure, but the variation is always present. Some sources of variation in the process cause very short-run piece-to-piece differences such as backlash and clearances within a machine and its fixtures, or in transactions: the accuracy of an accountant. Other sources of variation tend to cause changes in the output (Y) only over long periods of time (gradually with machine wear, tool-ramping of temperature, power surges, or similar/related problems). It is important to take into account time period for key measurements because large periods of time will account for present total variation.

From a standpoint of minimum requirements, out-of-specification measurements would be enough. To manage any process and reduce variation, the variation must be traced to its sources, and for this it is important to learn to make the distinction between common and special causes of variation (Section 15.8).

15.6   PREVENTION VERSUS DETECTION

To understand the concept of variation, we begin from the observation that *two product characteristics are not equal* because any process contains several sources of variation. Therefore, the difference among products, transactions, or services may range from very large to so small that it would be challenging to measure. No matter how small, variation is always present. For example, assume that the time required to process an invoice varies for different associates performing the task. This same principle applies to a call center for average speed to answer, equipment reliability, or associate training level. Here, the average speed to answer might depend on the complexity of the problem or invoice, the accuracy and legibility of the invoice or ticket, the procedures followed, and volume of work (in Lean = inventory).

It is always more effective to avoid "waste" (see previous chapters on Lean methodology) by not producing it—*Strategy of prevention*. A prevention strategy sounds obvious to most people and is often captured in slogans such as "Do it right the first time"; however, this approach is not sufficient. Some sources of variation in the process cause short-run piece-to-piece differences, such as the accuracy of a bookkeeper. Other sources of variation cause changes in the output over a long period of time. Time period and conditions, under which observations and measurements are made, affect the total process variation visible to the user. When minimum requirements are present, the issue of variation is simplified; everything within specification limits is acceptable and everything beyond specification limits is considered unacceptable. To control any process, the variation must be tracked to its sources before it can be reduced. The first step is to establish a difference between sources of variation(s). These are common variation and special cause (noncommon or assignable) variation.

The remaining subsections cover some of the important elements required to understand a statistical process control system. A standard manufacturing approach depends on production to make the product and on quality control to inspect final products to screen out those parts that do not meet specifications. Similarly, in the transactional world, the transactions are rechecked to catch errors. Both worlds require strategy and detection. However, in both worlds, waste is a non-value-added activity. The best strategy is prevention; do it right the first time. The second best strategy offers detection.

15.7   A PROCESS CONTROL SYSTEM DEFINITION

A process control system can be described as a feedback system. The four elements of this system are:

1. *The Process.* The process is the combination of people, equipment, materials, measurements, methods, and environment that together produce an output. The total performance of the process depends on the design and operation of the process. The process control system is useful if it contributes to improved overall performance of the process.

2. *Information about Performance.* Much information can be learned about the process by analyzing the process output (Y). The process output includes not only the products that are produced but also intermediate outputs that describe how the process is operating such as cycle time, temperature, and pressure. If this information is gathered and interpreted correctly, it will indicate when corrective action are required. If timely and appropriate corrective actions are not taken, any data gathered are considered waste.

3. *Action on the Process.* Action on the process is future-oriented because it is taken to prevent the creation of out-of-specification products. For example, actions might involve operator training, changes in the parts or components, or redesign of the process. The effect of these actions should be carefully monitored, and further actions taken when and if necessary.

4. *Action on the Output.* Action on the output is *past-oriented,* since it detects out-of-specification conditions after the output has already been produced. Unfortunately, out-of-specification products rarely meet customer requirements.

15.8   VARIATION

Variation is a measure of the difference between the values of characteristics describing a product. This section starts with observations such as that no two products or characteristics are exactly alike because any process contains many sources of variability. The difference(s) among products may be small or large but will always be present. In manufacturing these differences may be due to materials, machines, workstations, maintenance, or manufacturing environment (temperature, lighting, etc.). In the transactional area, examples include the time required for first time call resolution; this cycle time variation from system availability, operator/associate skills, operator's level of knowledge, or noise level in the office. Some sources of variation cause very short-run piece-to-piece differences such as in the transactional area; examples are accuracy of a bookkeeper's work and in manufacturing, piece-to-piece small color variations.

However, other sources of variation cause changes in output (*Y*) over a longer period of time, either gradually or irregularly. Time period and measurement conditions affect the total variation. When minimum requirements are present, situations such as parts within specification limits are acceptable while parts beyond specification limits are not acceptable. In the transactional and service area, on-time reports are acceptable, while late reports are not acceptable. For the call center first-time-call resolution, problems resolved during the first call are acceptable while problems not resolved during first-call are not acceptable. Therefore, to reduce variation in any process, this variation must be traced to its source(s). The first step is to differentiate between common causes of variation (natural cause of variation) and assignable cause of variation (special cause of variation).

15.8.1   Common Causes

Statistical control is a process that is statistically stable, meaning that the process has only natural variation and exhibits no patterns, cycles, or unusual points. Therefore, a process is in statistical control when the only source of variation is common causes. Common causes for random variation behave like a system of chance events. While individual measured values are all different, the set of values tends to form a pattern that can be described by a probability distribution. This distribution can be characterized by

- Location
- Spread
- Shape

15.8.2   Special Causes

Special causes, often called *assignable causes*, refer to any factors or causes of variation that cannot easily be explained if the process were in statistical control. Unless these causes are properly identified and timely corrected, they will continue to affect the process output in unpredictable ways.

Local Actions and Actions on the System.   Control charts have the unique ability to detect changes. First, the pattern is tested for evidence of unnaturalness. Unnatural patterns are then associated with causes. The causes that disturb the process are called *special or assignable causes* of variation and can be detected by simple statistical techniques. The discovery of a special cause of variation and its removal are usually the responsibility of someone in the operation; resolution of a special cause usually requires local action. Natural variation also results from causes, but these causes are unknown;

therefore, these causes are called "*nonassignable*" causes. Nonassignable causes are in general small in magnitude, numerous, closely intermingled, and statistically in balance. Thus, it is not possible to identify such causes without special efforts.

Section Summary.

- *Special cause of variation is* a source of variation that is intermittent, unpredictable, and unstable; thus, assignable variation results from causes that can be identified such as defective machinery or inadequately trained employees.
- *Random variation* is due to chance; it is the type of variation inherent in any process that is unable to produce every good or services exactly the same way every time.

15.9   PROCESS OUT OF CONTROL

A process is out of control when noncommon or assignable causes of variation force points on a control chart to exceed either of the control limits. It may also be deemed to be out of control if it fails one or more of a number of quantitative decision tests such as the Western Electric Tests. Figure 15.8 illustrates an example of an out-of-control process.

The following concepts will help determine when a process is out of statistical control:

1. *Upper control limit* (UCL)—the upper limit of process control is by convention set to the process mean plus three standard deviations.
2. *Centerline*—calculated as the process mean over the period being investigated.
3. *Lower control limit* (LCL)—the lower limit of process control is by convention set equal to the process mean minus three standard deviations.
4. One sigma line—a line calculated as being one standard deviation away from the centerline, or process mean.
5. *Two sigma line*—a line calculated as being two standard deviations away from the centerline, or process mean.

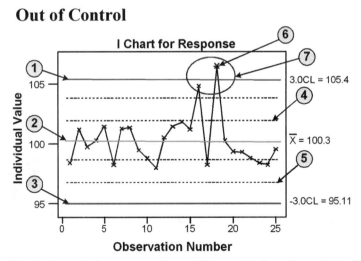

Figure 15.8 Graphical example of a process out of control. Observe point 6 out of control. Point *F* is outside the control limit $+ 3\sigma$.

6. *Out-of-control point*—a single process point showing the most obvious sign of an out-of-Control situation, that is, being beyond either the UCL or LCL.

7. *Out-of-control situation*—another example of an out-of-control condition, where the process has failed Western Electric Test 2, where a process is out of control when two out of any three successive points are on the same side of the central (center)line and simultaneously more than two standard deviations from the central line. There are many other tests to determine whether a process state of control, but the Western Electric rules are an excellent starting point and should be used whenever possible.

Likelihood of a Defect to Occur.   To determine the probability of defects occurring, various statistical distributions can be used for discrete and continuous data.

Discrete probability distribution.   A *probability distribution* is a graph, table, or formula used to assign probabilities to all of the possible outcomes or values of a characteristic measured in an experiment. A discrete probability distribution that forms/or shapes a graph describes the probabilities of occurrence of all possible results. In a discrete probability distribution, the characteristic measured can take on only discrete values, (0, 1, 2, 3, etc.). In this distribution the sum of probabilities is equal to 100%. Figure 15.9 graphically represents a discrete probability distribution.

The sum of the probabilities shown below equals 99.4%, due to rounding. However, the calculation is correct and the sum of the probabilities equals 100%. Examples of discrete distributions include the Poisson, binomial, and negative binomial.

Analysis techniques to determine system variability include several problem-solving techniques such as Pareto analysis, root cause analysis, and cause–effect analysis. However, the use of more advanced process analysis methods including statistical techniques such as DOE, response surface, and process simulation (Monte Carlo) may be necessary to achieve significant process variation reductions.

How can C_{pk} be good with data outside specifications?

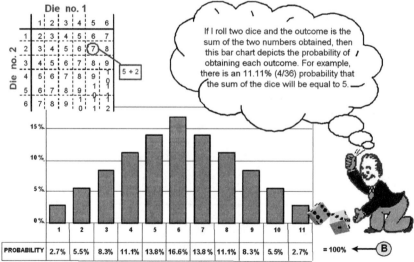

Figure 15.9   Figure showing a probability distribution for discrete data.

TABLE 15.1 d_2 Values

Subgroup Size	d_2
1	1.128
2	1.128
3	1.693
4	2.059
5	2.326
6	2.534
7	2.704

Example 15.5 A customer who recently called our technical support line could not understand why his calculated C_{pk} was above 1.0 when his data were not centered between the specifications and some of the data were outside the specification. How can you have a good C_{pk} when you have data outside the specification and/or data, that are not centered on the target/nominal value? To calculate C_{pk}, you need three pieces of information: the process average, the variation in the process, and the specification(s). First, determine whether the mean (average) is closest to the upper or lower specification. If the process is centered, then either z_{upper} or z_{lower} can be used. If you only have one specification, then the mean will be closest to that specification since the other one does not exist.

Solution To measure process variation, we use the estimated sigma (standard deviation). If you decide to use the standard deviation from the individual data, you should use the P_{pk} calculation, as it uses this sigma. To calculate the estimated sigma, divide the average range, $\bar{R}$ by d_2. The d_2 value depends on the subgroup size and will come from Table 15.1. If your subgroup size is one, you will use the average moving range, $M\bar{R}$.

You, or the customer, must provide the specifications. Now that you have these three pieces of information, the C_{pk} can be easily calculated. For example, if the process average is closer to the upper specification, then C_{pk} is calculated from the following equation:

$$C_{pk} = \frac{USL - mean}{3\sigma}$$

As you can see, the data are used indirectly to determine mean and average range, but the raw data are not used in the C_{pk} calculation. Suppose that you have the example of 14 subgroups with a subgroup size of 2 shown in Table 15.2.

The mean, $\bar{X}$, is 0.8057 and the average range, $\bar{R}$, is 0.220. For this example, the upper specification is 2.12, the target value is 1.12, and the lower specification is 0.12. For these values, more than 21% of the data are outside the specification, so you would expect C_{pk} to be low, right? As it turns out, C_{pk} is relatively healthy at 1.17. (Yes, for this example, we have ignored the first cardinal rule—before one calculates C_{pk}, the process must be in control.) Thus

$$\text{Mean} = \bar{X} = 0.8057$$

$$\bar{R} = 0.2200$$

$$\sigma_{estimated} = \bar{R}/d_2 = (0.2200/1.128) = 0.1950$$

$$C_{pk} = \frac{\text{smallest of } z_{upper} \text{ and } z_{lower}}{3}$$

$$z_{lower} = \frac{X - \bar{LSL}}{\sigma_{estimated}} = \frac{0.8057 - 0.12}{0.1950} = \frac{0.6857}{0.1950} = 3.516$$

TABLE 15.2   Sample Average and Range Dataset

Sample No.	Value	Value	Average	Range
1	0.03	0.06	0.045	0.030
2	0.10	0.20	0.150	0.100
3	0.05	0.10	0.075	0.050
4	1.00	0.00	0.500	1.000
5	1.50	1.50	1.500	0.000
6	1.10	1.50	1.300	0.400
7	1.10	1.00	1.050	0.100
8	1.10	1.01	1.055	0.090
9	1.25	1.20	1.225	0.050
10	1.00	0.30	0.650	0.700
11	0.75	0.76	0.755	0.010
12	0.75	0.50	0.625	0.250
13	1.00	1.10	1.050	0.100
14	1.20	1.40	1.300	0.200

We see that z_{upper} is larger, so for this example

$$C_{pk} = \frac{z_{\text{lower}}}{3} = \frac{3.516}{3} = 1.172$$

In this example, it appears that C_{pk} is good, yet the process is not centered and some data elements lie outside the specification limits. C_{pk} is good because the average range is understated, and therefore when we divide by the estimated sigma, which in this case uses the average range data, it inflates C_{pk}. When these values are shown on a control chart, we quickly observe that the process is not in control. Our fundamental goal is continuous improvement. However, in realistic terms, one should establish which processes should be improved first. In many businesses this is basically an economic decision and depends on the nature of the particular process in relation to the performance of other processes that could also be candidates for immediate improvement actions. To provide consistency in the improvement efforts, decisions should be based on guidelines. For example, certain customer requirements refer to across-the-board capability requirements of $\bar{X} \pm 3\sigma$ and further specify $\bar{X} \pm 4\sigma$ capability ($C_{p,\text{min}} \geq 1.33$) for new processes affecting significant product characteristics. The problems causing unacceptable process capability are usually common causes. When systematic process actions have been implemented, their effects are clearly visible on control charts. Therefore, run charts can verify the effectiveness of the action, and thus are a useful monitor of process stability. ∎

15.10   FUNDAMENTALS OF PROCESS CONTROL

A process control system can be described as a feedback system. Four elements of that system are important:

1. *The process*—the people, equipment, input(s) (x's), methods, environment, and all those elements that work together to produce the output.
2. *Information about the actual Performance of the Process*—this can be obtained by analyzing the process output (*Y*).
3. *Process output (Y)*—includes not only the products or services that are produced but any intermediate outputs (y's) that describe the operating state of the process. If this information is

gathered and interpreted correctly, it can show whether action is necessary to correct the output.

4. *Action of the output*—is past-oriented and involves detecting out-of-specification output already "done/produced." Corrective action(s) on the process must be taken and verified or service or product specifications must be changed.

5. *Action on the process*—is future-oriented because it is taken to prevent the production of out-of-specification products or services. These actions might consist of changes in the operator level of training to resolve the related problems during the customer call(s). The effects of actions taken should be carefully monitored, and further studies and actions should be taken when and if necessary.

15.11   CONTINUOUS STATISTICAL PROCESS CONTROL (SPC) TOOLS

Continuous SPC refers to a group of graphical tools called *control charts* that display prmocess input or output characteristics over time. These control charts serve three different purposes. They are used to control a CTP characteristic; we call this *statistical process control* (SPC). They can be used to monitor CTQ, CTC or CTD characteristics; we call this *statistical process monitoring* (SPM). Finally, these also can be used as diagnostic tools for any CT characteristic. Control charts offer other advantages such as grouping data into patterns that can be statistically tested, detection of assignable causes that affect the central tendency and/or the variability of the cause system, and determination of whether action is needed on the process. There are two categories of control charts for continuous data: charts for process location and variation. Generally, the two categories are combined. Typical control charts used in Six Sigma track individual values and moving ranges ($X–MR$), charts for averages and ranges (XBar-R), charts for Averages and Standard deviations ($\bar{X}–S$) and exponentially weighted moving-average (EWMA) chart.

On control charts, the different points represent statistical values of subgroup measurements ($X, \bar{X}$, R, S, etc.) over time. A centerline and two control limits are also displayed on control charts. The centerline is the average of subgroup statistical values. The control limits are set at a conventional distance of plus or minus three standard deviations ($\pm 3\sigma$) from the centerline. They represent the voice of the process and are independent of specification limits. We consider a CT characteristic to be in control when all points fall inside the control limits, and they display random variation only. There are several pattern types that can be found on control charts, which indicate different out-of-control situations such as cycles, shifts, or trends. The Western Electric rules and tests can help detect these out-of-control situations. When selecting control charts for practical application, we must consider criteria such as sample size, the desired sensitivity level for detecting small shifts in the process, and the allowable complexity level of the charts. Importance must be placed on planning and managing SPC and SPM implementation to ensure success. The general step-by-step implementation is to define the problem, establish the measurement system, design the control charts, prepare data collection, and implement control charts.

Example 15.6

1. What does *continuous SPC tools* mean, and how does it relate to Six Sigma?
2. What is the statistical foundation of control charts for continuous data?
3. What are the principal types of control charts for continuous data?
4. What are the decision criteria for matching a continuous control chart to a practical application?
5. What are the considerations for planning and implementing SPC for continuous data?
6. What are the considerations for sustaining the use of SPC for continuous data?

Solution

1. Continuous SPC tools are a group of graphical tools called *control charts*, which display process input or output characteristics measurements over time, to control or monitor continuous CTP characteristics.

2. Control charts display patterns that can be statistically tested and, thus provide information about the behavior of product and/or process characteristics. To evaluate this behavior, we use historical statistics of the process such as the mean, standard deviation, statistical control limits, and different tests to determine out-of-control situations.

3. There are two types of control charts for continuous data: charts for location, $\bar{X}$, the exponentially weighted moving average (EWMA), and charts for variation (R, S, MR, etc.). In many cases, these are combined to provide complete information ($\bar{X}-R$, $X-MR$, etc.).

4. Decision criteria for matching a continuous control chart to a practical application are sample size, sensitivity level for detecting small shifts in the process, and the complexity of the charts.

5. Important elements that should be considered when planning and implementing SPC include availability of resources (money, people, etc.), time constraints, risk and confidence requirements, impact on specific parties, and potential benefits derived from successful implementation.

6. To encourage continued use of SPC, the charts must be regularly reviewed, charts should be modified to improve process visibility by changing sampling interval, combining charts, eliminating charts found unnecessary, and corrective action must be taken when alerted to potential process problems. ∎

15.12   INTERPRETING PROCESS CONTROL

Several types of patterns that occur on control charts, which indicate different out-of-control situations such as cycles, shifts, or trends. The Western Electric rules and tests can help detect these out-of-control situations. When selecting control charts for practical application, criteria such as the sample size, the desired sensitivity level for detecting small shifts in the process and the allowable complexity level of the charts are considered. Importance must be placed in planning and managing SPC and SPM implementation in order to make it successful. The general step-by-step approach for the implementation is to define the problem, establish the measurement system, and determine the control charts, prepare data collection, implement control charts, use control charts and continuous improvement.

Solution

1. Continuous SPC tools are a group of graphical tools called control charts, displaying process input or output characteristics measurements over time to control continuous CTP characteristics, or/and monitor CTQ, CTC and CTD characteristics.

2. Control charts form data into patterns that can be statistically tested and, as a result, lead to information about the behavior of product and/or process characteristics. To evaluate this behavior, we use historical statistics of the process such as the mean, the standard deviation, statistical control limits, and different tests of out-of-control situations.

3. There are two categories of control charts for continuous data: charts for location ($\bar{X}$, EWMA, etc.) and charts for variation (R, S, MR, etc.). Generally, both categories are combined in order to have complete information ($\bar{X}-R$, $X-MR$, etc.).

4. Some of the decision criteria for matching a continuous control chart to a practical application are the sample size, the desired sensitivity level for detecting small shifts in the process, and the allowable complexity level of the charts.

5. Some important elements should be considered when planning and implementing SPC, such as availability of resources (money, people, etc.), time constraints, risk and confidence requirements, impact on specific parties, potential benefits to be derived, and potential for successful implementation.

6. In order to sustain the use of SPC, the charts have to be reviewed, such as changing sampling interval, combining charts, eliminating charts found unnecessary, and acting on the information provided by the charts.

15.13 STATISTICAL PROCESS CONTROL AND STATISTICAL PROCESS MONITORING

Six Sigma methodology focuses on the causal system, $Y = f(X)$. Individuals working with the Six Sigma approach often say that Y is a function of X to represent the idea that any output (or CT characteristic) is a function of some input factors (X), and that some of these x's constitute the CT process variables, or "vital few" variables. In the maskant dip process, the *maskant* mix viscosity is one of the vital few inputs, directly affecting the maskant uniform thickness (Y) on the aluminum parts. If we only collected data pertaining to the outcome (Y) in the Control Phase, we would be practicing Statistical Process Monitoring or SPM; if we collected data on the x's that influence Y, we would be practicing Statistical Process Control or SPC.

Example 15.7   A Black Belt asks, "How can we practice effective statistical process control if the vital few variables were identified during the improve phase, and if we made the process robust to variation?"

Solution   In an ideal world, the outcome of the improve phase should be a robust process whose performance approaches Six Sigma levels. In such a case, monitoring or controlling the process would not be required. However, we know that not everything in life works according to plan, so we implement control and monitor tools to ensure that the processes remain close to Six Sigma levels. An example of the application of the methodology of SPC is the following: ■

The Black Belt discusses with a champion the details of the project control phase related to the paint application process. During the improve phase, five CTPs were identified that had large influences on the variation of our CTQ characteristic (paint thickness). From these, three CTP characteristics are related to the application method including the method of using the paint gun to apply the paint. Since the paint application process is manual, the related CTPs are difficult to control, and some of these CTPs have discrete settings. We wrote a standard application method and presented it to the operators during training sessions. Given its manual nature and the existing conditions, we must try to optimize the process. Two other CTPs are related to the paint: pot life and paint viscosity. The *pot life* will be controlled following a standard operating procedure (SOP) and the *paint viscosity*, which is a *continuous variable*, can be controlled using statistical process control (SPC). Presently, the viscosity is measured after the paint mixing operation. We collect data, but do not use it to improve the process. With SPC, we measure viscosity and identify out-of-control situations. By understanding the causes of these out-of-control situations and finding solutions for them, we prevent future out-of-control situations of the CTQ characteristics such as paint thickness and viscosity.

The Behavior of Processes.   Process behavior is the result of superimposing random variation (due only to white noise) and variation due to special or assignable causes (black noise). White noise is due mainly to the level of technology used, and is related to process entitlement. Entitlement refers to the performance level a process expected by a customer; it should be achievable from prior investments.

The behavior of processes

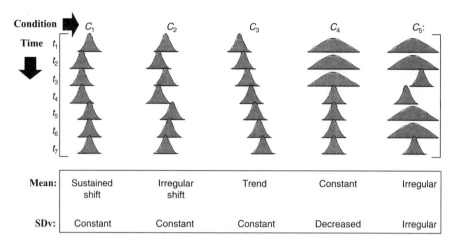

Mean:
| Sustained shift | Irregular shift | Trend | Constant | Irregular |

SDv:
| Constant | Constant | Constant | Decreased | Irregular |

What do your processes look like?

Figure 15.10   Figure illustrating the behavior of processes.

Black noise is the main target of Six Sigma improvements. We must eliminate these causes for our processes to meet Six Sigma performance targets. The five processes shown in Figure 15.10 exhibit different problems with the mean and/or standard deviation targets. In each case, we should decrease standard deviation (process variation) and center the mean on the target value. Statistical process control (SPC) enables us to control both process centering and spread to sustain the improvements. To avoid false alarms, SPC enables the operator to know if the process is behaving normally (i.e., it exhibits random variation only).

15.14   THE FOUNDATION OF SPC

Operating with minimum variance is achieved only when a process displays a reasonable degree of statistical control

—Walter A. Shewart

The word *statistical* means having to do with numbers; more specifically, it means drawing conclusions from numbers. The word *quality* as widely discussed in preceding chapters means much more than only goodness or badness of the product. It refers to the qualities or characteristics of the thing or process under study. The word *control* means to keep something within boundaries, or make that something behave the way we wasn't it to behave. Tying all these words together, the words *statistical quality control* mean

Statistical—with the assistance of numbers or data

Quality—the study of the characteristics of the process

Control—in order to make it behave the way we want it to behave

In general terms, process is a set of conditions, or causes, which work together to produce a given result. Walter A. Shewart from Bell Laboratories developed SPC in 1924. Dr. Shewart presented his theories in a series of lectures, and later published in the book, *Economic Control of Quality*

Manufactured Product, 1931. SPC became widely used in the1940s as a result of World War II manufacturing production efforts; it offers the following advantages:

- Groups data into patterns that can be statistically tested, and provides insight into the behavior of product or process characteristics
- Facilitates understanding of the underlying "cause system" of products and process characteristics
- Provides a graphical representation of product or process performance
- Detects assignable causes that affect the central tendency and/or the variability of the cause system (main use of SPC)
- Serves as a probability based decision tool based on probabilities
- Practical tool for detecting changes in product and/or process performance relative to historical performance or specified standards
- Enables inferences to be made on the basis of sample data.
- Points out when action is needed with known levels of risk and confidence

Dr. Shewart created the well-known control charts that are part of the much-discussed Japanese miracle and especially the Toyota production system discussed in Chapter 4 of this book.

15.15   TOOLS FOR PROCESS CONTROLS – CONTROL CHARTS

In addition, Dr. Shewart observed that when the data are plotted, the data formed a fluctuating zigzag pattern on a control chart. Generally, any series of observations or measurements obtained from a given process also formed a fluctuating pattern. Sampling the characteristics and estimating the parameters of the distribution can quantify the variation of any particular process characteristic. In addition, Dr. Shewart observed that a steady component, inherent to the process, could be attributed to chance and undiscoverable causes, known as *random variation*. If nothing disturbed the process, these fluctuating measurements remained within defined mathematical limits. In the long term, these measurements tend to form a predictable distribution. On the basis of these findings, Dr. Shewart proposed control charts as a simple and effective method to achieve statistical control.[2] Control charts provide reliable information on when action should be taken to control a process. When a process is in statistical control, its performance is predictable. Both manufacturer and customer can rely on stable costs to achieve the quality level and the process performance can be improved to reduce variation. Control charts provide a common language for describing process performance. By distinguishing special from common causes of variation, control charts indicate when problems are likely to be correctable locally or may require executive action. Choosing rational subgroups determines the effectiveness and efficiency of the control charts. Rational subgroups are discussed in Section 15.19.

Basic Components of a Control Chart. The vertical axis of a control chart represents the statistical values of the CT characteristic; the horizontal axis represents the subgroup number in chronological order. Each subgroup is characterized by its conditional distribution, which corresponds to the short-term variation (i.e., "white noise" only, or "within subgroup variation"). A point on the chart represents each subgroup statistic (e.g., mean, range, standard deviation), and these points form a marginal distribution (i.e., "black noise" or "between subgroup variation"). It is customary to connect the sample points on the control chart with straight-line segments, to highlight nonrandom patterns. A line that represents the process average divides the control chart horizontally. The two control limits (upper and lower) define three major zones. The first zone, between the control limits, represents random variation, while the two other zones, beyond the control limits, represent nonrandom variation.

Once we establish the control limits, they are utilized for any future samples taken from the same process, and running under the same conditions. Establishing the control limits is essential to determine whether the process is stable and remains in control.

15.16   CONTROL LIMITS

The *centerline* is the average of the plotted points for the statistic shown on the control chart (Fig. 15.11). The *upper control limit* (UCL) and the *lower control limit* (LCL) are horizontal lines at a conventional distance of three standard deviations ($\pm 3\sigma$) from the centerline. The control limits are chosen so that for a process in control, nearly all the points will fall between the limits. As long as the new points lie within the control limits, the process is assumed to be in control, and no action is necessary. Since these control limits make it possible to distinguish between intrinsic and special-cause variation, they minimize reaction to expected random variation (i.e., avoid false alarm). To create a control chart, we must have at least 20 points to calculate the trial control limits. This is sufficient to determine if the process was in control when the initial samples were selected. If all points fall inside the control limits and no systematic behavior is evident, then we conclude that the process was in control in the past, and the trial control limits are suitable for current and future production. Generally, the effective use of any control chart requires periodic revision of the control limits and centerlines. Some practitioners set regular periodic reviews of control charts, such as every week, month, or every 25, 50 or, 100 samples. Data corresponding to out-of-control points represent the effect of black noise; thus, when the cause is identified, the corresponding points must be excluded when calculating the new control limits.

Specifying control limits is a critical step in designing control charts and closely related to hypotheses testing. Moving the control limits further from the centerline decreases the alpha (α) risk or probability of committing type I error, encountering a point beyond the control limits. However, widening the control limits also increases the beta (β) risk or probability of committing type II error by placing a point within the control limits when the process is in fact, out of control.

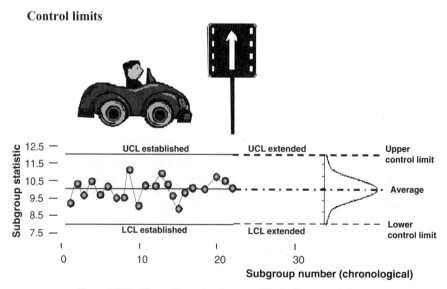

Figure 15.11   Figure illustrating the control limits in a control chart.

Out-of-control condition

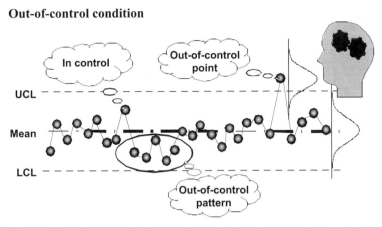

Figure 15.12   Illustration showing out-of-control point and an out-of-control pattern.

If we narrow the control limits: alpha (α) risk increases and beta (β) risk decreases. Using 3σ to establish control limits results in low probability of false alarm; a false alarm is generated 27 out 10,000 times. The advantage of using the 3 σ control limits is that it minimizes both α and β risks (i.e., types I and II errors).

15.17   PROCESS OUT-OF-CONTROL CONDITION

Sometimes we observe that the process is out of control, but all the parts produced are within specifications. Using specification limits in no way determines whether a process is in a state of statistical control. Being out of control does not mean that the units produced do not meet specifications. The CT characteristic distribution mean reflects the centering with respect to a target, and the standard deviation reflects its spread. For most distributions, more than 99% of the data lie within 3σ of the mean. The upper and lower control limits establish the band or region from -3σ to $+3\sigma$. If a process is in control, we expect to see that plotted points are randomly distributed within this region. On the other hand, if a process is out of control, we expect to see data points plotted outside this region, that is, above UCL or below the LCL and/or systematically distributed. Since the process is normally distributed, stable, and under control, there is only 0.135% of the points beyond one control limit; an thus out-of-control condition indicates a significant change in the process. Even if all the points plot inside the control limits, they may exhibit systematic or nonrandom patterns indicating an out-of-control process. Figure 15.12 shows out-of-control conditions. The upper and lower control limits used here should not be confused with the upper and the lower specification limits. The control limits are statistically driven, while the specification limits are customer-driven.

15.18   WESTERN ELECTRIC RULES

A process is out of control when noncommon or assignable causes of variation force points on a control chart to exceed either of the control limits. It may also be deemed to be out of control if it fails one or more of a number of quantitative decision tests such as the Western Electric rules.[3] Figure 15.13 shows an out-of-control point (6) that lies outside the upper control limit.

Out of Control

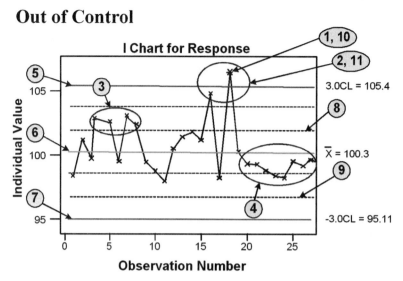

Figure 15.13 A control chart showing a process out of control—Rule 1 (point 6).

Criteria for Control Charts—Western Electric Rules

1. An out-of-control condition is indicated when a single point falls outside the 3σ ($\pm$) control limits.
2. At least two out of three successive points fall on the same side and are more than two sigma units away from the central line ($\bar{X}$).
3. At least four out of five successive points fall on the same side of and are more than one sigma unit value away from the central line ($\bar{X}$).
4. At least eight successive points fall on the same side of the central line ($\bar{X}$).
5. *Upper control limit* (UCL)—the upper limit of process control. The UCL is by convention set equal to the process mean plus three standard deviations.
6. *Centerline*—calculated as the process mean over the period being investigated.
7. *Lower control limit* (LCL)—the lower limit of process control. The LCL is by convention set equal to the process mean minus three standard deviations.
8. *One sigma line*—a line one standard deviation away from the centerline, or process mean.
9. *Two sigma line*—a line calculated two standard deviations away from the center line, or process mean.
10. *Out-of-control point*—a single process point showing the most obvious sign of an Out-of-control situation, i.e., being beyond either the UCL or LCL.
11. *Out-of-control situation*—this condition occurs when the process fails Western Electric rules test 2; a process is out of control when two out of any three successive points are on the same side of the central line and are simultaneously more than two standard deviations from the central line.

To determine whether a process is in or out of statistical control, apply quantitative tests to points on a control chart. Western Electric rules test 1 (Fig. 15.14) states that a process is out of control if any one point falls outside of either the LCL or the UCL.

Western Electric Rules - Test 1

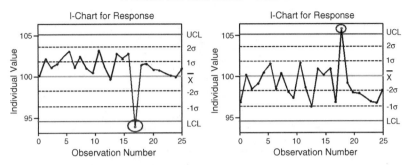

Western Electric Rules Test 1 :
A lack of control is indicated
whenever any one value
falls outside the three sigma limit.

Figure 15.14   Illustration showing example of process out of control in which Western Electric rules test 1 is clearly shown.

The use of formal tests such as the Western Electric rules, eliminates subjectivity as a factor in determining the state of process control.

Western Electric Rules Test 1.   One of the most common tests for the state of process control, and should always be applied first. MINITAB has the ability to test control charts for any of the Western Electric rules.

Interpretation of Figure 15.12–15.14; where the central line is the Mean or $\bar{X}$ or $\bar{\bar{X}}$.

- *Lower control limit* (LCL)—numerical value representing the lower limit of the expected variation if the process were in a state of statistical control; by convention, the LCL is set equal to the mean minus three standard deviations.
- *Centerline* (*central line*)—average value of the process parameter being plotted.
- *Upper control limit* (UCL)—line and numerical value representing the upper limit of the variation that could be expected if the process were in a state of statistical control, by convention equal to the mean plus three standard deviations.
- Out-of-control point—A single point above the UCL or below the LCL is sufficient for this process to fail test 1.

Western Electric Rules—Test 2.   Western Electric rule 2 (Fig. 15.15) states that a process is out of control if any two of three successive points are on the same side of the central line *and* are more than two standard deviations away from the central line.

Western Electric rule 2 is not used as frequently as rules 1 and 4, and is generally used when increased sensitivity is required.

Western Electric Rules—Test 3.   Western electric rule 3 states that a process is out of control if any four out of five successive points are on the same side of the central line *and* are more than one standard deviation away from the central line as is shown in Figure 15.16.

Interpretation of Figure 15.16

- *One sigma line*—line plotted on the control chart, corresponding to one standard deviation above or below the process mean.

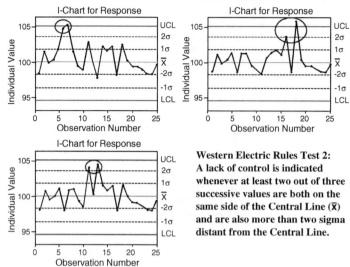

Western Electric Rules - Test 2

Western Electric Rules Test 2:
A lack of control is indicated whenever at least two out of three successive values are both on the same side of the Central Line ($\bar{x}$) and are also more than two sigma distant from the Central Line.

Figure 15.15 Illustration showing the application of Western Electric rule—test 2.

- *Two sigma line*—line plotted on the control chart, corresponding to two standard deviations above or below the process mean.
- *Out-of-control case*—four out of five successive points beyond the one sigma line constitute failure of Western Electric test 3.

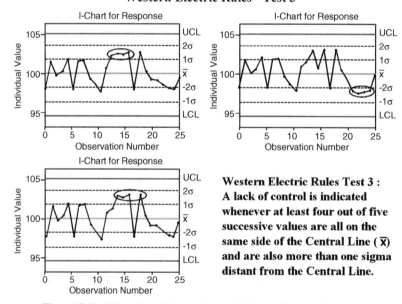

Western Electric Rules - Test 3

Western Electric Rules Test 3 :
A lack of control is indicated whenever at least four out of five successive values are all on the same side of the Central Line ($\bar{x}$) and are also more than one sigma distant from the Central Line.

Figure 15.16 Illustration of the application of Western Electric rule—test 3.

Western Electric Rules - Test 4

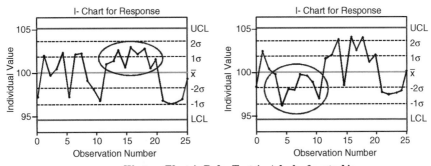

**Western Electric Rules Test 4 : A lack of control is
indicated whenever at least eight successive values are
all on the same side of the Central Line ($\bar{X}$).**

Figure 15.17   Drawing showing illustration of Western Electric rule—test 4.

The use of formal tests such as the Western Electric rules eliminates subjectivity as a factor in determining the state of process control. Western Electric rule 3 is not used as frequently as rules 1 and 4, and is generally used when increased sensitivity is required

Western Electric Rules—Test 4.   Western Electric rules 4 (Fig. 15.17) states that a process is out of control if any eight successive points are on the same side of the central line. This rule 4 is frequently used in conjunction with Western Electric test 1.

Interpretation of Figure 15.17

- *Lower control limit* (LCL)—line representing the lower limit of the variation that could be expected if the process were in a state of statistical control, by convention set equal to the mean minus three standard deviations
- *Central line*—average value of the process parameter being plotted, over the period of inspection being referenced.
- *Upper control limit* (UCL)—line representing the upper limit of the variation that could be expected if the process were in a state of statistical control, by convention set equal to the mean plus three standard deviations.
- *Out-of-control case*—eight successive points above the central line constitute failure of Western Electric rules—test 4.

15.19   CONTROL CHARTS AND HOW THEY ARE USED

Control charts are available for many variable types, and each serves a particular need. Therefore, we must carefully select the appropriate chart for our project. The most common control charts are

- *General average* ($\bar{X}$) *and individuals* (X)—To measure the central tendency of a process across time
- *Range* (R), *sigma* (s), *and moving range* (MR)—to measure the gain or loss of uniformity or variability of a process across time

- *Exponentially weighted moving average* (EWMA): Smoothes the data and emphasizes trends; this chart gives greatest importance to the most recent data collected.

Short-Run SPC Tools. The subgroups are chosen so that opportunities for variation among the units within a subgroup are small. If the variation within a subgroup represents the piece-to-piece variability over a short period of time, then any unusual variation between subgroups reflects changes in the process that should be investigated. For an individual process study, the subgroups typically consist of four or five consecutively produced pieces representing a single manufacturing tool. In transactions, these subgroups represent four or five consecutive transactions or service provided by one associate. Practitioners should remember that sample sizes should remain constant for all subgroups. The number of subgroups should satisfy two criteria: (1) from a process standpoint, enough subgroups should be gathered to ensure that the major sources of variation have the opportunity to appear and (2) from a statistical perspective, 25 or more subgroups containing about 100 or more individual "readings" provide a good test for stability and, if stable, good estimates of the process location and spread. (Usually the population from which the samples are selected is not normally distributed; in such cases, the shape of the sampling distribution of $\bar{X}$ is inferred from a very important theorem called the *central-limit theorem*.) Figure 15.18 reviews and confirms the key fundamental concepts for rational subgrouping.

(*Note:* Rational subgroups are usually selected in such a way that that the variation represented within each subgroup is small as feasible for the process—representing the variation for *common causes*—and so that any changes in the process performance due to *special causes* appears as differences between subgroups. Rational subgroups are typically statistically designed of consecutive product or workpieces, although occasionally random samples are sometimes used.)

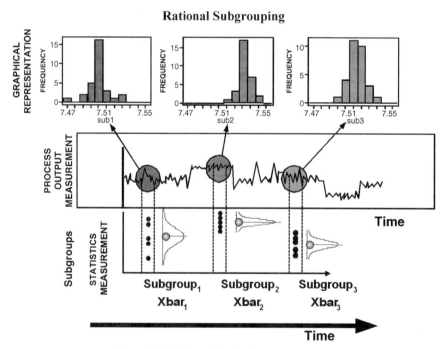

Figure 15.18   Illustration of rational subgrouping.

The shape of the sampling distribution $\hat{p}$ is inferred from the *central-limit theorem,* discussed in detail in previous chapters of this book. According to the central limit theorem, the sampling distribution of $\hat{p}$ is approximately normal for a sufficiently large sample size. In the case of the proportion, the sample size is considered to be sufficiently large if np and nq are both greater than 5; that is, if $np > 5$ and $nq > 5$, where $\hat{p} = $ sample proportion, p = corresponding population proportion, and $q = 1 - p$. Please observe that the sampling distribution of $\hat{p}$ will be approximately normal when $np > 5$ and $nq > 5$. This is the same condition that is required for the application of the normal approximation to the binomial approximation discussed in Section 11.8.3 (in Chapter 11).

The successful use of control charts, particularly $\bar{X}$ and R charts, depends on the proper selection of subgroups.

Issues to Remember about Subgroups

- Subgroups should be as homogeneous as possible.
- For process analysis and control, the individual items that constitute a subgroup should be produced as closely as possible to one another in time, which reduces within-subgroup variation and increases between-subgroup variation.
- This method gives the best estimate of the universe standard deviation if assignable causes of intersubgroup variation can be eliminated.
- This method increases sensitivity to process shifts.
- This method facilitates maintaining a given process average.
- In some cases, it is advisable to select individual random samples instead of sampling at fixed time intervals. In other cases sequential sampling is preferred.

Using production order to define subgroups tends to reveal assignable causes of variation, which are periodic in nature.

15.20   PRECONTROL METHOD

Traditionally precontrol has been perceived as an ineffective tool, and many quality practitioners remain skeptical of its benefits. This view originated because the limits of the three precontrol regions were usually based on process specifications, resulting in overreactions and increasing, rather than reducing, process variability. Six Sigma uses precontrol to monitor trends in the process center (precontrol assumes that the process is centered on target); by measuring two consecutive units of the response (Y), and comparing these measurements against three distinctive regions: green, yellow, or red. In the Six Sigma breakthrough strategy, precontrol is implemented after the *improve* phase. The zones are calculated after improvements are made; thus, the distribution is narrow compared to the specification band. Since specification limits are not used in calculating these zones, units may appear in the yellow or red zones before actual defects are produced.

Precontrol works like a traffic light by providing quick feedback to process operators. If a unit falls in the green region, we continue operating the process. If a unit falls in the yellow region, we suspect that an adjustment is needed. If two units fall in the same yellow band, we adjust the process, and if the units fall in opposite yellow bands, we conclude that variation has significantly increased. If a unit falls in a red zone, we stop the process and take corrective action. In the Six Sigma breakthrough strategy, precontrol is implemented after the improve phase.

The statistical basis for precontrol is found in probability theory; specifically, if A and B are two independent events, the probability that both A and B occur is $P(A \cap B) = P(A) \times P(B)$. From this, we know that the probability of encountering two units in the same yellow band is very small, and if we conclude that an adjustment is needed, we do so with a 99% confidence level. If the process distribution is not normally distributed, we use mathematical transformations to make it follow the normal distribution. The green zone is defined as $\mu \pm 1.5\sigma$, covering 86.64% of the area under the normal curve. The yellow zones are set from 1.5σ to 3.0σ, and from -1.5σ to -3.0σ. The yellow zones represent 6.55% of the area under each tail of the normal curve. Finally, the red zones are defined as the regions under $\mu -3.0\sigma$ and beyond $\mu +3.0\sigma$. Each red zone contains 0.13% of the area under the curve.

A process is qualified for precontrol when five consecutive units fall in the green region; we must requalify the process after an interruption or restart that modifies process conditions, such as change of shifts, or setups. If a process is not centered, it is very unlikely that it can be qualified. The preliminary activities associated with precontrol are:

1. Process characterization of short-term standard deviation (σ_{ST})
2. Process improvement (improve phase)
3. Definition of the precontrol regions
4. Process qualification
5. Process operation

There is no rule of thumb to establish the sampling frequency used in precontrol. Factors such as the nature of the process, production volume, and operator experience must be considered. Precontrol can be used on processes that have either one or two sided specifications, and is applicable to all areas of the business. Practical applications include manufacturing processes such as machining, forming, painting, assembly, and transactional applications such as customer service, order or form filling, and invoice collection. Figure 15.19 illustrates the fundamental concepts of precontrol.

15.20.1 The Foundations of Precontrol

Figures 15.19–15.21 illustrate the fundamental principles of the precontrol concept.

Assume a distribution A with a mean of 100 and a green zone ($+ - 1.5\sigma$) from 90 to 110, and that two units fall in the same yellow band. The probability of 2 units falling in the same yellow zone, either

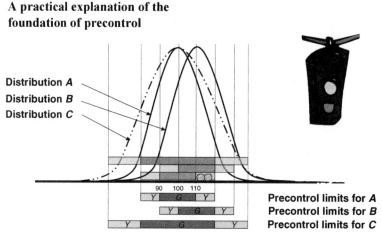

A practical explanation of the foundation of precontrol

Distribution A
Distribution B
Distribution C

90 100 110

Precontrol limits for **A**
Precontrol limits for **B**
Precontrol limits for **C**

Figure 15.19 Illustration of the fundamental principles of the precontrol concept.

What is precontrol?

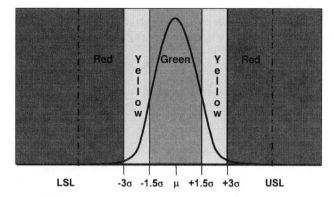

The limits of each zone are calculated from distribution of the characteristic measured, not the tolerances; therefore, units that fall in the yellow or red zones trigger an alarm before defects are produced.

Figure 15.20 Further Illustration of the fundamental principles of the precontrol concept.

left or right, is 0.858% (.00429 + .00429). Now imagine that the distribution changed to *B*, which has a mean of 110 and the same standard deviation as *A*, so that the two units found in the yellow zone of *A*, fall in the green area of *B*. The probability of finding two units in a green region is 75.06% (.8664 × .8664), and since this probability is larger than 0.858%, we conclude that the distribution shifted and that an adjustment back to *A* is required. In other words, our confidence that an adjustment

Precontrol chart

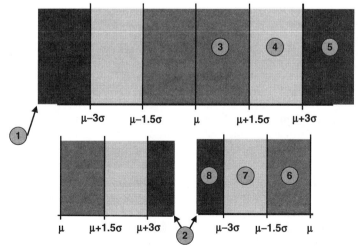

Figure 15.21 Further illustration of the fundamental principles of the precontrol concept.

back to A is needed is over 99% (1–0.00858). What if two subsequent units lie in opposite yellow bands? Assume that distribution A has changed to C, which, like A, is centered around a mean of 100, but has a much wider spread than A. Further, we assume that the two units that fall in the yellow zone of A are green for C. This clearly indicates that the variation has significantly increased, perhaps because of a technical problem, and that the process must be stopped and the variation reduced. What does "adjusting the process" mean? It means changing a process setting or resetting the process to its baseline condition. Often, these settings are specified during the improve phase of our project (DOE). Resetting a process as similar to performing a computer reboot.

15.20.2   Precontrol Charts

By comparing two successive units of process output against color-coded zones with respect to the existing process mean and standard deviation, the precontrol chart can determine whether a process has changed and if corrective action is required. The purpose of a precontrol chart is to provide a conceptual aid in determining the state of a process and the action to take in the event of a change in the process.

Major Considerations for Precontrol Charts

1. *Precontrol chart* for a two-sided process with output on either side of a mean, or target, value
2. *Precontrol chart* for a one-sided process with a single specification or target, value
3. *Green zone*—a three sigma wide band, centered on the process mean, and extending 1.5σ above and below. Process output in the green zone is considered acceptable and the process may continue to run.
4. *Yellow zone*—a pair of bands, extending from $\mu + 1.5\sigma$ to $\mu + 3.0\sigma$ and $\mu - 1.5\sigma$ to $\mu - 3.0\sigma$. Process output in the yellow zone may permit the process to continue, or may require process adjustment or stoppage.
5. *Red zone*—a pair of open-ended bands extending above the $\mu + 3.0\sigma$ point and below the $\mu - 3.0\sigma$ point. Process output in the red zone is cause for immediate process stoppage and investigation.
6. *Green zone* for a one-sided process – a band extending from the Mean to $\mu - 1.5\sigma$ for process with a lower specification (or $\mu + 1.5\sigma$ for a process with an upper spec).
7. *Yellow zone* for a one-sided process – a band extending from $\mu - 1.5\sigma$ to $\mu - 3\sigma$ (or $\mu + 1.5\sigma$ to $\mu + 3\sigma$).
8. *Red zone* for a one-sided process – an open-ended band extending beyond the 3σ limit.

The precontrol chart permits quick determination of the state of a process, and the potential need for corrective action. The use of precontrol requires that the process mean be centered on the target.

How to Construct a Precontrol Chart

1. Following the improve phase of your project, collect sufficient data from the process to determine the mean and the standard deviation.
2. Determine the red, yellow, and green zone limits by calculating the points $\pm 1.5\sigma$ and $\pm 3.0\sigma$ from the mean.
3. Qualify the process by taking five consecutive samples in the green zone.
4. Take two successive samples (A and B) from the process, and note their positions relative to the color-coded bands on the precontrol chart:
Case 1: continue	Both A and B inside green zone	Take no action
Case 2: continue	A is green, B is yellow (low)	Take no action
Case 3: continue	A is green, B is yellow (high)	Take no action

Case 4: continue	*A* is yellow (low),	*B* is green Take no action
Case 5: continue	*A* is yellow (high),	*B* is green Take no action
Case 6	Both *A* and *B* are yellow (low)	Adjust process
Case 7	Both *A* and *B* are yellow (high)	Adjust process
Case 8	*A* is yellow (low), *B* is yellow (high)	Stop process
Case 9	*A* is yellow (high), *B* is yellow (low)	Stop process

5. Continue selecting pairs of samples from the process using a predetermined sampling schedule and compare them per the control chart.

6. Requalify the process after any adjustment or stoppage.

7. Following the improve phase of a process change, collect sufficient data from the process to be able to determine the mean and the standard deviation.

8. Determine the red, yellow and green zone limits by calculating the points $\pm1.5\sigma$ and $\pm3.0\sigma$ from the mean.

9. Qualify the process by taking five consecutive samples in the green zone.

10. Take two successive samples (*A* and *B*) from the process, and note their positions relative to the color-coded bands on the precontrol chart.

11. Continue selecting pairs of samples from the process, per a predetermined sampling schedule, and compare them per the control chart.

12. Requalify the process after any adjustment or stoppage.

Example 15.8   MINITAB test results for zone chart average speed of answer (Fig. 15.22).

Detection Rule 3 −A lack of control is indicated whenever at least four out of five points/values fall on the same side of, and more than one StDev (1σ) from the central line $\bar{\bar{X}}$. In this case, six values/points fall on the lower side of $\bar{\bar{X}}$, and more than -1 StDev $= 2.709$. It should be noted that accordingly to the rules for defining lack of control, the process for average speed of answer is out of control because 6 points out of 14 points are in the same zone C and beyond (-1 StDev $= 2.78$; -2 StDev $= 1.344$; -3 StDev $= -0.020$). However, because this pattern indicates a downward $\bar{X}$ shift for the average speed of answer it means that the call center is averaging less time for answering the calls. In general, for call

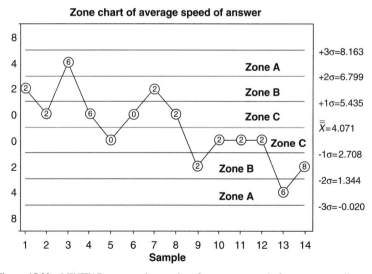

Figure 15.22   MINITAB generated zone chart for average speed of answer at a call center.

centers the fact that the average speed of answer is lowering is good as long as these observations are consistent over a large period of time.

15.21   CONTROL CHARTS FOR VARIABLES

Some data types are more sensitive than others in detecting the causes that tend to disturb a process. The common types of data listed in order of sensitivity are ranges (difference between highest and lowest value in a small group of measurements), averages, percentages and individual numbers (temperatures, pressures, records or earning, records of absences, etc.). The most sensitive data types are ranges and averages, and when plotted together in a control chart this form of graphical representation is known as the $\bar{X}$–$\bar{R}$ *chart*.

15.21.1 $\bar{X}$ Chart

In essence, control charts are a set of statistical limits applied to a sequence of points representing any process under observation or study. While each individual point is random, grouped points plotted in a preselected nonrandom arrangement highlight the most important performance changes. In a process capability study, the most important variable is "time." The $\bar{X}$ chart shows when a process is centered. If the center shifts, the $\bar{X}$ pattern also shifts. If the center of the distribution trends up or down, the $\bar{X}$ pattern follows the same trend. Therefore, the $\bar{X}$ chart assists in observing and evaluating the behavior of a process over time and signals the need for corrective action. The $\bar{X}$ chart plots the average values of each of a number of small sampled subgroups. The process variation must be in control before the process center is assessed by the use of the $\bar{X}$ chart because the control limits for this chart are calculated using the within-subgroup variation. If the process variation is not stable, then the chart control limits will be questionable and an out-of-control condition may be falsely signaled. Charts for individuals do not isolate the piece-to-piece repeatability of the process.

The conditions, which the $\bar{X}$ chart is intended to reflect, are shown in Figure 15.23.

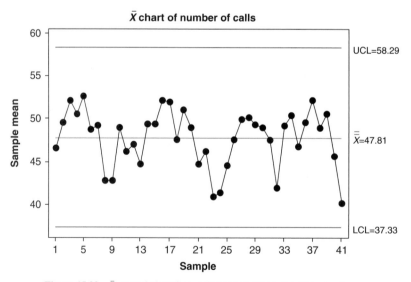

Figure 15.23 $\bar{X}$ control chart for a call center number of calls in one hour.

Control Chart - Xbar Chart

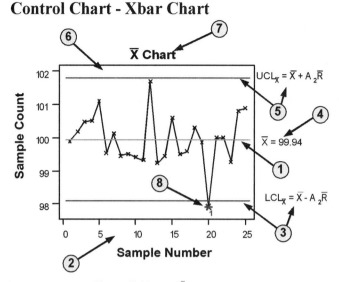

Figure 15.24 The $\bar{X}$ control chart.

In many specifications, is better to use the $\bar{X}-R$ chart with a small subgroup sample size (every subgroup size from three to four data points). Because of all of these reasons, the $\bar{X}$ chart is usually used in conjunction with the range (R) chart or the standard deviation (s) chart. The $\bar{X}$ chart, together with the R Chart, is a sensitive control chart used to identify assignable causes of product and process variation, and gives great insight into short-term variations. The control limits for the $\bar{X}$ chart depend on whether it is being plotted for use with the R chart or the S chart. In general, it is preferable to use an R chart or a combined $\bar{X}-R$ chart to assess whether the process variation is in control. An $\bar{X}$ control chart is shown in Figure 15.24.

Figure 15.24 Components

1. *Sample mean*—the means of the process subgroups collected in sequential or chronological, order from the process.
2. *Sample number*—the chronological index number for the sample or subgroup whose average value is being referenced.
3. *Lower control limit* (LCL)—Line representing the lower limit of the expected variation if the process is in a state of statistical control equal to the overall mean minus the average moving range multiplied by a conversion factor.
4. *Process average*—overall average value of the individual process readings over the period of inspection being referenced.
5. *Upper control limit* (UCL)—line and numerical value representing the upper limit of the variation that could be expected if the process were in a state of statistical control, equal to the overall mean $+ 3\sigma$.
6. Plot of the individual sample means versus sample number. Any excursion in this plot above the UCL or below the LCL represents an out-of-control condition and should be investigated.
7. $\bar{X}$ *chart*—the symbol $\bar{X}$ refers to the sample average value (X) being plotted.
8. *Out-of-control point*—by definition, any point that exceeds either the UCL or the LCL is out of control. MINITAB has a number of tests available for out-of-control conditions, and labels each point with a number corresponding to the test that the point fails.

Steps for construction of the $\bar{X}$ Chart

1. Determine purpose of the chart
2. Select data collection point
3. Establish basis for subgrouping
4. Establish sampling interval and determine sample size n
5. Set up forms for recording and charting data and write specific instructions on use of the chart.
6. Collect and record data. A minimum of 25 subgroups or samples of size n should be measured.
7. Compute the process average X.
8. Compute required upper control limit (UCL) $\bar{X}$.
9. Compute required lower control limit (LCL) $\bar{X}$.
10. Plot the data points.
11. Interpret the chart together with other pertinent sources of information on the process and take corrective action if necessary.

Mistakes to Avoid When Interpreting $\bar{X}$ Charts

- Interpret a $\bar{X}$ chart when an R chart is not in control.
- Attempt to relate the $\bar{X}$ chart to a specification while ignoring changes to the shape of the distribution. *Do not assume normality* in a distribution that may be truncated or screened.
- Assume that most of the data points are at or "near" the average.

15.21.2   *R* Chart (Range Chart)

Range is the simplest measure of dispersion to calculate; it is obtained by taking the difference between the largest and the smallest value of the dataset. *R* charts display variation of a process over time and compare it to the control limits.

The *R* chart plots the range values for a series of subgroups. The *R* chart is usually plotted in conjunction with the $\bar{X}$ chart. The range or (*R*) chart is a measure of the uniformity or consistency of a data set. It highlights changes in variation or spread. In general, we want the level shown on an *R* chart to be as low as possible. If all data points in a dataset receive the same treatment during production, the *R* chart will tend to stay in control. If the level shown on the *R* chart increases, this means that the data points corresponding to some units are receiving different treatment from the others. If the level on the *R* chart rises and then remains in control at a higher level, then some new element has entered the "cause system" and has become a regular part of the process. In general, an element that causes the *R* chart to rise is considered an undesirable element. The $\bar{X}$ chart and the *R* chart taken together quickly identify assignable causes of product and process variation and give insight into the causes of short-term variations. Figure 15.25 shows a sample *R* chart.

Figure 15.25 Components

1. Sample range
2. Sample number
3. Control limits (UCL and LCL)
4. Range average
5. Upper specification limit $+ 3\sigma = 6.445$
6. Range data point (high number − low number)
7. Chart type/name

R chart

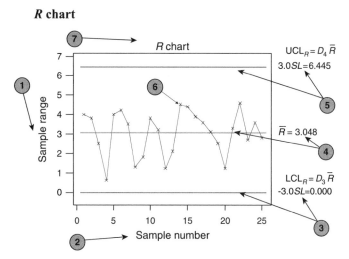

Figure 15.25   Example of an *R* chart.

The *R* chart is the best method for detecting mixtures, stratification, erratic conditions, interactions, and general statistical instability. Thus, it is considered by many to be the most important chart for process capability studies. Causes affecting an *R* chart include (1) poorly trained operator or inspector, (2) new operator or inspector, (3) tired operator or inspector, (4) nonuniform material, (5) fixture with loose or has excessive play, and (6) machine in need of repair. The following are some major considerations to take into account when interpreting an R chart:

- *Lower control limit* (LCL)—line and numerical value representing the lower limit of the variation that could be expected if the process were in a state of statistical control, equal to the average range over the period, multiplied by a conversion factor.
- *Process average range*—average value of the individual sample ranges, over the period of inspection being plotted
- *Sample number*—The chronological index number for the individual sample range value being referenced
- *Sample range*—range values calculated from sequential, or chronological, process subgroups
- *Upper control limit* (UCL)—line and numerical value representing the upper limit of the variation that could be expected if the process were in a state of statistical control (equal to the average range over the period, multiplied by a second conversion factor, different from the one used to calculate the LCL).

(*Note:* Plot of the range values vs. sample number.)

15.21.3 $\bar{X}$–*R* Chart

$\bar{X}$– *R* charts are also called "average and range" charts. The interpreted first; from this, it is possible to directly identify many of the causes of variation. The $\bar{X}$ pattern is then read in light of the *R* chart, and other causes are identified. Finally, the $\bar{X}$ pattern and *R* pattern are jointly interpreted, which gives still more information. The patterns that can be identified from an $\bar{X}$– *R* chart include

- Cycles
- Trends

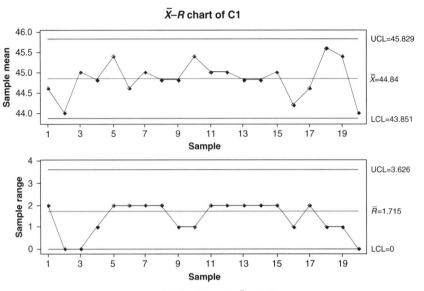

Figure 15.26   MINITAB $\bar{X}- R$ chart.

- Anomalies
- Mixtures
- Grouping of measurements
- Sudden shift in level
- Gradual change in level
- Instability (abnormal large fluctuations)
- Stratification (abnormal small fluctuations)
- Interactions (two or more variables acting together)
- Systematic variation
- Tendency of one chart to follow another.

The MINITAB chart in Figure 15.26 shows an $\bar{X} - R$ chart (C1 is the data column in MINITAB)
Three types of factors can affect the $\bar{X}$ chart indirectly but are not true $\bar{X}$ causes. These factors show on the R chart as well as the $\bar{X}$ chart and are in reality R causes. They appear on the $\bar{X}$ chart only as a reflection of the R chart. Finally, each pattern is associated with particular causes. Remember that the causes that affect the $\bar{X}$ chart are usually different from those that affect the R chart. When combined with experience and knowledge, it is possible to isolate the factors disturbing the process; remember, eliminate the R causes first, and the changes in $\bar{X}$ from these causes will also disappear.

15.21.4 Moving Range (MR) Chart

Moving range or *(MR)* is the difference between successive pairs of numbers in a series of numbers. In other words, it is the difference between the first and second numbers then between the second and the third number, then between the third number and the fourth, and so on. Each individual number is used to calculate two of the *Moving Ranges*. Therefore, M$\bar{R}$ is the average of a series of moving Ranges. The MR Chart displays process dispersion and is usually used in conjunction with the X (individual) Chart, which is used to visualize the process location.

R chart

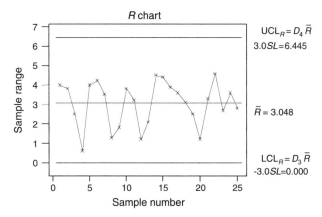

Figure 15.27   Illustrative example of the moving-range (*MR*) control chart.

Constructing the M*R* Chart

1. *Moving range*—the moving range values are calculated from sequential, or chronological, values from the process

2. *Observation number*—the chronological index number for the individual moving range value being referenced.

3. *Lower control limit* (LCL)—line and numerical value representing the lower limit of the variation that could be expected if the process were in a state of statistical control, equal to the average moving range over the period, multiplied by a conversion factor.

4. *Average of the moving range*—average value of the individual moving range values, over the period of inspection being referenced.

5. *Upper control limit* (UCL)—line and numerical value representing the upper limit of the variation that could be expected if the process were in a state of statistical control. It is equal to the average moving range over the period, multiplied by a second conversion factor, different from the one used to calculate the LCL.

6. *Plot of the moving range values vs. observation number*—any excursion in this plot above the UCL or below the LCL represents an out-of-control condition and should be investigated.

Moving range values are correlated, because each successive point has a point in common with the preceding value. Care must be exercised in interpretation. There will always be one less moving range value than individual values as shown in Figure 15.27.

15.21.5   Standard Deviation Chart

The purpose of this chart is to observe and evaluate the variation of a process over time, and against control limits, and take corrective action if necessary. The *S* chart plots the standard deviation of each of a number of sampled subgroups. The *s* chart is usually used in conjunction with the $\bar{X}$ chart. It is recommended for larger sample sizes (generally ≥ 10). It is less sensitive than the *R* chart in detecting causes that affect only a single value in a subgroup. An example of an s chart is shown in Figure 15.28.

Standard deviation chart

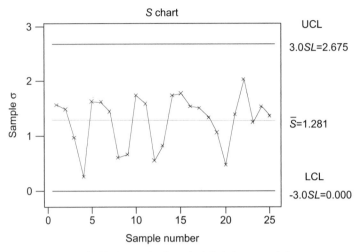

Figure 15.28   Example of standard deviation control chart.

15.22   CONTROL CHART FOR ATTRIBUTES

Control charts usually track variables; however, versions have also been developed to track attributes that have only two values. Control charts for attributes are useful for many technical and administrative processes. A major problem for attribute control charts is the operational definition of nonconformance.

15.22.1   *p* Chart

The purpose of the *p* chart, percentage or proportion chart, is to observe and evaluate the behavior of a process over time and take corrective action if necessary. The symbol *p* denotes proportion and often represents the proportion of bad components (percent defective). The *p* chart is less powerful than the $\bar{X}-R$ chart because it tracks only one variable. Note that the *p* chart forms the same kinds of patterns as the $\bar{X}-R$ chart and these patterns can be related to the variables in the process. The *p* chart relies more heavily on job knowledge than does the $\bar{X}-R$ chart. Therefore, it is used in situations where the operator and analyst know the important causes.

The *p* chart plots the proportion of defective units, not the proportion of defects. The *p* chart is preferred over the *np* chart because the rate of defective units is frequently more meaningful than using the actual number of defective units, and the subgroup, or sample, size varies from period to period. Large subgroup sizes should always be selected ($n > 50$ is considered normal), and the *np* value should always be greater than 5. See Figure 15.29.

15.22.2   Control Chart—*np* Chart

The *np* chart plots the number of defective units and is applicable to binomially distributed discrete defect data collected from subgroups of equal size. If the samples are the same size, it is simpler to plot the number of defective found in each sample instead of calculating the percentage. In this case, the chart is called an *np chart* instead of a *p chart*. Control limits for the *np* chart are calculated as follows;

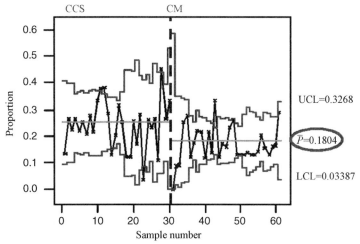

P chart for nonconformities for CCS and CM, Jan.–Aug. 2003
30 weeks for CCS and 30 weeks for CM, Jan.–July 2003

Figure 15.29   MINITAB-generated *p* chart (where CCS = corporate communications services; CM = corporate marketing). Observe the difference between the left hand side of the MINITAB fraction nonconforming control chart *p* chart (corporate communications services); and the right hand side if the *p* chart (corporate marketing).

for each sample, count the number of defectives, *np*, where *n* = number of units in a sample and $n\bar{p}$ = centerline for *np* chart, and

$$\text{Control limits} = n\bar{p} \pm 3\sqrt{n\bar{p}(1-\bar{p})}$$

$$\text{UCL}_{np} = n\bar{p} + 3\sqrt{n\bar{p}(1-\bar{p})}$$

$$n\bar{p} = \frac{\left(\sum_{i=1}^{k} np_i\right)}{k}$$

$$\text{LCL}_{np} = n\bar{p} - 3\sqrt{n\bar{p}(1-\bar{p})}$$

Except for plotting the number of defective instead of the percentage fraction, this series of equations is equivalent to the *p* chart. To construct the *np* chart, plot the points, connect them, mark the *X*s, and interpret patterns. Figure 15.30 shows an example *np* chart.

Figure 15.30 components

1. Determine purpose of the chart.
2. Select data collection point.
3. Establish basis for subgrouping.
4. Establish sampling interval and determine sample size.
5. Set up forms for recording and charting data and write specific instructions on use of the chart.
6. Collect and record data. It is recommended that at least 20 samples be used to calculate the control limits.
7. Count each np_i, the number of nonconforming units for each of the *i* subgroups.

Control chart—*np* chart

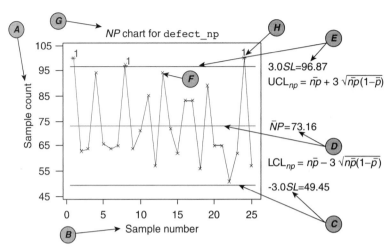

Figure 15.30 Figure showing the control chart—*np* chart.

8. Compute the process average number of units nonconforming *np*.
   - Compute upper control limit UCL*np*.
   - Compute lower control limit LCL*np*.
   - Plot data points.
   - Interpret chart together with other pertinent sources of information on the process and take corrective action if necessary.

15.22.3 *c* Chart

The *c* chart is a special type of attributes control chart that tracks the number of defects instead of the number of defective units. The distinction between "defect" and defectives" is as follows:

Defect—an individual failure to meet a single requirement
Defective unit—a production unit that contains one or more defects

It is possible for a product to contain many "defects" and still be counted as only one "defective." In cases where a single product unit is likely to contain many defects, a *c* chart is generally more appropriate to use than a *p* chart. When dealing with a complicated process in which the number of possibilities for defects increases very rapidly until it approaches, for practical purposes, an "infinite number" of possibilities. *c* charts are used only when the opportunity of defects occurring is kept constant. Thus, samples must grouped such that each is a certain fixed length, area, quantity, or other similar measure.

Example 15.9   All samples must of the same size. Thus, if the first sample consists of 1000 printed-circuit boards (PCBs) with multiple solder defects, then all samples must consist of 1000 PCBs. Typical places for using *c* charts are

- Number of loose solder connections on a PCB
- Number of failures in a satellite system per month.

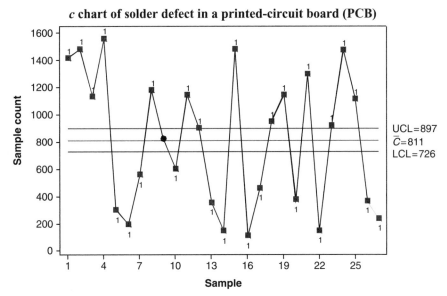

Figure 15.31   Illustration of process capability study using a c chart. Each point represents the number of solder defects found in a group of 100 printed-circuit boards. Each point represents the work of a different and new operator (change in workforce due to manufacturing cost effectiveness strategy).

To construct a c chart, it is necessary to count the number of defects c in each sample:

$$\bar{c} = \frac{c_1 + c_2 + c_3 + \cdots + c_n}{k} \qquad k = \text{subgroup}$$

where centerline on chart $= \bar{c}$ and control limits $= \bar{c} \pm 3\sqrt{\bar{c}}$ or UCL $= \bar{c} + 3\sqrt{\bar{c}}$ LCL $= \bar{c} - 3\sqrt{\bar{c}}$. The c chart measures the number of nonconformities (discrepancies or so-called defects) in the inspection lot of printed circuit boards(PCBs) (as opposed to the number of units found nonconforming as plotted on an np chart). C charts are interpreted in the same way as p-charts and can be constructed using MINITAB. See Figure 15.31.

15.22.4 u Chart

The u chart is a variation of the c chart. Each point is the average number of defects per unit in a sample of n units. Each unit in the sample must satisfy the requirements for a c chart, but the number of units that are averaged need not to be the same as for all samples. The u chart plots defects per unit collected from subgroups of equal or unequal size. u charts differ from c charts in that they plot the proportion of defects rather than the number of defects. This chart type is used in the aircraft industry, satellite and radar systems, and other complex assemblies.

Control limits for the u-chart are calculated as follows:

$$u = \frac{N}{n}$$

where N = number of defects and n = number of equal units. Also centerline on chart $= \bar{u}$; control limits $= \bar{u} \pm 3\sqrt{\bar{u}}/n$.

The u chart may be used when each sample includes more than one unit, and must be used when sample size varies from one period to the next. See Figure 15.32.

Control Chart - u Chart

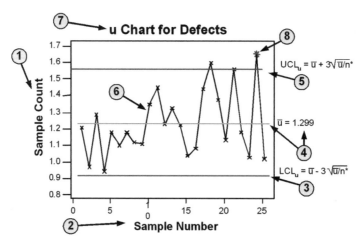

Figure 15.32   A *u* chart for defects.

Section Definitions

1. *Sample count*—number of defects per unit observed.

2. *Sample number*—chronological index number for the sample or subgroup whose number of defects per unit is being referenced.

3. *Lower control limit* (LCL)—represents the lower limit of the variation that could be expected if the process were in a state of statistical control; by convention, it is set equal to the Mean minus three times the square root of the process standard deviation

4. *Process average number of defects per unit*—average value of the number of defects per unit, over the period of inspection being referenced.

5. *Upper control limit* (UCL)—represents the upper limit of the variation that could be expected if the process were in a state of statistical control; by convention, it is set equal to the mean plus three times the square root of the process standard deviation.

6. *Plot of number of defects per unit versus sample number*—any excursion in this plot above the UCL or below the LCL represents an out-of-control condition and should be investigated.

7. *u chart*—the symbol *u* refers to the number of defects per unit in a subgroup.

8. *Out-of-control point*—by definition, any point that exceeds either the UCL or the LCL is out of control.

Figure 15.32 components

- Sample count (frequency)
- Sample number
- Lower control limit (LCL)—$3\sigma = 0.9109$
- $\bar{u}$, where $u = N/n = 1.229$ (where N = number of defects and n = number of equal units)
- Upper control limit (UCL) + $3\sigma = 1.548$
- *u*-chart title (*u* Chart for defects)
- Data point in *u* chart
- Data point outside the upper control limit (UCL) of *u* chart

How to Construct a *u* chart

1. Determine purpose of the chart.
2. Select data collection point.
3. Establish basis for subgrouping.
4. Establish sampling interval and determine sample size.
5. Create forms and instructions for recording and charting.
6. Collect and record data.
7. Count each c_i, the number of nonconformities for each of the i subgroups.
8. Compute the process average number of nonconformities per unit u.
9. Compute upper control limit UCL_u.
10. Compute lower control limit LCL_u.
11. Plot the data points.
12. Interpret chart together with other pertinent sources of information on the process and take corrective action if necessary

If the number of units is not the same from sample to sample, handle the varying sample sizes in the same manner as for a p chart; u charts are plotted, marked, and interpreted in the same way as p charts or c charts.

GLOSSARY

abscissa   Horizontal axis of a graph.

activity symbol   Rectangle used to designate an activity in a process flow diagram. A brief description of the activity is also included within this symbol.

affinity diagram   Tool used to organize and present large amounts of data. Ideas, issues, or solutions are grouped into logical categories.

alpha risk   Probability of accepting the alternative hypothesis (H_1), when, in reality the null hypothesis (H_0) is true.

alternative hypothesis   Tentative explanation for an event that does not follow a chance distribution.

attribute data   Numerical information at the nominal level.

c chart   Chart that displays the number of defects per sample.

cause   That which produces an effect or brings about a change.

centerline   Line on statistical process control charts that represents the central tendency of the subgroup.

characteristic   Measurable feature of process, product, or service.

champion   Senior manager who is responsible for the logistics and business aspects of Six Sigma projects.

classification   Placing factors or variables into different categories.

common cause   Random cause.

complex level   The first level of the process tree and the last process performed prior to delivery of the product/service to the customer.

confidence level   The probability that a random variable lies within a defined interval.

confidence limits   The two values that define the confidence level.

continuous data   Numerical information at the interval or ratio level; subdivision is conceptually meaningful.

continuous random variable   Random variable that can assume any value in some specified interval.

connections   Arrows indicating the flow and sequence of the key activities in a process flow diagram.

cnnector symbol   Circle used to illustrate a break and its continuation elsewhere on the same page or another page in a process flow diagram.

cntrol chart   Graphical presentation of a characteristic performance across time in relation to its natural limits and central tendency.

Cp **Index**   Process capability index.

C_{pk} **Index**   Measure of process dispersion and its centering about the average.

control limit   Line or lines on a control chart used to judge the significance of variation from subgroup to subgroup.

control specifications   Specifications called for by the product being manufactured or service being delivered.

cost of poor quality (COPQ)   Money spent or lost on non-value-added activities while producing a good service.

critical requirements for quality, delivery, and cost   Performance-level requirements to meet CTS characteristics.

critical-to-cost characteristics (CTC)   Product, service, and/or transactional characteristics that significantly influence one or more CTS in terms of cost.

critical-to-delivery characteristics (CTD)   Product, service, and/or transactional characteristics that significantly influence one or more CTS in terms of delivery.

critical-to-process characteristic (CTP)   Process parameters, that significantly influence a CTQ, CTD, and/or CTC.

critical-to-quality characteristic (CTQ)   A product, service, and/or transactional characteristic that significantly influences one or more CTS in terms of quality.

critical-to-satisfaction characteristic (CTS)   Expression of the customers' vital needs.

cycle time   Measure of time required for each step of a process or subprocess.

data   Factual information used as a basis for reasoning, discussion, or calculation. Often refers to quantitative information.

defect   The output of a process that does not meet customer specifications.

defective   Unit that contains at least one defect

defects per million opportunities (DPMO)   Number of defects divided by the total number of opportunities, then multiplied by 1,000,000; DPU × 1,000,000.

defects per unit (DPU)   (Total N defects)/(N units in group)

dependent variable   Response variable Y. $Y =$ output $=$ response variable from the transfer function $Y = f(x's)$.

(DFSS) design for Six Sigma   The process for designing new products or processes or redesigning ongoing processes.

detection   Past-oriented strategy that identifies unacceptable output after it has been produced and then separates it from the good output (*prevention*).

document symbol   Indicates a printed document pertinent to the process.

element level   This level represents a breakdown of the processes listed at the subsystem level. Critical-to-process (CTP) characteristics are identified directly below this level.

factors   Independent variables.

flowline   Arrow indicates the direction of the process and connects its elements in a process flow diagram.

independent variable   A variable whose value is independent of the value of another variable.

interval scale   Measure used to express numerical information on a scale with equal distance between categories, but no absolute zero. Examples of interval scales include temperature (°F and °C), a dial gage sitting on top of a gage block, and comparison of differences.

key process   List steps or subprocesses included in the process.

lower and upper specification limits   Actions based on the relationship between the specifications and the centering and dispersion of a controlled process depend on whether there are two specifications limits, a maximum upper limit USL and a minimum or lower limit LSL.

master black belt   Leader of a Six Sigma team responsible for applying the Six Sigma process.

mean   Measure of central tendency calculated by dividing the sum of all values by the number of values in the dataset.

measures of central tendency   Measures that describe the center of a distribution. The Mean, Median, and Mode are three such measurements.

measures of dispersion   Measures of the spread of a distribution.

median   Middle value in a group of data or measurements when arranged from lowest to highest if the number of values is odd. By convention the average of the middle of two values is used as a median when the number of samples is even.

Never-ending improvement in quality and productivity   Operational philosophy that makes the best use of the talents within the organization to produce products or services of increasing quality for their customers in an increasingly efficient manner.

nominal scale   Measure used to classify elements into categories without considering any specific property. Examples of nominal scales include "causes" on fishbone diagrams, yes/no, and pass/fail.

normal distribution   Continuous symmetrical density function characterized by a bell-shaped curve.

null hypothesis   Tentative explanation that an event was caused by random variation.

operation definition   Clear definition of quality expectations and performance.

opportunity   Potential defect, error, or failure mode.

ordinal scale   Measure used to order or rank nominal data in terms of a specific property. Examples of ordinal scales include relative height, Pareto diagrams, and customer satisfaction surveys.

outcome   Response dependent variable. The result of a trial with a given treatment is called a *response* or *outcome.*

outliers or extreme values   Values that are very small or very large relative to the majority of the values in a dataset.

p **chart**   Chart used to plot percent defectives in a sample.

pareto diagram   Chart used to plot percent defectives in a sample.

parameter   Summary measure calculated for population data.

percentile   Division of a ranked dataset into 100 equal parts.

process   Series of activities for doing something; combination of people, equipment, materials, and methods to produce an output for a given product or service.

process average   Central tendency of a given process characteristic.

process control   Application of statistical methods and procedures to ensure that a process meets a given set of standards.

process spread   Range of values for a process characteristic; this particular term is often applied to measure spread. The spread may be based on a set of data collected at a specific point in time or may reflect the variability across a given amount of time.

project   A problem usually calling for planned action

range   Difference between the highest and lowest values in a dataset or data subgroup.

random   Term applied in selecting a sample so that each member in the population has an equal chance of being selected; data that lack predictability or shows no pattern.

random variations   Variations in data resulting from causes that cannot be pinpointed or controlled.

ratio scales   Measure used to express numerical information on a scale with equal distance between categories and an absolute zero value in the range of measurement.

run chart   Graphical representation of a characteristic or sample statistic gathered from the process over time.

sample   In process control applications, a synonym for a subgroup. This use is different from the purpose of providing an estimate of a larger group of people or items.

sigma (σ)   Sigma, the Greek letter used to designate standard deviation.

spread   General concept for measuring the extent that values in a distribution differ from one another.

stable process   Process that is in statistical control.

standard deviation   Statistical measure of variability that describes the spread.

statistic   Summary measure calculated for sample data.

statistical control   The condition describing a process after all special causes of variation have been eliminated and only common causes remain.

subgroup   Logical grouping of objects or events that displays only random event-to-event variation. One or more events or measurements used to analyze the performance of a process.

symbol   Term indicating the nature of each step (see Section 7.9.1 for more details).

system   Something that is connected according to scheme

upper control limit   Horizontal line in a control chart that represents the upper limit of a process capability. In general at $+3\sigma$ of the centerline of the data distribution.

variable   Characteristic that may take on different values.

variance   Measure of spread

variation   Quantifiable difference between individual measurements due to common or special causes.

X and R charts   Control charts that represent the process capability over time.

yield   Amount of material that is processed by an operation less the scrap.

REFERENCES AND NOTES

1. Shigeo Shingo, Toyota's industrial engineer.
2. D. J. Wheeler, and D. S. Chambers, *Understanding Statistical Process Control*, SPC Press.
3. Western Electric Co., Inc., *Statistical Quality Control Handbook*, 1956.

16 Single-Population Hypothesis Tests

16.1 OVERVIEW

This chapter introduces the second topic in inferential statistics: tests of hypotheses. In a hypothesis test, we test a certain given theory or belief about a population parameter. We may want to find out, using some sample information, whether or not a given claim (or statement) about a population parameter is reasonable. This chapter discusses how to make such tests of hypotheses about the population mean μ and the population proportion p.

As an example, a soft-drink company may claim that, on average, its cans contain 12 ounces (oz) of soda. A government agency may want to test whether or not such cans contain, on average, 12 oz of soda. As another example, according to the US Bureau of Labor Statistics, 57.3% of married women in the United States were working outside their homes in 1991. An economist may want to check if this percentage is still reasonable for this year. In the first of these two examples we are to test a hypothesis about the population mean μ, and in the second example we are to test a hypothesis about the population proportion p.

16.2 INTRODUCTION TO HYPOTHESIS TESTING

Why do we need to perform a hypothesis test? Reconsider the example about soft-drink cans. Suppose that we take a sample of 100 cans of the soft drink under investigation. We then find out that the mean amount of soda in these 100 cans is 11.89 oz. On the basis of this result, can we state that, on average, all such cans contain less than 12 oz of soda and that the company is lying to the public? Not until we perform a hypothesis test can we make such an accusation. The reason is that the mean $\bar{x} = 11.89$ oz is obtained from a sample. The difference between 12 oz (the required average for the population) and 11.89 oz (the observed average for the sample) may have occurred only because of the variability from can to can. Another sample of 100 cans may give us a mean of 12.04 oz. Therefore, we make a hypothesis test to determine how large the difference between 12 oz and 11.89 oz is and whether or not this difference has occurred as a result of chance alone.

Now, if 11.89 oz is the mean for all cans, and not for only 100 cans, then we do not need to make a hypothesis test. Instead, we can immediately state that the mean amount of soda in all such cans is less than 12 oz. We perform a hypothesis test only when we are making a decision about a population parameter on the basis of the value of a sample statistic.

Consider a nonstatistical example of a person who has been indicted for committing a crime and is being tried in a court. Using the available evidence, the judge or jury will make one of two possible decisions:

Practitioner's Guide for Statistics and Lean Six Sigma for Process Improvements. By Mikel J. Harry, Prem S. Mann, Ofelia C. de Hodgins, Christopher J. Lacke, and Richard Hulbert
Copyright © 2010 John Wiley & Sons, Inc.

Two Decisions

1. The person is not guilty.
2. The person is guilty.

At the outset of the trial, the person is presumed innocent. The prosecutor's aim is to prove that the person has committed the crime and, hence, is guilty. In statistics, *the person is innocent* is called the *null hypothesis* and *the person is guilty* is called the *alternative hypothesis*. The null hypothesis is denoted by H_0 and the alternative hypothesis is denoted by H_1 (another common notation for the alternative is H_a). In the beginning of the trial it is assumed that the person is innocent. The null hypothesis is usually the hypothesis that is assumed to be true to begin with. The two hypotheses for the court case are written as follows (notice the colon after H_0 and H_1):

> Null hypothesis: H_0: The person is innocent
> Alternative hypothesis: H_1: The person is guilty

In a statistical example, the null hypothesis states that a given claim (or statement) about a population parameter is true. Reconsider the example of the soft-drink company's claim that, on average, its cans contain 12 oz of soda. In reality, this claim may or may not be true. However, we will initially assume that the company's claim is true (i.e., the company is innocent of making a false claim). To test the claim of the soft-drink company, the null hypothesis will be that the company's claim is true. Let μ be the mean amount of soda in all cans. The company's claim will be true if $\mu = 12$ oz. Thus, the null hypothesis will be written as

$$H_0: \mu = 12 \text{ oz}   \text{(the company's claim is true)}$$

In this example, the null hypothesis can also be written as $\mu \geq 12$ oz because the claim of the company will still be true if the cans contain, on average, more than 12 oz of soda. The company will be accused of cheating the public only if the cans contain, on average, less than 12 oz of soda. However, it will not affect the test whether we use an $=$ or a $\geq$ sign in the null hypothesis as long as the alternative hypothesis has a $<$ sign. Remember that in the null hypothesis (and in the alternative hypothesis also) we use the population parameter (such as μ or p) and not the sample statistic (such as $\bar{x}$ or $\hat{p}$).

The alternative hypothesis in our statistics example will be that the company's claim is false and its soft-drink cans contain, on average, less than 12 oz of soda, that is, $\mu < 12$ oz. The alternative hypothesis will be written as

$$H_1: \mu < 12 \text{ oz}   \text{(the company's claim is false)}$$

Thus, we can say that a *null hypothesis* is a statement about a population or population parameter that is assumed to be true. It typically implies a scenario of "no change," "no difference," or "no effect." Data are collected to determine the validity of the null hypothesis. An *alternative hypothesis* is a claim about a population parameter that will be concluded should the data contradict the null hypothesis. Sometimes the alternative hypothesis is referred to as the *research hypothesis*.

Let us return to the example of the court trial. The trial begins with the assumption that the null hypothesis is true, that is, that the person is innocent. The prosecutor assembles all the possible evidence and presents it in the court to prove that the null hypothesis is false and the alternative hypothesis is true (i.e., the person is guilty). In the case of our statistical example, the information obtained from a sample will be used as evidence to decide whether or not the claim of the company is true. In the court case, the decision made by the judge (or jury) depends on the quantity and quality of evidence presented by the prosecutor. At the end of the trial, the judge (or jury) will consider whether or not the evidence presented by the prosecutor is sufficient to declare the person guilty. The amount of evidence that will be considered to be sufficient to declare the person guilty depends on the discretion of the judge (or jury).

Rejection and Nonrejection Regions. In Figure 16.1, which represents the court case, the point marked "0" indicates that there is no evidence against the person being tried. The farther the point is to the right on the horizontal axis, the more convincing the evidence is that the person has committed the crime. We have arbitrarily marked a point C on the horizontal axis. Let us assume that a judge (or jury) considers any amount of evidence to the right of point C to be sufficient and any amount of evidence to the left of C to be insufficient to declare the person guilty. Point C is called the *critical value* or *critical point* in statistics. If the amount of evidence presented by the prosecutor falls in the area to the left of point C, the verdict will reflect that there is not enough evidence to declare the person guilty. Consequently, the accused person will be declared not guilty. In statistics, this decision is stated as *do not reject* H_0. It is equivalent to saying that there is not enough evidence to declare the null hypothesis false. The area to the left of point C is called the nonrejection region, that is, the region where the null hypothesis is not rejected. However, if the amount of evidence falls in the area to the right of point C, the verdict will be that there is sufficient evidence to declare the person guilty. In statistics, this decision is stated as reject H_0 or the null hypothesis is false. Rejecting H_0 is equivalent to saying that the alternative hypothesis is true. The area to the right of point C is called the rejection region, that is, this is the region where the null hypothesis is rejected.

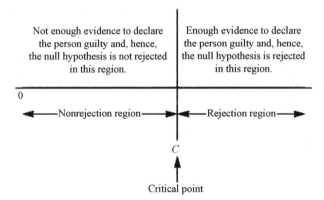

Figure 16.1   Nonrejection and rejection regions for the court case.

We all know that a court's verdict is not always correct. If a person is declared guilty at the end of a trial, there are two possibilities:

Two Types of Error

1. The person has not committed the crime but is declared guilty (because of what may be false evidence).
2. The person has committed the crime and is rightfully declared guilty.

In the first case, the court has made an error by punishing an innocent person. In statistics, this kind of error is called a *type I error*. In the second case, because the person committed the crime and has been found guilty, the court has made the correct decision. The bottom row in Table 16.1 shows these two cases. The last two columns of Table 16.1, "Innocent" and "Guilty," corresponding to *the person is innocent* and *the person is guilty*, respectively, give the two actual situations. Which one of these is true is known only to the person being tried. The two rows in this table,

TABLE 16.1   Possible Scenarios in a Court Trial

		Actual Situation	
		Innocent	Guilty
Court's	The person is not guilty	Correct decision	Type II error
decision	The person is guilty	Type I error	Correct decision

corresponding to *the person is not guilty* and *the person is guilty*, show the two possible court decisions.

In our statistical example, a type I error will occur when H_0 is actually true (i.e., the cans do contain, on average, 12 oz of soda), but it just happens that we draw a sample with a mean that is well below 12 oz and we wrongfully reject the null hypothesis H_0.

The value of α, called the *significance level* of the test, represents the probability of making a type I error. In other words, α is the probability of rejecting the null hypothesis H_0 when in fact it is true. Thus, a *type I error* occurs when a true null hypothesis is rejected. The value of α represents the probability of committing this type of error:

$$\alpha = P(H_0 \text{ is rejected}|H_0 \text{ is true})$$

The value of α represents the significance level of the test. The significance level is determined prior to performing the test. Although any value can be assigned to α, commonly used values of α are .01, .025, .05, and .10.

Now, suppose that in the court trial case the person is declared not guilty at the end of the trial. Such a verdict does not always indicate that the person is innocent. It is possible that the person is guilty but there is not enough evidence to prove the guilt. Consequently, in this situation there are again two possibilities:

1. The person has not committed the crime and is declared not guilty.

2. The person has committed the crime but, because of the lack of enough evidence, is declared not guilty.

In case 1, the court's decision is correct. But in case 2, the court has committed an error by setting a guilty person free. In statistics, this type of error is called a *type II error*. These two cases are shown in the top row of the court's decision portion of Table 16.1.

In our statistics example, a type II error will occur when the null hypothesis H_0 is actually false (i.e., the soda contained in all cans, on average, is less than 12 oz), but it happens by chance that we draw a sample with a mean that is close to or larger than 12 oz and we wrongfully conclude to fail to reject H_0. The probability of committing a type II error is represented by β. It represents the probability that H_0 is not rejected when actually H_0 is false. Thus, a *type II error* occurs when a false null hypothesis is not rejected. The value of β represents the probability of committing this type of error:

$$\beta = P(H_0 \text{ is not rejected}|H_0 \text{ is false})$$

The value of $1-\beta$ is called the *power of the test*. It represents the probability of rejecting H_0 when H_0 is false.

The two types of errors that occur in tests of hypotheses depend on each other. We cannot lower the values of α and β simultaneously for a hypothesis test for a fixed sample size. Lowering the value of α will raise the value of β, and lowering the value of β will raise the value of α. However, we can decrease both α and β simultaneously by increasing the sample size. Many statistical software packages have functions that will compute β based on α, the sample size, and the type of test used.

TABLE 16.2   Possible Scenarios in a Hypothesis Test

		Actual Situation	
		H_0 Is True	H_0 Is False
Decision	Fail to reject H_0	Correct decision	Type II error
	Reject H_0	Type I error	Correct decision

Table 16.2, which is similar to Table 16.1, is written for the statistical problem of a hypothesis test. In Table 16.2, *the person is innocent* (in "Innocent" column) is replaced by H_0 is true, *the person is guilty* (in "Guilty" column) by H_0 is false, and the *court's decision* by *decision.*

Tails of a Test. The statistical hypothesis testing procedure is similar to the trial of a person in the court but with two major differences:

1. In a statistical hypothesis test, the partition of the total region into rejection and nonrejection regions is not arbitrary. Instead, it depends on the value assigned to α (type I error). As mentioned earlier, α is the significance level of the test.

2. In the court case, the rejection region is on the right side of the critical point, as shown in Figure 16.1. However, in statistics, the rejection region for a hypothesis testing problem can be on both sides with the nonrejection region in the middle, or it can be on the left side or on the right side of the nonrejection region. These possibilities are explained in paragraphs 1–3, which follow. A test with two rejection regions is called a *two-tailed test,* and a test with one rejection region is called a *one-tailed test.* The one-tailed test is called a *left-tailed test* if the rejection region is in the left tail of the distribution curve, and it is called a *right-tailed test* if the rejection region is in the right tail of the distribution curve.

1. *A Two-Tailed Test.* According to the US Bureau of the Census, the mean family size in the United States was 3.17 in 1991. An economist wants to check whether or not this mean has changed since 1991. The keyword here is changed. The mean family size has changed if it has either increased or decreased during the period since 1991. This is an example of a two-tailed test. Let μ be the current mean family size for all families. The two possible decisions are

   a. The mean family size has not changed, that is, $\mu = 3.17$.

   b. The mean family size has changed, that is, $\mu \neq 3.17$.

We write the null and alternative hypotheses for this test as

$$H_0: \mu = 3.17 \quad \text{(the mean family size has not changed)}$$
$$H_1: \mu \neq 3.17 \quad \text{(the mean family size has changed)}$$

Whether a test is two-tailed or one-tailed is determined by the sign in the alternative hypothesis. If the alternative hypothesis has a not equal to ($\neq$) sign, as in this example, it is a two-tailed test. As shown in Figure 16.2, a two-tailed test has two rejection regions, one in each tail of the distribution curve. Figure 16.2 shows the sampling distribution of $\bar{x}$ for a large sample, assuming that the population standard deviation σ, is known. Assuming H_0 is true, $\bar{x}$ has a normal distribution with its mean equal to 3.17 (the value of μ in H_0). In Figure 16.2, the area of each of the two rejection regions is $\alpha/2$ and the total area of both rejection regions is α (the significance level). As shown in this figure, a two-tailed hypothesis test has two critical values that separate the two rejection regions from the nonrejection region. We will reject H_0 if the value of $\bar{x}$ obtained from the sample falls in either of the two rejection regions. We will fail to reject H_0 if the value of $\bar{x}$ lies in the nonrejection region. By rejecting H_0, we are

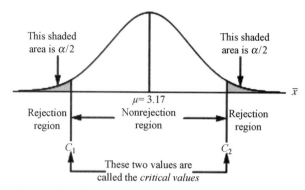

Figure 16.2   The regions and critical values for a two-tailed test.

saying that the difference between the value of μ stated in H_0 and the value of $\bar{x}$ obtained from the sample is too large to have occurred because of the sampling error alone. Consequently, this difference is real. By failing to reject H_0, we are saying that the difference between the value of μ stated in H_0 and the value of $\bar{x}$ obtained from the sample is small and it may have occurred because of the sampling error alone.

   2. *A Left-Tailed Test.* Reconsider the example of mean amount of soda in all soft-drink cans produced by a company. The company claims that these cans, on average, contain 12 oz of soda. However, if these cans contain less than the claimed amount of soda, then the company can be accused of cheating. Suppose that a consumer agency wants to test whether the mean amount of soda per can is less than 12 oz. Note that the key phrase this time is *less than*, which indicates a left-tailed test. Let μ be the mean amount of soda in all cans. The two possible decisions are

     a. The mean amount of soda in all cans is not less than 12 oz, that is, $\mu = 12$ oz.

     b. The mean amount of soda in all cans is less than 12 oz, that is, $\mu < 12$ oz.

The null and alternative hypotheses for this test are written as follows:

$$H_0: \mu = 12 \text{ oz} \quad \text{(the mean is not less than 12 oz)}$$
$$H_1: \mu < 12 \text{ oz} \quad \text{(the mean is less than 12 oz)}$$

Note that in this case, we can also write the null hypothesis as $H_0: \mu \geq 12$. This will not affect the result of the test as long as the sign in H_1 is *less than*($<$). When the alternative hypothesis has a *less than*($<$) sign, as in this case, the test is always left-tailed. In a left-tailed test, the rejection region is always in the left tail of the distribution curve, as shown in Figure 16.3, and the area of this rejection region is equal

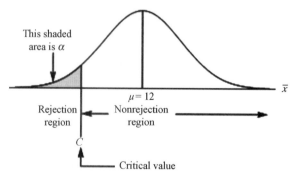

Figure 16.3   The regions and critical value for a left-tailed test.

to α (the significance level). We can observe from this figure that there is only one critical value in a left-tailed test. Assuming H_0 is true and that the population standard deviation σ is known, $\bar{x}$ has an approximate normal distribution for a large sample with its mean equal to 12 oz (the value of μ in H_0). We will reject H_0 if the value of $\bar{x}$ obtained from the sample falls in the rejection region; we fail to reject H_0 otherwise.

3. *A Right-Tailed Test.* To illustrate the third case, according to the US Bureau of the Census, the mean income of all households in the United States was $37,922 in 1991. Suppose that we want to test whether the current mean income of all households in the United States is higher than $37,922. The key phrase in this case is *higher than*, which indicates a right-tailed test. Let μ be the current mean income of all households in the United States. The two possible decisions this time are

   a. The current mean income of all households is not higher than $37,922, that is, $\mu = \$37,922$.

   b. The current mean income of all households is higher than $37,922, that is, $\mu > \$37,922$.

We write the null and alternative hypotheses for this test as

H_0: $\mu = \$37,922$ (the current mean income is not higher than $\$37,922$)
H_1: $\mu > \$37,922$ (the current mean income is higher than $\$37,922$)

In this case, we can also write the null hypothesis as H_0: $\mu \leq \$37,922$, which states that the current mean income is either equal to or less than $37,922. Again, the result of the test will not be affected whether we use an equal to $(=)$ or a less-than or equal-to $(\leq)$ sign in H_0 as long as the alternative hypothesis has a greater-than $(>)$ sign.

When the alternative hypothesis has a greater-than $(>)$ sign, the test is always right-tailed. As shown in Figure 16.4, in a right-tailed test, the rejection region is in the right tail of the distribution curve. The area of this rejection region is equal to α, the significance level. Like a left-tailed test, a right-tailed test has only one critical value.

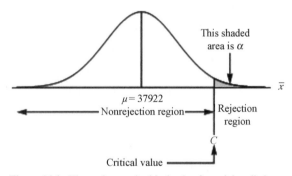

Figure 16.4   The regions and critical value for a right-tailed test.

Again, assuming that H_0 is true and the population standard deviation σ is known, $\bar{x}$ has an approximate normal distribution for a large sample with its mean equal to $37,922 (the value of μ in H_0). We will reject H_0 if the value of $\bar{x}$ obtained from the sample falls in the rejection region. Otherwise, we will fail to reject H_0.

The Test Statistic and the p-value.   Reconsider the scenario involving the mean income of all US households. We want to test if the current mean income of all households in the United States is higher than $37,922. The hypotheses for this test are

H_0: $\mu = \$37,922$
H_1: $\mu > \$37,922$

Suppose that we take a simple random sample of 50 US households, which produces a sample mean income of $\bar{x} = \$41,450$. Is this value far enough above $37,922 to fall in the rejection region for this test? We have two related methods for determining this. One method involves the *test statistic* and the *critical value* approach. The second method involves the *p-value* approach.

All hypothesis tests require a significance level. In this case, we will use a value of $\alpha = .05$. In addition, let us assume, for this example, that the standard deviation of the incomes for all US households is $\sigma = \$11,350$. The sampling distribution of $\bar{x}$ is approximately normal because of the central-limit theorem, so we can determine the probability of $\bar{x} \geq \$41,450$ when the null hypothesis is true. Specifically

$$P(\bar{x} \geq \$41,450 | \mu = \$37,922) = P\left(z \geq \frac{41,450-37,922}{11,350/\sqrt{50}}\right) = P(z \geq 2.20) = 1-.9861 = .0139$$

How do these values relate to our decision? Moreover, which value is the test statistic, and which value is the *p*-value?

A test statistic is calculated using data. Quite often, a test statistic identifies how far the value of the estimate is from the value in the null hypothesis. Whenever the sampling distribution of the estimator (such as the sample mean $\bar{x}$) is normal, the test statistic will be a z score, so in this case, the value of our test statistic is $z = 2.20$, which is shown in the calculation above. It is the z value for $\bar{x}$ obtained from the sample. Each type of hypothesis test has a specific method for calculating the test statistic, so it is important to choose the correct test. Thus, the *test statistic* is a statistic calculated from data. It is used to assess the relative likelihood that an alternative hypothesis is true when compared with a specific null hypothesis.

Using the test statistic to determine whether to reject or fail to reject the null hypothesis requires you to determine whether the test statistic falls in the rejection region. In order to do this, you need to determine the critical value in terms of z scores. This is equivalent to finding percentiles of the normal distribution, as we did in Chapter 12. In the preceding example, we decided that $\alpha = .05$, and the test is right-tailed, so we are looking for the 95th percentile of the standard normal distribution. If you recall from Chapter 14, the bottom row in the t table (Table III in Appendix A) identifies the z values for common right-tail areas, including .05. Looking in the column for .05, we see that the critical value is $z = 1.645$. Figure 16.4 showed that the rejection region for a right-tailed test includes all values that are *greater than or equal to the critical value*. As is shown in Figure 16.5, $z = 2.20$ falls in the rejection region, we reject the null hypothesis and conclude that the mean income of all US households has increased.

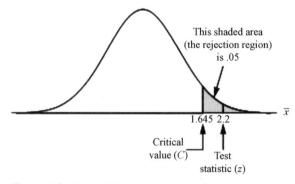

Figure 16.5 Test statistic and critical value for income example.

Remembering how to identify a *p*-value is easier when you remember that the *p* in this term stands for *probability*. This should tell you that the *p*-value in our example is (as calculated as above) .0139. The *p*-value identifies the probability of obtaining a value of the estimator that is *at least as extreme* as the sample produces. In this case, *extreme* means *greater than or equal to* because we are interested in determining whether the mean income has *increased*. If we were interested in determining whether the mean income has decreased, *extreme* would mean *less than or equal to* because we would be concerned only with values that are *far below* $37,922. Thus, the *p-value* identifies the probability of obtaining a value of the estimator that is *at least as extreme* as the sample produces. Graphically, the *p*-value for this example corresponds to the area of the region to the right of the test statistic, $z = 2.20$, which is shown in Figure 16.6.

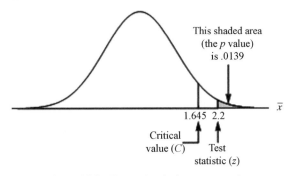

This shaded area (the *p* value) is .0139

1.645 2.2

Critical value (*C*)   Test statistic (*z*)

$\bar{x}$

Figure 16.6   The *p*-value for income example.

If we did not have the value of the test statistic, but we had the *p*-value, could we come to a conclusion? The answer is "Yes." Since the *p*-value for this example will always be the area to the right of the test statistic, any value of *z* that falls in the rejection region will have a right-tail area that is no more than .05. Also, any value of *z* that falls in the nonrejection region will have a right-tail area that is more than .05. Therefore, test statistic values in the rejection region will correspond to *p*-values that are less than or equal to the significance level α. This can be summarized in a very simple rule.

For any hypothesis test, if *p*-value $\leq \alpha$, reject H_0. If *p*-value $> \alpha$, do not reject H_0.

Is there a benefit to using the test statistic over the *p*-value, or vice versa? Before the advent of *user-friendly* statistical software, many nonstatisticians were taught how to use the standard normal table Table II of Appendix A, the *t* table (Table III of Appendix A), and so on. Typically, it was easier to use the test statistic in that case, even though tables were considered bulky and cumbersome. Statistical software calculates *p*-values, which allows the user to answer the simple question: "Is the *p*-value less than or equal to α?" Current statistical education stresses the use of the *p*-value.

Table 16.3 summarizes the foregoing discussion about the relationship between the signs in H_0 and H_1, the tails of a test, the rejection region(s), and the *p*-value.

TABLE 16.3   Relationship between H_1, the Rejection Region, and the *p*-value

	Two-Tailed Test	Left-Tailed Test	Right-Tailed Test
Sign in H_0	$=$	$=$ or $\geq$	$=$ or $\leq$
Sign in H_1	$\neq$	$<$	$>$
Rejection region	In both tails	In left tail	In right tail
p-value	Twice the area in the smaller tail beyond absolute value of test statistic	Area to left of test statistic	Area to right of test statistic

A hypothesis test involves the following five steps:

Steps in a Hypothesis Test

1. Select the type of test to use and check the underlying conditions.

2. State the null and alternative hypotheses.

3. Determine the significance level α.

4. Calculate the p-value/test statistic.

5. Make a decision.

16.3   TESTING A CLAIM ABOUT A POPULATION MEAN

In this section we will discuss how to perform a test of hypothesis about a population mean using the normal and the t distributions.

16.3.1   Hypothesis Test Using the Normal Distribution

As we just demonstrated in the income example, a hypothesis test for a population mean can be performed using a z statistic when specific conditions hold true. To use the normal distribution to perform a test of hypothesis about μ the following conditions must hold true:

1. The data were collected using a simple random sample (SRS).

2. The population standard deviation σ is known.

3. The variable of interest is normally distributed or the sample size is large enough for the central limit theorem to be applied.

These conditions are the same as those required for using the z procedure to make a confidence interval for a population mean. The test statistic for a test about μ using the normal distribution is defined by the following formula:

$$z = \frac{\bar{x} - \mu}{\sigma_{\bar{x}}} \quad \text{or} \quad z = \frac{\bar{x} - \mu}{\sigma / \sqrt{n}}$$

Examples 16.1 and 16.2 describe a five-step procedure for hypothesis testing about a population mean μ using the z statistic.

Example 16.1   A company that manufactures artificial fireplace logs advertises that the average burning time of their logs is at least 3 hours. A simple random sample of 42 of the company's artificial logs is selected, and the burning time is determined for each log. The average burning time for these 42 logs is found to be 2.96 hours. Using a significance level of .10, does this information contradict the company's claim? Assume that the standard deviation of all burning times for such logs is 0.23 hours.

Solution   Let μ be the mean burning time of all artificial logs made by this company and $\bar{x}$ be the corresponding mean for the sample. The following information is given:

$$n = 42, \; \bar{x} = 2.96 \text{ hours}, \quad \text{and} \quad \sigma = 0.23 \text{ hours}$$

Step 1. Select the type of test to use and check the underlying conditions.

The information provided corresponds to a z test. The problem states that the data came from a simple random sample, so the first condition is met. We are told to assume that the standard deviation of all burning times is 0.23 h, so we are assuming that σ is known. Nothing is stated about the distribution of burning times being normal, but the sample size is 42, which is large enough (30 or more) for the central limit theorem to be applied. Since the three conditions hold true, we can use a z test.

Step 2. State the null and alternative hypotheses.

Note that we are testing to determine whether or not the average burning time has been overstated. The average has not been overstated if it is 3 hours or more, that is, $\mu \geq 3$ hours. It would be overstated if the actual average burning time is less than 3 hours, which can be written as $\mu < 3$ hours. We write the null and alternative hypotheses as follows:

$$H_0: \mu \geq 3 \quad \text{(the average burning time is at least 3 hours)}$$
$$H_1: \mu < 3 \quad \text{(the average burning time is less than 3 hours)}$$

Step 3. Determine the significance level α.

The significance level is given to be .10; thus, $\alpha = .10$.

Step 4. Calculate the test statistic and the p-value:

$$z = \frac{\bar{x} - \mu}{\sigma/\sqrt{n}} = \frac{2.96 - 3}{.23/\sqrt{42}} = \frac{-.04}{.0355} = -1.13$$

Since the alternative hypothesis indicates that the test is left-tailed, the p-value is the probability of obtaining a test statistic that falls below -1.13. In other words, it is the area in the left tail of the normal curve to the left of $\bar{x} = 2.96$ or $z = -1.13$. Using the z table (Table II of Appendix A), we see that $P(z \leq -1.13) = .1292$. The p-value is shown in Figure 16.7.

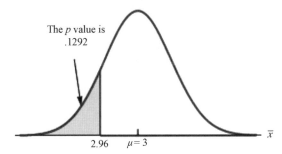

The p value is .1292

2.96 $\mu = 3$

Figure 16.7   The p-value for Example 16.1.

Step 5. Make a decision.

The significance level is .10. Using the p-value, we ask whether the p-value is $\leq \alpha$ or, more specifically, is p-value $\leq .10$? Since the p-value is .1292, it is greater than .10. Therefore, we fail to reject the null hypothesis. Our conclusion is that there is insufficient evidence to state that the average burning time for such logs is less than 3 hours, which implies that the company is not overstating the burning time of their artificial logs.   ∎

Example 16.2   Rob's Office Furniture makes and sells precut computer desks with sliding keyboard drawers. The drawers need to fit in a space that is 38 in. wide. If the drawers are too wide or too narrow, they will not work. The company's chief of production wishes to run a hypothesis test to determine whether the average drawer width is different from 38 inches. A simple random sample of 80 drawers is selected, and the drawer widths are measured. The average width is found to be 38.02 in. Past production data have shown the standard deviation of the drawer widths to be 0.098 in. Using a significance level of .05, determine whether the average drawer width differs from 38 in.

Solution Let μ be the mean width of all keyboard drawers made by Rob's and $\bar{x}$ be the corresponding mean for the sample. The given information is

$$n = 80, \bar{x} = 38.02 \text{ in.}, \quad \text{and} \quad \sigma = 0.098 \text{ in.}$$

Step 1. Select the type of test to use and check the underlying conditions.

The information provided corresponds to a z test. The problem stated that the data came from a simple random sample, so the first condition is met. We are given the standard deviation of all drawer widths as 0.098 in., so σ is known. Nothing is stated about the distribution of drawer width being normal, but the sample size is 80, which is large enough (30 or more) for the central limit theorem to be applied. Since the three conditions hold true, we can use the z test.

Step 2. State the null and alternative hypotheses.

Note that we are testing to determine whether or not the average drawer width is 38 in. If the average is 38 in., then $\mu = 38$ inches. If it is not, then we say that $\mu \neq 38$ in. We write the null and alternative hypotheses as follows:

$$H_0: \mu = 38 \quad \text{(on average, the drawer width is 38 in.)}$$
$$H_1: \mu \neq 38 \quad \text{(on average, the drawer width is not 38 in.)}$$

Step 3. Determine the significance level α.

The significance level is given to be .05; Thus $\alpha = .05$.

Step 4. Calculate the test statistic and the p-value:

$$z = \frac{\bar{x} - \mu}{\sigma/\sqrt{n}} = \frac{38.02 - 38}{.098/\sqrt{80}} = \frac{.02}{.011} = 1.82$$

Since the alternative hypothesis is two-sided, the p-value is the probability of obtaining a test statistic that falls below -1.82 or above 1.82. Using the z table, we see that $P(z \leq -1.82) = P(z \geq 1.82) = .0344$; therefore the p-value is $.0344 + .0344 = .0688$. Thus, the p-value is twice the area under the right tail beyond $\bar{x} = 38.02$ or $z = 1.82$. The p-value is shown in Figure 16.8.

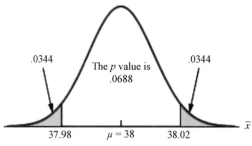

Figure 16.8 The p-value for Example 16.2.

Step 5. Make a decision.

The significance level is .05. Using the p-value, we ask whether the p-value is $\leq \alpha$ or more specifically, is p-value $\leq .05$? Since the p-value is .0688, it is greater than .05. Therefore, we fail to reject the null hypothesis. Our conclusion is that the average drawer width does not significantly differ from 38 in., which is the space in which the drawer has to fit. ■

16.3.2 Hypothesis Test Using the t Distribution

When we introduced confidence intervals for a population mean, we introduced the concept of the z interval, which we could potentially use as long as the population standard deviation σ was known. We then introduced the t interval for a population mean, which is used in practice because it does not require that σ be known. The same situation holds true for a hypothesis test for a mean. The following conditions must hold true to use the t distribution to test a hypothesis about μ:

1. The data are collected using a simple random sample (SRS).
2. The variable of interest (i.e., population) is normally distributed.

These are the same conditions that are required for using the t procedure to make a confidence interval for a population mean. The test statistic for the t test is defined by the formula:

$$t = \frac{\bar{x} - \mu}{s/\sqrt{n}} \qquad \text{or} \qquad z = \frac{\bar{x} - \mu}{s_{\bar{x}}}$$

Note that the same five steps are used for a t test using the t distribution as were used for a z test. Examples 16.3–16.5 illustrate the use of the five steps for performing a hypothesis test about the population mean μ.

Example 16.3   When a machine that is used to manufacture bolts at Sabana Steel Corporation is working properly, the mean length of these bolts is 2.5 in. However, from time to time this machine falls out of alignment and produces bolts that have a mean length of either less than 2.5 in. or more than 2.5 in. When this happens, the process is stopped and the machine is adjusted. To check whether or not the machine is producing bolts with a mean length of 2.5 in., the quality control department at the company takes a sample of bolts each week and performs a hypothesis test. One such simple random sample of 49 bolts produced a mean length of 2.49 in. and a standard deviation of 0.021 in. Using the 5% significance level, can we conclude that the machine needs to be adjusted? (*Note*: the sample did not contain any outliers.)

Solution   Let μ be the mean length of all bolts made on this machine and $\bar{x}$ be the corresponding mean for the sample. The following information is given:

$$n = 49, \quad \bar{x} = 2.49 \text{ in.}, \quad \text{and} \quad s = 0.021 \text{ in.}$$

Step 1. Select the type of test to use and check the underlying conditions.
   The information provided corresponds to a t test. The problem states that the data came from a simple random sample, so the first condition is met. Nothing is stated about the distribution of length bolt being normal, so we have to determine if the sample size is large enough to use the t test. The sample size is 49. According to the guidelines in Section 14.3, if the sample size is 40 or larger, a t test can be used as long as there are no outliers. Since we are told that there are no outliers, we can use a t test.

Step 2. State the null and alternative hypotheses.
   Note that we are testing to determine whether or not the machine needs to be adjusted. The machine will not need an adjustment if the mean length of the bolts produced is equal to 2.5 in., that is, $\mu = 2.5$ inches. The machine will need an adjustment if the mean length of these bolts is either less than 2.5 in. or more than 2.5 in., which can be written as $\mu \neq 2.5$ in. We write the null and alternative hypotheses as follows:

$$H_0: \mu = 2.5 \quad \text{(the machine does not need to be adjusted)}$$
$$H_1: \mu \neq 2.5 \quad \text{(the machine does need to be adjusted)}$$

Step 3. Determine the significance level α.
   The significance level is given to be .05; thus, $\alpha = .05$.

Step 4. Calculate the test statistic and the *p*-value:

$$t = \frac{\bar{x}-\mu}{s/\sqrt{n}} = \frac{2.49-2.5}{.021/\sqrt{49}} = \frac{-.01}{.003} = -3.33$$

Since the alternative hypothesis is two-sided, the *p*-value is the probability of obtaining a test statistic that falls below -3.33 or above 3.33. The sample size is 49, so we have df $= 48$. If we use the *t* table (Table III of Appendix A), we see that $P(t \geq 3.269) = P(t \leq -3.269) < .001$, therefore $P(t \geq 3.33) + P(t \leq -3.33) < .001 + .001 = .002$. So we know that the *p*-value is less than .002.

Step 5. Make a decision.

The significance level is .05. Using the *p*-value, we ask whether the *p*-value is $\leq \alpha$ or more specifically, is *p*-value $\leq .05$? Since the *p*-value is less than .002, it is less than or equal to .05. Therefore, we reject the null hypothesis. Our conclusion is that the mean length is not 2.5 in., so the company needs to shut down the machine in order to adjust it. ∎

Example 16.4   Grand Auto Corporation produces auto batteries. The company claims that its top-of-the-line Never Die batteries are good, on average, for at least 65 months. A consumer protection agency tested 15 such batteries to check this claim. It found the mean life of these 15 batteries to be 63 months with a standard deviation of 3 months. At the 5% significance level, can you conclude that the company's claim is true? A normal quantile plot shows that the data are approximately normally distributed.

Solution   Let μ be the mean length of all Never Die batteries and $\bar{x}$ be the corresponding mean for the sample. The given information is

$$n = 15, \ \bar{x} = 63 \text{ months, and } \ s = 3 \text{ months.}$$

Step 1. Select the type of test to use and check the underlying conditions.

The information provided corresponds to a *t* test. The problem states that the data came from a simple random sample, so the first condition is met. We are told that the data are approximately normally distributed, so we can use the *t* test regardless of the sample size.

Step 2. State the null and alternative hypotheses.

We are to test the claim that the mean life of Never Die batteries is at least 65 months. The null and alternative hypotheses are as follows:

$$H_0: \mu \geq 65 \quad \text{(the mean life is at least 65 months)}$$
$$H_1: \mu < 65 \quad \text{(the mean life is less than 65 months)}$$

Step 3. Determine the significance level α.

The significance level is given to be .05; thus, $\alpha = .05$.

Step 4. Calculate the test statistic and the *p*-value:

$$t = \frac{\bar{x}-\mu}{s/\sqrt{n}} = \frac{63-65}{3/\sqrt{15}} = \frac{-2}{.7746} = -2.582$$

Since the alternative hypothesis is left-sided, the *p*-value is the probability of obtaining a test statistic that falls below -2.582. The sample size is 15, so we have df $= 14$. If we use the *t* table, we see that $P(t \leq -2.624) = .01$ and $P(t \leq -2.145) = .025$ Since $-2.624 < -2.582 < -2.145$, $P(t \leq -2.582)$ will fall between $P(t \leq -2.624)$ and

$P(t \le -2.145)$. Therefore $.01 < P(t \le -2.582) < .025$. So we know that the p-value is between .01 and .025.

Step 5. Make a decision.

The significance level is .05. Using the p-value, we ask whether the p-value is $\le \alpha$ or, more specifically, $\le .05$. Since the p-value is between .01 and .025, it is less than or equal to .05. Therefore, we reject the null hypothesis. Our conclusion is that the mean lifetime of Never Die batteries is less than 65 months, which contradicts the manufacturer's claim.

This example reveals one of the benefits of using software for the calculation of a p-value instead of using a table. From the table, we knew that the p-value was less than .025, so it has to be less than or equal to .05, which concurs with the result using the p-value of .011. However, suppose that our significance level had been $\alpha = .02$, which is a very common significance level. Using the table, we know that the p-value is between .01 and .025, but so is the significance level, so we would be unable to come to a conclusion. Only with statistical software would we be able to obtain a more precise p-value, which would allow us to come to our conclusion.

Example 16.5   The management at Massachusetts Savings Bank is always concerned about the quality of service provided to its customers. With the old computer system, a teller at this bank could serve, on average, 22 customers per hour. The management noticed that with this service rate, the waiting time for customers was too long. Recently the management of the bank installed a new computer system in the bank expecting that it would increase the service rate and consequently make the customers happier by reducing the waiting time. To check if the new computer system is more efficient than the old system, the management of the bank took a simple random sample of 18 h and found the mean number of customers served by tellers during this time to be 23.7 per hour with a standard deviation of 4.2. Testing at the 1% significance level, would you conclude that the new computer system is more efficient than the old computer system? A normal quantile plot shows that the data are approximately normally distributed.

Solution   Let μ be the mean number of customers served per hour by a teller and $\bar{x}$ be the corresponding mean for the sample. The given information is

$$n = 18, \quad \bar{x} = 23.7 \text{ customers per hour}, \quad \text{and} \quad s = 4.2 \text{ customers per hour}.$$

Step 1. Select the type of test to use and check the underlying conditions.

The information provided corresponds to a t test. The problem states that the data came from a simple random sample, so the first condition is met. We are told that the data are approximately normally distributed, so we can use the t test regardless of the sample size.

Step 2. State the null and alternative hypotheses.

We are to test to see if the new computer system has increased the efficiency of serving customers. The null and alternative hypotheses are as follows:

$$H_0: \mu \le 22 \quad \text{(the new computer system has not increased efficiency)}$$
$$H_1: \mu > 22 \quad \text{(the new computer system has increased efficiency)}$$

Step 3. Determine the significance level α.

The significance level is given to be .01. Thus, $\alpha = .01$.

Step 4. Calculate the test statistic and the p-value:

$$t = \frac{\bar{x} - \mu}{s/\sqrt{n}} = \frac{23.7 - 22}{4.2/\sqrt{18}} = \frac{1.7}{.9899} = 1.717$$

Since the alternative hypothesis is right-sided, the p-value is the probability of obtaining a test statistic that falls at or above 1.717. The sample size is 18, so we have df $= 17$. If we use the t table, we see that $P(t \geq 1.740) = .05$ and $P(t \geq 1.333) = .10$ Since $1.333 < 1.717 < 1.740$, $P(t \geq 1.717)$ will fall between $P(t \geq 1.740)$ and $P(t \geq 1.333)$. Therefore $.05 < P(t \geq 1.717) < .10$. So we know that the p-value is between .05 and .10.

Step 5. Make a decision.

The significance level is .01. Using the p-value, we ask whether the p-value is $\leq \alpha$ or, more specifically, $\leq .01$. Since the p-value is between .05 and .10, it is greater than .01. Therefore, we fail to reject the null hypothesis. Our conclusion is that the new computer system does not significantly raise the average number of customers served per hour. ∎

16.3.3   Hypothesis Test About the Median

In Examples 16.3–16.5, we could use the t test because the data came from a simple random sample and either the underlying distribution was approximately normal or the sample size was large enough to compensate for the nonnormality of the data. In some cases, the sample size will not be large enough to compensate for the nonnormality of the data. In such a case, another method of analyzing the data will have to be used. Many different methods can be used. In this section we will describe the *median test*, which is a hypothesis test that can be used in such a situation, because the median is resistant to outliers, and is not substantially affected when the data are heavily skewed.

The median test falls in a group of tests that are referred to as *distribution-free* and/or *nonparametric* methods. Distribution-free tests are not based on any assumptions about the distribution from which the data are sampled. A t test is not distribution-free because it assumes that the data come from a normal distribution. Nonparametric tests, as the name suggests, do not involve any parameters, such as the mean μ or standard deviation σ. Tests that involve parameters and are not distribution-free are called *parametric* tests.

Before discussing the specifics of the median test, it is important to recognize that the five-step procedure we learned in performing the z test and the t test is used in every type of hypothesis test. Symbols in the hypotheses will change, as will the formula for the test statistic and/or p-value, but comparing the p-value with the significance level will remain the same, as will the need to reject or fail to reject the null hypothesis, as well as making a conclusion.

TABLE 16.4   Relationship between H_1, the Test Statistic, and the p-value in the Median Test

	Two-Tailed Test	Left-Tailed Test	Right-Tailed Test
Sign in H_1	$\neq$	$<$	$>$
Test statistic S	Count number of observations less than value of median in H_0 and number of observations greater than value of median in H_0; S is the larger of these two numbers	S is number of observations less than value of median in H_0	S is number of observations greater than value of median in H_0
p-value	Twice the probability $P(x \geq S)$, where x is a binomial random variable with n trials and $p = .50$	$P(x \geq S)$, where x is a binomial random variable with n trials and $p = .50$	$P(x \geq S)$, where x is a binomial random variable with n trials and $p = .50$

The population median is denoted by the lowercase Greek letter η (eta). Remember that the median divides a distribution into two equal parts with the lower 50% and the upper 50%, which implies that each observation is equally likely to fall on either side of η. The properties of the median test allow the p-value to be calculated using the binomial distribution, which was discussed in Chapter 11.

The only underlying condition in the median test is that the data come from a simple random sample. No assumption about the shape of the population distribution is necessary.

The value of n used in the p-value calculation is typically equal to the sample size. However, on occasion, one or more observations will be equal to the value in the null hypothesis. These observations are not counted in the value of n. For example, if a dataset has 24 observations, the null hypothesis is H_0: $\eta = 85$(i.e., median is equal to 85) and the dataset contains two observations equal to 85, then the calculation of the p-value is performed with $n = 22$, not $n = 24$.

The process of calculating the test statistic, which is denoted by S, involves counting the number of observations that fall on a specific side of the value in the null hypothesis. The appropriate side(s) are defined by the sign in the alternative hypothesis. Table 16.4 summarizes the calculation of the test statistic and the p-value in the median test.

Example 16.6   The records of a supermarket show that the median amount spent by its customers is $67.53 per visit at this store. Recently the management of the store initiated a promotional campaign according to which each customer receives points according to the total money spent at the store, and these points can be used to buy products at the store. The management expects that as a result of this campaign, the customers should be encouraged to spend more at the store. To check whether this is true, the manager of the store took a simple random sample of 12 customers who visited the store. The following data give the money (in dollars) spent by these customers at this supermarket during their visits:

97	41	53	74	69	119	78	83	65	203	161	79

Using the 10% significance level, can you conclude that the median amount of money spent by all customers at this supermarket after the campaign was initiated is higher than $67.53?

Solution   Let η be the median amount of money spent by all customers after the initiation of the campaign.

Step 1. Select the type of test to use and check the underlying conditions.
The information provided corresponds to a median test. The problem states that the data came from a simple random sample, so the underlying condition is met. Let's take a look at why we could not use a t test for the mean in this situation. Since the sample size is 12 (under 15), a t test can be used only when the data are close to being normally distributed. If the distribution is at least moderately skewed and/or has any outliers, we need to use the median test. Just by looking at the data, it seems that the data value 203 is an outlier. Therefore, we need to use the median test.

Step 2. State the null and alternative hypotheses.
We are to test to see if the promotional campaign has led to increased spending. The null and alternative hypotheses are as follows:

H_0: $\eta = \$67.53$ (the promotional campaign has not led to increased spending)
H_1: $\eta > \$67.53$ (the promotional campaign has led to increased spending)

Step 3. Determine the significance level α.
The significance level is given to be .10. Hence, $\alpha = .1$.

Step 4. Calculate the test statistic and the p-value.
Since the alternative hypothesis has a greater than ($>$) sign, S is the number of observations in the dataset that are larger than $67.53. As we can observe, 9 of the 12 observations are greater than $67.53, so $S = 9$. Since the alternative hypothesis is right-sided, the p-value is given by$P(x \geq 9)$, where x is a binomial random variable with $n = 12$ and $p = .50$. Using Table I in Appendix A, the

binomial probability table, we determine that the p-value is

$$P(x \geq 9) = P(x = 9) + \cdots + P(x = 12) = .0537 + .0161 + .0029 + .0002 = .0729$$

To use Table I of Appendix A, first we locate $n = 12$. Then in this section, we locate the column for $p = .50$. Finally, we add the probabilities of x from 9 to 12.

Step 5. Make a decision.

The significance level is .10. Using the p-value, we ask whether the p-value is $\leq \alpha$. Since the p-value is .0729, it is less than or equal to .1. Therefore, we reject the null hypothesis. Our conclusion is that the promotion seems to have increased the median amount spent. ∎

Often people ask why we worry about using a t test when the median test is less restrictive, which is a very good question. For a given sample size and significance level, the t test (which is a *parametric* test), has a lower probability of a type II error (or more *power*) than does the median test (which is *distribution-free*), when the normality assumption is reasonable. Since the lower probability of a type II error is preferred, one should use a t test over the median test when it is possible. However, when the normality assumption is unreasonable, we cannot trust the results of a t test because the significance level of the test is actually higher than whatever value we select. In general, the parametric tests have more power than nonparametric or (distribution-free) tests when the underlying conditions of the parametric test hold.

16.4   HYPOTHESIS TEST ABOUT A POPULATION PROPORTION

Often we want to conduct a hypothesis test about a population proportion. For example, a mail-order company claims that 90% of all orders placed with it are shipped within 72 hours. The quality control department of the company may want to check from time to time whether or not this claim is true. As another example, a company claims that 25% of all males use its after-shave products. A rival company may want to check whether or not this claim is true.

This section presents the procedure to perform tests of hypotheses about the population proportion p for large samples. The procedure includes the same five steps followed in the previous tests. Again, the test can be two-tailed or one-tailed. We know from Chapter 13 that when the sample size is large, the sample proportion $\hat{p}$ is approximately normally distributed with its mean equal to p and standard deviation equal to $\sqrt{p(1-p)/n}$. Hence, we use the normal distribution to perform a hypothesis test about the population proportion p for a large sample. As was mentioned in Chapters 13 and 14, in the case of a proportion, the sample size is considered to be large when np and $n(1-p)$ are both greater than 10.

The test statistic z for a large sample test of a population proportion is obtained as

$$z = \frac{\hat{p} - p}{\sqrt{\dfrac{p(1-p)}{n}}}$$

where p is the value given in the null hypothesis and $\hat{p}$ is the sample proportion derived from a simple random sample of size n.

Table 16.5 summarizes the relationship between H_1 and the p-value in this type of hypothesis test.

TABLE 16.5   Relationship between H_1 and the p-value in the One-Proportion Test

	Two-Tailed test	Left-Tailed test	Right-Tailed test
Sign in H_1	$\neq$	$<$	$>$
p-value	$P(Z \geq |z|) + P(Z \leq -|z|)$	$P(Z \leq z)$	$P(Z \geq z)$

Examples 16.7 and 16.8 describe the procedure for testing hypotheses about the population proportion p.

Example 16.7   According to Information Resources Inc., from studies based on the sales of teeth-cleaning products in supermarkets and drugstores during the period from October 1991 to September 1992, Crest toothpaste controlled a 31.2% share of the market. For convenience, assume that a 31.2% share of the market means that 31.2% of all people in the United States used Crest toothpaste. A researcher from a rival company wants to find whether or not the current market share controlled by Crest differs from 31.2%. She took a sample of 400 persons and found that 29% of them use Crest toothpaste. Using the .01 significance level, can you conclude that the current market share of Crest toothpaste is different from that for 1991–1992?

Solution   Let p be the proportion of all people in the United States who currently use Crest toothpaste and $\hat{p}$ be the corresponding sample proportion. Then, from the given information,

$$n = 400 \quad \text{and} \quad \hat{p} = .29$$

we note that we have changed all percentages to proportions in this example.

Step 1. Select the type of test to use and check the underlying conditions.

The information provided corresponds to a one-sample proportion test. The problem states that the data came from a simple random sample. In order to use the large sample test, np and $n(1-p)$ both must be greater than 10. The value of p for this test is $p = .312$. Thus

$$np = 400 \times .312 = 124.8 \quad \text{and} \quad n(1-p) = 400 \times 1-.312 = 275.2$$

Since 124.8 and 275.2 are both greater than 10, we can use the large-sample approximation.

Step 2. State the null and alternative hypotheses.

We are to test whether the market share of Crest toothpaste has changed. The null and alternative hypotheses are as follows:

$$H_0: p = .312 \quad \text{(Crest's market share has not changed)}$$
$$H_1: p \neq .312 \quad \text{(Crest's market share has changed)}$$

Step 3. Determine the significance level α.

The significance level is given to be .01; hence, $\alpha = .01$.

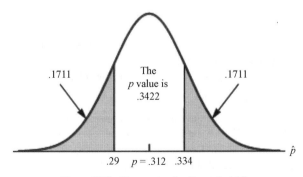

Figure 16.9   The p-value for Example 16.7.

Step 4. Calculate the test statistic and *p*-value.
The value of the test statistic is

$$z = \frac{\hat{p} - p}{\sqrt{\dfrac{p(1-p)}{n}}} = \frac{.29 - .312}{\sqrt{\dfrac{.312(1-.312)}{400}}} = \frac{-.022}{.0232} = -.95$$

Since the alternative hypothesis is two-sided, the *p*-value is $P(z \leq -.95) + P(z \geq .95)$. In other words, the *p*-value is the sum of the areas under the normal distribution curve to the left of $z = -.95$ and to the right of $z = .95$, which is twice the area to the left of $\hat{p} = .29$. Looking up $z = -.95$ in the standard normal table, we see that $P(z \leq -.95) = .1711$. Since $P(z \leq -.95) = P(z \geq .95)$, $P(z \leq -.95) + P(z \geq .95) = 2 \times .1711 = .3422$, which is shown in Figure 16.9.

Step 5. Make a decision.
The significance level is .01. Using the *p*-value, we ask whether the *p*-value is $\leq \alpha$. Since the *p*-value is .3422, it is greater than .01. Therefore, we fail to reject the null hypothesis. Our conclusion is that the Crest's market share has not changed significantly. ∎

Example 16.8   When working properly, a machine that is used to make chips for calculators does not produce more than 4% defective chips. Whenever the machine produces more than 4% defective chips, it needs an adjustment. To check whether the machine is working properly, the quality control department at the company often takes samples of chips and inspects them to determine whether they are good or defective. One such simple random sample of 300 chips taken recently from the production line contained 21 defective chips. Test at the 5% significance level whether or not the machine needs an adjustment.

Solution   Let *p* be the proportion of defective chips in all chips produced by this machine and $\hat{p}$ be the corresponding sample proportion. The given information is

$$n = 300 \quad \text{and} \quad \hat{p} = \frac{21}{300} = .07.$$

Step 1. Select the type of test to use and check the underlying conditions.
The information provided corresponds to a one-sample proportion test. The problem states that the data came from a simple random sample. In order to use the large sample test, np and $n(1-p)$ must both be greater than 10. The value of *p* for this test is $p = .04$; thus

$$np = 300 \times .04 = 12 \quad \text{and} \quad n(1-p) = 300 \times .96 = 288$$

Since 12 and 288 are both greater than 10, we can use the large-sample approximation.

Step 2. State the null and alternative hypotheses.
We are to test to see if the machine needs to be adjusted. The null and alternative hypotheses are as follows:

$$H_0: p = .04 \quad \text{(the machine does not need to be adjusted)}$$
$$H_1: p > .04 \quad \text{(the machine needs to be adjusted)}$$

Step 3. Determine the significance level α.
The significance level is given to be .05; hence, $\alpha = .05$.

Step 4. Calculate the test statistic and the *p*-value.
The value of the test statistic *z* is

$$z = \frac{\hat{p}-p}{\sqrt{\dfrac{p(1-p)}{n}}} = \frac{.07-.04}{\sqrt{\dfrac{.04(1-.04)}{300}}} = \frac{.03}{.0113} = 2.65$$

Since the alternative hypothesis is right-sided, the *p*-value is given by $P(z \geq 2.65)$. Looking up $z = 2.65$ in the standard normal table, we see that $P(z \leq 2.65) = .9960$. Thus, $P(z \geq 2.65) = 1 - .9960 = .0040$, which gives the *p*-value. See Figure 16.10.

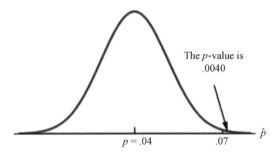

The *p*-value is
.0040

$p = .04$ .07 $\hat{p}$

Figure 16.10   The *p*-value for Example 17.8.

Step 5. Make a decision.
The significance level is .05. Using the *p*-value, we ask whether the *p*-value $\leq \alpha$. Since the *p*-value is .004, it is less than .05. Therefore, we reject the null hypothesis. Our conclusion is that the proportion of defectives exceeds .04 and hence the machine needs to be shut down and adjusted. ■

What if the sample size is not large enough to use the normal approximation?. As we discussed when computing the confidence interval for a population proportion, in some situations it is too expensive or impractical to obtain a large sample in order to use the normal approximation. The confidence interval based on the Wilson estimator $\tilde{p}$ provides just one of many alternate methods for estimating *p*.

Various methods are available for performing a hypothesis test for a population proportion when $np \geq 10$ and $n(1-p) \geq 10$ are not true or cannot be met. One of the more common methods is to calculate an exact *p*-value, using the binomial distribution, as opposed to using the normal approximation. This can be done quite easily with statistical software; however, the details are outside the scope of this book. As a note from our previous examples, the process in MINITAB is almost the same as the large-sample test. The only difference is that the "Use test and interval based on normal distribution" box in the "1 Proportion—Options" window should not be checked.

GLOSSARY

α   The significance level of a hypothesis test that denotes the probability of rejecting a null hypothesis when it actually is true. This gives the probability of committing a type I error.

alternative hypothesis   A claim about a population parameter that will be concluded should the data contradict the null hypothesis.

β   The probability of not rejecting a null hypothesis when it actually is false. This gives the probability of committing a type II error.

critical value or critical point   One or two values that divide the whole region under the sampling distribution of a sample statistic into rejection and nonrejection regions.

left-tailed test   A test in which the rejection region lies in the left tail of the distribution curve.

null hypothesis   A statement about a population or population parameter that is assumed to be true. It typically implies a scenario of "no change," "no difference," "no effect," "no increase," or "no decrease."

one-tailed test   A test in which there is only one rejection region, in either the left or right tail of the distribution curve.

***p*-value**   The *p*-value identifies the probability of obtaining a value of the estimator that is *at least as extreme* as the sample produces.

right-tailed test   A test in which the rejection region lies in the right tail of the distribution curve.

significance level   The value of α that gives the probability of committing a type I error.

test statistic   A statistic calculated from data. It is used to assess the relative likelihood that an alternative hypothesis is true when compared with a specific null hypothesis.

two-tailed test   A test in which there are two rejection regions, one in each tail of the distribution curve.

type I error   An error that occurs when a true null hypothesis is rejected.

type II error   An error that occurs when a false null hypothesis is not rejected.

EXERCISES

16.1   Write the null and alternative hypotheses for each of the following examples. Determine whether each is a case of a two-tailed, a left-tailed, or a right-tailed test.

   (a) To test whether or not the mean price of houses in Connecticut is greater than $2,103,000.

   (b) To test if the mean number of hours spent working per week by college students who hold jobs is different from 15 hours.

   (c) To test whether the mean life of a particular brand of auto batteries is less than 45 months.

   (d) To test whether the mean amount of time taken by all workers to do a certain assembly job is more than 35 min.

   (e) To test whether the mean age of all managers of companies is different from 50 years.

16.2   According to the Hertz Corporation, the mean cost of owning and operating a car was $3002 in 1986. Suppose this estimate to be true for the population of all cars for 1986. A simple random sample of 45 cars showed that the sample mean cost of owning and operating these cars was $3550 in 1993 with a sample standard deviation of $475. Using the 2.5% significance level, can you conclude that the mean cost of owning and operating a car in 1993 was greater than $3002? Assume that there were no outliers in the dataset.

16.3   Are we enjoying more or less leisure time now than in the past? A survey conducted by Louis Harris and Associates showed that the mean time that Americans spent "to relax, watch TV, take part in sports or hobbies, go swimming or skiing, go to the movies, theater, concerts, or other forms of entertainment, get together with friends, and so forth" was 16.6 h per week in 1988. A recent poll of 200 Americans showed that they spend an average of 17.2 h a week on these leisure activities with a standard deviation of 3.9 h.

   (a) Testing at the 5% significance level, do you think the mean number of hours spent per week on leisure activities by all Americans is now different from 16.6? Assume that there were no outliers in the dataset.

(**b**) What is the type I error in this case? Explain in words. What is the probability of making this error?

16.4 According to the National Agricultural Statistics Service, the mean yield of potatoes per acre was 292 cwt (hundredweight) (1 cwt = 100 lb) in 1986. A random sample of 20 acres gave a mean yield of potatoes to be 301 cwt for 1993 with a standard deviation of 22 cwt. A normal quantile plot of the data revealed that the data are approximately normally distributed. Using the 5% significance level, can you conclude that the mean yield of potatoes for 1993 is different from 292 cwt? Explain your conclusion in words.

16.5 A business school claims that students who complete a 3-month typing course can type, on average, at least 1200 words an hour. A random sample of 25 students who completed this course typed, on average, 1130 words an hour with a standard deviation of 85 words. A normal quantile plot of the data revealed that the data are not heavily skewed and that there are no outliers.

(**a**) Suppose that the probability of making a type I error is selected to be zero. Can you conclude that the claim of the business school is true? Answer without performing the five steps of a hypothesis test.

(**b**) Using the 5% significance level, can you conclude that the claim of the business school is reasonable?

16.6 A past study claims that adults in America spend an average of 18 hours a week on leisure activities. A researcher wanted to test this claim. She took a sample of 10 adults and asked them about the time they spend per week on leisure activities. Their responses (in hours) are as follows:

14	25	22	38	16	26	19	23	41	33

Using the 5% significance level, can you conclude that the claim of the earlier study is reasonable?

16.7 A computer company that recently introduced a new software product claims that at least half of those people who are somewhat familiar with computers should be able to learn how to use this software within 2 hours. A random sample of 12 such persons was selected. The following data give the time taken (in hours) by these persons to learn how to use this software:

1.75	2.15	2.25	2.25	2.40	1.80
1.90	2.20	1.50	3.25	2.75	2.60

Test at the 1% significance level if the company's claim is reasonable.

16.8 According to a survey, the median amount of time that adults spend with their family per day is 86 min. Suppose this is true of all adults at the time of that survey. A recently taken sample of 25 adults showed that 18 spent less than 86 min/day with their families. None of the 25 adults spent exactly 86 min/day with his/her family. Using a 1% significance level, test whether the median time currently spent by all adults with their families is less than 86 min/day.

16.9 According to a 1992 survey conducted by the Roper Organization for Shearson Lehman Brothers, 45% of the persons polled said that they would change careers if they could (*The Wall Street Journal*, Sept. 1, 1992). Suppose this percentage to be true for the 1992 population of all workers. A researcher wanted to check whether this percentage is still true. A sample of 350 workers taken recently by this researcher showed that (approximately) 39% of them would change careers if they could. Using the 2.5% significance level, can you conclude that the current percentage of workers who would change careers if they could is less than 45%?

16.10   Providing health insurance coverage to all people was one of the major issues in the recent presidential election. According to a 1992 survey, 45% of the American people would support higher taxes to pay for health insurance for people who cannot afford it (*US News & World Report*, Aug. 10, 1992). A consumer agency wanted to check whether or not this percentage is true. The agency recently took a sample of 400 people and found that 50% of them would support higher taxes to pay for health insurance for people who cannot afford it. Using the 1% significance level, can you conclude that the percentage of people who would support higher taxes to pay for health insurance for people who cannot afford it is greater than 45%?

16.11   Two years ago, 75% of the customers of a bank said that they were satisfied with the services provided by the bank. The manager of the bank wants to know if this percentage of satisfied customers has changed since then. She assigns this responsibility to you. Briefly explain how you would conduct such a test.

17 Estimation and Hypothesis Tests: Two Populations

17.1 OVERVIEW

Chapters 14 and 16 discussed the estimation and hypothesis testing procedures for μ and p involving a single population. This chapter extends the discussion of estimation and hypothesis testing procedures to the difference between two population means and the difference between two population proportions, as well as nonparametric and/or distribution-free tests that can be used when certain underlying conditions are not fulfilled. For example, we may want to make a confidence interval for the difference between the mean prices of houses in California and New York. Or we may want to test the hypothesis that the mean price of houses in California is different from that in New York. As another example, we may want to make a confidence interval for the difference between the proportions of defective items in all items manufactured on each of two machines. Or we may want to test the hypothesis that the proportion of defective items in all items manufactured on machine 1 is different from the proportion of defective items manufactured on machine 2. Constructing confidence intervals and testing hypotheses about population parameters are referred to as *making inferences*.

17.2 INFERENCES ABOUT DIFFERENCES BETWEEN TWO POPULATION MEANS FOR INDEPENDENT SAMPLES

Let μ_1 be the mean of the first population and μ_2 be the mean of the second population. Suppose that we want to make a confidence interval and test a hypothesis about the difference between these two population means, that is, $\mu_1 - \mu_2$. Let $\bar{x}_1$ be the mean of a sample taken from the first population and $\bar{x}_2$ be the mean of a sample taken from the second population. Then, $\bar{x}_1 - \bar{x}_2$ is the sample statistic that is used to make an interval estimate and to test a hypothesis about $\mu_1 - \mu_2$. This section discusses how to make confidence intervals and test hypotheses about $\mu_1 - \mu_2$ when the two samples are independent. The concept of independent and dependent samples is explained next.

Independent versus Dependent Samples. Two samples are *independent* if they are drawn from two different populations and the elements of one sample are unrelated to the elements of the second sample. If the elements of the two samples are somehow related, then the samples are said to be *dependent*. Thus, in two independent samples, the selection of one sample has no effect on the selection of the second sample. Examples 17.1 and 17.2 illustrate independent and dependent samples, respectively.

Example 17.1   Suppose that we want to estimate the difference between the mean salaries of all executives in finance companies and all executives in chemical companies. To do so, we draw two samples, one from the population of finance executives and another from the population of chemical

Practitioner's Guide for Statistics and Lean Six Sigma for Process Improvements. By Mikel J. Harry, Prem S. Mann, Ofelia C. de Hodgins, Christopher J. Lacke, and Richard Hulbert
Copyright © 2010 John Wiley & Sons, Inc.

executives. These two samples are independent because they are drawn from two different populations and the samples have no effect on each other. ∎

Example 17.2   Suppose that we want to estimate the difference between the mean number of customers served per hour by all tellers working at a bank before and after the tellers take a new training course. To accomplish this, we take a sample of 20 tellers and observe the mean number of customers served per hour by them before and after this course. Note that these two samples include the same 20 tellers. This is an example of two dependent samples. Such samples are also called *paired* or *matched* *samples*. ∎

This section and Section 17.4 discuss how to make confidence intervals and test hypotheses about the difference between two population parameters when samples are independent. Section 17.3 discusses how to make confidence intervals and test hypotheses about the difference between two population means when samples are dependent.

17.2.1   Two-Sample *t* Test

Suppose that we draw independent samples from two normally distributed populations and call them populations 1 and 2. Let

μ_1 = the mean of population 1
μ_2 = the mean of population 2
σ_1 = the standard deviation of population 1
σ_2 = the standard deviation of population 2
$\bar{x}_1$ = the mean of the sample drawn from population 1
$\bar{x}_2$ = the mean of the sample drawn from population 2
s_1 = the standard deviation of the sample drawn from population 1
s_2 = the standard deviation of the sample drawn from population 2
n_1 = the size of the sample drawn from population 1
n_2 = the size of the sample drawn from population 2

The sampling distribution of $\bar{x}_1 - \bar{x}_2$ has the following properties:

1. The mean of $\bar{x}_1 - \bar{x}_2$, denoted by $\mu_{\bar{x}_1 - \bar{x}_2}$, is

$$\mu_{\bar{x}_1 - \bar{x}_2} = \mu_1 - \mu_2$$

2. The standard deviation of $\bar{x}_1 - \bar{x}_2$, denoted by $\sigma_{\bar{x}_1 - \bar{x}_2}$, is

$$\sigma_{\bar{x}_1 - \bar{x}_2} = \sqrt{\frac{\sigma_1^2}{n_1} + \frac{\sigma_2^2}{n_2}}$$

3. The shape of the sampling distribution of $\bar{x}_1 - \bar{x}_2$ is normal with the mean and standard deviation stated in properties 1 and 2. Therefore, the two-sample z statistic

$$z = \frac{(\bar{x}_1 - \bar{x}_2) - (\mu_1 - \mu_2)}{\sqrt{\dfrac{\sigma_1^2}{n_1} + \dfrac{\sigma_2^2}{n_2}}}$$

is normally distributed with mean 0 and standard deviation 1.

As mentioned in Chapters 14 and 16, very rarely do we know the population standard deviations. Therefore, we need to consider the sampling distribution when we have s_1 and s_2 as estimates for σ_1 and σ_2. However, the sampling distribution also depends on whether we can assume that the two population standard deviations are equal, that is, whether $\sigma_1 = \sigma_2$. If this assumption is reasonable, we will develop a single estimate of the standard deviation, called the *pooled standard deviation*, which is denoted s_p and is given by the formula

$$s_p = \sqrt{\frac{(n_1-1)s_1^2 + (n_2-1)s_2^2}{n_1+n_2-2}}$$

In this situation, the two-sample t statistic is defined as

$$t = \frac{(\bar{x}_1-\bar{x}_2)-(\mu_1-\mu_2)}{s_p\sqrt{\dfrac{1}{n_1}+\dfrac{1}{n_2}}}$$

and has a t distribution with n_1+n_2-2 degrees of freedom, provided that the assumption that each population is normally distributed holds true.

In many cases, the equal-variance assumption is inappropriate. In that case, the corresponding two-sample t statistic is defined as

$$t = \frac{(\bar{x}_1-\bar{x}_2)-(\mu_1-\mu_2)}{\sqrt{\dfrac{s_1^2}{n_1}+\dfrac{s_2^2}{n_2}}}$$

and t has an approximate t distribution. Calculation of the degrees of freedom for this case is more complicated, which are calculated as follows:

$$df = \frac{\left(\dfrac{s_1^2}{n_1}+\dfrac{s_2^2}{n_2}\right)^2}{\dfrac{1}{n_1-1}\left(\dfrac{s_1^2}{n_1}\right)^2+\dfrac{1}{n_2-1}\left(\dfrac{s_2^2}{n_2}\right)^2}$$

This calculation is included in statistical software. However, if hand calculations need to be performed, the degrees of freedom can be approximated by the formula

$$df = \min(n_1-1, n_2-1)$$

which means that df are equal to the minimum of the two numbers given by n_1-1 and n_2-1.

The methods using the approximate t distribution work well under a number of circumstances. Here are the situations in which the t methods can be used for a hypothesis test or confidence interval for $\mu_1-\mu_2$ for two independent samples:

1. The two populations are both normally distributed.
2. The two populations have similarly shaped distributions, and the sample sizes are both five or more.
3. If the populations do not have similarly shaped distributions, then you will need to use the robustness conditions provided in Chapter 14. In such a case, checking the normality of each sample will have to be done separately. However, the sample size criteria are based on the total sample size. For example, if you have samples of size 20 and 25, the total sample size is 45. Therefore, as long as neither sample contains any outliers, the two-sample t procedures could be used in this situation.

Since σ_1 and σ_2 are seldom known, we will not discuss the normal distribution procedures here to make confidence intervals for and to test hypotheses about $\mu_1 - \mu_2$. We will discuss only the t distribution procedures.

Confidence Interval Estimation of $\mu_1 - \mu_2$. By constructing a confidence interval for $\mu_1 - \mu_2$, we find the difference between the means of two populations. For example, we may want to find the difference between the mean salaries of finance and chemical executives. The difference between the two sample means $\bar{x}_1 - \bar{x}_2$ is the point estimator of the difference between the two population means $\mu_1 - \mu_2$. As with the t statistic, the formula used for calculating the two-sample t interval for $\mu_1 - \mu_2$ depends on whether the equal standard deviations assumption is reasonable. If it is reasonable, then the formula for the confidence interval for $\mu_1 - \mu_2$ is

$$(\bar{x}_1 - \bar{x}_2) \pm t\, s_p \sqrt{\frac{1}{n_1} + \frac{1}{n_2}}$$

with df $= n_1 + n_2 - 2$. Note that s_p is calculated as explained earlier. This interval is commonly referred to as the *pooled two-sample t interval*. If the equal standard deviation assumption is not reasonable, then the formula for the confidence interval is

$$(\bar{x}_1 - \bar{x}_2) \pm t \sqrt{\frac{s_1^2}{n_1} + \frac{s_2^2}{n_2}}$$

with df $= \min(n_1 - 1, n_2 - 1)$ if statistical software is not being used. This interval is commonly referred to as the *nonpooled two-sample t interval*.

Note that the value of t for these formulas is obtained from Table III of Appendix A for the given confidence level and the appropriate degrees of freedom.

Determining Whether the Equal Standard Deviation Assumption Is Reasonable. There are many formal procedures for determining whether the equal standard deviation assumption is reasonable. Many of these procedures have a variety of underlying assumptions that must be checked. Moreover, if the underlying assumptions are violated, the methods should not be used. Fortunately, there is a conservative *ad hoc* method that is very simple. Specifically, calculate the ratio of the smaller of the two sample standard deviations to the larger of the two sample standard deviations as follows:

$$\frac{\max(s_1, s_2)}{\min(s_1, s_2)}$$

If this ratio is less than or equal to 2, then the pooled two-sample t interval can be used. If this ratio is greater than 2, then the nonpooled two-sample t interval should be used.

Examples 17.3 and 17.4 illustrate the procedure for constructing a confidence interval for $\mu_1 - \mu_2$.

Example 17.3   A consumer agency wanted to estimate the difference in the mean amounts of caffeine in two brands of coffee. The agency took a sample of 15 one-pound jars of Reglof's coffee that showed the mean amount of caffeine in these jars to be 80.7 mg per jar with a standard deviation of 5.3 mg. Another sample of 12 one-pound jars of Niravas coffee gave a mean amount of caffeine equal to 77.2 mg per jar with a standard deviation of 6.7 mg. Construct a 95% confidence interval for the difference between the mean amounts of caffeine in one-pound jars of these two brands of coffee. Assume that the two populations are normally distributed.

Solution   Let μ_1 and μ_2 be the mean amounts of caffeine per jar in all one-pound jars of Reglof's and Niravas, respectively, and let $\bar{x}_1$ and $\bar{x}_2$ be the means of the two respective samples. The given information is as follows:

$$\bar{x}_1 = 80.7 \text{ mg} \qquad s_1 = 5.3 \text{ mg} \qquad n_1 = 15$$
$$\bar{x}_2 = 77.2 \text{ mg} \qquad s_2 = 6.7 \text{ mg} \qquad n_2 = 12$$

The confidence level is $1-\alpha = .95$.

In determining whether the equal standard deviation assumption is reasonable, we use

$$\frac{\max(s_1, s_2)}{\min(s_1, s_2)} = \frac{6.7}{5.3} = 1.26 < 2$$

Since the ratio is less than 2, we can use the pooled confidence interval procedure. The value of the estimated pooled standard deviation is

$$s_p = \sqrt{\frac{(n_1-1)s_1^2 + (n_2-1)s_2^2}{n_1+n_2-2}} = \sqrt{\frac{14(5.3)^2 + 11(6.7)^2}{15+12-2}} = \sqrt{\frac{887.05}{25}} = \sqrt{35.482} = 5.957$$

Thus,

$$s_p\sqrt{\frac{1}{n_1}+\frac{1}{n_2}} = 5.957\sqrt{\frac{1}{15}+\frac{1}{12}} = 2.307$$

Since we are using the pooled confidence interval, the number of degrees of freedom is

$$\text{df} = n_1+n_2-2 = 15+12-2 = 25$$

Since the confidence level is 95%, the area in each tail is $.05/2 = .025$. The t value corresponding to a tail area of .025 and df $= 25$ is 2.060. The 95% confidence interval for $\mu_1-\mu_2$ is

$$(\bar{x}_1-\bar{x}_2) \pm t\, s_p\sqrt{\frac{1}{n_1}+\frac{1}{n_2}} = (80.7-77.2) \pm 2.060 \times 2.307 = 3.5 \pm 4.752 = (-1.252,\ 8.252)$$

Thus, with 95% confidence we can state that based on these two sample results, the difference in the mean amounts of caffeine in one-pound jars of these two brands of coffee lies between -1.252 and 8.252 mg. Because the lower limit of the interval is negative, it is possible that the mean amount of caffeine is greater in the second brand than in the first brand of coffee. Note that the value of $\bar{x}_1-\bar{x}_2$, which is $80.7-77.2=3.5$, gives the point estimate of $\mu_1-\mu_2$. ∎

Example 17.4   According to the Bureau of Labor Statistics, in June 2006, construction workers earned an average hourly wage of $20.03, and manufacturing workers earned an average hourly wage of $16.78. Assume that these mean earnings have been calculated for samples of 500 and 700 workers taken from the two populations, respectively. Further assume that the sample standard deviations of the hourly earnings of the two populations are $3.26 and $1.42, respectively.

1. What is the point estimate of $\mu_1-\mu_2$?
2. Construct a 90% confidence interval for the difference between the mean hourly earnings of the two populations.
3. Assuming that the employees work a 40-hour week with no overtime, construct a 90% confidence interval for the difference between the mean weekly earnings of the two populations.

Solution   Refer to all construction workers as population 1 and all manufacturing workers as population 2. The respective samples, then, are samples 1 and 2. Let μ_1 and μ_2 be the means of populations 1 and 2, and let $\bar{x}_1$ and $\bar{x}_2$ be the means of the respective samples. The given information is

$$\begin{array}{ccc} \bar{x}_1 = \$20.03 & s_1 = \$3.26 & n_1 = 500 \\ \bar{x}_2 = \$16.78 & s_2 = \$1.42 & n_2 = 700 \end{array}$$

The confidence level is $1-\alpha = .90$. The sample sizes are very large. As long as there are no outliers in the data, the t procedures can be used. In determining whether the equal standard deviation assumption is reasonable, we use

$$\frac{\max(s_1, s_2)}{\min(s_1, s_2)} = \frac{3.26}{1.42} = 2.30 > 2$$

Since the ratio is greater than two, we should use the nonpooled confidence interval procedure:

1. The point estimate of $\mu_1 - \mu_2$ is given by

$$\bar{x}_1 - \bar{x}_2 = \$20.03 - \$16.78 = \$3.25$$

2. Since we are not using software, we will use the df $= \min(n_1 - 1, n_2 - 1)$ criterion. Therefore, df $= \min(500 - 1700 - 1) = 499$, so there are 499 degrees of freedom. Since the confidence level is 90%, the area in each tail is $.10/2 = .05$. The t table does not have a line corresponding to df $= 499$, so we will use the line corresponding to df $= 200$, which is the closest to 499 without going over. The t value corresponding to a tail area of .05 and df $= 200$ is 1.653. The 90% confidence interval for $\mu_1 - \mu_2$ is

$$(\bar{x}_1 - \bar{x}_2) \pm t\sqrt{\frac{s_1^2}{n_1} + \frac{s_2^2}{n_2}} = (20.03 - 16.78) \pm 1.653\sqrt{\frac{3.26^2}{500} + \frac{1.42^2}{700}}$$
$$= 3.25 \pm 1.653(.1554) = 3.25 \pm .257 = (\$2.993, \$3.507)$$

Thus, we are 90% confident that the average hourly wage of construction workers is between $2.99 and $3.51 higher than the average hourly wage of manufacturing workers.

3. The process of converting the interval to the difference in the average weekly wages requires multiplying the endpoints of the difference in the average hourly wages by 40. Thus, the 90% confidence interval for the difference in the average weekly wage of construction and manufacturing workers is

$$(40(\$2.993), 40(\$3.507) = (\$119.72, \$140.28)$$

Thus, we are 90% confident that average weekly wage of construction workers is between $119.72 and $140.28 higher than the average hourly wage of manufacturing workers. ∎

Hypothesis Testing about $\mu_1 - \mu_2$. When the underlying conditions regarding the normality of the data are satisfied or compensated for by the sample size and distribution of the data, an approximate t test about $\mu_1 - \mu_2$ can be performed. The possible test statistics are the two-sample t statistics that were mentioned at the beginning of this chapter. The choice of the test statistic depends on the assumption whether or not the two population standard deviations are equal, using the same criterion that was used in choosing the appropriate form for the confidence interval:

If $\max(s_1, s_2)/\min(s_1, s_2) \le 2$, then the test statistic is the pooled t statistic given by

$$t = \frac{(\bar{x}_1 - \bar{x}_2) - (\mu_1 - \mu_2)}{s_p\sqrt{\dfrac{1}{n_1} + \dfrac{1}{n_2}}}$$

TABLE 17.1   Relationship between H_1, the Rejection Region, and the p-value

	Two-Tailed Test	Left-Tailed Test	Right-Tailed Test
H_1	$\neq$	$<$	$>$
Rejection region	In both tails	In left tail	In right tail
p-value	Twice the area to right of absolute value of test statistic	Area to left of test statistic	Area to right of test statistic

If $\max(s_1, s_2)/\min(s_1, s_2) > 2$, then the test statistic is the nonpooled t statistic given by

$$t = \frac{(\bar{x}_1 - \bar{x}_2) - (\mu_1 - \mu_2)}{\sqrt{\dfrac{s_1^2}{n_1} + \dfrac{s_2^2}{n_2}}}$$

Table 17.1 summarizes the foregoing discussion about the relationship between H_1, the tails of a test, the rejection region(s), and the p value.

As you may have noticed, Table 17.1 is the same as the table for the single-sample tests mentioned in Chapter 16.

The null hypothesis can be regarded as H_0: $\mu_1 - \mu_2 = $ a specific difference. However, in many cases, the null hypothesis is the assumption that the two means are equal. In this case we have $H_0 : \mu_1 = \mu_2$ or $H_0 : \mu_1 - \mu_2 = 0$.

Table 17.2 demonstrates the equivalent alternative hypotheses when the null hypothesis is $H_0 : \mu_1 - \mu_2 = 0$.

TABLE 17.2   Equivalent Versions of H_1 When H_0: $\mu_1 - \mu_2 = 0$

Stated in Terms of $\mu_1 - \mu_2$	Equivalent Version
$H_0 : \mu_1 - \mu_2 = 0$	$H_0 : \mu_1 = \mu_2$
$H_1 : \mu_1 - \mu_2 \neq 0$	$H_1 : \mu_1 \neq \mu_2$
$H_1 : \mu_1 - \mu_2 < 0$	$H_1 : \mu_1 < \mu_2$
$H_1 : \mu_1 - \mu_2 > 0$	$H_1 : \mu_1 > \mu_2$

Examples 17.5 and 17.6 illustrate how a test of hypothesis about the difference between two population means, $\mu_1 - \mu_2$, is conducted using the t distribution.

Example 17.5   A simple random sample of 14 cans of Liquid Plutonium Energy Drink gave the mean vitamin B_{12} content as 125 μg per can (approximately 5200% of the recommended daily allowance) with a standard deviation of 4.7 μg per can. Another simple random sample of 16 cans of Ultimate Insomnia gave the mean vitamin B_{12} content as 119μg per can with a standard deviation of 3.9 μg per can. At the 1% significance level, is the mean vitamin B_{12} content for Liquid Plutonium different from the mean vitamin B_{12} content for Ultimate Insomnia? Assume that the vitamin B_{12} contents are normally distributed for each of the two brands.

Solution   Refer to all cans of Liquid Plutonium as population 1 and all cans of Ultimate Insomnia as population 2. Let μ_1 and μ_2 be the mean vitamin B_{12} levels for populations 1 and 2, and let $\bar{x}_1$ and $\bar{x}_2$ be the means of the respective samples. The given information is

$$\begin{array}{lll} \bar{x}_1 = 125\,\mu g & s_1 = 4.7\,\mu g & n_1 = 14 \\ \bar{x}_2 = 119\,\mu g & s_2 = 3.9\,\mu g & n_2 = 16 \end{array}$$

Step 1. Select the type of test to use and check the underlying conditions.

The information provided corresponds to a two independent samples t test. The problem states that the data came from two simple random samples, which appear to be independent. The vitamin B_{12} content levels are assumed to be normal for both drinks, so the normality condition is met. Given that both conditions have been satisfied, we can use a t test. Last, we need to determine whether we should use the pooled t test or the nonpooled t test. Specifically

$$\frac{\max(s_1, s_2)}{\min(s_1, s_2)} = \frac{4.7}{3.9} = 1.21 < 2$$

So we can use the pooled t test.

Step 2. State the null and alternative hypotheses.

Note that we are testing to see if the mean vitamin B_{12} content for Liquid Plutonium is different from the mean vitamin B_{12} content for Ultimate Insomnia. This will be the case if $\mu_1 - \mu_2 \neq 0$. We write the null and alternative hypotheses as follows:

$$H_0 : \mu_1 - \mu_2 = 0 \quad \text{(the average } B_{12} \text{ contents are the same)}$$
$$H_1 : \mu_1 - \mu_2 \neq 0 \quad \text{(the average } B_{12} \text{ contents are different)}$$

Step 3. Determine the significance level α.

The significance level is given to be .01; hence, $\alpha = .01$.

Step 4. Calculate the test statistic and the p-value.

Since we are using the pooled t test, we need to calculate the value of s_p.

$$s_p = \sqrt{\frac{(n_1-1)s_1^2 + (n_2-1)s_2^2}{n_1 + n_2 - 2}} = \sqrt{\frac{13(4.7)^2 + 15(3.9)^2}{14 + 16 - 2}} = \sqrt{\frac{515.32}{28}} = \sqrt{18.404} = 4.290$$

$$t = \frac{(\bar{x}_1 - \bar{x}_2) - (\mu_1 - \mu_2)}{s_p \sqrt{\dfrac{1}{n_1} + \dfrac{1}{n_2}}} = \frac{(125 - 119) - 0}{4.29 \sqrt{\dfrac{1}{14} + \dfrac{1}{16}}} = \frac{6}{1.57} = 3.822$$

Since the alternative hypothesis is two-sided, the p-value is the probability of obtaining a test statistic that falls below -3.822 or above 3.822. The sample sizes are 14 and 16, so we use df $= 14 + 16 - 2 = 28$. If we use the t table, we see that $P(t \geq 3.674) = .0005$, so $P(t \geq 3.822) < .0005$. Therefore $P(t \leq -3.822) + P(t \geq 3.822) < .001$. So we know that the p-value is less than .001 as is shown in Figure 17.1.

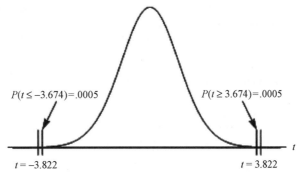

Figure 17.1   The p-value for Example 17.5.

Step 5. Make a decision.

The significance level is .01. Using the p-value, we ask whether the p-value is $\leq \alpha$. Since the p-value is less than .001, it is less than .01. Therefore, we reject the null hypothesis. Our conclusion is that the mean vitamin B_{12} content levels are different for Liquid Plutonium and Ultimate Insomnia.   ■

Example 17.6   The management at a supermarket wants to investigate whether or not a promotional campaign increased the sales of a product. A sample of 28 days during the promotional campaign showed that an average of 316 units of this product were sold per day with a standard deviation of 29 units. A sample of 24 days before the promotional campaign showed that an average of 282 units of this product were sold per day with a standard deviation of 13 units. Test at the 5% significance level if the promotional campaign increases the mean number of units sold per day. Assume that the number of units sold per day is approximately normally distributed for each population.

Solution   Let the number of units sold per day during the promotional campaign be referred to as population 1 and the number of units sold per day before the promotional campaign be referred to as population 2. Let μ_1 and μ_2 be the mean numbers of units sold per day for populations 1 and 2, and let $\bar{x}_1$ and $\bar{x}_2$ be the means of the respective samples. The given information is

$$\begin{array}{lll} \bar{x}_1 = 316 \text{ units} & s_1 = 29 \text{ units} & n_1 = 28 \\ \bar{x}_2 = 282 \text{ units} & s_2 = 13 \text{ units} & n_2 = 24 \end{array}$$

Step 1. Select the type of test to use and check the underlying conditions.

The information provided corresponds to a two-independent-sample t test. The problem states that the data came from two simple random samples, which we are assuming to be independent. The numbers of units sold each day are approximately normal for both periods, so the normality condition is met. Given that both conditions have been satisfied, we can use a t-test procedure. Last, we need to determine whether we should use the pooled t test or the nonpooled t test. Specifically

$$\frac{\max(s_1, s_2)}{\min(s_1, s_2)} = \frac{29}{13} = 2.23 > 2$$

So we need to use the nonpooled t test.

Step 2. State the null and alternative hypotheses.

Note that we are testing to see if the mean the number of units sold per day during the promotional campaign is higher than the number of units sold per day before the promotional campaign. This will be the case if $\mu_1 - \mu_2 > 0$. We write the null and alternative hypotheses as follows:

H_0: $\mu_1 - \mu_2 = 0$ (average sales are the same for the two periods)
H_1: $\mu_1 - \mu_2 > 0$ (average sales increased during the promotional period)

Step 3. Determine the significance level α.

The significance level is given as .05; thus, $\alpha = .05$.

Step 4. Calculate the test statistic and the p-value.

Since we are using the nonpooled t test, calculate t as follows:

$$t = \frac{(\bar{x}_1 - \bar{x}_2) - (\mu_1 - \mu_2)}{\sqrt{\dfrac{s_1^2}{n_1} + \dfrac{s_2^2}{n_2}}} = \frac{(316 - 282) - 0}{\sqrt{\dfrac{29^2}{28} + \dfrac{13^2}{24}}} = \frac{34}{6.089} = 5.584$$

Since the alternative hypothesis is one-sided, the p-value is the probability of obtaining a test statistic that falls above 5.584. The sample sizes are 28 and 24, so we use df $= \min(28 - 1, 24 - 1) = 23$. If we

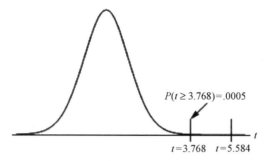

Figure 17.2   The p value for Example 17.6.

use the t table, we see that $P(t \geq 3.768) = .0005$, so $P(t \geq 5.584) < .0005$. So we know that the p-value is less than .0005 as is shown in Figure 17.2.

Step 5. Make a decision.

The significance level is .05. Using the p-value, we ask whether the p-value is $\leq \alpha$. Since the p-value is less than .0005, it is less than .05. Therefore, we reject the null hypothesis. Our conclusion is that the mean number of units sold is higher during the promotional period. ■

17.2.2 Mann–Whitney Test

The two-sample t tests described above rely on the assumption that the data are selected from a population that is normally distributed. However, there are many situations in which this assumption is not reasonable, such as when a dataset contains one or more outliers. In such a case, we must have another method for comparing two independent populations. One such method is a distribution-free test known as the *Mann–Whitney Test*.

The Mann–Whitney Test can be used for two different types of comparisons. The most popular, which is discussed here, is to compare the population medians of two independent populations. A less frequent comparison is to examine the spreads of two independent populations, assuming that they have similar medians.

The underlying assumptions of the Mann–Whitney Test are not as strict as the assumptions of the independent samples t test. These assumptions are

1. The two samples are independent and random.
2. The variable observed is a continuous random variable.
3. The data are recorded using an ordinal or continuous scale.
4. The distribution functions of the two populations differ, at most, in their location.

Assumptions 1 and 2 are equivalent to assumptions of the independent samples t test. Assumption 3 allows for the observations to be rescaled or recoded, as long as the order of the size of the observations is maintained. In other words, if the smallest observation is 1.7 and the second smallest is 8.3, the observations could be recoded to 1 and 2. The last assumption is similar to the assumption that the variances of the two populations are equal although their means or medians may not be equal.

Let η_1 and η_2 represent the population medians for populations 1 and 2. Table 17.3 contains the possible hypotheses for the Mann–Whitney Test.

The Mann–Whitney test presents an example of a test based on *ranks*. In a test that is based on ranks, we are not concerned with the individual values of the observations. Instead, we are concerned with where they fall in an ordered list of the data. To begin the calculation of the test statistic, we need to combine the observations from the two samples, order them from smallest to largest, assign ranks to the observations, and sum the ranks for each of the two samples. Example 17.7 demonstrates this procedure.

TABLE 17.3   **Hypotheses for the Mann–Whitney Test**

$H_0: \eta_1 - \eta_2 = 0$	$H_0: \eta_1 = \eta_2$
$H_1: \eta_1 - \eta_2 \neq 0$	$H_1: \eta_1 \neq \eta_2$
$H_1: \eta_1 - \eta_2 < 0$	$H_1: \eta_1 < \eta_2$
$H_1: \eta_1 - \eta_2 > 0$	$H_1: \eta_1 > \eta_2$

Example 17.7   A consumer testing agency is interested in testing two popular MP3 players to determine whether their battery lifetimes differ. Random samples of five units of model A and six units of model B were selected. Each unit was run until the battery died. The following data give the battery lifetimes in hundreds of hours:

Model A	9.5	10.8	12.9	16.0	9.7	
Model B	12.9	11.1	13.0	16.4	8.9	4.6

Assign ranks to the data for performing the Mann–Whitney test and calculate the sums of the ranks for the two models.

Solution   In order to assign ranks, we need to combine all of the data into one set and sort them. We also need to keep track of which model each observation comes from. The following table contains the combined and sorted data:

Lifetime	4.6	8.9	9.5	9.7	10.8	11.1	12.9	12.9	13.0	16.0	16.4
Model	B	B	A	A	A	B	A	B	B	A	B

When we assign ranks, the smallest value is assigned a rank of 1, while the largest value will be assigned a rank equal to the total number of observations. If two or more observations have the same value, we assign ranks equivalent to the average of the ranks that would have gone in those locations had there been no ties. In the case of these data, observations 7 and 8 in the list are both equal to 12.9. Therefore, they will be assigned a rank of 7.5, which is equal to the average of 7 and 8. The observations and their ranks are shown in the following table:

Lifetime	4.6	8.9	9.5	9.7	10.8	11.1	12.9	12.9	13.0	16.0	16.4
Model	B	B	A	A	A	B	A	B	B	A	B
Rank	1	2	3	4	5	6	7.5	7.5	9	10	11

The next step involves calculating the sums of the ranks for the two samples. Let S_1 and S_2 be the sums of the ranks for samples 1 and 2. Then we find the values of S_1 and S_2 as follows:

$S_1 = $ sum of ranks for model A $= 3 + 4 + 5 + 7.5 + 10 = 29.5$

$S_2 = $ sum of ranks for model B $= 1 + 2 + 6 + 7.5 + 9 + 11 = 36.5$

The calculation of the test statistic is dependent on which population is considered to be population 1. Let n_1 and n_2 be the sample sizes from populations 1 and 2, respectively. Then the test statistic, U, is defined as

$$U = S_1 - \frac{n_1(n_1 + 1)}{2}$$

Example 17.8   Using the data provided in Example 17.7, determine the value of the Mann–Whitney test statistic for the null hypothesis that the median lifetimes are the same for battery models A and B, assuming that model A represents population 1. Do the same assuming that model B represents population 1.

Solution   Assuming that model A represents population 1, we have $S_1 = 29.5$ and $n_1 = 5$. Then, the value of the test statistic U is

$$U = S_1 - \frac{n_1(n_1+1)}{2} = 29.5 - \frac{5(5+1)}{2} = 29.5 - 15 = 14.5$$

Assuming that model B represents population 1, we have $S_1 = 36.5$ and $n_1 = 6$. In this case, the value of the test statistic U is

$$U = S_1 - \frac{n_1(n_1+1)}{2} = 36.5 - \frac{6(6+1)}{2} = 36.5 - 21 = 15.5$$

It is important to note that the critical value for the Mann–Whitney test is a function of the alternative hypothesis and the two sample sizes, as well as the significance level, therefore the fact that the value of the test statistic is different depending on which model represents population 1 is not a problem.

Determining the rejection region for the Mann–Whitney test depends, as do many other tests, on the alternative hypothesis. Table VI in Appendix A gives critical values for samples sizes of 2–20. The calculation of the critical value is broken down by the version of the alternative hypothesis.

- If $H_1 : \eta_1 < \eta_2$; go to the section of the Table VI that corresponds to the value of n_1, which is shown in the leftmost column of the table. Find the row corresponding to the significance level. Last, go to the column corresponding to the value of n_2. For example, Table 17.4 contains the part of Table VI of Appendix A corresponding to $n_1 = 7$ and $n_2 = 5$.

- If the significance level is .05, the critical value occurs at the intersection of the .05 row and the $n_2 = 5$ column, thus the critical value, denoted by is U_c is $U_c = 7$. If the significance level is .01, the critical value is $U_c = 4$. The decision rule is to reject the null hypothesis if the test statistic U is less than the critical value U_c.

- If $H_1 : \eta_1 > \eta_2$, use the same set of instructions for locating the value at the intersection of the row and column corresponding to the values of n_1, n_2, and α as explained above. Note that this value obtained from the table is not the critical value in this case. Let us label it U'. The critical value U_c is calculated as

$$U_c = n_1 n_2 - U'$$

TABLE 17.4   **Portion of Mann–Whitney Table VI for $n_1 = 7$ and $\leq n_2 \leq 6$**

n_1	probability	$n_2 = 4$	$n_2 = 5$	$n_2 = 6$
	.10	7	9	12
	.05	5	7	9
7	.025	4	6	7
	.01	2	4	5
	.005	1	3	4

The decision rule is to reject the null hypothesis if the test statistic U is greater than the critical value U_c. If the significance level is .05, then $U_c = n_1 n_2 - U' = (7 \times 5) - 7 = 28$. For Example 17.8, if the significance level is .01, the critical value is $U_c = n_1 n_2 - U' = (7 \times 5) - 4 = 31$.

- If $H_1 : \eta_1 \neq \eta_2$, we need to find two critical values. To find the lower critical value, which we will call U_{lower}, follow the instructions for the $H_1 : \eta_1 < \eta_2$ case, but use the row in Table VI that corresponds to $\alpha/2$. The upper critical value, denoted by U_{upper}, is calculated as

$$U_{\text{upper}} = n_1 n_2 - U_{\text{lower}}$$

The decision rule is to reject the null hypothesis if U is less than U_{lower} or U is greater than U_{upper}. If the significance level is .05, use the row that corresponds to a probability of .025. Then from Table VI, $U_{\text{lower}} = 6$, and $U_{\text{upper}} = n_1 n_2 - U_{\text{lower}} = (7 \times 5) - 6 = 29$. Now, if the significance level is .01, use the row corresponding to a probability of .005. Then, $U_{\text{lower}} = 2$ and $U_{\text{upper}} = n_1 n_2 - U_{\text{lower}} = 7 \times 5 - 2 = 33$. ■

Example 17.9   Using the data on battery lifetime given in Example 17.7, determine whether there is a significant difference in the median lifetimes between model A and model B batteries. Use a 5% significance level.

Solution
Step 1. Select the type of test to use and check the underlying conditions.

As was stated in Example 17.7, the data come from independent simple random samples. The variable, the battery lifetime, is a continuous random variable. For the purpose of this example, we will assume that the only possible difference in the distributions is the location, not the spread.

Step 2. State the null and alternative hypotheses.

Note that we are to test whether or not there is a difference between the median lifetimes for the two brands of battery. Let

$$\eta_1 = \text{median } \textit{battery lifetime of all model A batteries}$$
$$\eta_2 = \text{median } \textit{battery lifetime of all model B batteries}$$

If there is no difference between the median lifetimes, $\eta_1 = \eta_2$. If there is a difference, $\eta_1 \neq \eta_2$. We write the null and alternative hypotheses as follows:

$$H_0: \eta_1 = \eta_2 \quad \text{(there is no difference between the median lifetimes)}$$
$$H_1: \eta_1 \neq \eta_2 \quad \text{(there is a difference between the median lifetimes)}$$

Step 3. Determine the significance level α.

The significance level is given to be .05; thus, $\alpha = .05$.

Step 4. Calculate the test statistic.

Let model A be classified as population 1. From Examples 17.7 and 17.8 we have

$$S_1 = 29.5, \quad n_1 = 5, \quad n_2 = 6, \quad \text{and} \quad U = 14.5.$$

Table 17.5 contains the portion of Table VI of Appendix A corresponding to $n_1 = 5$ and $n_2 = 6$. Since the alternative hypothesis is two-tailed, we need to obtain U_{lower} and U_{upper}, using the row for $\alpha/2 = .025$. The corresponding entry in the $n_2 = 6$ column gives $U_{\text{lower}} = 4$. Thus

$$U_{\text{lower}} = 4 = U_{\text{upper}} = n_1 n_2 - U_{\text{lower}} = 5 \times 6 - 4 = 26$$

TABLE 17.5 Portion of Mann–Whitney Table for $n_1 = 5$ and $2 \leq n_2 \leq 6$

n_1	Probability	$n_2 = 2$	$n_2 = 3$	$n_2 = 4$	$n_2 = 5$	$n_2 = 6$
	.10	2	3	5	6	8
	.05	1	2	3	5	6
5	.025	0	1	2	3	4
	.01	0	0	1	2	3
	.005	0	0	0	1	2
	.001	0	0	0	0	0

Step 5. Make a decision.

The decision rule calls for us to reject the null hypothesis if $U < 4$ or $U > 26$. Since the value of $U = 14.5$ falls between 4 and 26, we fail to reject the null hypothesis. Our conclusion is that there is insufficient evidence to state that the median lifetimes of model A and model B batteries are different. ∎

When both samples are large, an approximate test based on the normal distribution can be performed for the difference between two medians. The test statistic for the approximate z test, in the case presented above, is

$$z = \frac{S_1 - n_1(n_1 + n_2 + 1)}{\sqrt{\dfrac{n_1 n_2(n_1 + n_2 + 1)}{12}}}$$

The p-values for the approximate z test are calculated in the same manner as are p-values for any other test based on the standard normal distribution:

1. If $H_1 : \eta_1 \neq \eta_2$, p-value $= 2P(Z \geq z)$
2. If $H_1 : \eta_1 < \eta_2$, p-value $= P(Z \leq z)$
3. If $H_1 : \eta_1 > \eta_2$, p-value $= P(Z \geq z)$

17.3 INFERENCES ABOUT DIFFERENCES BETWEEN TWO POPULATION MEANS FOR PAIRED SAMPLES

In this section we will learn about inference making for paired samples using the t distribution and the nonparametric procedure called the Wilcoxon signed-rank test.

17.3.1 Paired t Test

Section 17.2 discussed estimation and hypothesis tests regarding the difference between two population means when the two samples were drawn independently from two different populations. This section describes estimation and hypothesis testing procedures for the difference between two population means when the samples are dependent.

In a case of two dependent samples, two data values—one in each sample—are collected from either the same source (or element) or two similar sources; hence, these are also called *paired* or *matched samples*, respectively. For example, we may want to make inferences about the mean number of units of a product sold by salespersons before and after taking a course in sales management. To do so, we select a sample of 15 salespersons and record their sales before and after taking this course. In this example, both sets of data are collected from the same 15 salespersons, once before and once after the course. Thus, although there are two samples, they contain the same 15 salespersons. This is an example of paired (or dependent or matched) samples.

Another example may involve determining whether the size of a tip, as a percentage of the restaurant bill, differs between male and female servers. We would obtain data from two servers, one male and one female, at the same restaurant. We may take a sample of 20 restaurants and calculate the difference in tip percentage for each of the 20 pairs of male and female servers. The procedures for making confidence intervals and test hypotheses in cases of paired samples differ from the ones for independent samples discussed in earlier sections of this chapter.

As another example of paired samples, suppose that an agronomist wants to measure the effect of a new brand of fertilizer on the yield of potatoes. To do so, he selects 10 pieces of land and divides each piece of land into two portions. Then he randomly assigns one of the two portions from each piece of land to grow potatoes without using fertilizer (or using some other brand of fertilizer). The second portion from each piece of land is used to grow potatoes using the new brand of fertilizer. Thus, he will have 10 pairs of data values. Then, using the procedure to be discussed in this section, he will make inferences about the difference in the mean yields of potatoes with the new fertilizer and without it.

The question arises as to why the agronomist does not choose 10 pieces of land on which to grow potatoes without using the new brand of fertilizer and another 10 pieces of land to grow potatoes by using the new brand of fertilizer. If he does so, the effect of the fertilizer might be confused with the effects due to soil differences at different locations. Thus, he will not be able to identify the effect of only the new brand of fertilizer on the yield of potatoes. Consequently, the results will not be reliable. By choosing 10 pieces of land and then dividing each of them into two portions, the researcher decreases the possibility that the difference in the productivities of different pieces of land affects the results. Even with this precaution, there is no guarantee that the other issues, such as drainage, disease, and infestation, to name only a few, may not cause differences within the same plot of land. Similar arguments can be made with regard to the male and female servers, but it is impossible to control for every possible influence.

In paired samples, the difference between the two data values for each element of the two samples is denoted by d. This d value is called the *paired difference*. We then treat all the values of d as one sample and make inferences applying procedures similar to the ones used for one-sample cases in Chapters 14 and 16. Note that as each source (or element) gives a pair of values (one for each of the two datasets), each sample contains the same number of values; that is, both samples are of the same size. Therefore, we denote the (common) *sample size* by n, which gives the number of paired difference d values. The *degrees of freedom* for the paired samples are $n-1$. Let

μ_d = the mean of the paired differences for the population

σ_d = the standard deviation of the paired differences for the population

$\bar{d}$ = the sample mean of the paired differences

s_d = the sample standard deviation of the paired differences

n = the number of paired difference values

The sample mean and sample standard deviation of the paired differences are calculated using the same methods that were used to calculate the sample mean and standard deviation for any other dataset in Chapter 8. The only difference in the formulas is the fact that x will be replaced by d. The formulas are

$$\bar{d} = \frac{\sum d}{n} \quad \text{and} \quad s_d = \sqrt{\frac{\sum (d-\bar{d})^2}{n-1}} = \sqrt{\frac{\sum d^2 - \frac{(\sum d)^2}{n}}{n-1}}$$

The standard error of $\bar{d}$ is denoted by se_d and is calculated as

$$se_d = \frac{s_d}{\sqrt{n}}$$

In paired samples, instead of using $\bar{x}_1 - \bar{x}_2$ as the sample statistic to make inferences about $\mu_1 - \mu_2$, we use the sample statistic $\bar{d}$ to make inferences about μ_d. Actually the value of μ_d is always equal to $\mu_1 - \mu_2$, and the value of $\bar{d}$ is always equal to $\bar{x}_1 - \bar{x}_2$. However, se_d is not the same as $s_{\bar{x}_1 - \bar{x}_2}$.

As with all of the other inferential methods we have discussed, it is unreasonable to assume that the population standard deviation σ_d is known. As a result, inferences for μ_d involve the sample standard deviation of the differences s_d, and the methods are based on the t procedures. The procedures are subject to a set of assumptions and conditions similar to those used in the one-sample t procedures discussed in Chapters 14 and 16. These assumptions are

1. The sample of differences is based on a simple random sample.
2. The population of paired differences is normally distributed.

As in Chapters 14 and 16, if the normality assumption is not reasonable, the paired t procedures can still be used if the sample size is large enough. The criteria were partitioned into three scenarios in those chapters in which the sample size is less than 15, from 15 up to 40, and 40 or more.

Constructing a Confidence Interval for μ_d. Here, the mean $\bar{d}$ of paired differences for two paired samples is the point estimator of μ_d. The following formula is used to construct a confidence interval for μ_d provided that the previously stated underlying conditions hold true

$$\bar{d} \pm t\,\frac{s_d}{\sqrt{n}}$$

where the value of t is obtained from the t distribution table for the given confidence level and $n-1$ degrees of freedom. Note that in this formula, $s_d / \sqrt{n}$ is se_d. Example 17.10 illustrates the procedure to construct a confidence interval for μ_d.

Example 17.10   Sham Corporation is considering installing a new assembly procedure. The company randomly selected seven employees. Each employee had to assemble one unit of the product using the old procedure and another unit using the new procedure. The order of the procedures was assigned at random for each employee. Table 17.6 gives the assembly times (in minutes) for these seven employees for each procedure.

TABLE 17.6   Data for Example 17.10

Employee	1	2	3	4	5	6	7
Old procedure	64	71	68	66	73	62	70
New procedure	60	66	66	69	63	57	62

Let μ_d be the population mean of the differences between the assembly times for the two procedures. Construct a 95% confidence interval for μ_d. Assume that the population of paired differences is (approximately) normally distributed.

Solution   Because the information obtained is from paired samples (i.e., each pair of data belongs to one employee), we will construct the confidence interval for the paired difference mean μ_d of the population by using the paired difference mean $\bar{d}$ of the sample. Let d be the difference in the times taken by an employee to assemble one unit of the product by using the old procedure and by using the new procedure. The values of d are obtained by subtracting the assembly time for the new procedure from the assembly time for the old procedure for the same employee. The fourth column of Table 17.7 lists the values of d for the seven sampled employees. The fifth column of the table records the values of $(d-\bar{d})^2$, which are needed for calculating the sample standard deviation of the d values. The first value

TABLE 17.7   Calculation of Paired Differences for Example 17.10

Employee	Old Procedure	New Procedure	d	$(d-\bar{d})^2$
1	64	60	4	.1837
2	71	66	5	.3265
3	68	66	2	5.8980
4	66	69	-3	55.1837
5	73	63	10	31.0408
6	62	57	5	.3265
7	70	62	8	12.7551
			$\Sigma d = 31$	$\Sigma(d-\bar{d})^2 = 105.7143$

of d is 4, which is obtained as $64-60$. Other values of d are obtained the same way. The mean of differences $\bar{d}$ is 4.4286 as calculated below. The first value in the column of $(d-\bar{d})$ is obtained as $(4-4.4286)^2$. Other values in this column are calculated the same way.

from the given information, we have

$$\bar{d} = \frac{31}{7} = 4.4286 \quad \text{and} \quad s_d = \sqrt{\frac{\Sigma(d-\bar{d})^2}{n-1}} = \sqrt{\frac{105.7143}{7-1}} = 4.198$$

Since the confidence level is 95%, the area in each tail is $(1-.95/2) = .05/2 = .025$. The number of degrees of freedom is $n-1 = 7-1 = 6$. The t value corresponding to a tail area of .025 and df $= 6$ is 2.447. The 95% confidence interval for μ_d is

$$\bar{d} \pm t\frac{s_d}{\sqrt{n}} = 4.4286 \pm 2.447\frac{4.198}{\sqrt{6}} = 4.4286 \pm 4.1937 = (.2349, 8.6223)$$

Thus, we can state with 95% confidence that the mean difference between the assembly times for the two procedures is between .2349 and 8.6223 min. In other words, the old procedure takes, on average, .2349 to 8.6223 min longer than the new procedure to assemble one unit of the product. ∎

Hypothesis Testing about μ_d. A hypothesis about μ_d is tested by using the sample statistic $\bar{d}$. The assumptions and underlying conditions for this test are the same as those for the one-sample t test for a population mean μ, discussed in Chapter 16. The following formula is used to calculate the value of the test statistic t when testing the null hypothesis about μ_d

$$t = \frac{\bar{d} - \mu_d}{s_d/\sqrt{n}}$$

where μ_d is the hypothesized average difference among all pairs in the population. In many cases, the null hypothesis assumes that there is no difference (i.e., the hypothesized difference is zero), making $H_0 : \mu_d = 0$.

The critical value of t is found from the t distribution table for the given significance level and $n-1$ degrees of freedom. The rejection regions and p-values are described in Table 17.8, which is comparable to Tables 16.1 and 17.1.

Examples 17.11 and 17.12 illustrate the hypothesis testing procedure for μ_d.

TABLE 17.8 Relationship between H_1, Rejection Region, and p-value for Paired t Test

	Two-Tailed Test	Left-Tailed Test	Right-Tailed Test
H_1	$\neq$	$<$	$>$
Rejection region	In both tails	In the left tail	In the right tail
p-value	Two times the area to the right of the absolute value of the test statistic	The area to the left of the test statistic	The area to the right of the test statistic

Example 17.11 A company wanted to know whether attending a course on *how to be a successful salesperson* can increase the average sales of its employees. The company sent a simple random sample of six of its salespersons to attend this course. Table 17.9 gives the one-week sales of these salespersons before and after they attended this course.

TABLE 17.9 Data for Example 17.11

Before	18	9	14	16	12	25
After	24	14	19	20	18	24

Using the 1% significance level, can you conclude that the mean weekly sales for all salespersons increased as a result of attending this course? Assume that the population of paired differences has a normal distribution.

Solution We use our five-step procedure to make this test as follows:

Step 1. Select the type of test to use and check the underlying conditions.

The information provided corresponds to paired samples. The problem states that the data came from a simple random sample, so the first condition is met. We are told that the distribution of paired differences has a normal distribution, which satisfies the second condition. It is important to note that the sample size of 6 is very small, so the population distribution would have to be very close to a normal distribution in order to justify the use of the t procedure. Nevertheless, the two conditions hold, so we will use a paired t test.

Step 2. State the null and alternative hypotheses.

Note that we are to test whether or not the average one-week sales level for all salespersons has increased as a result of taking the course. Suppose that

$$d = (\text{one week sales after taking course}) - (\text{one week sales before taking course})$$

If the course does not help to increase sales, then $\mu_d \leq 0$. If it does help, then $\mu_d > 0$. We write the null and alternative hypotheses as follows:

$$H_0: \mu_d \leq 0 \quad (\text{average sales level has not increased})$$
$$H_1: \mu_d > 0 \quad (\text{average sales level has increased})$$

Step 3. Determine the significance level α.

The significance level is given to be .01; thus, $\alpha = .01$.

Step 4. Calculate the test statistic and the p-value.

The values of the paired differences are calculated by taking the difference (after− before). All calculations are shown in Table 17.10.

Thus $\quad \bar{d} = \dfrac{\Sigma d}{n} = \dfrac{25}{6} = 4.167 \quad$ and $\quad s_d = \sqrt{\dfrac{\Sigma(d-\bar{d})^2}{n-1}} = \sqrt{\dfrac{34.8334}{6-1}} = 2.639$

TABLE 17.10   Calculation of d and $(d-\bar{d})^2$

Salesperson	Before	After	d	$(d-\bar{d})^2$
1	18	24	6	3.3599
2	9	14	5	.6939
3	14	19	5	.6939
4	16	20	4	.0279
5	12	18	6	3.3599
6	25	24	−1	26.6979
			$\Sigma d = 25$	$\Sigma(d-\bar{d})^2 = 34.8334$

The value of the test statistic is

$$t = \frac{\bar{d}-\mu_d}{s_d/\sqrt{n}} = \frac{4.167-0}{2.639/\sqrt{6}} = \frac{4.167}{1.077} = 3.869$$

Since the alternative hypothesis is right-tailed, the p-value is the probability of obtaining a test statistic that falls above 3.869. Using the t table for df $= 5$, we see that $P(t \geq 3.365) = .01$ and $P(t \geq 4.032) = .005$. Therefore, $.005 < P(t \geq 3.869) < .01$. The p-value is shown in Figure 17.3.

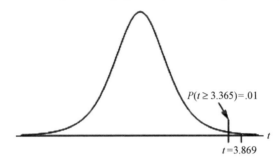

$P(t \geq 3.365) = .01$

$t = 3.869$

Figure 17.3   The p-value for Example 17.11.

Step 5. Make a decision.

The significance level is .01. The p-value is less than .01. Therefore, we reject the null hypothesis. Our conclusion is that there is enough evidence to state that the average increase in weekly sales after taking the course is positive, which implies that the course is effective in increasing weekly sales. ■

Example 17.12   Refer to Example 17.10. The table that gives the assembly times for the two procedures is reproduced below:

Employee	1	2	3	4	5	6	7
Old procedure	64	71	68	66	73	62	70
New procedure	60	66	66	69	63	57	62

Test at the 5% significance level whether the mean of the paired differences μ_d is different from zero. Assume that the population of paired differences is (approximately) normally distributed.

Solution   From Example 17.10 we have

$$\bar{d} = 4.4286 \quad \text{and} \quad n = 7, \; s_d = 4.198.$$

Step 1. Select the type of test to use and check the underlying conditions.

As we already discussed in Example 17.10, the data correspond to paired samples and the underlying conditions are met. Once again, it is important to note that the sample size of 7 is very small, so the distribution would have to be very close to normal in order to justify the use of the t procedure. Nevertheless, the two conditions hold true, so we will use a paired t test.

Step 2. State the null and alternative hypotheses.

Note that we are to test whether the average difference in the time it takes to assemble one unit is different for the two methods. Suppose that

$$d = \text{(time to assemble one unit using old procedure)}$$
$$-\text{(time to assemble one unit using new procedure)}$$

If there is no difference between the methods, then $\mu_d = 0$. If there is a difference, then $\mu_d \neq 0$. We write the null and alternative hypotheses as follows:

$$H_0: \mu_d = 0 \quad \text{(average difference in assembly times is zero)}$$
$$H_1: \mu_d \neq 0 \quad \text{(average difference in assembly times is not zero)}$$

Step 3. Determine the significance level α.

The significance level is given to be .05; thus, $\alpha = .05$.

Step 4. Calculate the test statistic and the p-value.

From Example 17.10 we have

$$\bar{d} = 4.4286 \quad \text{and} \quad s_d = 4.198.$$

The value of the test statistic is

$$t = \frac{\bar{d} - \mu_d}{s_d / \sqrt{n}} = \frac{4.4286 - 0}{4.198 / \sqrt{7}} = \frac{4.4286}{1.587} = 2.791$$

Since the alternative hypothesis is two-tailed, the p-value is the probability of obtaining a test statistic that falls above 2.791 or below -2.791, which is the same as $2P(t \geq 2.791)$. Using the t table for df $= 6$, we see that $P(t \geq 2.447) = .025$ and $P(t \geq 3.143) = .01$. Therefore, $.01 < P(t \geq 2.791) < .025$, so $.02 < 2P(t \geq 2.791) < .05$. The p-value is shown in Figure 17.4.

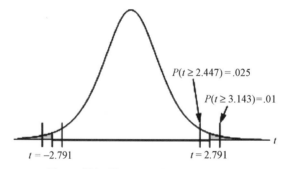

Figure 17.4   The p-value for Example 17.12.

Step 5. Make a decision.

The significance level is .05. The p-value is less than .05. Therefore, we reject the null hypothesis. Our conclusion is that there is enough evidence to state that the average difference in the time to assemble one unit of the product using the two methods is different from zero. ∎

17.3.2 Wilcoxon Signed-Rank Test

As with the independent-samples t test, the paired samples t test discussed above relies on the fact that the data come from a normal distribution. Furthermore, as the Mann–Whitney test can be used in place of the independent-samples t test when the underlying normality assumption is unreasonable, the *Wilcoxon signed-rank test* can be used in place of the paired samples t test (discussed above) when the assumption that the population differences have a normal distribution is unreasonable.

As the term suggests, the *signed-rank test* is a test that also involves ranking the data. However, just like the paired t test, we are interested in the differences of the pairs of data, so the ranking does not take place until after the differences have been calculated. Thus, the underlying assumptions of the Wilcoxon signed-rank test are based on the differences of the paired observations. The following are the assumptions of the Wilcoxon signed-rank test:

1. The paired differences are based on a simple random sample selected from all possible paired differences.
2. The paired differences represent a continuous random variable.

The hypotheses are based on the value of η_d, which is the median of the paired differences. The null hypotheses for such a case is $H_0: \eta_d = 0$, and the alternative hypothesis can be one of the three possibilities: $H_1: \eta_d \neq 0$, $H_1: \eta_d > 0$, $H_1: \eta_d < 0$.

Finding the value of the test statistic. The procedure for calculating the test statistic for the Wilcoxon signed-rank test is as follows:

1. Calculate the paired differences between the paired observations. In order to standardize the calculations, subtract the first observation from the second observation for each pair.
2. Do not include any pairs for which the difference is zero.
3. Rank the absolute values (i.e., without considering the negative sign) of the differences from smallest to largest. If two or more pairs have the same difference, assign them the average of the ranks that would have been assigned to their placeholders if there were no ties.
4. For each difference, assign a plus $(+)$ or minus $(-)$ sign, depending on whether the corresponding difference is positive or negative, respectively.
5. Calculate W_+ and W_-, where $W_+ = $ sum of ranks with $+$ sign and $W_- = $ sum of ranks with $-$ sign.
6. The test statistic depends on the alternative hypothesis. To simplify the notation, the test statistic will be labeled W. Specifically, the test statistic will be as mentioned in Table 17.11.

TABLE 17.11   Test Statstic for Each Alternative Hypothesis for Wilcoxon Signed-Rank Test

Alternative Hypothesis	Test Statistic W
$H_1: \eta_d \neq 0$	Minimum (W_+, W_-)
$H_1: \eta_d > 0$	W_-
$H_1: \eta_d < 0$	W_+

Critical Value and Decision Rule for Wilcoxon Signed-Rank Test. As with the Mann–Whitney test, the p-value calculation for Wilcoxon signed-rank test is very cumbersome without statistical software, so we will use only the critical value approach here and will use Table VII of Appendix A to determine the critical value for the decision rule. Table 17.12 provides a section of Table VII of Appendix A.

Denoting the critical value by W_c, the decision rule is to reject the null hypothesis if the test statistic is less than or equal to the critical value, that is, if $W \leq W_c$. Note that here n is the number of paired differences that are nonzero.

TABLE 17.12   Portion of Wilcoxon Signed-Rank Table for $5 \leq n \leq 11$

Probability	$n = 5$	$n = 6$	$n = 7$	$n = 8$	$n = 9$	$n = 10$	$n = 11$	$n = 12$
.10	2	3	5	8	10	14	17	21
.05	0	2	3	5	8	10	13	17
.025	$-^a$	0	2	3	5	8	10	13
.01	—	—	0	1	3	5	7	10
.005	—	—	—	0	1	3	5	7

^{a}Blank cell (—) implies that it is impossible to reject H_0 at this combination of significance level and sample size.

Example 17.13   A manufacturing company is interested in purchasing a new type of filter in order to reduce emissions. The company selects a random sample of 10 of their manufacturing plants. They compare emission levels, in parts per million, from the last time the filter was replaced, with emission levels with the new filter. See Table 17.13.

TABLE 17.13   Data for Example 17.13

Plant:	1	2	3	4	5	6	7	8	9	10
Current filter	87	102	95	73	99	84	91	110	106	65
New filter	84	72	89	75	90	88	80	101	95	65

Perform a hypothesis test to determine whether the new filter is effective in reducing the median emission level. Use a 10% significance level.

Solution   We perform the test of hypothesis using the same five steps as follows:

Step 1. Select the type of test to use and check the underlying conditions.

As stated in the problem, the plants were selected using a simple random sample. The random variable, the emissions level, is a continuous random variable. The population distribution is unknown. The conditions for the Wilcoxon signed-rank test are met.

Step 2. State the null and alternative hypotheses.

Note that we are testing to see if the new filter reduces the median emission discharge. We will calculate the change in emission level using the paired differences as follows:

$$d = (\text{emission level using new filter}) - (\text{emission level using current filter})$$

If the new filter is effective in reducing emission levels, the new filter readings will be less than the current filter readings and the differences will be negative. This would imply that $\eta_d < 0$. If the new filter is not effective, then $\eta_d = 0$. We write the null and alternative hypotheses as follows:

H_0: $\eta_d = 0$ (the median difference in emission levels is zero)
H_1: $\eta_d < 0$ (the median difference in emission levels is negative)

Step 3. Determine the significance level α.

The significance level is given as .10; hence, $\alpha = .10$.

Step 4. Calculate the test statistic.

First we need to calculate the differences and their absolute values as shown below:

Plant:	1	2	3	4	5	6	7	8	9	10
Current filter	87	102	95	73	99	84	91	110	106	65
New filter	84	72	89	75	90	88	80	101	95	65
d = new − current	−3	−30	−6	2	−9	4	−11	−9	−11	0
Absolute value of d	3	30	6	2	9	4	11	9	11	0

Next, we assign the ranks and the $+/-$ signs depending on the signs of the differences. Remember that plant 10 will not be assigned a rank because the difference is zero for this plant.

Plant:	1	2	3	4	5	6	7	8	9	10
Current filter	87	102	95	73	99	84	91	110	106	65
New filter	84	72	89	75	90	88	80	101	95	65
$d = $ new $-$ current	-3	-30	-6	2	-9	4	-11	-9	-11	0
Absolute value of d	3	30	6	2	9	4	11	9	11	0
Ranks of d	2	9	4	1	5.5	3	7.5	5.5	7.5	
Signs of d	$-$	$-$	$-$	$+$	$-$	$+$	$-$	$-$	$-$	

The values of W_+ and W_- are given by the sums of differences with $+$ and $-$ signs, respectively. Thus

$$W_+ = 1 + 3 = 4$$
$$W_- = 2 + 4 + 5.5 + 5.5 + 7.5 + 7.5 + 9 = 41$$

Since the alternative hypothesis is $\eta_d < 0$, the value of the test statistic comes from W_+. Hence

$$\text{Value of test statistic} = W = 4.$$

Step 5. Make a decision.

The significance level is .10. From Table VII of Appendix A, for $\alpha = .10$ and $n = 9$ (recall that we have nine nonzero differences), the critical value is $W_c = 10$. The decision rule is to reject the null hypothesis if $W \le W_c$. Since $4 \le 10$, we reject H_0. Our conclusion is that there is enough evidence to state that the median reduction in emission is less than zero, that is the new filter is effective in reducing emissions. ∎

In Example 17.13 the sample size was very small. When the number of paired differences is large, an approximate test based on the normal distribution can be performed. The test statistic for the approximate z test is calculated as

$$z = \frac{W - \dfrac{n(n+1)}{4}}{\sqrt{\dfrac{n(n+1)(2n+1)}{24}}}$$

Where n is the number of paired differences that are not zero, and W is calculated as above in the Wilcoxon Signed-Rank Test. The p-values for the approximate z test are calculated in the same manner as the p-values for any other test based on the standard normal distribution as follows:

1. If H_1: $\eta_d \ne 0$, p value $= 2P(Z \ge z)$
2. If H_1: $\eta_d < 0$, p value $= P(Z \le z)$
3. If H_1: $\eta_d > 0$, p value $= P(Z \ge z)$

17.4   INFERENCES ABOUT DIFFERENCES BETWEEN TWO POPULATION PROPORTIONS

Quite often we need to construct a confidence interval and test a hypothesis about the difference between two population proportions. For instance, we may want to estimate the difference between the proportion of defective items produced on two different machines. If p_1 and p_2 are the proportions of defective items produced on the first and second machines, respectively, then we are to construct a confidence interval for $p_1 - p_2$. We may want to test the hypothesis that the proportion of defective items produced on machine 1 is different from the proportion of defective items produced on machine 2. In this case, we are to test the null hypothesis $p_1 - p_2 = 0$ against the alternative hypothesis $p_1 - p_2 \neq 0$.

In Section 17.4.1, we discuss how to construct a confidence interval and test a hypothesis about $p_1 - p_2$ for two independent samples. The large sample methods are contained in all the common statistical software packages.

17.4.1   Large-Sample Procedure

The sample statistic used to make inferences about $p_1 - p_2$ is $\hat{p}_1 - \hat{p}_2$, where $\hat{p}_1$ and $\hat{p}_2$ are the proportions for two independent samples. As discussed in Section 13.3, we define a sample proportion by dividing the number of elements in the sample with a given attribute by the sample size. Thus

$$\hat{p}_1 = \frac{x_1}{n_1} \qquad \text{and} \qquad \hat{p}_2 = \frac{x_2}{n_2}$$

where x_1 and x_2 are the number of elements with a given characteristic in the two samples and n_1 and n_2 are the sizes of the two samples, respectively. As discussed in Chapter 13, for a large sample the sample proportion $\hat{p}$ is (approximately) normally distributed with mean p and standard deviation $\sqrt{p(1-p)/n}$. Hence, for two large and independent samples of sizes n_1 and n_2, respectively, their sample proportions $\hat{p}_1$ and $\hat{p}_2$ are (approximately) normally distributed with means p_1 and p_2 and standard deviations $\sqrt{p_1(1-p_1)/n_1}$ and $\sqrt{p_2(1-p_2)/n_2}$ respectively. Using these results, we can make the following statements about the shape of the sampling distribution of $\hat{p}_1 - \hat{p}_2$ and its mean and standard deviation. The mean and standard deviation of the sampling distribution of $\hat{p}_1 - \hat{p}_2$ are denoted by $\mu_{\hat{p}_1 - \hat{p}_2}$ and $\sigma_{\hat{p}_1 - \hat{p}_2}$ respectively.

For two large and independent samples, the sampling distribution of $\hat{p}_1 - \hat{p}_2$ is (approximately) normal with its mean and standard deviation as

$$\mu_{\hat{p}_1 - \hat{p}_2} = p_1 - p_2$$

and

$$\sigma_{\hat{p}_1 - \hat{p}_2} = \sqrt{\frac{p_1(1-p_1)}{n_1} + \frac{p_2(1-p_2)}{n_2}}$$

Thus, to construct a confidence interval and test a hypothesis about $p_1 - p_2$ for large and independent samples, we use the normal distribution. As was indicated in Chapter 13, in the case of proportion, the sample is large if np and $n(1-p)$ are both greater than or equal to 10. However, in the case of two samples, both sample sizes will be large if $n_1\hat{p}_1$, $n_2\hat{p}_2$, $n_1(1-\hat{p}_1)$, and $n_2(1-\hat{p}_2)$ are all greater than 5.

Interval Estimation of $p_1 - p_2$.   The difference between two sample proportions $\hat{p}_1 - \hat{p}_2$ is the point estimator for the difference between two population proportions $p_1 - p_2$. Because we do not know p_1 and p_2 when we are constructing a confidence interval for $p_1 - p_2$, we cannot calculate the value of $\sigma_{\hat{p}_1 - \hat{p}_2}$. Therefore, we use $s_{\hat{p}_1 - \hat{p}_2}$ as the point estimator of $\sigma_{\hat{p}_1 - \hat{p}_2}$ in the interval estimation. We construct the confidence interval for $p_1 - p_2$ using the formula

$$(\hat{p}_1 - \hat{p}_2) \pm z\, s_{\hat{p}_1 - \hat{p}_2}$$

where z is obtained from the standard normal table for the given confidence level and the standard error of $\hat{p}_1 - \hat{p}_2$, $s_{\hat{p}_1 - \hat{p}_2}$, is calculated as

$$s_{\hat{p}_1 - \hat{p}_2} = \sqrt{\frac{\hat{p}_1(1-\hat{p}_1)}{n_1} + \frac{\hat{p}_2(1-\hat{p}_2)}{n_2}}$$

Example 17.14 describes the procedure to make a confidence interval for the difference between two population proportions for large samples.

Example 17.14   A business consultant agency wanted to estimate the difference between the percentages of users of two toothpastes who will never switch to another toothpaste. In a sample of 500 users of toothpaste A, 20% said that they will never switch to another toothpaste. In another sample of 400 users of toothpaste B, 13% said that they will never switch to another toothpaste.

1. Let p_1 and p_2 be the proportions of all users of toothpastes A and B, respectively, who will never switch to another toothpaste. What is the point estimate of $p_1 - p_2$?
2. Construct a 95% confidence interval for the difference between the proportions of users of the two toothpastes who will never switch.

Solution   Let p_1 and p_2 be the proportions of all users of toothpastes A and B, respectively, who will never switch to another toothpaste, and let $\hat{p}_1$ and $\hat{p}_2$ be the respective sample proportions. The following information is given:

For toothpaste A:	$n_1 = 500$	$\hat{p}_1 = .20$
For toothpaste B:	$n_2 = 400$	$\hat{p}_2 = .13$

1. The point estimate of $p_1 - p_2$ is $\hat{p}_1 - \hat{p}_2 = .20 - .13 = .07$.
2. The values of $n_1 \hat{p}_1$, $n_2 \hat{p}_2$, $n_1(1-\hat{p}_1)$, and $n_2(1-\hat{p}_2)$ are

$$n_1 \hat{p}_1 = 500(.20) = 100 \qquad n_1(1-\hat{p}_1) = 500(.80) = 400$$
$$n_2 \hat{p}_2 = 400(.13) = 52 \qquad n_2(1-\hat{p}_2) = 400(.87) = 348$$

All of these values are 5 or more, so we can use the normal distribution to make a confidence interval for $p_1 - p_2$.

The standard error of $\hat{p}_1 - \hat{p}_2$ is

$$s_{\hat{p}_1 - \hat{p}_2} = \sqrt{\frac{\hat{p}_1(1-\hat{p}_1)}{n_1} + \frac{\hat{p}_2(1-\hat{p}_2)}{n_2}} = \sqrt{\frac{(.20)(.80)}{500} + \frac{(.13)(.87)}{400}} = .0246$$

The z value for a 95% confidence interval is 1.96. The 95% confidence interval for $p_1 - p_2$ is

$$(\hat{p}_1 - \hat{p}_2) \pm z\, s_{\hat{p}_1 - \hat{p}_2} = (.20 - .13) \pm 1.96(.0246) = .07 \pm .0482 = (.0218, .1182)$$

Thus, we can say that we are 95% confident that the difference between the two population proportions is between .0218 and .1182 or between 2.18% and 11.82%.   ■

Hypothesis Testing About $p_1 - p_2$.   In this section we learn how to test a hypothesis about $p_1 - p_2$ for two large and independent samples. The procedure involves the same five steps that we used previously. Once again, we calculate the standard deviation of $\hat{p}_1 - \hat{p}_2$ as

$$\sigma_{\hat{p}_1 - \hat{p}_2} = \sqrt{\frac{p_1(1-p_1)}{n_1} + \frac{p_2(1-p_2)}{n_2}}$$

When a test of hypothesis about $p_1 - p_2$ is performed, usually the null hypothesis is $p_1 = p_2$ and the values of p_1 and p_2 are not known. Assuming that the null hypothesis is true and $p_1 = p_2$, a common value of $\hat{p}_1$ and $\hat{p}_2$, denoted by $\hat{p}_c$, is calculated by using one of the following formulas:

$$\hat{p}_c = \frac{x_1 + x_2}{n_1 + n_2} \quad \text{or} \quad p_c = \frac{n_1 \hat{p}_1 + n_2 \hat{p}_2}{n_1 + n_2}$$

Which of these formulas is used depends on whether the values of x_1 and x_2 or the values of $\hat{p}_1$ and $\hat{p}_2$ are known. Note that x_1 and x_2 are the number of elements in each of the two samples that possess a certain characteristic. This value of p_c is called the *pooled sample proportion*. Using the value of the pooled sample proportion, we compute the standard error of $\hat{p}_1 - \hat{p}_2$, under the assumption $p_1 = p_2$, as

$$s_{\hat{p}_1 - \hat{p}_2} = \sqrt{\hat{p}_c(1 - \hat{p}_c)\left(\frac{1}{n_1} + \frac{1}{n_2}\right)}$$

The value of the test statistic z for $\hat{p}_1 - \hat{p}_2$ is calculated as

$$z = \frac{(\hat{p}_1 - \hat{p}_2) - (p_1 - p_2)}{s_{\hat{p}_1 - \hat{p}_2}}$$

The value of $p_1 - p_2$, which is typically zero, is given in the null hypothesis. Examples 17.15 and 17.16 illustrate the procedure to test hypotheses about the difference between two population proportions for large samples.

Example 17.15   Reconsider Example 17.14 regarding the percentages of users of two toothpastes who will never switch to another toothpaste. At the 1% significance level, can we conclude that the proportion of users of toothpaste A who will never switch to another toothpaste is greater than the proportion of users of toothpaste B who will never switch to another toothpaste?

Solution   Let p_1 and p_2 be the proportions of all users of toothpastes A and B, respectively, who will never switch to another toothpaste and let $\hat{p}_1$ and $\hat{p}_2$ be the corresponding sample proportions. The given information is

For toothpaste A:	$n_1 = 500$	$\hat{p}_1 = .2$
For toothpaste B:	$n_2 = 400$	$\hat{p}_2 = .13$

Step 1. Select the type of test to use and check the underlying conditions.

As we discussed in Example 17.14, these data correspond to two independent samples. Furthermore, we showed that the sample sizes are large enough to use the normal distribution for performing the test.

Step 2. State the null and alternative hypotheses.

Note that we are testing to find whether the proportion of users of toothpaste A who will never switch to another toothpaste is greater than the proportion of users of toothpaste B who will never switch to another toothpaste. If there is no difference between the two proportions, then $p_1 - p_2 = 0$. If the proportion for toothpaste A is larger, then $p_1 > p_2$ or $p_1 - p_2 > 0$. We write the null and alternative hypotheses as follows.

$$H_0: p_1 - p_2 = 0 \quad \text{(there is no difference in the proportions)}$$
$$H_1: p_1 - p_2 > 0 \quad \text{(the proportion for toothpaste A is larger)}$$

Step 3. Determine the significance level α.

The significance level given to be .01; hence, is $\alpha = .01$.

Step 4. Calculate the test statistic and the *p*-value.

The pooled sample proportion is

$$\hat{p}_c = \frac{n_1 \hat{p}_1 + n_2 \hat{p}_2}{n_1 + n_2} = \frac{500(.20) + 400(.13)}{500 + 400} = .1689$$

The value of the standard error is

$$s_{\hat{p}_1 - \hat{p}_2} = \sqrt{\hat{p}_c(1 - \hat{p}_c)\left(\frac{1}{n_1} + \frac{1}{n_2}\right)} = \sqrt{.1689(.8311)\left(\frac{1}{500} + \frac{1}{400}\right)} = .0251$$

The value of the test statistic z for $\hat{p}_1 - \hat{p}_2$ is

$$z = \frac{(\hat{p}_1 - \hat{p}_2) - (p_1 - p_2)}{s_{\hat{p}_1 - \hat{p}_2}} = \frac{(.20 - .13) - 0}{.0251} = 2.79$$

Since the alternative hypothesis is right-tailed, the *p*-value is $P(z \geq 2.79) = .0026$, as is shown in Figure 17.5.

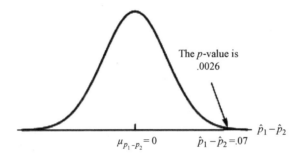

The *p*-value is .0026

$\mu_{p_1 - p_2} = 0$ $\hat{p}_1 - \hat{p}_2 = .07$ $\hat{p}_1 - \hat{p}_2$

Figure 17.5   The *p*-value for Example 17.14.

Step 5. Make a decision.

The significance level is .01. The *p*-value, .0026, is less than .01. Therefore, we reject the null hypothesis. Our conclusion is that there is enough evidence to state that the proportion of toothpaste A users who will never switch is larger than the proportion for toothpaste B.   ■

Example 17.16   A company is planning to buy a few machines. It is considering two types of machines but will buy all machines of the same type. The company selected one machine of each type and used it for production for a few days. A sample of 800 items produced on machine 1 showed 48 of them to be defective. Another sample of 900 items produced on machine 2 showed that 45 of them were defective. Testing at the 1% significance level, can we conclude, from the information on these samples, that the proportions of defective items produced on the two machines are different?

Solution   Let p_1 and p_2 be the proportions of defective items in all items produced by machines 1 and 2, respectively. Let $\hat{p}_1$ and $\hat{p}_2$ be the corresponding sample proportions. The given information is

For machine 1:	$n_1 = 800$	$x_1 = 48$
For machine 2:	$n_2 = 900$	$x_2 = 45$

Step 1. Select the type of test to use/check underlying conditions.

The data come from two independent samples. Furthermore, the check for large samples results in the values 48, 752, 45, and 855, for $n_1\hat{p}_1$, $n_1(1-\hat{p}_1)$, $n_2\hat{p}_2$, and $n_2(1-\hat{p}_2)$, respectively, which are all greater than 5, so the sample sizes are large enough to use the normal distribution for performing the test.

Step 2. State the null and alternative hypotheses.

Note that we are testing to find whether the proportion of defective items manufactured on machine 1 is different from the proportion for machine 2. If there is no difference between the proportions, then $p_1-p_2 = 0$. If the proportions are different, then $p_1 \neq p_2$, or $p_1-p_2 \neq 0$. We write the null and alternative hypotheses as follows:

H_0: $p_1-p_2 = 0$ (there is no difference in the two proportions)
H_1: $p_1-p_2 \neq 0$ (the proportions of defectives are different)

Step 3. Determine the significance level α.

The significance level is given to be .01; hence, $\alpha = .01$.

Step 4. Calculate the test statistic and the p-value.

The values of the sample proportions are

$$\hat{p}_1 = \frac{48}{800} = .06 \quad \text{and} \quad \hat{p}_2 = \frac{45}{900} = .05$$

The pooled sample proportion is

$$\hat{p}_c = \frac{x_1+x_2}{n_1+n_2} = \frac{48+45}{800+900} = .0547$$

The value of the standard error is

$$s_{\hat{p}_1-\hat{p}_2} = \sqrt{\hat{p}_c(1-\hat{p}_c)\left(\frac{1}{n_1}+\frac{1}{n_2}\right)} = \sqrt{.0547(.9453)\left(\frac{1}{800}+\frac{1}{900}\right)} = .0110$$

The value of the test statistic z for $\hat{p}_1-\hat{p}_2$ is

$$z = \frac{(\hat{p}_1-\hat{p}_2)-(p_1-p_2)}{s_{\hat{p}_1-\hat{p}_2}} = \frac{(.06-.05)-0}{.0110} = .91$$

Since the alternative hypothesis is two-tailed, the p-value is $2P(z \geq .91) = 2(.1814) = .3628$, as is shown in Figure 17.6.

Step 5. Make a decision.

The significance level is .01. The p-value, .3628, is greater than .01. Therefore, we fail to reject the null hypothesis. Our conclusion is that there is not enough evidence to state that the proportion of defective items in all items made by machine 1 is different from the proportion for machine 2.   ∎

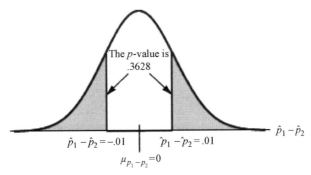

Figure 17.6   The p-value for Example 17.15.

GLOSSARY

d   The difference between two matched values in two samples collected from the same source. It is called the *paired difference*.

$\bar{d}$ The mean of the paired differences for a sample.

independent samples   Two samples drawn from two populations such that the selection of one does not affect the selection of the other.

Mann–Whitney test   A distribution-free test that is used for comparing the medians of two independent populations. The Mann–Whitney test is commonly used when the normality assumption of the independent samples t test is unreasonable.

nonpooled t test   A test used to compare the means of two independent simple random samples. The nonpooled t test is used when the equal variance assumption is unreasonable.

paired or matched samples   Two samples drawn in such a way that they include the same elements and two data values are obtained from each element, one for each sample. Also called *dependent samples*.

pooled standard deviation s_p   Estimate of the common population standard deviation when the assumption $\sigma_1 = \sigma_2$ is reasonable. The assumption is considered to be reasonable when $\max(s_1, s_2)/\min(s_1, s_2) \leq 2$.

pooled t test   A test used to compare the means of two independent simple random samples. The pooled t test is used when the equal variance assumption is reasonable.

ranks   Values denoting the relative ordering of the observations in one or more datasets. The smallest observation is assigned a rank of 1, while the largest observation is assigned a rank equal to the number of observations in the dataset(s).

signed ranks   Values denoting the sign and relative ordering of the absolute value of the differences in paired samples. Differences that are closer to zero, regardless of sign, will have lower ranks, while differences that are farther from zero, regardless of sign, will have higher ranks.

S_1 and S_2   The sums of the ranks of the observations in samples 1 and 2, respectively.

U   The test statistic used in the Mann–Whitney test.

U_c   The critical value in a Mann–Whitney test with a one-sided alternative hypothesis.

U_{lower} and U_{upper}   The lower and upper critical values in a Mann–Whitney test with a two-sided alternative hypothesis

Wilcoxon signed–rank test   A distribution-free test that is used for making an inference about the median of the paired differences in a populations. The Wilcoxon signed-rank test is commonly used when the normality assumption of the paired t test is unreasonable.

W_+ **and** W_- The sums of the ranks of the positive paired differences and negative paired differences, respectively, in a Wilcoxon signed-rank test.

W The test statistic in the Wilcoxon signed-rank test. W will be the value of W_+ or W_-, depending on the alternative hypothesis.

W_c The critical value in the Wilcoxon signed-rank test.

μ_d The mean of the paired differences for the population.

s_d The standard deviation of the paired differences for a sample.

σ_d The standard deviation of the paired differences for the population.

EXERCISES

17.1   A consulting firm was asked by a large insurance company to investigate whether business majors were better salespersons. A sample of 40 salespersons with a business degree showed that they sold an average of 10 insurance policies per week with a standard deviation of 1.80. Another sample of 45 salespersons with a degree other than business showed that they sold an average of 8.5 insurance policies per week with a standard deviation of 1.35. Neither sample of data contained any outliers.

  (a) Construct a 99% confidence interval for the difference between the two population means.

  (b) Using the 1% significance level, can you conclude that persons with a business degree sell more insurance policies, on average, than those who have a degree in another area?

17.2   A business consultant wanted to investigate whether providing daycare facilities on premises by companies reduces the absentee rate of working mothers with 6-year-old or younger children. She took a sample of 45 such mothers from companies that provide daycare facilities on the premises. These mothers missed an average of 6.4 days from work last year with a standard deviation of .97 days. Another sample of 50 such mothers taken from companies that do not provide daycare facilities on the premises showed that these mothers missed an average of 9.3 days last year with a standard deviation of 1.96 days.

  (a) Construct a 98% confidence interval for the difference between the two population means.

  (b) Test at the 2.5% significance level if the mean number of days missed per year by mothers working for companies that provide daycare facilities on the premises is less than the mean number of days missed per year by mothers working for companies that do not provide daycare facilities on the premises.

17.3   According to a recent survey, the average yield of corn per acre is 132.8 bushels (bu) for Ohio and 140.6 bu for Iowa. Assume that these two results are based on random samples of 25 acres from Ohio and 28 acres from Iowa. Further assume that the sample standard deviations for the two states are 5.3 and 7.7 bu, respectively, and that the datasets are neither extremely skewed nor contain any outliers.

  (a) Construct a 95% confidence interval for the difference between the two population means.

  (b) Test at the 2.5% significance level assuming the mean yield of corn per acre for Ohio is lower than that for Iowa.

17.4   A grocery store chain decided to use two different discount styles for advertising the same discount. Style 1 was "50% discount" and style 2 was "2 for the price of 1." The "50% discount" promotion was run for one month at a randomly selected 11 store locations, while the "2 for the price of 1" promotion was run for one month at nine randomly selected store locations. The variable measured was the increase in profit from the sale of the product compared to the month prior to the promotion, in hundreds of dollars. The following table shows the increase in profit (in thousand dollars) for each style.

50% discount	2.7	3.1	2.5	3.3	2.1	3.5	3.0	2.6	2.5	3.1	3.3
2 for the price of 1	2.4	2.8	2.2	2.5	2.4	2.2	2.9	2.5	1.9		

Perform the Mann–Whitney test to determine whether the median increase in profit for the 50% discount wording style is different from the median increase in profit for the 2 for the price of 1 wording style. Use a significance level of 5%.

17.5 A local bank usually has three tellers working at any given time. They are interested in determining the effect of adding another teller on the amount of time the first person in line has to wait to be called to begin service. A consulting firm measured the amount of time that the first person in line had to wait to begin service for eight randomly selected customers while three tellers were working, as well as the amount of time for eight randomly selected customers while four tellers were working. The times, given in the following table, were measured in seconds:

3 tellers	13	32	27	65	21	40	52	44
4 tellers	25	9	36	24	44	18	16	31

Perform the Mann–Whitney test to determine whether the median waiting time is shorter when there are four tellers working. Use a significance level of 5%.

17.6 A random sample of eight employees was selected to test for the effectiveness of hypnosis on their job performances. The following table gives the job performance ratings (on a scale of 1–4, where 1 is the lowest and 4 is the highest) before and after these employees tried hypnosis. Assume that the population of paired differences is (approximately) normally distributed.

Before	2.3	2.8	3.1	2.7	3.4	2.6	2.8	2.5
After	2.6	3.2	3.0	3.5	3.7	2.4	2.9	2.9

(a) Construct a 99% confidence interval for the mean μ_d of the population paired differences where a paired difference is given by the job performance rating of an employee before trying hypnosis minus the job performance rating of an employee after trying hypnosis.

(b) Test at the 5% significance level if there is an improvement in the average job performance of employees due to hypnosis.

(c) Perform the Wilcoxon signed-rank test on the data to answer the question with regard to the median improvement. Are the conclusions consistent?

17.7 Many students suffer from math anxiety. A professor who teaches business statistics offered her students a 2-hour lecture on math anxiety and ways to overcome it. The following table gives the scores of seven students in business statistics before and after they attended this lecture; assume that the population of paired differences is (approximately) normally distributed.

Before	56	69	48	74	65	71	58
After	62	73	44	85	71	70	69

(a) Construct a 99% confidence interval for the mean μ_d of the population paired differences where a paired difference is given by the score before attending this lecture minus the score after attending this lecture.

(b) Test at the 1% significance level whether attending this lecture increases the average score in business statistics.

(c) Perform the Wilcoxon signed-rank test on the data to answer the question with regard to the median improvement. Are the conclusions consistent?

17.8 According to a recent survey, 66% of households headed by single women and 81.9% of households headed by single men own cars. Assume that these estimates are based on random samples of 1640 households headed by single women and 1800 households headed by single men.

(a) Determine a 99% confidence interval for the difference between the two population proportions.

(b) At the 1% significance level, can you conclude that the proportion of households headed by single women who own cars is less than the proportion of households headed by single men who own cars?

17.9 A company has two restaurants in two different areas of New York City. The company wants to estimate the percentages of patrons who regard the food and service at each of these restaurants as excellent. A sample of 200 patrons selected from the restaurant in area A showed that 114 of them consider the food and service to be excellent at this restaurant. Another sample of 250 patrons selected from the restaurant in area B showed that 155 of them consider the food and service excellent at this restaurant.

(a) Construct a 97% confidence interval for the difference between the two population proportions.

(b) Testing at the 2.5% significance level, can you conclude that the proportion of patrons at the restaurant in area A who consider the food and service to be excellent is lower than the corresponding proportion at the restaurant in area B?

17.10 The management of a supermarket wanted to investigate whether the percentages of men and women who prefer to buy national-brand products over the store-brand products are different. A sample of 600 men shoppers at the company's supermarkets showed that 246 of them prefer to buy national-brand products over the store-brand products. Another sample of 700 women shoppers at the company's supermarkets showed that 266 of them prefer to buy national-brand products over the store brand products.

(a) What is the point estimate of the difference between the two population proportions? Construct a 95% confidence interval for the difference between the proportions of all men and all women shoppers at these supermarkets who prefer to buy national-brand products over the store-brand products.

(b) Testing at the 5% significance level, can you conclude that the proportions of all men and all women shoppers at these supermarkets who prefer to buy national-brand products over the store-brand products are different?

17.11 Airbags for automobiles manufactured by two different companies were tested to determine whether these bags deploy when these autos are involved in a collision at 20 mph (miles per hour). Of the 50 cars tested with airbags made by company A, 37 deployed when involved in a collision at 20 mph. Of the 60 cars tested with airbags made by company B, 49 deployed when involved in a collision at 20 mph.

(a) Make a 99% confidence interval for the difference between the two population proportions.

(b) Testing at the 5% significance level, determine whether there is significant evidence of a difference in the proportion of all airbags made by company A that deploy in a 20-mph collision and the same proportion for company B.

18 Chi-Square Tests

18.1 OVERVIEW

In Section 14.5 we introduced the chi-square distribution for the purpose of developing a confidence interval for a population variance. Like many other distributions, the chi-square distribution is the basis for the sampling distribution for a variety of test statistics. As you may recall, the chi-square distribution has only one parameter, the degree(s) of freedom (df). The shape of a specific chi-square distribution depends on the degrees of freedom, which are calculated by using different formulas for different tests. The random variable χ^2 assumes nonnegative values only. Figure 18.1 re-creates Figure 14.5, demonstrating the shapes of chi-square distributions with 2, 5, and 10 degrees of freedom, respectively.

As we can see from Figure 18.1, the shape of a chi-square distribution curve is skewed to the right for very small degrees of freedom and changes drastically as the degrees of freedom increase. Eventually, for large degrees of freedom, the chi-square distribution curve looks like a normal distribution curve. The peak (or mode) of a chi-square distribution curve with 1 or 2 degrees of freedom occurs at zero and for a curve with $df \geq 3$ at $df - 2$. For instance, the peak of the chi-square distribution curve with $df = 2$ in Figure 18.1 occurs at zero. The peak for the curve with $df = 5$ occurs at $5 - 2 = 3$. Finally, the peak for the curve with $df = 10$ occurs at $10 - 2 = 8$. Like all other continuous distribution curves, the total area under a chi-square distribution curve is 1.0.

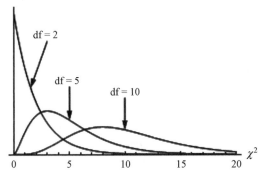

Figure 18.1 Three chi-square density curves.

If we know the degrees of freedom and the area in the right tail of a chi-square (χ^2) distribution, we can find the value of χ^2 from Table IV of Appendix A. Section 14.5 of Chapter 14 discussed how to read this table. Readers can refer to that section and Example 14.10.

In this chapter, we will study two types of tests a goodness-of-fit-test and a test of independence and homogeneity. Both these tests are performed by using the χ^2 distribution.

Practitioner's Guide for Statistics and Lean Six Sigma for Process Improvements. By Mikel J. Harry, Prem S. Mann, Ofelia C. de Hodgins, Christopher J. Lacke, and Richard Hulbert
Copyright © 2010 John Wiley & Sons, Inc.

18.2   A GOODNESS-OF-FIT TEST

This section explains how to make tests of hypotheses about experiments with more than two possible outcomes (or categories). Such experiments, called multinomial experiments, possess four characteristics. Note that a binomial experiment is a special case of a multinomial experiment.

An experiment with the following characteristics is called a *multinomial experiment*:

1. It consists of n identical trials (repetitions).
2. Each trial results in one of the k possible events.
3. The trials are independent.
4. The probabilities of the various events remain constant for each trial.

An experiment of many rolls of a die is an example of a multinomial experiment. It consists of many identical rolls (trials); each roll (trial) results in one of the six possible outcomes; each roll is independent of the other rolls; and the probabilities of the six outcomes remain constant for each roll.

As a second example of a multinomial experiment, let us select a random sample of people and ask them whether or not the quality of American cars is better than that of Japanese cars. The response of a person can be "Yes," "No," or "I don't know." Each person included in the sample can be considered as one trial (repetition) of the experiment. There will be as many trials for this experiment as the number of persons selected. Each person can belong to any of the three categories: "Yes" (Y), "No" (N), or "[I] don't know" (D). The response of each selected person is independent of the responses of other persons. Given that the population is large relative to the sample size, the probabilities of a person belonging to the three categories effectively remain the same. Consequently, this is an example of a multinomial experiment.

The frequencies obtained from the actual performance of an experiment are called the *observed frequencies*. In a *goodness-of-fit test*, we test the null hypothesis that the observed frequencies for an experiment follow a certain pattern or theoretical distribution. The test is called a goodness-of-fit test because the hypothesis tested is how well the observed frequencies fit a given pattern.

For our first example involving the experiment of many rolls of a die, we may test the null hypothesis that the given die is fair. The die will be fair if the observed frequency for each outcome is close to one-sixth of the total number of rolls.

For our second example involving opinions of people on the quality of American cars, suppose that such a survey was conducted in 2004 and in that survey 41% were Y, 48% were N, and 11% were D. We want to test if the current percentages match the 2004 results. Suppose that we take a random sample of 1000 adults and observe that 536 of them think that the quality of American cars is better than that of Japanese cars, 362 say it is worse, and 102 have no opinion. The frequencies 536, 362, and 102 are called the observed frequencies. These frequencies are obtained by actually performing the survey. Now, assuming that the 2004 percentages are still true (which will be our null hypothesis), in a sample of 1000 adults we will expect 410 Y, 480 N, and 110 D responses. These frequencies are obtained by multiplying the sample size (1000) by the 2004 proportions. These frequencies are called the *expected frequencies*. Then, we will make a decision to reject or not to reject the null hypothesis according to how large the difference between the observed frequencies and the expected frequencies is. To perform this test, we will use the chi-square distribution. Note that in this case, we are testing the null hypothesis that all three percentages (or proportions) are unchanged. However, if we want to make a test for only one of the three proportions, we use the procedures learned in Section 16.4. For example, if we are testing the hypothesis that the percentage of people who think the quality of American cars is better than that of the Japanese cars is different from 41%, then we will test the null hypothesis H_0: $p = .41$ against the alternative hypothesis H_1: $p \neq .41$. This test will be conducted using the procedures discussed in Section 16.4.

As mentioned earlier, the frequencies obtained from the performance of an experiment are called the observed frequencies. They are denoted by O_i. To make a goodness-of-fit test, we calculate the expected frequencies for all categories of the experiment. The expected frequency for category i, denoted by E_i, is given by the product of n and p_i, where n is the total number of trials and p_i is the probability for category i.

The degrees of freedom for a goodness-of-fit test are

$$df = k-1,$$

where k equals the number of possible categories in the experiment.

The test statistics for a goodness-of-fit test is χ^2, and its value is calculated as

$$\chi^2 = \sum \frac{(O_i - E_i)^2}{E_i}$$

where O_i =observed frequency for category i and E_i = expected frequency for category $i = np_i$.

The procedure for performing a goodness-of-fit test involves the same five steps that were used in the preceding chapters. *The chi-square goodness-of-fit test is always a right-tailed test.*

Whether or not the null hypothesis is rejected depends on how much the observed and expected frequencies differ from each other. To find how large the difference between the observed frequencies and the expected frequencies is, we do not look only at $O_i - E_i$ because some of the $O_i - E_i$ values will be positive and others will be negative. The net result of the sum of these differences will always be zero. Therefore, we square each of the $O_i - E_i$ values to obtain $(O_i - E_i)^2$ and then weight them according to the reciprocals of their expected frequencies. The sum of the resulting numbers gives the computed value of the test statistic χ^2.

To perform a goodness-of-fit test, *the sample size should be large enough so that the expected frequency for each category is at least 5*. If there is a category with an expected frequency of less than 5, we can increase the sample size or combine two or more categories to make each expected frequency at least 5.

Examples 18.1 and 18.2 describe the procedure for performing goodness-of- fit tests using the chi-square distribution.

Example 18.1   A large shopping mall in New York City has five department stores. A business consultant to one of these stores was asked by the management to investigate whether the proportion of visitors to this shopping mall who prefer each of the five department stores is the same. The consultant selected a random sample of 1000 visitors to the shopping mall and obtained the frequencies listed in the following table. The frequency for each store gives the number of persons in the sample who expressed a preference for that store.

Store	FloweringEarnhardts	KD Nickel	Old King's	Parades 'R' Us	Prophets
Frequency	214	231	182	219	154

Using the 1% significance level, test the null hypothesis that the proportion of visitors to this shopping mall who prefer each of the five department stores is the same.

Solution   To perform this test, we proceed as follows:

Step 1. Select the type of test to use and check the underlying conditions.

In order to perform a goodness-of-fit test, we need to ensure that the conditions of a multinomial experiment are satisfied. In this case

  a. Here, $n = 1000$ persons, and these persons were asked the same question.

  b. Each person may respond with one of the five stores, so that $K = 5$ possible outcomes.

c. All trials (persons) are independent.

d. The population is large. Hence, we can assume that the probabilities of the five outcomes are constant.

The conditions of a multinomial experiment have been met. Next, in order to use the goodness-of-fit test, the expected counts must all be greater than or equal to 5. Since we are testing to see if the five stores are equally preferable, then $p_i = \frac{1}{5} = .20$ for each store. Thus, the expected frequency/count for each store will be $np_i = 1000 \times .20 = 200$. Since the expected count requirement has been met, we can use a goodness-of-fit test for this experiment.

Step 2. State the null and alternative hypotheses.

We are to test whether the five stores are equally preferable. The null and alternative hypotheses are as follows:

$$H_0 : p_1 = p_2 = p_3 = p_4 = p_5 = .20 \quad \text{(all five stores are equally preferable)}$$
$$H_1 : \text{at least two of the } p_i \neq .20 \quad \text{(not all five stores are equally preferable)}$$

Step 3. Determine the significance level α.

The significance level is given to be .01; hence, $\alpha = .01$.

Step 4. Calculate the test statistic and the p-value.

Since there are five categories (stores), we have

$$\text{df} = k-1 = 5-1 = 4$$

The value of the test statistic is calculated as follows:

$$\chi^2 = \sum \frac{(O_i - E_i)^2}{E_i} = \frac{(214 - 200)^2}{200} + \frac{(231 - 200)^2}{200} + \frac{(182 - 200)^2}{200}$$
$$+ \frac{(219 - 200)^2}{200} + \frac{(154 - 200)^2}{200} = 19.790$$

All the required calculations to find the value of the test statistic χ^2 are shown in Table 18.1. The p value is given by $P(\chi^2 \geq 19.790)$. From the chi-square distribution table (Table IV of Appendix A), $P(\chi^2 \geq 14.860) = .005$, therefore the $P(\chi^2 \geq 19.790) < .005$. The region corresponding to the p-value is shown in Figure 18.2.

TABLE 18.1   Detailed Calculations of χ^2 for Example 18.1

Store	Observed Frequency O_i	p_i	Expected Frequency $E_i = np_i$	$O_i - E_i$	$(O_i - E_i)^2$	$\dfrac{(O_i - E_i)^2}{E_i}$
FloweringEarnhardts	214	.20	1000(.20) = 200	14	196	.980
KD Nickel	231	.20	1000(.20) = 200	31	961	4.805
Old King's	182	.20	1000(.20) = 200	−18	324	1.620
Parades 'R' Us	219	.20	1000(.20) = 200	19	361	1.805
Prophets	154	.20	1000(.20) = 200	−46	2116	10.580
Total	1000	1.00	1000			19.790

Step 5. Make a decision.

The significance level is .01. Using the p value, we ask whether the p value is $\leq \alpha$. Since the p value is less than .005, it is less than .01. Therefore, we reject the null hypothesis. Our conclusion is that the stores are not equally preferable.

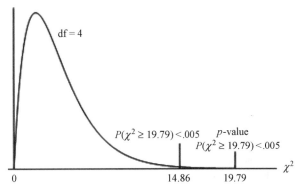

Figure 18.2   The p-value for Example 18.1.   ■

Example 18.2   The following table gives the January–July 2006 US auto market shares held by various auto companies. (*Source*: http://www.automotivedigest.com/research/research_results.asp? sigstats_id=1176.) *Market shares* are usually defined as percentages of the total revenue earned by various companies. Suppose for convenience that in this case the market shares are synonymous with percentages of the total new autos sold by various auto companies in 2006.

Company	DaimlerChrysler	Ford	GM	Honda	Toyota	Others
Market share	14.4%	18.1%	24.8%	14.1%	14.9%	13.7%

A business organization wanted to investigate if the current US auto market shares of these companies are the same as in July 2006. A sample of 2000 recently sold new autos showed that 270 were manufactured by DaimlerChrysler, 382 by Ford, 467 by General Motors, 317 by Honda, 288 by Toyota, and 276 were manufactured by other companies. Testing at the 2.5% significance level, will you reject the null hypothesis that the current US auto market shares held by these companies are the same as for the year-to-date figures for July 2006?

Solution   To perform this test, we proceed as follows:

Step 1. Select the type of test to use and check then underlying conditions.

In order to perform a goodness-of-fit test, we need to check to see if the conditions of a multinomial experiment are satisfied. In this case, the following conditions apply.

a. Here, $n = 2000$, which gives the number of new autos sold.

b. Each car will fall under one of the five named companies or the "*others*" group. Thus, $k = 6$ possible outcomes.

c. All 2000 trials (autos) are independent.

d. Because the population of autos sold is large, probabilities can be considered constant.

The conditions of a multinomial experiment have been met. Next, in order to use the goodness-of-fit test, each of the expected counts must be greater than or equal to 5. Since the assumed probabilities vary for the different companies, we need to make sure that the smallest expected count is greater than or equal to 5. The group with the lowest market share is "others," for which $p_i = .137$. The expected count for "others" is $np_6 = 2000 \times .137 = 274$. This is well above 5, so all of the other expected counts will also be well above 5. Since the expected count requirement has been met, we can use a goodness-of-fit test for this experiment.

Step 2. State the null and alternative hypotheses.

We are to test whether or not the current market shares are the same as the ones for January–July 2006. Labeling the five brands 1–5, in alphabetical order, and labeling the "others" group as category 6, the values in the null hypothesis correspond to the given market shares. The null and alternative hypotheses are as follows:

$$H_0 : p_1 = .144, \; p_2 = .181, \; p_3 = .248, \; p_4 = .141, \; p_5 = .149, \; p_6 = .137$$
$$H_1 : \text{At least two of the equalities in } H_0 \text{ do not hold true}$$

Step 3. Determine the significance level α.

The significance level is given as 2.5%, hence, $\alpha = .025$.

Step 4. Calculate the test statistic and the p-value.

Since there are six categories,

$$df = k-1 = 6-1 = 5.$$

The value of the test statistic is calculated as follows:

$$\chi^2 = \sum \frac{(O_i - E_i)^2}{E_i} = \frac{(270 - 288)^2}{288} + \frac{(382 - 362)^2}{362} + \frac{(467 - 496)^2}{496}$$
$$+ \frac{(317 - 282)^2}{282} + \frac{(288 - 298)^2}{298} + \frac{(276 - 274)^2}{274}$$
$$= 8.6210$$

The p-value is given by $P(\chi^2 \geq 8.621)$. From the chi-square distribution table (Table IV of Appendix A), $P(\chi^2 \geq 9.236) = .10$; therefore $P(\chi^2 \geq 8.621) > .10$. The region corresponding to the p-value is shown in Figure 18.3.

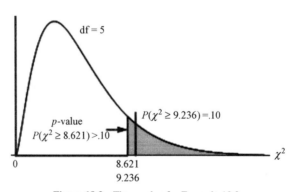

Figure 18.3   The p-value for Example 18.2.

All the required calculations to find the value of the test statistic χ^2 are shown in Table 18.2.

Step 5. Make a decision.

The significance level is .025. Using the p-value, we ask whether the p-value is $\leq \alpha$. Since the p-value is greater than .10, it is greater than .025. Therefore, we fail to reject the null hypothesis. Our conclusion is that the current market shares are comparable to the ones for January–July 2006.

TABLE 18.2 Detailed Calculations of χ^2 for Example 18.2

Store	Observed Frequency O_i	p_i	Expected Frequency $E_i = np_i$	$O_i - E_i$	$(O_i - E_i)^2$	$\dfrac{(O_i - E_i)^2}{E_i}$
DaimlerChrysler	270	.144	2000(.144) = 288	−18	324	1.125
Ford	382	.181	2000(.181) = 362	20	400	1.105
GM	467	.248	2000(.248) = 496	−29	841	1.696
Honda	317	.141	2000(.141) = 282	35	1225	4.344
Toyota	288	.149	2000(.149) = 298	−10	100	0.336
Others	276	.137	2000(.137) = 274	2	4	0.015
Total	2000	1.00	2000			8.621

18.3 CONTINGENCY TABLES

We often may have information on more than one variable for each element. Such information can be summarized and presented using a two-way classification table, which is also called a *contingency table* or *cross-tabulation*. Table 18.3 is an example of a contingency table. It gives information on a sample of male and female workers selected from a large company who are in favor of approving a new labor management contract, are against it, or have no opinion. Table 18.3 has two rows (one for male workers and the other for female workers) and three columns (corresponding to opinions representing the approval of the contract, against it, or no opinion, respectively). Hence, it is also called a 2 × 3 (read as "two by three") contingency table.

TABLE 18.3 Opinions of Workers on a Labor-Management Contract

	Support	Against	No Opinion
Female	87	32	6
Male	93	70	12

A contingency table can be of any size. For example, it can be 2 × 3, 3 × 2, 3 × 3, 4 × 2, or of any other dimension. Note that in these notations, the first digit refers to the number of rows in the table and the second digit refers to the number of columns. For example, a 3 × 2 table will contain three rows and two columns. In general, an $R \times C$ (row–column) table contains R rows and C columns.

Each of the six boxes that contain numbers in Table 18.3 is called a *cell*. The number of cells for a contingency table is obtained by multiplying the number of rows by the number of columns. Thus, Table 18.3 contains 2 × 3 = 6 cells. The subjects that belong to a cell of a contingency table possess two characteristics. For example, 12 workers listed in the third cell of the second row in Table 18.3 are *males* and *have no opinion*. The numbers entered in the cells are usually called the *joint frequencies*. For example, 12 workers belong to the joint category of *males* and *have no opinion*. Hence, it is the joint frequency of this category.

18.4 TESTS OF INDEPENDENCE AND HOMOGENEITY

This section is concerned with tests of independence and homogeneity, which are performed using the contingency tables. Except for a few modifications, the procedure used for such tests is almost the same as the one applied in Section 18.2 for a goodness-of-fit test.

18.4.1 Test of Independence

In a *test of independence* for a contingency table, we test the null hypothesis that the two attributes (characteristics) of the elements of a given population are not related (i.e., are independent) against the alternative hypothesis that the two characteristics are related (i.e., are dependent). For example, we may want to test whether the gender and opinions of workers about the labor management contract mentioned in Table 18.3 are dependent. We perform such a test by using the chi-square distribution. As another example, we may want to test whether or not an association exists between the job satisfaction index and the absentee rate of employees.

As we mentioned earlier, the methods used in a test of independence are similar to the methods used in a goodness-of-fit test. The three differences involve the hypotheses, the calculation of the number of degrees of freedom, and the method for calculating the expected frequencies. In the case of a test of independence, the null and alternative hypotheses are of the form

H_0: The two attributes are independent

H_1: The two attributes are dependent

The following formula is used to calculate the number of degrees of freedom for a test of independence

$$df = (R-1)(C-1)$$

where R and C are the numbers of rows and columns, respectively, in the contingency table.

In order to calculate the *expected frequencies*, we must first calculate the row and column totals for the contingency table. The procedure that is used to calculate the expected frequencies is described in Example 18.3.

Example 18.3 The labor union at a large auto manufacturing company has drafted a new labor management contract. Before this contract is presented to and negotiated with management, the labor union leaders want to seek opinions of workers on this proposed contract. The union leaders selected a random sample of 300 workers and asked their opinions. The two-way classification of the responses of these 300 workers was presented earlier in Table 18.3. Calculate the expected frequencies for this table assuming that the gender and opinions of workers are independent.

Solution Table 18.3 is reproduced here as Table 18.4. Note that we have added the row and column totals in Table 18.4. It is important to recognize that the row and column of totals do not count in the dimension of the table. Table 18.4 is still a 2 × 3 contingency table.

TABLE 18.4 Opinions of Workers on a Labor Management Contract

	Support (S)	Against (A)	No Opinion (N)	Row Totals
Female (F)	87	32	6	125
Male (M)	93	70	12	175
Column totals	180	102	18	300

The numbers 87, 32, 6, 93, 70, and 12 in Table 18.4 are called the *observed frequencies* for the six cells.

As mentioned earlier, the null hypothesis in a test of independence is that the two attributes (or classifications) are independent. In an independence test of hypothesis, to begin with we assume that the null hypothesis is true and that the two attributes are independent. Assuming that the null hypothesis is true and that the gender and opinions of workers are not related, the expected frequency for the cell corresponding to *female* and support is calculated as follows. From Table 18.4, we have

$$P(\text{a worker is a female}) = P(\text{F}) = \frac{125}{300}$$

$$P(\text{a worker supports the proposed contract}) = P(\text{S}) = \frac{180}{300}$$

Because we are assuming that F and S are independent (by assuming that the null hypothesis is true), the joint probability of these two events is

$$P(\text{F and S}) = P(\text{F}) \times P(\text{S}) = \frac{125}{300} \times \frac{180}{300}$$

Then, assuming that F and S are independent, the number of workers expected (E) to be female and support the proposed contract in a sample of 300 is

$$E(\text{F and S}) = 300 \times P(\text{F and S})$$
$$= 300 \times \frac{125}{300} \times \frac{180}{300} = \frac{125 \times 180}{300}$$
$$= \frac{(\text{row total})(\text{column total})}{\text{sample size}}$$

Thus, the rule for obtaining the expected frequency for a cell is to divide the product of the corresponding row and column totals by the sample size. To provide notation for the calculations, let

E_{ij} = expected frequency for the cell at the intersection of row i and column j

r_i = total for row i

c_j = total for column j

n = sample size

Then

$$E_{ij} = \frac{r_i \times c_j}{n}$$

Using this rule, we calculate the expected frequencies of the six cells as follows:

E for female *and* support = $E_{11} = (125 \times 180)/300 = 75$
E for female *and* against = $E_{12} = (125 \times 102)/300 = 42.5$
E for female *and* no opinion = $E_{13} = (125 \times 18)/300 = 7.5$
E for male *and* support = $E_{21} = (175 \times 180)/300 = 105$
E for male *and* against = $E_{22} = (175 \times 102)/300 = 59.5$
E for male *and* no opinion = $E_{23} = (175 \times 18)/300 = 10.5$

The expected frequencies are usually written in parentheses below the observed frequencies within the corresponding cells, as shown in Table 18.5.

Like a goodness-of-fit test, a test of independence is always right-tailed. To apply a χ^2 test of independence, the sample size should be large enough so that the expected frequency for each cell is at least 5. If the expected frequency for a cell is not at least 5, we either increase the sample size or combine some categories. Examples 18.4 and 18.5 describe the procedure for performing tests of independence using the chi-square distribution.

TABLE 18.5   Opinions of Workers on a Labor Management Contract

	Support (S)	Against (A)	No Opinion (N)	Row Totals
Female (F)	87 (75.0)	32 (42.5)	6 (7.5)	125
Male (M)	93 (105)	70 (59.5)	12 (10.5)	175
Column totals	180	102	18	300

∎

Example 18.4   Reconsider the two-way classification of 300 workers based on gender and opinions about the proposed labor management contract given in Table 18.3. Using the 2.5% significance level, can you conclude that the gender and opinions of workers are dependent?

Solution   The test involves the following five steps:

Step 1. Select the type of test to use and check the underlying conditions.

The data were collected using a simple random sample of 300 employees, analyzing two categorical variables. As shown in Table 18.5, the expected counts are all greater than or equal to 5, so we can perform a chi-square test of independence.

Step 2. State the null and alternative hypotheses:

$$H_0: \quad \text{Gender and opinions are independent}$$
$$H_1: \quad \text{Gender and opinions are dependent}$$

Step 3. Determine the significance level α.

The significance level is given as .025; hence, $\alpha = .05$.

Step 4. Calculate the test statistic and the *p*-value.

Since $R = 2$ and $C = 3$, it follows that

$$\text{df} = (R-1)(C-1) = (2-1)(3-1) = 2$$

Let

O_{ij} = observed frequency for the cell at the intersection of row i and column j
E_{ij} = expected frequency for the cell at the intersection of row i and column j

The formula for the test statistic is

$$\chi^2 = \sum \frac{(O_{ij} - E_{ij})^2}{E_{ij}}$$

Using this formula, the value of the test statistic is

$$\chi^2 = \frac{(87-75)^2}{75} + \frac{(32-42.5)^2}{42.5} + \frac{(6-7.5)^2}{7.5} + \frac{(93-105)^2}{105} + \frac{(70-59.5)^2}{59.5} + \frac{(12-10.5)^2}{10.5}$$
$$= 1.920 + 2.594 + .300 + 1.371 + 1.853 + .214 = 8.252$$

The *p*-value is given by $P(\chi^2 \geq 8.252)$. From the chi-square distribution table (Table IV of Appendix A), $P(\chi^2 \geq 7.378) = .025$ and $P(\chi^2 \geq 9.210) = .010$. Therefore $.01 < P(\chi^2 \geq 8.252) < .025$. The region corresponding to the *p*-value is shown in Figure 18.4.

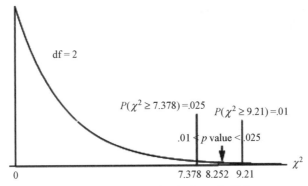

Figure 18.4   The p-value for Example 18.4.

Step 5. Make a decision.

The significance level is .025. Using the p-value, we ask whether the p-value is $\leq \alpha$. Since the p-value is between .01 and .025, it is less than .025. Therefore, we reject the null hypothesis. Our conclusion is that gender and employee opinion regarding support of the new contract are dependent. ∎

Example 18.5   Windham Electronics Company manufactures DVDs (digital versatile disks). The company has two machines that are used to manufacture DVDs. From time to time the quality control inspector at the company takes a sample of DVDs and checks them for being good or defective. A recent such sample of 200 DVDs produced the two-way classification presented in the Table 18.6.

TABLE 18.6   DVD Quality Outcomes by Machine

	Good (G)	Defective (D)
Machine 1 (A)	109	11
Machine 2 (B)	66	14

Does the sample provide sufficient evidence to conclude that the two attributes, the machine type and DVDs being good or defective, are dependent? Use $\alpha = .01$.

Solution   We perform this test of hypothesis using the usual five steps as follows:

Step 1. Select the type of test to use/check underlying conditions.

The data were collected using a simple random sample of 200 DVDs, analyzing two categorical variables. We need to ensure that the expected counts are all greater than or equal to 5, in order to perform a chi-square test of independence. Table 18.7 re-creates Table 18.6, with the row and column totals included.

TABLE 18.7   DVD Quality Outcomes by Machine

	Good (G)	Defective (D)	Row Totals
Machine 1 (A)	109	11	120
Machine 2 (B)	66	14	80
Column totals	175	25	200

The expected counts are calculated as follows, using the formula $E_{ij} = (r_i \times c_j)/n$ and the symbols from Tables 18.6 and 18.7:

$$E(A \text{ and } G) = E_{11} = \frac{120 \times 175}{200} = 105$$

$$E(A \text{ and } D) = E_{12} = \frac{120 \times 25}{200} = 15$$

$$E(B \text{ and } G) = E_{21} = \frac{80 \times 175}{200} = 70$$

$$E(B \text{ and } D) = E_{22} = \frac{80 \times 25}{200} = 10$$

As we observe, the smallest expected frequency is 10. Thus, all the expected frequencies are 5 or more. Hence, we perform a chi-square test of independence.

Step 2. State the null and alternative hypotheses:

H_0:   The machine and DVD quality are independent
H_1:   The machine and DVD quality are dependent

Step 3. Determine the significance level α.
    The significance level is given to be .01; hence, $\alpha = .01$.

Step 4. Calculate the test statistic and the p-value.
    Since $R = 2$ and $C = 2$, it follows that

$$df = (R-1)(C-1) = (2-1)(2-1) = 1$$

The value of the test statistic is

$$\chi^2 = \frac{(109-105)^2}{105} + \frac{(11-15)^2}{15} + \frac{(66-70)^2}{70} + \frac{(14-10)^2}{10}$$
$$= .152 + 1.067 + .229 + 1.600 = 3.048$$

The p-value is given by $P(\chi^2 \geq 3.048)$. From the chi-square distribution table (Table IV of Appendix A), $P(\chi^2 \geq 2.706) = .10$ and $P(\chi^2 \geq 3.841) = .05$; therefore $.05 < P(\chi^2 \geq 3.048) < .10$. The region corresponding to the p-value is shown in Figure 18.5.

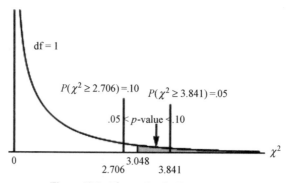

Figure 18.5   *The p*-value for Example 18.5.

Step 5. Make a decision.

The significance level is .01. Using the *p*-value, we ask whether the *p*-value is $\leq \alpha$. Since the *p*-value is between .05 and .10, it is greater than .01. Therefore, we fail to reject the null hypothesis. Our conclusion is that there is not enough evidence to conclude that production machine and DVD quality are dependent. In other words, there is insufficient evidence to conclude that the defective rates for the two machines differ.   ∎

18.4.2   Test of Homogeneity

In a *test of homogeneity*, we determine whether two (or more) populations are homogeneous (similar) with regard to the distribution of a certain characteristic. For example, we might be interested in testing the null hypothesis that the proportions of households that belong to different income groups are the same in California and Wisconsin. Or we may want to test whether the preferences of people in Florida, Arizona, and Vermont are similar with regard to Coke, Pepsi, and 7-Up. Thus, a test of homogeneity involves testing the null hypothesis that the proportions of elements with certain characteristics in two or more different populations are the same against the alternative hypothesis that these proportions are not the same.

The economic recession and budget deficits in the early 1990s led many states to cut benefits and freeze salaries of state employees. Combined with an increase in taxes, this may have affected the morale of many state employees. A citizens group wanted to test the null hypothesis that the distribution of the job satisfaction index of state employees in California and New York is the same. (*Note*: In a test of homogeneity the null hypothesis will always be that the proportions of elements with certain characteristics are the same in two or more populations. The alternative hypothesis will be that these proportions are not the same.) Suppose that the researcher, who was assigned this job by the citizens group, took one sample of 500 state employees from California and another sample of 400 state employees from New York State. The information that she collected from these employees is tabulated and recorded in Table 18.8.

TABLE 18.8   Job Satisfaction Ratings in California and New York

	California	New York	Row Totals
Very satisfied	60	75	135
Somewhat satisfied	100	125	225
Somewhat dissatisfied	184	140	324
Very dissatisfied	156	60	216
Column totals	500	400	900

Note that in this example the column totals are fixed; that is, we decide in advance to take a stratified random sample of 500 state employees from California and 400 from New York. However, the row totals of 135, 225, 324, and 216 are determined randomly by the outcomes of the two samples. If we compare this example to Example 18.4 regarding the sample of men and women workers and their opinions on the proposed contract, we will note that neither the column nor the row totals were fixed in that example. Instead, the union took just one sample of 300 workers, collected the information on their gender and opinions, and prepared the contingency table. Consequently, in that example, the row and column totals were all determined randomly by the outcome of the survey. Thus, when both the row and column totals are determined randomly, we perform a test of independence. However, when either the column totals or the row totals are fixed, we perform a test of homogeneity. In the case of state employees in California and New York, we will perform a test of homogeneity to test for the similarity of job satisfaction levels for workers in the two states.

The procedure for performing a test of homogeneity is similar to the one used for a test of independence discussed earlier. Like a test of independence, a test of homogeneity is also right-tailed. Example 18.6 illustrates the procedure to make a homogeneity test.

Example 18.6 Consider the data on the job satisfaction of state employees in California and New York given in Table 18.8. At the 2.5% significance level, test the null hypothesis that the distribution of state employees with regard to job satisfaction is similar (homogeneous) for the two states.

Solution We perform this test using the five steps as follows.

Step 1. Select the type of test to use and check the underlying conditions.

The data were collected using a stratified random sample of 500 California state employees and 400 New York state employees, analyzing job satisfaction using the same set of categories for each state. We need to ensure that the expected counts are all greater than or equal to 5 in order to perform a chi-square test of homogeneity.

The expected counts are calculated as follows (where CA = California, NY = New York, SD = Somewhat dissatisfied, SS = Somewhat satisfied, VD = Very dissatisfied, VS = Very satisfied):

$$E(\text{CA and VS}) = E_{11} = (135 \times 500)/900 = 75$$

$$E(\text{CA and SS}) = E_{21} = (225 \times 500)/900 = 125$$

$$E(\text{CA and SD}) = E_{31} = (324 \times 500)/900 = 180$$

$$E(\text{CA and VD}) = E_{41} = (216 \times 500)/900 = 120$$

$$E(\text{NY and VS}) = E_{12} = (135 \times 400)/900 = 60$$

$$E(\text{NY and SS}) = E_{22} = (225 \times 400)/900 = 100$$

$$E(\text{NY and SD}) = E_{32} = (324 \times 400)/900 = 144$$

$$E(\text{NY and VD}) = E_{41} = (216 \times 400)/900 = 96$$

The smallest expected frequency is 60, so all expected frequencies are 5 or more, and thus we perform a chi-square test of homogeneity.

Step 2. State the null and alternative hypotheses:

H_0: The sets of state employee job satisfaction proportions are the same for California and New York

H_1: The sets of state employee job satisfaction proportions are different for California and New York

Step 3. Determine the significance level α.

The significance level is given to be .025; thus, $\alpha = .025$.

Step 4. Calculate the test statistic and the *p*-value.

Since $R = 4$ and $C = 2$, it follows that

$$\text{df} = (R{-}1)(C{-}1) = (4{-}1)(2{-}1) = 3$$

Table 18.9 reproduces Table 18.8, including the expected frequencies that were calculated in step 1 above.

Note that we have dropped the row and column of totals.

TABLE 18.9   Job Satisfaction Ratings in California and New York

Rating	California	New York
Very satisfied	60	75
	(75)	(60)
Somewhat satisfied	100	125
	(125)	(100)
Somewhat dissatisfied	184	140
	(180)	(144)
Very dissatisfied	156	60
	(120)	(96)

The value of the test statistic is

$$\chi^2 = \frac{(60-75)^2}{75} + \frac{(100-125)^2}{125} + \frac{(184-180)^2}{180} + \frac{(156-120)^2}{120}$$

$$= \frac{(75-60)^2}{60} + \frac{(125-100)^2}{100} + \frac{(140-144)^2}{144} + \frac{(60-96)^2}{96}$$

$$= 3 + 5 + .089 + 10.8 + 3.75 + 6.25 + .111 + 13.5 = 42.5$$

The p-value is given by $P(\chi^2 \geq 42.5)$. From the chi-square distribution table (Table IV of Appendix A), $P(\chi^2 \geq 12.838) = .005$; therefore, $P(\chi^2 \geq 42.5) < .005$. The actual region corresponding to the p-value is so small (it is approximately .000000003) that it cannot be shown in a figure.

Step 5. Make a decision.

The significance level is .025. Using the p-value, determine whether the p-value $\leq \alpha$. Since the p-value is less than .005, it is less than .025. Therefore, we reject the null hypothesis. Our conclusion is that the sets of job satisfaction proportions are different for California and New York.   ∎

GLOSSARY

chi-square distribution   A distribution, with degrees of freedom as the only parameter, that is skewed to the right for small df and resembles a normal curve for large df.

expected frequencies   The frequencies for categories of a multinomial experiment or for cells of a contingency table that are expected to occur when a given null hypothesis is true.

goodness-of-fit test   A test of the null hypothesis that the observed frequencies for an experiment follow a certain pattern or theoretical distribution.

multinomial experiment   An experiment with n trials for which (1) the trials are identical, (2) there are more than two possible outcomes per trial, (3) the trials are independent, and (4) the probabilities of various outcomes remain constant for each trial.

observed frequencies   The frequencies actually obtained from the performance of an experiment.

test of homogeneity   A test of the null hypothesis that the proportions of elements that belong to different groups in two (or more) populations are similar.

test of independence   A test of the null hypothesis that two attributes of a population are not related.

EXERCISES

18.1 Find the value of χ^2 for 4 degrees of freedom and

(a) .005 area in the right tail of the chi-square distribution curve

(b) .05 area in the left tail of the chi-square distribution curve

18.2 Determine the value of χ^2 for 13 degrees of freedom and

(a) .025 area in the left tail of the chi-square distribution curve

(b) .995 area in the right tail of the chi-square distribution curve

18.3 The following table gives the percentage distribution of the June 2006 US labor force by educational attainment:

Educational Attainment	Percentage of Workers
Less than a high school diploma	9.94
High school graduate, no college	29.97
Some college, no degree or associate degree	27.16
Bachelor's degree or higher	32.93

Source: US Bureau of Labor Statistics.

A recent random sample of 1000 workers showed that 108 have less than a high school diploma, 323 are high school graduates with no college, 283 have some college with an associate degree or no degree, and 286 hold a bachelor's degree or higher. Test at the 1% significance level whether the percentage distribution of the US labor force by educational attainment has changed since June 2006.

18.4 DOM Corporation owns department stores in almost all states. Of all persons who possess credit cards for these department stores, 48% usually make payments on their credit card bills within 15 days of the issue of the statements, 44% pay within 16–30 days, and 8% take more than 30 days to pay. The management of DOM Corporation wants to know whether the payment patterns of the credit card–holders in a specific state are different from those of credit card–holders in other states. The research department at the company took a random sample of 800 credit card–holders from this state and found that 397 of them made the payments on their credit card bills within 15 days of the issue of the statement last month, 342 paid within 16–30 days, and 61 took more than 30 days to pay. Test at the 2.5% significance level if the credit card–holders in this state are different from all credit card holders with respect to making the payments on their credit card bills.

18.5 Home Mail Corporation sells products by mail. The company's management wants to find out if the number of orders received at the company's office on each of the 5 days of the week is the same. The company took a sample of 400 orders received during a 4-week period. The following table lists the frequency distribution for these orders by the day of the week.

Day of the week	Monday	Tuesday	Wednesday	Thursday	Friday
Number of orders received	92	71	65	83	89

Test at the 5% significance level if the null hypothesis that the orders are evenly distributed over all 5 days of the week should be rejected.

18.6 A management consultant thinks that job satisfaction affects the absentee rate of employees. He took a sample of 400 employees of companies and collected information on job satisfaction (on a scale of 1–10, where 1 is the lowest and 10 is the highest) and the number of days they were absent for last year. The following table gives the tabulated results:

	Job Satisfaction Index		
Number of Days absent Last Year	< 4	4–7	> 7
< 6	12	61	107
6–12	22	80	50
> 12	41	18	9

Do you think that the sample information provides sufficient evidence to reject the null hypothesis that job satisfaction and absentee rate of employees are not related? Use $\alpha = .025$.

18.7 A random sample of 200 jurors was selected and asked whether or not each of them had ever been a victim of crime. The jurors were also asked whether they are strict, fair, or lenient regarding punishment for crime. The following table gives the results of the survey:

	Strict	Fair	Lenient
Have been a victim	40	16	6
Have never been a victim	44	76	18

Test at the 5% significance level if the two attributes for all jurors are dependent.

18.8 A company owns four restaurants in four different parts of New York City. The management wanted to determine the opinions of its patrons on the quality of food and service at each of the four restaurants. A random sample of 100 patrons was selected for each restaurant and asked to rate the quality of food and service at that restaurant. The results of the survey are recorded in the following table:

	Opinions of Patrons		
	Excellent	Average	Below Average
Restaurant 1	59	32	9
Restaurant 2	60	28	12
Restaurant 3	64	26	10
Restaurant 4	50	40	10

From these sample surveys, can you conclude that the distributions of opinions are homogeneous for all four restaurants? Use $\alpha = .025$.

18.9 A random sample of 100 persons was selected from each of four regions in the United States. These people were asked whether or not they support a certain farm subsidy program. The results of the survey are summarized in the following table.

Region	Favor	Oppose	Uncertain
Northeast	56	33	11
Midwest	65	23	12
South	67	20	13
West	53	30	17

Using the 1% significance level, test the null hypothesis that the percentages of people with different opinions are similar for all four regions.

18.10   The following table gives the two-way classification of a sample of 1000 students selected from certain colleges and universities:

	Business Major	Nonbusiness Major
Male	77	389
Female	52	482

Test at the 1% significance level whether being a male or a female and being a business or nonbusiness major are related.

18.11   The president of a bank selected a sample of 200 loan applications to check whether the approval or rejection of an application depends on which one of the two loan officers, Thurow or Webber, handles that application. The information obtained from the sample is summarized in the following table:

	Approved	Rejected
Thurow	57	38
Webber	69	36

Test at the 2.5% significance level if the approval or rejection of a loan application depends on which loan officer handles the application.

19 Analysis of Variance

19.1 OVERVIEW

Chapter 17 described the procedures that are used to test hypotheses about differences between two population means using the t distribution. Also described in that chapter were the hypothesis testing procedures for differences between two population proportions using the normal distribution. Then, Chapter 18 explained the procedures to test hypotheses about the equality of more than two population proportions using the chi-square distribution.

This chapter explains how to test a null hypothesis that the means of more than two populations are equal. For example, suppose that a company that produces detergents is planning to buy new machines to fill 64-oz detergent jugs. The company is considering three types of such machines. Before the company decides which type of machine to buy, it puts all three machines to a test. One of each type of machine is randomly selected and used to fill the 64-oz jugs for a few hours. Let μ_1, μ_2, and μ_3 be the mean number of 64-oz jugs filled per hour by each of these three machines, respectively. To test whether the three types of machines fill the same number of jugs, on average, we test the null hypothesis

$$H_0: \quad \mu_1 = \mu_2 = \mu_3 \quad \text{(all three population means are equal)}$$

against the alternative hypothesis

$$H_1: \quad \text{Not all three population means are equal}$$

We use the analysis of variance, usually referred to as ANOVA, procedure to perform this test of hypothesis.

Note that the analysis of variance procedure can be used to compare two population means in the case of a two-sided alternative hypothesis. However, the procedures learned in Chapter 18 can be used for performing tests of hypotheses about the difference between two population means, regardless of the alternative hypothesis.

The *analysis of variance* (ANOVA) tests are performed using the F distribution. First, the F distribution is described in Section 19.2. Then, Section 19.3 discusses the application of the one-way analysis of variance procedure to perform tests of hypotheses.

19.2 THE *F* DISTRIBUTION

Like the t and chi-square distributions, the shape of a particular *F distribution* curve depends on the number of degrees of freedom. However, the F distribution has two numbers of degrees of freedom:

Practitioner's Guide for Statistics and Lean Six Sigma for Process Improvements. By Mikel J. Harry, Prem S. Mann, Ofelia C. de Hodgins, Christopher J. Lacke, and Richard Hulbert

degrees of freedom for the numerator and degrees of freedom for the denominator. These two df numbers are the parameters of the F distribution. Each set of degrees of freedom for the numerator and for the denominator gives a different F distribution curve. The units of an F distribution are denoted by F, which assumes only nonnegative values. Like the normal, t, and chi-square distributions, the F distribution is also a continuous distribution. The shape of an F distribution curve is skewed to the right, but the skewness decreases as the number of degrees of freedom increases.

Properties of the F Distribution

1. The F distribution is continuous and skewed to the right.
2. The F distribution has two numbers of degrees of freedom: df for the numerator and df for the denominator.
3. The units of an F distribution, denoted by F, are nonnegative.

For an F distribution, degrees of freedom for the numerator and degrees of freedom for the denominator are usually written as follows:

$$df = (8,14)$$

The first number denotes the degrees of freedom for the numerator and the second number denotes the degrees of freedom for the denominator, so the example above refers to an F distribution with 8 df for the numerator and 14 df for the denominator.

Figure 19.1 gives three F distribution curves for three sets of degrees of freedom for the numerator and for the denominator. In the figure, the first number gives the degrees of freedom associated with the numerator and the second number gives the degrees of freedom associated with the denominator. We can observe from this figure that as both of the degrees of freedom increase, the peak of the curve moves to the right; that is, the skewness decreases.

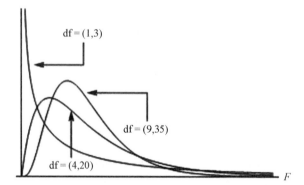

Figure 19.1   Three F distribution curves demonstrating the effect of increasing both degrees of freedom.

Figure 19.2 gives three F distribution curves having the same degrees of freedom for the numerator, which is 3. In the figure, the first number gives the degrees of freedom associated with the numerator and the second number gives the degrees of freedom associated with the denominator. As in Figure 19.1, the peak of the curve moves to the right as the number of degrees of freedom for denominator increases. However, the shift is not as substantial as when both numbers of degrees of freedom are increased. Furthermore, the three curves have 3, 10, and 305 degrees of freedom for the denominator, respectively. The shift that occurs when changing from 3 to 10 degrees of freedom for the denominator is comparable to the shift that occurs when changing from 10 to 305 degrees of freedom for the denominator. This demonstrates that, for a fixed number of degrees of freedom for the

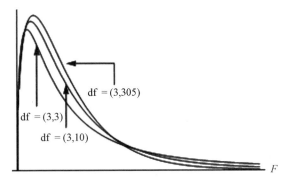

Figure 19.2   Three F distribution curves demonstrating the effect of increasing the degrees of freedom for the denominator.

numerator, the shift of the *F* distribution curve is limited as the degrees of freedom for the denominator becomes substantially large.

It is extremely rare, in practice, for the degrees of freedom for the numerator to be extremely large, so we will not examine a graph in which the number of degrees of freedom for the denominator is fixed and the number of degrees of freedom for the numerator gets large.

Table V in Appendix A lists the values of *F* for the *F* distribution. To read Table V, we need to know three quantities: the degrees of freedom for the numerator, the degrees of freedom for the denominator, and an area in the right tail of an *F* distribution curve. Note that the *F* distribution table (Table V) is read only for an area in the right tail of the *F* distribution curve. Also note that Table V has four parts. These four parts give the *F* values for an area of .01, .025, .05, and .10, respectively, in the right tail of the *F* distribution curve. Example 19.1 illustrates how to read Table V.

Example 19.1   Find the *F* value for 8 degrees of freedom for the numerator, 14 degrees of freedom for the denominator, and .05 area in the right tail of the *F* curve.

Solution   To find the required value of *F*, we consult the portion of Table V of Appendix A that corresponds to .05 area in the right tail of the *F* distribution curve. The relevant portion of that table is shown as Table 19.1. To find the required *F* value, we locate 8 in the row for the degrees of freedom for the numerator (at the top of Table V) and 14 in the column for the degrees of freedom in the denominator (the first column on the left side in Table V). The entry where the column for 8 and the row for 14 intersect gives the required *F* value. This value of *F* is 2.70, as shown in Figure 19.3. The *F* value taken from this table for a test of hypothesis is called the *critical value* of *F*.

TABLE 19.1

Denominator df	Numerator df					
	1	2	. . .	**8**	. . .	100
1	161.5	199.5	. . .	238.9	. . .	253.0
2	18.51	19.00	. . .	19.37	. . .	19.49
. . .	. . .	. . .	. . .	. . .	. . .	. . .
14	4.60	3.74	. . .	**2.70**	. . .	2.19
. . .	. . .	. . .	. . .	. . .	. . .	. . .
. . .	. . .	. . .	. . .	. . .	. . .	. . .
100	3.94	3.09	. . .	2.03	. . .	1.39

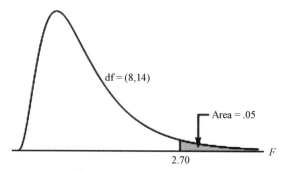

Figure 19.3   The critical value of F for 8 df for the numerator, 14 df for the denominator, and .05 area in the right tail.   ■

19.3   ONE-WAY ANALYSIS OF VARIANCE

As mentioned in the beginning of this chapter, the analysis of variance procedure is used to test the null hypothesis that the means of three or more populations are equal against the alternative hypothesis that not all these population means are equal.

Reconsider the example of the company that produces detergents and is planning to buy a new machine to fill 64-oz detergent jugs. The company is considering three types of such machines. Before it decides which type of machine to buy, the company puts all three machines to a test. One of each type of machine is randomly selected and used to fill the 64-oz jugs for a few hours. Let μ_1, μ_2, and μ_3 be the mean numbers of 64-oz jugs filled per hour by each of these three machines, respectively. To test whether the three types of machines fill the same number of jugs, on average, we test the null hypothesis

$$H_0: \mu_1 = \mu_2 = \mu_3 \quad \text{(all three population means are equal)}$$

against the alternative hypothesis

$$H_1: \text{ Not all three population means are equal}$$

We use the ANOVA procedure to perform this test of hypothesis. Note that the alternative hypothesis above states that at least one population mean is different from the other two.

One method to test such a hypothesis is to test the three hypotheses: $H_0: \mu_1 = \mu_2$, $H_0: \mu_1 = \mu_3$, and $H_0: \mu_2 = \mu_3$ separately using the procedure discussed in Chapter 17. Besides being time-consuming, such a procedure has other disadvantages: (1) if we reject even one of these three hypotheses, then we must reject the null hypothesis $H_0: \mu_1 = \mu_2 = \mu_3$; and (2) combining the type I errors for the three tests (one for each test) will give a very large type I error for the test $H_0: \mu_1 = \mu_2 = \mu_3$. Hence, we should prefer a procedure that can test the equality of three means in one test. The ANOVA procedure provides such a procedure. It is used to compare three or more population means in a single test.

This section discusses the *one-way ANOVA* procedure to make tests comparing the means of several populations. By using a one-way ANOVA test, we analyze only one *factor* (or grouping variable). For instance, in the example of testing for the equality of means of three filling machines, we are considering only one factor, which is the effect of different machines on the number of jugs filled. Sometimes we may analyze the effects of two factors. For example, if the three filling machines are operated by different persons, we can analyze the effects of machines and operators on the number of jugs filled. This is done using a *two-way ANOVA*. The procedure under discussion is called *analysis of variance* because the test is based on the analysis of variation in the data obtained from samples. The application of one-way ANOVA requires that the following assumptions hold true:

1. The populations from which the samples are drawn are (approximately) normally distributed.
2. The populations from which the samples are drawn have the same variance (or standard deviation).
3. The samples drawn from different populations are random and independent.

For instance, in the example of three types of machines being considered by a company to fill 64-oz jugs, we first assume that the number of jugs filled per hour by each machine is (approximately) normally distributed. We can use a normal quantile plot to analyze the data from each machine to determine whether the assumption is reasonable. Then, the means of the distributions of the number of jugs filled per hour for three machines may or may not be the same, but all three distributions have the same variance. As with the two sample t test, we can use the ratio of the largest and smallest sample standard deviations to determine whether the equal variance assumption is reasonable. Finally, when we take samples to perform an ANOVA test, these samples are drawn independently and randomly from three different populations.

The ANOVA test is applied by calculating two estimates of the variance σ^2 of population distributions: the *variance between samples* and the *variance within samples*. The variance between samples is commonly called the *mean square for samples* (some texts and software also call it the *mean square for groups*) (MSG). The variance within samples is also called the mean square for error or *MSE*.

The variance between groups MSG gives an estimate of σ^2 based on the variation among the means of samples taken from different populations. For the example of three filling machines, MSG will be based on the values of the means for three samples, which represent the number of jugs filled per hour by three machines, respectively. If the means of all populations under consideration are equal, the means of respective samples will still be different but the variation among them is expected to be small and, consequently, the MSG value is expected to be small. However, if the means of populations under consideration are not all equal, the variation among the means of respective samples could be large. Consequently, the value of MSG could be large.

The within-sample variance MSE (mean square error) gives an estimate of σ^2 based on the variation within the data in each sample. For the example of three filling machines, MSE will be based on the measurements (the number of jugs filled per hour) included in the three samples taken from three populations. The concept of MSE is similar to the concept of the pooled standard deviation s_p for two samples discussed in Section 17.2.

The one-way ANOVA test is always right-tailed with the rejection region in the right tail of the F distribution curve. The hypothesis testing procedure using ANOVA involves the same five steps that were used in earlier chapters. The next subsection explains how to calculate the value of the test statistic F for an ANOVA test.

19.3.1   Variance between Groups

In calculating the variance between groups, we are measuring how far apart the group means are. As with many measures of spread, this measure is based on the squared distance between each group mean and the mean of all the observations, which is called the *grand mean*. Before proceeding with this calculation, it is necessary to introduce some notation. Let

$\bar{x}_i = $ the sample mean for group (or treatment) i

$k = $ the number of groups (or treatments)

$n_i = $ the sample size of group i

$\bar{x} = $ the average (the grand mean) of all of the observations in all groups

$n = $ the sum of the k sample sizes $= n_1 + n_2 + n_3 + \cdots + n_k$

$s_i^2 = $ the sample variance for group (or treatment) i

As mentioned earlier, MSG is based on the squared distance between each group mean and the grand mean. The quantity

$$\text{SSG} = n_1(\bar{x}_1 - \bar{x})^2 + n_2(\bar{x}_2 - \bar{x})^2 + n_3(\bar{x}_3 - \bar{x})^2 + \cdots + n_k(\bar{x}_k - \bar{x})^2$$

is called the *sum of squares for groups* (or *treatments*), and is denoted by SSG. If all the group means are equal to the grand mean, the SSG value will be zero. As the group means move farther away from the grand mean, the SSG value will increase.

To calculate a mean square, one must divide the relevant sum of squares by its number of degrees of freedom. In equation form, this relationship is

$$\text{Mean square} = \frac{\text{sum of squares}}{\text{degrees of freedom}}$$

In the case of the MSG, the number of degrees of freedom is the number of groups minus one, or $k-1$. Thus

$$\text{MSG} = \frac{\text{SSG}}{k-1}$$

19.3.2   Variance within Groups

While the MSG represents the distance between the group means, the mean square error (MSE) represents the distance between observations within each group. As with the *between group* measure, the within group measure also includes a sum of squares. The calculation of SSE, the sum of squared error, is

$$\text{SSE} = (n_1-1)s_1^2 + (n_2-1)s_2^2 + (n_3-1)s_3^2 + \cdots + (n_k-1)s_k^2$$

Thus, the calculation of SSE involves multiplying the sample variance and the sample size minus one for each group and taking the sum of the products. As you may recall from Chapter 17, the number degrees of freedom for the pooled t test was equal to $(n_1-1) + (n_2 - 1) = n - 2$. Similarly, the degrees of freedom for error in ANOVA is $(n_1-1) + (n_2-1) + (n_3 - 1) + \cdots + (n_k - 1) = n - k$. Thus, the mean square error (MSE), which represents the variance within groups, is

$$\text{MSE} = \frac{\text{SSE}}{n - k}$$

19.3.3   Total Sum of Squares (SST)

One other measure of variability in the data is called the *total sum of squares* (SST). The formula for SST is

$$\text{SST} = \Sigma(x - \bar{x})^2$$

If you think that this formula looks familiar, you are right. Recall from Chapter 8 the formula for the sample variance of a dataset

$$s^2 = \frac{\Sigma(x - \bar{x})^2}{n-1}$$

Thus, SST is the numerator in the formula for the sample variance. Unlike those for SSG and SSE, the calculation of SST does not include a group distinction. For each data point, take the difference between each observation and the grand mean, square each difference, and sum the squared differences to obtain SST.

While it is mathematically feasible to divide SST by its number of degrees of freedom, we do not refer to this quantity as the total mean square (MST). The primary reason will be explained later in this chapter. However, it is important to note that the number of degrees of freedom for SST is $n-1$. Thus, as shown in the previous paragraph, the result of dividing SST by its number of degrees of freedom is the sample variance.

19.3.4 Relationships within Sums of Squares and Degrees of Freedom

The ANOVA process results in additive relationships among the sums of squares and the degrees of freedom. The relationship is easy to see for the numbers of degrees of freedom. From Sections 19.4.1 through 19.4.3, it follows that

	Groups	Error	Total
df	$k-1$	$n-k$	$n-1$

It is the case that

$$(k-1) + (n-k) = k - 1 + n - k = n - 1$$

Thus

$$\text{Group df} + \text{error df} = \text{total df}$$

A similar relationship exists among the sums of squares. Specifically

$$\text{SSG} + \text{SSE} = \text{SST}$$

The proof of this result is beyond the scope of this book.

Earlier we mentioned that we do not talk about the concept of a "total mean square." One primary reason for this is that

$$\frac{\text{SSG}}{k-1} + \frac{\text{SSE}}{n-k} \neq \frac{\text{SST}}{n-1}$$

19.3.5 Equal Sample Sizes

Analysis of variance is a very computationally intense procedure. Some older texts provide multiple sets of formulas for MSG and MSE. One set is based on the assumption that the group sample sizes are equal, that is, $n_1 = n_2 = \cdots = n_k$, while the other set contains generalized formulas that will work regardless of the group sample sizes. Moreover, conceptual and *calculation-friendly* formulas have been provided in the past. The availability and use of statistical software packages dramatically reduces the need for calculation-friendly formulas. Moreover, functions such as the mean and variance are readily available in spreadsheet packages, so the conceptual formulas given for SSE and SST can be used quite easily.

19.3.6 Calculating the Value of the Test Statistic

The value of the test statistic F for a test of hypothesis using ANOVA is given by the ratio of the mean square for groups (MSG) and the mean square error (MSE):

$$F = \frac{\text{MSG}}{\text{MSE}} = \frac{\text{SSG}/(k-1)}{\text{SSE}/(n-k)}$$

19.3.7 The One-Way ANOVA Table

Whether one uses statistical software or performs an ANOVA *by hand*, the standard convention for reporting the results is to use an ANOVA table. The format of an ANOVA table with one factor is as follows:

Source	Sum of Squares	df	Mean Square	F	P
Groups	SSG	$k-1$	$MSG = \dfrac{SSG}{k-1}$	$\dfrac{MSG}{MSE}$	p-value
Error	SSE	$n-k$	$MSE = \dfrac{SSE}{n-k}$		
Total	SST	$n-1$			

Some common statistical software packages label the p-value column using "Sig.," which is short for the term "observed significance level." The terms p-value and *observed significance level* are synonymous. Example 19.2 demonstrates the ANOVA process by hand. The remaining examples utilize statistical software to obtain the ANOVA table.

Example 19.2 One of the products made at Abe Chemicals Company is detergents. Because of increased sales, the company is planning to buy a few new machines that will be used to fill the 64-oz detergent jugs. The company is considering three types of such machines but will eventually buy only one type of machine. Before making such a decision, the company wanted to test the three machines for the number of jugs filled per hour. To do so, the company used each of the three types of machines for 5 h and recorded the number of jugs filled during each hour by these machines. Table 19.2 gives the number of jugs filled by these machines during each of the 5 h.

TABLE 19.2 Data for Example 19.2

Machine 1	Machine 2	Machine 3
54	53	49
49	56	53
52	57	47
55	51	50
48	59	54

At the 1% significance level, can we reject the null hypothesis that the mean number of jugs filled per hour by each of these three machines is the same? Assume that the distribution of the number of jugs filled per hour is approximately normal for each of the three machines.

Solution In ANOVA terminology, *machine used* is called the *factor*. There are three *levels* of this factor, corresponding to the three machines; thus $k = 3$. The response variable (x) is *the number of jugs filled during a one-hour period*. Each machine was used for 5 one-hour periods, so the sample size for each of the three machines is 5.

We calculate the following summary statistics that are necessary for completing the ANOVA table. Here, means and variances for the three machines are calculated using the formulas of Chapter 8. The grand mean $\bar{x}$ and the grand variance s^2 are calculated for all 15 values belonging to all three machines.

- For machine 1:

$$n_1 = 5$$

$$\bar{x}_1 = \frac{\sum x_1}{n_1} = \frac{54 + 49 + 52 + 55 + 48}{5} = 51.6$$

$$s_1^2 = \sqrt{\frac{\sum (x - \bar{x}_1)}{n_1 - 1}} = 9.3$$

- Similarly for machines 2 and 3:

$$n_2 = 5, \ \bar{x}_2 = 55.2, \ s_2^2 = 10.2$$
$$n_3 = 5, \ \bar{x}_3 = 50.6, \ s_3^2 = 8.3$$

- For machines 1–3 together:

$$n = 15, \quad k = 3, \quad \bar{x} = 52.6667, \quad s^2 = 169.7333$$

The sums of squares can now be calculated as follows:

$$\begin{aligned}
\text{SSG} &= n_1(\bar{x}_1 - \bar{x})^2 + n_2(\bar{x}_2 - \bar{x})^2 + n_3(\bar{x}_3 - \bar{x})^2 \\
&= 5(51.6 - 52.4667)^2 + 5(55.2 - 52.4667)^2 + 5(50.6 - 52.4667)^2 \\
&= 5(.7512) + 5(7.4709) + 5(3.4846) \\
&= 3.756 + 37.3545 + 17.423 = 58.5335
\end{aligned}$$

$$\begin{aligned}
\text{SSE} &= (n_1 - 1)s_1^2 + (n_2 - 1)s_2^2 + (n_3 - 1)s_3^2 \\
&= (5 - 1)9.3 + (5 - 1)10.2 + (5 - 1)8.3 \\
&= 37.2 + 40.8 + 33.2 \\
&= 111.2
\end{aligned}$$
$$\text{SST} = \text{SSG} + \text{SSE} = 58.5335 + 111.2 = 169.7335$$

The numbers of degrees of freedom are

$$\text{Group df} = k - 1 = 3 - 1 = 2$$
$$\text{Error df} = n - k = 15 - 3 = 12$$
$$\text{Total df} = n - 1 = 15 - 1 = 14$$

The values of the mean squares and the test statistic F are

$$\text{MSG} = \frac{\text{SSG}}{k - 1} = \frac{58.5335}{2} = 29.26675$$

$$\text{MSE} = \frac{\text{SSE}}{n - k} = \frac{111.2}{12} = 9.26667$$

$$F = \frac{\text{MSG}}{\text{MSE}} = \frac{29.26675}{9.26667} = 3.1583$$

We assumed that each distributions of the number of jugs filled per hour is approximately normal. In order to determine whether the assumption that the three populations have the same population variance is valid, we can use the same rule that we used in the two-sample t test. Specifically, the assumption is reasonable if

$$\frac{\max(s_i)}{\min(s_i)} \leq 2$$

From our summary statistics, we have

$$\max(s_i) = \sqrt{s_2^2} = \sqrt{10.2} = 3.19, \quad \min(s_i) = \sqrt{s_3^2} = \sqrt{8.3} = 2.88$$

Therefore,

$$\frac{\max(s_i)}{\min(s_i)} = \frac{3.19}{2.88} = 1.11 \le 2$$

Thus, the equal variance assumption is reasonable.

Unless there is a learning process that would allow people operating the filling machines to become more efficient over time, there is no reason to believe that the independent-samples assumption is unreasonable.

One cannot determine the exact p-value using Table V of Appendix A. Instead, one needs to look at the F value for numerator df $= 2$, denominator df $= 12$, and the significance level corresponding to each part of Table V. Table 19.3 summarizes this information.

TABLE 19.3   Specified F Distribution Values for df $=$ (2,12)

α	.10	.05	.025	.01
F	2.81	3.89	5.10	6.93

As calculated above, the value of F observed is 3.1583. Since $2.81 < 3.1583 < 3.89$, $.05 < P(F \ge 3.1583) < .10$; thus the p-value is between .05 and .1. Table 19.4 is the complete ANOVA table for this problem.

TABLE 19.4   ANOVA Table for Example 19.2

Source	Sum of Squares	df	Mean Square	F	P
Group	58.5335	2	29.26775	3.1583	$.05 < p$-value $< .10$
Error	111.2	12	9.26667		
Total	169.7335	14			

The null and alternative hypotheses for this problem are

$$H_0: \mu_1 = \mu_2 = \mu_3 \quad \text{(all three population means are equal)}$$
$$H_1: \text{Not all three population means are equal}$$

Since the p-value is between .05 and .10, it is greater than .01, which is the significance level. We fail to reject the null hypothesis. Thus, there is not significant evidence to conclude that at least two of the machines differ in the average number of jugs filled per hour.

Figure 19.4 gives the ANOVA table for this problem as produced by MINITAB. Note that MINITAB uses the term `Factor` instead of `Group`.

```
One-way ANOVA: Machine I, Machine II, Machine III

Source DF     SS    MS    F      P
Factor  2  58.53 29.27 3.16  0.079
Error  12 111.20  9.27
Total  14 169.73
```

Figure 19.4   ANOVA table for Example 19.2 as produced by MINITAB.

Two items about the ANOVA table in Figure 19.4 must be noted: (1) the columns for degrees of freedom and sums of squares have been switched (statistical software packages and texts are inconsistent on the order of these columns, so remember to look at the headings), and (2) the use of software has provided us with a specific p-value instead of a p-value range. Specifically, the p-value for this test is .079, which, as was noted, falls between .05 and .1. Figure 19.5 displays the p-value for this test. ∎

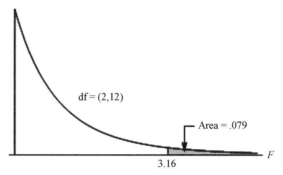

Figure 19.5   The p-value for Example 19.2.

Example 19.3   From time to time, unbeknownst to its employees, the research department at Post Bank observes various employees for work productivity. Recently this department wanted to check whether the four tellers at a branch of this bank serve, on average, the same number of customers per hour. The research manager observed each of the four tellers for a certain number of hours. Table 19.5 gives the number of customers served by the four tellers during each of the observed hours.

TABLE 19.5   Teller Data for Example 19.3

Teller A	Teller B	Teller C	Teller D
19	14	11	24
21	16	14	19
26	14	21	21
24	13	13	26
18	19	16	20
—	11	18	—

At the 5% significance level, test the null hypothesis that the mean number of customers served per hour by each of these four tellers is the same. Assume that the number of customers served per hour by each teller is approximately normal and that the data come from independent samples.

Solution   We perform this test as follows:

Step 1. Check the underlying assumptions.

In an ANOVA test, we assume that the samples are selected independently from distributions that are approximately normally distributed. We still need to verify that the equal variance assumption is reasonable. Figure 19.6 contains the standard deviations for the four tellers, as calculated using MINITAB.

```
Variable          StDev
Teller A           3.36
Teller B           2.74
Teller C           3.62
Teller D           2.92
```

Figure 19.6   Teller standard deviations for Example 19.3.

The largest of the four standard deviations is 3.62, while the smallest is 2.74. Thus

$$\frac{\max(s_i)}{\min(s_i)} = \frac{3.62}{2.74} = 1.32 \leq 2$$

Thus, the equal variance assumption is reasonable.

Step 2. Select the test to use.

We are testing for the equality of four means. The assumptions of the ANOVA procedure are met, so we will use the ANOVA procedure for this analysis.

Step 3. State the null and alternative hypotheses.

Let $\mu_1, \mu_2, \mu_3,$ and μ_4 be the mean number of customers served per hour by tellers A, B, C, and D, respectively. The null and alternative hypotheses are

$$H_0: \quad \mu_1 = \mu_2 = \mu_3 = \mu_4 \quad \text{(all four population means are equal)}$$

$$H_1: \quad \text{Not all four population means are equal}$$

Step 4. Determine the test statistic and *p*-value.

Figure 19.7 provides the ANOVA table as generated by MINITAB.

```
One-way ANOVA: Teller A, Teller B, Teller C, Teller D

Source  DF    SS     MS    F      P
Factor   3   255.6  85.2  8.42  0.001
Error   18   182.2  10.1
Total   21   437.8
```

Figure 19.7   ANOVA table for Example 19.3.

The value of the test statistic is $F = 8.42$. It has 3 degrees of freedom in the numerator $(k - 1 = 4 - 1 = 3)$ and 18 degrees of freedom in the denominator $(n - k = 22 - 4 = 18)$, and the *p*-value is .001.

Step 5. Make a conclusion.

The significance level is .05. The *p*-value is .001, which is less than the significance level, so we will reject the null hypothesis. Thus, we will conclude that not all tellers serve the same number of customers, on average. The *p*-value for this test is shown in Figure 19.8.

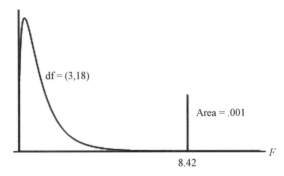

Figure 19.8   The *p*-value for Example 19.3.

19.4   PAIRWISE COMPARISONS

If the result of an ANOVA is to reject the null hypothesis, the conclusion is that not all groups have the same mean. However, the process of rejecting the null hypothesis does not identify which group means are significantly different. Another procedure is needed to determine which means are different. The type of procedure is called a *pairwise comparison* (or *post hoc comparison*).

Many types of pairwise comparison procedures are available in statistical software packages. As computationally intensive as a one-way ANOVA procedure is, a pairwise comparison procedure is even more computationally intensive. Because different software packages will include many of the procedures, we will not discuss specific tests. Instead, using the data from Example 19.3, we will examine the output from one pairwise comparison procedure and discuss how to analyze it. Figure 19.9 contains the output from the Tukey HSD (Honestly Significant Differences) pairwise comparison procedure.

Pairwise comparison methods calculate a confidence interval for the difference of every unique pair of means. The differences can be taken in either order, but taking both "C–D" and "D–C" is unnecessary, because the same conclusion will be derived from each interval.

To determine whether two means are significantly different, check to see whether zero falls in the interval. If zero does not fall in the interval, the two means in the interval are significantly different. If zero does fall in the interval, the two means are significantly different.

Examining Figure 19.9, we see that the interval comparing tellers A and B extends from -12.549 to -1.651. Since this interval does not contain zero, we can conclude that the mean numbers of customers served per hour are different for tellers A and B. Similarly, the interval comparing tellers A and C goes from -11.549 to $-.651$. It does not contain zero, so we can conclude that the mean numbers of customers served per hour are different for tellers A and C. Similar conclusions are reached for tellers B and D and tellers C and D.

On the other hand, the interval comparing B and C goes from -4.195 to 6.195. Since zero falls in the interval, it is feasible that there is no difference between the average service rates for tellers B

```
Tukey Simultaneous Confidence Intervals
All Pairwise Comparisons

Teller A subtracted from:
        Lower Center  Upper
Teller B -12.549 -7.100  -1.651  (-------*-------)
Teller C -11.549 -6.100  -0.651   (------*-------)
Teller D  -5.291  0.400   6.091        (--------*-------)
                        --------+---------+---------+---------+-
                          -7.0      0.0       7.0      14.0
Teller B subtracted from:
        Lower Center  Upper
Teller C  -4.195  1.000   6.195             (------*-------)
Teller D   2.051  7.500  12.949               (-------*------)
                        --------+---------+---------+---------+-
                          -7.0      0.0       7.0      14.0
Teller C subtracted from:
        Lower Center  Upper
Teller D   1.051  6.500  11.949               (------*-------)
                        --------+---------+---------+---------+-
                          -7.0      0.0       7.0      14.0
```

Figure 19.9   Pairwise comparison output for Example 19.3.

and C. Thus, we would conclude that there is not a significant difference in the average service rates for tellers B and C. A similar conclusion is made regarding tellers A and D.

19.5   MULTIFACTOR ANALYSIS OF VARIANCE

Often, we will be interested in the impact of two or more categorical variables on the average of a response variable. Provided that a set of assumptions similar to those for a one-way ANOVA is reasonable, we can use a multiple-factor ANOVA. The reader should understand that following descriptions of *two-way ANOVA* and *N-way ANOVA* are at a very introductory level. The mathematics of multifactor ANOVA are well above the level of this book. It is best to consult a professional statistician prior to doing any data collection to make sure that the methodology one wishes to use is actually the proper methodology.

19.5.1   Two-Way ANOVA

In two-way ANOVA, we are interested in answering two sequential questions:

1. Do the two categorical variables (factors) act together to impact the averages for the various groups?
2. If the two factors do not act together to impact the averages, does at least one of the factors have an impact on the averages for the various groups?

Consider the following situation. A company has two production lines that produce the same product. Over time, some of the equipment on each line has had to be repaired or replaced. In addition, each production line has three teams of employees who work on different 8-h shifts. The company is interested in comparing the average output rates for the various shift/production line combinations. Given this setting, our two general questions become the following:

1. If we compare shifts across production lines, is there an inconsistent difference in the average shift production rates from line 1 to line 2? One such example would be that the morning shift on line 1 outproduces the morning shift on line 2, but the afternoon shifts have the same average production rates. If this is the case, we conclude that there is an *interaction* between the shift and the production line.
2. If there is no interaction, does the production rate of one production line consistently differ from that of the other production line? (for example, the three shifts on line 1 each have a production rate that is five units higher than their corresponding shifts on line 2.) Or, do certain shifts on both production lines consistently outperform other shifts by the same amount? (for instance, both evening shifts have an average production rate that is 6.2 units higher than the afternoon shifts on both production lines.) If the answer to either of these questions is affirmative, then we conclude that there is a *main effect*. Note that it is possible to have two main effects while not having an interaction.

19.5.2   *N*-Way ANOVA

An *N*-way ANOVA is similar to a two-way ANOVA because we are looking for evidence of interaction of factor or for the presence of one or more main effects. However, the addition of more categorical variables increases the number of possible conclusions. To understand the difference, let us compare the possible conclusions of a two-way ANOVA to the possible conclusions of a three-way ANOVA. To do this, we will label the factors A, B, and C:

Two-Way Anova (Factors A and B)	Three-Way Anova (Factors A, B, and C)
A and B interact	A, B, and C interact
No interaction; A and B have main effects	Only A and B interact
No interaction; A is the only main effect	Only A and C interact
No interaction; B is the only main effect	Only B and C interact
No effects (all group means are the same)	No interaction; A, B, and C have main effects
	No interaction; only A and B have main effects
	No interaction; only A and C have main effects
	No interaction; only B and C have main effects
	No interaction; A is the only main effect
	No interaction; B is the only main effect
	No interaction; C is the only main effect
	No effects (all group means are the same)

The list of possibilities has grown from 5 to 12 by adding one more factor, so you can just imagine the realm of possibilities with seven factors.

In addition to the complexity of the problem, adding more factors typically results in the need for greater data collection. Suppose that a statistical consultant informs a company that they need to obtain 10 observations for each group in a multifactor ANOVA. Using the aforementioned scenario with the factory that has two production lines and three shifts, the company would need to collect 10 observations from each of the six shift/production-line combinations. However, imagine that this company needed to compare all of its factories that have this setup. A brewing company that might have 10 regional factories would need to collect 600 data points.

These examples are provided to stress the need for proper planning when performing a statistical experiment. They are not provided to dissuade one from performing an important analysis simply because of the need to collect a large dataset.

19.6   WHAT TO DO WHEN THE ASSUMPTIONS ARE UNREASONABLE

Recall that the ANOVA procedure required that three conditions be reasonable in order to perform the analysis: (1) the data are acquired via independent samples, (2) the population distribution for each group is normal, and (3) each group has the same variance. As with other statistical methods, if the underlying assumptions are unreasonable, the ANOVA will produce unreliable results. At this point we will briefly discuss how to handle situations in which any of these assumptions are unreasonable.

The independent-sample assumption is not really an assumption, but a method used when collecting the data. If, in the process of designing the study, it is determined that the samples are dependent, other common statistical procedures can be used. Two of the more common sets of procedures are classified as *randomized block* or *repeated measures* procedures.

Violation of the normality assumption is a major problem, unless the sample sizes are large and there are no outliers in any of the k independent samples. As mentioned earlier, a normal quantile plot can be used to determine whether any of the distributions of the sample data differs substantially from a normal distribution. If so, one procedure that can be used is the *Kruskal–Wallis test*, which is an extension of the Mann–Whitney test, and is commonly found in most software packages.

The equal variance assumption is mentioned last for two reasons: (1) if the normality assumption is unreasonable, an ANOVA procedure cannot be used, regardless of the equal variance assumption; and (2) if the normality assumption holds, there is a variation on the pooled ANOVA that can be used in its place. The nonpooled ANOVA, which is formally known as the *Welch ANOVA*, is an extension of the nonpooled t test to be used when comparing three or more means. In addition, post hoc tests assuming unequal variances also exist and should be used to determine which pairs of means are different should the Welch ANOVA lead to the conclusion that at least two of the group means are different.

GLOSSARY

analysis of variance (ANOVA) A statistical technique that is used to test whether the means of three or more populations are equal.

F distribution A continuous distribution that has two parameters: df for the numerator and df for the denominator.

grand mean Average of the data across all samples

interaction Occurs when two or more factors have a multiplicative effect on group means.

main effect Occurs when the levels of one or more factors have different group means. Main effects are not discussed in one-way ANOVA.

mean square for groups (MSG) A measure of the variation among means of samples taken from different populations.

mean square for error (MSE) A measure of the variation within data of all samples taken from different populations.

N-way ANOVA The analysis of variance technique that uses N grouping variables, or factors.

one-way ANOVA The analysis of variance technique that uses one grouping variable, or factor.

paired comparisons Method that determines which pairs of means are significantly different. Used only when the null hypothesis of equal means is rejected. Also called *post hoc* tests.

SSE The sum of squares for error.

SST The total sum of squares given by the sum of SSG and SSE.

SSG The sum of squares for groups.

two-way ANOVA The analysis of variance technique that uses two grouping variables, or factors.

EXERCISES

19.1 Find the critical value of F for the following:
    (a) df $= (5,12)$ and area in the right tail $= .05$
    (b) df $= (4,18)$ and area in the right tail $= .10$
    (c) df $= (12,7)$ and area in the right tail $= .01$

19.2 The following ANOVA table, based on information obtained for three samples selected from three independent populations that are normally distributed with equal variances, has a few missing values:

Source	Sum of Squares	df	Mean Square	F	P
Group		2	19.2813		
Error	89.3677				
Total		12			

    (a) Find the missing values to complete the ANOVA table.
    (b) Using $\alpha = .01$, what is your conclusion for the test with the null hypothesis that the means of the three populations are all equal against the alternative hypothesis that not all three population means are equal?

19.3 A large manufacturer of copying machines recently hired three new salespersons with degrees in marketing, mathematics, and sociology. The company wants to check if the fields of study have any effect on the mean number of sales made by salespersons. The following table lists the number of sales made by these three salespersons during certain randomly selected days:

Salesperson with Marketing Degree	Salesperson with Mathematics Degree	Salesperson with Sociology Degree
9	2	4
10	1	1
3	3	1
7	2	3
4	5	6
12	3	8
8	1	1

Using the 5% significance level, can you reject the null hypothesis that the mean numbers of copying machines sold per day by all salespersons with degrees in each of these three areas are the same?

19.4 A consumer agency wanted to investigate whether four insurance companies differed with regard to the premiums charged for auto insurance. The agency randomly selected a few auto drivers who were insured by each of these four companies and had almost similar driving records, autos, and insurance policies. The following table gives premiums paid per month by these drivers insured with these four insurance companies:

Company A	Company B	Company C	Company D
65	48	57	62
73	69	61	53
54	88	89	45
43		77	51
		69	

Using the 1% significance level, test the null hypothesis that the mean auto insurance premiums paid per month by all drivers insured by each of these four companies are the same.

19.5 Three new brands of fertilizer that a farmer can use to grow crops just came on the market. Before deciding which brand he should use permanently for all crops, a farmer decided to experiment for one season. To do so, he randomly assigned each fertilizer to eight 1-acre tracts of land that he used to grow wheat. The following table gives the production of wheat (in bushels) for each acre for three brands of fertilizer:

Fertilizer 1	Fertilizer 2	Fertilizer 3
72	58	61
69	42	58
75	53	63
59	47	68
64	45	55
68	52	65
71	47	59
67	57	63

At the 5% significance level, can you conclude that the mean yield of wheat for each of these three brands of fertilizer is the same?

19.6 COPE Inc. owns a chain of department stores, including three in three different sections of New York City. The management wants to determine whether the mean gross sales per day are the same for these three stores. The research department at the company collected data on the gross sales for randomly selected days for each of these three stores. The following table lists the gross sales (rounded to thousands of dollars) for these stores for the selected days:

Store 1	Store 2	Store 3
43	52	38
37	49	45
57	59	40
39	57	49
46	39	36
51	58	42
47	63	39

At the 1% significance level, will you reject the null hypothesis that the mean gross sales per day for each of these three stores are the same?

20 Linear and Multiple Regression

20.1 OVERVIEW

Regression analysis is the process of developing the estimate of a mathematical model for the purpose of describing the relationship between a variable, called a *response variable,* and a set of one or more variables called *predictor* or *explanatory variables.* (*Note*: Some books refer to these variables as *dependent* and *independent variables*, respectively, which often leads to confusion with dependent and independent events.) By using the regression equation, we can evaluate the magnitude of change in one variable due to a certain change in another variable. For example, an economist can estimate the amount of change in food expenditure due to a certain change in the income of a household by using the regression equation. A company's research department can estimate the increase in gross sales due to a particular increase in its advertising expenditure by using the regression equation.

The variables in regression models can be quantitative or categorical. For the purpose of this book, we will discuss regression models having one quantitative response variable. These models will be divided into two groups. The first group will contain models having only one explanatory variable, which will also be quantitative. The second group will have two or more explanatory variables, at least one of which will be quantitative. The first group of models consists of *simple regression models*, while the second group of models consists of *multiple regression models.*

In addition to answering the preceding questions, a regression equation helps in predicting the value of one variable for a set of given values of the explanatory variables. For example, by using the predictive regression equation, we can predict the (approximate) food expenditure of a household with a given income.

20.2 SIMPLE REGRESSION MODEL

Let us return to the example of an economist investigating the relationship between food expenditure and income. What factors or variables does a household consider when deciding how much money should be spent on food every week or every month? Certainly, household income is one factor. However, many other variables also affect food expenditure. For instance, the assets owned by the household, the size of the household, the preferences and tastes of household members, and any special dietary needs of household members are some of the variables that will influence a household's decision about food expenditure. These variables are called *explanatory variables* because they explain the variation in food expenditure among different households. In other words, these variables explain why different households spend different amounts of money on food. Food expenditure is called the *response variable* because we are examining how the amount of money spent on food responds to changes in the explanatory variables. Studying the effect of two or more explanatory variables on a response variable using regression analysis is called *multiple regression.* However, if we choose only one explanatory variable and study the effect of that single variable on a response variable, it is called a *simple regression.* Thus, a simple regression includes only two variables: one response

Practitioner's Guide for Statistics and Lean Six Sigma for Process Improvements. By Mikel J. Harry, Prem S. Mann, Ofelia C. de Hodgins, Christopher J. Lacke, and Richard Hulbert
Copyright © 2010 John Wiley & Sons, Inc.

variable and one explanatory variable. Note that whether it is a simple or a multiple regression analysis, it always includes one and only one response variable. It is the number of explanatory variables that changes in simple and multiple regressions.

20.3   LINEAR REGRESSION

The relationship between two variables in a regression analysis is expressed by a mathematical equation called a *regression equation*. A regression equation, when plotted, may assume one of many possible shapes, including that of a line. A regression equation that gives a linear relationship between two variables is called a *linear regression equation*; otherwise, it is called a *nonlinear regression equation*. [*Note*: the term "regression" was first used by Sir Francis Galton (1822–1911), who studied the relationship between the heights of children and the heights of their parents.] Figure 20.1 shows linear and nonlinear relationships between the response variable food expenditure and the explanatory variable income.

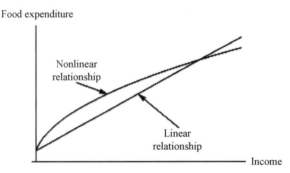

Figure 20.1   Linear and nonlinear relationships between food expenditure and income.

A *linear relationship* between income and food expenditure indicates that as income increases, the food expenditure always changes (increases, in this case) at the same rate. However, a *nonlinear relationship* between income and food expenditure shows that as income increases, the rate at which the food expenditure changes (increases, in this case) is not always the same. The nonlinear relationship shown in Figure 20.1 demonstrates that the rate of increase in food expenditure is lower for every subsequent increase in income.

The equation of a linear relationship between two variables x and y is written as

$$y = a + bx$$

Where each set of values of a and b gives a different line. For instance, when $a = 50$ and $b = 5$, then the above equation becomes

$$y = 50 + 5x$$

To plot a line, we need to know two points that lie on that line. We can find two points on a line by assigning any two values to x and then calculating the corresponding values of y. For the equation $y = 50 + 5x$, when

1. $x = 0$, then $y = 50 + 5(0) = 50$
2. $x = 10$, then $y = 50 + 5(10) = 100$

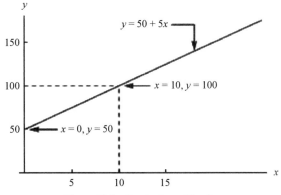

Figure 20.2 Graph of $y = 50 + 5x$.

These two points are plotted in Figure 20.2. By joining these two points we obtain the line representing the equation $y = 50 + 5x$.

Note that in Figure 20.2, the line intersects the y (vertical) axis at 50. Consequently, 50 is called the y *intercept*. The y intercept is given by the constant term in the equation. It is the value of y when x is zero. In the equation $y = 50 + 5x$, the 5 is called the *coefficient* of x or the *slope* of the line. It gives the amount by which y changes when there is a one unit increase in the value of x. For example, if

1. $x = 10$, then $y = 50 + 5(10) = 100$
2. $x = 11$, then $y = 50 + 5(11) = 105$

Hence, as x increases by one unit (from 10 to 11), y increases by five units (from 100 to 105). This is true for any value of x. Such changes in x and y are shown in Figure 20.3.

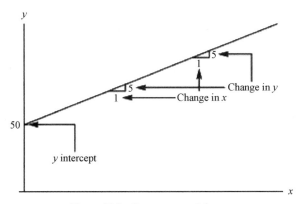

Figure 20.3 Constant rate of change.

In general, when an equation is written in the form

$$y = a + bx$$

a represents the y intercept and b represents the slope of the line. In other words, a represents the point where the line intersects the y axis and b gives the amount of change in y due to an increase of one unit in x. Here b is also called the *coefficient* of x. (*Note*: Those of you who remember your algebra classes from high school might recall using m for the slope and b for the y intercept. Statisticians have been using the a–b notation for years.)

20.3.1   Simple Linear Regression Analysis

In a regression model, the explanatory variable is usually denoted by x and the response variable is usually denoted by y. The x variable, with its coefficient, is written on the right side of the equality ($=$) sign, whereas the y variable is written on the left side of this $=$ sign. The y intercept and the slope, which we earlier denoted by a and b, respectively, can be represented by any of the many commonly used symbols. Let us denote the y intercept (which is also called the *constant term*) by A, and the slope (or the coefficient of the x variable) by B. Then, our simple linear regression model is written as

$$y = A + Bx \tag{20.1}$$

where $y = $ response variable, $A = y$-intercept, $B = $ slope, and $x = $ explanatory variable.

In model (20.1), A gives the value of y for $x = 0$, and B gives the change in y due to a one unit increase in x. Model (20.1) is called a *deterministic model*. It gives an exact relationship between x and y. This model simply states that y is determined exactly by x and for a given value of x there is one and only one (unique) value of y. If we plot such a (deterministic) model on a graph, all the points that will satisfy this equation will fall on the straight line.

However, in many cases the relationship between variables is not exact. For instance, if y is food expenditure and x is income, model (20.1) would state that food expenditure is determined by income only and that all households with the same income will spend the same amount on food. But as mentioned earlier, food expenditure is determined by many variables, only one of which is included in model (20.1). In reality, different households with the same income spend different amounts of money on food because of differences in the size of the household, the assets they own, and their preferences and tastes. Hence, to take these variables into consideration and to make our model complete, we add another term to the right side of model (20.1). This term is called the random error term. It is denoted by ε (cursive Greek letter epsilon). The complete regression model is written as

$$y = A + Bx + \varepsilon \tag{20.2}$$

The regression model 20.2 is called a *probabilistic model* or a *nondeterministic model*. The random error term ε is included in the model to represent the following two phenomena:

1. *Missing or Omitted Variables.* As mentioned earlier, food expenditure is affected by many variables other than income. The random error term ε is included to capture the effect of all those missing or omitted variables that have not been included in the model.

2. *Random Variation.* Human behavior is unpredictable. For example, a household may have many parties during one month and may spend more than usual on food during that month. The same household may spend less than usual during another month because it spent quite a bit of money to buy furniture. The variation in food expenditure for such reasons may be called *random variation*.

In model (20.2), A and B are the *population parameters*. The regression line obtained for model (20.2) by using the population data is called the *population regression line*. The values of A and B in the population regression line are called the *true values of the y intercept and slope*.

However, population data are difficult to obtain. As a result, we almost always use sample data to estimate model (20.2). The values of the y intercept and slope calculated from sample data on x and y are called the *estimated values of A and B* and are denoted by a and b. Using a and b, we write the *estimated regression equation* as

$$\hat{y} = a + bx \tag{20.3}$$

where $\hat{y}$ is the *estimated or predicted value of y* for a given value of x.

**TABLE 20.1   Monthly Incomes and Food Expenditures of
Seven Households**

Income (hundreds of dollars)	Food Expenditure (hundreds of dollars)
63	16
88	25
38	13
70	19
27	9
51	15
44	16

20.3.2   Scatterplots

Suppose that we take a sample of seven households and collect information on their incomes and food expenditures for the past month. The information obtained (in hundreds of dollars) is given in Table 20.1.

In Table 20.1, we have a pair of observations for each of the seven households. Each pair consists of one observation on income and a second on food expenditure. For example, the first household's income for the past month was $6300 and its food expenditure was $1600. By plotting all seven pairs of values, we obtain a *scatterplot*. Figure 20.4 gives the scatterplot for the data of Table 20.1. Each dot in this diagram represents one household. A scatterplot is helpful in detecting a relationship between two variables. For example, by looking at the scatterplot of Figure 20.4, we can observe that there exists a strong linear relationship between food expenditure and income. If a straight line is drawn through the points, the points will be scattered closely around the line.

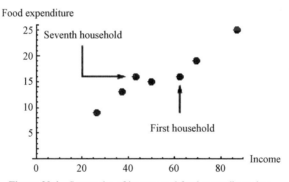

Figure 20.4   Scatterplot of income and food expenditure data.

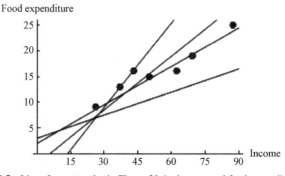

Figure 20.5   Lines for scatterplot in Figure 20.4—income and food expenditure data.

As shown in Figure 20.5, a large number of lines can be drawn through the scatterplot of Figure 20.4. Each of these lines will give different values for *a* and *b* of the estimated regression equation (20.3).

In regression analysis, we try to find a line that *best* fits the points in the scatterplot. While it is relatively easy to discern whether a specific line is a lousy choice or a good choice, visually attempting to determine which line is the best is quite difficult. One commonly used method for determining the "best" line utilizes the *method of least squares*, which is discussed next. The line obtained by using the method of least squares is called the *least squares regression line*.

20.3.2.1 Least Squares Line

The value of *y* obtained for a member from the survey is called the *observed* or *actual value of y*. As mentioned in Section 20.2, the value of *y*, denoted by $\hat{y}$, obtained for a given *x* by using the regression line is called the *predicted value of y*. The random error ε denotes the difference between the actual value of *y* and the predicted value of *y* for population data. For example, for a given household, ε is the difference between what this household actually spent on food during the past month and what is predicted using the population regression line. The term ε is also called the *residual*, as it measures the surplus (positive or negative) of actual food expenditure over what is predicted by using the regression model. If we estimate model (20.2) by using sample data, the difference between the actual *y* and predicted *y* on the basis of this estimation cannot be denoted by ε. The error for the predicted regression equation is denoted by *e*. Thus, *e* is an estimator of ε. If we estimate model (20.2) using sample data, then the value of *e* is given by

$$e = (\text{actual food expenditure}) - (\text{predicted food expenditure}) = y - \hat{y}$$

In Figure 20.6, *e* is the vertical distance between the actual position of a household and the point on the least-squares regression line. Note that in such a diagram, we always measure the response variable on the vertical axis and explanatory variable on the horizontal axis.

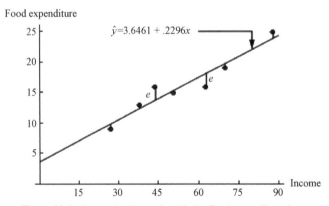

Figure 20.6   Regression line and residuals–Food expenditure data.

The value of an error is positive if the point that gives the actual food expenditure is above the regression line and negative if it is below the regression line. The sum of these errors is always zero. In other words, the sum of the actual food expenditures for seven households included in the sample will be the same as the sum of the food expenditures predicted from the regression equation. Thus

$$\Sigma e = \Sigma(y - \hat{y}) = 0$$

Hence, to find the line that best fits the data, we cannot minimize the sum of the errors (residuals). Instead, we minimize the *error sum of squares*, denoted by SSE, which is obtained by adding the squares of errors. Thus

$$\text{SSE} = \Sigma e^2 = \Sigma(y - \hat{y})^2$$

The least squares method gives the values of a and b for model (20.3) such that the sum of squared errors (SSE) is minimized. These values of a and b are called the *least squares estimates* of A and B. The least-squares values of a and b are computed using the following formulas

$$b = \frac{SS_{xy}}{SS_{xx}} \quad \text{and} \quad a = \bar{y} - b\bar{x}$$

where

$$SS_{xy} = \Sigma(x - \bar{x})(y - \bar{y}) = \Sigma xy - \frac{(\Sigma x)(\Sigma y)}{n} \quad \text{and} \quad SS_{xx} = \Sigma(x - \bar{x})^2 = \Sigma x^2 - \frac{(\Sigma x)^2}{n}$$

The least-squares regression line $\hat{y} = a + bx$ is also called the *least squares regression of y on x*.

Calculating the least squares regression line without the use of statistical software is extremely cumbersome. Nevertheless, we will present a simple example to demonstrate the calculations. The results in all other examples will be provided using MINITAB software. Example 20.1 illustrates how to estimate a regression line for sample data.

Example 20.1   Find the least squares regression line for the data on incomes and food expenditures of seven households given in Table 20.1. Use income as the explanatory variable and food expenditure as the response variable.

Solution   We are to find the values of a and b for the estimated regression equation $\hat{y} = a + bx$. Table 20.2 (based on Table 20.1) shows the calculations required for the computations of a and b. We denote the explanatory variable (income) by x and the response variable (food expenditure) by y.

The following steps are performed to compute a and b:

Step 1. Compute $\bar{x}$ and $\bar{y}$:

$$\bar{x} = \frac{\Sigma x}{n} = \frac{381}{7} = 54.4286 \quad \text{and} \quad \bar{y} = \frac{\Sigma y}{n} = \frac{113}{7} = 16.1429$$

TABLE 20.2

Income (x)	Food Expenditure (y)	xy	x^2
63	16	(63)(16) = 1008	3969
88	25	(88)(25) = 2200	7744
38	13	(38)(13) = 494	1444
70	19	(70)(19) = 1330	4900
27	9	(27)(9) = 243	729
51	15	(51)(15) = 765	2601
44	16	(44)(16) = 704	1936
$\Sigma x = 381$	$\Sigma y = 113$	$\Sigma xy = 6744$	$\Sigma x^2 = 23,323$

Step 2. Compute SS_{xy} and SS_{xx}:

$$SS_{xy} = \Sigma xy - \frac{(\Sigma x)(\Sigma y)}{n} = 6744 - \frac{(381)(113)}{7} = 593.5714$$

$$SS_{xx} = \Sigma x^2 - \frac{(\Sigma x)^2}{n} = 23323 - \frac{(381)^2}{7} = 2585.7143$$

Step 3. Compute a and b:

$$b = \frac{SS_{xy}}{SS_{xx}} = \frac{593.5714}{2585.7143} = .2296$$

$$a = \bar{y} - b\bar{x} = 16.1429 - .2296(54.4286) = 3.6461$$

Thus, our predicted regression equation $\hat{y} = a + bx$ is

$$\hat{y} = 3.6461 + .2296x$$

This least squares regression line gives the *regression of food expenditure on income.*   ■

Note that we have rounded all calculations to four decimal places. Different statistical software packages will round these values to different numbers of decimal places. However, it is important to remember that we will use this regression line for prediction and estimation purposes later on, so rounding to fewer decimal places early in the process will lead to more roundoff error in our predicted values and residuals.

Using this regression equation, we can find the predicted value of y for a specific value of x. For instance, suppose that we randomly select a household whose monthly income is $6300 so that $x = 63$ (recall that x denotes income in hundreds of dollars). The predicted value of food expenditure for this household is

$$\hat{y} = 3.6461 + .2296(63) = \$18.1109 \text{ hundred}$$

In other words, on the basis of our regression line, we predict that a household with a monthly income of $6300 is expected to spend $1811.09 per month on food. This value of $\hat{y}$ can also be interpreted as the mean value of y for $x = 63$. Thus, we can state that, on average, all households with a monthly income of $6300 spend $1811.09 per month on food.

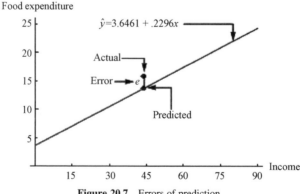

Figure 20.7   Errors of prediction.

In our data on seven households, there is one household whose income is $6300. The actual food expenditure for that household is $1600 (see Table 20.1). The difference between the actual and predicted values gives the estimated residual. Thus, the estimated residual for this household, which is shown in Figure 20.7, is

$$e = y - \hat{y} = 16 - 18.11094 = -\$2.1109 \text{ hundred}$$

Therefore, the residual is $-\$2.1109$ hundred dollars (or $-\$211.09$). The negative error indicates that the predicted value of y is greater than the actual value of y. Thus, if we use the regression equation, this household's food expenditure is overestimated by $211.09.

20.3.2.2 Interpretations of a and b

How do we interpret $a = 3.6461$ and $b = .2296$ obtained in Example 20.1 for the regression of food expenditure on income? A brief explanation of the y intercept and slope of a regression line was given earlier in this section. The next two parts of this subsection explain the meaning of a and b in more detail.

Interpretation of a. Consider a household with zero income. Using the predictive equation obtained in Example 20.1, the predicted value of y for $x = 0$ is

$$\hat{y} = 3.6461 + .2296(0) = \$3.6461 \text{ hundred}$$

Thus, we can state that a household with no income is expected to spend $364.61 per month on food. Alternatively, we can also state that the average monthly food expenditure for all households with zero income is $364.61. Note that here we have used $\hat{y}$ as a point estimate of $\mu_{y|x}$. Thus, $a = 3.6461$ gives the predicted or mean value of y for $x = 0$, based on the regression model estimated for the sample data.

However, we should be very careful while making this interpretation of a. In our sample of seven households, the incomes vary from a minimum of $2700 to a maximum of $8800. (Note that in Table 20.1, the minimum value of x is 27 and the maximum value is 88.) Hence, our regression line is valid only for the values of x between 27 and 88. If we predict y for a value of x outside this range, a process called *extrapolation*, the prediction is often questionable. The reason for this is that the pattern of expenditures may change for monthly incomes that fall outside this range. Thus, since $x = 0$ is outside the range of household incomes that we have in the sample data, the prediction that a household with zero income spends $364.61 per month on food does not carry much credibility. Many families with low incomes receive government-issued food stamps, which allow them to receive more food than approximately $365 per month.

The same is true if we try to predict y for an income greater than $8800, which is the maximum value of x in Table 20.1. Someone making $50,000 per month ($600,000 per year), would have a predicted monthly food expenditure of

$$\hat{y} = 3.6461 + .2296(500) = \$118.4461 \text{ hundred}$$

or $11,844.61, which equates to an average of $131.61 per meal. Although someone could be this extravagant, it is highly unlikely.

Interpretation of b. The value of b in a regression equation gives the change in y (response variable) due to an increase of one unit in x (explanatory variable). For example, by using the regression line obtained in Example 20.1, when $x = 50$, then $\hat{y} = 3.6461 + .2296(50) = \$15.1261$, and when $x = 51$, then $\hat{y} = 3.6461 + .2296(51) = \$15.3557$. Hence, when x increased by one unit, from 50 to 51, $\hat{y}$ increased by $15.3557 - 15.1261 = .2296$, which is the value of b. Because our unit of measurement is hundreds of dollars, we can state that, on average, a $100 increase in income will cause a $22.96 increase in food expenditure. We can also state that, on average, a $1 increase in income of a household will increase the food expenditure by $.2296. Note the phrase "on average" in these statements. The regression line is seen as a measure of the mean value of y for a given value of x. If one household's income is increased by $100, that household's food expenditure may or may not increase by $22.96. But if the incomes of all households are increased by $100 each, the average increase in their food expenditures will be very close to $22.96.

Note that when b is positive, an increase in x will lead to an increase in y and a decrease in x will lead to a decrease in y. In other words, when b is positive, the movements in x and y are in the same direction. Such a relationship between x and y is called a *positive linear relationship*. The regression line in this case slopes upward from left to right. On the other hand, if the value of b is negative, an increase in x will cause a decrease in y and a decrease in x will cause an increase in y. The changes in x and y in this case are in opposite directions. Such a relationship between x and y is called a *negative linear relationship*. The regression line in this case slopes downward from left to right. The two diagrams in Figure 20.8 show these two cases.

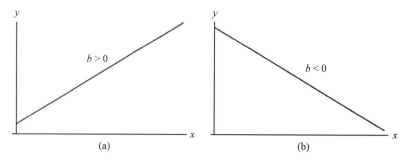

Figure 20.8   Positive (a) and negative (b) linear relationships between x and y.

Remember that b is computed as $b = SS_{xy}/SS_{xx}$. The value of SS_{xx} is always positive, and that of SS_{xy} can be positive or negative. Hence, the sign of b depends on the sign of SS_{xy}. If SS_{xy} is positive (as in our example on incomes and food expenditures of seven households) then b will be positive. On the other hand, b will be negative if SS_{xy} is negative.

20.3.3   Assumptions of the Regression Model

Like any other theory, the linear regression analysis is also based on a set of underlying conditions or assumptions. Consider the population regression model:

$$y = A + Bx + \varepsilon \tag{20.4}$$

There are four assumptions regarding this model, which are described next. These assumptions are explained with reference to the example regarding incomes and food expenditures of households. Note that these assumptions are made about the population regression model and not about the predictive (sample) regression equation.

Assumption 1.   The random error term ε has a mean equal to zero for each x. In other words, among all households with the same income, some spend more than the predicted food expenditure (and, hence, have positive errors) and others spend less than the predicted food expenditure (and, consequently, have negative errors). This assumption simply states that the sum of the positive errors is equal to the sum of the negative errors so that the mean of errors for all households with the same income is zero. Thus, when the mean value of ε is zero, the mean value of y for a given x is equal to $A + Bx$ and is written as

$$\mu_{y|x} = A + Bx$$

The quantity $\mu_{y|x}$ is read as the *mean value of y for a given value of x*. When we find the values of A and B for model (20.4) using the population data, the points on the regression line give the average values of y, denoted by $\mu_{y|x}$ for the corresponding values of x for the population.

Assumption 2. The errors associated with different observations are independent. According to this assumption, the errors for any two households in our example are independent. In other words, all households decide independently how much to spend on food.

Assumption 3. For any given x, the distribution of errors is normal. The corollary of this assumption is that the food expenditures for all households with the same income are normally distributed.

Assumption 4. The distribution of population errors for each x has the same (constant) standard deviation, which is denoted by σ_ε. This assumption indicates that the spread of points around the regression line is similar for all x values.

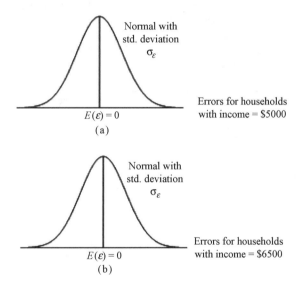

Figure 20.9   Errors for households with incomes of (a) $5000 and (b) $6500 per month.

Figure 20.9 illustrates the meaning of the first, third, and fourth assumptions for households with incomes of $5000 and $6500 per month. The same assumptions hold true for any other income level. In the population of all households, there will be many households with a monthly income of $5000. Using the population regression line, if we calculate the errors for all these households and prepare the distribution of these errors, it will resemble the distribution given in Figure 20.9a. Its standard deviation will be σ_ε. Similarly, Figure 20.9b gives the distribution of errors for all those households in the population whose monthly income is $6500. Its standard deviation is also σ_ε. Both of these distributions are identical. Note that the mean of both of these distributions is $E(\varepsilon) = 0$.

A Note on the Use of Simple Linear Regression.   We should apply linear regression with caution. When we use simple linear regression, we assume that the relationship between two variables is

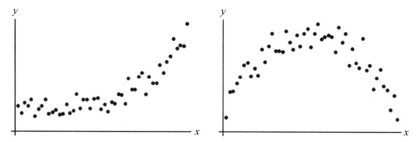

Figure 20.10 Nonlinear relationships between x and y.

described by a straight line. In the real world, the relationship between variables may not be linear. Hence, before we use a simple linear regression, it is better to construct a scatterplot and look at the plot of the data points. We should estimate a linear regression model only if the scatterplot indicates such a relationship. The scatterplots of Figure 20.10 give two examples where the relationship between x and y is not linear. Consequently, fitting linear regression in such cases would be wrong.

As you may have noticed, the calculation of the least-squares regression line is rather cumbersome when done by hand. The calculations are usually performed using statistical software, so the examples that occur throughout the remainder of this chapter will contain MINITAB output. Detailed directions for obtaining the output can be found in the accompanying MINITAB supplement. The basic command for accessing the set of regression commands in MINITAB is STAT > REGRESSION > REGRESSION. Figure 20.11 contains the MINITAB output generated for the income and food expenditure data given in Example 20.1.

```
Regression Analysis: Food Expenditure versus Income

The regression equation is                        (1)
Food Expenditure = 3.65 + 0.230 Income

Predictor    Coef SE Coef    T      P             (2)
Constant    3.648    1.802 2.02 0.099
Income    0.22956 0.03122 7.35 0.001

S = 1.58733   R-Sq = 91.5% R-Sq(adj) = 89.8% (3)

Analysis of Variance                              (4)
Source         DF     SS     MS     F       P
Regression      1 136.26 136.26 54.08 0.001
Residual Error 5  12.60    2.52
Total           6 148.86
```

Figure 20.11 MINITAB regression output for Example 20.1.

As you can see, the predictive regression equation is written at the top of the output with the names of the variables inserted, as opposed to the basic terms x and $\hat{y}$. The next section of the output contains the values of a and b to more decimal places. These values appear in the Coef column, which stands for "Coefficient". The intercept will always appear in the row labeled Constant, while the slope appears in the row labeled with the name of the explanatory variable. The remaining portions of the output will be discussed as we progress through the chapter.

20.3.4 Standard Deviation of Random Errors

When we consider income and food expenditures, all households with the same income are expected to spend different amounts on food. Consequently, the random error ε will assume different values for these households. The standard deviation σ_ε measures the spread of these errors around the population regression line. The *standard deviation of errors* tells us how widely the errors and, hence, the values of y are spread for a given x. In Figure 20.10, the points on the vertical line through $x = 50$ give the monthly food expenditures for all households with a monthly income of \$5000. The distance of each dot from the point on the regression line gives the value of the corresponding error. The standard deviation of errors σ_ε measures the spread of such points around the population regression line. The same is true for $x = 65$ or for any other value of x.

Note that σ_ε denotes the standard deviation of errors for the population. However, usually σ_ε is unknown. In such cases, it is estimated by s_e, which is the standard deviation of errors for the sample data. The following is the basic formula for calculating the standard deviation of errors s_e

$$s_e = \sqrt{\frac{\text{SSE}}{n-2}} \quad \text{where} \quad \text{SSE} = \Sigma(y-\hat{y})^2$$

where SSE means the *sum of the squared errors* and $n-2$ represents the *degrees of freedom for the simple linear regression model*. The rationale for df $= n-2$ is that we lose one degree of freedom to calculate $\bar{x}$ and one for $\bar{y}$. Example 20.2 demonstrates the calculation of s_e and SSE $= \Sigma(y-\hat{y})^2$ using the food expenditure data, then shows how to determine the values using the MINITAB output.

Example 20.2   Compute the standard deviation of errors s_e for the data on monthly incomes and food expenditures of seven households given in Table 20.1. Identify the relevant values in the MINITAB output.

Solution   To compute s_e, we need to know the value of $(y-\hat{y})^2$ for each observation. Using the data from Example 20.1, we prepare Table 20.3.

Since SSE $= \Sigma(y-\hat{y})^2 = 12.5981$, we obtain

$$s_e = \sqrt{\frac{\text{SSE}}{n-2}} = \sqrt{\frac{12.5981}{7-2}} = \sqrt{2.51962} = 1.5873$$

Hence, the standard deviation of errors s_e is 1.5873.   ∎

Figure 20.12 reproduces the MINITAB output from Example 20.1, as was shown in Figure 20.11.

TABLE 20.3   Residuals and Squared Residuals for Food Expenditure Data

Income (x)	Food Expenditure (y)	$\hat{y} = 3.6461 + .2296x$	$e = y-\hat{y}$	$(y-\hat{y})^2$
63	16	18.1109	−2.1109	4.4559
88	25	23.8509	1.1491	1.3204
38	13	12.3709	0.6291	0.3958
70	19	19.7181	−0.7181	0.5157
27	9	9.8453	−0.8453	0.7145
51	15	15.3557	−0.3557	0.1265
44	16	13.7485	2.2515	5.0693
				$\Sigma(y-\hat{y})^2 = 12.5981$

```
┌─────────────────────────────────────────────────────────────────┐
│ Regression Analysis: Food Expenditure versus Income               │
│                                                                   │
│ The regression equation is                           (1)          │
│ Food Expenditure = 3.65 + 0.230 Income                            │
│                                                                   │
│ Predictor    Coef  SE Coef     T      P              (2)          │
│ Constant    3.648    1.802  2.02  0.099                           │
│ Income    0.22956  0.03122  7.35  0.001                           │
│                                                                   │
│ S = 1.58733   R-Sq = 91.5%  R-Sq(adj) = 89.8%        (3)          │
│                                                                   │
│ Analysis of Variance                                 (4)          │
│ Source          DF     SS     MS      F     P                     │
│ Regression       1 136.26 136.26  54.08 0.001                     │
│ Residual Error   5  12.60   2.52                                  │
│ Total            6 148.86                                         │
└─────────────────────────────────────────────────────────────────┘
```

Figure 20.12   MINITAB regression output for Example 20.1.

Note the section labeled `Analysis of Variance` in Figure 20.12. The row that belongs to `Residual Error` corresponds to output related to the estimated residuals, or $y-\hat{y}$ values. The specific entries in this row are shown in the following table:

Source	DF	SS	MS
Residual Error	$n-2 = 5$	$\text{SSE} = \Sigma(y-\hat{y})^2 = 12.60$	$s_e^2 = \text{MSE} = \dfrac{\text{SSE}}{n-2} = 2.52$

Thus, we can obtain the value of s_e by taking the square root of s_e^2 or MSE as follows:

$$s_e = \sqrt{\text{MSE}} = \sqrt{\frac{\text{SSE}}{n-2}} = \sqrt{2.52} = 1.5875$$

20.4   COEFFICIENT OF DETERMINATION AND CORRELATION

We may ask the question, "How good is the regression model?" In other words, "How well does the explanatory variable explain the response variable in the regression model?" The *coefficient of determination* is one concept that answers this question.

For a moment, assume that we possess information only on food expenditures of households and not on their incomes. Hence, in this case, we cannot use the regression line to predict the food expenditure for any household. As we did in earlier chapters, in the absence of a regression equation, we use $\bar{y}$ (mean food expenditure) to estimate or predict every household's food expenditure. Consequently, the error of prediction for each household is now given by $y-\bar{y}$, which is the difference between the actual food expenditure of a household and the mean food expenditure. If we calculate such errors for all households, then square and add them, the resulting sum is called the *total sum of squares* and is denoted by SST. Actually SST is the same as SS_{yy} and is defined as

$$\text{SST} = SS_{yy} = \Sigma(y-\bar{y})^2 \quad \text{or} \quad \Sigma y^2 - \frac{(\Sigma y)^2}{n}$$

The value of SST can be found in the `Analysis of Variance` section of the MINITAB output. It is the value located at the intersection of the row labeled `Total` row and the column labeled `SS`. From the MINITAB solution given in Figure 20.12, we have SST = 148.86. From Example 20.1, $\bar{y} = 16.1429$. Figure 20.13 shows the total errors for each of the seven households in our sample.

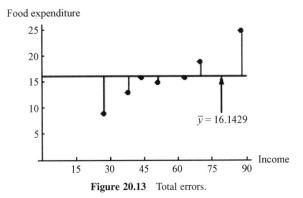

Figure 20.13 Total errors.

In Example 20.2 we used the simple linear regression equation to calculate the predicted food expenditure, the estimated residual, and the squared estimated residual for each of the seven households in our sample. Table 20.4 includes this information, as well as the values of $y-\bar{y}$ and $(y-\bar{y})^2$.

TABLE 20.4

x	y	$y-\bar{y}=y-16.1429$	$e=y-\hat{y}$	$(y-\bar{y})^2$	$(y-\hat{y})^2$
63	16	−.1429	−2.1109	0.0204	4.4559
88	25	8.8571	1.1491	78.4482	1.3204
38	13	−3.1429	0.6291	9.8778	0.3958
70	19	2.8571	−0.7181	8.1630	0.5157
27	9	−7.1429	−0.8453	51.0210	0.7145
51	15	−1.1429	−0.3557	1.3062	0.1265
44	16	−.1429	2.2515	0.0204	5.0693
				$\Sigma(y-\bar{y})^2=148.8570$	$\Sigma(y-\hat{y})^2=12.5981$

The sums of the last two columns in Table 20.4 represent the values of SST and SSE, respectively. As one can see, the value of SST is almost 12 times as large as the value of SSE. Thus, the least squares regression line has a much smaller level of errors associated with it than it would be when using $\bar{y}$ for predicting the average food expenditure.

The *errors of prediction* for the regression equation for seven households are shown in Figure 20.14.

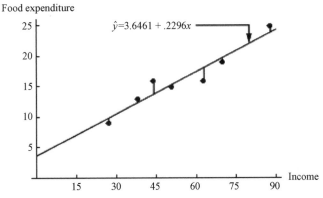

Figure 20.14 Errors of prediction when regression model is used.

As mentioned previously, the sum of squared errors decreased from 148.8570 (see Table 20.4) to 12.5981 when we used $\hat{y}$ to predict food expenditures in place of $\bar{y}$. This reduction in squared errors is called the *regression sum of squares* (SSR). Consequently

$$SSR = SST - SSE = 148.8570 - 12.5981 = 136.2589$$

The value of SSR can also be computed by using the formula

$$SSR = \Sigma(\hat{y} - \bar{y})^2$$

Thus, SSR is that portion of SST that is explained by the use of the regression model, and SSE is that portion of SST that is not explained by the use of the regression model. The sum of SSR and SSE is always equal to SST. Thus

$$SST = SSR + SSE$$

The ratio of SSR to SST gives the *coefficient of determination*. The coefficient of determination calculated for population data is denoted by ρ^2 (ρ is the Greek letter rho) and the one calculated for sample data is denoted by R^2. The coefficient of determination gives the proportion of SST that is explained by the use of the regression model. The value of the coefficient of determination always lies in the range from 0 to 1. The coefficient of determination can be calculated by using the formula

$$R^2 = \frac{SSR}{SST} \quad \text{or} \quad R^2 = 1 - \frac{SSE}{SST}$$

Most statistical software packages provide the value of R^2 with the basic set of regression output. Example 20.3 illustrates the calculation of the coefficient of determination using the preceding formulas and the MINITAB output from Example 20.2.

Example 20.3   For the data of Table 20.1 on monthly incomes and food expenditures of seven households, calculate the coefficient of determination.

Solution   The MINITAB output previously provided in Figure 20.11 is reproduced here as Figure 20.15.

```
Regression Analysis: Food Expenditure versus Income

The regression equation is                      (1)
Food Expenditure = 3.65 + 0.230 Income

Predictor    Coef   SE Coef    T      P        (2)
Constant    3.648    1.802   2.02   0.099
Income     0.22956  0.03122  7.35   0.001

S = 1.58733   R-Sq = 91.5%  R-Sq(adj) = 89.8% (3)

Analysis of Variance                            (4)
Source          DF    SS      MS     F      P
Regression       1 136.26  136.26 54.08  0.001
Residual Error   5  12.60    2.52
Total            6 148.86
```

Figure 20.15   MINITAB regression output for Example 20.1.

The SS column in the *Analysis of Variance* section of Figure 20.15 gives the values of SSR, SSE, and SST, respectively, from top to bottom:

$$\text{SSR} = 136.26, \qquad \text{SSE} = 12.60, \qquad \text{and} \qquad \text{SST} = 148.86$$

This provides us with the necessary information to calculate R^2:

$$R^2 = \frac{\text{SSR}}{\text{SST}} = \frac{136.26}{148.86} = .915 \qquad \text{or} \qquad R^2 = 1 - \frac{12.60}{148.86} = 1 - .085 = .915$$

Sometimes R^2 is expressed as a percentage instead of a proportion. In this case, we would have $R^2 = 91.50\%$. Many people find the percentage easier to use because it is easier to understand that SST is reduced by approximately 92% (from 148.86 to 12.60) when we use $\hat{y}$, instead of $\bar{y}$, to predict the food expenditures of households. ∎

The total sum of squares (SST) is a measure of the total variation in food expenditures, the regression sum of squares (SSR) is the portion of total variation explained by the predictive equation (or by incomes), and the error sum of squares (SSE) is the portion of total variation not explained by the predictive equation. Hence, for Example 20.3 we can state that 91.5% of the total variation in food expenditures of households occurs because of the variation in their incomes, and the remaining 8.5% is due to randomness and other variables.

Usually, the higher the value of R^2 is, the better the regression model. This is so because if R^2 is larger, a greater portion of the total errors is explained by the included explanatory variable, and a smaller portion of errors is attributed to other variables and randomness. However, the value of R^2 is irrelevant if the four assumptions discussed earlier in this section are not met.

Correlation. Another measure of the relationship between two variables is the *correlation coefficient*. This section describes the *Pearson product–moment correlation coefficient*, which is more commonly referred to simply as the *correlation coefficient*. It measures the strength of the linear association between two quantitative variables. (*Note*: There are many different types of correlation, depending on the type of variables being analyzed.) In other words, the linear correlation coefficient measures how closely the points in a scatterplot are spread around the regression line. The correlation coefficient calculated for the population data is denoted by ρ, and the one calculated for sample data is denoted by r. In the case of simple linear regression, the square of the correlation coefficient r^2 is equal to the coefficient of determination R^2. However, outside of simple linear regression, the values of r^2 and R^2 are not related. In those cases, the value of R^2 will be the relevant value to use.

Unlike R^2, which identifies only how well a specific regression equation fits a set of data, the correlation coefficient r identifies whether the association is positive, negative, or neutral. The value of the correlation coefficient always lies in the range -1 to 1:

$$-1 \leq r \leq 1 \qquad \text{and} \qquad -1 \leq \rho \leq 1$$

Although we can explain the linear correlation using the population correlation coefficient ρ, we will do so using the sample correlation coefficient r.

If $r = 1$, it is said to be a case of perfect positive correlation. In such a case, all points in the scatterplot lie on a line that slopes upward from left to right, as shown in Figure 20.16a. If $r = -1$, the correlation is said to be a perfect negative linear correlation. In this case, all points in the scatterplot fall on a line that slopes downward from left to right, as shown in Figure 20.16b. If the points are scattered throughout the diagram, as shown in Figure 20.16c, then there is no linear correlation between the two variables and consequently $r \approx 0$.

We seldom encounter an example with perfect positive or perfect negative correlation. What we observe in real-world problems is either a positive linear correlation with $0 < r < 1$ (i.e., the correlation

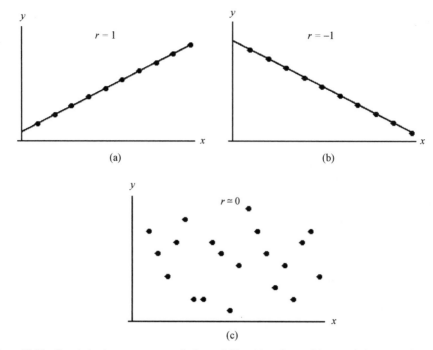

Figure 20.16   Correlation between two quantitative variables: (a) perfect positive correlation, $r = 1$; (b) perfect negative correlation, $r = -1$; (c) no correlation, $r \approx 0$.

coefficient is greater than zero but less than 1) or a negative linear correlation with $-1 < r < 0$ (i.e., the correlation coefficient is greater than -1 but less than zero).

If the correlation between two variables is positive and close to 1, we say that the variables have a *strong positive correlation*. If the correlation between two variables is positive but close to zero, then the variables have a *weak positive correlation*. On the other hand, if the correlation between two variables is negative and close to -1, then the variables are said to have a *strong negative correlation*. Also, if the correlation between variables is negative but close to zero, there exists a *weak negative correlation* between the variables. Graphically, a strong correlation indicates that the points in the scatterplot are very close to the regression line, and a weak correlation indicates that the points in the scatterplot are widely spread around the regression line. These four cases are shown in Figure 20.17, parts (a), (b), (c), and (d), respectively.

There are many algebraic versions of the formula for calculating the Pearson product–moment correlation coefficient. Since statistical software will calculate the correlation, we mention two versions that assist in understanding some of the properties of correlation

$$r = b\frac{s_x}{s_y} \quad \text{or} \quad r = \frac{SS_{xy}}{\sqrt{SS_{xx}SS_{yy}}}$$

where b = slope of regression line
s_x = standard deviation of explanatory variable, x
s_y = standard deviation of response variable, y

The first version of the equation reveals that there is a relationship between the slope of the least squares regression line and correlation. The standard deviation of a variable is almost always positive (as opposed to zero), thus the sign of r will always be the same as the sign of the slope b.

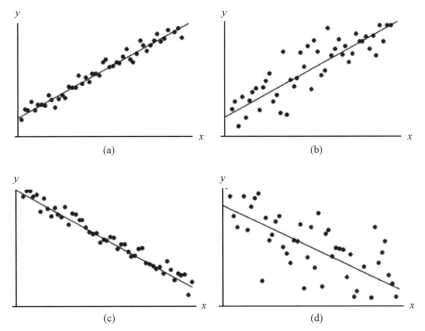

Figure 20.17   Linear correlation between variables: (a) strong positive correlation (r close to 1); (b) moderate positive correlation (r closer to .5 than 1); (c) strong negative correlation (r close to -1); (d) weak negative correlation (r negative and closer to 0).

The second version of the formula helps us understand what would happen if we were to switch the explanatory and response variables. Using our example of income (x) and food expenditure (y), if we were to predict income based on food expenditure, the values of SS_{xx} and SS_{yy} would be switched, but the value of SS_{xy} would remain the same. The value of SS_{xy} would not change because the only change in the formula involves reversing the order of the multiplication. Therefore, the value of r will be the same, regardless of the assignment of the two variables.

If you are using a software package that gives the value of the correlation coefficient in the basic output, you will not need to use the formula. If you are using software that only gives the value of R^2 in the basic output, then you can calculate the value of r by using the following rule:

$$r = \begin{cases} \sqrt{R^2} & \text{if} \quad b > 0 \\ -\sqrt{R^2} & \text{if} \quad b < 0 \end{cases}$$

If you are using the previously stated rule, make sure that the value of R^2 is given as a proportion, not as a percentage. Example 20.4 illustrates the calculation of the linear correlation coefficient r.

Example 20.4   Calculate the correlation coefficient for the example on incomes and food expenditures of seven households.

Solution   From earlier calculations made in Examples 20.1 and 20.2, we have

$$SS_{xx} = 2585.7143, \qquad SS_{yy} = 148.8571, \qquad \text{and} \qquad SS_{xy} = 593.5714$$

Substituting these values in the formula for r, we obtain

$$r = \frac{SS_{xy}}{\sqrt{SS_{xx}SS_{yy}}} = \frac{593.5714}{\sqrt{(2585.7143)(148.8571)}} = \frac{593.5714}{620.4047} = .9587$$

Thus, the linear correlation coefficient is .9587.

Using the MINITAB output most recently referenced in Example 20.3, we had $R^2 = .9150$ and $b = .2642$. Since the slope is positive, we have

$$r = \sqrt{.9150} = .9567$$

The linear correlation coefficient simply tells us how strongly the two variables are (linearly) related. The correlation coefficient of .9567 for incomes and food expenditures of seven households indicates that income and food expenditure are very strongly and positively correlated. This correlation coefficient does not, however, provide us with any more information.   ∎

Sometimes the calculated value of r indicates that the two variables are very strongly linearly correlated but they do not have a meaningful relationship. For example, if we calculate the correlation coefficient between the price of Coke and the size of families in the United States using data for the last 30 years, we will find a strong negative linear correlation. Over time, the price of Coke has increased and the size of families has decreased. This finding does not mean that family size and price of Coke are related. As a result, before we calculate the correlation coefficient, we must seek help from a theory or from common sense to postulate whether or not the two variables have a causal relationship.

20.5 MULTIPLE REGRESSION

A simple regression model includes one independent variable and one response variable, and it presents a very simplified scenario of real-world situations. In the real world, a response variable is usually influenced by a number of explanatory variables. For example, the sales of a company's product may be determined by the price of the product, the quality of the product, and advertising expenditure incurred by the company to promote that product. Therefore it makes more sense to use a regression model that includes more than one explanatory variable. Such a model is called a *multiple regression model*. Remember, whether it is a simple or a multiple regression model, it always includes one response variable.

The simple linear regression model discussed in Section 20.3 was written as

$$y = A + Bx + \varepsilon$$

A multiple regression model with y as a response variable and $x_1, x_2, x_3, \ldots, x_k$ as explanatory variables is written as

$$y = A + B_1x_1 + B_2x_2 + B_3x_3 + \cdots + B_kx_k + \varepsilon \tag{20.5}$$

where A represents the constant term and $B_1, B_2, B_3, \ldots, B_k$ are the regression coefficients of explanatory variables $x_1, x_2, x_3, \ldots, x_k$, respectively. The variable ε represents the random error term. Note that this model contains k explanatory variables: $x_1, x_2, x_3, \ldots, x_k$. Here, A is the constant term, which gives the value of y when all explanatory variables assume zero values. The coefficients $B_1, B_2, B_3, \ldots, B_k$ are called the *partial regression coefficients*. For example, B_1 is the partial regression coefficient of x_1. It gives the change in y due to a one-unit increase in x_1 when all other explanatory variables included in the model are held constant. In other words, if we increase x_1 by one unit but keep $x_2, x_3, \ldots, x_k$ unchanged, then the resulting change in y is measured by B_1.

Similarly, the value of B_2 represents the change in y due to a one-unit increase in x_2 when all other explanatory variables are held constant. In model (20.5) above, $A, B_1, B_2, B_3, \ldots, B_k$ are called the *true regression coefficients* or *population parameters*.

A positive value for a particular B_i in model (20.5) will indicate a positive relationship between y and the corresponding x_i variable. A negative value for a particular B_i in that model will indicate a negative relationship between y and the corresponding x_i variable.

Note that regression model (20.5) is linear. The relationship between each x_i and y is a linear relationship. In this model, $A + B_1 x_1 + B_2 x_2 + B_3 x_3 + \cdots + B_k x_k$ is called the *deterministic portion* and ε is called the *stochastic portion* of the model. We will discuss only multiple *linear* regression models in this chapter. Multiple *nonlinear* regression models will not be discussed here.

In a situation where a multiple regression model includes only two explanatory variables, model (20.5) reduces to

$$y = A + B_1 x_1 + B_2 x_2 + \varepsilon \qquad (20.6)$$

and in this case $k = 2$.

A multiple regression model that includes three explanatory variables is written as

$$y = A + B_1 x_1 + B_2 x_2 + B_3 x_3 + \varepsilon \qquad (20.7)$$

and in this case $k = 3$.

If model (20.5) is estimated using sample data, which is usually the case, the predictive regression equation is written as

$$\hat{y} = a + b_1 x_1 + b_2 x_2 + b_3 x_3 + \cdots + b_k x_k \qquad (20.8)$$

In equation (20.8), $a, b_1, b_2, b_3, \ldots, b_k$ are the sample statistics, and the point estimators of $A, B_1, B_2, B_3, \ldots, B_k$ respectively.

In model (20.5), y denotes the actual values of the response variable. In model (20.8), $\hat{y}$ denotes the predicted or estimated values of the response variable. The difference between a pair of y and $\hat{y}$ values gives the error of prediction. For a multiple regression model, we obtain

$$\text{SSE} = \Sigma e^2 = \Sigma (y - \hat{y})^2$$

where SSE denotes the error sum of squares or sum of squared errors.

As in Section 20.3, the predictive regression equation (20.8) is obtained by minimizing the sum of squared errors:

$$\text{Minimize } \Sigma (y - \hat{y})^2$$

The predictive regression equation (20.8) obtained by minimizing the sum of squared errors is called the *least-squares regression model*.

The calculations for a multiple regression are much more complex than the calculations for a simple linear regression, and these calculations are made using statistical software. We will describe the process using MINITAB, instead of using the formulas manually. The solutions obtained by using other statistical software packages such as JMP, SAS, S-plus, SPSS, and STATA can be interpreted in the same way.

20.5.1   Assumptions of the Multiple Regression Model

Like the simple linear regression model, the multiple (linear) regression analysis is also based on certain assumptions. Consider the multiple regression model (20.5), which is

$$y = A + B_1 x_1 + B_2 x_2 + B_3 x_3 + \cdots + B_k x_k + \varepsilon$$

Six assumptions regarding this model are described next. Note that these assumptions pertain to the population regression model, not the predictive (sample) regression equation.

Assumption 1. The random error term ε has a mean equal to zero for each set of values for the explanatory variables $x_1, x_2, x_3, \ldots, x_k$. In other words, if we calculate errors for all measurements for a given set of values of explanatory variables for a population dataset, the mean of these errors will be zero. When this assumption holds true, the mean value of y is given by the deterministic part of the regression model (20.5):

$$E(y) = A + B_1 x_1 + B_2 x_2 + B_3 x_3 + \cdots + B_k x_k$$

Here, $E(y)$ is the expected or mean value of y. The mean value of y is also denoted by $\mu_{y|x_1, x_2, \ldots x_k}$.

Assumption 2. The errors associated with different observations are independent.

Assumption 3. The probability distribution of errors is normal.

Assumption 4. The distribution of population errors for each set of values of explanatory variables $x_1, x_2, x_3, \ldots, x_k$, has the same (constant) standard deviation, which is denoted by σ_ε.

Assumption 5. The explanatory variables are not linearly related. However, they can have a nonlinear relationship. When explanatory variables are highly linearly correlated, this condition is referred to as *multicollinearity*. This assumption is about the nonexistence of the multicollinearity problem. For example, consider the following multiple regression model:

$$y = A + B_1 x_1 + B_2 x_2 + B_3 x_3 + \varepsilon$$

All of the following relationships (and other similar relationships) between x_1, x_2, and x_3 are linear and, consequently, should be invalid for this model with three explanatory variables:

$$\begin{aligned} x_1 &= x_2 + 4x_3 \\ x_2 &= 5x_1 - 2x_3 \\ x_1 &= 3.5x_2 \end{aligned}$$

If any of these linear relationships is true, by making a substitution, we can eliminate one variable from our model and reduce the number of explanatory variables to two. However, nonlinear relationships between explanatory variables are acceptable. For example, the relationships of the type $x_1 = x_2^2$ are acceptable.

Assumption 6. There exists a zero correlation between the random error term ε and each explanatory variable x_i.

20.5.2   Standard Deviation of Errors

The *standard deviation of errors* (also called the *standard deviation of the estimate*) for the multiple regression model (20.5) is denoted by σ_ε and is a measure of the variation among errors. However, when sample data are used to estimate multiple regression model (20.5), the estimated standard deviation of errors is denoted by s_e, which is also called the *standard error of the estimate*. The formula for calculating s_e is as follows:

$$s_e = \sqrt{\frac{SSE}{n-k-1}} \qquad \text{where} \qquad SSE = \Sigma(y-\hat{y})^2$$

Note that here, k is the number of explanatory variables in the model and $(n - k - 1)$ is the degrees of freedom (df) for this multiple regression model. We will not use this formula to calculate SSE manually; rather, we will obtain it from the computer solution.

20.5.3   Coefficient of Multiple Determination

In Section 20.4, we denoted the coefficient of determination for a simple linear regression model by R^2 and defined it as the proportion of the total sum of squares (SST) that is explained by the regression model. We also noted that the coefficient of determination for a simple linear regression model is equal to the square of the correlation between the explanatory and response variables:

$$\text{For simple linear regression:} \quad R^2 = r^2$$

The coefficient of determination for a multiple regression model, usually called the *coefficient of multiple determination*, is also denoted by R^2 and is defined as the proportion of SST that is explained by the predictive multiple regression equation. It tells us how good the predictive equation is and how well the explanatory variables included in the model explain the response variable.

The properties of R^2 and the components in the calculation of R^2 are the same in the simple linear regression and multiple regression situations. Specifically, SST is the total sum of squares, SSR is the regression sum of squares, and SSE is the error sum of squares. SST is always equal to the sum of SSE and SSR. They are calculated as follows:

$$SSR = \Sigma(\hat{y}-\bar{y})^2$$
$$SSE = \Sigma e^2 = \Sigma(y-\hat{y})^2$$
$$SST = SS_{yy} = \Sigma(y-\bar{y})^2$$

SSR is the portion of SST that is explained by use of the regression model and SSE is the portion of SST that is not explained by use of the regression model. The *coefficient of multiple determination* is given by the ratio of SSR and SST as follows:

$$R^2 = \frac{SSR}{SST} \qquad \text{or} \qquad R^2 = 1 - \frac{SSE}{SST}$$

and

$$0 \leq R^2 \leq 1$$

The coefficient of multiple determination R^2 has one major shortcoming. Its value either increases or remains the same, but never decreases, as we add more and more explanatory variables to the regression model. Therefore, by adding a large number of explanatory variables to our regression model (even if they do not belong in the model), we might be able to make the value of R^2 very close to 1. Such a value of R^2 will be misleading, and it will not represent the true explanatory power of the regression model. To eliminate this shortcoming of R^2, it is preferable to use the *adjusted coefficient of multiple determination*, which is denoted by $\bar{R}^2$. Note that $\bar{R}^2$ is the coefficient of multiple determination adjusted for degrees of freedom. The value of $\bar{R}^2$ may increase, decrease, or stay the same as we add more explanatory variables to our regression model. If a new variable added to the regression model contributes significantly to explain the variation in y, then $\bar{R}^2$ increases; otherwise it decreases or changes very little. The value of $\bar{R}^2$ is calculated as follows:

$$\bar{R}^2 = 1 - \left(1 - R^2\right) \frac{n-1}{n-k-1} \quad \text{or} \quad 1 - \frac{\text{SSE}/(n-k-1)}{\text{SST}/(n-1)}$$

Thus, if we know R^2, we can find the value of $\bar{R}^2$. Almost all statistical software packages give the values of both R^2 and $\bar{R}^2$ for a regression model.

Another property of $\bar{R}^2$ to remember is that whereas R^2 can never be negative, $\bar{R}^2$ can be negative.

20.6 REGRESSION ANALYSIS

The inferential methods in simple linear regression and multiple regression utilize the same procedures and, hence, we will discuss them here together. We will discuss three different inferential procedures:

1. Performing a hypothesis test of overall significance on the set of regression coefficients $B_1, B_2, B_3, \ldots, B_k$.
2. Performing a hypothesis test on an individual regression coefficient B_i.
3. Developing a confidence interval for an individual regression coefficient B_i.

We will explain these inferential methods before taking examples and applying them.

20.6.1 Testing for Overall Significance of Multiple Regression Model

In regression analysis, we can test for the overall significance of the model. We perform a test of hypothesis with the null hypothesis that the coefficients of all explanatory variables in the regression model are equal to zero and the alternative hypothesis that one or more of the coefficients of all explanatory variables are not equal to zero. In other words, we are testing to determine whether at least one of the explanatory variables is useful in predicting the value of the response variable. For the multiple regression model

$$y = A + B_1 x_1 + B_2 x_2 + B_3 x_3 + \cdots + B_k x_k + \varepsilon$$

the two hypotheses for such a test are written as

$$\begin{aligned} H_0: \quad & B_1 = B_2 = B_3 = \cdots B_k = 0 \\ H_1: \quad & \text{At least one of the } B_i \text{ is not equal to zero} \end{aligned}$$

What exactly does the null hypothesis mean here? If H_0 is true, then our model becomes

$$y = A + 0x_1 + 0x_2 + 0x_3 + \cdots + 0x_k + \varepsilon = A + \varepsilon$$

Thus, if the null hypothesis is true, the values of $x_1, x_2, x_3, \ldots, x_k$ have no influence on the calculation of y. However, if one or more of the B_i is not equal to 0, at least one of the $B_i x_i$ terms will remain in the equation.

A test of hypothesis for the overall significance of a multiple regression model is performed by using the F distribution, which was covered in Chapter 19. Although the value of the test statistic F is obtained from the computer solution, it can also be calculated by using the following formula

$$F = \frac{\text{SSR}/k}{\text{SSE}/(n-k-1)} = \frac{\text{MSR}}{\text{MSE}}$$

where MSR stands for the *mean square regression* and MSE for the *mean square error*.

A test of hypothesis for the overall significance of a multiple regression model is always a right-tailed test. The value of F for the given significance level is obtained from Table V of Appendix A for k degrees of freedom for the numerator and $(n - k - 1)$ degrees of freedom for the denominator.

A regression model in which $k = 1$ is expressed as follows

$$y = A + B_1 x_1 + \varepsilon$$

which is the same as

$$y = A + Bx + \varepsilon$$

Thus, a simple linear regression model is simply a special case of a multiple regression model with $k = 1$.

To conduct a test of overall significance for a multiple regression model, we will use the information in the area labeled `Analysis of Variance` in the MINITAB output.

20.6.2   Inferences about a Single Regression Coefficient, B_i

This section is concerned with estimation and tests of hypotheses about a single regression coefficient, B_i. We can also construct confidence intervals and test hypotheses about the y intercept A of the population regression line. However, making inferences about A is beyond the scope of this text. As with testing the significance of an overall regression model, the results for simple linear regression are a special case of the results for multiple regression. Therefore, we shall discuss them together.

20.6.2.1   *Sampling Distribution of b*

One of the main purposes for determining a regression line is to find the true value of the slope B of the population regression line. However, in almost all cases, the regression line is estimated using sample data. Then, on the basis of the estimated regression equation, inferences are made about the population regression model. The slope b of an estimated regression equation is a point estimator of slope B of the population regression line. The different estimated regression lines resulting from different samples taken from the same population will give different values of b. If only one sample is taken and the regression line for that sample is estimated, the value of b will depend on which elements are included in the sample. Thus, b is a random variable and it possesses a probability distribution that is more commonly called its *sampling distribution*.

Because of the assumption of normally distributed random errors, the sampling distribution of b is normal. The mean and standard deviation of b, denoted by μ_b and σ_b, respectively, are

$$\mu_b = B \quad \text{and} \quad \sigma_b = \frac{\sigma_\varepsilon}{\sqrt{\text{SS}_{xx}}}$$

However, the standard deviation of population errors σ_ε is rarely known. Hence, the sample standard deviation of errors s_e is used to estimate σ_ε In such a case, when σ_ε is unknown, the standard deviation of b is estimated by the standard error of b, denoted by s_b, which is calculated as

$$s_b = \frac{s_e}{\sqrt{\text{SS}_{xx}}}$$

We use the t distribution to make inferences about B.

20.6.2.2   Estimation of B for a Simple Linear Regression

The $(1-\alpha) \times 100\%$ *confidence interval for B* is given by

$$b \pm (t \times s_b) \quad \text{where} \quad s_b = \frac{s_e}{\sqrt{SS_{xx}}}$$

If the calculations are being performed by hand, the value of t is obtained from the t distribution table for the given confidence level and $n-2$ degrees of freedom. We will demonstrate the procedure through the use of MINITAB output.

20.6.2.3   Estimation of B_i for a Multiple Linear Regression

The values of $a, b_1, b_2, b_3, \ldots, b_k$ obtained by estimating model (20.5) using sample data give the point estimates of $A, B_1, B_2, B_3, \ldots, B_k$, respectively, which are the population parameters. Using the values of sample statistics $a, b_1, b_2, b_3, \ldots, b_k$, we can construct confidence intervals for the corresponding population parameters $A, B_1, B_2, B_3, \ldots, B_k$, respectively.

As with simple linear regression, the errors for a multiple linear regression are assumed to be normally distributed. Thus, the sampling distribution of each b_i is normal with its mean equal to B_i and standard deviation equal to σ_{b_i}. For example, the sampling distribution of b_1 is normal with its mean equal to B_1 and standard deviation equal to σ_{b_1}. However, σ_ε is typically unknown; therefore we use s_{b_i} as an estimator of σ_{b_i} and is obtained from MINITAB output.

The formula for a *confidence interval for a parameter B_i* is

$$b_i \pm (t \times s_{b_i})$$

This is similar to the formula we used to formulate a confidence interval for B for a simple regression. The only difference is that to construct a confidence interval for a particular B_i for a multiple regression model, the *number of degrees of freedom is $(n-k-1)$*.

20.6.2.4   Hypothesis Testing about B for a Simple Linear Regression

Testing a hypothesis about B when the null hypothesis is $H_0: B = 0$ (i.e., the slope of the regression line is zero) is equivalent to testing the hypothesis that x does not determine y and that the regression line is of no use in predicting y for a given x. However, we should remember that we are testing for a linear relationship between x and y. It is possible that x may determine y nonlinearly. Hence, a nonlinear relationship may exist between x and y.

To test the hypothesis that x does not determine y linearly, we will test the null hypothesis that the slope of the regression line is zero, that is, $H_0: B = 0$. The alternative hypothesis can be one of the following:

1. There is a significant linear relationship between the explanatory variable x and the response variable y, that is, $H_1: B \neq 0$
2. There is a significant positive linear relationship between the explanatory variable x and the response variable y, that is, $H_1: B > 0$
3. There is a significant negative linear relationship between the explanatory variable x and the response variable y, that is, $H_1: B < 0$

The procedure used to make a hypothesis test about B is similar to the one used in earlier chapters. It involves the same five steps. It is based on the t distribution, and has the following test statistic with $n-2$ degrees of freedom:

$$t = \frac{b-0}{s_b}$$

Note that the test $H_0: B = 0$ versus $H_1: B \neq 0$ is equivalent to testing for the overall significance of a simple linear regression model.

20.6.2.5 *Hypothesis Testing about Individual Coefficients for a Multiple Linear Regression*

We can formulate a test of hypothesis about any of the B_i coefficients of model (20.5) using the same procedure that we used to make a test of hypothesis about B for a simple regression model above. The only difference is the degrees of freedom, which are equal to $(n-k-1)$ for a multiple regression model where k is the number of explanatory variables in the model.

Again, because of the assumption that the errors are normally distributed, the sampling distribution of each b_i is normal with its mean equal to B_i and standard deviation equal to σ_{b_i}. Once again, σ_ε is seldom known; therefore we use s_{b_i} as an estimator of σ_{b_i} and base the test on the t distribution. The value of the test statistic t for b_i is calculated as

$$t = \frac{b_i - 0}{s_{b_i}}$$

As with the simple linear regression case, the null hypothesis to test for a single coefficient in a multiple regression is $H_0: B_i = 0$ and the alternative hypothesis can be one of the following three alternative hypotheses:

1. $H_1: B_i \neq 0$
2. $H_1: B_i > 0$
3. $H_1: B_i < 0$

However, a test of $H_0: B_i = 0$ in a multiple regression assumes that all the other variables belong to the model. For example, consider the multiple regression model:

$$y = A + B_1 x_1 + B_2 x_2 + B_3 x_3 + \varepsilon$$

Suppose that we are to test the null hypothesis $H_0: B_2 = 0$. If we fail to reject the null hypothesis, then we are literally eliminating the $B_2 x_2$ term from the model. Prior to running any further hypothesis tests, one must develop a predictive equation for the adjusted model:

$$y = A + B_1 x_1 + B_3 x_3 + \varepsilon$$

20.7 USING THE REGRESSION MODEL

Example 20.5   A random sample of eight auto drivers insured with a company and having similar auto insurance policies was selected. The following table lists their driving experience (in years) and the monthly auto insurance premium (in dollars) paid by them:

Driving experience (years)	5	2	12	9	15	6	25	16
Monthly auto insurance premium (dollars)	92	127	73	104	65	82	62	87

1. Does the insurance premium depend on driving experience, or does the driving experience depend on insurance premium? Do you expect a positive or a negative relationship between these two variables?
2. Find the least squares regression line by choosing appropriate response and explanatory variables on the basis of your answer in part 1.

3. Interpret the meaning of the values of a and b calculated in part 2.
4. Plot the scatterplot and the regression line.
5. Calculate r and R^2 and explain what they mean.
6. Predict the monthly auto insurance premium for a driver with 10 years of driving experience.
7. Compute the standard deviation of errors.
8. Construct a 90% confidence interval for B.
9. Test at the 5% significance level if B is negative.

Solution

1. According to theory and intuition, we expect the insurance premium to depend on driving experience. Consequently, the insurance premium would be a response variable and driving experience an explanatory variable in the regression model. A new driver is considered a high risk by insurance companies and has to pay a higher premium for auto insurance. On average, the insurance premium is expected to decrease with an increase in the years of driving experience. Therefore, we expect a negative relationship between these two variables. In other words, both the population correlation coefficient ρ and the population regression slope B are expected to be negative.

2. The remaining answers will be generated using MINITAB. Figure 20.18 contains the MINITAB output resulting from the command STAT > Regression > Regression and placing Monthly Premium in the Response box and Driving Experience in the Predictor (or explanatory) box.

   Using part (1) in Figure 20.18, our estimated regression line $\hat{y} = a + bx$ is

$$\hat{y} = 101 - 2.01x$$

where x represents the driving experience and y the monthly premium. If more precision is needed, one can use the values given in the Coef column in part (2) of the output. In this case, the predictive regression equation is

$$\hat{y} = 101.044 - 2.0150x$$

```
Regression Analysis: Monthly Premium versus Driving Experience

The regression equation is
Monthly Premium = 101 - 2.01 Driving Experience      (1)

Predictor              Coef    SE Coef     T       P    (2)
Constant             101.044     9.170  11.02   0.000
Driving Experience   -2.0150     0.6942  -2.90   0.027

S = 13.5940   R-Sq = 58.4%  R-Sq(adj) = 51.5%         (3)

Analysis of Variance                                  (4)
Source          DF      SS      MS      F      P
Regression       1   1557.1  1557.1   8.43   0.027
Residual Error   6   1108.8   184.8
Total            7   2665.9
```

Figure 20.18 MINITAB output for Example 20.5.

3. The value of $a = 101.044$ gives the value of y for $x = 0$, that is, gives the monthly auto insurance premium for a driver with no driving experience. However, as mentioned earlier in this chapter, this is an example of extrapolation because the sample contains drivers with only 2 or more years of experience. Therefore, we should not attach much importance to this statement. The value of b gives the change in y due to a change of one unit in x. Thus, $b = -2.0150$ indicates that, on average, for every extra year of driving experience the monthly auto insurance premium decreases by $2.015. Note that when b is negative, y decreases as x increases.

4. Figure 20.19 shows the scatterplot and the regression line for the data on eight auto drivers. Note that the regression line slopes downward from left to right. This result is consistent with the negative relationship we anticipated between driving experience and insurance premium. The scatterplot can be created in MINITAB by using the GRAPH > Scatterplot > With Regression command.

Figure 20.19   Scatterplot of monthly premium versus driving experience.

5. The value of R^2 is given as a percentage in section (3) of Figure 20.18, which is 59.0%. As a decimal, the value is $R^2 = .584$. Since the slope of the predictive regression equation is negative, the value of r is the negative square root of .584. Thus, $r = -\sqrt{.584} = -.76$ indicates that the driving experience and the monthly auto insurance premium are negatively related. The linear relationship is strong but not very strong. The value of $R^2 = .584$ states that 58.4% of the total variation in insurance premiums is explained by years of driving experience, and 41.6% is not. The value of R^2 indicates that there may be many other important variables that contribute to the determination of auto insurance premiums. For example, the premium is expected to depend on the driving record of a driver and the type and age of the car. As an additional note, recall that the value of R^2 can be calculated by using the values in the SS column in section (4) of Figure 20.18:

$$R^2 = \frac{\text{SSR}}{\text{SST}} = \frac{1557.1}{2665.9} = .584$$

6. Using the estimated regression line, the predicted value of y for $x = 10$ is

$$\hat{y} = 101.044 - 2.0150(10) = 80.894$$

Thus, we expect the monthly auto insurance premium of a driver with 10 years of driving experience to be $80.89.

7. The standard deviation of errors s_e is given in section (3) of Figure 20.18 as S. This value is 13.5940. We can also derive the standard deviation of errors using the information given in section (4) of Figure 20.18. The value is $s_e = \sqrt{\text{MSE}} = \sqrt{184.8} = 13.594$.

8. To construct a 90% confidence interval for B, we need s_b, the standard deviation of b. This value of s_b is located in section (2) of Figure 20.18. The value can be found at the intersection of the SE Coef column and the Driving Experience row. Specifically, $s_b = .6942$. The degrees of freedom for this interval is $n-2 = 8-2 = 6$. From the t distribution table, the value of t for a 90% confidence level with 6 degrees of freedom is 1.943. Then, the 90% confidence interval for B is

$$b \pm (t \times s_b) = -2.0150 \pm (1.943 \times .6942) = -2.0150 \pm 1.3488 = -3.3638 \text{ to } -.6662$$

Thus, we can state with 90% confidence that B lies in the interval -3.36 to $-.66$. That is, on average, the monthly auto insurance premium of a driver decreases by an amount between $.66 and $3.36 for every extra year of driving experience.

9. We perform the following five steps to test the hypothesis about B.

Step 1. Select the type of test to use and check the underlying conditions.
A hypothesis test for the slope of a population regression line utilizes the t test. The data came from a simple random sample. Other underlying assumptions pertain to the distribution of residuals, which will be covered in Section 20.8. For now we will assume that the assumptions are reasonable, although one would typically verify them before proceeding to the test.

Step 2. State the null and alternative hypotheses:

$$H_0: B = 0 \quad \text{(the population slope is not negative)}$$
$$H_1: B < 0 \quad \text{(the population slope is negative)}$$

Note that the null hypothesis can also be written as $H_0: B \geq 0$.

Step 3. Determine the significance level α:
The significance level is $\alpha = .05$.

Step 4. Calculate the test statistic and the p-value.
The value of the test statistic is

$$t = \frac{b-0}{s_b} = \frac{-2.0150-0}{.6942} = -2.90$$

The test statistic and the information regarding the p-value are provided in the Driving Experience row of section (2) of Figure 20.18. The value .027 in the P column in section (2) of Figure 20.18 is the p-value for a two-sided alternative, $H_1: B \neq 0$. To determine the p-value for a one-sided test, use the following rule. If the sign of the test statistic t is the same as the direction of the alternative hypothesis, that is, if t is negative and $H_1: B < 0$, or if t is positive and $H_1: B < 0$, divide the value in the P column by 2. If the sign of the test statistic is not the same as the direction of the alternative hypothesis, divide the value in the P column by 2 and subtract the result from 1. In this problem, t is negative and $H_1: B < 0$, so the p-value for the test is $.027/2 = .0135$.

Step 5. Make a decision.
The significance level is .05. Using the p value, we ask whether the p-value is $\leq \alpha$? Since the p-value is .0135, it is less than or equal to .05. Therefore, we reject the null hypothesis. Our conclusion is that there is a significant negative linear association between driving experience and monthly premium.

■

TABLE 20.5

Monthly Premium (dollars)	Driving Experience (years)	Number of Driving Violations (past 3 years)
102	5	2
51	14	0
71	6	1
90	10	3
163	4	6
74	8	2
79	11	3
57	16	1
168	3	5
64	9	1
47	19	0
85	13	3

Example 20.6 A researcher wanted to find the effect of driving experience and the number of driving violations on auto insurance premiums. A random sample of 12 drivers from a low-risk neighborhood insured with a company and having similar auto insurance policies was selected. Table 20.5 lists the monthly auto insurance premiums (in dollars) paid by these drivers, their driving experience (in years), and the number of driving violations that each of them has committed during the past 3 years.

1. Estimate the regression model $y = A + B_1 x_1 + B_2 x_2 + \varepsilon$.
2. Explain the meaning of the estimated regression coefficients.
3. What are the values of the standard deviation of errors, the coefficient of multiple determination, and the adjusted coefficient of multiple determination?
4. What is the predicted auto insurance premium paid per month by a driver with 7 years of driving experience and three driving violations?
5. What is the point estimate of the expected (or mean) auto insurance premium paid per month by all drivers with 12 years of driving experience and four driving violations?
6. Determine a 95% confidence interval for B_1 (the coefficient of driving experience) for the multiple regression of auto insurance premium on driving experience and the number of driving violations.
7. Using the 1% significance level, can you conclude that the slope of the number of driving violations in regression model of part a is positive?
8. Using the 2.5% significance level, can you conclude that the coefficients of all explanatory variables in the regression model of part a are equal to zero?

Solution We are to estimate the regression model

$$y = A + B_1 x_1 + B_2 x_2 + \varepsilon \tag{20.9}$$

where $y =$ monthly auto insurance premium (in dollars) paid by a driver
$x_1 =$ driving experience (in years) of a driver
$x_2 =$ number of driving violations committed by a driver during past 3 years

Figure 20.20 contains the MINITAB output resulting from the command STAT > Regression > Regression and placing Premium in the Response box and Experience and Driving Violations in the Predictor (or explanatory) box.

```
Regression Analysis: Premium versus Experience,  Driving Violations

The regression equation is
Premium = 73.5 - 2.21 Experience + 15.9 Driving Violations (1)

Predictor              Coef   SE Coef      T       P        (2)
Constant              73.47     15.83   4.64   0.001
Experience           -2.213     1.058  -2.09   0.066
Driving Violations   15.945     2.830   5.63   0.000


S = 13.1543    R-Sq = 91.0%   R-Sq(adj) = 89.1%            (3)

Analysis of Variance                                       (4)
Source            DF       SS      MS      F       P
Regression         2  15827.6  7913.8  45.73   0.000
Residual Error     9   1557.3   173.0
Total             11  17384.9


Source             DF    Seq SS
Experience          1   10335.8
Driving Violations  1    5491.8
```

Figure 20.20   MINITAB output for Example 20.6.

1. From part (1) in Figure 20.20, the predictive (sample) regression equation is

$$\hat{y} = 73.5 - 2.21x_1 + 15.9x_2 \tag{20.10}$$

If more precision is needed, one can use the values given in the Coef column in part (2) of the output. In this case, the predictive regression equation is

$$\hat{y} = 73.47 - 2.213x_1 + 15.945x_2 \tag{20.11}$$

2. The value of $a = 73.47$ in the estimated regression equation (20.11) gives the value of $\hat{y}$ for $x_1 = 0$ and $x_2 = 0$. Thus, a driver with no experience and no driving violations is expected to pay an auto insurance premium of $73.47 per month. Again, this is the technical interpretation of a. In reality that may not be true because none of the drivers in our sample has both zero experience and zero driving violations. The value of $b_1 = -2.213$ in the predictive regression equation gives the change in $\hat{y}$ for a one-unit change in x_1 when x_2 is held constant. Thus, we can state that a driver with one extra year of experience but the same number of driving violations is expected to pay $2.213 (or $2.21) less for the auto insurance premium per month. Note that because b_1 is negative, an increase in experience decreases the premium paid. In other words, $\hat{y}$ and x_1 have a negative relationship. The value of $b_2 = 15.945$ in the predictive regression equation gives the change in $\hat{y}$ for a one-unit change in x_2 when x_1 is held constant. Thus, a driver with one extra driving violation but with the same years of driving experience is expected to pay $15.945 (or $15.95) more per month for the auto insurance premium. Note that because b_2 is positive, an increase in violations increases the premium paid. In other words, $\hat{y}$ and x_2 have a positive relationship.

3. The values of the standard deviation of errors, the coefficient of multiple determination, and the adjusted coefficient of multiple determination are given in section (3) of Figure 20.20. These values are

$$s_e = 15.1543, \quad R^2 = .910, \quad \text{and} \quad \bar{R}^2 = .891.$$

Thus, the standard deviation of errors is 15.1543. The value of $R^2 = .910$ tells us that the two explanatory variables, years of driving experience and the number of driving violations, included in our model explain 91.0% of the variation in the response variable. The value of $\bar{R}^2 = .891$ is the value of the coefficient of multiple determination adjusted for degrees of freedom. It states that when adjusted for degrees of freedom, the two explanatory variables explain 89.1% of the variation in the response variable.

4. To find the predicted auto insurance premium paid per month by a driver with 7 years of driving experience and three driving violations, we substitute $x_1 = 7$ and $x_2 = 3$ in the predictive regression equation (20.11):

$$\hat{y} = 73.47 - 2.213(7) + 15.945(3) = 73.47 - 15.491 + 47.835 = 105.814$$

Thus, the predicted premium is $105.81.

5. To obtain the point estimate of the expected (mean) auto insurance premium paid per month by all drivers with 12 years of driving experience and four driving violations, we substitute $x_1 = 12$ and $x_2 = 4$ in the estimated regression equation (20.11):

$$\hat{y} = 73.47 - 2.213(12) + 15.945(4) = 73.47 - 26.556 + 63.780 = 110.694$$

Thus, the predicted premium is $110.69.

6. To make a confidence interval for B_1, we use the Experience row from part (2) of Figure 20.20. From Figure 20.20 we have

$$n = 12, \quad b_1 = -2.213, \quad \text{and} \quad s_{b_1} = 1.058.$$

The confidence level is 95%. The number of degrees of freedom is $n - k - 1 = 12 - 2 - 1 = 9$, where k is the number of explanatory variables in the model. From Table III of Appendix A, the appropriate value of t is 2.262. The 95% confidence interval for B_1 is

$$b_1 \pm (t \times s_{b_1}) = -2.213 \pm (2.262 \times 1.058) = -2.213 \pm 2.393 = -4.606 \text{ to } .16$$

Thus, the 95% confidence interval for B_1 is -4.606 to .16. Thus, we can state with 95% confidence that for one extra year of driving experience, the monthly auto insurance premium will change by an amount between an increase of $.16 and a decrease of $4.61. Using this procedure, we can make a confidence interval for any of the coefficients of a multiple regression model, such as A and B_2 in model (20.9). For example, the 95% confidence intervals for A and B_2, respectively, are

$$a \pm (t \times s_a) = 73.47 \pm (2.262 \times 15.83) = 37.66 - 109.28$$

$$b_2 \pm (t \times s_{b_2}) = 15.945 \pm (2.262 \times 2.830) = 9.54 - 22.35$$

7. We are to test whether B_2 is positive. We perform the following five steps to test this hypothesis about B_2:

Step 1. Select the type of test to use and check underlying conditions.

A hypothesis test about a single coefficient of a multiple regression coefficient is performed using the t distribution. In this example, the data came from a simple random sample. Other underlying assumptions pertain to the distribution of residuals, which will be covered in Section 20.8. For now we will assume that the assumptions are reasonable, although one would typically verify them before proceeding to the test.

Step 2. State the null and alternative hypotheses:

$$H_0: B_2 = 0 \quad \text{(the population slope is not positive)}$$
$$H_1: B_2 > 0 \quad \text{(the population slope is positive)}$$

Step 3. Determine the significance level α:
The significance level is $\alpha = .01$.

Step 4. Calculate the test statistic and the p-value.
The test statistic and information needed to calculate the p-value are provided in the `Driving Violations` row of part (2) of the MINITAB solution given in Figure 20.20. These values are

$$t = \frac{b_2 - 0}{s_{b_2}} = \frac{15.945 - 0}{2.830} = 5.63$$
$$\text{P} = .000$$

Recall from Example 20.5 part 9 that we need to perform an additional calculation to obtain the p-value when the alternative hypothesis is one-sided. In this case, the test statistic is positive and the alternative hypothesis is $H_1: B_2 > 0$, so the p-value is half of the value in the P column of MINITAB solution. The value P $= .000$ is not really zero, but a number that is less than .0005. MINITAB rounds .0005 to .001, and a number less than .0005 (such as .0003) to .000. Thus, for this example, the p-value is half of the value in the P column, so it is also less than .0005.

Step 5. Make a decision.
The significance level is .01. Using the p-value, we ask whether the p-value is $\leq \alpha$. Since the p-value is less than .0005, it is less than or equal to $\alpha = .01$. Therefore, we reject the null hypothesis. Our conclusion is that there is a significant positive association between the number of violations in the last three years and the monthly premium.

Note that the observed value of t in step 4 of part 7 is obtained from the MINITAB solution only if the null hypothesis is $H_0: B_2 = 0$. However, if the null hypothesis is that B_2 is equal to a number other than zero, then the t value obtained from the MINITAB solution is no longer valid. For example, suppose the null and alternative hypotheses in part 7 are:

$$H_0 : B_2 = 10$$
$$H_1 : B_2 > 10$$

In this case the observed value of t will be calculated as

$$t = \frac{b_2 - 10}{s_{b_2}} = \frac{15.945 - 10}{2.830} = 2.101$$

Using the t table (Table III of Appendix A), we see that $P(t \geq 1.833) = .05$ and $P(t \geq 2.262) = .025$. Thus

$$.025 < P(t \geq 2.101) < .05$$

The p-value for this test is greater than the significance level of .01, so we fail to reject $H_0 : B_2 = 10$.

8. We are to test whether or not the coefficients of all explanatory variables are equal to zero. This is a test for the overall significance of the regression model. To performs this test, we will use part (4) of the MINITAB output given in Figure 20.20:

Step 1. Select the type of test to use and check underlying conditions.
A hypothesis test for the set of coefficients of all explanatory variables utilizes the F test. The data came from a simple random sample. Other underlying assumptions relate to the distribution of residuals, which will be covered in Section 20.8. For now we will assume that the assumptions are reasonable, although one would typically verify them before proceeding to the test.

Step 2. State the null and alternative hypotheses:

$$H_0: B_1 = B_2 = 0 \quad \text{(all coefficients are zero)}$$
$$H_1: \text{at least one of the } B_i \neq 0 \quad \text{(at least one coefficient is not zero)}$$

Step 3. Determine the significance level α:
The significance level is $\alpha = .025$

Step 4. Calculate the test statistic and the p-value.
The test statistic and p-value for this test are provided in part (4) of the MINITAB solution in Figure 20.20. Their values are

$$F = \frac{\text{MSR}}{\text{MSE}} = \frac{7913.8}{173.0} = 45.73$$

$$P = .000$$

The degrees of freedom for the test statistic are

$$\text{Numerator } df = k = 2$$
$$\text{Denominator } df = n-k-1 = 9$$

The p-value is less than .0005, which MINITAB rounds to .000.

Step 5. Make a decision.
The significance level is .025. Using the p-value, we ask whether the p-value is $\leq \alpha$. Since the p-value is less than .0005, it is less than or equal to .025. Therefore, we reject the null hypothesis. Our conclusion is that at least one of the B_i is not zero, which implies that at least one of the explanatory variables is useful in predicting the monthly insurance premium. ∎

20.8   RESIDUAL ANALYSIS

In Sections 20.3 and 20.5, we mentioned a series of assumptions regarding the residuals in simple linear and multiple regression. A basic method for assessing the validity of these assumptions is to examine a *residual plot*, which is a scatterplot that involves the residuals. In simple linear regression, the most common residual plot involves the explanatory variable on the horizontal axis and the residuals on the vertical axis. In multiple regression, the predicted ($\hat{y}$) values are on the horizontal axis and the residuals are on the vertical axis.

The rule for assessing a residual plot is simple. If the residual plot has no pattern to it (i.e., it shows a random scatter), the model used is a good choice for representing the data. If the residual plot has a pattern, then the model is inappropriate. There are many reasons why a model could be inappropriate, such as leaving out an explanatory variable or choosing the wrong mathematical function. A study of these reasons is beyond the scope of this book. Examples 20.7–20.9 demonstrate examples of residual plots for three different datasets.

Example 20.7   Figures 20.21a and 20.21b are the scatterplot and corresponding residual plot for a dataset that is best modeled by a line. Note that the residual plot shows a random scatter. The points (residuals) are scattered throughout the plot, with no obvious pattern. The horizontal line running through the middle of the residual plot occurs where a residual equals zero. Points above the line

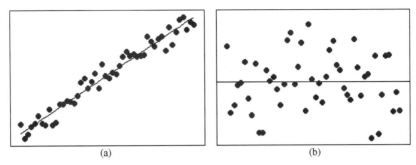

Figure 20.21   (a) Scatterplot of two linearly related variables; (b) residual plot resulting from fitting the least-squares regression line to the data in (a).

correspond to residuals with positive values, while those below the horizontal line have negative values.   ■

Example 20.8   Figure 20.22a is the scatterplot for a dataset that is best modeled by a curved function. Figure 20.22b is the residual plot that results when a straight line is fitted to the data in Figure 20.22a and corresponding residual plot for a dataset that is best modeled by a curved function. Note that the residual plot has a pattern, which happens to be curved. Figure 20.22c is the residual plot when the best curved function is used to model the data in Figure 20.22a. Note that the residual plot is now a random scatter. Thus, a random scatter will occur when the appropriate model is fitted to a dataset.   ■

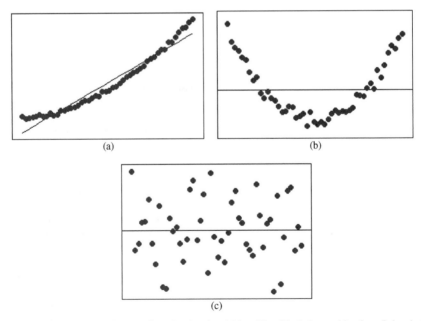

Figure 20.22   (a) Scatterplot of two nonlinearly related variables; (b) residual plot resulting from fitting the least-squares regression line to the data in (a); (c) residual plot resulting from fitting the least-squares regression curve to the data in (a).

Example 20.9   Figure 20.23a is the scatterplot for a dataset in which the variance of the errors increases as the value of the explanatory variable increases. Note that for low values of x, the observed y values are close to the least-squares regression line. However, as x increases, the data become more spread out. The residual plot, shown in Figure 20.23b, also demonstrates this pattern. Whenever this type of pattern exists, it is called a "fan". Fans can also go in the opposite direction, with the residuals getting closer to zero as the explanatory variable gets larger. Typically, this type of plot identifies that at least one more explanatory variable is needed in order to adequately predict the response variable. (*Note*: The technical term for this type of situation is *heteroskedasticity*, which means that the variance of the residuals is different for different values of the explanatory variable. It is important to recognize that there are shapes other than a fan that imply the existence of heteroskedasticity, but the fan shape is one of the more common forms of heteroskedasticity.)

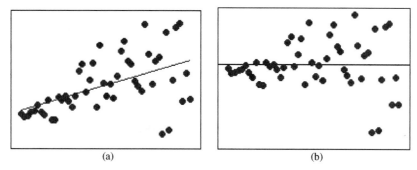

(a)   (b)

Figure 20.23   (a) Scatterplot representing nonconstant variance; (b) residual plot resulting from fitting the least-squares regression line to the data in (a). The shape of the plot is called a "fan." ∎

20.9   CAUTIONS IN USING REGRESSION

Throughout this chapter, we have mentioned potential issues that can arise when using regression. In this section we will examine a few other potential issues such as relying solely on the values of r, R^2, and the p-value to determine whether a model is good or bad, the impact of outliers and influential observations, multicollinearity, extrapolation, and causality.

20.9.1   Determining whether a Model is Good or Bad

Too often, people who use statistics rely on numbers to make conclusions, without looking at graphs to confirm their results. Example 20.10 provides one set of results in which using only the numbers would result in a bad decision.

Example 20.10   Figure 20.24 contains output from a simple least-squares regression procedure. The value of R^2 for this model, which is shown in section (3) of Figure 20.24, is .831. Since the slope is positive, the value of the correlation is $r = \sqrt{.831} = .912$. These pieces of information imply that the model is very good, and that there is a very strong correlation between x and y. Moreover, the p-value for the test $H_0 : B = 0$ versus $H_1 : B \neq 0$ is .000, so one would once again conclude that x contributes linearly to the prediction of y.

Figures 20.25a and 20.25b give the scatterplot of the data and the residual plot, respectively. Here, the plot of residuals results from fitting the least-squares regression line to the data. As is evident from the scatterplot, the relationship between x and y is not linear, but curved. However, this curved relationship oscillates around the line. The regression formulas and the hypothesis test do not "see"

```
The regression equation is
y = 121 + 19.0 x                                            (1)

Predictor   Coef   SE Coef    T      P
Constant   121.15     18.72  6.47  0.000
x           19.040     1.241 15.35  0.000                   (2)

S = 63.2916   R-Sq = 83.1%  R-Sq(adj) = 82.7%              (3)

Analysis of Variance
Source            DF      SS       MS      F      P
Regression         1   943711   943711  235.58  0.000      (4)
Residual Error    48   192280     4006
Total             49  1135991
```

Figure 20.24   Output from a simple least squares regression procedure.

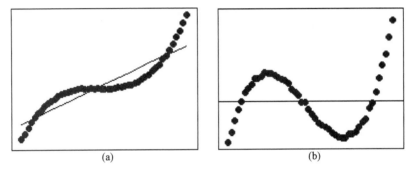

(a)   (b)

Figure 20.25   (a) Scatterplot and (b) residual plot for data in Example 20.10.

this pattern. Instead, all that the formulas recognize is that the points are fairly close to a line. However, the residual plot shows a pattern in the residuals, not a random scatter. Therefore, a curved model would be more appropriate for representing the relationship between these variables.   ∎

20.9.2   Outliers and Influential Observations

As we just saw, the formulas for r and R^2 can result in misleading values and conclusions. Two other causes of misleading values and conclusions are outliers and influential observations. Figure 20.26a is a scatterplot of a dataset that has an outlier, while Figure 20.26b has two influential observations.

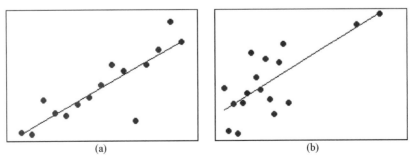

(a)   (b)

Figure 20.26   (a) Scatterplot involving one outlier; (b) scatterplot involving two influential observations.

In regression, an *outlier* is a data point that falls far from the general pattern of the data. In Figure 20.26a, the observation that falls at the lower right of the graph is much farther away from the regression line than all of the other data points, so it appears to be an outlier.

In Figure 20.26b, most of the data are widely spread around the regression line, with what appears to be a very weak linear relationship. However, the two observations in the upper right fall close to this weak pattern, even though they are far away from the rest of the observations. These two observations are called *influential observations* because their positions will influence the regression line by pulling it toward them.

To understand the impact of outliers and influential observations, we can look at the values of r and R^2 when they are included in the regression process, as well as when they are ignored. For the data in above two figures, we obtain

With outlier:	$R^2 = .733$ and	$r = .856$
Without outlier:	$R^2 = .915$ and	$r = .957$
With influential observations:	$R^2 = .570$ and	$r = .755$
Without influential observations:	$R^2 = .195$ and	$r = .442$

Thus, outliers reduce the value of R^2 and move the value of r toward zero. Essentially, they can cause a relationship to appear weaker. Whenever an outlier exists, it is important to determine the source of the outlier to determine whether it is an anomaly, a recording error, or a sign of a production flaw. If an outlier is due to something other than a recording error, it is important to include the value in the analysis.

Influential observations increase the value of R^2 and move the value of r away from zero. They can make a relationship seem stronger. Although the effects of outliers and influential observations are opposite in nature, the recommendations for handling them are similar.

20.9.3 Multicollinearity

Multicollinearity occurs when two or more explanatory variables in a multiple regression model are strongly correlated. When multicollinearity exists, including both explanatory variables in the multiple regression model will usually lead to a very small increase in R^2, but a substantial decrease in $\bar{R}^2$. This occurs because the two variables provide redundant information.

A simple way to determine if multicollinearity exists is to create a scatterplot and calculate the correlation for each pair of explanatory variables. Many statistical software packages include a command that will create a set of graphs and a table of correlations for all explanatory variables at the same time. This is especially useful when the number of explanatory variables is large. A set of just six explanatory variables requires 15 comparisons, so it is easy to understand how using a single command once instead of a command 15 times can be a great time-saver.

20.9.4 Extrapolation

The regression line estimated for the sample data is valid only for the range of x values observed in the sample. For example, the values of x in our example on incomes and food expenditures vary from a minimum of 15 to a maximum of 49. Hence, our estimated regression line is only applicable for values of x between 15 and 49, that is, we should use this regression line to estimate the mean food expenditure or to predict the food expenditure of a single household only for income levels between $1500 and $4900. If we estimate or predict y for a value of x either less than 15 or greater than 49, it is called *extrapolation*. This does not mean that we should never use the regression line for extrapolation. Instead, we should interpret such predictions cautiously and not attach much value to them.

Similarly, if the data used for the regression estimation are time-series data, the predicted values of y for periods outside the time interval used for the estimation of the regression line should be interpreted very cautiously. When using the estimated regression line for extrapolation, we are assuming that the same linear relationship between the two variables holds true for the values of x outside the given range. The relationship between the two variables might not be linear outside that range. Nonetheless, even if it is linear, adding a few more observations at either end will probably give a new estimation of the regression line.

20.9.5 Causality

Note that a regression line does not prove causality between the two variables; that is, the regression equation does not predict that a change in y is caused by a change in x. The information about causality is based on theory or common sense. A regression line only describes whether a significant quantitative relationship between x and y exists. Significant relationship means that we reject the null hypothesis $H_0 : B = 0$ at a given significance level. The estimated regression line gives the change in y due to a change of one unit in x. Note that it does not indicate that y has changed because x has changed. In our example on incomes and food expenditures, it is economic theory and common sense, not the regression line, that tell us that food expenditure depends on income. The regression analysis simply helps us determine whether this dependence is significant.

GLOSSARY

adjusted coefficient of multiple determination   Denoted by $\bar{R}^2$, it gives the proportion of SST that is explained by the multiple regression model and is adjusted for the degrees of freedom.

coefficient of determination   A measure that gives the proportion (or percentage) of the total variation in a response variable that is explained by a given explanatory variable.

coefficient of multiple determination   Denoted by R^2, it gives the proportion of SST that is explained by the multiple regression model.

degrees of freedom for a multiple regression model with k explanatory variables $n-k-1$

degrees of freedom for a simple linear regression model   Sample size minus 2, that is, $n-2$.

deterministic model   A model in which the explanatory variable determines the response variable exactly. Such a model gives an exact relationship between two variables.

estimated or predicted value of y   The value of the response variable, denoted by $\hat{y}$, that is calculated for a given value of x (or set of x values) using the estimated regression equation.

explanatory (predictor/independent) variable   The variable(s) included in a model to explain the variation in the response variable.

extrapolation   Predicting values of y using values of predictor variables that are outside the range of the existing data.

influential observation   A data point that is far away from most of the data, but close to the general pattern.

least-squares estimates of A and B   The values of a and b that are calculated by using the sample data.

least-squares method   The method used to fit a regression line through a scatterplot such that the error sum of squares is minimized.

least-squares regression line   A regression line obtained by using the least-squares method.

Pearson product-moment correlation coefficient   A measure of the strength of the linear relationship between two variables.

least-squares regression model   The estimated regression model obtained by minimizing the sum of squared errors.

linear regression model   A regression model that gives a straight-line relationship between two variables.

mean square error (MSE)   Given by $\text{SSE}/(n-k-1)$.

mean square regression (MSR)   Given by SSR/k.

multicollinearity   Exists when two or more predictor variables are strongly correlated.

multiple regression model   A regression model that contains two or more explanatory variables.

negative relationship between two variables   As the value of predictor variable increases, the value of the response variable decreases. The value of the slope in the regression line and the correlation coefficient between two variables are both negative.

nonlinear (simple) regression model   A regression model that does not give a straight-line relationship between two variables.

outlier   An observation that is far away from the general pattern between two or more variables. An outlier will have a residual that has a large absolute value.

partial regression coefficients   The coefficients of explanatory variables in a multiple regression model are called the *partial regression coefficients* because each of them gives the effect of the corresponding explanatory variable on the response variable when all other explanatory variables are held constant.

population parameters for a simple regression model   The values of A and B for the regression model $y = A + Bx + \varepsilon$ that are obtained by using population data.

positive relationship between two variables   As the value of predictor variable increases, the value of the response variable decreases. The value of the slope in the regression line and the correlation coefficient between two variables are both positive.

probabilistic or statistical model   A model in which the explanatory variable(s) does/do not determine the response variable exactly.

random error term (e)   The difference between the actual and predicted values of y.

response variable   The variable to be predicted or explained.

scatterplot   A plot of the paired observations of x and y.

simple linear regression   A regression model with one dependent variable and one explanatory variable that assumes a straight-line relationship.

slope   The coefficient of x in a regression model that gives the change in y for a one unit increase in x.

standard deviation of errors   Also called the *standard deviation of estimate*, it is a measure of the variation among errors.

SSE (error sum of squares)   The sum of the squared differences between the actual and predicted values of y. It is that portion of the SST that is not explained by the regression model.

SSR (regression sum of squares)   That portion of the SST that is explained by the regression model.

SST (total sum of squares)   The sum of the squared differences between actual y values and y.

standard deviation of errors   A measure of spread for the random errors.

EXERCISES

20.1   A car rental company charges \$30 a day and 10 cents per mile for renting a car. Let y be the total rental charges (in dollars) for a car for one day and x be the miles driven. The equation for the relationship between x and y is

$$y = 30 + .10x$$

(a) How much will a person pay who rents a car for one day and drives it 100 mi.?

(b) Suppose each of 20 persons rents a car from this agency for one day and drives it 100 mi. Will each of them pay the same amount for renting a car for a day or is each person expected to pay a different amount? Explain.

(c) Is the relationship between x and y exact? Will all the points for this relationship fall on a straight line?

20.2 A researcher took a sample of 25 electronic companies and found the following relationship between x and y, where x is the amount of money (in millions of dollars) spent on advertising by a company in 2008 and y represents the total gross sales (in millions of dollars) of that company for 2008:

$$y = 3.4 + 11.55x$$

(a) An electronics company spent \$2 million on advertising in 2008. What are its expected gross sales for 2008?

(b) Suppose that four electronics companies spent \$2 million each on advertising in 2008. Do you expect these four companies to have the same actual gross sales for 2008? Explain.

(c) Is the relationship between x and y exact? Will all the points for this relationship fall on a straight line?

20.3 An economist wanted to determine whether the amount of phone bills and incomes of households are related. The following table gives information on the biweekly incomes (in hundreds of dollars) and monthly telephone bills (in dollars) for a random sample of 10 households:

Income	16	45	36	32	30	13	41	15	36	40
Phone bill	35	78	102	56	75	26	130	42	59	85

(a) Find the regression line with income as an explanatory variable and the amount of the phone bill as a response variable.

(b) Give a brief interpretation of the values of a and b calculated in part (a).

(c) Estimate the amount of the monthly phone bill for a household with a biweekly income of \$5000.

(d) Find the values of the following: standard deviation of errors, SST, SSE, SSR, coefficient of determination, and correlation coefficient.

(e) Make a 98% confidence interval for B.

(f) An earlier study claims that for the relationship between incomes and amounts of phone bills, $B = 1.50$. Test at the 1% significance level if B is less than 1.50.

(g) Create and examine a scatterplot and residual plot to determine whether a line is appropriate for modeling the relationship between *biweekly income* and *monthly telephone bill*. Comment on your reasoning.

20.4 A consumer welfare agency wants to investigate the relationship between the sizes of houses and rents paid by tenants in a small town. The agency collected the following information on the sizes (in hundreds of square feet) of houses and the monthly rents (in dollars) paid by tenants for six houses:

Size of the house	21	16	19	27	34	23
Monthly rent	1700	1580	1720	1850	2050	1800

(a) Construct a scatterplot for these data. Does the scatterplot show a linear relationship between the sizes of houses and monthly rents?

(b) Find the predictive regression line $\hat{y} = a + bx$ with the size of a house as an explanatory variable and monthly rent as a response variable.

(c) Give a brief interpretation of the values of a and b calculated in part (b).

(d) Plot the regression line on the scatterplot of part (a) and show the errors by drawing vertical lines between the scatter points and the regression line.

(e) Predict the monthly rent for a house with 2500 square feet.

(f) One of the houses in our sample is 2700 square feet and its rent is $1850. What is the predicted rent for this house? Find the error for this observation.

(g) Compute the standard deviation of errors.

(h) Calculate the coefficient of determination. What percentage of the variation in monthly rents is explained by the sizes of the houses? What percentage of this variation is not explained?

(i) Construct a 99% confidence interval for B.

(j) Testing at the 5% significance level, can you conclude that B is different from zero?

20.5   The American Manufacturing Company makes baseball bats. The management of the company wants to know the relationship between the number of bats produced and the total cost of production. The manager of the company gathered data on the number of baseball bats (in thousands) manufactured and the total cost (in thousands of dollars) of production for 8 selected months. These data are given in the following table:

Baseball bats	56	40	44	48	54	51	46	59
Total cost	795	658	664	682	770	747	692	806

(a) Determine which of the two variables is the explanatory variable and which is the response variable. Do you expect a positive or a negative relationship between these two variables?

(b) Find the least-squares regression line.

(c) Interpret the meaning of the values of a and b calculated in part (b).

(d) Plot the scatterplot, the regression line, and the residual plot. Is a linear relationship appropriate for representing these data? Explain why.

(e) Calculate r and R^2 and briefly explain what they mean.

(f) Predict the total cost of production for 49,000 baseball bats.

(g) Compute the standard deviation of errors.

(h) Construct a 95% confidence interval for B.

(i) Test at the 1% significance level if B is positive.

20.6   The salaries of workers are expected to depend, among other factors, on the number of years that they have spent in school and their work experience. The following table gives information on the biweekly salaries (in hundreds of dollars) for 12 workers, the number of years each of them spent in school, and the total number of years of work experience:

Salary	32	24	28	57	48	28	39	43	18	41	17	49
Schooling	16	12	13	20	18	16	14	18	12	16	12	16
Experience	6	10	15	8	11	2	12	4	6	9	2	8

Using MINITAB (or any other statistical software package), find the regression of salary on schooling and experience. Using the solution obtained, answer the following questions:

(a) Write the estimated regression equation.

(b) Explain the meaning of the estimates of the constant term and the regression coefficients of explanatory variables.

(c) What are the values of the standard deviation of errors, the coefficient of multiple determination, and the adjusted coefficient of multiple determination?

(d) What is the value of the total sum of squares? What portion of SST is explained by our regression model? What portion of SST is not explained by our regression model?

(e) How much salary is a person with 18 years of schooling and 7 years of work experience expected to earn?

(f) What is the point estimate of the expected (mean) biweekly salary for all people with 16 years of schooling and 10 years of work experience?

(g) Determine a 99% confidence interval for the coefficient of schooling.

(h) Using the 1% significance level, test whether the coefficient of experience is positive.

(i) Using the 2.5% significance level, can you conclude that the coefficients of both explanatory variables in the population regression model are equal to zero?

(j) Determine the simple linear correlation coefficient between each pair of the three variables.

20.7 The CTO Corporation has a large number of chain restaurants in the Midwest and Southern states. The research department at the company wants to know whether the sales of restaurants depend on the size of the population within a certain area surrounding the restaurants and the mean income of households in those areas. The company collected information on these variables for 11 restaurants. The following table gives information on the weekly sales (in thousands of dollars) of these restaurants, the population (in thousands) within 5 mi. of the restaurants, and the mean annual income (in thousands of dollars) of the households for those areas:

Sales	19	29	17	21	14	30	33	22	18	27	24
Population	21	15	32	18	47	69	29	43	75	39	53
Income	57	73	44	48	41	54	61	39	29	51	42

Using MINITAB (or any other statistical software package), find the regression of sales on population and income. Using the solution obtained, answer the following questions:

(a) Write the estimated regression equation.

(b) Explain the meaning of the estimates of the constant term and the regression coefficients of population and income.

(c) What are the values of the standard deviation of errors, the coefficient of multiple determination, and the adjusted coefficient of multiple determination?

(d) What is the value of the total sum of squares? What portion of SST is explained by our regression model? What portion of SST is not explained by our regression model?

(e) What are the predicted sales for a restaurant with 50,000 people living within a 5-mi area surrounding it and $52,000 thousand mean annual income of households in that area?

(f) What is the point estimate of the expected (mean) sales for all restaurants with 45,000 people living within a 5-mi area surrounding them and $60,000 mean annual income of households living in those areas?

(g) Determine a 95% confidence interval for the coefficient of income.

(h) Using the 1% significance level, test whether the coefficient of population varies from zero.

(i) Using the 1% significance level, can you conclude that the coefficients of both explanatory variables in the population regression model are equal to zero?

(j) Determine the simple linear correlation coefficient between each pair of the three variables.

21 Measurement Analysis

21.1 OVERVIEW

The objective of this chapter is to understand the importance of and language of good measurements, the types of variation involved in measurement systems, and how to conduct and interpret a measurement system analysis for different types of data.

Measurement system analysis (MSA) is a key tool/method for assessing whether the measurement systems of an organization are providing true and accurate information. This chapter provides information in the use of MSA as one of the tools for validating and maintaining the capability and effectiveness of the measurement systems, equipment, and processes used in the manufacturing and support of the products and services provided by an organization.

21.2 INTRODUCTION

No two things are alike, but even if they were, we would still get different values when we measure them.
                                                                                        —Anonymous

When you can measure what you are speaking about, and express it in numbers, you know something about it; but when you cannot measure it, when you cannot express it in numbers, your knowledge is of a meager an unsatisfied kind.

Even if when you can measure it, you may still not know anything if the variation in the measurement due to the measuring system is large compared to that of the variation of the items under scrutiny.
                                                                                        —Lord Kelvin, 1883

Measurement is perhaps the most important Six Sigma concept. Without measurement we only have opinions about the efficiency of our processes. One must measure the correct parameters to determine cause–effect relationships. Good data are often rendered useless because some important factor was not recorded. Likewise, if a certain statistical tool requires data in a specific format, one needs to ensure that the data are collected in that format. To improve our processes, we must collect measurement data on the "critical to" characteristics. Any variation in these measurement data can be attributed to either the characteristic that is being measured or the way that measurements are being taken. When the measurement error is large, it may lead to inaccurate decisionmaking.

Measurement error is defined as the effect of all the sources of measurement variability that cause an observed value to deviate from the true value being measured. The measurement system is the complete process used to obtain measurements, such as the procedures, gages, and human inputs employed to obtain measurements. Each component of this system represents a potential source of error. We must identify these sources of error. Characterizing measurement error is important but is a poorly understood and frequently overlooked aspect of the measurement process. Several types of measurement error affect the location and the spread of the distribution; accuracy, linearity, and stability affect location.

Practitioner's Guide for Statistics and Lean Six Sigma for Process Improvements. By Mikel J. Harry, Prem S. Mann, Ofelia C. de Hodgins, Christopher J. Lacke, and Richard Hulbert
Copyright © 2010 John Wiley & Sons, Inc.

Measurement accuracy describes the difference between the observed average and reference value of the measurements. A linearity problem describes a change in accuracy across the expected operating range of the measuring instrument. A stability problem suggests that there is a lack of consistency in the measurement over time.

Precision is the standard deviation between replicated measurements. For accuracy and precision, multiple measurements of one single characteristic must be taken. Primary contributors to measurement system error affect the spread of the distribution (measurement precision) and therefore affect the measuring system's repeatability (precision) and reproducibility. *Repeatability* is the variation in measurements obtained by one operator measuring the same characteristic on the same parts with the same measuring instrument. *Reproducibility* refers to the variation in the average of measurements of an identical characteristic taken by different operators using the same instrument. A basic knowledge of stability, bias, and linearity is required to understand measurement accuracy (central location).

Given that reproducibility and repeatability are important contributors to error, they are the objects of a specific study called *gage repeatability and reproducibility* (gage R&R). This study can be performed on both attribute-based and variable-based measurement systems. It enables evaluation of measurement consistency among operators after having at least two operators measure several parts at random on a few trials. If consistency is poor, then the measurement system must be improved.

Indeterminate statistical measurements may require consideration of respond-to-control variable correlation. A variable-gage R&R study can be done with MINITAB software. When precision and accuracy measurements are assessed in combination, the analysis is referred to as *measurement capability analysis or measurement system analysis* (MSA). When measuring the output of the process, two sources of variation are found: (1) part-to-part variation and (2) measurement system variation. Two methods can be used to estimate repeatability and reproducibility. $\bar{X}$ and R charts are easy to use and provide a graphical representation. ANOVA, a more powerful tool, is a statistical technique used to estimate and analyze the components of variance, including the interaction between different parts and different operators. When precision and accuracy measurements are assessed in combination, the graphical output of MINITAB software provides information on both aspects. With this software, it is possible to determine whether the measurement system is reliable enough to collect data for a Six Sigma project.

21.3   MEASUREMENT

Measurement is a process that compares data against a standard to gain information and knowledge about a part, component, product, feature, service, or transaction. This process is usually an estimate as sample data are used rather than the entire data population. The objective of the measurement process is to learn as much as possible in a short amount of time and to identify required improvements. For the measurement process, as with any other process, we are concerned with the level and variability of the process over time as the measurement process may shift or drift. To improve a measurement process, the transfer function $Y = f(x)$ and the roles that the input variables play in driving variation in the measurement process must be clearly understood. Thus we must identify and quantify the different sources of variation that affect measurement. One of the main sources of variation is measurement uncertainty.

Measurement error (*uncertainty*) is a variation in the measurement system that can be attributed to what is being measured or to the measurement system itself:

$$\sigma^2 \text{ total} = \sigma^2 \text{ item} + \sigma^2 \text{ measurement} + \varepsilon$$
$$\sigma^2 \text{ measurement} = \sigma^2 \text{ individual} + \sigma^2 \text{ process} + \sigma^2 \text{ other}$$

The second source of variation is *part-to-part variation*. The measurement system must identify differences between repeated measurements in order to assess the capability and therefore control of the process. In other words, the measurement system must be able to discriminate.

21.4   MEASUREMENT ERROR

Finally, if measurements are used to guide decisions, then it follows that the more error there is in a measurement, the more error there will be in the decisions that are based on these measurements. Measurement error describes the sources of measurement variability that cause an observed value to deviate from the "true" value of the characteristic that is being measured. A large measurement error affects the data and may lead to inaccurate decisions. A measurement system is the complete process used to obtain measurements. Each component of the process represents a potential source of error, such as procedures, measuring instruments, personnel inputs (operators), and environmental factors and materials.

Measurement precision refers to the ability of a measuring instrument to reproduce the same measurements. *Measurement accuracy* refers to the ability of a measuring instrument to give measurements that are close to the "true" value.

For estimating precision and accuracy, multiple measurements of one single characteristic must be taken. Precision is quantified as the standard deviation between these replicated measurements. *Accuracy* is the difference between the observed average of measurements and the reference value. For estimating reproducibility and repeatability, gage R&R studies should be performed. Figure 21.1 illustrates some tools used for measurements, a person doing the measurements and the occurrence of "measurement errors" taking place during the process of measuring. Example 21.1 describes the scenario depicted in Figure 21.1.

For example, assume that during a Six Sigma training session, a Black Belt asks: "Is it really necessary to verify the measurement system used to collect our data?" The instructor would reply something like: "Yes, in fact even when collecting data on the critical-to-quality (CTQ) characteristics of your project, you should verify and analyze your measurement system." Experience has shown that the presence of measurement error is inevitable. Measurement error is the effect of all sources of measurement variability that cause an observed value to deviate from the real value (true value) of the characteristic being measured, and there is always variability in the way we measure. Consider a figure-skating competition where judges rarely score exactly the same way.

Measurement error

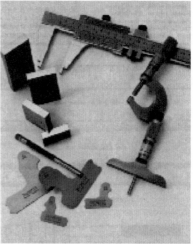

Figure 21.1   Illustration showing some well-known tools used for measuring.

Example 21.1   While observing two quality auditors during a quality system auditing process verifying the organization's documents, does the practitioner think that he/she will come up with the same number of errors? Further, what does the practitioner think would happen if we gave one of these auditors the same document to unknowingly review for a second time? Would this auditor record the same number of errors?

Maybe; however, measurement process experience shows that this does not tend to be the case. When measurements are repeated by a second person/auditor, we rarely obtain the same results. So, how would measurement error affect a project?

Solution   Erroneous results such as incorrect defect numbers or misclassified products will affect graphical outputs, such as control charts. Since we rely heavily on these tools for decisionmaking, this output may lead the practitioner to make the wrong decisions.

Measurement system errors can be classified into two categories: (1) accuracy and (2) precision. *Accuracy* describes the difference between the measurement and the part's actual value. *Precision* describes the variation that is observed when the same part is measured repeatedly with the same device. Within any measurement system, one or both errors could occur.

21.5   ACCURACY AND PRECISION

There are some inconsistencies in measurement system analysis (MSA) terminology. Therefore we have include the Automotive Industry Action Group (AIAG)[1] definitions here:

- *Accuracy* is the closeness of agreement between an observed and the accepted reference value.
- *Precision* is the net effect of discrimination, sensitivity, and repeatability over the operating range (size, range, and time) of the measurement system. The practitioner should be aware that some authors, some organizations, and some Six Sigma training material use *precision* interchangeably with *repeatability*. Actually precision is most often used to describe the expected variation of repeated measurements over time and measurement range. Be aware that MINITAB's definition for precision differs from the AIAG definition.[1] MINITAB divides precision or measurement of variation into two components:

    Repeatability according to MINITAB is the variation due to measuring device, and is observed when the same operator uses the same measurement device to repeatedly measure the same part. In other words, is the variation between successive measurements of the same characteristic, on the same part (or service) by the same person (associate, operator) using the same measuring instrument.

    Reproducibility is the variation due to the measurement system itself. Reproducibility is also the variation observed when different operators measure the same part using the same measurement device. Figure 21.2 illustrates this principle. In other words, when different operators/associates on different measuring instruments obtain the *same* characteristic on the same part.

Some authors define accuracy as the ability to produce an average measured value that agrees with the true value of the "set" standard. Precision is the ability to measure with repetition a part, a product, a service, or a transaction. Figures 21.3 and 21.4 graphically illustrate the differences between the terms *accuracy* and *precision*.

In the other hand, *precision* is the ability to measure with repetition a part, a product, a service, or a transaction and as is illustrated in terms of process control in Figure 21.4.

Some of the measurements must be requirements for capable measurement systems are as follows:

- The statistically stable over time.
- Variability must be small compared to that of the process variability.

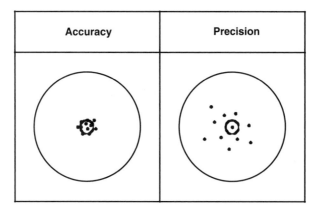

Figure 21.2   Schematic representations of the difference between accuracy and precision.

- Variability must be small compared to the specification limits (tolerance, service level agreements).
- Resolution or discrimination of the measurement device must be small relative to the smaller of either the specification tolerance or the process spread (variation). As a rule of thumb, the measurement system should have a resolution of at least one-tenth the smaller of either the specification tolerance or the process spread. If the resolution is not fine enough, process variability will not be recognized by the measurement system.

Figure 21.5 shows an example of a manufacturing application of *accuracy* and *precision*.

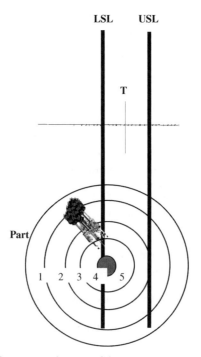

Figure 21.3   Drawing illustrating accuracy in terms of the measurement process. The T (target) in the figure denotes target measurement.

LSL   USL

Part

Figure 21.4 Drawing illustrating precision in terms of the measurement process.

Example 21.2   A manufacturer wants to determine whether the bore diameter measurements for car cylinders are consistent. The company has determined that the critical-to-quality (CTQ) characteristic for this part is the measurement of the cylinder bore diameter. Communication with another company recommended the use of a capacitance measuring system. The documentation states

Evaluating accuracy and precision of an instrument
Repeated readings on 3 master ring gages

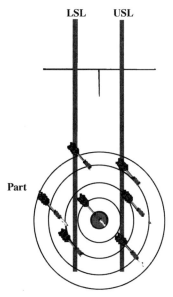

Reading #	Ring 1	Ring 2	Ring 3
1	0.245013	0.312508	0.436999
2	0.245013	0.312508	0.436993
3	0.245013	0.312509	0.437001
4	0.245013	0.312507	0.436997
5	0.245013	0.312508	0.436996
.	.	.	.
.	.	.	.
.	.	.	.
28	0.245013	0.312507	0.436987
29	0.245013	0.31251	0.436986
30	0.2450127	0.312508	0.436985
Average	0.245013	0.312509	0.436992
True value	0.245	0.3125	0.437
Accuracy (Average-true value)	1.34E-5	8.7E-6	-7.8E-6
Precision (Standard deviation)	8.5E-7	2.07E-6	4.83E-6

Figure 21.5 Measurement listings for three different diameter ring gages. Note that measurements were performed with accuracy in the first column and performed with precision in the second and third columns.

Measurement system as a process

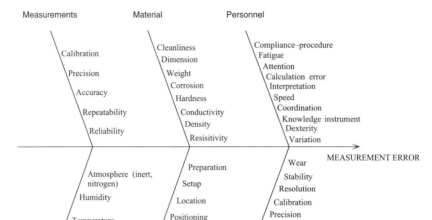

Figure 21.6   Illustration showing a CE diagram assisting in the process measurement system.

that the measuring system is quite easy to use and is very reliable. Since the system is quite expensive and has never been used in this company, manufacturing borrows the instrument so that it can evaluate it with the help of its metrology department. To confirm the instrument reliability, the metrology department will use this instrument to measure different cylinder bore diameters. It is explained to the project team that it choose three of the most common produced cylinder bore sizes. Using the master ring gages, that correspond to these bore sizes: (1) 30 repeated measurements should be taken by one operator (to determine consistency of the operator measuring a single part and at the same time assess the ability of the operators to use the measuring system), and (2) the manufacturer will also be able to determine the accuracy of the instrument by quantifying the difference between the average of the 30 repeated measurements against the reference value of the master ring gages measured with metrology equipment. The precision of the instrument from the sample standard deviation of the 30 measurements is illustrated in Figure 21.6. From the results recorded in Figure 21.5, what can be concluded?

Solution   The team, in conjunction with the metrology department, concluded that the instrument was quite reliable and corresponded to the quoted accuracy and precision for this type of instrument. The assigned operator consistently measured a single part and was able to use the measuring system.

21.6   MEASUREMENT SYSTEM AS A PROCESS

Just as there is variability in the products or services being measured, there is also variability in the way that they are being measured. Measurement system analysis is a key component for many quality improvement initiatives because is important to assess the ability of the measurement system to detect the differences of the process variables. This section covers the ways that variability of services and products are being measured. In fact, there is a complete process used to obtain measurements on a specific part or service characteristic that is referred to as the measurement system. Measurement system analysis (MSA) characterizes the accuracy and variation of a measurement system and provides ways to validate that the measurement system meets the measurement requirements for the product or service being measured. Some examples of measurement systems as mentioned in above sections include operator visual inspection, manual gages, automated inspection processes, measurement systems, process measurement equipment, test measurement equipment and other. MSA is used to validate the integrity and therefore quality of the data of a measurement system with the goal to reduce measurement variation due to measurement error.

Example 21.3   Let us assume that a machine component is the measuring instrument in which cleanliness, temperature, design, and calibration can impact the measurement error.

Solution   It is easier to visualize the components of this system with the cause-and-effect [cause–effect (CE)] diagram in Figure 21.6. Each of these components, *m*easurement, *m*aterial, personnel/*m*anpower, environment/*m*other nature, *m*ethods, and *m*achines/equipment (the "6 ms"), represents a potential cause of measurement error. We understand that a measurement system can be influenced by many factors that will directly affect the quality of the measurements taken. This could have a big impact on any project. What could be done to prevent this impact?

An important first step is to identify the factor or factors (variables) that specifically influence the measurement system (in fact, this may even be a Lean, a DFFS (design for Six Sigma), or a Six Sigma project. In past chapters, we have used tools in the breakthrough strategy that can be applied to study and improve the measurement process. Therefore, the practitioner could select the CE diagram as first step of a measurement system process to visualize and assist with the analysis of how factors such as cleanliness, temperature, design and calibration impact the measurement error.

The system shown in Figure 21.7 can help the practitioner revise the fundamental concepts of accuracy and precision.

The combination of the CE diagram and knowledge of the fundamental concepts of accuracy and precision provide educated information on how important each component of the machine is for manufacturing a defect-free product. By brainstorming each one variable mentioned in this example (cleanliness, temperature, design, and calibration) with the assistance of the CE diagram, it is easier to visualize the impact of measurement errors.

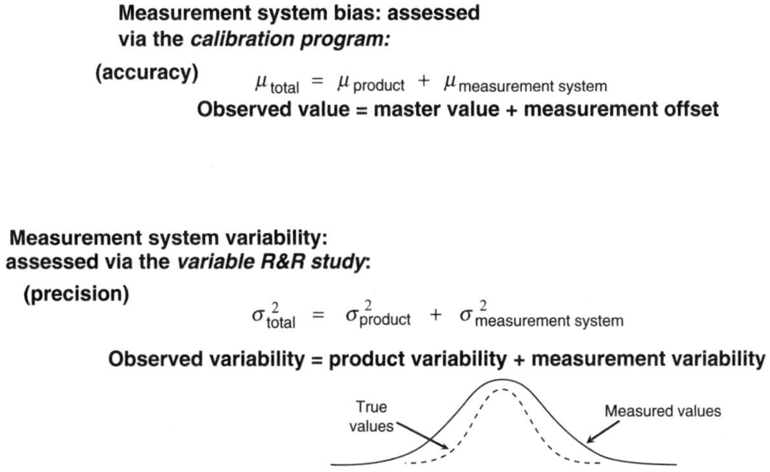

**Measurement system bias: assessed
via the *calibration program*:**

(accuracy) $\mu_{total} = \mu_{product} + \mu_{measurement\ system}$

Observed value = master value + measurement offset

**Measurement system variability:
assessed via the *variable R&R study*:**

(precision) $\sigma_{total}^2 = \sigma_{product}^2 + \sigma_{measurement\ system}^2$

Observed variability = product variability + measurement variability

Figure 21.7   Drawing showing the output of the different components of a measurement system.

In general, a measurement system can be described in terms of the following parameters:

1. *Location*   Average measurement value versus actual value:
   - *Stability*—capacity of a measurement system to produce the same values over time when measuring the same sample. With respect to statistical process control (SPC) charts, stability means the absence of special cause(s) of variation, leaving only common causes of variation (random variation).
   - *Bias* (accuracy)—measure of the distance between the average value of the measurements and the true or actual value of the sample or part.
   - *Linearity*—measure of consistency of bias (accuracy) over the range of the measurement device.
2. *Variation*   Spread of measurement values (precision):
   - *Repeatability*—assesses whether the *same appraiser* (operator, associate, etc.) can measure the same part, sample, or document multiple times with the same measurement device and obtain with consistency the same value(s).
   - *Reproducibility*—assesses whether *different appraisers* (operators, associates, etc.) can measure the same sample, part, or document with the same measurement device and obtain the same value(s).

21.7   CATEGORIES OF MEASUREMENT ERROR THAT AFFECT LOCATION

How do we characterize variation in the measurement system? As with every process, the distribution used to describe the measurement system's variation can be characterized by the location and the width or spread. There are several categories of measurement that affect this location. Assume that you measure one characteristic with a measuring instrument, and that you note a systematic difference between the true or reference value of the product or service characteristic and the measurement that you have obtained.

The general basic parameters of measurement system analysis (MSA) are (1) bias, (2) stability, and (3) linearity. Some types of error that affect location are insufficient accuracy (bias), linearity, and stability:

1. *Bias* (*accuracy*) refers to the difference between the observed average value of the measurements and the accepted standard.
2. *Stability* refers to processes that are normally free from special cause variations.
3. *Linearity* refers to different statistical results from measurements when subjected to different metric spaces. Linearity in a system is determined using higher levels of calibration in measurement standards that are often guided by inferences drawn from several interaction factors influencing s system. See also Figure 21.8.

The bias (accuracy) of the measuring instrument is characterized by the difference between the average of the repeated measurements on the same product and the (*true*) or reference value of the product or service. You may have heard accuracy described as bias. *Accuracy* is also a measure of the distance between the average value of the measurements and the *true* or actual value of the sample or part.

Several notes are in order here:

- A rule of thumb for acceptance of a measurement system (discussed in more detail in following sections) is a total gage R&R of 30% or the lesser of total variation or the specification tolerance. Over 30% error is considered unacceptable.[1]

Categories of measurement error that affect location

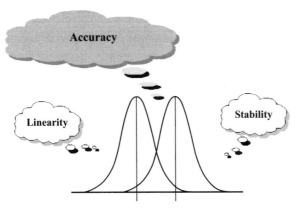

Figure 21.8   Figure showing categories of measurement error that affect location.

- An error of 10–30% error suggests that the system is acceptable depending on the importance of the application, cost of measurement device, cost of repair, and other factors.[1]
- If the measurement system error has an error in excess of 30%, the first step in improving results is to analyze the breakdown of the error source. If the largest contributor to error is repeatability, then the equipment must be revised, calibrated, and improved. If reproducibility is the largest source of error, appraiser/operator training and adherence to procedures can enhance improvement.[1]

21.8   CATEGORIES OF MEASUREMENT THAT AFFECT SPREAD

Repeatability and reproducibility are important contributors to measurement error. The practitioner might ask: "What is the *difference* between *repeatability* and *reproducibility*?" The answer is that both affect the spread of the distribution; however, they focus on different factors (see Fig. 21.9).

Repeatability of the measurement process implies that the measurement process variability itself is consistent. It is the variation in measurements obtained with one measuring instrument when one operator uses the same gage to measure identical characteristics of the same parts. Precision is actually the same as repeatability (AIAG 2002),[2] but instead of focusing on the measurement system, it usually refers to evaluating a measuring instrument in a metrology environment. It is important to observe that for this brief discussion, focus is on the system. AIAG also states that the number of distinct categories into which the measurement system divides a process should be ≥ (greater than, or equal to) 5. (*Note*: In addition to percent error and the number of distinct categories, the practitioner is advised to review graphical analysis over time to decide acceptability of a measurement system.)

It is important to focus on the reproducibility of the measurement system. *Reproducibility* of the measurement system implies that the operators are consistent. Operator variability represents the incremental bias that can be attributed to each operator. Thus, reproducibility is the variation in the average of measurements made by different operators using the same gage when measuring identical characteristics of the same parts. Reproducibility is the variation in the average of the measurements made by different operators/associates using the same measuring instrument when measuring the same identical characteristic on the same part—a measure of precision on the same part.

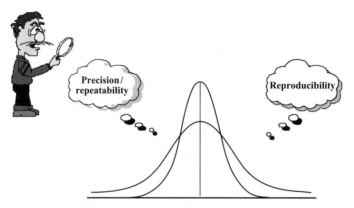

Categories of measurement error that affect spread

Figure 21.9   Illustration showing some of the different categories of measurement error that affect the process spread.

For example, with measurement systems that produce quantitative measurements such as weight, concentration of a solution, or strength of a component, it is important to determine the magnitude of any error in the resulting measurements. If the error is large, it may be impossible to determine whether an individual sample is within specifications (specs). In addition, *design of experiments* (DOE) relies on the ability to separate the real effects of making changes from the background noise, and these DOE methods can be of no value if an inadequate measurement system is used. When quantifying measurement error, it is common practice to separate the error into *repeatability* (error due to the instrument or measurement procedure) and *reproducibility* (error due to the appraiser).

21.9   GAGE ACCURACY AND PRECISION

How is the true value of the product or deliverable good known? The *true value* serves as an agreed-on reference(s) for comparison. This information can be obtained through research that includes studies by various organizations or by simply taking an average of several measurements that have been obtained with more accurate measuring equipment in the metrology laboratory. The key characteristic of gages is that they should give the same true measurements each time. Gages should be able to distinguish one item from another. In other words, a gage should be accurate, precise, and provide adequate resolution. In general, some of the preferred characteristics for gages are

1. Resolution
   - Number of distinct categories
2. Accuracy (bias)
   - *Linearity*
   - *Stability*
   - Correlation
3. Precision (variability)
   - Repeatability
   - Reproducibility

In general, the bias (accuracy) of a measurement system is usually divided into two components: linearity and stability. Also

- *Bias*—the difference between the observed average measurement and a master value.
- *Linearity*—a measure of how size of the part affects the bias of the measurement system. It is the difference in the observed bias values through the expected range of measurements.
- *Stability*—a measure of how a measurement system performs over time. It is the total variation obtained using a particular measurement device to measure a single characteristic on the same part over time.

Example 21.4   A Six Sigma project team member asks the organization's Black Belt: "When we evaluate our measurement system, what happens if there is a change in accuracy through the expected operating range of the instrument?"

Solution   This is an excellent question; it means that there is a *linearity problem*. At the upper limit of the measuring instrument's range, the error is proportionally larger than at its middle range. For example, when comparing a 0–6-in. caliper with the appropriate standard, the instrument is relatively accurate between 0.0 and 4.55 in.; however it is observed that errors occur between 4.551 and 6 in. This change in accuracy throughout the range of the caliper is known as a *linearity problem*. *Stability* refers to the consistency of measurements over time. If there is change in the measurements over time, then stability is a problem.

How do we evaluate accuracy, linearity, and stability? Usually multiple measurements are taken of a master or standard part. The procedure is to take multiple measurements on the same part and then calculate the difference between the observed average and the reference value and then calculate *accuracy*.

Linearity is calculated in a similar way. However, for linearity, the practitioner should compare measurements of different master parts throughout the operating range of the measurement instrument. As for *stability*, a single characteristic on the same master part(s) is measured over an extended period of time. The practitioner should monitor the results using graphical outputs such as control charts.

21.10   EXPLORING LINEARITY ERROR

While repeatability and reproducibility studies concentrate on the variability or precision of a measuring system, linearity and accuracy studies quantify the bias. In these studies, multiple measurements are made on reference samples. One important source of error associated with a measurement system is reproducibility, which describes the variation introduced by different operators in the measurement process.

At this point, the practitioner should wonder whether a gage has the same accuracy for all the sizes of objects being measured. Could there be a linearity error in the instrument? If measurement system variation or bias change depends on the magnitude of the measurement, then, linearity is an issue because it is a measure of the consistency of bias over the range of the measurement device. For example, if a laboratory scale is off by 0.5 lb when measuring a product of 100 lb, but is off by 1.5 lb when measuring a 150 lb product, the scale bias is nonlinear in the sense that the degree of bias changes over the range of use. As can be seen, linearity can also be an issue when correlating gages. For example, the magnitude of the differences between gages might vary over the measurement range. Also, gages can be nonlinear in terms of variation. In some measurement systems error is a percentage of the mean value. (*Note*: Linearity can be evaluated using the gage facility in MINITAB. The graphical output of MINITAB will indicate whether the error is uniform throughout the measurement range.)

Exploring linearity error

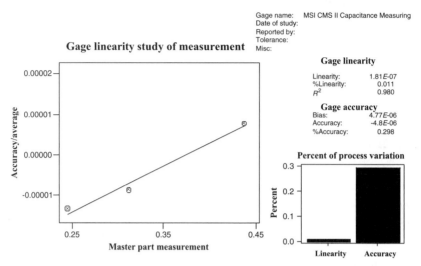

Figure 21.10   Illustration showing exploration of linearity error with MINITAB.

Example 21.5   Using the data from the first study that consists of 30 repeated measurements on each of three master-ring gages; it is observed that these measurements don't represent the complete range of the measuring device. We want to know whether there is a difference in accuracy for the different cylinder bore diameters. Visualize the linearity of a gage in MINITAB as it shows a best-fit line relating the average deviation to the master measurements. This analysis will indicate whether the gage has the same accuracy for all sizes of objects being measured. Generally, the closer the slope is to zero, the better the gage linearity.

Solution   Refer to Figure 21.10.

We suspect that a linearity effect is present because the accuracy increases linearly with diameter size. However, when we examine the MINITAB output, we observe that the variation due to linearity and accuracy for this gage is 0.01% and 0.3% for the overall measurement system variation. We conclude that the linearity error for this instrument is not *overly* important.

21.11   GAGE REPEATABILITY AND REPRODUCIBILITY (R&R)

There are statistical methods to estimate the repeatability and reproducibility (R&R) of parts in destructive scenarios such as tensile strength, compression tests, impact test, hardness test, and other parameters. Once the measurement is obtained for the part, this part is no longer available for additional measurements by other operators. However, there are statistical methods available to estimate the repeatability and reproducibility (R&R) of the parts. The theoretical assumption is that all parts from the same process produced with the same machines/tools are alike. Thus, the measurement characteristics of interests are assumed to be identical for each part of the group (batch). This assumption means that the batch is homogeneous in nature (all parts of the batch/group are equal); this assumption is important because the observed within-batch variability is used to estimate the repeatability of the measurement system. This assumption is fundamental for considering conducting gage R&R studies. Even some practitioners are interested in the percent variation of the measurement system; the practitioner should also be interested in the proportion of the measuring system variation due to repeatability and reproducibility. MINITAB displays graphical and tabular data that quantify the variation in the measuring system as a proportion to the total variation. MINITAB and JMP identify the sources of variation within the measurement system. Both statistical software programs show the measurement error as a percent of total variance and compare it with historical process variations.

Gage R&R identifies the different types of variation that may occur in a measurement system such as

- Repeatability due to equipment variation (within variation)
- Reproducibility due to appraiser/operator variation (between variation)
- Residual or pure error
- Interaction effects variation

Measurement Repeatability. An important source of error associated with a measurement system is repeatability, which describes the variation obtained in measurements made by one instrument over repeated trials; it is the variation in measurements obtained when an operator measures the same characteristic several times with the same instrument under the following conditions:

Repeatability Criteria

- On the same part
- In the same location on the part
- Under the same condition(s) of use
- Over a short period of time

Repeatability is the inherent variation of the measurement system. Repeatability error comes from the instrument itself and the position of the part in relation to the instrument. Note that measurements may show good repeatability without being accurate.

In contrast to *precision*, where the measuring instrument is generally evaluated by the metrology function, repeatability is evaluated through a gage R&R study of the measurement system.

Measurement Reproducibility. *Reproducibility* is the variation between the averages of measurements obtained by different operators who measure the same characteristic on the same parts under the following conditions:

Criteria for Reproducibility

- Use of the same measuring instrument/device
- Use of the same method
- In the same location on the parts
- Under the same conditions of use
- Over a short period of time

Reproducibility represents the incremental bias that can be attributed to each operator. If there is variability between the operators, then the reproducibility of the measurements represents the variation between operators. Reproducibility can be evaluated through a gage R&R study. Figure 21.11 graphically illustrates the fundamental concept of reproducibility.

[*Note*: For measurement systems that result in quantitative measurements (weight, size, or diameter of a part; concentration or strength of a solution; etc.), it is important to determine the magnitude of any error in the resulting measurements. If the error is large, it may be impossible to determine whether an individual part is within specs. The practitioner should always remember that when quantifying measurement error, it is common practice to separate the error into

Reproducibility

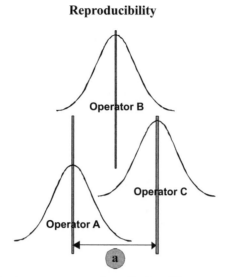

Figure 21.11   Illustration showing measurement performed by three different operators lacking of reproducibility (**a** shows the difference in the measurement between operator A and operator C).

repeatability—error due to the measuring instrument—or error of procedure; and reproducibility which is error due to the appraiser/operator.]

21.11.1   Variable Gage R&R

Variable gage R&R analyzes errors produced when a variable-based measurement system interacts with its environment. The study focuses on two types of error with the greatest effect on the measurement system. Repeatability and reproducibility measure the variation introduced by one measuring instrument and different operators. A variable gage R&R is a carefully planned way to study a measurement system. In addition, it should be noted that as there are two major categories of data— (1) continuous data and (2) attribute data—so there are two major categories of gages. Gages that are used for *continuous data* include stopwatches for time measurements, voltmeters, thermometers, pressure gages, gas monitors for measuring gas concentration, graduated cylinders for liquids, scales for weights, and strain gages; and *attribute gages* include pass/fail, good/bad, and go/no go measurement devices. Attribute gage studies assess the amount of bias and repeatability of a gage when the response is a binary attribute variable. Unlike a variable gage, an attribute gage does not give information on the quality of a part but only indicates whether a part is accepted or rejected according to a set of specified standards. To conduct an attribute gage study, you must have part names or numbers, a unique reference value for each part, and the number of acceptances for each reference value. The practitioner should select reference values that are at nearly equidistant intervals. Also the practitioner should specify either a lower or upper tolerance limit.

In addition, there are different techniques for analyzing the measurement systems. The gage R&R method studies the variation in measurement data employing one measuring instrument and several operators. If we were required to analyze other factors, such as different equipment, and measuring methods a design of experiment (DOE) method would have to be in place.

Whenever the results of a process are measured, some variation is observed. There are two main sources of this variation: (1) inevitable differences between parts made by any process and (2) an imperfect measurement system. Thus, measuring the same part repeatedly does not always result in identical measurements.

Measurement—variable gage R&R study

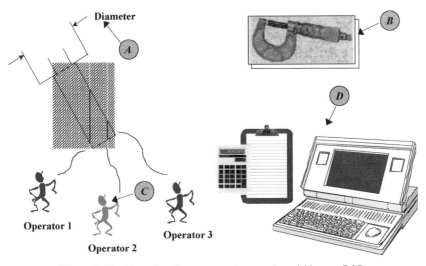

Figure 21.12   Examples of measurement sources in variable gage R&R.

Criteria for Variable Gage R&R

1. Determine the continuous characteristic (CTQ, CTP, etc.) to be measured with an instrument.
2. Carefully select the measuring instrument that is to be verified.
3. Select the representative group of personnel who normally operate the instrument.
4. Correlate data collection and analysis of the results with MINITAB, JMP, or other statistical software.
5. The data structure in MINITAB is very important in order to perform this study.

Figure 21.12 graphically illustrates the fundamental concept of variable gage R&R.

Although the numbers of operators, trials, and parts may vary, the following steps are common for most gage R&R studies:

1. The instrument must have an adequate resolution that allows at least one-tenth of the expected process variation of the characteristic to be read directly. For example, if the characteristic's variation is .001, the equipment should be able to read a change of .0001.
2. Identify three operators to participate in the study. These operators must be familiar with the instrument and the measuring method.
3. Obtain 10 samples (e.g., parts) that represent the actual or expected range of process variation. Number the parts and mark the area to be measured on each part.
4. Ensure that the gage has been calibrated.
5. Have the first operator measure all the samples once in random order (another technique is to reduce human bias by "blind sampling," where the operator does not know the identity of the parts being measured).
6. Record the values obtained on a data collection sheet or directly in a computer file.
7. Have the second and then the third operators measure all the samples once in random order.
8. Repeat steps 5–7 thrice (3 times).
9. Check for outliers in the data and if possible, repeat these readings (wrong adjustments, handling errors, dirt on the samples or instruments, reading errors, etc.). A good way to find outliers is out-of-control points on the range chart.

10. Analyze results using MINITAB software (`Stat > Quality Tools > Gage R&R Study`).

11. Determine follow-up actions if necessary. For example, improve or clarify measuring method, organize training for the operators, or find another instrument.

[*Note*: The preferred method of analysis is ANOVA. To use the gage R&R function, MINITAB needs three column variables: *part numbers* (sample names or numbers for each observed measurement), *operators* (operator names or numbers for each observed measurement), and *measurement data* (observed measurements). To generate part and operator numbers, the `Simple Set of Numbers` function can be used (`Calc > Make Patterned Data > Simple Set of Numbers`).]

21.11.2 Crossed Gage R&R

Whenever you measure the results of a process, you will see some variation. There are two sources of this variation: (1) there are always differences between parts made by any process, and (2) any method of taking measurements is imperfect—thus, measuring the same part repeatedly does not result in identical measurements. In general, crossed gage R&R and nested gage R&R are associated with alternative methods such as $\bar{X}-R$ and ANOVA analysis. The study steps are the same for both, and it is up to the practitioner to select the method that he/she feels more comfortable with. Some practitioners with manufacturing background might feel more at ease with the $\bar{X}-R$ method. One advantage of this method is that it derives from the Automotive Industry Task Force (AIAG) report (1994), which can be used as a standard method.

The gage R&R method is used to determine what portion of the variability in measurements may be due to the measurement system. Measurement system variability includes variations due to both the gage and operator-to-operator variability. The crossed method involves collecting data from all combinations of *n* units, *m* appraisers, and *w* trials each. This method is defined for the case $2 \leq n \leq 10$, $2 \leq m \leq 3$, and $2 \leq w \leq 3$. In general, it is desirable that the total number of evaluations be > 20 (i.e., $n \times m \times w \geq 20$). This method is suitable for nondestructive testing; however, it can also be used for destructive tests when the following assumptions have been met:

- All parts within a single batch are similar enough to be identified as the same part.
- All operators measure from each batch.

For some practitioners, the calculations used in the $\bar{X}-R$ method are simpler; however, the ANOVA method is more accurate. In a gage R&R study, it is important that the measurements be done in a random order and that all the parts selected provide a representative sample across the possible range of responses.

Example 21.6   Nine parts representing the expected range of the process variation were selected. Two operators measured the thickness of each of the nine parts, 2 times per part, in a random order.

Solution   Use the MINITAB `Stat > Quality Tools > Gage Study > Gage R&R Study (Crossed)` Function. Enter data in MINITAB as shown in Figure 21.13.

Interpretation.   The graphs in Figure 21.14 indicate the contributions of different components of variance, any differences between operators for each part, the mean difference between operators, and the variation of measurements for each part:

The $\bar{X}$ chart points inside the control limits in this example are almost indistinguishable from each other. For an ideal gage, the control limits should be on top of the grand mean, showing

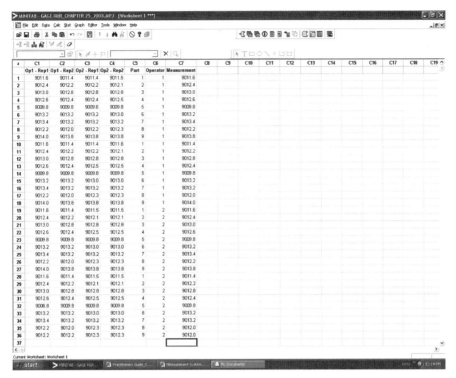

Figure 21.13   MINITAB's captured screen showing data worksheet for performing a crossed gage R&R study.

Gage R&R (ANOVA) for Meas2

Gage name:
Date of study:

Reported by:
Tolerance:
Misc:

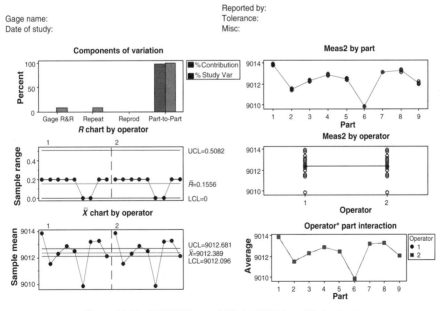

Figure 21.14   MINITAB gage R&R (ANOVA) graphical output.

perfect resolution. The `Xbar Chart by Oper` graph displays the measurements in relation to the overall mean for each operator, allowing to compare the operators to each other, and to the mean.

$\bar{X} - R$ charts assess the repeatability of the measuring system for each operator, as well as the discrimination of the measuring system. *Discrimination* is the ability to detect small changes in the measured characteristics. The $\bar{X}$ chart compares the part-to-part variation top repeatability.

The `R Chart by Oper` graph displays the variation in the measurements done by each operator, allowing the practitioner to compare different operators. In this chart it appears that both operators had almost identical measurements.

The `Oper*Part Interaction` graph displays the operator by part interaction, enabling the practitioner to see how the relationship between operator and part changes depending on the operator. It also graphs the average part measurement—by part—for each operator. When the operators measure the same sequence of parts, these two lines are almost identical.

The `Meas2 by Oper` graph displays the main effects for the operator; thus we can compare mean measurements between operators.

See also Figure 21.15.

MINITAB analyzes variance (ANOVA) to calculate variance components. These are then used to estimate the percent variation due to the measuring system. The analysis of variance tables (ANOVA) are used to calculate variance components, which are then used to estimate the percent variation due to the measuring system. The percent variation appears in the gage R&R table. The two-way ANOVA table includes terms for the part and operator. The gage R&R table shows how the total variability is divided among the following sources:

- *Repeatability*, which is the variability in measurements obtained when the same part is measured multiple times by the same operator
- *Reproducibility* which can be further divided into operator and operator-by-part components, which is the variability in measurements obtained when the same part is measured by different operators
- *Part-to-part* variability, which is the variability in measurements across the different parts

Ideally, very little variability should be due to repeatability and reproducibility. Instead, difference between parts should account for most of the variability. In the example session window in Figure 21.15 we find:

1. Two-way ANOVA table with interaction. The example ANOVA table contains the terms `Operator`, `Part`, and `Oper*Part Interaction`.
2. Two-way ANOVA without interaction. When the *p* value for the `Oper*Part Interaction` term is ≥ 0.25, MINITAB generates a second ANOVA table that omits the interaction term from the model.

In addition, MINITAB calculates a column of variance components (`VarComp`) that are the base values for calculating `%Gage R&R` using the ANOVA method. While an ANOVA table compares each term (`Operator`, `Part`, `Oper*Part`) to error—repeatability—we really need to compare `error + Operator + Operator*Part` variation to `Total Variation`. Estimates for a Gage R&R study are obtained by calculating the variance components for each term and for error repeatability, the `Operator` and `Operator*Part` components are summed to obtain total variability of the measuring system. The `%Contribution` column, based on variance components, is calculated by dividing each value in `VarComp` by the total variation then multiplying the result by 100. The `Part-to-Part %Contribution` value of 99.26 means that 99.26% of the variation in the measurement is contributed by the difference between parts. `%Study` variation compares measurement system variation to the total variation in the parts measurements for this study. Finally,

SESSION WINDOW-

Gage R&R Study-ANOVA Method

Two-Way ANOVA Table With Interaction

Source	DF	SS	MS	F	P
Part	8	46.5156	5.81444	*	*
Oper	1	0.0000	0.00000	*	*
Part * Oper	8	0.0000	0.00000	0	1.000
Repeatability	18	0.2800	0.01556		
Total	35	46.7956			

Alpha to remove interaction term = 0.25

Two-Way ANOVA Table Without Interaction

Source	DF	SS	MS	F	P
Part	8	46.5156	5.81444	539.913	0.000
Oper	1	0.0000	0.00000	0.000	1.000
Repeatability	26	0.2800	0.01077		
Total	35	46.7956			

Gage R&R

Source	VarComp	%Contribution (of VarComp)
Total Gage R&R	0.01077	0.74
Repeatability	0.01077	0.74
Reproducibility	0.00000	0.00
Oper	0.00000	0.00
Part-To-Part	1.45092	99.26
Total Variation	1.46169	100.00

Source	StdDev (SD)	Study Var (6 * SD)	%Study Var (%SV)
Total Gage R&R	0.10377	0.62265	8.58
Repeatability	0.10377	0.62265	8.58
Reproducibility	0.00000	0.00000	0.00
Oper	0.00000	0.00000	0.00
Part-To-Part	1.20454	7.22725	99.63
Total Variation	1.20900	7.25402	100.00

Number of Distinct Categories = 16
Gage R&R for Meas2

Figure 21.15 MINITAB's ANOVA method session window.

the column $StDev\ (SD)$ is the standard deviation of each variance component (square root of the variance), and the column $Study\ Var$ followed by the second line $(6 * SD)$ represents 99.7% of the variation. For example, the column $\%Study\ Var = 8.58 = 8.6$ was obtained by

$$\frac{P}{SV} = \frac{\sigma_{MS}}{\sigma_{total}} = \frac{\text{total gage R\&R}}{\text{total variation}} = \frac{0.104}{1.21} = 8.6$$

$\%\ Contribution$ is calculated by dividing each value in $VarComp$ by the total variation, then multiplying the result by 100. For example, to calculate the $\%Contribution\ for$ part-to-part variation, we divide the $Part$-To-$Part$ in the $VarComp$ column by the total variation in the same

`VarComp` and then multiply by 100 as follows:

$$\frac{1.45092}{1.46169} = 0.9926 \times 100 = 99.26$$

21.11.3   Attribute Gage R&R Study

The objective of the attribute Gage R&R study is to evaluate an attribute-based measurement system. The attribute gage R&R is based on a go/no-go, pass/fail, or good/bad assessment. Thus, we can say that a "critical to" (CT) characteristic is measured in attribute data when it is compared to a specified limit. Data are then classified into categories: accepted or rejected. An attribute gage R&R evaluates the consistency between measurement decisions to accept or reject. An attribute gage R&R determines the consistency of agreement and/or correctness of subjective ratings or classifications. An attribute gage R&R should be used when two or more appraisers rate the same items or one appraiser rates the same items more than once. The data for this study can be binary, nominal, or ordinal. Thus, attribute gage R&R studies are divided into two categories: (1) attribute gage R&R study with nominal data and (2) attribute gage R&R study with ordinal data. An attribute measurement system is a system that compares the part being measured to a standard of the part and accepts the part as pass, go, good-if the standard is met.

When measuring with attribute data, the issue of measurement error arises because attribute data are being measured in terms that rely on human judgment even when the operational definition for the measurement exists. One example of these measurements exists today in the semiconductor industry when operators assess the pass/fail results of a wafer with thousands of chips with the simple aid of an optical microscope. *Attribute gage R&R* represents repeatability and reproducibility. Again, a review of these terms is provided. *Repeatability* means that the same operator measuring the same part, using the same gage, should obtain the same reading every time. *Reproducibility* means that different operators, measuring the same thing, using the same gage, should obtain the same measurement reading every time. Attribute gage R&R shows the percentage of repeatability and the percentage of reproducibility. In an ideal study, both percentages should be 100%—in general, anything above 90% is considered acceptable. To perform this study, there is no need for sophisticated software, nevertheless MINITAB has a module called *attribute agreement analysis* (*attribute gage R&R* in MINITAB 13). Attribute gage studies assess the amount of bias and repeatability of a gage when the response is a binary attribute variable.

Example 21.7   Perform the following steps for an attribute gage R&R:

1. Select a minimum of 30 samples (parts) to test to represent a minimum range of variation.
2. Mix 50% of good parts (good against standard) and 50% of bad parts (parts that did not meet standard).
3. Identify at least three operators (inspectors).
4. Select a master appraiser to categorize each test part into its true attribute category.
5. Have each operator independently and in random order, assess the parts to determine whether they pass or fail.
6. Enter data into a worksheet to report the effectiveness of the attribute measurement system (see Figs. 21.16 and 21.17).
7. For each operator, count the number of times that they agreed with the *master expert* categorization. This is the individual repeatability of that operator that MINITAB calls "within appraiser."
8. Results (see Fig. 21.18). The three operators matched 16 of the 2 trials for a 53.33% match. An `SE Kappa` value of $<.7$ indicates that the measurement system may require improvement.

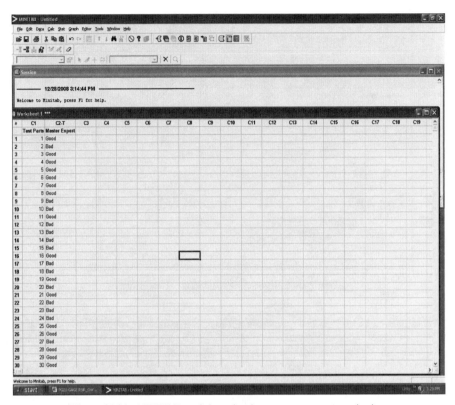

Figure 21.16   MINITAB worksheet showing master expert categorization.

Between Appraisers

Assessment Agreement

# Inspected	# Matched	Percent	95 % CI
30	16	53.33	(34.33, 71.66)

\# Matched: All appraisers' assessments agree with each other.

Fleiss' Kappa Statistics

Response	Kappa	SE Kappa	Z	P (vsY>Y0)
Fail	0.581395	0.0471405	12.3333	0.0000
Pass	0.581395	0.0471405	12.3333	0.0000

All Appraisers versus Standard

Assessment Agreement

# Inspected	# Matched	Percent	95 % CI
30	0	0.00	(0.00, 9.50)

\# Matched: All appraisers' assessments agree with known standard.

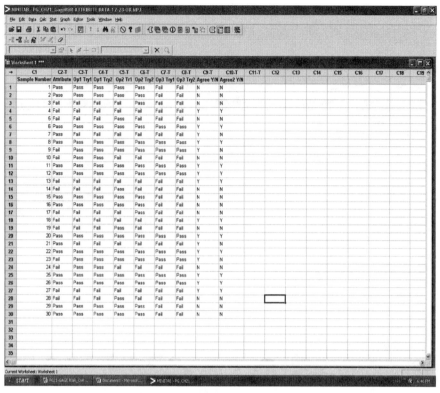

Figure 21.17 MINITAB worksheet after operators categorized the parts.

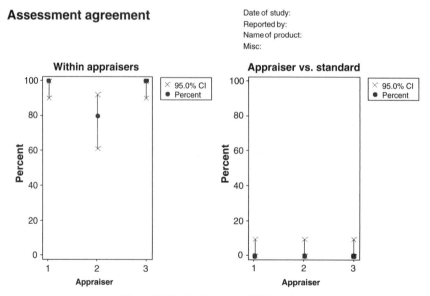

Figure 21.18 Attribute gage R&R results.

9. Implement appropriate actions to fix the inspection process if necessary.

10. Rerun the study to verify the fix.

A 30-sample test should yield and estimate of operator efficiency and capability with a fair amount of uncertainty. Typically a larger sample is not needed (cost effectiveness, cycle time).

Example 21.8   A Black Belt works with her project team to improve an automatic riveting machine process. She says that a critical-to-quality (CTQ) characteristic to measure is rivet bucking height. First, the system must be evaluated. Although this characteristic is continuous by nature, the team is currently measuring it with a go/no-go gage. So, in order to evaluate this measurement system, it is necessary to perform an attribute gage R&R.

Solution   Select 20 rivets that have been installed on sample parts. Referring to the specifications of these parts, identify those that have been installed slightly below or above the limits. Take note of the rivets that are good and bad and enter them into a data spreadsheet in Excel under the attribute column. This enables us to determine how consistently operators evaluate a set of samples against a known standard, which we refer to as *attribute*. A team member could ask, "How do you evaluate the consistency between them?" The answer is "to work with both day and night shift operators for this machine." They will both measure the 20 rivets of this test sample, which is provided to them in random order, twice. Meanwhile, the team leader needs to record all the measurements for both trials (in either an Excel spreadsheet or a MINITAB worksheet) so that afterward, the team can compare these scores to the standard scores that have already been entered. Figure 21.19 shows an example of a Excel spreadsheet. The selected attribute is "go" (G) and "NG" (no-go), and "SBU" is sector business unit.

Review of Attribute Data and Related Features

- *Attribute data*—discrete (go/no go) data that can be tallied for recording and analysis
- *Attribute measurement system*—a measurement system that compares each part to a standard and accepts the part if this standard is met
- *Screen*—100% evaluation of product using inspection techniques (an attribute measurement system)
- *Screen effectiveness*—the ability of the attribute measurement system to properly differentiate good from bad

Review of Purpose of Attribute Data

1. To assess your inspection standards against your customer's requirements
2. To determine whether inspectors use the same criteria to distinguish "good" from "bad"
3. To quantify the ability of inspectors to accurately repeat their inspection decisions
4. To identify how well these inspectors are conforming to a "known master," which includes

    - How often operators decide to ship a truly defective product
    - How often operators do not ship a truly acceptable product

5. Discover areas where

    - Training is needed
    - Procedures are lacking
    - Standards are not defined

Major Considerations in Interpreting Attribute Gage Study with Excel

1. If % appraiser score is less than 100%, training in specific areas is required.
2. "% score vs. attribute" is an error against a known population as deemed by experts.
3. 100% is the target for screen % effectiveness score.
4. "Screen % effective vs. attribute" is an error against a known population as deemed by the experts; 100% is the target.
5. Attribute legend allows Excel to use a macro to count the number of occurrences of the legend text.

On the spreadsheet in Figure 21.19, the appraiser's score shows how consistently the operators were able to repeat their own measurements. The score was obtained by comparing the operator's trials for all samples to determine the proportion of consistent scores. The value referred to as the "score versus attribute" shows how consistent the operators' measurements are with the known standard. Here, the attribute is taken, and two trials are selected: trial 1 and trial 2 to calculate the proportion of scores that are consistent across the samples. The results of the first operator revealed a score of 95% (19/20) between the two trials, and 90% (18/20) for the score versus the attribute value. For operator 2, the appraiser score was 100% and the "%score vs. attribute" is 95%. It is important to also know the overall consistency of the measurement decisions. In this case, the effective scores were compared for both operators. Then, a comparison between the effective score and the standard score is performed. The target value is 100%, meaning that all measurement decisions should be consistent.

Attribute Gage Study with Excel

SCORING REPORT

Attribute Legend
1 G
2 NG

DATE: 4/8/97
NAME: Dep.282 Boeing
PRODUCT: Rivets
SBU: 148T2610-238
TEST CONDITIONS: GT996645-GTOO01(Mod)

Known Population		Operator 1		Operator 2		Y/N	Y/N
Sample#	Attribute	Try 1	Try 2	Try 1	Try 2	Agree	Agree
1	G	G	G	G	G	Y	Y
2	G	G	G	G	G	Y	Y
3	G	G	G	G	G	Y	Y
4	G	G	G	G	G	Y	Y
5	G	G	G	G	G	Y	Y
6	G	NG	G	G	G	N	N
7	G	G	G	G	G	Y	Y
8	G	G	G	G	G	Y	Y
9	NG	G	G	NG	NG	N	N
10	NG	NG	NG	G	G	N	N
11	G	G	G	G	G	Y	Y
12	G	G	G	G	G	Y	Y
13	NG	NG	NG	NG	NG	Y	Y
14	G	G	G	G	G	Y	Y
15	G	G	G	G	G	Y	Y
16	G	G	G	G	G	Y	Y
17	NG	NG	NG	NG	NG	Y	Y
18	G	G	G	G	G	Y	Y
19	G	G	G	G	G	Y	Y
20	G	G	G	G	G	Y	Y

% Appraiser score (1) ->	95.00%		100.00%
% Score vs. attribute (2) ->	90.00%		95.00%

Screen % effective score (3) ->	85.00%	
Screen % effective score vs. attribute (4) ->		85.00%

Figure 21.19   Excel spreadsheet of data for an attribute Gage R&R study. The attribute selected for this study is go/no-go.

For this example case, both scores were 85%. These numbers were obtained by comparing all of the operators' scores for both trials in all 20 samples and determining the proportion of consistent samples. When measurement decisions are not consistent, the measurement should be improved by improving the measuring method and/or the gage itself, or by providing better training to the operators.

Example 21.9   A manufacturer wants to determine whether the operators answers are correct. Each operator provides the same answer to each part across both trials. All operators provide the same answers to each part.

Solution   The solution was found using MINITAB: Stat > Quality Tools > Attribute Gage R&R Study. This score is obtained by comparing the operator's trials for all samples to determine the proportion of consistent scores. The value of the score versus attribute shows us how consistent the operators' measurements are with the known standard, under the attribute column. For this example we took the attribute, trial 1 and trial 2, and calculated the proportion of scores that are consistent across the samples. The results of the first operator reveal a score of 100% (114/14) between the two trials, and 78.57% (11/14) for the score versus the attribute value. For operator 2, the appraiser score is 78.57% and the % score vs. attribute is 64.29%. In addition, it is important to know how consistent our measurement decisions are overall. In this case, we need the effective scores for both operators. We then compare the effective score with the standard score. The target value is 100%, meaning that all measurement decisions are consistent.

In this case, % of appraiser scores is 100% for operators 1 and 3 and 78.57 for operator 2. The score vs. attribute is 78.57% for operator 1, 64.29% for operator 2, and 71.43% for operator 3. We obtain these numbers by comparing all of the operators' scores for both trials in all 14 samples and determining the proportion of consistent samples. When measurement decisions are not consistent, we should improve our measurement system by improving the measuring method and/or the gage itself, or by providing better training to the operators.

See Figures 21.20 and 21.21.

Results Within Appraisers – Confidence Interval (CI) = 95%

```
AGREEMENT

Appraiser #    Inspected #    Matched    Percent      95 % CI
1              14             14         100.00    (80.74, 100.00)
2              14             11          78.57    (49.20, 95.34)
3              14             14         100.00    (80.74, 100.00)
```

\# Matched: Appraiser agrees with him/herself across trials.

Major Considerations.   An attribute-based measurement system, such as a go/no-go or pass/fail gage, cannot indicate how good or how bad a part is. It can only indicate whether the part was accepted, was rejected, passed, or failed. Although this gage R&R permits an evaluation of the measurement system used to obtain data, we recommend using a measurement system that provides continuous data.

Although the number of operators, trials, and parts may vary, the following steps are commonly followed in performing an attribute gage R&R:

1. Identify two operators who will participate in the study. These operators should be selected from those who normally take the data.

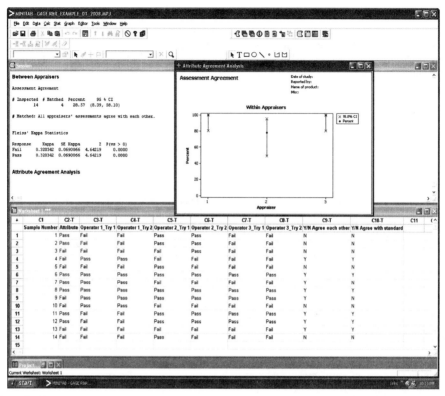

Figure 21.20   Illustration showing MINITAB worksheet, MINITAB session window, and MINITAB graph for attribute agreement analysis.

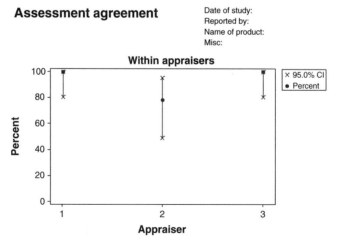

Figure 21.21   Illustration showing MINITAB graph for assessment agreement.

2. Obtain at least 30 samples (e.g., parts) that fall within the range of possible outcomes of the process. For example, when selecting parts for evaluating a go/no-go gage, it is desirable that some of the parts be slightly below or above the standard.

3. Have the first operator measure all the samples once in random order (blind sampling, where the operator does not know the identity of the sample, as is often done to reduce human bias).

4. Record the values obtained on a data collection sheet or directly in a computer file.

5. Have the second operator measure all the samples once in random order.

6. Repeat steps 3–5 until finishing two trials.

7. Analyze the results and determine follow-up action if necessary.

An Excel spreadsheet, JMP, or MINITAB can be used to perform and analyze the results. The measurement system is acceptable if the majority of the measurement decisions agree. If the measurement decisions do not agree, the measurement system must be improved and reevaluated. Some improvement methods are (1) establish a standardized measuring procedure, (2) provide training to the operators, (3) change the go/no-go gage, or (4) find an alternative measurement system that can obtain continuous data.

Example 21.10   A transactional Black Belt asks a master Black Belt, "In my transactional project, I don't have any measuring equipment, go/no-go or variable gages. What can I do to evaluate the way I collect my data? As you know, my project is about documentation errors on a production order form."

Solution   The master Black Belt replies, "In the world of transactions, although the same measurement error principles apply, measuring instruments tend to be less clear-cut and sometimes are not economically viable to evaluate. But, *they do exist*." Consider a survey that measure perceptions. Developing this instrument requires the professional expertise. In the meantime, apply the same principles that you have seen with the attribute gage R&R study. Remember, the objective is to better understand your measurement system and verify its adequacy. A sample of 30 document sheets was selected and two auditors were asked to verify them. Questions were asked about the verification method and how the number of errors was counted. The auditors explained that after they verified each sample (each sheet) separately, the number of errors was completely different! The master Black Belt said, "It seems that different people have different interpretations of what an error is." Establish a measurement procedure to standardize operational definitions and classify the different error types. You must do this before you continue the study.

It is also obvious that there is a lack of consistency between the auditors; therefore, make sure that each one of them can repeat their measurements.

21.12   ANOVA METHOD VERSUS $\bar{X}$–R METHOD

An ANOVA method and an $\bar{X}-R$ method are available for the crossed gage R&R study. The calculations used in the $\bar{X}-R$ method are simpler; however, the ANOVA method is more accurate. In a gage R&R study, it is important that the measurements be done in a random order and that the selected parts provide a representative sample across the possible range of responses. Of the two methods used to estimate repeatability and reproducibility, ANOVA is the more powerful method. ANOVA is a statistical technique that is used to estimate and analyze the variance, whereas the $\bar{X}-R$

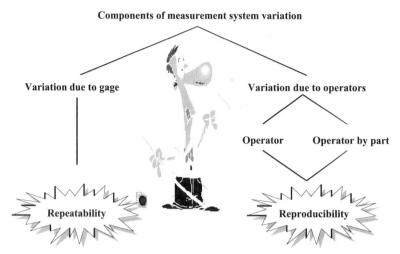

**Evaluating repeatability and reproducibility
with ANOVA**

Figure 21.22   Scenario depicting the evaluation of repeatability and reproducibility with the ANOVA method.

chart is a graphical method known as the *control chart method*. We have explored the differences between these two methods in previous chapters of this book. The $\bar{X}-R$ method was developed first because the calculations stemming from control charts are simpler. However, the ANOVA method is more accurate because

- The ANOVA method accounts for the possible interaction between operators and parts, while the $\bar{X}-R$ method does not.
- The variance components used by the ANOVA method are better estimates of variability than are the ranges used by the $\bar{X}-R$ method. See Figure 21.22.

Example 21.11   An organization master Black Belt says, "It is important to note that the ANOVA method provides a more accurate assessment of the measurement system study than does the $\bar{X}-R$ chart method." The Black Belt asks, "I know that I will be learning ANOVA later on in my training, but I still want to know what ANOVA is, and why it is a more precise method." ANOVA is a standard statistical technique used to analyze the variation between measurement observations and then to identify the important contributing factors. When we do a gage R&R, both the ANOVA and the $\bar{X}-R$ chart method break down the measurement system variation into reproducibility and repeatability. However, the ANOVA method allows us to go one step further in the analysis by breaking down reproducibility into operator and operator-by-part components.

Operator-by-Part Component. This component explores the interaction between each operator and part. This quantifies the variation between average part measurements for each operator. It enables us to understand different situations where, for example, one operator may obtain more variation when measuring smaller parts rather than larger parts, or vice versa. Another reason for the greater accuracy of ANOVA is that the estimators that are calculated are of higher quality. With the $\bar{X}-R$ chart, the range is used to estimate the variation, but with ANOVA, the variance is used. These *variance components* are shown in Figure 21.23.

ANOVA/variance component analysis

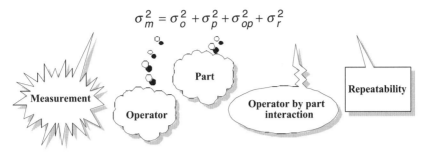

Figure 21.23   Illustration of the fundamental concepts behind ANOVA analysis.

21.13   ANOVA/VARIANCE COMPONENT ANALYSIS

The ANOVA method partitions the total measurement variance into different components. For a traditional gage R&R study, the variation is broken down into four categories of components: (1) operator, (2) part-to-part, (3) operator-by-part, and (4) repeatability. From the *operator* component, we quantify the variation observed between different operators who are measuring the same set of parts. From the *part* component, we quantify the variation observed for a characteristic measured on different parts, regardless of the operator. From the *operator-by-part* component, we explore the interaction between the operators and the parts. Finally, the *repeatability* quantifies the variation due to the instrument itself and the position of the parts in the instrument. Calculation of variance components; the session window of MINITAB shows the ANOVA table. This information is used to calculate the variance components.

Interpretation of MINITAB Graphical Output.   The available information allows us to perform the analysis and to reach conclusions. An example of MINITAB's graphical method is shown in Figure 21.24.

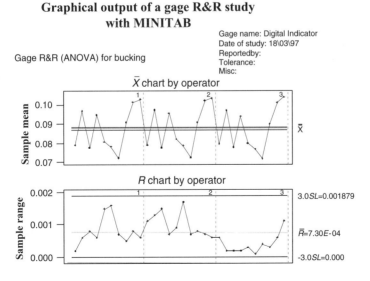

Figure 21.24   Graphs showing MINITAB's gage R&R output.

Interpretation. The first two graphs represent the $\bar{X}-R$ charts. Observe that all of the points on the $\bar{X}$ chart represent the averages of the repeated measurements on the parts for the three operators. The range chart represents the range, which is the maximum value minus the minimum value of the repeated measurements on the parts for these three operators. From the $\bar{X}$ charts it is observed that the measurements lie/are outside the control limits; we interpret this as an indication that variation is due mainly to differences between parts.

The range chart is satisfactory mainly because none of the range values are too large and they are within the control limits. The *control limits* are variation limits for repeatability error. On the $\bar{X}$ chart they are compared with the part averages, and on the range chart, they limit the repeatability error for each part. Therefore, *the patterns observed on these charts indicate that there is no problem.* What can we say about the operators when we look at the $\bar{X}-R$ chart? On the $\bar{X}$ chart, the averages of the repeated measurements on each part for the three operators seem quite similar. The pattern is virtually the same. Since the operators have obtained the same measurements for the same parts, it may be concluded that the interoperator variability is consistent.

From the range chart for operator 3, we see that these ranges are lower than those of the other two operators. This makes sense because operator 3 is actually an experienced quality inspector and therefore makes more consistent measurements than the others do. Although it is very important to have a reliable measuring instrument, how the instrument is used is equally important. The quality and metrology groups can help an organization standardize the measurement procedure.

Operator–Part Components. In Figure 21.25 we can see the results and interaction between operators and part components.

Besides the $\bar{X}-R$ charts, the graphical MINITAB output of a gage R&R study also provides graphs representing operator and part components. The operator–part interaction graph shows the effect of different parts on each operator's measurements. By looking at the average measurements of each part for each operator, we can learn how differently configured parts affect an operator's measurements. Since the lines are almost parallel, we can see that the operators perform equally well for each part. Is there any *difference among the operators*? On the measurement-by-operator graph, we see that the line connecting the average measurements between the operators is not horizontal. Thus, the averages between these operators is not the same.

Gage R&R (ANOVA) for measurement

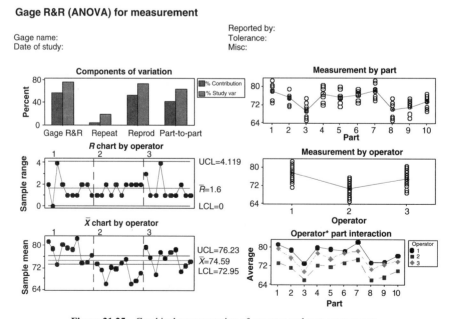

Figure 21.25   Graphical representation of operator and part components.

Next we want to know is if there is any variation between the parts. The by-part graph represents the variation in average measurements taken on the same characteristic for each part. It shows us that there are in fact large differences among the parts. The nonhorizontal line reveals that we must look at the parts themselves as there is large variation among the measurements of the characteristic.

Graphical Representation of Components of Variation Indices. The *components of variation* chart is very useful for visualizing the different ratios calculated with a gage R&R. It shows the percentage of the measurement's system variation that is due to the combined and separated effects of repeatability and reproducibility and the parts themselves. The meanings of the three columns in the chart are as follows: The `%Total` Var column represents the value of the variance components divided by the total variation, then multiplied by 100. This is known as the *ratio of variances*. Remember that the standard deviation is easier to interpret than variance because it has the same unit of measurement as the original observation. So, in practice, the standard deviation is used more often than the variance. The `%Study Var` column represents the standard deviation for each component divided by the total standard deviation. Finally, the `%Toler` column shows ratios of the standard deviation for all components divided by the tolerance of the product that we want to measure with the instrument.

Knowing these definitions, it becomes easier to understand the graph. For the three columns, we see that the differences among the parts account for more variation than do both the gage Reproducibility and repeatability combined. However, the measurement system itself accounts for very little of the variation in the measurements. Thus most of the observed variation comes from the parts, not from the measurement system.

21.14 RULES OF THUMB

Different indices can be calculated using the data from the gage R&R data. In the MINITAB session window, the following information is presented in table format. The table columns are as follows:

1. The `% Contribution` column indicates the percent contribution to the total variation made by each variance component. The sum of the percentages in this column is 100. In the case of the preceding example, the percent contribution from the `Part-To-Part` is 99.51%. In this case, most of the variation was due to differences between parts, and very little was due to measurement system error.

2. The `%StudyVar` column shows the standard deviation for each component divided by the total standard deviation. The sum of these percentages does not equal 100. In the case of the example, it can be observed that the standard deviation for repeatability and reproducibility errors represents only 6.97 % of the total standard deviation observed in the study. The total standard deviation accounts for repeatability and reproducibility plus the part-to-part variation.

3. The `%Tolerance` column is 5.15 times the standard deviation for each component divided by the tolerance spread of the measured product. MINITAB uses 5.15 because the software finds 99% of the area under a normal curve within an interval; this is a wide 5.15 standard deviation. In this case, repeatability and reproducibility errors consume 10.56% of the product tolerance.

The value of 6 times the standard deviation represents the full (99.7%) spread of the measurement system variation. Historically, some organizations have used the value of 5.15 rather than 6. For a normal distribution, 5.15 times the standard deviation captures an even 99% of the distribution. The practitioner can select whichever is needed for compatibility with standards. The precision: tolerance ratio *P/T* is then a gage metric based on quality. Knowing *P/T* is essential to determining whether the parts are inside the specification limits, and also for reducing the product tolerances to allow gage variation. For example, if a product specification is 65 ± 1 mm, and the *P/T* is 5%, the parts must be produced to a measured tolerance of 65 ± 0.32 mm. The precision/tolerance ratio can be expressed as follows:

$$\text{Tolerance} = \text{USL} - \text{LSL}$$

$$\frac{P}{T} = \frac{6 \times \sqrt{\sigma^2 \text{MS}}}{\text{USL} - \text{LSL}}$$

This fraction is usually expressed as a percentage. The best case is $P/T < 10\%$; an acceptable case is $P/T < 30\%$.

Number of Distinct Categories. Finally, the number of distinct categories represents the number of groups within data that the measurement system can distinguish. For this study, 20 categories were a good number. However, the suggested sample size is 30. The number of distinct categories is calculated by dividing the standard deviation for *parts* by the standard deviation for *gage*, then multiplying by 1.41 and rounding down to the nearest integer. This number represents the number of nonoverlapping confidence intervals that will span the range of product variation.

The most frequently used gage R&R formulas are as follows (where P = precision, T = tolerance, and V = variation):

- P/T (% tolerance)—the most common estimate of measurement system precision, this metric compares gage precision to the tolerance requirements. Use P/T if the gages being used are to classify production samples.
- P/SV (% R&R)—measures how the gage performs compared to the gage study variation and best estimate for performing process improvement. (*Note*: Gage study variance is not necessarily representative of the true process variance.)
- P/TV (% R&R)—measures how the gage performs compared to the process variation. Gage study variance is not necessarily representative of the true process variance.

(*Note*: $P/TV = P/TV$ when the variation in the gage samples represents true process variation.)
Figure 21.26 illustrates an example of an application of the rules of thumb.

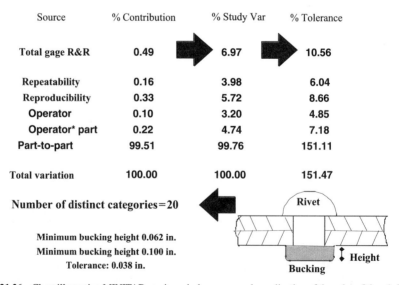

Rules of thumb

Source	% Contribution	% Study Var	% Tolerance
Total gage R&R	0.49	6.97	10.56
Repeatability	0.16	3.98	6.04
Reproducibility	0.33	5.72	8.66
Operator	0.10	3.20	4.85
Operator* part	0.22	4.74	7.18
Part-to-part	99.51	99.76	151.11
Total variation	100.00	100.00	151.47

Number of distinct categories = 20

Minimum bucking height 0.062 in.
Minimum bucking height 0.100 in.
Tolerance: 0.038 in.

Rivet

Height

Bucking

Figure 21.26 Chart illustrating MINITAB session window—example application of the rules of thumb for gage R&R.

21.15   ACCEPTABILITY CRITERIA

What is acceptable? The AIAG provides guidelines based on standard deviation for acceptable gage variation, as compared to total variation, which are given as a percentage of tolerance. These are

- *Ideal*—10% or less
- *Acceptable*—10–20%
- *Marginal*—20–30% depending on the application

To arrive at % gage R&R as a percentage of tolerance, MINITAB's ANOVA method calculates the variance component for the gage R&R term. Then it calculates the estimated standard deviation (StdDev) for the measuring system by 5.15 and divides it by the tolerance. The tolerance is the difference between specification limits. In the session window MINITAB shows this value at the top of an output column entitled %Tolerance. A disadvantage of this MINITAB method for calculating percentages is that the sum of these columns never total 100%. Therefore, it is suggested that you use the %Contribution that measures variance rather than standard deviation. See Figure 21.27.

In other words, for indices such as the %Study Variation or %Tolerance, the measurement system is considered acceptable if the R&R% is less than 10%. If the R&R% falls between 10 and 30%, it may be acceptable under certain considerations. If it is greater than 30%, the measurement system is considered to be unacceptable, and it will need to be improved before collecting data for our Six Sigma projects. The values of your R&R% for the %Study Variation and for the %Tolerance columns in the MINITAB output are 6.97% and 10.56%.

Number of Distinct Categories. A value ≤ 1 means that the measurement system cannot distinguish between parts. A value $= 2$ means that the data can be divided into two groups, for example, high and low. A value $= 3$ means that the data can be divided into 3 groups, say, high, middle, and low (and so on). But to a value > 5 is recommended. Because the value is 20, this means that the measurement system has an excellent discrimination between the parts. However, the final acceptance of a measurement system should not come down to a single set of indices. The performance of the measurement system should also be reviewed using the ANOVA technique and graphical analysis.

Acceptability criteria

% R&R Indices	
<10%	Acceptable measurement system
10–30%	May be acceptable, based on importance of application, cost of measurement device, cost of repair, etc.
>30%	Considered unacceptable; measurement system needs improvement

Number of Distinct Categories Indices	
1	Unacceptable; one part can not be distinguished from another
2–4	Generally unacceptable
>=5	Recommended

Figure 21.27   Summary of acceptability criteria.

21.16   CHAPTER REVIEW

1. It is important to define and validate the measurement system before collecting data.

2. The measurement system is the complete process used to obtain measurements.

3. Experience has shown that measurement error is inevitable. We must identify, evaluate, and control the sources of measurement error. This could even be the object of a Six Sigma project itself.

4. The common types of measuring instrument errors are accuracy, linearity, stability, and precision.

5. The common types of measuring system errors are repeatability and reproducibility.

6. To estimate precision and accuracy, multiple measurements of one single characteristic must be taken. *Precision* is quantified as the standard deviation between these replicated measurements. *Accuracy* is quantified as the difference between the observed average of measurements and the reference value.

7. Attribute and variable gage studies allow us to evaluate reproducibility and repeatability.

8. In transactional areas, with a little imagination, the same measurement evaluation studies apply. Establishing and communicating measurement procedures will reduce errors.

9. A word of advice for transactional projects. Since measuring instruments are not always clearly defined; avoid spending too much time evaluating the system. It may not be economically viable.

10. In the usual method of performing an attribute gage R&R, at least two operators measure 20 parts at random, twice each. If there is little consistency among the operators, then the measurement system must be improved, by defining a measurement method, training the operators, etc. A significant improvement can be achieved by changing the measuring system to obtain continuous data.

11. In the usual method of performing a variable gage R&R, three operators measure 10 parts with the same nominal dimension in a random order, 3 times each.

12. A gage R&R, like any study, requires careful planning.

13. A variable gage R&R can be analyzed using two methods. $\bar{X}$ and R charts are easy as they provide graphical representations of the variation. ANOVA is a more powerful method because it gives better estimators of the variation and considers operator and part interactions. These tools will be covered in more detail in other phases of Black Belt training.

14. The graphical output of MINITAB software provides both methods of analysis:
    - *Control charts*—show discrimination, stability and variation in the range of measurements for each part
    - *ANOVA*—helpful for estimating error sources and their contribution to overall variability
    - *Linear regression*—helpful for estimating the linearity of system response
    - *Charts and scatterplots*—can also be used to study variation between and across operators and parts

15. As a rule of thumb, if percentage R&R indices are >30% and the number of distinct categories indices is < 5, then the measurement system is considered to be unacceptable. In this case, it must be improved before the practitioner begins to collect data for his/her Six Sigma projects.

16. The percentages of variation are determined as follows:
    - (Repeatability variation)/(total variation) = % variation that gage R&R contributes to total variation
    - (Reproducibility variation)/(total variation) = % variation that operator contributes to total variation

- (Reproducibility variation (EV) + Reproducibility variation (AV))/(total variation) = % gage R&R variation (where EV = equipment variation, AV = appraiser variation)

GLOSSARY

abscissa   Horizontal axis of a graph.

activity symbol   Rectangle used to designate an activity. A brief description of the activity is also included within this symbol.

accuracy   Items are correctly categorized. Requires knowledge of true value. Also defined as (number of times correctly classified by all)/(total number of evaluations by all).

affinity diagram   Tool used to organize and present large amounts of data (ideas, issues, solutions) into logical categories on the basis of perception.

alpha risk   Probability of accepting the alternative hypothesis (H_1), when in reality the null hypothesis (H_0) is true.

alternative hypothesis   Tentative explanation indicating that an event does not follow a chance distribution; a contrast to the null hypothesis.

attribute data   Numerical information at the nominal level.

Bias   The proportion of items in a given category is correct. Overall average proportion in a given category for all inspectors/operators) minus correct proportion in a given category, averaged over all categories. Requires knowledge of true value.

brainstorming   Technique that a group of people use to generate ideas on a particular issue.

c chart   Chart that displays the number of defects per sample.

cause   That which produces an effect or brings about a change.

centerline   Line of statistical process control charts that represents the characteristics of central tendency.

champion   Senior manager who is responsible for the logistics and business aspects of a project or projects.

characteristic   Measurable feature of a process, product, or service.

classification   Differentiation of factors or variables.

common cause   Random cause.

common cause variation   Changes in the system outputs or quality characteristic values under usual circumstances.

complex level   First level of the process tree. It is the last process performed to deliver the product/ service to the customer.

confidence level   The probability that a random variable x lies within a defined interval.

confidence limits   The two values that define the confidence level.

continuous data   Numerical information at the interval of ratio level; subdivision is conceptually meaningful.

continuous random variable   Random variable that can assume any value continuously in some specified interval.

connections   Arrows indicating the flow and sequence of the key activities.

connector symbol   Circle used to illustrate a break and its continuation elsewhere on the same page or another page.

control chart   Graphical representation of a characteristic performance across time in relation to its natural limits and central tendency.

C_p **index**   Process capability index.

C_{pk} **index**   Measure of process dispersion and its centering about the average.

control limit   Line(s) on a control chart used as a basis for judging the significance of variation from subgroup to subgroup.

control specifications   Specifications called for by the product being manufactured or service being delivered.

cost of poor quality (COPQ)   Money spent or lost on non-value added activities while producing a good service.

critical requirements for quality, delivery, and cost   The needs translated from the CTS characteristics.

critical-to-cost (CTC) characteristics   Product, service, and/or transactional characteristics that significantly influence one or more CTS in terms of cost.

critical-to-delivery (CTD) characteristics   Product, service, and/or transactional characteristics that significantly influence one or more CTS in terms of delivery.

critical-to-process (CTP) characteristics   Process parameters that significantly influence a CTQ, CTD, and/or CTC.

critical-to-quality (CTQ) characteristics   Product, service, and/or transactional characteristics that significantly influence one or more CTS in terms of quality.

critical-to-satisfaction (CTS) characteristic   Expression of the customers' vital needs.

customer   Person or organization that receives a service or finished product.

cycle time   Measure of each process step or subprocess.

data   Factual information used as a basis for reasoning, discussion, or calculation. Often refers to quantitative information.

defect   Output of a process that does not customer specifications.

defective   Unit with one or more defects. Unit that contains at least one defect.

defects per million opportunities (DPMO)   Number of defects divided by the total number of opportunities, then multiplied by 1,000,000; DPU $\times 1,000,000$.

defects per unit (DPU)   The total number of defects (N) divided by the total number of units in the group (n): N/n.

dependent variable   Response variable Y: $Y =$ output $=$ response variable from transfer function $Y = f(x\text{'s})$.

design for Six Sigma (DFSS)   The process for designing to develop new products or processes or redesigning ongoing processes.

detection   Past-oriented strategy that attempts to identify unacceptable output after it has been produced and then separate it from the good output (*prevention*).

discrimination   Ability of the measurement system to adequately differentiate between repeated measurements.

document symbol   Term indicating a printed document pertinent to the process.

element level   This level represents a breakdown of the processes listed at the subsystem level. Critical-to-process (CTP) characteristics are identified directly below this level.

factors   Independent variables, or x's, or input factors.

flowline   Arrow indicating the direction of the process and connects its elements.

gage   Instrument used to measure a part to generate a measurement number.

gage reliability and reproducibility (gage R&R)   Analysis to determine the portion of the observed process variation caused by measurement system variation.

independent variable   A controlled variable; a variable whose value is independent of the value of another variable.

interval scales   Measure used to express numerical information on a scale with equal distance between categories, but no absolute zero. Examples of interval scales include temperature (°F and °C), a dial gage sitting on top of a gage block, and comparison of differences.

key process   List steps or subprocesses included in the process.

linearity   When an inspector/operator evaluates items covering a full set of range measurements, his/her classifications are to be consistent across all range measurements.

lower inner fence   Value in a dataset that is $1.5 \times$ IQR below the first quartile.

lower outer fence   Value in a dataset that is $3.0 \times$ IQR below the first quartile.

lower and upper specification limits   Actions based on the relationship between the specifications and the centering and dispersion of a controlled process depend on whether there are two specifications limits. A maximum upper limit USL and a minimum or lower limit LSL.

master Black Belt   Leader of a Six Sigma team responsible for applying the Six Sigma process.

mean   Measure of central tendency calculated by dividing the sum of all values by the number of values in the dataset.

measurement system   A systematic study of all aspects of data measurements produced by the system being measured with the objective of determining the extent to which the measurement system introduces a measurement error.

measures of central tendency   Measures that describe the center of a distribution. The mean, median, and mode are three of such measurements.

measures of dispersion   Measures that give the spread of a distribution.

median   Middle value in a group of data or measurements when arranged from lowest to highest if the number of values is even. By convention, the average of the middle of two values is used as a median.

never-ending improvement in quality and productivity   Operational philosophy that makes the best use of the talents within the organization to produce products or services of increasing quality for our customers in an increasingly efficient manner that protects the return on investment of stockholders.

nominal scales   Measures used to classify elements into categories without considering any specific property. Examples of nominal scales include "causes" on fishbone diagrams, yes/no, and pass/fail.

nonconforming unit   A unit that does not conform to one or more customer specifications or standards.

normal distribution   Continuous, symmetrical density function characterized by a bell-shaped curve.

null hypothesis   Tentative explanation indicating that a chance distribution is operating; a contrast to the null hypothesis.

operation definition   Means of clearly communicating quality expectations and performance.

opportunity   Potential defect or error that could result in failure to meet customer requirements.

ordinal scales   Measure used to order or rank nominal data with respect to a specific property. Examples of ordinal scales include relative height, Pareto charts, and customer satisfaction surveys.

outcome   Response-dependent variable. The result of a trial with a given treatment is called *response*.

outliers or extreme values   Values that are very small or very large relative to the majority of the values in a dataset.

pareto diagram   Chart used to plot percent defectives in a sample.

parameter   Summary measure calculated for Population data.

***p* chart** Chart to plot percent defectives in a sample.

percentiles Ninety-nine values that divide a ranked dataset into 100 equal parts.

population Collection of all elements whose characteristics are being studied.

process Series of activities for doing something-Combination of people, equipment, materials, methods, and environment to produce an output for a given product or service.

process average Central tendency of a given process characteristic.

process control Application of statistical methods and procedures relative to process and a given set of standards.

process spread Range of values that gives process characteristic displays; this particular term most often applies to the range but may also encompass the variance. The spread may be based on a set of data collected at a specific point in time or may reflect the variability across a given amount of time.

project A problem usually calling for planned action.

quartiles Three summary measures that divide a ranked dataset into four equal parts.

random Term applied for selecting a sample so each item in the population has an equal chance of being selected; lack of predictability; without pattern.

random variations Variations in data that result from causes that cannot be pinpointed or controlled.

range Difference between the highest and lowest values in a dataset or data subgroup.

ratio scales Measure used to express numerical information on a scale with equal distance between categories, but with an absolute zero in the range of measurement.

reliability Ability of one gage to consistently return the same results for the same activity.

repeatability Variation due the measuring device—the difference in repeated measurements on the same part, by the same operator/inspector, using the same device; referred to as *equipment variation* (EV). (*Note*: EV in gage R&R should not be confused with the acronym EV used for *earned* value in project management.)

replication Recording data under identical controlled conditions.

reproducibility Variation due to the measurement system. The differences in measurements on the same part, by different operators, using the same device, referred to as *operator/inspector effect* or *appraiser variation* (AV).

run chart Simple graphical representation of a characteristic of a process showing plotted values of some statistic gathered from the process.

sample In process control applications, a synonym with sub-group. This use is totally different from the purpose of providing an estimate of a larger group of people or items.

sigma (σ) Greek letter used to designate standard deviation.

special cause Cause of variation that are not common to all events involved.

spread General concept for the extent by which values in a distribution differ from one another. Data dispersion.

stability Variability between attribute R&R studies at different times.

stable process Process that is in statistical control.

standard deviation Statistical index of variability which describes the spread.

statistic Summary measure calculated for sample data.

statistical control The condition describing a process from which all special causes of variation have been eliminated and only common causes remain.

subgroup Logical grouping of objects or events which displays only random event to event variations. One or more events of measurements used to analyze the performance of a process.

symbols Terms indicating the nature of each step (see Section 7.9.1 for more details).

system   Something which is connected according to scheme.

upper control limit   Horizontal line in a control chart that represents the upper limits of a process capability. In general at $+3\sigma$ of the data distribution centerline.

variable   Characteristic that may take on different values.

variance   Measure of spread.

variation   Quantifiable difference between individual measurements such as differences that can be classified as being due to common causes or special causes.

X and R charts   Control charts that represent the process capability over time.

yield   Amount of material that is processed by an operation less the scrap.

REFERENCES

1. Automotive Industry Action Group, *Measurement System Analysis* (MSA). (http://www.aiag.org/publications/quality/dcxfordgm.html/.

2. AIAG, *Measurement Systems Analysis, Reference Manual*, 3rd ed., Automotive Industry Action Group, Southfield, MI, 2002.

22 Design of Experiments

22.1 OVERVIEW

As mentioned earlier, to apply statistical methods, we need access to data. Design of experiments is a large and complex topic; therefore, this chapter will cover only introductory concepts and some of the most important aspects of this topic. The authors assume that the practitioner is familiar with process capability principles and understands the basics concepts of variation (e.g., that apparent differences in data may or may not be significant), and that data must be statistically tested to determine whether an effect is significant. The theoretical basis for process capability is described in Chapters 9 and 15.

By changing one or more elements in a process and observing the effects of these changes, the analyst can learn much about the process. Therefore, after carefully reading this chapter, the practitioner will be able to create a simple experimentation strategy using the DOE process.

22.2 INTRODUCTION

An important issue in statistical process control (SPC) is monitoring, controlling, and understanding process variation. In some cases, process capability could be a major concern, while in others it could be judgment errors introduced when recording or classifying causes during process monitoring. A *process capability* is a single-factor experiment. In contrast, a *designed experiment* may include a number of factors that can be studied simultaneously. Data can be arranged and rearranged for study according to these factors. A designed experiment is successful if the data have been collected in such a way as to make all these rearrangements possible. By using formal experimental techniques, the effect of several variables (factors) can be studied simultaneously, and thus we can observe interactions between them. This technique is useful to identify the effects of hidden variables in an ongoing process when the effect of one factor is offset by another factor. Therefore, design of experiments is a technique to be used with reserve and never substituted for the broader process capability study. Control charts are used to confirm these results and allocate resources between business areas or between shifts.

To redesign an existing process (process improvement) or design a new process, some experimentation is needed. Sometimes, we need to control process input variables and change their levels, while at other times, we may need to redesign training or data collection methods. When the number of factors or variables that strongly influence the process is known, it is easier to select the appropriate design of experiment. For example, in a call center, assume that there are three key factors: (1) training duration; (2) associate level of analytical skills; and (3) associate knowledge of the product, service, or/and transaction. To determine the call center performance under low and high call rates, the practitioner can assign two levels (low and high) to each of the three variables and the call volume. The two levels, low and high, are to be coded as -1 and $+1$). In this particular case, the practitioner has a three-factor (three variables), two-level design (-1 and $+1$). Referring to the DOE table, the practitioner learns that for three factors and two levels a full factorial design is recommended.

Practitioner's Guide for Statistics and Lean Six Sigma for Process Improvements. By Mikel J. Harry, Prem S. Mann,
Ofelia C. de Hodgins, Christopher J. Lacke, and Richard Hulbert
Copyright © 2010 John Wiley & Sons, Inc.

The DOE table also indicates that the sample size (n reps) is ≥ 9 with a 95% confidence level in $\bar{s}$ and 99.99% in the sample mean. Similar information is provided by MINITAB. An example, using MINITAB, will be shown at the end of this chapter.

22.3   DESIGN OF EXPERIMENTS (DOE) DEFINITION

An experiment is any test in which the inputs are controlled or directly manipulated according to a plan. A *design of experiment* (DOE) is a scientific process for planning an experiment that will yield statistically useful results. DOE or DoE (design of experiments) is also a test in which the experimenter makes purposeful changes to the input variables (x's of the input–process–output diagram/transfer function graphical representation) of the process and then observes the output (Y). Well-designed experiments (DOEs) can produce a significant amount of good information compared to unplanned experiments. In addition, well-planned experiments will identify the factors that are most important for the output or response (Y). For example, if it is believed that there is an interaction between two factors (two input x variables), the experimenter must include both factors in the design rather than doing "one factor at a time type of experiment." Interaction occurs when the level of one factor $x_1 =$ input variable influences the effect of second input variable x_2. To provide valid results, a few basic steps should be followed when designing an experiment:

1. *Identify the objective.* Identify the relevant population and the exact question to be answered. For example, is technical knowledge important to resolve calls rapidly and correctly during the first contact?

2. *Collect Sample data.* Sample data must be representative of the population, and sample size must be large enough so the effects of the treatment can be known.

3. *Use a random procedure to avoid bias.*

4. *Analyze the data and form conclusions.* Analyzing the survey results might suggest a great opportunity for improvement.

Example 22.1   The analysis data and creation of a DOE for all centers or help desks came from a presentation to the 2002 Six Sigma Meeting by O. de Hodgins, R. Hulbert, and B. Salyer of Bank of America entitled "Performance Improvement for call centers." De Hodgins et al. reported an interesting event. The better Bank of America became in closing calls at first time customer contact, the number of calls increased. The selected variables (factors) for this DOE example were $A =$ number of calls, $B =$ first-call resolution (number of problem resolution during the first call), and $C =$ associate/analyst training level. Before redesigning the process, Bank of America's call closure performance was poor. After redesign, when customers called, analysts from a call control group greeted them. These analysts had less training than the technically trained group but occasionally provided answers to their questions. This group functioned as a phone reception and semitechnical resources group for all incoming calls ($\approx 10{,}000{,}000$ calls per week).

Bank of America based its Florida call center redesign on real-time support. The intent was to place a customer directly in contact with an analyst with the proper technical skills on the first call, with the purpose of solving the customer's problem during the call. The expected outcome was to attain 40–50% real-time closures from the 18% (problem resolution calls that the bank received, meaning that 18% of the 10,000,00 calls per week were classified as problem resolution calls only). This outcome was reasonable as it was based on historical data (patterns), a review of queuing calls, a review of typical calls (requests), and staffing levels during "rush hours." The good news was that the new processes and process redesign worked well and call closure jumped from 8% to 38%. The bad news was that the redesign worked too well, and the call volume increased by 30–50%. The large impact of the increased call volume became a new problem. Therefore, the process redesign was reviewed one more time using the DOE approach.

The results of the DOE provided enough information to guide a combination of changes in policies, procedures [use of an improved voice recognition unit (IVRU)] and rotation of call center staff. After tracking the improvements for some time, it was observed that real-time call closure remained the same 50% while the volume of calls subsided. It is well known that best-practice organizations use a simple approach to ensure that their organization understands the link between performance measures and business strategies. This example suggests an interesting and predictable dynamic. As performance improved, the staff was able to handle more calls; therefore, an increased volume of calls was recorded and fewer customers abandoned their calls

A possible reason behind this response is that if something works, customers will like it, use it more, and tell others. Before redesigning a process it is important to create a thought map to eliminate some of the factors such as how many calls could have been eliminated if something else were done right the first time. When conducting an experiment, we often encounter interference from factors that are not relevant (noise variables) to the problem being studied. These effects can be controlled through good experimental design. When designing an experiment to test the effectiveness of one or more treatments, it is important that subjects (often called *experimental units*) be placed in different groups or blocks. For example, if we have two shifts, we can call the first shift block 1 and the second shift block 2. A *block* is a group of similar subjects. Note that they need to be similar only in the ways that might affect the outcome of the experiment.

When testing one or more *different treatments*—different blocks—it is important that each block consist of subjects that are similar. When assigning elements to the different blocks, random selection should be used. One approach is a *randomized experimental design*, in which subjects are placed into different blocks by a *random selection process*. An example of a randomized experimental design for the call center example is call center first-time call resolution. Different call centers are assigned to test the new process; these call centers are randomly selected. Another approach is a rigorously controlled design, in which experimental units are *carefully* selected to ensure that subjects placed in each block are similar in the ways that are important for the experiment. With a rigorously controlled design, it is easier to start with a carefully selected call center site (e.g., one in Atlanta, GA) from the organization. The practitioner must carefully define beyond geographic sites and site size to identify other key relevant factors. When conducting experiments, the results are sometimes ruined because of *confounding*. Some definitions are needed here[1]:

Confounding occurs in an experiment when the effects from two or more variables cannot be distinguished from each other.

Aliasing occurs when two factors that interact with each other interaction are held constant throughout the entire experiment.

Two-level design occurs when an experiment is set at two levels denoted as low (-1) and high $(+1)$.

Three-level design occurs when an experiment is set at three levels denoted as low (-1), medium (0), and high $(+1)$.

A *blocking variable* is a factor (variable), that cannot be randomized. The DOE is usually run in "blocks" for each level of the blocking variable and randomization is calculated within the blocks.

Capability (of a process) is a measure of quality, usually expressed as *Cpk*, or defects per million (DPM) and is obtained by comparing the ongoing process with the specification limits.

An example of confounding in a bank call center in Manhattan, NY, is when the bank implements a new attendance policy, Attendance average will drop one point for each absent day. After this policy is implemented, a mild winter improves call center attendance by reducing commuting, transportation, and weather-related delays that in the past have reduced the call center staff (analyst) attendance.

When attendance improves, it is not very clear whether the improvement is attributable to the new attendance policy or to the mild winter because the effects of the attendance policy combined with weather conditions (mild winter) are confounded. Thus, even a well-planned experiment is susceptible to some pitfalls and dangers. Among the most serious are

1. *Experiment planning*—unless the experiment is planned with care and the results studied statistically, conclusions may be wrong or misleading.
2. *Experimental strategy.*
3. *Analysis*—improper analysis will not lead to proper and productive courses of action.
4. *Wrong variables*—studying the wrong variables or gathering too much or too little data may waste time and effort.
5. By carefully planning experiments and analyzing them, the experimenter will be able to
   - Save time and money
   - Statistically remove the effects of unwanted variables
   - Evaluate the results when experiments fail to repeat
   - Reconcile new results with previous knowledge
   - Plan scientifically how much data to collect, what variables to use, and how many factors and levels to include

Some of the Terminology Most frequently Used in DOE

1. *Block*—the component of the experimental event of the service or transaction whose members are likely to be more homogeneous than the whole.
2. *Confounding*—allowing two or more factors to vary together precludes our ability to separate their unique effects.
3. *Data*—the result, for example, the magnitude of the response variable.
4. *DOE*—design of experiment, or "an invitation for an information event to occur."
5. *Effect of a factor*—the change in response produced by a change in the level of the factor.
6. *Experimental error*—occurs when two identically treated experimental units fail to give the same value.
7. *Experimental unit*—one item to which a single treatment is applied in one replication of the basic experiment.
8. *Factor*—input to a process that can be manipulated during experimentation. A process variable that will be purposely changed in value during the experiment. An independent variable is a feature of the experimental conditions in which may be varied from one observation to another. This factor or independent variable (x) may be qualitative or quantitative, fixed or random. Also *factor*—a process variable that will be purposely altered during the experiment.
9. *High level*—high setting [e.g., 40 psi. (lb/in.^{2}), new, shift 1]. High-level coded $= +1$.
10. *Interaction*—occurs when the effect of one factor varies for different levels of another factor; if this occurs, the two factors are said to interact or have interaction.
11. *Level*—factor setting.
12. *Levels of a factor*—the various values of a factor considered in the experiment.
13. *Low level*—low setting. High setting $= 200\,°C$, low setting $= 50\,°C$. Low-level coded $= -1$.
14. *Main effect*—the average effect of a factor.
15. *Noise*—input variable that could be discrete or continuous in nature.
16. *Order*—sequence or run order for the experiment.
17. *Output (response)*—dependent variable. The result of a trial with a given treatment.
18. *Replication*—the number of repetitions of an experiment

19. *Response*—the numerical result of a trial based on a given treatment combination.
20. *Response variable*—the subject of the experiment (CTQ, CTD, CTC).
21. *Treatment combination*—combination of levels for this element of the experiment.

Experimental Strategy

- State the practical *problem*—*clear problem statement.*
- Clearly define *experiment objective*—this objective is more specific than the problem statement.
- Ensure careful selection of output/response variable $= Y$.
- Ensure careful selection of input variables—some organizations use the thought map approach.
- Careful selection of levels for the *input variables* $(x_1, x_2, x_3, x_4, \ldots, x_n)$
- Select the experimental design.
- Collect data.
- Execute the experiment.
- Perform data and result analysis.
- Reach statistical conclusions.
- Replicate and validate experimental results.
- Draw practical conclusions.
- Obtain agreement from all stakeholders.
- Create documentation (SOPs, etc.).
- Implement proposed improvements.

22.4   ROLE OF EXPERIMENTAL DESIGN IN PROCESS IMPROVEMENT

Experimental design consists of changing the x's or inputs or factors of the process. It is important to review the process control system from previous chapters as a feedback system.

1. *The Process.* The whole combination of people, equipment, input materials, methods, and environment that work together to produce an output. The total performance of the process, the quality of its output, and its productivity efficiency depend on the way the process has been designed and built, and on the way the process is operated. The rest of the process control and improvement system is useful only if it contributes to improved performance of the process.

2. *Information about Performance.* Much information on the process can be obtained by studying the process output. Process output includes not only the products that are produced but also any intermediate output that describe the operating state of the process, such as average speed of answer or call duration or number of abandoned calls.

3. *Action on the Process.* Action on the process is future-oriented because it prevents the production of out-of-specification services (products). This action might include changes in the operations, improved analyst training, and changes to the policies and/or procedures or the more basic elements of the process itself (such as redesign of the process as a whole.) The effects of new actions should be monitored, and further analysis and action(s) undertaken when necessary.

4. *Action on the Output.* Action on the output is past-oriented, because it involves detecting out-of-specification output already produced. If current output does not consistently meet customer requirements, it may be necessary to sort all products and to scrap and rework any non-conforming transactions. This must continue until the necessary corrective and preventive actions on the process has been implemented and verified, or the service/transaction specifications have been changed.

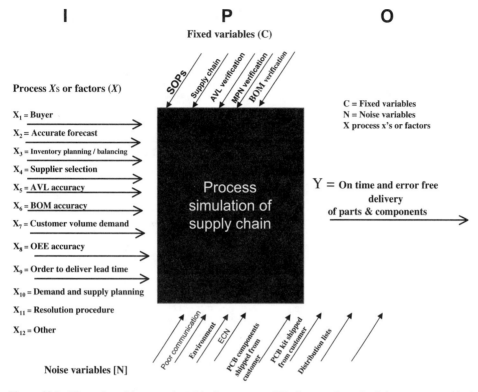

Figure 22.1   Illustration of the general model of a process or IPO diagram of supply chain process (graphical representation of the transfer function).

Figure 22.1 shows the graphical representation of the Transfer Function for planning a supply chain process. This input–process–output (IPO) diagram shows the *x*'s or input factors of the process and the outputs of the factors [*y*'s ("small *y*'s")] and finally the process output or response [*Y*] or "big *Y*."

Figure 22.1 highlights the factors of the DOE (*x*'s) that together result in the single effect *Y*; in this illustration the transfer function appears as

$$Y = f(x_1, x_2, x_3, \ldots, x_{12})$$

Remember that the primary reason for using statistically designed experiments is to obtain the maximum amount of information at a minimum cost and determine a course of action. Many aspects of successfully designing and optimizing a process require efficient, accurate experiments:

1. *Scientific method*—all scientific information is based on experimentation.
2. *Optimization of process parameters*—process parameters of a transaction or service are the different elements of the service or elements of the transaction.
3. *Cost savings*—result from fewer experiments, increased investigator efficiency, and improved output from an optimal process are usually substantial.
4. *Statistical methods*—designing experiments to simultaneously gather information on all factors, offers additional savings compared to single-factor experiments.

Figure 22.2 shows the general model of a process system and the different types of experimental factors.

General model of a process or system

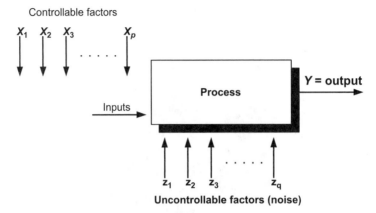

Controllable factors

X_1 X_2 X_3 . . . . . X_p

Inputs

Process

Y = output

z_1 z_2 z_3 . . . . . z_q

Uncontrollable factors (noise)

Figure 22.2   Graphically representation of two types of experimental factors (noise and controllable).

22.5   EXPERIMENT DESIGN TOOLS

Most organizations collect data to monitor the "health" of their processes; however, frequently, there no corrective actions because the data do not contain meaningful information. When these organizations decide to perform an experiment in an effort to produce meaningful data on a specific CT (critical-to) characteristic, they design a one-factor-at-a-time test, and progress is slow and expensive. Organizations that use design of experiments (DOE) leverage their efforts by conducting well-thought-out experiments where several factors are altered simultaneously in one efficient series of tests. DOE provide more information in a shorter period of time than obtained from multiple single-factor experiments. The role of a statistically designed experiment is to identify the most influential factors or "vital few" associated with a particular CT characteristic and to define their relationships using analytical quantities. Since interactions between various factors are addressed, a single DOE can guide the experimenters to quickly improve their process. The purposes of DOE are to determine those variables that have more influence on the output. Choosing input levels to produce the most desirable output level and reduce output variability. Organizations frequently use the *best-guess* approach to select the combination of factors and then run a test and compare the output with expectations. If deficient, factor levels are changed, retest takes place, and then this procedure is repeated as necessary. If the first guess is wrong, more trials will be necessary, which translates into rework and inefficient use of time, which in turn adds costs to the DOE. On the other hand, if the first guess is acceptable, experimentation stops, and the DOE solution is never found. It would be more effective to conduct a carefully planned DOE early in the design process experimenting with the effect of tolerances and failure modes, so that when the product or service is implemented, customer satisfaction will be high and warranty claims will be low. The same experimental techniques can then be used throughout the lifecycle of the product or service to continuously improve process performance.

DOE Considerations[2]   There are two components to a DOE: (1) experimental design and (2) analytical procedure, which should be carefully considered before conducting the DOE. Depending on the level of understanding available about a process, various types of DOE may be employed. *Screening designs* are typically used early in the process where many factors are involved. *Characterization designs* narrow the number of factors down to only a few, while *optimization designs* focus on a few significant factors in much more depth. To ensure an effective DOE, we must consider such topics as Yates standard order, factor levels using the $-1, +1$ notation, column contrast, balance, orthogonality, confounding issues, and interactions between factors. Other aspects of the design, such as replications and the basic design

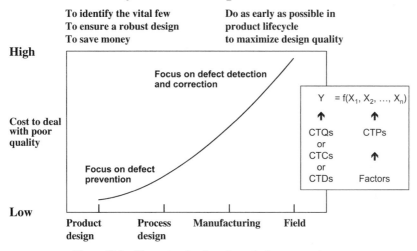

Figure 22.3 Illustration showing why and when to experiment.

type, influence the size and therefore the cost of the DOE. The two-level two-factor design is relatively common and is a good entry point for most problems. When replications are considered, meaningful results can be achieved with as few as 8–16 runs. When one does not know the factors or factor levels that have substantial effect on a response variable, a factorial or fractional factorial design is the preferred design approach. Fractional factorial experiments are commonly applied when five or more factors are to be considered. In this case, one only conducts runs, which represent a portion of the matrix, because, for example, a six-factor two-level experiment with no replications would involve 64 runs. The advantage, therefore, of a factorial experiment is that factors are varied together instead of one at a time and the experiments are a complete set of experimental runs and are ready to be analyzed when all the runs are completed. An advantage of a fractional factorial experiment is that fewer tests are required.

The statistical significance of the data from a DOE is evaluated immediately after conclusion of the DOE. *One-way* and *N-way ANOVA* are the favored methods for this phase of the analysis. Practical significance can be evaluated through the study of sums of squares, pie charts, Pareto diagrams, main-effects plots, and normal probability plots. Note that factors may be statistically significant, but have no practical significance. In any analysis, the use of cross-functional teams in the selection of factors, design of the experiment, conducting the experiment, analysis of the data, recommending actions, and implementing DOE results all require careful planning and attention to detail to achieve success. Figure 22.3 places the DOE in the design process. A thought map of proposed experiments could be created during experiment definition. In the definition part it is necessary to state the practical problem, state the experiment objective, select the output (Y = response), and carefully select the input factors ($x_1, x_2, \ldots, x_n$). The execution part is to select the experiment design and sample size, run the experiment, collect data, analyze the data, draw statistical and practical conclusions, and translate the results and conclusions into practical business actions.

22.6 PRINCIPLES OF AN EXPERIMENTAL DESIGN

Four fundamental principles are essential for a statistically well-designed experiment matrix:

1. Understanding process relationships
   - Between-process variables
   - Between-response variables

2. Randomization
   - Select experimental units randomly
   - Randomizing run sequence
3. Replication
   - Increased precision of effect estimates
   - Providing measure of experimental error
4. Experimental control
   - Active control of process variables
   - Control of other sources of variation

We will now explore each of these principles:

Understanding Process Relationships. A statistically designed experiment is a test or a series of tests in which purposeful changes are made to the input variables (x's) of the process or system to identify reasons for changes in the output response (Y). See Figure 22.1.

Randomization. If separate experimental runs are run in random order, statistical analysis of responses will be free of bias. The danger of not randomizing could be exemplified by considering an experimental run. For the experiment shown in Figure 22.2, assume that the training level of the staff (skills and knowledge) is not considered to be a factor, although in reality it strongly affects the variation of the process and therefore the output of the process (Y). If all runs at one level were completed before runs at the second level are completed, the results would be biased. Randomization provides validity to the statistical tests.

Replication: Experimental Error. If two identically treated experimental units give different results, replication provides the means to estimate experimental error and shed light on potential interactions between input variables. Significant process factors produce large changes compared to the difference between replicates. Replication is a repetition of the entire experiment to estimate experimental error, increase precision (detect smaller changes), and increase sensitivity. Replication makes statistical tests possible.

Local Controls. Local controls means experimental planning, referring to the amount of blocking or balancing of grouped experimental runs among different settings of variables being studied. The purpose of local controls is to achieve a more efficient experimental design, while guarding against unforeseen surprises nullifying experimental adjustments. In summary, local controls make the experiment more efficient. Together, each of these principles can produce good experiments, which can be analyzed using the statistical analysis of variance methods discussed in Chapter 19.

Summary of the Role of Experimentation Once results are obtained, we must choose the settings or levels of the factors to improve yield, quality, or the CT characteristic. This discovery would then lead to another round of experiments. Although it is not always possible, the goal is to strive to find factors that will show independent effects. On the other hand, the objective in many cases may be to develop a robust process, that is, a process affected minimally by the many sources of variability, namely, white noise; which cannot be controlled (see Fig. 22.1). In general, DOE is the controlled, active variation of process inputs to identify better settings or procedures to improve the product quality. Specifically, *DOE* is a systematic method where a number of factors are simultaneously changed following a predetermined pattern to investigate the effect of these changes on the response or output. Experimentation provides a sound, effective, and economical method for determining the best way to implement changes and improve results.

22.7 DIFFERENT TYPES OF EXPERIMENTS

Figure 22.4 lists different types of experiments:

Hierarchy of Experiment Types The first four items in the list in Figure 22.4 are familiar. The remaining headings that describe the Six Sigma approach may be unfamiliar, but we will be surprised that up to 20 factors may be addressed in a single DOE experiment. A *factorial experiment* is used to determine the process response for a specific combination of process inputs and levels. A number of multiple runs (repetitions) of an experiment depend on the number of factors and the type of experimental design selected.

Several types of DOE can be applied to improve a process, such as

- *Screening designs*, which are used to identify the "vital few" process factors. Usually, such designs involve a large number of factors and are limited to only two levels.
- *Characterization designs*, which are generally used to study the effects of a small number of factors (three or four). Such designs are usually based on full factorial models using two or three levels.
- *Optimization designs* (also referred to as *modeling designs*), which are used to study complicated effects and interactions involving one or two factors. Such designs are employed to define a mathematical model of the process under investigation and may involve several levels (to capture nonlinearities).

Standard order (2^k) designs, as defined by Yates, are designs that are easily generated and analyzed. The *y* (number of variables/factors) metric is used to determine the minimum number of trials needed to run and the number of initial columns in a design or Yates standard order (2^2). Coding used for these designs is as follows:

$(-) =$ low level of a factor

$(+) =$ high level of a factor

Using the $(-1)/(+1)$ designation will be helpful in understanding how designs are generated.

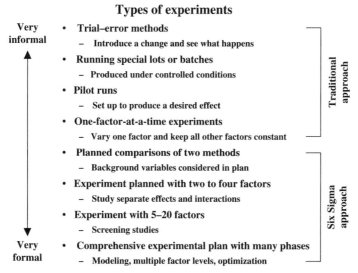

Figure 22.4 Schematic representation of the different types of experiments from very informal to very formal.

Example 22.2   A Six Sigma team conducts an experiment on a heat treatment process. Three factors are identified:

Factor	Level of Factor	
A = quenching speed (factor)	(−) low: 1 m/s;	(+) high: 2 m/s
B = aging time (factor)	(−) low: 4 min;	(+) high: 8 min
C = bath type (factor)	(−) low: water;	(+) high: glycol

Solution   When designing the experiment, the Yates standard order is most often used. We used a two-level three-factor design. For each run, the corresponding factor level is identified as −1 when the factor is set at "low," and as +1 when it is set at "high." The first combination shown in the Yates standard order will be a treatment with quenching speed set at −1 (low, or 1 m/s), aging time at −1 (low, or 4 min). The type of bath is set at −1 (low, or water). This run is represented in the first row by the numbers: −1, −1, −1. All other runs are set in the same fashion, up to the eighth run, where the settings are +1, +1, +1. It is important to randomize the Yates standard order used in the experiment to average out the noise caused by nuisance variables. The same table can be used to plan an experiment for a two-level three-factor design. The team recognizes that using this Yates standard order helps avoid errors in preparing an experiment. This experiment is a three-factor two-level design or 2^3, 8 runs.

Full Factorial Design

Factors: 3

Runs: 8

Blocks: 1

Base design: 3, 8

Replicates: 1

Center points (total): 0

All terms are free from aliasing.

To create factorial design using MINITAB; select `Stat > DOE > Factorial > Create Factorial Design` (see Fig. 22.5).

MINITAB automatically stores the design in the worksheet. Be careful when changing design elements in the worksheet. If changes corrupt the design, the experiment must be redesigned using the `Define Custom Factorial Design` option. The column standard order is low–high low–high for the first factor, high–two lows–high for the second factor, and low–high–two lows–high for the third factor. MINITAB assigns a random order to the experimental runs when the experiment is performed. By default, MINITAB displays the design in run order. The `Center Point` column indicates the type of design point. Centerpoints are coded as 0; corner points are coded as 1. The design points in this experiment were selected to be corner points. The block concept, discussed later in this chapter, indicates a particular combination of settings that run the first and second repetition. When more than one block is selected, each block is randomized independently.

22.7.1.1   Main Effects

The change in response observed when a factor goes from low to high level is referred to as the *main effect* of the factor. An *interaction* exists between two or more factors when the effect of one factor depends on the level of other factor(s).

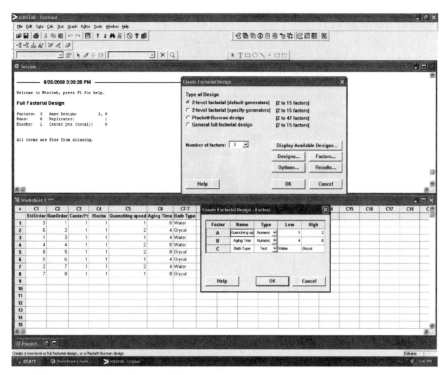

Figure 22.5   Screenshot of MINITAB `create factorial` design.

Example 22.3: One-Way Main-Effects Plot   An engineer is working on a project to maximize the strength of the adhesion bond of the circuitry to a printed-circuit board (PCB). Three different types of solder paste were suggested as possible solutions to the low-tensile-strength problem. Fifteen measurements were taken, with five measurements for each of the different types of solder paste. To analyze the data, the engineer created a one-way main-effects plot. On the chart he depicted the average response for the three different levels, that is, the three different types of paste. Having completed the analysis, the engineer presented the results. As you can see by this main-effects plot, the solder paste at level 3 provides the maximum tensile strength of the three types tested; the engineer recommends selection of this solder paste for future production. This looks very good and appears to support the selection that you have made. However, do you have statistical evidence to support this selection?

"Statistical evidence?"   The main-effects plot (see Fig. 22.6) is a graphical tool used to verify conclusions; however, it should be used to complement other tools. With this tool alone, we cannot be sure that the variation presented in the plot is significant. We need to know if the differences between the averages could have resulted from chance random sampling variations. To determine this, we need a tool such as one-way ANOVA.

Interaction Plot   The *interaction plot* (see Fig. 22.7) is used to visually assess the nature of the interactions between two or more factors. The interaction plot is created by plotting the response variable on the Y (vertical) axis (ordinate) and one of the independent variables on the X (horizontal) axis (abscissa). Then the second independent variable is plotted on the graph. Parallel lines indicate weak or no interaction between the factors. Strong interaction between factors in

Plotting the main effects (2^2)

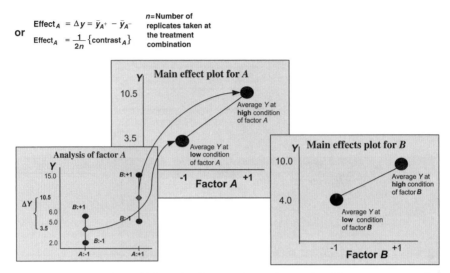

Figure 22.6 Main-effects plot for two factors, A and B.

indicated by nonparallel lines. The strength of the interaction is indicated by the slopes of the two lines.

Example 22.4 A bank wants to know which type of customer (premier, private, commercial, small-business, etc.) showed interaction with the invoice value (factor B). A DOE was performed, and the interaction plots are analyzed here. When factor A changes from its low level to a high level and factor B is kept at its low level, the output [Y] changes by an average of 4.0 units (from 2.0 to 6.0). When B is kept at its high level and factor A varies from its low level to a high level, the output changes by an average of 10.0 units (from 5.0 to 15.0). This indicates interaction between factors A and B. If the

Plotting the interaction effect (2^2)

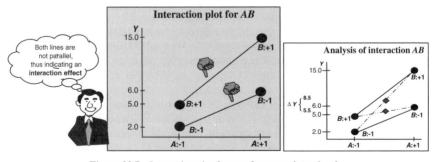

Figure 22.7 Interaction plot for two factors and two levels.

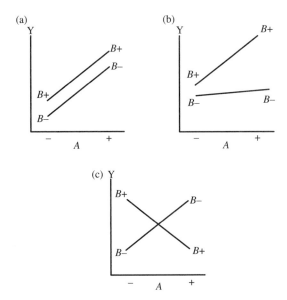

Figure 22.8 Graphical examples of interaction plots of two factors, A and B, at two levels [low (−1) and high (+1)]. Note that in these graphs, the relationship between variable A and Y changes as the level of B changes. Plot (a) shows no interaction (lines are parallel), while plot (b) shows strong interaction. In this figure, when B is at it high (+) Level, variable A has a strong effect on Y. When B is low (−), A has some effect on Y. Plot (c) shows a full reversal. The distinguishing feature of strong interactions is the degree of nonparallelism between the two lines.

lines are not parallel, an interaction exists between the factors. Graphs are useful for visualizing significant interactions between factors, but they should not be the only means of interpreting data. Interactions also occur in transactional processes. Consider a project analyzing overdue invoices. The first factor is A, the type of customer; the second factor is B, invoice value. We would be surprised to discover that there is an important interaction effect between the two factors (invoice value and type of customer). Two factor interactions are common. However, three-factor and higher-factor interactions are rarely seen in practice. Figure 22.8 shows some examples of interactions.

Figure 22.9 graphically depicts interactions and main effects.

22.8 INTRODUCTION TO FACTORIAL DESIGNS

Full factorial designs are often used when a few input factors (x's) need to be studied. If the study is performed one factor at a time, the factor interactions will not be observed.

Example 22.5   As part of its marketing campaign, a hotel chain would like to test whether offering customers delightful showers at optimal water temperature will create more business for the organization. For this, the hotel management needs to know the optimal knob adjustment for cold and hot water.

Solution   By adjusting hot and cold knobs, we will change the water temperature of the shower, which represents the response of the process Y. For this example, we will set the knobs at high and low levels. The two knobs (hot and cold) represent the two input variables (x_1 and x_2) or factors ($A = x_1$ and $B = x_2$), where $x_1 = $ *flowrate of cold water* and $x_2 = $ *flowrate of hot water*. This two-level two-factor

Main effects and interactions

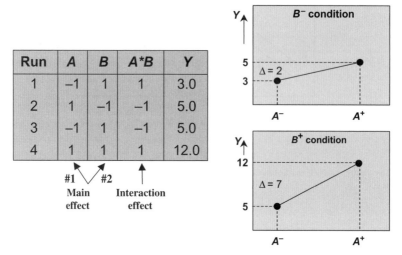

Run	A	B	A*B	Y
1	−1	1	1	3.0
2	1	−1	−1	5.0
3	−1	1	−1	5.0
4	1	1	1	12.0

#1 #2
Main   Interaction
effect   effect

Figure 22.9   Illustration of main effects and interactions.

matrix, $2^2 = 4$, requires four combinations of input variables. The experiment itself still needs to be done (see Figs. 22.10 and 22.11).

When conducting an experiment, at least two types of resources are required: (1) human and (2) materials. To achieve worthwhile results, each DOE should be carefully planned to maximize the amount of information while requiring minimal resources. To perform a factorial design, the experimenter identifies the factors and corresponding resource levels.

Executing the Experimental Strategy

1. Define the problem—when taking a shower, find the most comfortable water temperature setting at minimum cost.

The shower experiment

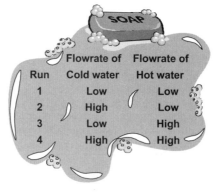

Run	Flowrate of Cold water	Flowrate of Hot water
1	Low	Low
2	High	Low
3	Low	High
4	High	High

$$Y = f(X_1, X_2)$$

Water temperature $= f$ **(flow of hot water, flow of cold water)**

Figure 22.10   Matrix representation for DOE: two factors, two levels, four runs.

Other factorial design notation matrix (2^2)
two-level two-factor design

Run No.	Sign Notation		Alternative Notation		Japanese Notation		Classical Notation	
	A	*B*	*A*	*B*	*A*	*B*	*A*	*B*
1	-	-	0	0	1	1	(1)	
2	+	-	1	0	3	1	a	
3	-	+	0	1	1	2	b	
4	+	+	1	1	2	2	a b	

Figure 22.11   Notation for a two-level, two-factor design.

2. Establish the experimental objective—determine the best knob settings.

3. Select the output (response)—minimum cost.

4. Identify critical x's values.

5. Select the input(s) (independent variables)—$A = x_1$ and $B = x_2$ (where $A =$ cold-water flowrate, $B =$ hot-water flowrate).

6. Select optimal settings for x values to achieve the identified objective.

7. Select the low and high levels for the input variables. The low and high levels are coded as -1 and $+1$. The range of these levels must be sufficiently wide to affect the output.

8. Blocking—there are no blocking factors.

9. Centerpoints—type of design points; centerpoints are coded as 0.

10. Corner points—corner points are coded as 1.

For this example, a full factorial design is selected consisting of all possible factor and level combinations. Full factorial designs are referred to as 2^k factorial design or 2^2 (two factors, each factor at two levels). This type of design investigates the individual effects of each factor (main effects) and determines whether these factors interact. The 2^2 designs are easy to use, and the analysis can be graphical. A factorial design for four factors at two levels requires represented by $2^4 = 16$ runs. A factorial design for four factors at three levels is represented by $3^4 = 81$ runs. A factorial design for four factors at four levels is represented by $4^4 = 256$ runs. The combination of factors and levels that make up the factorial design is called the *design matrix*. When factors are studied at two levels and the numerical value is not known, the convention is to enter these two levels as high $(+1)$ and low (-1). However, these levels can be assigned in any arbitrary manner.   ■

22.9   FEATURES OF FACTORIAL DESIGNS—ORTHOGONALITY

Orthogonality   The sum should be balanced, meaning that this column should be equal to 0. We obtain this column by multiplying the signs of each of the variable columns; for example, by multiplying columns A and B, we obtain the AB column. This column represents the interaction between A and B. The property of orthogonality ensures that the plus and minus signs are arranged in the main-effects columns A and B such that their product column is balanced. This, in turn,

Run	Order	*A*	*B*	**AB**
1	(1)	-1	-1	+1
2	*a*	+1	-1	-1
3	*b*	-1	+1	-1
4	*ab*	+1	+1	+1

Figure 22.12   Tabular representation of two-factor, full factorial design in a design matrix display.

ensures that each of the factors will be independent of the others. In the design shown in Figure 22.12, the variable *A* is set high for two runs and at the low level for two runs (see also Fig. 2.13).

Advantages of factorial designs are as follows:

- Simple design
- Three 2-factor interactions
- No confounding (confusion among factors)

Features of Factorial Designs—Balance   Each variable in the experiment contains the same number of runs at both high and low levels. In the above design; the variable *A* has two runs in which it is at its high level and two runs in which the variable is at its low level. When we sum the *A* variable, we obtain: $(-1) + (+1) + (-1) + (+1) = 0$. This property simplifies the mathematical analysis of the design by ensuring that each level of the variable has equal *impact* on the final result.

Three-Variable Full Factorial   The matrix in Figure 22.14 shows factors *A*, *B*, and *C*. These factors are the *main effects*. Column *AB*, a *two-way interaction*, is obtained by multiplying the content of columns *A* and *B*. Column *ABC*, a *three-way interaction*, is obtained by multiplying columns *A*, *B*, and *C*. All full factorial experiments can be expanded using this method.

Factorial experimental runs are shown in Figure 22.15.

A disadvantage of three-factor designs is that we have only eight tests. Advantages are its simple design; we have three, two-factor interactions, and there is no confounding, and we have one three-factor interaction.

A MINITAB cube plot for a three-factor, two-level DOE 2^k is shown in Figure 22.16.

A *cube plot* is a graph generally used by the process design engineer to visualize how the response data are distributed across the experiment space.

A	**B**	**C**
-1	-1	-1
+1	-1	-1
-1	+1	-1
+1	+1	-1
-1	-1	+1
+1	-1	+1
-1	=1	+1
+1	+1	+1

Figure 22.13   Design matrix display of three-factor (*A,B,C*), two-level $(+1, -1)$ design.

A	B	AB	C	AC	BC	ABC
-1	-1	+1	-1	+1	+1	-1
+1	-1	-1	-1	-1	+1	+1
-1	+1	-1	-1	+1	-1	+1
+1	+1	+1	-1	-1	-1	-1
-1	-1	+1	+1	-1	-1	+1
+1	-1	-1	+1	+1	-1	-1
-1	+1	-1	+1	-1	+1	-1
+1	+1	+1	+1	+1	+1	+1

Figure 22.14 Design matrix display of three factors (variables), two levels, and three way-interactions.

22.10   FULL FACTORIAL DESIGNS

The full factorial design tests every possible combination of factors at a given number of levels. Full factorial designs are most useful for quantifying interactions and optimizing process settings. Full factorial designs simultaneously estimate factor and interaction effects with equal precision. However, the resources required to run a full factorial design are often impractical. Full factorial designs are effective when properly used in the overall process improvement strategy. Some characteristics of full factorial standard order designs are:

- 2^k (where 2 = levels, k = number of factors)
- Balanced (orthogonal)
- Useful for quantifying interactions

Figure 22.17 shows a full factorial standard order design for two-factor two-level design.

Example: 2^3 Design

- three factors at two levels
- $2^3 = 2 \times 2 \times 2 = 8$ runs

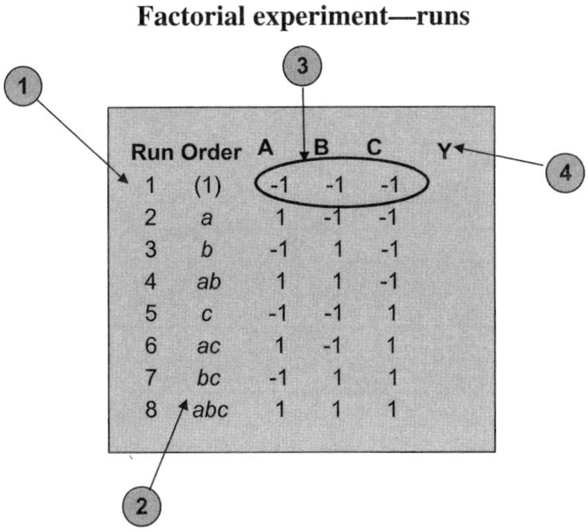

Factorial experiment—runs

Figure 22.15 Factorial experiment runs [1—experimental run number; 2—yates designation for the run; 3—experimental settings (high or low) for the main process factors; 4—response/output of the experimental run].

A	B	C
-1	-1	-1
+1	-1	-1
-1	+1	-1
+1	+1	-1
-1	-1	+1
+1	-1	+1
-1	+1	+1
+1	+1	+1

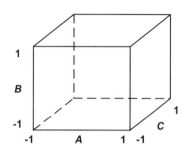

Pros

➢ Simple design
➢ Three 2-factor interactions
➢ No confounding

Cons

➢ Only three factors allowed
➢ Eight tests to control

Figure 22.16   Geometric representation of a three-factor two-level design.

- Eight runs minimum
- Three factor columns (*A, B,* and *C*)

To create full factorial models in MINITAB, select Stat > DOE > Create Factorial Design.

22.11   RESIDUAL ANALYSIS (2^2)

While analyzing her first 2×2 factorial design, a Black Belt notices that MINITAB can graph residuals. In real life, we rarely find a perfect normal distribution. Often, the factor or factors of interest are impacted by other factors that are not part of the data.

MINITAB standardizes residuals by placing them on a common scale. How this is done is not within the scope of this chapter. However, we know that if a model fits the data perfectly, the residuals

Run Order	A	B	AB	C	AC	BC	ABC
1	-1	-1	+1	-1	+1	+1	-1
2	+1	-1	-1	-1	-1	+1	+1
3	-1	+1	-1	-1	+1	-1	+1
4	+1	+1	+1	-1	-1	-1	-1
5	-1	-1	+1	+1	-1	-1	+1
6	+1	-1	-1	+1	+1	-1	-1
7	-1	+1	-1	+1	-1	+1	-1
8	+1	+1	+1	+1	+1	+1	+1

Figure 22.17   Matrix representation showing variables *A*, *B*, and *C*, referred to as the *main effects* as well as the interactions that are computed from the main effects. Column *AB* is referred to as *two-way interaction*. Column *ABC* is referred to as *three-way interaction*.

are all equal to 0; we also know that analysis of residuals effectively assesses how well the model fits the data. Therefore, it is important to study a variety of residual plots and search for patterns and trends. In general, the absence of any patterns or trends (i.e., a random scatter of points) indicates that the model is adequate. When the model is not adequate, it cannot account for curvature in Y, and the residuals contain both random variation and curvature. An interesting pattern would warn the practitioner that something of importance should not be overlooked. Therefore, the experimental process is not complete until the residuals are thoroughly evaluated. Remember that when estimating a statistical model, the practitioner often assumes that the residuals are sampled from a normal distribution with constant variance and no correlation exists between residuals. These assumptions are not made in all experimental analysis, but be aware that these assumptions are frequently made. Under residual plots, MINITAB creates histograms of the residuals and normal probability plots of the residuals, residuals versus fitted values, and residuals versus the order of the data.

When conducting a DOE, the objective is to test the appropriate hypotheses and estimate the effects of treatment. Errors are assumed to be independent random variables, normally distributed with mean equal to zero, and are assumed to remain constant for all levels of the factors. We must

1. Verify that residuals are normally distributed. The normal plot shows dots aligned on a straight line indicating that residuals are indeed normally distributed.
2. Verify that residuals are normally distributed and centered on zero.
3. Review the individual chart of residuals. Since the points are randomly distributed, no points lie outside the control limits, and no particular pattern exists, we conclude that the residuals are randomly distributed. Remember, we must analyze residuals prior to drawing any conclusions about experiment result. See Figure 22.18.

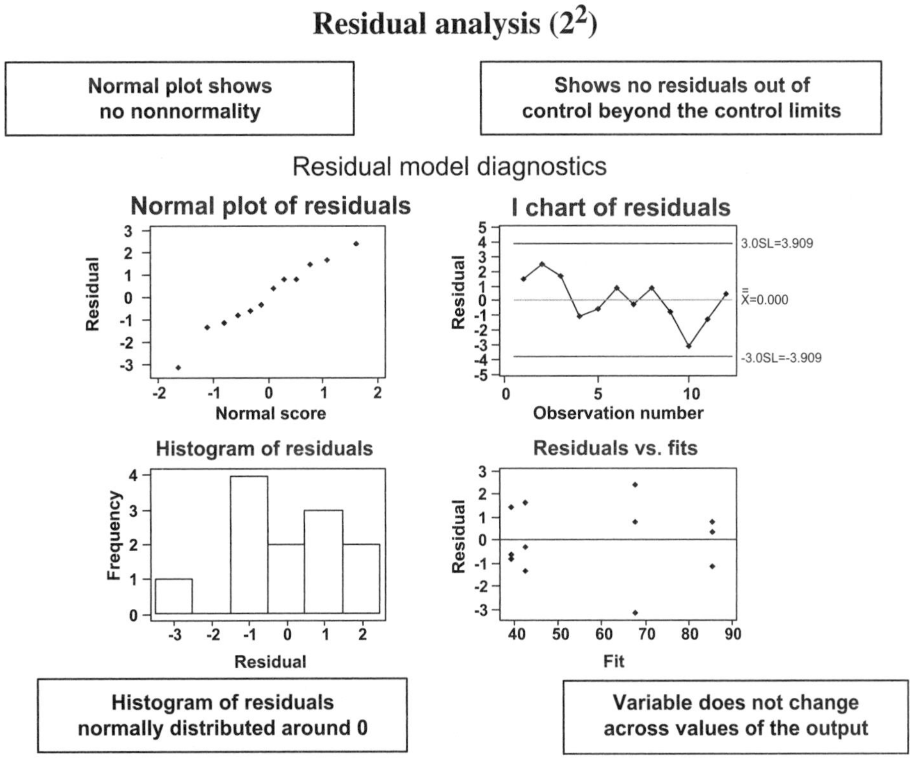

Residual analysis (2^2)

Figure 22.18   Residual analysis plot.

Modeling - 2^2

Estimated effects and coefficients for yield

Term	Effect	Coef	StDevCoef		T	P
Constant		58.017	1.341		43.25	0.000
A	34.367	17.183	1.341		12.81	0.000
B	7.800	3.900	1.341		2.91	0.020
A*B	4.600	2.300	1.341		1.71	0.125

Analysis of variance for yield

Source	DF	Seq SS	Adj SS	Adj MS	F	P
Main effects	2	3725.72	3725.72	1862.86	86.29	0.000
2-way interactions	1	63.48	63.48	63.48	2.94	0.125
Residual error	8	172.71	172.71	21.59		
Pure error	8	172.71	172.71	21.59		
Total	11	3961.92				

$$\hat{Y} = 58.017 + 17.183 * A + 3.9 * B$$

Figure 22.19   MINITAB graphical display showing experiment results for estimated effects and coefficients for yield and analysis of variance for yield.

22.12   MODELING (2^2)

The results of the DOE quantify the effect that two factors and the interaction have on the process response (Y). Are we limited to these two factor settings? Or, is it possible to know more about the process behavior? Is it correct to say that the DOE analysis provides useful information about the factor effects and their interactions?

The average response output yield (Y) when temperature (A) and Pressure (B) are set to their *high levels* is computed using the following model: $Y = 58.017 + 17.183*1 + 3.9*1 = 79.1$. With a similar calculation, we can determine that when A is *high* and B is *low*, the average yield will be $Y = 58.017 + 17.183*1 - 3.9*(-1) = 71.3$. We usually assume a linear response; in that case, this mathematical model is valid. We use other techniques, such as centerpoints or multilevel factorial design, to confirm this linearity. See also Figure 22.19.

DOE analysis provides useful information about the factor effects and their interaction. However, from the mathematical model of this example, we can observe that if no change is made, the process will have an average yield of 58.01%. Also, we observe that the level of factor A (temperature) is key in controlling the yield of the process as compared to factor B (pressure). The average response output yield when temperature (A) and pressure (B) are set to their high levels is computed using the model $Y = 58.017 + 17.183*1 + 3.9*1 = 79.1$. Similarly, when A is high and B is low, the average yield will be $Y = 58.017 + 17.183*1 - 3.9*(-1) = 71.3$.

22.13   MULTIFACTOR EXPERIMENT

Example 22.6   A Black Belt from a manufacturing organization is discussing a project with a master Black Belt. During the brainstorming session, the team identified several factors that could influence the CTQ characteristic, angularity of the parts formed on the fluid cell press. They must design an experiment with at least five factors, but a full factorial experiment would be very large. Although the factors have been identified for some time, the high leverage factors have not been isolated. So far, we have calculated that a five-factor two-level experiment requires 32 runs. For six

Multifactor experiment

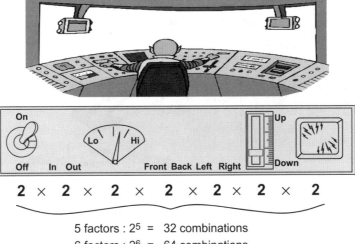

5 factors : 2^5 = 32 combinations
6 factors : 2^6 = 64 combinations
7 factors : 2^7 = 128 combinations

Figure 22.20 Graphical representation of a multifactor experiment.

factors, 64 runs are needed, and seven factors will require 128 runs—and we have not included any replications. The Black Belt asks: "What do you recommend?"

Solution   The Black belt asked, "if there are many factors in the experiment, a full factorial experiment will require a large number of tests, even if only two levels of each factor are being investigated." The master Black Belt replied, "to identify leverage variables from a large number of variables, the fractional factorial design is the preferred analytical tool. In fact, it is common practice to do a fractional factorial experiment whenever we have five or more factors." See Figure 22.20.

22.14   FRACTIONAL FACTORIAL DESIGNS

Although MINITAB and JMP create fractional factorial designs, it is important to understand how they are constructed. Our example is a three-factor two-level design. In most industries, third-order interactions are statistically negligible, so we can neglect the **ABC** interaction when creating our fractional design. First, we sort the entire matrix by the highest-order interaction; this is the **ABC** column for our example. The fractional design is formed by selecting all those treatment combinations where the **ABC** effect is either + 1 or −1. The sign of the **ABC** effect constitutes a "block." Therefore, the design generator can be **ABC** = −1 or **ABC** = + 1. We choose the block where **ABC** = + 1, and then retain runs 2, 3, 5, and 8 shown in Figure 22.21. Look what happens to the matrix. First, the **ABC** effect is lost since it was used to generate the half fraction of the full factorial matrix. Also, the unique effect of **A** cannot be separated form the unique effect of the **BC** interaction because both columns have the same sign for each experimental run. This also occurs for **B** with **AC**, and **C** with **AB**. This is called *confounding* or *aliasing*. If we had chosen **ABC** = −1 to generate the design, we would have reached the same conclusions. The design is thus reduced to these four runs and we only need the three columns for **A**, **B**, and **C**. For the three-factor design, the two-level interactions are confounded with main effects.

The fundamental idea for a fractional factorial design are better illustrated in Figure 22.22.

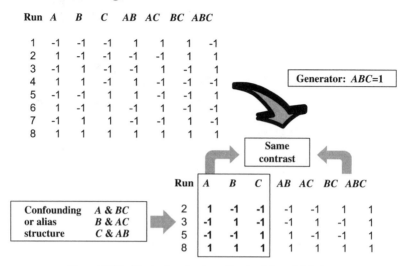

Forming a half-fraction of a full three-factor design

Run	A	B	C	AB	AC	BC	ABC
1	-1	-1	-1	1	1	1	-1
2	1	-1	-1	-1	-1	1	1
3	-1	1	-1	-1	1	-1	1
4	1	1	-1	1	-1	-1	-1
5	-1	-1	1	1	-1	-1	1
6	1	-1	1	-1	1	-1	-1
7	-1	1	1	-1	-1	1	-1
8	1	1	1	1	1	1	1

Generator: $ABC=1$

Same contrast

Run	A	B	C	AB	AC	BC	ABC
2	1	-1	-1	-1	-1	1	1
3	-1	1	-1	-1	1	-1	1
5	-1	-1	1	1	-1	-1	1
8	1	1	1	1	1	1	1

Confounding or alias structure: A & BC, B & AC, C & AB

Figure 22.21   Illustration showing a fractional factorial design.

A single fraction of the 2^k design contains 2^{k-1} runs and is known as the 2^{k-1} *fractional factorial design*. Table 22.1 shows plus and minus signs for the 2^3 design.

The 2^{3-1} design is formed by selecting only those runs that yield a plus on the ABC effect. Therefore, ABC is the generator of this fraction. The identity element I is also plus for the four runs, so $I = ABC$, which is the defining relation for the design.

Fractional factorial designs

Full factorial   **Fractional factorial**

Example with 4 factors

Main Effects	Interactions	Main Effects	Interactions
A	AB, AC, AD, BC, BD, CD	A	AB, AC, AD, BC, BD, CD
B	ABC, ABD, ACD, BCD	B	
C	ABCD	C	
D		D	

Figure 22.22   Illustrations of the fundamental concept of fractional factorial designs.

TABLE 22.1   Plus and Minus Signs for the 2^3 Design

Run	Identity Element (I)	A	B	C	AB	AC	BC	ABC
a	+	+	−	−	−	−	+	+
b	+	−	+	−	−	+	−	+
c	+	−	−	+	+	−	−	+
abc	+	+	+	+	+	+	+	+
ab	+	+	+	−	+	−	−	−
ac	+	+	−	+	−	+	−	−
bc	+	−	+	+	−	−	+	−
(1)	+	−	−	−	+	+	+	−

Fractional Factorial Design

Factors: 3

Runs: 4

Blocks: 1

Base design: 3, 4

Replicates: 1

Centerpoints (total): 0

Resolution: III

Fraction: $\frac{1}{2}$

(*Note*: Main effects are confounded with two-way interactions.)
Design generators are $C = AB$. Alias structure is as follows:

$I + ABC$

$A + BC$

$B + AC$

$C + AB$

If one of more factors from a $\frac{1}{2}$ fraction of a 2^k can be dropped, the design becomes a full factorial design. If we believe that at most two of the three factors are important, then the 2^{3-1} design is likely to identify the significant factors. When practitioners seek to identify a few significant factors from a larger number of factors, these experiments are called *screening experiments*. The design resolution concept catalogs fractional factorial designs according to the alias patterns that they produce. Designs of resolution III, IV and V, listed below, are particularly important:

Resolution III. In these designs, each main effect is not aliased with any other main effect, but main with two-factor interactions, and these two-factor interactions may be aliased with each other. The 2^{3-1} design with $I = ABC$ is of resolution III as was found using MINITAB. In general, a roman numeral represents design resolution. Thus, a 2^{3-1} half fraction is represented by 2^{3-1}_{III}.

Resolution IV. In these designs, main effects are not aliased with each other or with two-factor interactions. Two-factor interactions are aliased with each other. The 2^{4-1} design with $I = ABCD$ is of resolution IV 2^{2-1}_{IV}.

Resolution V. In these designs no main effect or two-factor interaction is aliased with any other main effect or two-factor interaction. However, two-factor interactions are aliased with three-factor interactions. A 2^{5-1} design with $I = ABCDE$ is of resolution 2^{5-1}_{V}.

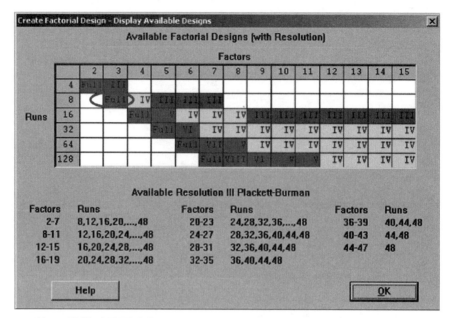

Figure 22.23   MINITAB illustration showing available factorial designs (with resolution).

Resolution III and Resolution IV. These designs are particularly useful in factor screening experiments. A resolution IV design provides good information on main effects and some information on two-factor interactions.

A 2^{k-1} design is useful because it reduces the number of runs required for an experiment. However, smaller fractions will provide almost as much useful information and are more cost-effective. Figure 22.23 illustrates this principle.

The principal advantage of fractional designs is the reduction in number of runs. Thus, they provide effective screening experiments. Depending on the number of factors and runs, resolution defines the alias or confounding structure of a design. For example, a resolution III design confounds the main effects with interactions between factors. Even these designs don't seem to be very useful, we must remember that these designs are the so called *screening designs,* and their advantage is that a small number of runs are required to use them.

As the quantity of information is limited, another experiment will be needed to study, in more detail, the important factors that have been found from the screening experiments.

In resolution IV and resolution V designs, main effects are not confounded with interactions but interactions are confounded together. We consider these designs *characterization designs,* and they are generally used to study relatively uncomplicated effects and interactions from a smaller number of factors compared to those designs of resolution III. Resolution V and higher designs are called *optimization designs* and are often used to study more complicated effects and relationships involving two or three factors. Such designs are employed to define a mathematical model describing the relationship between the CT characteristic and the factors $[Y = f(x_1, \ldots, x_n)]$.

Example 22.7: Variables and Factors   The design of experiments (DOE) plan identified the factors that affect the quality of the parts formed in the fluid cell press. The CTQ characteristic that needed improvement was the angularity of the parts. The analysis indicated that two types of parts are produced: *curved* and *straight*. A separate statistical analysis will be done for each part type.

Objective, response variables, and factors

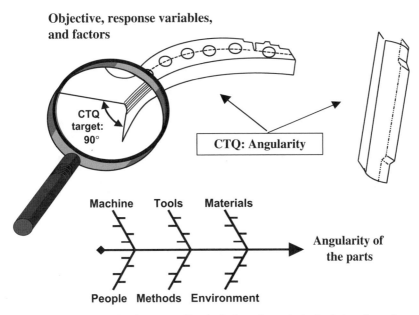

Figure 22.24   Illustration showing cause effect Analysis used to assist in the design of experiments.

Improvements are required because rework is regularly performed on the parts produced by this process. Can you clarify how using the CTQ characteristic can solve the problem that is being experienced? The parts produced have an angle that almost never meets the 90° target. Thus, each part must be checked and straightened to meet the CTQ target.

After the review of the DOE results, the Six Sigma champion asked the quality department, "Do you have a good measuring system to obtain accurate data for your DOE?" "Yes, we have defined a location where all the parts will be measured, and a gage R&R study has been conducted with the team members who will measure the parts. The results of the gage R&R study detected a problem with the measuring system, but it has been corrected."

With the information gathered during the project and a recent brainstorming session, more than 60 causes have been identified that could affect the angularity of the parts. The cause–effect diagram (Fig. 22.24) was used to organize potential causes.

Example 22.8: Selecting Factors and Levels   During the presentation of a DOE to a group of Six Sigma champions, a Black Belt described another aspect of the fluid cell project, the selection of factors and levels. The team identified seven factors that influence the CTQ (critical-to-quality) characteristic, the angularity of the parts. "With such a complex process, there are more than seven factors," said the method's champion. Initially, more than 60 causes were identified but these were reduced to the seven factors studied in this experiment. Initially, the causes were placed in three different categories: (1) controllable, (2) noise, and (3) fixed (held constant). From the group of the controllable factors, the nominal group technique was used where each team member ranked the factors by importance. Finally, seven factors were identified as a reasonable number of factors for additional study since the number of factors is usually set to be eight or less for industrial experiments.

How were these levels selected? "Some factors such as forming speed have just two options," said the master Black Belt. Other levels come from operators' experience or by choosing factor levels at reasonable extremes. In general, the levels should be set "wide," but not so wide that they extend beyond the operational range of the factor to improve our chances of detecting effects.

ANOVA table

Analysis of Variance for Angularity

Source	DF	Seq SS	Adj SS	Adj MS	F	P
A	1	9.753	9.753	9.753	41.06	0.000
B	1	6.722	6.722	6.722	28.30	0.000
C	1	10.889	10.889	10.889	45.84	0.000
D	**1**	**375.836**	**375.836**	**375.836**	**1582.26**	**0.000**
E	1	2.170	2.170	2.170	9.14	0.008
F	1	1.389	1.389	1.389	5.85	0.027
G	**1**	**583.679**	**583.679**	**583.679**	**2457.27**	**0.000**
A*B	**1**	**134.070**	**134.070**	**134.070**	**564.43**	**0.000**
A*C	1	13.133	13.133	13.133	55.29	0.000
A*D	1	3.445	3.445	3.445	14.50	0.001
A*E	1	3.897	3.897	3.897	16.40	0.001
A*F	1	5.147	5.147	5.147	21.67	0.000
A*G	1	3.230	3.230	3.230	13.60	0.002
B*C	1	1.188	1.188	1.188	5.00	0.039
Error	17	4.038	4.038	0.238		
Total	31	1158.586				

Figure 22.25   MINITAB session window showing analysis of variance for the factor angularity.

22.15   THE ANOVA TABLE

The results of the DOE appear in the session window of MINITAB (Fig. 22.25). To obtain the sums of squares of each term in a model, the function General Linear Model can be used. The Six Sigma champion asked, "Is it really important to calculate all the sums-of-squares?" "Yes, the sums of squares are the basic measure of variation of each term in the model. They are also used to evaluate the practical significance of the effects."

If the design is a fractional factorial 2^{7-3} of resolution IV, how do we choose the model to use in MINITAB? Generally for a resolution IV design alias structure, we select the main effects and the two-factors interaction. The ANOVA table indicates that all effects are statistically significant at an alpha level of 0.05. Although they are statistically significant, all are not practically significant. This is evident if we consider the magnitude of each sum of squares compared to the total sum of squares. It appears that factors $D*G$ and $A*B$ have the largest sums of squares compared to the total. *(Note: A*B is a label; it does not define the method used to calculate the data.)*

Analyzing the alias structure of the design, it is observed that the $A*B$ interaction is confounded with D^*G and E^*F. Since factors D and G are both important as main effects, it is more likely that the D^*G interaction is present here. We next graph the residuals to verify the assumptions associated with ANOVA, and to detect the presence of outliers that could contaminate the results.

22.16   NORMAL PROBABILITY PLOT OF THE EFFECTS

What is the difference between the graph shown in Figure 22.26 and a Pareto chart of the sums of squares? The difference is that instead of using the sums of squares, the figure compares values with the normal probability plot, but each graph can be used to identify the important effects. How is this plot interpreted? If the comparisons represent merely random variation (i.e., if there are no real differences between the averages at -1 compared to $+1$), then the contrast effects form a reasonably straight line on the normal probability plot. However, if some comparisons represent real differences rather than random variation, then these will be plotted on the lower left or upper right of the straight line formed by the other contrasts. In other words, the normal probability plot of effects is used to compare the relative magnitude and the statistical significance of both main and interaction effects. MINITAB

Normal probability plot of the effects

Normal probability plot of the standardized effects
(response is Angulari, α=.10)

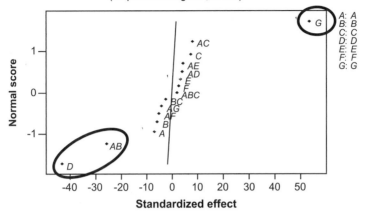

Figure 22.26   Normal probability plot of standardized effects and $\alpha = 0.05$.

draws a line to indicate where the points would be expected to fall if there were no effects. Significant effects are larger and fall farther from the line than do the nonsignificant effects. In addition, MINITAB uses $\alpha = 0.05$ as default, and labels significant effects only. This plot also indicates the direction of the effect; for example, in Figure 22.26, the factors A, C, and B have positive effects because they appear on the right side of the plot, meaning that when the low level changes to the high level of the factor, the response increases. In the figure, D appears on the left side of the plot, meaning that this factor has a negative effect. This indicates that when the low level changes to high, the response decreases.

22.17   MAIN-EFFECTS PLOT

A Black Belt feels that the ANOVA table is not user-friendly because it displays many numbers arranged in a tabular form. She asks, "Is there a tool that creates a visual representation of the results?" The *main-effects plot* is a very useful tool because it plots the mean of the response variable for the levels of each factor. It helps visualize and confirm the important main effects. The main-effects plot shows a large difference in the mean of the response variable between the low (-1) and high (1) levels of the D and G factors in Figure 22.26.

Discussion

1. *Where do these data come from?* The data are the result of a fractional factorial design related to the project.
2. *What is the CT characteristic and the type of design you used?* The CT characteristic is the angularity of the parts, and the design is a fractional factorial 2^{7-3} of resolution IV with two replications. Since we wish to minimize the response variable (CT characteristic), we should set factor D at its high level ($+1$) and factor G at its low level (-1). If there is an interaction between these two factors, we should do an interaction plot to find the best setting for both factors.

22.18   BLOCKING VARIABLE

Blocking is a design technique used to improve the precision of comparisons for the factors being studied. Blocking is frequently used to reduce or eliminate the variability from nuisance factors. When using blocking variables, experimental runs are divided into groups such that the runs within

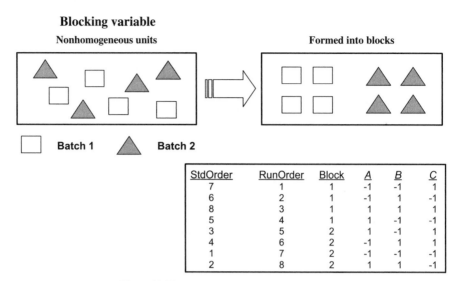

Figure 22.27 Drawing showing blocking approach.

a group are relatively homogeneous; this reduces the impact of experimental errors or noise on the runs within the group. However, when creating a design matrix, the practitioner should not forget that blocking variables are usually confounded with higher-order interactions. To remove the effect of this added noise, a "randomized block design" is to determine the run order of the block members. Blocking variables remove the noise from the error term by creating a separate term. Thus, blocking provides two benefits: (1) it reduces the experimental error, which makes it easier to detect effects; and (2) it tests the significance of the blocking variable. Figure 22.27 shows this blocking approach (*Note:* The three basic principles of experimental design are randomization, replication, and blocking.)

22.19 STATISTICAL SIGNIFICANCE

To illustrate this concept, consider that a Six Sigma team conducting a DOE related to the curing process of composite parts. Two factors are being studied: pressure (Factor A) and temperature (Factor B). The team conducts two replicates for this *two-factor three-level experiment*. After recording the responses, an ANOVA table is analyzed in MINITAB to determine whether there is a statistical difference between the output for different input levels of the factors (pressure and temperature). There are various ANOVA functions in the MINITAB menu; which one should we choose? For this two-factor experiment, *balanced ANOVA* was selected because it highlights the interaction between factors A and B. From the table in Figure 22.28, we conclude that pressure (Factor A) is not statistically significant with a risk of 5%. The p value is 0.466, indicating that there is a 46.6% probability that the change in the output is due to chance. On the other hand, temperature (Factor B) has considerable influence on the output. With a p value of .008, it is highly improbable that the change in output is due to chance alone. The *interaction* between factors A and B ($A*B$) has a significant effect on the CTQ, as shown by a p value smaller than 0.05. Thus, we conclude that the change in the output is not random. We recognize that even if the main effect of factor A is not statistically significant, it affects the output through its interaction with factor B.

22.20 PRACTICAL SIGNIFICANCE

In some situations certain factors are statistically significant, but are not practically significant. Assume that we have conducted an experiment with two factors: type of coolant (Factor A) and

A	B	Y
1	1	40
1	2	50
1	3	60
1	1	20
1	2	30
1	3	50
2	1	50
2	2	70
2	3	50
2	1	10
2	2	60
2	3	60
3	1	30
3	2	20
3	3	70
3	1	40
3	2	10
3	3	80

Statistically significant?
multilevel full factorial

Analysis of variance for _Y_   **The ANOVA table**

Source	DF	SS	MS	F	P
A	2	277.8	138.9	0.83	0.466
B	2	2877.8	1438.9	8.63	0.008*
A * B	4	2788.9	697.2	4.18	0.035*
Error	9	1500.0	166.7		
Total	17	7444.4			

Figure 22.28   The ANOVA table and the results of the analysis of variance for output Y: two-factor three-level experiment.

type of end mill (Factor _B_). Each factor was set at three levels. Assume that the ANOVA table, similar to the table shown in Figure 22.28, shows a risk of 0.05, and a _p_ value for factor _A_ of .014. However, in the analysis of the sum of squares, it is observed that this factor _A_ accounts for just 4.1% _of the total sum of squares_; thus, we conclude that this factor is not of practical significance. In fact, the effect caused by factor _B_ represents the largest percentage of the total sum of squares, 43.0%.

Does this mean that factor A should not to be considered? No, this means that we must base our decision not only on main effects. As we saw, the _interaction A*B_ is both _statistically_ and _practically significant_. To understand this interaction, we must study both factors _A_ and _B_.

22.21 FUNDAMENTALS OF RESIDUAL ANALYSIS

A _residual_ is simply the difference between the observed value of _y_ (response variable) and the value of _y_ predicted output by the model.

$$\text{Residual} = y_{\text{observed}} - y_{\text{predicted}}$$

There is one residual for each observation. MINITAB normalizes residuals and shows them on a common scale. Residuals from any model are helpful in evaluating the adequacy of the model relative to the data and assumptions made during the analysis.

So far, we have conducted the DOE, interpreted the ANOVA table, and identified the factors that have statistical and practical significance. Does this mean that we have finished the DOE analysis? Not yet; there is a very important statistical analysis remaining. We must analyze the residuals.

Figure 22.29 shows the results of the MINITAB residual analysis.

**Residual analysis
multilevel full factorial**

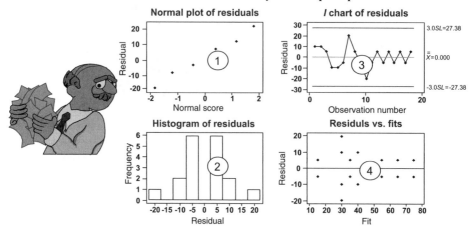

Figure 22.29 MINITAB graphs showing residual analysis for a multilevel full factorial DOE.

Figure 22.29 Graphs

- The first graph (upper left) shows that the residuals are normally distributed.
- The second graph (upper right) is the histogram. It indicates that the residuals are grouped around zero, as they should be when nothing special is affecting the residuals.
- The third graph (lower left) is the chart of the residuals in run order. The residuals are randomly distributed and no points occur outside the control limits.
- The fourth graph (lower right) indicates that the residuals are randomly distributed around the fitted values.

From these observations, the project team concludes that the residuals are normally distributed around zero and are not affected by special causes.

22.22 CENTERPOINTS

Sometimes it is desirable to study quantitative factors at more than two levels. Examples that would lead to three or more levels include

- Checking for quadratic effects (curvature)
- Estimating the error
- Including the current level of the factors as well as the high and the low levels.

By selectively adding centerpoints to a 2^k factorial design, we may address the above mentioned concerns without significantly increasing the number of experimental runs by adding centerpoints to the design by performing n replicates at the midlevel of each factor in the factorial design. In this case, the ANOVA table shows a curvature at the centerpoint. Since its p value is less than 5%, the curvature cannot be assigned to a particular factor, but adding centerpoints indicates whether it is appropriate to interpolate between the factorial points. Note that the magnitude of the curvature can be directly compared to the factor effects for relative importance if the number of centerpoints is equal to one-half the number of factorial points. Also, the number of centerpoints does not affect the other effects. The organization's Six Sigma champion recognizes that, with only two centerpoints, the error is calculated and the curvature is detected, necessitating only six experiments (e.g., four runs for the 2^2 with one replicate plus two runs at centerpoint) compared to a 3^2 full factorial design, which needs nine runs.

22.23 NOISE FACTORS

What to do about factors that can't be controlled and held constant during the experiment? These factors are called noise (N), or nuisance factors. In many experiments, the variation due to background or nuisance variables will equal or exceed the variation because of the chosen factors. Generally, noise is a factor that we think would have an effect on the CT characteristic, and we classify it into two types as follows: (1) factors those that are known but uncontrollable, and (2) factors that are unknown and therefore, uncontrollable. We can use the blocking technique if we suspect that noise is known and controllable, and may have a significant effect on the CT characteristic. It is always important to minimize the effects from all types of noise throughout the experimental sequence; this is usually achieved through a complete randomization of the experimental sequence.

22.24 STRATEGY OF GOOD EXPERIMENTATION

Can you outline the steps to take to ensure that it is successful? A 10-step process can be followed to ensure that the practitioner covers the main topics of good experimental design. Steps 1 and 2 define the problem and the objective for the experiment. A well-constructed problem statement, combined with a clear objective statement, is key to the improvement cycle. The objective, preferably with a statement of planned actions, should be stated such that it provides guidance to those designing the experiment. The objective statement should clarify whether the experiment requires screening. The 10 steps are as follows:

Step 1. Define the problem.
Step 2. Establish the objective.
Step 3. Select the response variable(s).
Step 4. Select the independent variable(s).
Step 5. Choose the variable levels.
Step 6. Select the experiment design.
Step 7. Collect the data.
Step 8. Analyze the data.
Step 9. Draw conclusions.
Step 10. Achieve the objective.

The objective should be specific. For example, the objective of an experiment might be to identify factors that could be used to improve yield by 5%, or identify factors that have the greatest effect on the variation of the parts. The objective should help identify the response variables and the appropriate factors to be studied. Variables that are found by the experiment to be significant are probably important variables. These can be used to improve the processes, reduce costs, and explain effects that were previously unexplainable. On the other hand, variables that are not statistically significant may be important. The amount of data required by the DOE process establish certain variables as significant but require more data to conclude that they are insignificant.

22.25 SELECTING THE VARIABLE LEVELS

Once the factors for the experiment are selected, are there any tips for selecting the factor levels for the experiment? In step 5 (in Section 22.24), the range of factor levels selected for the

Selecting the experimental design

Current Knowledge	Type of Experiments
Little knowledge	Fractional factorials (screening studies)
Some knowledge	Factorial studies (new levels, new factors)
Much knowledge	Confirmatory full factorial studies

Figure 22.30   Simple guideline showing level of knowledge and selection of a design of experiments (DOE).

experiment will ultimately determine our confidence in the result of the experiment. For a quantitative variable with specifications, the high setting for the factor can be determined by adding approximately 20% of the difference between the specification limits to the upper specification limit. For the low level, the 20% value is subtracted from the lower specification limit. If we have qualitative variables, with many possible levels, we should select levels using extreme values whenever possible. In most cases, the factor levels must differ enough to increase our chances of observing experimental results. "A key consideration in selecting factor levels is the safe operation of the experiment across the experimental conditions," replies the master Black Belt.

What about factors that are held constant throughout the experiment and nuisance factors?

Where possible, constant factors should be kept at their nominal settings and tightly controlled throughout the experiment. Nuisance or noise factors should be held constant throughout the experiment. Finally, where possible, if you can define levels for a nuisance factor, then it may be possible to assess its impact by using blocking.

22.26   SELECTING THE EXPERIMENTAL DESIGN

Once all factors and their associated levels are established, what is the next step in planning an experiment? Well...! We now proceed to step 6, by selecting an experimental design that will meet the stated objective of your study. The initial phases of DOEs will usually be screening studies to determine the key factors affecting a process. These fractional factorial designs examine five or more variables. As we gain knowledge, experiments can be repeated using new levels for existing factors and adding new factors. If we wish to further study the relationships between the factors, two to four factors can be studied in a full factorial design. As experience is gained, confirmatory experiments are usually run to increase our confidence in the results from the previous experiments.

See Figure 22.30.

22.27   REPLICATION

At this point, we know that replication refers to repeating aspects of an experiment, but what benefit does it give us? Replication of a well-designed experiment is the primary means for analyzing the stability of factor effects and increase our confidence in the results of the experiment.

Replication measures experimental variation caused by nuisance variables. It minimizes the impact of noise by ensuring that nuisance variables are averaged out. The amount of replication in an experiment increases the resources needed for the study. As a guideline, the initial experiment should consume less than 25% of the resources allocated to the total DOE study. However, most studies will have budgetary constraints. Remember that the most important consideration in conducting a DOE is not the number of replications, but the breadth of conditions under which the comparisons are made. Replication over a small range of factor settings will not provide confidence in the experimental results. As a *rule of thumb,* two to five replications of each experiment should be done.

22.28   ANALYZING THE DATA (ANOVA)

What steps should be followed when analyzing experimental data?   Analyzing the data is step 8 of conducting a good experiment. The first task is to construct an ANOVA table for the full model in MINITAB. We then store the "residuals" and "fits" using the function `Stat > ANOVA > Residual Plots` and ensure that the residuals model satisfies the normality assumptions. After creating the ANOVA table and checking the residuals, we verify that results of the ANOVA analysis have practical significance. It is possible for an effect to be statically significant, but have little value in improving the process. From analyzing "normal probability plot for effects" charts, we can determine what practical significance each main effect, and interaction effect, has on the process.

Are there any more steps in analyzing the data?   After producing the main-effects and interaction plots in MINITAB using the function Stat > ANOVA > Main Effect, we select `Stat > ANOVA > interaction Plot` to graphically interpret the two-way interactions.

22.29   RECOMMENDATIONS

Once data have been analyzed, and we have determined the practical significance of each factor and interaction, and the combination of factors, it is good practice to replicate the experiment to ensure that the results can be reproduced. When we have confirmed our findings, we make our recommendations to senior management.

What steps should be followed in drawing our conclusions?   Verify that you have considered all the observed data. Base initial conclusions on the experimental evidence. Explain the analysis in both graphical and numerical terms. Explain conclusions in practical terms. It is important to summarize the results of the study, implications of the results, and, if required, future studies.

Your report should clearly describe the experiment work, give the background of the study, and present a clear description of the problem and the business meaning of the results. Always supply sufficient information to permit readers of the report to verify results and draw their own conclusions.

22.30   ACHIEVING THE OBJECTIVE

What to do if the process has not been optimized by our experiment?   It is seldom feasible to optimize a process on the basis of only one experimental sequence. Remember that it was recommended that 25% of the budget and resources be allocated to the first experiment. Most experiments lead to follow-on studies.

When we have confirmed that the process has been improved and operating at the optimum level, what comes next? When this stage has been reached in the experimental process, we are ready to implement step 10. We can improve and optimize the process, but if we do not institutionalize the recommended changes, it may, with time, revert back to a less than perfect process. This is the main reason for using control techniques, such as SPC, to ensure that critical to process parameters (CTPs) continue to meet objectives.

22.31   CHAPTER SUMMARY

1. To design an experiment, we must first clearly understand the process and determine the features that we need to investigate. When the experiment is completed and additional understanding of important issues has been gained, additional experiments may be required. We repeat this cycle until the desired results are achieved.

2. Each DOE has two elements, *experimental design* and *analytical approach.* The two elements must be well thought through to ensure that we collect appropriate data that can be meaningfully analyzed.

3. Compared to the traditional one-factor-at-a-time approach to experimentation, a DOE can provide a wealth of knowledge on multiple factors and their interactions in a very cost-effective manner.

4. The most effective approach to design an experiment is to use Yates standard order. This notation $(-1, +1)$ represents factor levels. The *Contrast* concept is used to determine whether a design is balanced and orthogonal.

5. The ANOVA table is initially used to assess statistical significance and determine practical significance of factor effects and interaction effects on the output.

6. Pie charts and Pareto diagrams are used to display the relative importance of several factors. We often plot the main effects on a graph to visualize the difference in the response output between the low and high settings of each important factor. Graphs are used to visualize interaction effects between factors and to find the optimal settings for both factors.

7. Even though the main effect of a factor is not statistically significant, it may affect the output through an interaction with another factor.

8. We should consider both the main-effects and interaction plots when specifying optimal settings.

9. Two-factor interactions are common, but three-factor and higher-factor interactions are rarely seen in industry.

10. It is common practice to do a fractional factorial experiment when five or more factors must be evaluated.

11. In general, factor levels should be set wide, but they should not extend beyond the operational range of the factors.

12. The general steps for conducting a statistical analysis are ANOVA (with residual analysis), practical significance (sum of squares, pie chart, etc.), and main effect plots (with interactions).

13. A successful experiment does not occur by chance; it requires careful planning and execution to ensure that an experiment produces meaningful results. There is no such thing as a poor experiment, only an experiment that is poorly designed or executed.

14. Remember to review your plan with all those concerned to ensure a successful experiment.

TABLE 22.2   Design of Experiments Factors and Levels

Factor	Low Level	High Level	Response $Y =$ Call Closure
Skill level (A)	5	10	
Staff number (B)	30	50	

22.32   CHAPTER EXAMPLES

Real-Time Call Closure

- State the practical problem—first-time call resolution (call closure). During 2004, a bank call center received 2 million calls for problem resolution. The bank call center wants to close/solve most of the problems in real time and needs to know which factors must be improved to achieve this objective.
- State the experiment objective—first-time call resolution.
- Select the output response—$Y =$ number of problems solved in real time.
- Select the input factors—$A =$ staff skill level, $B =$ staff number.
- Select the input factor levels for A—low $= 5$ training days—generalist $= 30$; $B =$ high $= 10$ training days—technical level $= 50$.
- Select the experiment design and the sample size.
- Full factorial: two factors, two levels $= 2^2$.
- Simple design, two-factor interactions with no confounding.
- Choose sample size (number of runs).
- Run the experiment and collect data.
- Analyze data.
- Analyze results.
- Draw statistical and practical conclusions.
- Implement solutions.

See Tables 22.2 and 22.3.
See Figure 22.31.
To choose the experiment sample size with MINITAB, go to Stat > Power and Sample Size > Two Level Factorial Design.

8/26/2008 5:58:33 PM ————————————————— ————

Power and Sample Size

    Alpha = 0.05
    Assumed standard deviation = 1

TABLE 22.3   Two-Variable Full Factorial Design

Run	A	B	AB
1	−1	−1	+1
2	+1	−1	−1
3	−1	+1	−1
4	+1	+1	+1

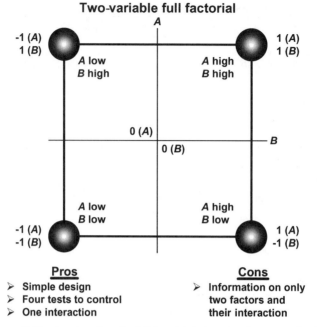

Two-variable full factorial
A
-1 (A)
1 (B)
A low
B high
A high
B high
1 (A)
1 (B)
0 (A)
0 (B)
B
A low
B low
A high
B low
-1 (A)
-1 (B)
1 (A)
-1 (B)
Pros
➢ Simple design
➢ Four tests to control
➢ One interaction
Cons
➢ Information on only two factors and their interaction

Figure 22.31 An other view of a full factorial design—two factors, two levels.

Generating a Full Factorial Model in MINITAB

Stat > DOE > Factorial > Create Factorial Design
(see Figure 22.32).

Figure 22.32 Components

Blocks: none

Center Points	Reps	Total Runs	Power	Effect
0	2	8	0.9	3.10834

Full Factorial Design

Factors:	2	Base Design:	2, 4
Runs:	8	Replicates:	2
Blocks:	1	Center pts (total):	0

All terms are free from aliasing.

- The first column factor is alternating as 5, 5, 10, 5, 5, 10, 10.
- The second column factor is alternating as: 50, 50, 30, 30, 50, 30, 30, 50.
- This is called the *Yates standard order* developed for ease of computation.
- The *base design* is a full factorial design with two factors and eight runs because there are two replicates of each run.

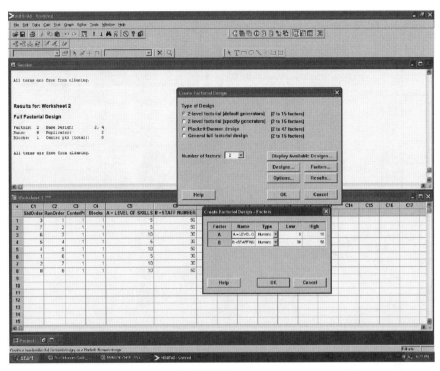

Figure 22.32   Screen capture showing MINITAB Create Factorial Design command.

- *Block Generators*—because the design is replicated, the first replicate of each treatment will be placed in one block.
- *Aliasing*—because this design is a full factorial, all the terms are free from aliasing.

Executing the Experiment Strategy

- Select Stat > DOE. Display Design.
- Select Standard Order for Design.
- Select uncoded units.
- Click OK.

MINITAB stores the information necessary to set up experimental runs.

Analyzing Factorial Designs—Initial Analysis, Model, and Effects Plots

- Select Stat > DOE > Factorial > Analyze Factorial Design.
- In response, enter Response.
- Click Terms.
- Complete dialog box.
- Click OK.
- Click Graphs.
- In effect plots: select Normal, select Pareto.
- In residual plots: select Regular.
- In residual plots: select 4 in one.
- Click OK.

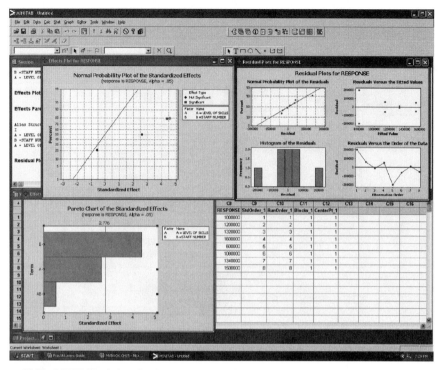

Figure 22.33   MINITAB window showing normal probability plot, residual plots for response, and Pareto chart for standardized effects.

(*Note*: To select appropriate graphs, use normal and Pareto charts to determine which terms are important in the model and should be used in subsequent analyses. Remember that a term is important when the measurements differ at the low and high settings.) See Figure 23.33.

Additional Diagrams and Related Parameters

Pareto diagram—if the bars on the Pareto chart extend past the red dashed line, this indicates a significant effect. In this example, the effect extends past the line, indicating that the number of staff has a significant effect on the residue remaining after first time call resolution.

Staff number—the second term involving skill level is not significant. In the effects plot the default α level is 0.05. In a 2^k design, α is the risk of being wrong if a term is significant. Because many designs are screening designs used to identify factors that contribute to the variability in the response, MINITAB sets the α level (risk of being wrong) higher so that "borderline" terms will be retained and considered for further analysis.

Normal probability plot—much of the information in the normal probability plot is found on the Pareto chart; the standardized effects on the normal probability plot can be positive or negative. An effect is calculated by subtracting the average response at the low (-1) level from the average response at the high ($+1$) level. MINITAB labels significant terms:

- The *staff number* has a significant positive effect on the response. When the number of staff increases, the call center closes more problems during the first call.

- In the normal probability plot it is important to verify that the residuals are normally distributed. In our example, the straight line alignment indicates that the residuals are likely to be normally distributed.

Histogram of the residuals—the histogram graph reveals that the residuals are randomly distributed around the fitted values.

Residuals versus the order of the data—the dots are randomly distributed and the overall pattern is random; thus we conclude that the residuals are randomly distributed relative to experiment run order.

—— **8/26/2008 6:24:00 PM** ——————————

Welcome to MINITAB, press F1 for help.

Full Factorial Design

Factors:	2	Base Design:	2, 4
Runs:	8	Replicates:	2
Blocks:	1	Center pts (total):	0

All terms are free from aliasing.

Results for: Worksheet 1

Factorial Fit: RESPONSE versus A = LEVEL OF SKILLS, B =STAFF NUMBER

Estimated Effects and Coefficients for RESPONSE (coded units)

Term	Effect	Coef	SE Coef	T	P
Constant		1205000	53735	22.42	0.000
A = LEVEL OF SKILLS	280000	140000	53735	2.61	0.060
B = STAFF NUMBER	470000	235000	53735	4.37	0.012
A = LEVEL OF SKILLS*B =STAFF NUMBER	−60000	−30000	53735	−0.56	0.606

S = 151987   R−Sq = 86.77%   R−Sq(adj) = 76.84%

Analysis of Variance for RESPONSE (coded units)

Source	DF	Seq SS	Adj SS	Adj MS	F	P
Main Effects	2	5.98600E+11	5.98600E+11	2.99300E+11	12.96	0.018
2-Way Interactions	1	7200000000	7200000000	7200000000	0.31	0.606
Residual Error	4	92400000000	92400000000	23100000000		
Pure Error	4	92400000000	92400000000	23100000000		
Total	7	6.98200E+11				

Estimated Coefficients for RESPONSE using data in uncoded units

Term	Coef
Constant	−515000
A = LEVEL OF SKILLS	104000
B = STAFF NUMBER	32500.0
A = LEVEL OF SKILLS*B =STAFF NUMBER	−1200.00

Alias Structure

I

A = LEVEL OF SKILLS

B = STAFF NUMBER

A = LEVEL OF SKILLS*B =STAFF NUMBER

Factorial Fit: RESPONSE versus A = LEVEL OF SKILLS, B =STAFF NUMBER

Estimated Effects and Coefficients for RESPONSE (Coded Units)

Term	Effect	Coef	SE Coef	T	P
Constant		1205000	53735	22.42	0.000
A = LEVEL OF SKILLS	280000	140000	53735	2.61	0.060
B = STAFF NUMBER	470000	235000	53735	4.37	0.012
A = LEVEL OF SKILLS*B =STAFF NUMBER	-60000	-30000	53735	-0.56	0.606

S = 151987   R-Sq = 86.77%   R-Sq(adj) = 76.84%

Analysis of Variance for RESPONSE (Coded Units)

Source	DF	Seq SS	Adj SS	Adj MS	F	P
Main Effects	2	5.98600E+11	5.98600E+11	2.99300E+11	12.96	0.018
2-Way						
Interactions	1	7200000000	7200000000	7200000000	0.31	0.606
Residual Error	4	92400000000	92400000000	23100000000		
Pure Error	4	92400000000	92400000000	23100000000		
Total	7	6.98200E+11				

Estimated Coefficients for RESPONSE using data in uncoded units

Term	Coef
Constant	-515000
A = LEVEL OF SKILLS	104000
B = STAFF NUMBER	32500.0
A = LEVEL OF SKILLS*B =STAFF NUMBER	-1200.00

Effects Plot for RESPONSE

Alias Structure

I

A = LEVEL OF SKILLS

B = STAFF NUMBER

A = LEVEL OF SKILLS*B =STAFF NUMBER

Statistically Significant Factor $B =$ (staff number) is the factor that shows statistical significance. Observe that the p value of factor B is less than .05. The p value of factor B is .012. For factor A, level of skills, is not statistically significant, for a risk of 5%. A p value equal to .60 means that there is a 60% probability that the change in the output due to factor A is due to chance.

An interaction plot is given in Figure 22.34.

A main-effects plot for response is presented in Figure 22.35.

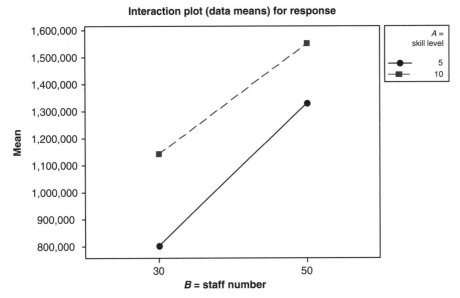

Figure 22.34 Note that in this MINITAB generated integration plot the relationship between factors B and Y changes as the level of A changes. When A is at its high $(+)$ level, variable B has a strong effect on $Y =$ mean. The distinguishing feature of interactions is the degree of nonparallelism between the two lines.

MINITAB Cube Plot The cube plot shows the mean value for each combination of factors. This graph is seldom a useful presentation plot; however, it is used by the experimenter to visualize distribution of the response data across the experiment space. From Figure 22.36 we conclude that to maximize first-time call resolution, it is best to have a highly skilled call center combined with a good level of staffing.

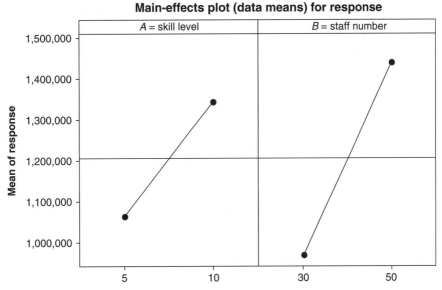

Figure 22.35 The main-effects plot generated by MINITAB is the primary graphical output of a factorial experiment. As MINITAB displays them, all of the variables are plotted side-by-side with a common Y axis. This graph shows that factor B is the major contributor to the experiment response.

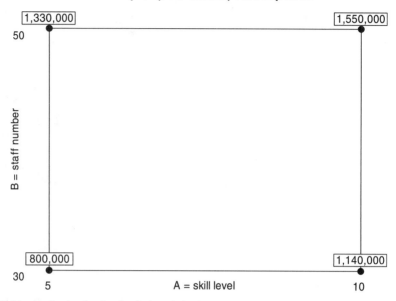

Figure 22.36 A cube plot showing that the best choice is to run the call center with associates with high level of skills and high staff number. When the staff number is high (+) and this staff is highly skilled (+), a large number of problems are solved during the call is (Y = real-time call resolution).

GLOSSARY

abscissa   Horizontal axis of a graph.

acceptance region alpha risk   Region of values for which the null hypothesis is accepted.

accuracy   Deviation of the measured or observed value from the true value.

alias   When the estimate of an effect includes the influence of one or more other effects—usually high-order interactions—the effects are said to be *aliased*.

alternate hypothesis   Probability of accepting the alternate hypothesis when, the null hypothesis is true.

analysis of variance (ANOVA)   Mathematical process for separating the variability of a group of observations into assignable causes and setting up various significance tests.

assignable cause   Source of variation that is nonrandom.

assignable variations   Variations in a dataset that can be attributed to specific causes.

attribute   Characteristic that may take on only one value, for example, either 1 or 0.

attribute data   Numerical information at the nominal level. Data that represent the frequency of occurrence within some discrete category, for example, 42 solder defects in the X box.

average   Sum of values divided by the number of values.

awareness   Personal understanding of the interrelationship between quality and productivity, directing attention to the requirement for management commitment and statistical thinking to achieve *never-ending improvement*.

background variables   Variables that are not of experimental interest and are not held constant. Their effects are frequently assumed negligible or are randomized to ensure that contamination of the primary response does not occur.

balanced design   Experimental design where all cells (treatment combinations) have the same number of observations.

beta risk   Probability of accepting the null hypothesis when the alternate hypothesis is true.

blocking   Schedule for conducting treatment combinations in an experimental study such that any effects on the experimental results due to a known change in raw materials, operators, machines, or other elements become concentrated in the levels of the blocking variable. The reason for blocking is to isolate a systematic effect and prevent it from obscuring the main effects. Blocking is achieved by restricting randomization.

blocking variables   Relatively homogeneous set of conditions within which different conditions of the primary variable are compared.

capability   Can be determined only after the process is in statistical control. When the process average $\pm 3\sigma$ spread of the distribution of individuals ($\bar{\bar{x}} + 3$) is contained within the specification tolerance (variables data), or when at least 99.73% of individuals are within specification (attributes data), a process is said to be capable.

causality   The principle that every change implies the operation of a cause.

causative   Effective as a cause.

cause   That which produces an effect or brings about a change

cause–effect diagram   Graphical representation/diagram for individual or group problem solving that uses visual description of the various process elements to analyze potential sources of the variations of the process. Also called *fishbone diagram* or *Ishikawa diagram.*

c charts   Charts that display the number of defects per sample.\

centreline   Line on a statistical process control chart that represents the characteristic's central tendency.

centerpoints   Points at the center value of all factor ranges.

central tendency   Numerical average (mean, median, and mode). Centerline on a statistical process control chart.

champion   In Six Sigma, a member of senior management who is responsible for the logistics and business aspects of the organization program.\

characteristic   Definable or measurable feature of a process, product, service, transaction, or variable.

classification   Differentiation of variables.

coding factor levels   Transforming the scale of measurement for a factor so that the high value becomes $+1$ and the low value becomes -1. After coding all factors in a two-level factorial experiment, the design matrix has all *orthogonal* columns.

coefficient of determination r^2 represents the adequacy of the regression model or the amount of variation explained by the regression equation.

common cause   Source of variation that affects all the individual values of the process output being studied in a control chart analysis. It appears as part of the random process variation.

common cause variation   Source of variation that is random. Inherent natural source of variation.

comparative experiment   An experiment whose objective is to compare the treatments rather than to determine absolute values.

confidence interval   Range in which a parameter is expected to perform in terms of the amount of data collected and the desired confidence level.

confidence level   Probability that a random variable X is within a defined interval.

confidence limits   The two values that define the confidence interval.

consecutive   Describing units of output produced in succession. A basis for selecting subgroup samples.

continuous data   Numerical information at the interval ratio level; subdivision is conceptually meaningful.

continuous random variable   Random variable that can assume any value in some specific interval.

control chart   Graphical representation of characteristic over time in relation to its natural limits and central tendency.

control limit   Line(s) on a control chart used to determine the significance of variation between subgroups.

control specifications   Specifications for the product being manufactured or service being delivered.

cost of poor quality (COPQ)   Resources consumed by non-value-added activities while producing a good or service.

critical Y   Effect that a series of subprocesses might have on a CTQ characteristic.

CTQ   Critical-to-quality characteristic used to describe an element of a design, a characteristic of a part, a transaction, a product, or attribute of a service that is critical in the eyes of the customer.

cube plot   Graph used by the process design engineer to visualize how response data are distributed across the experiment space.

customer   Person or organization that will receive or use the product or service. Customers may be internal or external.

data   Information used for reasoning or calculations.

decision variables   Defined by upper and lower range bounds and value type (discrete or continuous) and increment.

defect   Failure to meet a critical-to-quality characteristic.

defective   Unit with one or more defects.

defects per million opportunities (DPMO)   (Number of defects)/(number of units) $\times$ (number of opportunities) $\times$ 1 million.

defect of experiment (DOE or DoE)   Technique that changes experimental factors simultaneously while retaining the ability to separate the effect from each factor

defects per million opportunities (DPMO) $D/O \times 1,000,000$.

defects per unit (DPU)   (Number of defects)/(number of units).

dependent variable   Response variable.

design   Set of experimental runs that allows fit to a particular model and estimation of desired effects.

design for six sigma (DFSS)   Methodology for developing new or redesigned products and processes that deliver Six Sigma performance.

design matrix   Matrix description of an experiment that is useful for constructing and analyzing experiments.

detection   Past-oriented strategy that attempts to identify unacceptable output after it has been produced and then separates it from the good output.

discrete variable   Random variable that can have only two possible values.

distribution   Tendency of a large number of observations to group around some central value with a certain amount of variation—scatter—above and below the central value.

DMAVD   An alternative/substitute abbreviation to DFSS, and like DFSS, DMADV is central to Six Sigma initiatives. DMADV more specifically describes a method comprising linked steps—*Define, measure, analyze, design, verify*—for ensuring that products and processes are designed at the outset to meet Six Sigma requirements.

effect   That which was produced by a cause.

effect of a factor   Change in response produced by a change in the level of the factor.

effectiveness   Focuses on results, such as turning the right crank versus turning the crank right.

error   Unexplained variation in a collection of observations. (*Note*: DOEs typically require understanding of both random error and lack of fit error.)

experiment   Planned set of operations leading to a corresponding set of observations.

experimental unit   One item to which a single treatment is applied in one replication of the basic experiment.

factor   Independent variable that may be varied from one observation to another. Variable may be qualitative, quantitative, fixed, or random.

factor-independent variable   Feature of the experimental conditions that may be varied from one observation to another. May be quantitative, qualitative, fixed, or random.

failure mode–effects analysis (FMEA)   Process in which each potential failure mode in every subitem of an item is analyzed to determine its effect and impact on customer requirements or other aspects of the process.

fractional factorial DOE   A design of experiment where a limited number of factor combinations are tested.

frequency distribution   Shape formed by the group of measurements.

hidden factory   Those costs that cannot be broken down and traced to specific operations or are not made known to management.

Histogram   Graphical representation of the sample frequency distribution that describes the occurrence of grouped items.

individual   Single unit, or single measurement of a characteristic.

interaction   If the effect of one factor differs from effects at different levels of another factor, the two factors are said to *interact* or to have *interaction*.

levels of a factor   Various values of a factor considered in the experiment are called *levels*.

line charts   Charts used to track performance over time without relationship to process capability or control limits.

local controls   *Local controls* means experimental planning considering the amount of blocking or balancing of grouped experimental runs.

location   General concept for the typical values or central tendency of a distribution.

main effect   Average effect of a factor is called the main effect of the factor.

master Black Belt   Leader of a Six Sigma team responsible, among other things, for applying the Six Sigma methodology.

mean   Average of values in a group of measurements.

median   Middle value in a group of measurements when arranged from lowest value to highest value. By convention, when the dataset is even, the median is the average of the two values in the center. When the dataset is odd, the median is the value in the center.

never-ending improvement in quality and productivity   Operational philosophy that makes the best use of talents and resources within an organization to produce products or services of increasing quality for customers in an increasingly efficient manner to ensure return on investment by stockholders.

noncomparative experiments   An experiment whose objective is the determination of the properties or characteristics of a population.

nonconforming units   Units that do not conform to a specification or other inspection standard.

nonconformities   Specific occurrences of a condition that does not conform to specifications or other inspection standards. Sometimes called *discrepancies* of *defects*.

normal distribution   Continuous, symmetric, bell-shaped frequency distribution for variables data that underlies the control charts for variables. When measurements have a normal distribution, about 68.26% of all individuals lie within $\pm 1\sigma$ unit of the mean. About 95.44% lie within $\pm 2\sigma$ of the mean and about 99.73% lie within $\pm 3\sigma$ units of the mean. These percentages are the basis for control limits and control chart analysis.

operational definition   Means of clearly communicating quality expectations and performance.

opportunity   Potential defect or error that could result in failure to meet customer requirements.

outcome   Response–dependent variable. The result of a trial with a given treatment is called response (Y).

Outcome–Response-dependent variable   Result of a trial with a given treatment

optimization   Optimization examples include production scheduling, project/strategy selection and prioritization, workforce queuing, and inventory optimization.

pareto diagram   Simple graphical representation (tool) for problem solving that involves ranking of all potential problem areas or sources of variation according to their contribution to cost or to total variation.

pareto priority index (PPI)   There is an alternative approach for feasibility evaluation based on a known procedure PPI calculation procedure. The feasibility estimate can be calculated as the geometrical mean of values of all feasibility- related dimensions (cost, time consumption, chance of success, etc.)

p-**chart**   Chart that plots percent defectives in a sample.

poisson distribution   Discrete probability distribution for attributes data that applies to nonconformities and underlies the c and U control charts.

precision   Measurement precision is related to its repeatability in terms of the deviation of a group of observations from a mean value.

prevention   Future-oriented strategy that improves quality and productivity by directing analysis and action toward correcting the *process* itself.

problem solving   Process of moving from symptoms to causes (special or/and common) to actions that improve performance.

process   Series of activities to accomplish a task. Combination of people, equipment, materials, methods, and environment (mother nature) that produce an output—a given service, a product, a transaction, etc. A process can involve any aspect of a business. A key tool/method for managing processes is *statistical process control* (SPC).

process average   Central tendency of a given process characteristic. Location of the distribution of measured values of a particular process characteristic, usually designated as overall average $\bar{X}$.

process control   Application of statistical methods and procedures to a process. See **statistical process control**.

process spread   Extent to which the distribution of individual values of the process characteristic vary. Often shown as the process average plus or minus some number of the standard deviation (e. g., in MINITAB and JMP found as a default as $\bar{\bar{X}} + 3\sigma$).

project   A projects is the work unit of the Six Sigma methodology.

project management   Application of knowledge, skills, tools, methods, and techniques to a range of activities to meet objectives of a project.

project management plan (output/input)   Formal approved document that defines how a project is to be executed, monitored, and controlled.

project management professional   Person certified as a PMP by the Project Management Institute.

*p*value   The probability of making an alpha-type error.

quality functional deployment (QFD)   Structured method in which customer requirements are translated into the appropriate technical requirements for each stage of product development and production.

randomness   Condition in which individual values are not predictable, although they may come from a definable *distribution*.

range   Difference between the highest and the lowest values in a subgroup or sample.

replication   Performing the same treatment combination more than once.

response   Numerical result of a trial based on a given treatment combination.

response surface design   DOE that fully explores the process window and models the responses.

robust design   Term generically used to describe the ability of a product or process design to be tolerant of sources of variation

rotatability   Characteristic of design that is rotatable if the variance of the predicted response at any point x depends only on the distance of x from the design centerpoint.

run   Consecutive number of points consistently increasing or decreasing, or above or below the *central line*.

run chart   Simple graphical representation of a characteristic of a process showing plotted values of some *statistic* gathered from the process.

sample   A synonym for *subgroup*.

scaling factor levels   Transforming factor levels so that the high value becomes $+1$ and low value becomes -1.

scatterplot   Two-dimensional plot to assess the relationship between two variables.

screening designs   DOE that identifies which of many factors have a significant effect on the response.

shape   General concept for the overall pattern formed by a distribution of values.

sigma (σ)   Greek letter used to designate a *standard deviation*.

simulation   Technique effectively applied to solve a wide variety of problems, including strategic analysis, value stream analysis, tolerance design, cost estimation, financial analysis, market forecasting, and resource allocation.

specification   Engineering requirement for "judging"/calculating acceptability of a particular product characteristic. A *specification limit* should *Never* be *confused* with a *control limit*.

spread   General concept for the extent by which values in a distribution differ from one another—*dispersion*.

stability for control charts   Absence of special causes of variation.

stable process   Process that is in statistical control.

standard deviation   Measure of the spread of the process output or the spread of a sampling statistic from the process of subgroup averages)—denoted by the Greek letter sigma (σ)

statistic   Value calculated from or based on sample data (subgroup average or range) used to make inferences about the process that produced the output from which the sample came.

statistical process control (SPC)   Use of statistical techniques to analyze a process to ensure that the process remains under control.

stochastic optimization   Combination of simulation and optimization.

strategic planning   Process used by organizations to envision its future and develop appropriate goals, objectives, and therefore action plans.

subgroup   Logical grouping of objects or events that displays only random event-to-event variations; one or more events of measurements used to analyze the performance of a process.

system   Something that is connected according to scheme.

treatment   Treatment is a specific combination of factor levels whose effect is to be compared with other treatments.

treatment combination   Combination of the settings of several factors in a given experimental trial.

upper control limit (UCL)   Horizontal line on a control chart that represents the upper limits of process capability.

variable   Characteristic that may take on different values.

variation   Quantifiable difference between individual measurements that can be classified as being due to common causes or special causes.

X **and** R **charts**   Control charts that graphically present process capability over time.

x's values   Critical elements (factors, variables, inputs) that determine an output Y.

y's values   Main responses being measured in a project (lowercase y).

REFERENCES

1. G. E. P. Box, W. G. Hunter, and S. Hunter, *Statistics for Experimenters*, Wiley, New York, 1978.
2. R. D. Moen, T. W. Nolan, and L. P. Provost, *Quality Improvement through Planned Experimentation*, 2nd ed., McGraw-Hill, New York, 1999.

23 Design for Six Sigma (DFSS), Simulation, and Optimization

23.1 OVERVIEW

After reading this chapter, the practitioner will be familiar with a proactive, systematic method for designing new products, services, or/and processes. This methodology is a disciplined approach/strategy to discover problems during product design and to avoid the high costs related to making changes after product launch. Design for experiments (DFSS) incorporates important customer requirements during the design into and the related aspects of the product or service development process that can be measured, verified, and therefore optimized. Thus, DFSS focuses on planned and measurable quality.

23.2 INTRODUCTION

What do they mean when organizations say "we are using Six Sigma." Most are using the DMAIC methodology, because resources are being wasted with their existing processes (this approach led to the large savings claimed by General Electric, Allied Signal, Sony Electronics, and others since 1998). Some Six Sigma practitioners may be using design for Six Sigma (DFSS) to design new Six Sigma quality products. Since its inception in 1984, Six Sigma theory has evolved amid continued debate. Central to this debate is the idea of process capability, where operational bandwidth is compared with theoretical or expected performance bandwidth. Quality professionals have long recognized that existing metrics for tracking process capability can be analyzed using well-established statistical and application practices. To better understand the practical application of Six Sigma, we will explore its key ideas and statistical concepts. This chapter focuses on the theoretical underpinnings, analytical rationale and supporting simulation practices that offer technically sound assessments of design qualification. Doing so reveals that a valid assessment of process capability is highly dependent on items such as (1) the inherent nature of variation, (2) rational sampling, (3) analysis of variance, and (4) control chart theory. These concepts underlie the quality science field and should be familiar to most Six Sigma practitioners. However, embedded within the discussion are unique twists and turns that depart from conventional usage. In other words, we will lay a conventional foundation and then construct the unconventional arguments that differentiate Six Sigma practice from other well-known but less effective initiatives. While the DMAIC methodology is based on the define, measure, analyze, improve and control phases, almost all companies apply define for Six Sigma (DFSS) differently. Often a company modifies DFSS to suit its business needs and culture; at other times it will implement the version of DFSS favored by the organization assisting them with the deployment. Thus, DFSS is more of an approach than a defined methodology.

The principal DFSS goal is to design or redesign a product or service from the ground up. The expected sigma level for a DFSS product or service is at least 4.5σ—no more than ≈ 1 defect per 1000 opportunities. The Six Sigma practitioner should determine whether the organization can attain a 6σ level for its product or service. To achieve such a low defect level that meets customer expectations

Practitioner's Guide for Statistics and Lean Six Sigma for Process Improvements. By Mikel J. Harry, Prem S. Mann, Ofelia C. de Hodgins, Christopher J. Lacke, and Richard Hulbert
Copyright © 2010 John Wiley & Sons, Inc.

and needs, all critical-to-quality (CTQ) factors must be completely understood before a design is implemented. The purpose of this chapter is to provide the readers with a conceptual understanding of producibility assessment in the context of Motorola's Six Sigma initiative. Although our focus is on product design, we shall also show its relevance to process design and manufacturing control. More specifically, this chapter will develop a fundamental understanding of why producibility must be studied prior to the release of a design for production and the role that process, material, and component variation play in producibility. Because of its broad scope, emphasis is placed on the development of a conceptual understanding of design for Six Sigma (DFSS). Design for Six Sigma differs from the DMAIC methodology in that the phases of DFSS are not universally recognized, as almost every organization defines DFSS differently. For this reason, DFSS is more of an approach than a defined methodology. DFSS tools can be used in the design of new products or services and their supporting processes to meet customer requirements utilizing fewer steps and components to minimize opportunities for defects, improve sigma levels, and ensure that the outcome is both simple and cost-effective while meeting customer needs. One of the most widely used DFSS methods is the five-phase DMADV(define, measure, analyze, design, verify).

Therefore, DFSS is used in industrial and commercial organizations to help create superior products that can be exchanged for a different form of value. To enhance the efficacy of product creation and the likelihood of realizing value entitlement (for the customer and provider), we formally or informally "design" the ways and means used to achieve these ends. For example, many organizations have used DFSS to map, document, and govern their design processes. Design processes, which have been formalized, take on many different names, such as *new-product introduction* (NPI) process, *integrated product development system* (IPDS), and *strategic design system* (SDS). However, such design processes are most often exclusively used to design products. They do not support the informal design activities, which do not directly affect the product. For example, few organizations have a formal design system for creating or improving an accounting system or a quality information system. Supporting systems, processes, and even lower-level activities are most often designed on an ad hoc basis. In most organizations, the idea of design is being employed on a daily basis. Even at the personal level, we design our schedules, work activities, as well as many other things. Perhaps the Pareto effect is alive and well with respect to the idea of design—85 % of the total design effort in an organization is unknowingly apportioned to the ad hoc creation of those things not connected to the principal product. Given this, we conclude that organizations have many "hidden design departments" in operation on any given workday. Needs are to be related to utility, availability, and worth. Is in this context that this chapter presents design for Six Sigma (DFSS).

It is clear to most that the DFSS approach has many practical applications and economic benefits. Both Six Sigma and DFSS use the same advanced design tools that include but are not limited to (1) benchmarking, (2) parameter characterization, (3) parameter optimization, (4) system design, (5) detail design, (6) design of experiments, (7) failure modes, (8) cause–effect analysis, (9) statistical optimization, (10) error proofing, (11) reliability analysis, (12) product simulation, (13) process simulation, and (14) cost-effectiveness. The ideas and methods set forth in this chapter can greatly extend the breadth and scope of such applications by providing the bedrock on which a network of unique knowledge can be built.

23.3   SIX SIGMA AS STRETCH TARGET

During the mid-1980s Motorola Inc. was the first to set an extremely aggressive quality goal—the implementation of Six Sigma. In the 1990s, other major corporations such as Allied Signal, General Electric, and Sony Electronics implemented Six Sigma and greatly benefited from its use. Today Six Sigma is no longer just another quality program to simply reduce defects. It has matured into a management system capable of achieving concurrent optimization of customer and stakeholder satisfaction. Fortunately, the fundamental concept of Six Sigma is relatively simple—it is founded on the idea that a defect opportunity can be directly linked to many of the economic fundamentals that underlie a business, such as cycle time, work in process, labor cost, and capital equipment utilization.

An ability to use this approach depends on a thorough understanding of the basic statistical principles on which Six Sigma is based. Six Sigma is the stretch target for a serious quality aim.

At its heart, Six Sigma is a world class expectation applied to a single defect opportunity (CTQ) related to any type of product or service. The short-term goal of Six Sigma thus translates into approximately two onconformities per billion opportunities (NPBO). This performance level is relative to a design or process characteristic that has been assigned a bilateral performance tolerance. In the instance of a unilateral performance specification, we would observe NPBO = 1. To determine a long-term goal, this short-term goal is increased to 3.4 nonconformities per million opportunities (NPMO) to reflect the influence of process centering error over many cycles of operation. Given the random performance variable (X) in the context of a symmetric bilateral specification (two-sided tolerance with a centered nominal specification), the short-term performance goal of 2 NPBO can be described by a $\pm 6\sigma$ distribution, assuming that $X \sim$ Normally and independently Distributed (Statistics) (NID) ($\mu\sigma$) and $\mu = T$, where T is the target value (nominal specification). To account for dynamic perturbations in process centering over many cycles of operation, the performance of X is decreased or otherwise diminished via a compensatory static offset in the distribution mean (μ) such that $\delta = 1.5\sigma$. Applying this compensatory correction results in a long-term performance expectation of $6\sigma - 1.5\sigma$ 4.5σ. These two statistical conditions are graphically displayed in Figure 23.1.

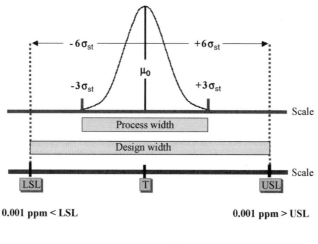

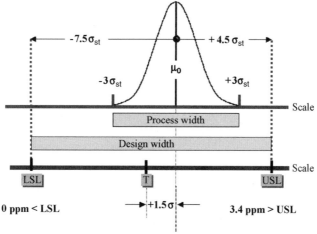

Figure 23.1 Graphs showing process in two statistical conditions and compensatory correction revealing a long-term performance expectation of $6\sigma - 1.5\sigma = 4.5\sigma$.

Converting the 4.5σcapability level into a defect rate provides the long-term expectation of 3.4 NPMO, or 3.4 defective parts per million (PPM). Thus, when we refer to a product, process, or service as being "Six Sigma," we say that the capability of any given CTQ related to that product, process, or service maintains a short-term capability of $\pm 6\sigma$ and will exhibit no more than 3.4 NPMO over the long term (i.e., after many cycles or iterations of the process).

23.4   PRODUCIBILITY

Bob Galvin[1] once told Dr. Mikel Harry that designing a phenomenally fast electronic device is one thing, but figuring out how to make a million of them is another issue. As many practitioners are well aware, the notion of *producibility* encompasses a myriad of seemingly complex topics and analytical methods. However, at the root of this complexity lie several common ideas and underlying principles. In other words, the complexity factor begins to quickly evaporate once the "vital few" principles and practices are understood. When this happens, we see the forest, not the trees. When the global art of producibility analysis and optimization is rationally segregated, it can be compartmentalized and subsequently analyzed using scientific methods. Once this occurs, the related ideas, methods, and tools can be consistently, effectively, and efficiently disseminated to a wide array of practicing designers. When knowledge transfer has occurred, meaningful application can begin. We are then free to confidently design a reliable high-speed circuit, produce a million of them in a quality manner, deliver them to the customer on time, and realize economic prosperity for all who are involved. This is best accomplished by first studying the individual compartments and then integrating their key ideas and grounding principles. In this manner, the higher-level topic can be understood, mastered, communicated, and practiced on a larger scale. Before we venture too far into this forest of technical detail, we will outline the key aspects of producibility. In this manner, we will create a conceptual topology of the forest before exploring its vital few trees.

In 1989 the United States Department of the Navy sponsored a working producibility measurement conference. Their stated purpose was to define and study producibility, and to devise ways to improve producibility. Producibility is concerned with how easily a design concept can be moved from development to production of a quality product, on time, at the lowest possible cost (better product or service, faster at a lower cost than the competition). Thus, we seek to create a product design concept that not only is viable in terms of its physical mission (*product viability*) but also can be easily replicated (*process viability*). Given a viable design concept, we then work to realize its existence in the real world by creating detail drawings, assigning tolerances, and selecting processes. Throughout the design process, we continually strive to identify and adjudicate functional performance issues related to, for example, reliability, cost, and timing.

Process viability determines producibility. Essentially, we divide this into two distinct, but interrelated elements—process economics and process capability. The first, process economics, is concerned with financial implications. This aspect of producibility analysis is not the primary focus of this document; although it will be considered and discussed from time to time. The second, process capability, is concerned with fostering or constraining the repeatability of a design. Here, we are concerned with the repeatability of a design in terms of process capability. We understand that a design must be "Six Sigma–qualified" where this dimension of producibility is concerned and that such qualification must be done prior to full-scale production. In other words, we will discuss several of the key ideas, principles, practices, methods, and tools necessary to ensure (and assure) that the performance specifications of a design are fully compatible with the process capabilities. Inherently, this optimization must be compatible with economic constraints and organizational controls. Expressed differently, a design's performance goals must be compatible with the capabilities and economics of those processes that will ultimately realize its *physical—real—existence*. Another key step for a product physical and real existence is the clear understanding of the concept of *design qualification*.

23.5 STATISTICAL TOLERANCES

In this chapter, we review the symmetric bilateral case discussed in Chapter 9 because this case is the most complex in terms of its constituents, widely recognized, and most frequently applied. Regardless of organizational complexity, a process owner is often forced (by default) to assume that T and Δ are optimal (see Fig. 23.1). On the other hand (flip side), the designer assumes that the process owner will maintain design margin integrity. Even though this chapter will not address the "ins and outs" of how to determine statistical performance standards, we will provide a cursory discussion and illustrative examples. Operationally, the main goal of any process owner (and dream of a designer) is to set $\mu = T$, and minimize the extent and pattern of error for μc, and σ. Naturally, this is done to minimize $p(d)$, where d is a nonconformance to standard (defect). Thus, the minimization of d, results in the maximization of process yield and related efficiencies. Although the aforementioned goal is seductively simple, the seasoned practitioner understands that it is anything but simple to achieve. If it were, the world would have surpassed the Six Sigma quality threshold long ago. Although standards centering, and error are interrelated, we will treat each one separately to simplify our discussion. Since any given CTQ specification must be realized by the selected process, the CTQ is inherently bilateral and variant over time; thus, the boundary specifications (i.e., USL and LSL) are somewhat artificial. Deferring the pros and cons of the arguments, we shall assume certain pervasive and simple facts. In reality, these specifications are frequently established by physical considerations and limitations. A process of some form must be utilized to replicate the CTQ of concern, all distributions are bilateral by nature, the various parameters of all distributions generally change over time, many types of distributions can be mathematically transformed to adequately emulate a normal density function, simply tightening a specification will not reduce the variations that are naturally inherent to the process, it is generally more beneficial to operate a process in the center of a specification regardless of where the target value is located within the tolerance zone, the assignment of an asymmetric target value is most often done on the basis of some physical or performance rationale, and the list of facts goes on, perhaps forever.

The full consideration of such issues can be daunting. Thus, many designers (and process owners) simply throw their hands in the air and continue to rely on crude *thumb rules*. Many do not realize that methodologies exist to handle such considerations in a very easy and convenient way in this day and age of high technology (powerful software for desk and laptops). To address the concept of statistical tolerancing, we must recognize that it may be desirable to "statistics"ize a performance specification. By doing this sort of thing, the physical constraints and requirements of the designer can be fully realized while ensuring the full *operational*ization of the design by the process owner. In an oversimplified manner, such a tolerance takes the form

$$S = T^{+z_\sigma}_{-z_\sigma}$$

where T is the nominal specification, z is some number of standard normal deviates (i.e., Z value), and σ is the specified standard deviation. The design is harmonized with the intended process to create its reality in the physical world. In some cases, σ should be bounded. Although there may be some benefit to establishing a maximum value of σ, such benefit is much less obvious for a minimum specification (but highly relevant in some instances). To cover such cases, we amend our previous example and show

$$S = T : \sigma^{+z_{max}}_{-z_{min}}$$

To illustrate the need for a lower bound on σ, we will consider a product design example. We assume that there exists a certain design problem in which the symmetric bilateral specifications associated with several independent variables must *play* together in a linear combination such that a *gap* will be intentionally formed. In other words, we define an *envelope* in which some number of *parts* must fit. To address process error, also provide an *assembly gap* between the envelope and the collective set of parts. We govern the gap *size* by establishing a symmetrical bilateral specification for the assembly space (i.e., clearance). Furthermore, we recognize that the error contribution of each variable is fully independent and additive and that T is constant and additive in all cases. To simplify discussion, we assume that $\mu = T$ in all cases.

Given this simple scenario, reproducibility errors may be accumulated or *stacked* and *paired* to the gap specification. In this manner, the designer is free to compute the probability of exceeding either gap constraint and, on the basis of such an analysis, react accordingly. Perhaps a superior design strategy would first establish a desired gap probability by specifying σ_{gap}, then reverse propagate the error by way of a rational weighting scheme. In this manner, optimal tolerances could be assigned to each subordinate variable that contribute to gap size. This, of course, assumes an a priori knowledge of the process capabilities that will produce the parts.

Next, we will reduce the total error component. In this case, μ_{gap} remains unchanged; however, σ_{gap} is reduced by deliberate improvement efforts on behalf of the process owner. Given such a reduction in σ_{gap}, the typical unit will now exhibit too much freedom relative to the gap limits and, as a consequence, fewer units will possess a near perfect fit, but fewer units would exhibit a poor fit. There would be a corresponding increase in the number of units that exhibit an average fit. This holds because the overall range for natural random mating of parts is significantly diminished; thereby, increasing the probability that the typical unit will display too much freedom in terms of "rattle space." The corrective action for this situation is to modify one or more of the nominal specifications related to the subordinate elements (i.e., the envelope and/or parts). Although this scenario is clearly related to specific engineering problems, it may be extended to a broad range of business and operations management applications. For example, in business we often study the positive or negative margin, the gap between sales price and total cost. When considering each line item that aggregates to form the total cost, we recognize the idea of cost variance, or σ. Here again, the errors described by each σ accumulate, thereby providing the opportunity for previously described statistical analysis. In operations management, we are frequently concerned with total cycle time, as this impacts delivery times, cost and profitability. Here again, there is a ceiling related to total cycle time. By the same reasoning, we can say that the cycle time associated with each process operation can be described in terms of μ and σ. Hence, the process cycle time errors can be aggregated and subsequently optimized. From the previous discussions, we can define a statistically based specification as

$$T \pm \Delta : \sigma_{\max \, \alpha}^{n}$$

where Δ is a range within which we expect to locate the true process center μ, $\sigma_{\max}$ is the maximum allowable process standard deviation for the given sampling requirement n, and α is the allowable risk. The specified range of statistical expectation related to μ, for any given value of n, is defined as

$$T - \frac{t_{\alpha/2}\hat{\sigma}}{\sqrt{n}} \leq \mu \leq T + \frac{t_{\alpha/2}\hat{\sigma}}{\sqrt{n}}$$

where $\hat{\sigma}$ is an unbiased estimator of σ, $t_{\alpha/2}$ is the t distribution value corresponding to $\alpha/2$, α is the estimated probability of type I decision error, and n is the sample size on which $\hat{\sigma}$ is estimated. The upper limit for T is simply labeled $T_{\max}$, and the lower limit is labeled $T_{\min}$. To compensate for the uncertainties associated with natural and expected process variations, we use the same α and compute the interval width related to σ using the chi-square distribution. This operation is relationally given as

$$\hat{\sigma}\sqrt{\frac{n-1}{\chi_{\alpha/2}^2}} \leq \sigma \leq \hat{\sigma}\sqrt{\frac{n-1}{\chi_{1-\alpha/2}^2}}$$

where $\chi^2_{1-\alpha/2}$ is the upper tail of the chi-square distribution corresponding to the area $1-(\alpha/2)$ For the sake of brevity, we shall refer to the minimum as $\sigma_{\min}$ and the maximum condition as $\sigma_{\max}$. After accounting for the interval width (i.e., statistical confidence interval) associated with μ and σ (for any given n and α), we may now seek to establish the "statistical specification limits." Such limits may be generally described by

$$\text{USL} = T_{\max} + \frac{3\sigma^{n/\alpha}_{\max}}{1-M}$$

$$\text{LSL} = \frac{3\sigma^{n/\alpha}_{\min}}{1-M} - T$$

where n is the specified sampling requirement, 3 is the constant used to define unity (i.e., $U = 1.0 = \pm 3\sigma$), and M is the extent of design margin considered desirable for a particular application. Theoretically, the idea of design margin exists in the range $0 < M < 1$; however, in practice, M is most always established in the lower half of this range. Conventionally speaking, we most often observe $M = 0.25$; however, in Six Sigma practice $M = 0.5$ is frequently used.

23.6   DESIGN APPLICATION

A *defect* has previously been defined as a failure to conform to requirements. These requirements are interpreted and incorporated into the specification or tolerance of the product, service, or process. Achieving these tolerances ensures that these products, services, or processes can be easily assembled, with minimum adjustments, to perform their required functions. Properly designed specifications reduce DPMOs (defects per million opportunities). Tight tolerance specifications ensure functional requirements but are not cost-effective, while loose tolerance specifications will likely affect assembly processes through frequent rework to maintain product, service, or process performance. Different methods can be used to determine tolerance specifications. One conventional method depends on the designer's experience and perception; a second follows a disciplined procedure. To improve our understanding of the DFSS qualification process, we will consider a simple application that emphasizes the characterization and optimization stages. To limit complexity, we have chosen a design consisting of a single product requirement and its related process distribution. Using the deductive design approach, we explore how to match product performance requirements to a candidate process. The reader should be able to apply these methods to more complex cases after the example has been fully presented.

For this example, we assume that an organization is currently designing a new product that is somewhat different from its previous products. Thus, it is believed that much of the existing process technology can be effectively utilized to produce the new product. Because market analysis has predicted reasonably strong demand for this product, the management of this organization believes that the new product concept will deliver the revenues necessary to support the investment required to develop and produce the new product. Consequently, production planning anticipates a moderate volume over the next several years. After several months in development, the design project proceeded through the *new-product development process* (NPDP) to the detail design stage. At this point a designer configured a particular assembly contained within the larger system. Following this, the designer declared that characteristic "4" was very critical to the overall producibility of the assembly. In this instance, the critical characteristic was defined as CTQ_4. As the designer has had no previous experience with such a CTQ, little was known about its potential producibility. Therefore, *it was determined that the CTQ_4 characteristic should be Six Sigma–qualified before releasing the assembly design to production.*

The performance standard implied by CTQ_4 is a symmetric bilateral specification where the lower limit has been declared as $\text{LSL} = 70$, the nominal requirement has been centered at $T = 100$, and the

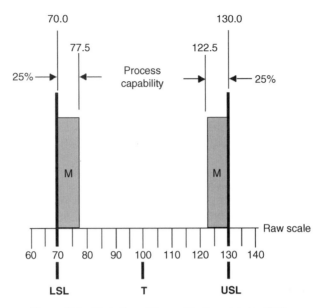

Figure 23.2 Illustration of the specification related to CTQ$_4$.

upper limit has been established at USL $= 140$. In addition, the designer used historical convention and set the design margin at $M = 0.25$, or simply 25%. Figure 23.2 presents the specification and design margins assigned to CTQ$_4$. Given these performance requirements, the designer recognized that a process must eventually be identified, selected, and subsequently employed to achieve the CTQ$_4$ targets. He believed that the process capability should extend between the limits of 122.5 and 77.5 to ultimately realize the desired reproducibility level. Given that the process could operate within this region, the designer asserted that the assigned design margin would be sufficient to contain anticipated operational uncertainties.

By convention, M is usually implied and not specified. Unless specified during requirements planning or called for in a *standard operating procedure* (SOP), the extent of design margin is generally accepted as being $M = 0.25$, and is not presented on the design documents (i.e., engineering drawings). From this perspective, we see that by letting $M = 25\%$, the designer is indirectly specifying a $\pm 4\sigma$ process capability requirement, since the inner boundary limits of M (e.g., 77.5 and 122.5) defines the outer boundary limits of unity (e.g., -3σ and $+3\sigma$). Thus the designer is allowing a 1σ margin of error on each side of the specification. Figure 23.3 is a visualization of this important concept.

The choice of a $\pm 4\sigma$ level capability is reasonable if it is consistent with the complexity (number of CTQs) of the process. However, few designers realize that the assumed limits of unity (related to such a producibility model) are predicated on infinite degrees of freedom. In other words, the $\pm 4\sigma$ process capability specified in Figure 23.3 describes a theoretical distribution consisting of an unlimited number of operational experiences. Thus, to verify reproducibility, we require an infinite number of operational experiences. Since this is not practical, the candidate process must be judiciously specified and evaluated to ensure that we can achieve the given design objectives. Naturally, this will involve some form of analysis based on highly limited degrees of freedom.

The heart of Six Sigma design qualification is selection of a process without knowing its true capability yet capable of ensuring a high level of producibility. Through the judicious application of inferential statistics, the designer can address the myriad of operational uncertainties that will confront the design over its production life. To specify a process capability for CTQ$_4$, the designer must recognize the two types of risk resulting from random error: (1) a process may be selected that would

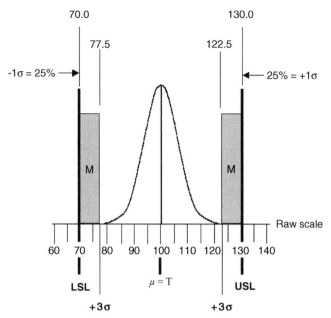

Figure 23.3   Illustration showing the implied process capability requirement related to CTQ₄.

ultimately prove insufficient in terms of capability—this is called *alpha risk*, and (2) , the designer may reject a process that would ultimately prove sufficient—this is called *beta risk*. Each of these risk types carries practical and theoretical consequences. By addressing these two forms of risk prior to full-scale production, a designer can significantly enhance the overall design producibility. Absence of such consideration often leads to inappropriate and costly changes in the design and production system. Although our discussion has been limited to the upper specification limit of this symmetric bilateral example, the reader should be able to extrapolate our discussion to include the asymmetric bilateral and the unilateral cases.

23.7   DESIGN MARGIN

In this context, a single point-in-time measurement of a given CTQ can be fully explained or otherwise accounted for by the momentary settings associated with each of the independent variables related to the process $(x_1, \ldots, x_n)$. Using determinism as a framework, we begin to form a scientific understanding of process capability. To fully comprehend the implications of an increase in design margin, it is important to consider the meaning of margin. By convention, the limits of a process are declared as plus or minus three sigma $(\pm 3\sigma)$ and the resulting range is defined as unity. However, when establishing tolerance specifications, we would extend these limits by a sufficient amount to encompass room/space for the process to wiggle. In this manner the design is configured to allow for the many types and forms of uncertainty that the design will inevitably experience. In other words, the design is made robust to Murphy's law.

Many designers know how to design but do not know how to statistically enhance the realization of that design (i.e., *producibility*). Usually, the first step establishes a design concept based on customer needs, performance requirements, and market research. Next, a design concept is agreed to and then proof of principles is accomplished—often by way of computer modeling, prototypes. After this, the system level design is established followed by detail design, at which point tolerances are assigned.

Most often, tolerances are established by historical analysis of previous designs, worst-case analysis, experience based instinct, or worse, left unassigned (default specifications). In many cases, the final design is realized and studied during a limited production run. During this process test (pilot run), errors are discovered and subsequently addressed. Finally, the design is put into full production—most often resulting in a flood of design changes. Eventually, the design (and process) reaches a state consistent with a performance level of about $3.5\sigma - 4.5\sigma$. Essentially, this classic design process boils down to putting the idea on paper and then throwing it over the wall to be realized. Given this gross but reality-based generalization, the process owner must ensure design margin integrity. When the selected process is fired up and it is discovered that the process can't meet the specifications (specs), the process owners often resort to *detect and sort* to ensure margin integrity, or in those instances of profound courage seek a modification of the design, which that often results in *political battles*, cost overruns, and missed delivery commitments. In those cases that utilize the age-old method of detect and sort as a means to ensure margin integrity, we recognize that no inspection or test is perfect—some defects will inevitably escape. In fact, there is theoretical and empirical evidence indicating that the escape rate is frequently proportional to the rate at which defects are created during processing. Consequentially, this particular strategy often results in reliability problems that the process owner (or designer) never considers and hence, the driving need behind design for Six Sigma (DFSS).

Now that we understand the idea of design margin and some of its implications, we move on to its mathematical form. In this context, we may describe the idea of design margin as

$$M = 1 - \frac{\pm\Delta}{|USL - LSL|}$$

where M is the design margin, Δ is the process bandwidth (i.e., $\pm3\sigma$), and $|USL - LSL|$ is the specification bandwidth, or some predetermined extent of guard banding. If Δ is not established on the basis of process capability, we should consider

$$M = 1 - \frac{|P - T|}{|SL - T|}$$

where P is a defined setpoint, most often symmetrically bilateral and established by pragmatic considerations. By convention, the upper USL and lower LSL specification limits would each be set at their respective four sigma points. When normalized to the process standard deviation, the design margin is given as

$$M = \frac{\pm Z_{unity}}{\pm Z_{spec}} = 1 - \frac{\pm3}{\pm4} = 1 - \frac{3}{4} = .25$$

or simply, 25%. By convention, the standard normal deviation associated with the upper specification limit is defined as

$$Z_{USL} = \frac{USL - T}{\sigma_{st}}$$

and that of the lower specification limit is characterized by

$$Z_{LSL} = \frac{T - LSL}{\sigma_{st}}$$

where σ_{st} is the short-term standard deviation of the process. Designing to a 4.0σ–4.5σ level of reproducibility has historically been considered quite normal and conventional for most applications. From a quality engineering perspective, the design margin is

$$M = 1 - \frac{1}{C_p}$$

where C_p is the short-term process capability ratio. C_p usually lies in the range of $1.33 < C_p < 1.67$ (C_p is widely discussed in Chapters 9 and 15 of this book).

23.8 DESIGN QUALIFICATION

As mentioned above, one of the most widely used DFSS approaches is DMADV, which stands for: *d*efine the project goals and customer (internal and external) requirements, *m*easure and determine needs and specifications, benchmark competitors and organization, *a*nalyze the process options to meet customer needs, detail *d*esign the process, and *v*erify the design performance and ability to meet customer needs. However, in the literature the practitioner will find other flavors of DFSS such as DCCDI, IDOV, and DMEDI:

- DCCDI is an acronym for define, customer, concept/ideas, design, and implementation; in further detail, *d*efine the project goals; *c*ustomer analysis completed; *c*oncepts developed, reviewed, and selected; *d*esign performed to meet customer and business specifications; and *i*mplementation completed to develop and commercialize products or services.
- IDOV is a design approach (Fig. 23.4) commonly used in the manufacturing world. IDOV is the acronym for identify, design, optimize, and validate: *i*dentify customer and customer specifications (CTQs); *d*esign translates the customer CTQs into functional requirements and into solution alternatives. A selection process gradually reduces or eliminates potential solutions until the "best solution" is obtained. *Optimize* uses the advanced statistical tools and modeling to

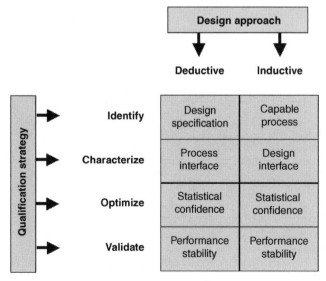

Figure 23.4   Illustration showing the IDOV design approach.

predict and optimize the design and performance. The *validate* phase ensures that the design meets customer CTQs.

- DMEDI stands for define, measure, explore, develop, and implement; it is used by Price Waterhouse Coopers.

The DFSS approach may utilize any of the many methodologies. Each of these DFSS approaches use the same advanced design tools such as quality function, deployment, cause–effect (CE) analysis, failure mode–effects analysis (FMEA), root cause analysis (RCA), benchmarking, advanced statistics, optimization, errorproofing (referred to as "mistakeproofing" in other chapters), design of experiments, and Monte Carlo simulation. Each approach differs mainly in the number of phases, their names, and of course, the acronym, but they all ensure that a thorough design evaluation procedure is followed.

At the very core of Six Sigma is the idea of design qualification. A Six Sigma–qualified design is a formalized product or service that has been fully evaluated and optimized for both instantaneous (i.e., short-term) repeatability, and that is sustainable (i.e., long-term) over time. To meet this criteria, the final product and all supporting critical levels, segments, and nodes must also be Six Sigma–qualified designs. Naturally, we use a scientific process to accomplish this aim. Most experienced practitioners of Six Sigma will recognize that the design qualification strategy is identical to the breakthrough strategy that this researcher developed many years ago. Not only does this scientifically grounded strategy solve operational problems; it also ensures that a design is properly analyzed and validated with respect to design producibility. As we continue, the reader should recognize that each phase of the breakthrough strategy represents a distinct stage of qualification and will be referred to accordingly (stages 1–4).

In terms of the deductive approach to design qualification, we first *identify* each performance characteristic or quality attribute that is believed to have a strong influence on producibility. We identify such contributory factors as critical-to-quality characteristics and label each of them as a CTQ. Thus, we may assert that $P = f(CTQ_1, \ldots, CTQ_n)$, where P represents the producibility of a design and n is the last provisional contributing design characteristic that exerts an undue influence on P. For each CTQ, we identify a functional process (real or postulated) that has the operative potential to create its respective CTQ. Note that a provisional process is frequently called a "candidate process." To the extent possible, each candidate process selected for stage 1 qualification should complement the overall economic aims of the business, realize the performance specifications, and be robust to operational constraints. The identification of a candidate process is based on operational experience and involves many qualitative judgments. As a consequence, this task should not be left to the novice or an inexperienced team.

In stage 2, we establish a theoretical design model (i.e., performance distribution). Naturally, such a model describes the expected producibility level for each CTQ. We then *characterize* or otherwise "design" a model sampling distribution to account for the potential "worst case" influence of random sampling error. To accomplish this task, the standard Six Sigma design qualification sampling criteria should be utilized to build a theoretical sampling distribution that accounts for the effect of random error in σ (and μ in some cases). The sampling model is then used to guide process selection and adoption. We recommend that the designer use $n = 30$ and $\alpha/2 = .005$ as the sampling criterion for model construction during stage 2 of the design qualification. Following the development of a theoretical sampling model for each CTQ, the resulting estimates of capability must be translated into a producibility measure (e.g., first-time yield expectation). In turn, each of the related CTQ producibility measures is hierarchically pooled and subsequently normalized to form an overall measure of producibility for each level, segment, or system of interest. If an unacceptable discrepancy exists between the producibility objective and projected value, this difference must be reconciled or otherwise eliminated during stage 3 of the design qualification.

At this point, we are ready to hierarchically *optimize* the producibility measures related to critical nodes, segments, and levels of the design. This understanding defines the overriding task in stage 3.

Given that a higher-order producibility measure (e.g., first-time yield expectation) does not conform to its idealized value, we must first determine the magnitude of departure and then reverse and propagate the positive or negative solution back down (and across) the hierarchy to each critical design–process node interface. Following this, the relative differential between the "is condition" and the "ought to be condition" must be judiciously appraised and properly adjudicated. After adjusting or improving the targeted CTQs, we begin the second producibility analysis iteration following the same approach. Such vertical (and horizontal) iterations continue until an overall solution is realized. Of course, such a solution considers not only the statistical aim of confidence maximization but also that of constraint minimization as related to certain operational considerations such as cycle time and cost. Given the myriad of possibilities that could likely exist during the course of these vertical (and horizontal) iterations, this writer is compelled to limit further discussion in the interest of space and continuity of reasoning. On the basis of the outcomes of this stage, we provisionally adopt the candidate processes deemed worthy of validation.

Last, we execute the fourth stage of design qualification. This particular stage of qualification consists of two phases. During the first phase, we *validate* that the candidate processes adhere to their respective sampling model, primarily in terms of σ. The second validation phase addresses the longitudinal aspects of process control. We must confirm that the capability of each selected process can be controlled over time in terms of μ and σ, without an appreciable loss of system level confidence (i.e., overall producibility). In this manner, we ensure that the composite design is reasonably robust to both expected and unanticipated sources of variation that might exist at the node level, within one or more segments of the design. Considering the design qualification outcomes, we validate that the design and each of its related processes is "Six Sigma–qualified."

23.9 DESIGN FOR SIX SIGMA (DFSS) PRINCIPLES

Six Sigma employs a set of tools and methodologies to design new products, services, and supporting processes that ensure a simple and cost-effective outcome while serving customer needs. Products/ services are defined at a Six Sigma level using processes that are operating at their optimum technological capability level (level of entitlement). Empowered cross-functional teams are the most effective users of these tools. These guiding principles ensures that product/service definition decisions are "data-based," judgment guided, and experience guarded, rather than the traditional mostly "judgment-based" approach. Since customer requirements are rigorously identified, prior-itized, and translated into design requirements rather than just making broad-based assumptions, proper cost and performance tradeoff decisions can be achieved. Then, specific component or process step targets can be established for the product, service, or process definition. A partial list of these primary application tools include quality function deployment (QFD), CT matrix, FMEA, design for manufacturing and assembly (DFMA), robust design, risk analysis, and key critical characteristics (key characteristics in engineering). The DFSS tools can be used to design products, services, and manufacturing or transactional processes to meet customer requirements that result in fewer steps and parts to minimize defect opportunities.

23.9.1 DFSS Leverage in Product Design

In Figure 23.5 we find that while design typically is the smallest actual product cost element, it has the largest cost influence. Therefore, it will likely be the very top item on the vital few list. It can also be deduced from Figure 23.5, that any incremental improvement in the design, through the use of DFSS methodology will have a large direct impact on actual product cost. Then, *"Why is it so important to develop a DFSS methodology?"* Carefully observe the illustration in Figure 23.5.

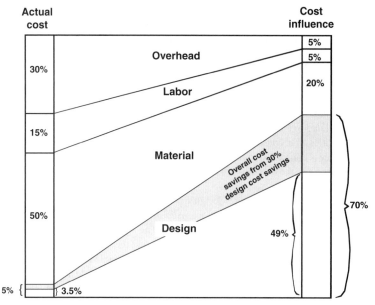

Figure 23.5   Simple graphical thought map for design for Six Sigma (DFSS) leverage in product design.

Example 23.1   Achieve a 30% savings through design simplification.

Solution   The solution can be found by using Figure 23.5. A 30% reduction in design cost translates into over 20% overall cost savings (cost influence)—a phenomenal achievement, indeed! Yet, the same 30%, applied to labor or overhead only, results in only 5% each or less of overall savings. Elemental improvement optimizes a part of the process, while DFSS may eliminate one or more parts of the process, or simplify it. This often results in a direct cost reduction in material and labor, and in an indirect reduction of overhead cost. Yet, any improvement in one of the other three elements remains within that element. For example, a more efficient way of nesting sheetmetal on a two-axis router machine will reduce the raw-material requirement. If this process change is carried out well, the process redesign will decrease machine utilization time and its related cost. The savings from this redesign carries over into the overall cost of the manufacturing cost of the product. This principle can be applied to all types of transactions and services.

23.9.2   Importance of DFSS for Product Design[2]

How does DFSS differ from the classical approach to product design, and why is it better?   There are several major differences that fall into two distinct categories:

1. *Collect and torture*—analyze relevant data to facilitate making *knowledge-based* design decisions.
2. Implement concurrent product development throughout the definition process. See Figure 23.6.

Another example is the use of *statistical tolerancing* as opposed to the current practice of worst-case *tolerancing techniques*. To apply statistical tolerancing, process capability data must be developed and available to the design team. A further example is the setting of component or system-level cost targets to ensure adequate end-product profitability. A rapid and accurate method of end-product cost estimating is required to validate each option. As better estimating methods are

Why DFSS for product design?

Concurrent design saves time

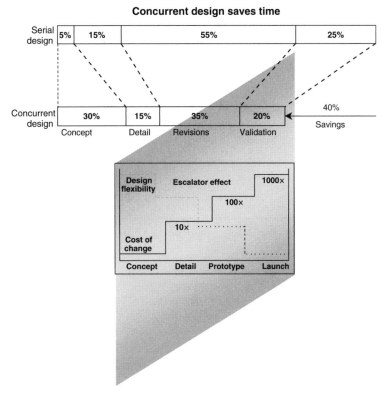

Figure 23.6   Illustration of why DFSS is used for product design.

developed and coupled with risk analysis techniques, better design decisions can be made. At this point the following question might arise.

Are the process phases in DFSS identical to the phases of the breakthrough strategy—measure, analyze, improve, and control (MAIC)? Because of the unique nature of DFSS, the appropriate phasing is IDOV. IDOV is a design process that can occur separately within each of the company's engineering system processes, (i.e., concept, preliminary, and product definition), or it can occur as an offshoot of MAIC, where technology entitlement has been achieved in the process and we seeking to further improve the process sigma. This is best illustrated in Figure 23.7, which presents a side-by-side comparison between MAIC and DFSS.

Again, the IDOV acronym is defined as follows:

Identify represents market needs, customer requirements (CTS), regulatory requirements, process optimization requirements, subsystem design requirements, etc. It can occur in a variety of situations and phases. Data should be presented in a ranking format (ordinal scale) to permit some cost and performance tradeoff decisions to occur.

Design represents either the initial concept generation or any redesign required to accommodate technology limitations.

Opotimize represents a variety of activities and utilizes the majority of the DFSS tools. Here, the design concept is repeatedly adjusted and modified to produce a simple robust design with verifiable cost and quality levels.

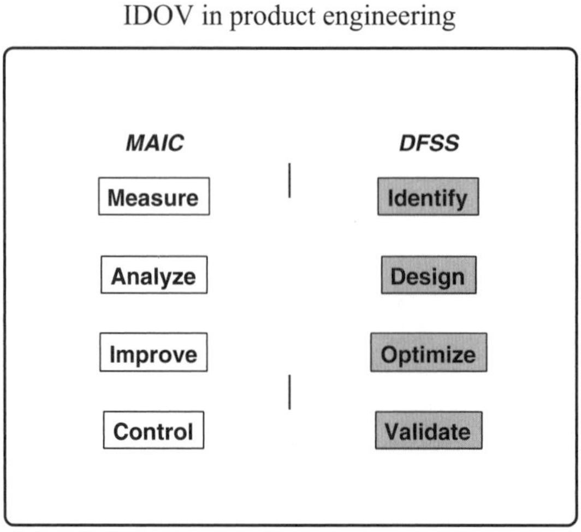

IDOV in product engineering

Figure 23.7   Comparison between IDOV and design for Six Sigma.

Validate represents all the detail, product, and prototype testing as well as verification of the product/process interaction.

23.10   DECISION POWER

A frequent and important *risk type* frequently overlooked by a designer is most often called a type II decision error. Generally speaking, type II decision errors result whenever we *falsely* assert that some state of a system is in equilibrium. From a Six Sigma design qualification perspective, the design model distribution parameters T and σ are in tune with the corresponding process parameters; that is, the *process capability is properly aligned* with the *design intent*. The statistical probability of falsely asserting that such an alignment exists when, in reality, the process and design is not a good match is also referred to as *beta risk* and summarized as β. Thus β is the statistical likelihood of accepting a premised H_0 when in reality, it is false. This difference is of practical concern if δ did, in fact, exist at the time of sampling, but was not detected or it was falsely declared to be an artifact of random sampling error. On the flip side, $1-\beta$ is called the *power of the test*. In this sense, $1-\beta$ is the statistical likelihood of being able to detect some δ of practical concern if such a magnitude of difference is really present. For example, if a given design qualification circumstance is such that $\mu-T = \delta$ is greater than 0, and that δ is of practical concern, then $1-\beta$ is the statistical probability of detecting δ with n observations (sample size). Thus $1-\beta$ is often called *test sensitivity*. $1-\beta$ is analogous to the power of a metal detector. This device is intended to signal or otherwise "ping" in the presence of a metallic object. As the power of a metal detector is increased, the device can reliably find progressively smaller pieces of metal; however, such an increase in detection power necessarily requires a larger power source. The power source of a metal detector is analogous to sample size and the detection device is much like a statistical test; (T test, ANOVA, F test, etc). When α and δ are treated as constants, β will vary inversely with n. As n is increased, β is reduced in value; hence, the overall test sensitivity $(1-\beta)$ is increased and the probability of detecting δ increases. From a different perspective, we are able to detect progressively smaller values of δ as the sample size n increases, given that α and δ are constant across the range of n being considered. Therefore, the statistical power of the hypothesis test is highly dependent on the available degrees of freedom (df), given that α and δ are established a priori. In the case of a stage 2 design qualification, this researcher standardized on $\beta = 0.10$, or 10%.

This particular value of type II risk was selected and standardized because it falls within the range of conventional practice, and it represents a rational level of risk when qualifying a design–process interface at the node level. In other words, $(1-\beta) = 0.9$ or 90% is sufficient confidence to detect most practical design differences. Practitioners with substantial design qualification experience will immediately recognize that the operational consequences of a β error are often not as profound as the consequences associated with an α error. Therefore, β is most often established at a level less than that of α. Whereas the realization of β risk has the potential to extend the designer's search for a suitable process, the realization of α risk is manifested in the form of operational inefficiency—across the entire lifespan of process operation. In short, the consequence of realizing β risk is most often a one-time thing, whereas the consequences associated with the realization of α risk are likely to be repetitive and ongoing. While this may not be true in all applications, it is certainly this researcher's experience where design qualification trials are concerned. Here again, the reader should recognize that this last statement is a generality. We recognize that the blind application of such general thumb rules carry the risk of inappropriate application, particularly by a novice practitioner. However, the advantages associated with the mass application of a robust thumb rule outweighs the negative consequence that may occasionally be experienced as a result of faulty or inappropriate application. In this investigator's humble opinion, the Six Sigma design qualification guidelines provide reasonable robustness because the guidelines promote the use of relatively high levels of confidence and power. Naturally, this assumes that the sample size is sufficient.

Since a great many process qualifications are undertaken with limited observations, we recognize the conventional sampling range $30 < n < 60$ and specify that our design qualification test should be based on $n = 30$. This particular value is frequently sited as the ideal sample size because it represents a point of diminishing return. When the general range $30 < n < 60$ is considered, it can be mathematically demonstrated that the quantity $1/\sqrt{n}$ represents a theoretical point where the statistical precision is relatively high, but n is relatively low. As should be apparent, this point represents a reasonable and rather pragmatic tradeoff in terms of theory and practice. In terms of statistical precision, the standardized qualification test of $n = 30$ random samples provides a relatively precise estimate of capability. Justification for this choice is shown in Figure 23.8.

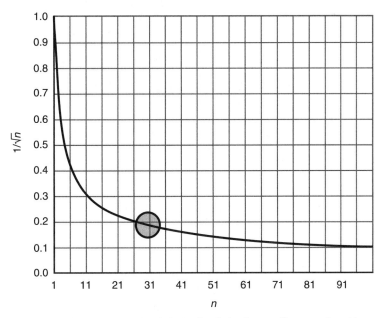

Figure 23.8 Illustration of analytical rationale for the sampling case of $n = 30$.

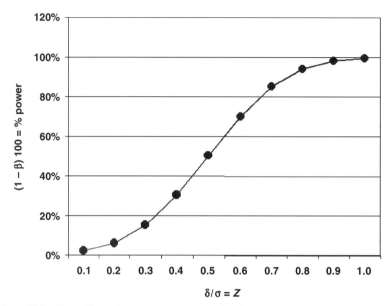

Figure 23.9 Figure illustrating operating characteristic (OC) curve related to the case example.

When sample size is $n = 30$, operational convenience and cost are often perceived as reasonable and acceptable. In addition, $1-\beta$ is relatively high when considering that $\alpha/2 = 0.005$, given that the expected δ is rational. Solely for the purpose of design qualification in a moderate to high-volume environment, the recommended degrees of freedom was established as df $= n - 1 = 30 - 1 = 29$. As discussed previously, this particular value of n represents a relatively sound tradeoff between maximizing statistical precision while minimizing operational inefficiency and waste. Therefore, this researcher standardized on the sampling practice $n = 30$. Meanwhile back at the design ranch, we recall that the Six Sigma design qualification guidelines specifies that the probability of type I decision error should be set at $\alpha/2 = .005$, under the sampling constraint df $= n-1 = 29$. With these criteria, it may be statistically demonstrated that when $\delta/\sigma = 0.75$, then $1-\beta = 0.90$. In other words, we would be 90% certain of detecting a 0.75σ change if such a difference (or larger) were actually present during the time of sampling. If we set $\delta/\sigma = 1.0$, we would discover the test sensitivity to be 99.5%. The advanced reader may recognize that such estimates of statistical power are based on the mathematics underlying the classic one-sample t test. However, when considering the case $\delta/\sigma = 0.70$, we would compute $1-\beta = 0.85$, or 85%. Thus, we would have 85% certainty in our ability to detect a 0.70σ difference if such a difference was actually present. On the flip side, if after testing for a 0.70σ difference we failed to find such a magnitude of change, we would consequentially be exposed to a 15% risk of falsely accepting H_0. Thus, we find a high level of power (test sensitivity) within the practical range $0.75 < \delta/\sigma < 1.0$ for the case $n = 30$ and $\alpha/2 = 0.005$. Thus the design qualification guidelines (related to the deductive approach) offer a reasonable balance between risk minimization and maximum operational efficiency. For the reader's convenience, this writer has constructed an operating characteristic (OC) curve to display the relative power that is inherent in the standard Six Sigma design qualification sampling criteria (Fig. 23.9).

23.11 EXPERIMENTATION

Many of us perform experiments seeking clues or evidence that may eventually solve personal or worldly puzzles. We consciously (and perhaps unconsciously) manipulate our environment in this quest for enlightenment. These experiments may lead us to specific solutions, yet at other times, to a deeper understanding of the question rather than the answer that we sought. This acquisition of knowledge through experimentation is an old concept. What is relatively new (at least to general industry) is how experiments are conducted and the results analyzed. Without a full retelling of history, it is evident that the age-old *art* of haphazard trial and error is evolving into a more disciplined pseudoscientific approach. This has fostered the idea of manipulating one factor at a time while holding all other variables constant.

While the single-factor experiment is wonderfully simple and highly useful, it falls short when addressing those cases involving two or more variables. Today, experimentation is focused on the creation of more efficient test plans to understand multiple variables and their impact on a process or problem. Such test plans are characterized by concurrent and combinatorial arrangement of several test variables (and even multiple dependent variables). Thus, experimentation can no longer rely on common sense reasoning supported by single-factor analysis but has evolved into a mature science using mathematics and statistics to view the world in all of its multidimensional character. Specifically, this author is referring to a branch of applied mathematical statistics named *design of experiments* (DOE). Although DOE is based on the application of mathematical statistics, the practical side is relatively independent of such esoteric knowledge. In fact, it is eloquently simple and fairly intuitive to apply through the use of computers. As previously stated, the aim of DOE is related to the combinatorial manipulation of two or more independent variables (x_1, x_2, ..., x_n) in the context of some dependent variable Y. The selected test combinations and means of data manipulation ensure that the outcomes can be usefully interpreted. Without the a priori application of DOE, the normal and expected variations (which are inevitable during testing) have a far greater potential to mask reality. Thus, the judicious application of DOE ensures that the input variables are configured such that the variations of the outcome variable Y can be properly organized and analytically manipulated during data analysis. When an experiment has been properly designed, the resulting analyses can provide greater insight than other forms of experimentation. The reader must always remember that the most powerful mathematics cannot rescue one from faulty conclusions drawn from poorly designed experiments.

23.12   EXPERIMENT DESIGN

To initiate a DOE project, the investigator must first articulate the problem and purpose of experimentation. Then the researcher must choose a response characteristic (i.e., dependent variable) and select one or more experimental factors (i.e., independent variables) that are believed to influence the response characteristic. Next, two or more settings are established for each of the experimental factors, to provide a basis for comparative analysis. After assigning performance-testing levels to each of the experimental factors, the investigator prepares test patterns prescribing the combinations to be explored. Then the analytical fundamentals must be considered to establish the appropriate levels of risk and confidence. These decisions determine sample size (i.e., the number of repetitions in the experiment). At this point, tradeoffs may be required to balance practical considerations of implementation against the statistical requirements. Following the execution of each test combination (i.e., experimental run), the researcher "interrogates" the resulting response data using an array of statistical tools, methods, charts, and graphs. This allows the researcher to "separate" the effect of each experimental factor. When the *experimental test pattern* and *data analysis scheme* have been properly designed, the investigator can translate the analytical outcomes into useful information. In this sense, the raw data that fall out of an experiment cannot be viewed as *information*—until it has been analytically "tortured" to reveal what it "knows."

Perhaps in bad taste, it may be said that the torture chamber (i.e., test pattern) must be sufficiently configured and equipped with the proper interrogation tools (i.e., statistics) if we want the prisoners (data) to reveal how the enemy (i.e., variation) may attack our position (i.e., design). From this perspective, we recognize that variation is the enemy of producibility and our special mission is to locate, study, and destroy its fortified positions in the forest of design details—before the product configuration crosses the forward edge of the battle zone and engages itself in full-scale production. Once a design has been released for production, a battle with variation will be fought. The only question is, How big will the battle be, how long will it last, and what are our chances of winning?

In this sense, the application of DOE (and the other methods discussed so far) is a way to stack the statistical odds in our favor. Gathering relevant intelligence information and conducting preemptive strikes within our design rooms, laboratories, and processes best realize such probabilistic enhancements. In this manner, we are able to wear down the enemy of variation and reduce its numbers before we go on the battlefield of variation (within our industrial and commercial processes). We execute this strategy so that not one single enemy soldier (i.e., defect) reaches the populated cities of our customers and has the opportunity to decimate their rightful levels of performance expectation. When this unthinkable action occurs, the hubs of economic activity within our own cities will eventually collapse. Therefore, we must learn the skills that can better prevent war, rather than honing our process improvement swords and strengthening our field service activities. Perhaps this is message is best communicated by saying that our *design pens* are more powerful than our *process improvement swords.* When properly developed and exercised, a DOE has the ability to help us (1) draw meaningful conclusions with known degrees of risk and confidence, (2) study the effect of each experimental factor, and (3) establish the factor's impact on the total observed variation. Beyond any shadow of scientific doubt, DOE should be included in each designer's "A team" list when configuring, analyzing, and/or optimizing a product, service, or process design. However, in spite of its proven track record, DOE is frequently not employed to yield a producible design. Many designers are simply unaware that such a powerful tool exists, or they simply overlook its potential application (due to insufficient training and exercise). Here again, we must reconsider one of the key Six Sigma principles: We can't do what we don't know, and we won't know until we learn.

Figures 23.10 and 23.11 illustrate how the interaction between two variables can impact process performance (signal to noise ratio in this case).

From a cursory examination of Figure 23.10, the similar slopes for factors A and B indicate similar impact on SNR. As a consequence, we would most likely reason that they should be given equal consideration in the context of our improvement effort. It would appear that when factors A and B are controlled to their respective low setting, the overall performance of Y is optimized. In the interests of extending the reader's perspective, Figure 23.11 graphically illustrates the best and worst settings for factors A and B in terms of the response Y.

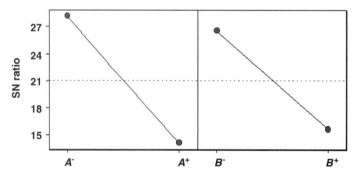

Figure 23.10 Main-effects plots for factors *A* and *B* in terms of signal-to-noise ratios (SNRs).

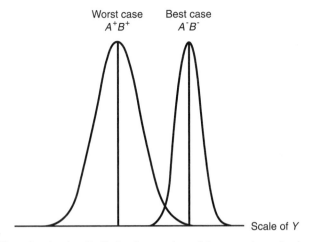

Figure 23.11   Illustration showing distributional comparison of the test settings related to factors A and B.

23.13   RESPONSE SURFACE DESIGNS

Response surface design is an extension of a full factorial design of experiment. It is also a special case of multiple regression analysis. In a response surface design, the experimental factor levels are varied to determine which combination of settings produces an optimum or desired response. Response surface design is used to model the relationship between the response variable and two or more factors. For a two-factor model, is represented by the following equation

$$Y = \beta_0 + \beta_1 {}^* A + \beta_2 {}^* B + \beta_3 {}^* A {}^* B + \beta_4 {}^* A^2 + \beta_5 {}^* B^2 + \varepsilon$$

where A and B are the factors of interest. This equation allows us to model curvature (the quadratic term). The difference between the equation for a response surface as compared to the equation for a factorial design is the addition of the quadratic terms that can produce the curvature in the response. The response surface method is used when the inputs are all continuous variables when (1) curvature is found in a DOE (sequential experimentation—central composite design), (2) curvature is known (Box-Behnken design), and (3) when the optimum operating point is in a region of curvature. The most important reasons for using the response surface method are

1. Understanding the effect of a set of factors on a response of interest: $Y = f(X)$
2. Knowing which settings of these factors ensure that the process satisfies desired specifications
3. Evaluation of interactions and curvature
4. Knowing which settings of these factors optimize the response
5. Using response surface method (RSM) experiments, which are statistically designed examples of multiple regression modeling

Example 23.2   Identify the combination of coffee bean acidity and sugar that optimize coffee flavor (see Fig. 23.12).

Design points for a central composite design

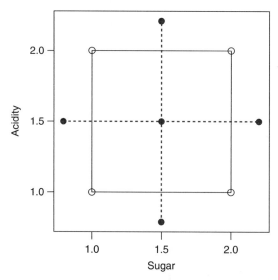

Figure 23.12   MINITAB diagram showing design points of the central composite design for the coffee taste satisfaction experiment. This plot contains four cube points located on the corners of the square, five centerpoints (all located in the exact center of the cube), and four axial points extending outside the cube through the center of the four sides of the cube.

Solution   For a two-factor (corn syrup and sugar) design we can select between the following:

- A 13-run design with one block
- A 14-run design with two blocks

The solution of the problem is to identify the combination of coffee acidity and sugar that optimize the taste of corn syrup.

Using MINITAB, go to `Stat > DOE > Response Surface > Create Response Surface Design > OK`.

- Screen "Create Response Surface Design" (Figure 23.14).
- In number of factors, type/enter 2. The two factors are: factor 1 = coffee beans acidity, and factor 2 = sugar (Figure 23.13).
- Click OK.

(*Note:* MINITAB provides default α values to ensure that the design exhibits the desirable properties of orthogonal blocks and, whenever possible, rotatability. Remember that with an orthogonally blocked design, the linear and interaction terms in the model can be estimated independently of one another and that with a rotatable design, all points equidistant from the design center have the same prediction variance.)

MINITAB's session window is interpreted as follows:

—— **10/27/2008 5:22:37 PM** ——————————

Welcome to Minitab, press F1 for help.

Central Composite Design

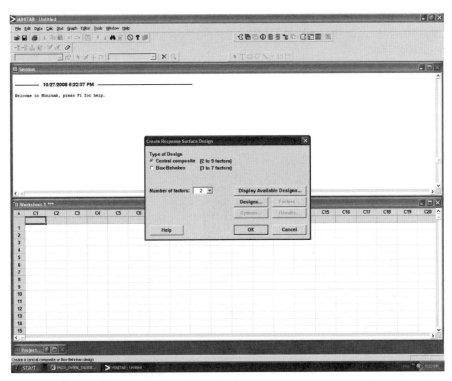

Figure 23.13   MINITAB captured screen showing how to create a response surface design.

Factors: 2 (coffee bean acidity and sugar) Replicates: 1
Base runs: 13 Total runs: 13
Base blocks: No blocks Total blocks: 0
Two-level factorial: Full factorial
Cube points: 4
Centerpoints in cube: 5
Axial points: 4
Centerpoints in axial: 0
Alpha α: 1.41421

For this coffee delight taste experiment, we have created a central composite design with the characteristics listed above. MINITAB displays the experimental conditions or settings for each of the factors in coded form in the session window. When performing the experiment, use the order that is shown to determine the conditions of each run. Run the uncoded factor levels as follows:

Coffee bean acidity (factor A)	1 low and 2 high
Sugar (factor B)	1 low and 2 high

We have selected in MINITAB to randomize the order of the runs. Randomization is recommended because it scrambles the order of the runs so that this external source of variation is not confounded with one or more factor levels. When repeating these instructions to create RSM, do not worry whether the order of the data matrix looks different; just remember that MINITAB scrambles the order of the runs—randomizing:

Data Matrix

Run Order	Coffee Bean Acidity	Sugar
1	0.00000	1.41421
2	−1.41421	0.00000
3	0.00000	0.00000
4	1.00000	1.00000
5	0.00000	0.00000
6	−1.00000	−1.00000
7	1.00000	−1.00000
8	0.00000	0.00000
9	0.00000	0.00000
10	−1.00000	1.00000
11	0.00000	−1.41421
12	1.41421	0.00000
13	0.00000	0.00000

The cube points allow estimation of linear and interaction effects, but not for curvature. These points are comparable to the corner points of a 2^k factorial design. The centerpoints allow for detection

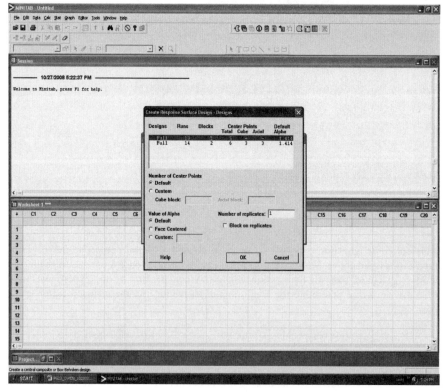

Figure 23.14   MINITAB captured screen showing selected type of design.

of curvature, but not estimation of the individual quadratic terms. The point in the middle of the cube represents the center points for both the cube and axial blocks, and finally, the axial points allow for the estimation of the quadratic terms. The points outside the cube and joined by dotted lines represent the axial block. The objective of this exercise is to identify the combination of corn syrup and sugar that optimize the taste of the cereal. In this experiment we measured the taste satisfaction for each sample tasted by 10 testers. The response variable is taste.

```
S=2.153 R-sq=22.2% R-sq(adj)=12.8%
Analysis of Variance for Taste
```

Source	DF	Seq SS	Adj SS	Adj MS	F	P
Blocks	9	54.338	54.338	6.038	1.30	0.243
Regression	5	98.143	98.143	19.629	4.23	0.001
Linear	2	6.362	6.362	3.181	0.69	0.506
Square	2	10.556	10.556	5.278	1.14	0.324
Interaction	1	81.225	81.225	81.225	17.52	0.000
Residual						
Error	115	533.242	533.242	4.637		
Lack-of-Fit	75	364.442	364.442	4.859	1.15	**0.317**
Pure Error	40	168.800	168.800	4.220		
Total	129	685.723				

The residual error of the full model is split into two components, lack of fit and pure error. *Lack of fit* is composed of higher-order terms omitted from the model. For a two-factor model, the higher-order terms include the interactions with squared terms and all interactions with the blocking variable. For your central composite design, these terms are

```
Sugar * C-Acidity
Sugar1*C-Acidity²
Sugar2*C-Acidity²
```

(where C-Acidity = coffee bean acidity).

All Possible Interactions Involving Blocks. The P value for interaction $P = 0.0000$ suggests that there is at least one significant interaction. Therefore, we need to include the linear terms in the model. In this cereal experiment the following groups of terms are nonsignificant:

- Blocks, where $p = .243$
- All linear terms, where $p = .506$
- All squared terms, where $p = .324$

The nonsignificant lack-of-fit test $p = 0.317$ indicates that none of these terms need to be in the model. For the central composite design all possible interactions involving blocks are part of the terms of lack of fit. Therefore, blocks will be excluded and the model will be refit.

Removing Blocks from the Model. Because there is no effect due to the difference in the 10 testers, the blocking variable used for this variable is not considered significant. Removing blocks reduces the model.

```
Stat > DOE > Response Surface > Analyze Response Surface Design > Click
Terms > Uncheck include blocks in the model > click OK
S = 2.177  R-Sq = 14.3%  R-Sq(adj) = 10.9%
Analysis of Variance for Taste
```

Source	DF	Seq SS	Adj SS	Adj MS	F	P
Regression	5	98.143	98.143	19.6285	4.14	0.002
Linear	2	6.362	6.362	3.1810	0.67	0.513
Square	2	10.556	10.556	5.2778	1.11	0.332
Interaction	1	81.225	81.225	81.2250	17.14	0.000
Residual Error	124	587.580	587.580	4.7386		
Lack-of-Fit	3	1.300	1.300	0.4335	0.09	**0.966**
Pure Error	121	586.280	586.280	4.8453		
Total	129	685.723				

There is no significant lack of fit of P = 0.966. Therefore we conclude that removing blocks had no effect on the model fit.

Analysis of Variance (ANOVA) for Taste

Source	DF	Seq SS	Adj SS	Adj MS	F	P
Regression	3	87.587	87.587	29.196	6.15	0.001
Linear	2	6.362	6.362	3.181	0.67	**0.513**
Interaction	1	81.225	81.225	81.225	17.11	0.000
Residual Error	126	598.136	598.136	4.747		
Lack-of-Fit	5	11.856	11.856	2.371	0.49	**0.784**
Pure Error	121	586.280	586.280	4.845		
Total	129	685.723				

and:

(*Note*: The analysis was done using coded units.)

Estimated Regression Coefficients for Taste

Term	Coef	SE Coef	T	P
Constant	5.04615	0.1911	26.407	0.000
C-Acidity	-0.07589	0.2436	-0.312	0.756
Sugar	0.27160	0.2436	1.115	**0.267**
C-Acid*Sugar	-1.42500	0.3445	-4.136	0.000

```
S = 2.179  R-Sq = 12.8%  R-Sq(adj) = 10.7%
```

The ANOVA table allows us to conclude that there is no significant lack of fit P = 0.784. Therefore removing the squared terms did not affect the model fit. For model terms, all linear terms are nonsignificant because P = 0.513 and there is at least one significant interaction: P = 0.0000. C-Acidity * Sugar shows that the effect of coffee bean acidity on taste depends on the amount of sugar because for sugar P = 0.267. (*Note:* Remember that when $p \geq .5$, there is no significant effect.) The R^2 and adjusted R^2 values represent the proportion of variation in the response data explained by the model. R^2 describes the amount of variation in the observed response values that is explained by the model. $R^2 = 12.8\%$ is the coefficient of determination that indicates that 12.8% of the variation in taste is explained by the amount sugar, and 87.2% of the variation in taste remains unexplained. We can reduce the model further by removing all square terms.

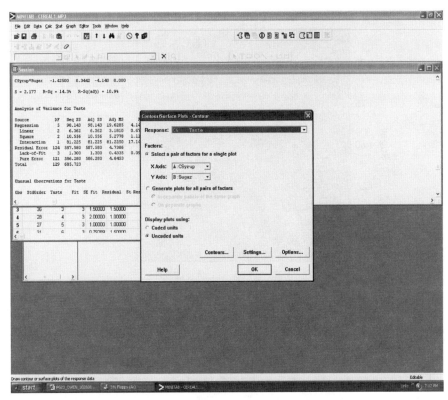

Figure 23.15 MINITAB captured screen with settings for creating the contour plot for Example 23.2.

Finding Optimal Settings Go to. `Stat > DOE > Response Surface > Contour/Surface` `(wire-frame) plots.` Then

- Check contour plot (See Fig. 23.15).
- Click `setup`.
- Complete the dialog box with the response and a pair of factors > uncoded units > `OK` or coded units.
- Clicking OK in each dialog box.

The contour plot in Figure 23.16 shows two optimal regions: high sugar and low corn syrup and low sugar and high coffee bean acidity.

Finding the Optimal Settings. `Go to Stat > DOE > Response Surface Optimizer.` Then

- Move `taste` from `available` to `selected`.
- Click `setup`.
- Complete the dialog box.

MINITAB employs a numerical algorithm to find the factor settings that will optimize the response (see Figs. 23.17 and 23.18). For the cereal data, taste is optimized when corn syrup is low and sugar is high. The predicted maximum response for satisfaction for this example was selected to be 7.9964.

Contour Plot of Taste vs Sugar, C-Acidity

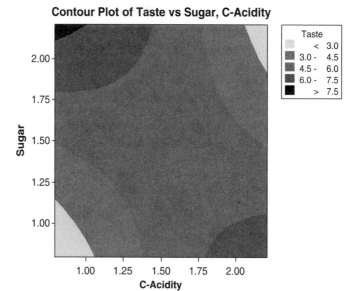

Figure 23.16 MINITAB graph showing contour plot of Taste vs Sugar, C-acidity.

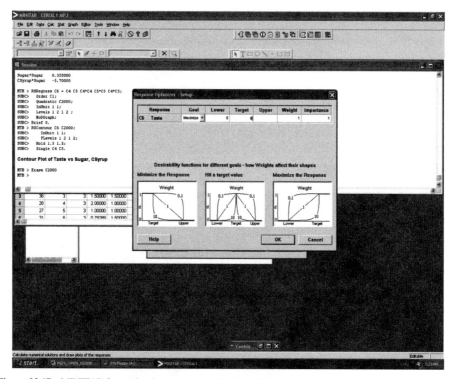

Figure 23.17 MINITAB figure showing response optimizer settings as follows: Response: Taste, Goal: Maximize, Lower: 5, Target: 9, Weight: 1, and Importance: 1.

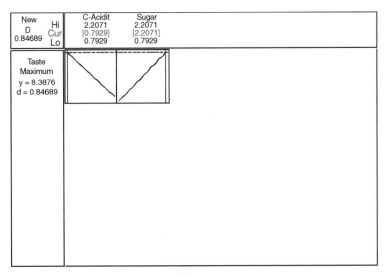

Figure 23.18 MINITAB's response optimizer graph.

MINITAB displayed the maximum values of the response Taste $= 8.3876$ and the settings needed to achieve these results: C-Acidity (coffee bean acidity), Sugar $= 2.2071$.

While the response optimizer finds a single combination of factor settings, you can use a contour plot to determine whether there are multiple regions that optimize the response. The practitioner can use this information to find reasonable alternatives to the default settings obtained by the response optimizer.

23.14 FACTORIAL PRODUCIBILITY

In many design and process situations, we are unwilling or unable to generate *live* data to analyze and guide our improvement efforts. We recognize that such situations are usually the rule, not the exception for most product and process designs. *Specifically, the authors propose the use of computer simulation for these situations.* To effectively employ simulation, we must be able to emulate the underlying function of a performance characteristic. For example, if we consider two factors (*A* and *B*) in the context of a given *Y*, it might be possible to simulate a factorial experiment if we are able to model the response. We offer a simple application to illustrate how a (computer) simulation can be leveraged in the context of a full factorial experiment. Here, it is assumed that the designers confronted the problem of how to study and subsequently optimize producibility. More specifically, we assumed that the designers are concerned with the upfront optimization of process reproducibility during the concept design phase. The team believes they can better ensure design replication (during process operation) by identifying the key elements of reproducibility and then optimizing these factors minimizes risks relative to certain other design goals. Needless to say, they agree that the task should be undertaken, but they differ on how it should be done. One team member insists that the key to producibility is a simple design (low complexity in terms of parts, connections etc.). Another team member agrees, but also asserts that the influence of process capability cannot be underestimated and, as a consequence, should be optimized before the design is released for full-scale production. Yet another team member, who previously worked in production, recognized the power of process control (in terms of stability). It was this team's member position that process centering (over time) strongly affects reproducibility. Still another engineer stated, "I believe that each of you have valid points, but I do not know how to choose between the alternatives. There is just too much at stake to be grabbing at straws and too little time to be

TABLE 23.1   Testing Combinations for the Producibility Analysis Example

Run	Capability	Complexity	Stability	TDPU[a]
1	3.0	500	0.5	3.11
2	4.5	500	0.5	0.02
3	3.0	1000	0.5	6.23
4	4.5	1000	0.5	0.03
5	3.0	500	2.0	86.38
6	4.5	500	2.0	3.11
7	3.0	1000	2.0	172.75
8	4.5	1000	2.0	6.23

[a] Total defects per unit.

guessing. " After a brainstorm meeting, the Six Sigma design Black Belt came into the room and, overhearing the argument, announced that he knew how to bring clarity to the matter.

For this case, the Black Belt had the team describe the anticipated future performance of the process:

1. They first specified the general range of process capability that a typical design CTQ would likely experience on the shop floor. The team believed that a liberal estimate of process capability would be 4.5σ and that a conservative estimate would be 3.0σ. For the most part, the team recognized that a 4.0σ capability level was typical since most of their designs reflected 25% margins. Although they knew a few of their operations exhibited world-class capability of about 5.0σ (as a result of capital-intensive automation), they also recognized that some operations were as low as 2.0σ, from use of low-quality materials and tools that required frequent human intervention.

2. Then, the simulations Black Belt had them choose the *best-case* and *worst-case* scenarios with respect to design complexity. After some friendly debate on complexity, the team decided to scale this concept by considering the total number of CTQs. This was fairly easy since the total number of CTQs is frequently highly correlated to the total number of key design features, which, in turn, is determined largely by the total number of sensitive parts. Following some more lively debate, the team agreed that (on the basis of their engineering judgment), the complexity factor should be sized between 500 and 1000 CTQs. This was based on the belief that a revolutionary design configuration (relative to their situation) would yield about 500 CTQs, while similar configurations have historically exhibited about 900. Allowing for new—bells, whistles, and buzzers—the team decided that they should bump the historical estimate by 100 CTQ's, thereby raising the upper estimate of complexity to 1000.

3. The Black Belt successfully stimulated a discussion about process stability (i.e., control over centering). This was done by personally injecting the idea that a typical process will shift and drift over time, thereby degrading the initial capability estimate. Because of this phenomenon, he believed that the team should select two competing levels of process stability. Here again, the team chose the extreme values by considering the *best-case* and *worst-case* scenarios. The team also agreed that the typical process will shift and drift on the order of 1.5σ, but that this estimate will vary by as much as 0.5σ, given the nature of their particular production facility. Thus, the team determined that the expected lower limit should be defined as 0.5σ, while the upper limit should be 2.0σ. Armed with this information, the Black Belt formed a 2^3 full factorial matrix and substituted the specified level values. The design matrix is illustrated in Table 23.1.

23.15   TOOLBOX OVERVIEW

To address the needs of a specific task, the DFSS toolbox can be selectively equipped from the large array of tools in the master toolbox. The ultimate aim of the tools is to facilitate the development and delivery of Six Sigma products and services, on time, at the lowest possible cost. Naturally, we all recognize that DFSS is not a replacement for sound engineering and management. The tools suitable for conducting market surveys such as the Likert scale would have an unlikely application in the preliminary design phase. An analogy would be would be trying to loosen a six-point nut using a Phillips head screwdriver. Therefore, as the practitioner can appreciate, a substantial amount of detailed training will be required to expand the knowledge base of the tools and their applicability and selection criteria. In general, phase and process appropriate *tool trays* can be prepared ahead of time to address most needs. These tools include the following:

- Cause–effect (CE) analysis
- Failure mode–effects analysis (FMEA)
- Mistakeproofing
- Tolerance analysis
- Highly accelerated life testing (HALT©)
- Quality function deployment in engineering
- Design simplification
- Robust analysis
- Multiple response optimization
- Monte Carlo simulations

23.16   MONTE CARLO SIMULATIONS

Among the organizations practicing Six Sigma, there is no single established method for project selection. Practitioners have not yet agreed on which approach is superior, and no standard has appeared. This section describes a powerful project selection methodology using a combination of Monte Carlo simulation and stochastic optimization. Monte Carlo simulation is an increasingly popular technique for Six Sigma professionals because, with the advent of faster personal computers, it has become a flexible, accessible, and visually informative tool. Monte Carlo simulation is a random sampling technique that requires probability distributions as inputs to demonstrate the impact of variability on the process outputs. Crystal Ball®, a Microsoft Excel–based software suite from Oracle, is used in this chapter to demonstrate Monte Carlo simulation to show how variation or uncertainty affects project outcomes. Simulation can also be used to guide strategic decisions.

We will use these tools to analyze two distinct projects: *bank loan process improvement*, a transactional example of simulation; and *simulation with DOE*, a DMAIC project that begins in MINITAB and progresses to simulation and optimization in Crystal Ball. These projects will use both normal and other distributions based on observed data, theoretical knowledge, or expert opinion.

23.16.1   Monte Carlo Simulation Defined

Monte Carlo simulation is a random sampling experiment. The technique was named after Monte Carlo, Monaco, where the primary casino attractions are games of chance such as roulette. These games all exhibit random behavior. When you roll a die, you know that a 1, 2, 3, 4, 5, or 6 will come up equally given a fair die, but you don't know which number will appear for any particular roll. One of the earliest applications of Monte Carlo simulation occurred in the late 1940s, when scientists at the Manhattan Project at Los Alamos (NM) National Laboratory used the method to predict the range of possible nuclear explosion results.

Monte Carlo simulation is a proven, efficient technique that requires only a random-number table or a computer random-number generator.[3] In practical applications, simulation is applied to a "model," defined here as a combination of inputs and outputs that capture the behavior and performance of a business process or product. A model could be a transfer function, cost estimate procedure, or back-office process. A random number is defined as a mathematically selected value generated to conform to a probability distribution assigned around a model input.

With increasingly powerful personal computers and sophisticated software products, Monte Carlo simulation has quickly become a staple in the desktop analytic toolkit. The most effective way to use Monte Carlo simulation is through a commercial tool such as Crystal Ball that offers the ability to define distributions and run simulations on the familiar Microsoft Excel platform. For each simulation trial, Crystal Ball randomly selects a value from a defined distribution and enters that value into the spreadsheet, which then recalculates the affected formulas. Crystal Ball saves and displays the results of each simulation trial and calculates descriptive statistics for the entire group of trials. Multiple scenarios created through simulation can be analyzed via interactive charts and tables to give more insight into the mechanisms and drivers of the spreadsheet model. When used correctly, Monte Carlo simulation can provide valuable insights not available through deterministic models.

23.16.2 When Simulation is an Appropriate Tool

As with any Six Sigma tool, simulation can be effectively applied to solve a wide variety of problems, including strategic analysis, value stream analysis, tolerance design, cost estimation, financial analysis, market forecasting, and resource allocation. Six Sigma professionals often apply Monte Carlo analysis to two broad applications:

- *Project management*, where inputs such as costs and potential revenue are uncertain (or unknown)
- *Product and process design*, where inputs such as part dimensions and task cycle time display variation

Table 23.2 lists some of the more common concerns and conditions within project management and design that indicate the need for Monte Carlo simulation.

TABLE 23.2   Reasons for Using Simulation in Project Management and Product and Process Design

Issue	Concern/Condition
Project management	*Financial uncertainty*: When financial uncertainties exist for a proposed or existing process or product, and financial estimates or forecasts can be described in a spreadsheet model.
	Schedule uncertainty: When scheduling uncertainties (task cycle time) exist for a proposed or existing process or product, and the schedule is such that activities can be defined along a simple critical path.
	Cost controls: When it is prohibitively expensive to acquire data for an input or output of a process or product, and simulation can create realistic virtual data.
	High risk: The term *risk* is most often used to describe a financial loss. When you are making decisions about a high-risk project (e.g., one that has the potential for a huge negative revenue impact), simulation can be used to estimate the certainty of success, magnitude of risk, and key drivers impacting success.
Product and process design	*Robust design*: When there is a need to determine whether a product or process design is robust to variation of inputs and validation is prohibitively expensive or time-consuming. Robust design requires stochastic optimization.
	Tolerance analysis: When engineering variation predictions need to be quantified.
	Relatively simple process: When a business process is linear or can be viewed at a high level; the process can be quickly modeled in spreadsheet form and simulated. This can help measure baseline performance and test the effect of process improvements prior to implementation.

TABLE 23.3 Data Format and Model Conditions that Require Monte Carlo Simulation

Issue	Condition
Data format	*Data are infrequent*: When significant results or data take a long time to obtain because of the nature of the project variables
	Data DOEs do not exist: When there is a notable absence of data for an input or output of a process or product (e.g., competitive benchmarking analysis)
	Data are estimated: When there is an estimated mathematical relationship (between inputs and outputs) with estimated input variations
	Data are costly: When the cost of obtaining adequate samples is prohibitively expensive
	Variation is nonnormal: When nonnormal variation occurs on process or product inputs, this can invalidate closed-form solutions based on the normal assumptions like RSS[2].
Data model	*Mathematical relationship exists*: When there is a known mathematical relationship (between product or process inputs and outputs) with known input variations
	Spreadsheet model already exists: When you already have a process or product modeled in an existing spreadsheet (e.g., an inventory model), and this model contains inputs with variations and outputs as formulas
	Equation is nonlinear: When the transfer function is highly nonlinear and nonintuitive
	Physical models are impractical or impossible to create: when physical models of the process or product are impossible/expensive or impossible/impractical to create because of technological limitations

[a]Root sum square.

Of course, the decision to use Monte Carlo simulation will be impacted most by the quality and quantity of the available data and the format of the model that describes the process or product. Table 23.3 provides a checklist for data and model conditions that indicate the need for Monte Carlo simulation.

Practitioners should recognize that every Six Sigma software tool has strengths and weaknesses. In a Monte Carlo simulation, the quality of the results is directly related to the quality of the simulation model. Prior to simulation, each model must be verified and validated as representing the process or product as closely as possible. The maxim of "garbage in, garbage out" (GIGO) is nowhere more appropriate than in regard to the use of simulation. Practitioners should be able to defend their reasoning for applying specific probability distributions within a model. Simulation itself can often be used as a testing tool to ferret out poor model design and flaws in defining assumptions; poorly built models will cause a simulation tool to shut down because of calculation errors, and poorly defined assumptions will lead to inaccurate and outrageous forecasts and sensitivity charts. Inexperienced modelers and those with particularly narrow agendas can easily misrepresent and misuse simulation results.

23.16.3 Defining Distributions and Outputs in Crystal Ball

Crystal Ball offers 20 predefined probability distributions and an additional custom distribution used to represent unique situations that cannot be described with the existing distributions. Crystal Ball distributions, referred to as "assumptions," are available in two formats, as graphical objects and as Excel functions. The graphical selection method is the most popular means of defining assumptions given the strong visual impact of the software interface. Once a distribution is selected from the distribution gallery (Fig. 23.19), the analyst can shape the distribution from the broad set of parameters provided.

Any model should first be validated to ensure that the *base case*, which represents the deterministic or most likely solution, is as realistic and accurate as possible. Once the team or Black Belt has signed off on the base case, distributions should be added via Crystal Ball. Figure 23.20 shows the assumption definition dialog for the normal distribution. Assumptions can also be created through a distribution fitting to existing data using one of three goodness-of-fit methods: chi squared, Anderson–Darling,

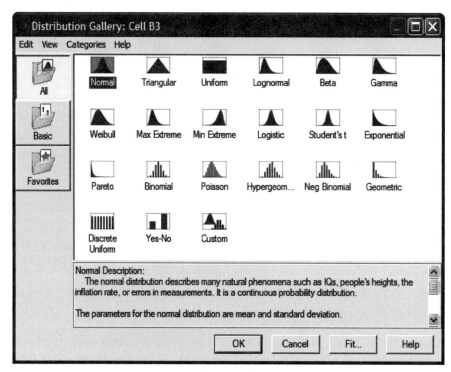

Figure 23.19   Crystal Ball distribution gallery.

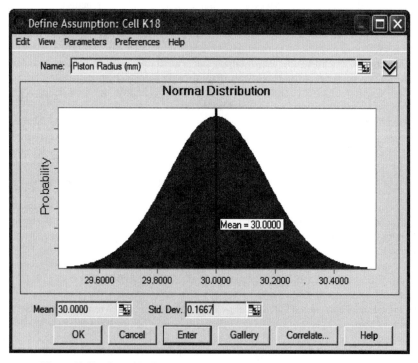

Figure 23.20   Crystal Ball assumption definition dialog for normal distribution.

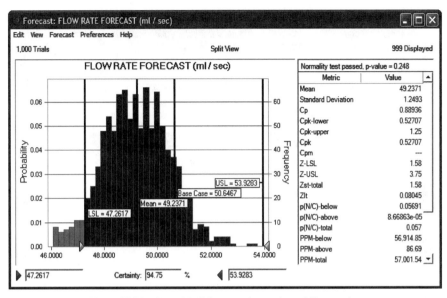

Figure 23.21   Crystal Ball forecast chart and capability metrics.

and Kolmogorov–Smirnov. A complex model can include hundreds of assumptions; Crystal Ball can also define dependencies between assumptions given a known or expected correlation coefficient.

Once all variable or uncertain inputs have been defined, Crystal Ball requires analysts to define one or more outputs, referred to in Crystal Ball as "forecasts." These forecasts, which are the responses, Ys, or effect of the model, are displayed as histograms, tables, or a combination of both following a successful simulation (Fig. 23.21). With one or both specification limits defined for a forecast, Crystal Ball (version 7.2 or later) will calculate the long- or short-term capability of the forecast as well as the descriptive statistics and percentiles.

Crystal Ball's sensitivity analyses helps modelers to quickly and easily judge the influence of each assumption cell on a particular forecast cell. During simulation, Crystal Ball calculates sensitivity by computing rank correlation coefficients between each assumption and every forecast. These correlation coefficients provide a meaningful measure of the degree to which assumptions and forecasts change together. If an assumption and a forecast have a high correlation coefficient, then the assumption has a larger impact on the forecast (both through its uncertainty and its model sensitivity), relative to the other assumptions. Crystal Ball ranks the assumptions according to their importance to each forecast cell. The sensitivity chart (Fig. 23.22) displays these rankings as a bar or pie chart, indicating which of the critical factors (Xs) in the analysis cause(s) the predominance of variation in the response variable of interest (Y). The insights provided through sensitivity analysis can improve the quality and accuracy of the model, help the project team understand what's driving defects, and define where the team should focus (or where *not* to focus) their energies and resources. After reducing the variation for these few critical Xs, a modeler can rerun the simulation and examine the effects on the output.

Example 23.3   A healthcare software company used Lean Sigma and Monte Carlo simulation to reduce the lengthy installation and implementation cycle time of its enterprise software. This simulation forecast also reinforced management confidence in the proposed process solution. This software company produces software used by physicians and hospitals to manage the complexities of healthcare. Products, some acquired from other companies, included physician practice software, laboratory systems, radiology systems, and enterprise software systems. When the process study

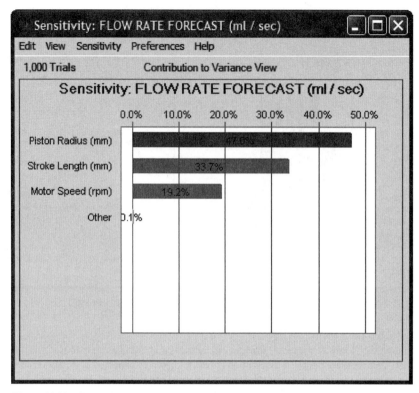

Figure 23.22   Crystal Ball sensitivity chart showing flowrate forecast in milliliters per second.

began, the company learned that the installation and implementation of the enterprise software could take 18–36 months. This result was deemed unacceptable, and so they began to redesign the implementation process using the DMADV Lean Sigma (LS) methodology.

Gathering data for the project was a problem because the long cycle time meant that years would pass before a statistically significant sample of new installs could be completed. The only viable option was to determine the anticipated cycle time through simulation. Crystal Ball software was used to simulate the high-level uncertain components of the new process (e.g., the assignment of sufficient resources). Additional model uncertainties included the scheduling and staffing of concurrent implementation projects. The primary simulation output was the cycle time of the proposed process where the mean, standard deviation, and distribution of values revealed just how successful the new design would be. The solution validated the intuition of the LSS team that the cycle time could actually be reduced by the goal of 50%. This enabled the company to recognize revenue in half the time. The simulation of the new process design and the resulting critical company reorganization provided direct evidence to senior management that the LSS team had met their project goals.

Example 23.4: Bank Loan Process Improvement   A financial organization wishes to make its loan process "best-in-class". The current loan process, from initial customer inquiry through loan disbursement, took an average of 91 h. Given a cycle time target of 96 h, the process would seem to be solid. However, loan specialists complained (quite loudly) that the entire loan process can take over 130 h and that the process should be improved. The goals of the Six Sigma Black Belt are to measure the current state, understand what is driving the variation, reduce the standard deviation of the process, and quantify the process capability and Six Sigma quality level.

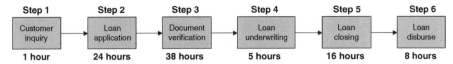

Step 1	Step 2	Step 3	Step 4	Step 5	Step 6
Customer inquiry	Loan application	Document verification	Loan underwriting	Loan closing	Loan disburse
1 hour	24 hours	38 hours	5 hours	16 hours	8 hours

Figure 23.23 Macro map of loan process.

Solution   In the *define* phase, the Black Belt and her team defined six broad transactional steps. Figure 23.23 shows the macro map for the loan process with the average time for each major process step. How stable is this process? By tracking 100 loans through the application process, the team calculated the range and variation for each step using a combination of observed data, theoretical knowledge, and expert opinion. The details of this research are presented in Table 23.4.

TABLE 23.4   Description of Steps for Loan Process

Process Step	Step Description	Probability Distribution
Customer inquiry	Occurs via a phone call, office visit, the Internet, or home visit by a mortgage officer; includes creation of an initial rate quote for the customer.	Observed data indicate that this cycle time is lognormal with a mean of one hour and a standard deviation 0.25 h.
Loan Application	With the inquiry completed, the applicant must complete all necessary forms. The elapsed time from the distribution of blank forms and the collection of completed forms is difficult to estimate.	From expert opinion it is estimated that it takes no more than one week (40 h), no less than one day (8 h), and most often takes 3 days (24 h). This is a triangular distribution.
Document verification and processing	Completed application is reviewed by a loan specialist, who contacts the applicant to verify the information, present the best loan alternatives, and help the applicant to choose the best option for his/her situation. The loan specialist obtains other information, such as credit score and history, from the credit bureaus, and independently verifies the information in the application.	Observed data show that this step usually takes 2–4 days (16–32 h), but 20% of the time, due to suspended loans, the step takes between 4–6 days (32–48 h). This situation is best described by a custom distribution with 80% of the values generated form the 16–32 h range, and 20% of the values generated from the 32–48 h range.
Loan underwriting	Here the application is sent to an underwriter for review. The underwriter either approves it outright, approves it with conditions to be met, or declines it.	Expert opinion indicates that the loan will receive preapproval in no more than 8 h but no less than 1 h. All values within 1–8 h have the same likelihood of occurrence. This is described by a uniform distribution.
Loan closing	Closing of the loan includes locking in of the interest rate, preparation of the final documentation, and arranging for depositing the funds.	You have historical data (100 sample values) for this step. Using distribution fitting, you determine that the data are normally distributed with a mean of 16 h and a standard deviation of 4 h.
Loan disbursement	Moving the funds to the applicant's bank usually takes 2 days (16 h).	Disbursement time is normally distributed with a standard deviation of 4 h.

	A	B	C	D	E	F	G	H	I
9		Performance Target:			96				
10									
11		Process	Simulated						
12		Step	Cycle Time			Assumption Parameters			
13									
14		Step 1:	1		1	0.25		lognormal (mean, st dev)	
15									
16		Step 2:	24		8	24	40	triangular (min, likely max)	
17									
18		Step 3:	29		16	32	80%	custom (two conditions)	
19					32	48	20%		
20		Step 4:	5		1	8		uniform (min, max)	
21									
22		Step 5:	16		16	4		normal (mean, st dev)	
23									
24		Step 6:	16		16	4		normal (mean, st dev)	
25									
26		Cycle Time	91						
27									

Figure 23.24 Figure showing simple Excel model summing the process steps.

The Black Belt created a simple Excel model that sums the process steps (Fig. 23.24). Using Crystal Ball, she made the appropriate assumptions described in Table 23.4 for each of the six process steps. The process cycle time formula sums these steps and represents the model forecast. The target and the upper specification limit (USL) for this forecast were both set at 96 h.

After a Monte Carlo simulation of 1000 trials—the equivalent of 1000 virtually processed loans—the Black Belt determined that the current baseline performance was as poor as the loan specialists had originally described. The cycle time forecast chart (Fig. 23.25) indicated that the process mean was 89 h with a standard deviation of 12 h, a C_{pk} of 0.19, and a sigma level of .61. The process was nonnormal,

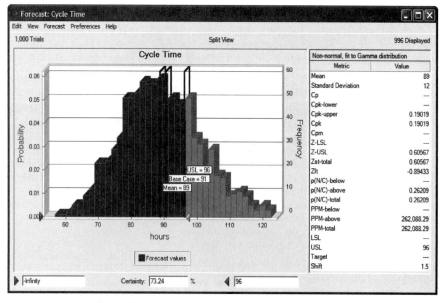

Figure 23.25 Cycle time forecast of simulated loan process.

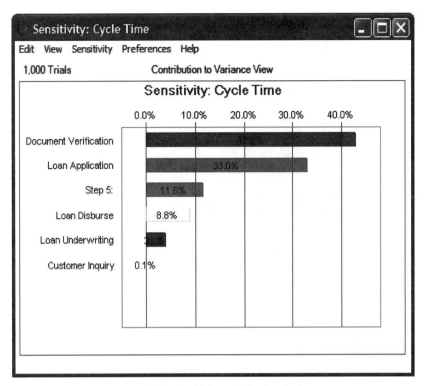

Figure 23.26   Sensitivity analysis of cycle time.

slightly skewed to the right (resulting in longer processing hours), and fit to a gamma distribution. The probability of achieving the 96-h USL and target was roughly 73%, or 732 of the 1000 simulation trials.

Using sensitivity analysis, the Black Belt learned that step 3, document verification (see Table 23.4), was strongly and positively correlated with the cycle time forecast, contributing nearly 43% of the variation (Fig. 23.26). Given this result, the team returned to the loan specialists to break down the substeps of this task and investigate the root causes for the associated delays. By automating several of the subtasks, the team calculated that they could reduce the variation in the step 3 assumption so that 95% of loans would be verified in 8–16 h and 5% would be verified in 16–24 hours. The team knew that this one change would make substantially improve the capability of the process, but implementing the change would take several weeks.

Using the simulation model of the process, the Black Belt could predict the effects of the proposed solution prior to actual implementation. The improved solution (Fig. 23.27) passed the normality test (p value of .223) and displayed a mean of 74 h, standard deviation of 9 h, a C_{pk} of 0.78, and a sigma (z) level of 2.34. In this proposed state, just 0.83% of loans exceeded the USL, with a maximum loan process time of 100 h. With these data in hand and with the approval of the process owner, the Black Belt and her team worked with the loan specialists to implement the changes required to achieve the improved solution.

As demonstrated in Example 23.4, simulation can help identify and reduce the causes of variation, and answer critical "what if" questions. A more difficult—and pertinent—question is "What's best?" This is often where using simulation alone fails to provide the best solution. The *combination* of Monte Carlo simulation and optimization, referred to as *stochastic optimization*, helps modelers find the proper settings of controllable input variable values that result in the best solution for the forecast consistent with the inherent variation of the system or process.

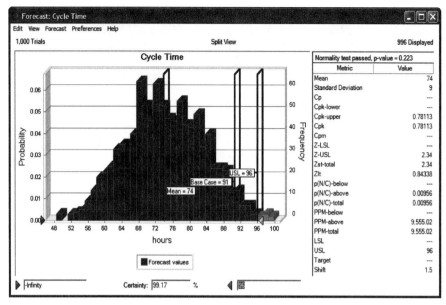

Figure 23.27 Improved solution of loan process cycle time.

In Crystal Ball, the controllable inputs, called "decision variables," are defined by upper- and lower-range bounds, value type (discrete or continuous), and increment. The classic application of stochastic optimization is in defining robust designs that optimize the tradeoff between the highest possible material quality and the lowest possible material cost. Other examples include production scheduling, project/strategy selection and prioritization, workforce queuing, and inventory optimization. Crystal Ball utilizes OptQuest® for optimization. OptQuest is a computer software system that allows users to automatically search for solutions to complex systems. OptQuest enhances Crystal Ball by identifying optimal solutions to simulation models. A single OptQuest run is the equivalent of a Crystal Ball simulation (e.g., 1000 trials) at specific settings of the decision variables. OptQuest runs sequences of Crystal Ball simulations and uses a mixture of techniques, including scatter search and advanced "tabu" search to find the right combination of decision variables for the best possible results. The program uses adaptive and neural network technologies to learn from past optimizations to quickly achieve the desired optimum solution. Stochastic optimization is especially useful for models with several decision variables and because of many of the complex conditions cited in Tables 23.2 and 23.3.

Example 23.5: Simulation and Optimization with Design of Experiments (DOE)   *A* Six Sigma team needs to improve a process that involves manufacturing plastic injected-molded parts. These parts have length specifications (LSL 59.0 and USL 67.0), but variation in the process has resulted in incomplete filling of the mold and different part lengths. The inputs that allow control of the process are mold temperature, cycle time, and hold pressure. In the measure phase, short-term measurements (30 samples) were made of the variation of the input factors, and they were found to be normal (passed the normality test). All attempts to control the variation of those factors have failed, but the nominal (mean) value of the inputs can be controlled by operational controls on the molding equipment. Using design of experiments and Monte Carlo simulation methods, the team first predicted the variation of part length in terms of C_{pk} and then optimized the nominal factor settings for the best quality available with the current process.

TABLE 23.5   Input factors and Levels for the DOE

Input Factor	Low	High
Mold temperature (x_1)	100	200
Cycle Time (x_2)	60	140
Hold Pressure (x_3)	120	140

Solution   Using MINITAB, the team created, completed, and analyzed a full factorial 2^5 DOE with five replicates using the three input factors at the levels shown in Table 23.5.

In their analysis, the team determined that all the potential factors were significant, as was the interaction between mold temperature and cycle time. The reduced model and final analysis are shown in Figure 23.28.

The team analyzed data on factorial fit from the following MINITAB output:

Factorial Fit: Length versus Mold Temp, Cycle Time, HoldPres

Estimated Effects and Coefficients for Length (coded units)

Term	Effect	Coef	SE Code	T	P
Constant		63.5850	0.2039	311.87	0.000
MoldTemp	2.8300	1.4650	0.2039	7.19	0.000
CycleTime	3.8300	1.9150	0.2039	9.39	0.000
HoldPres	9.1300	4.5650	0.2039	22.39	0.000
MoldTemp*CycleTime	-1.1300	(0.5650	0.2039	(2.77	0.009

S = 1.28946   R-Sq = 94.88%   R-Sq(adj) = 94.30%

Analysis of Variance for Length (coded units)

Source	DF	Seq SS	Adj SS	Adj MS	F	P
Main Effects	3	1066.11	1066.11	355.369	213.73	0.000
2-Way						
Interactions	1	12.77	12.77	12.769	7.68	0.009
Residual Error	35	58.20	58.20	1.663		
Lack of Fit	3	5.75	5.75	1.916	1.17	0.337
Pure Error	32	52.45	52.45	1.639		
Total	39	1137.07				

Estimated Coefficients for Length using data in uncoded units

Term	Coef
Constant	-9.18000
MoldTemp	0.0575500
CycleTime	0.0902500
HoldPres	0.456500
MoldTemp*CycleTime	-2.82500E-04

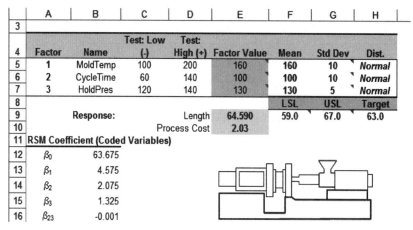

	A	B	C	D	E	F	G	H
3								
4	Factor	Name	Test: Low (-)	Test: High (+)	Factor Value	Mean	Std Dev	Dist.
5	1	MoldTemp	100	200	160	160	10	Normal
6	2	CycleTime	60	140	100	100	10	Normal
7	3	HoldPres	120	140	130	130	5	Normal
8						LSL	USL	Target
9		Response:		Length	64.590	59.0	67.0	63.0
10				Process Cost	2.03			
11	RSM Coefficient (Coded Variables)							
12	β_0	63.675						
13	β_1	4.575						
14	β_2	2.075						
15	β_3	1.325						
16	β_{23}	-0.001						

Figure 23.28 Microsoft Excel model for plastic injection-molded part lengths.

The transfer function for part length generated from the analysis was as follows:

```
Part  Length  =  0.0575500*MoldTemp  +  0.0902500*CycleTime  +
0.456500*HoldPres - 2.82500E-04*(MoldTemp*CycleTime) - 9.18000
```

Even with a successful DOE, the team still needed to determine the quality of the transfer function, the capability of the process, and the probability of noncompliance. Monte Carlo simulation provided them with a method for quickly testing the quality of the response equation. The team entered the response equation into Excel (Fig 23.29) and then used Crystal Ball to define normal distributions around the three inputs on the basis of short-term measurements collected prior to the DOE. They then used Crystal Ball to define the part length equation as a forecast and ran a simulation of 1000 trials, the virtual equivalent of 1000 plastic parts.

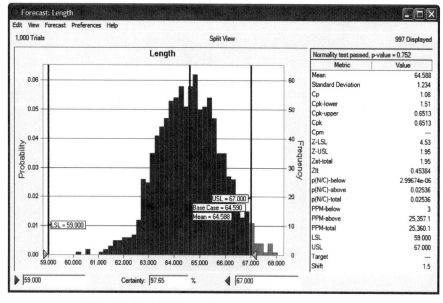

Figure 23.29 Initial simulation forecast of part length response.

Analyzing the results, they determined that the part length process passed the normality test (p value = .752) but that the defect rate was 2.35% (a sigma level of $\sim$ 2%). The process, shown in Figure 23.29, was off center to the right ($C_{pk} = 0.65$), indicating that the defective part lengths were long rather than short. Sensitivity analysis revealed that the part length variation was greatly influenced ($\sim$50%) by the variation in the mold temperature. Studying the effects of mold temperature, they found that the standard deviation could be reduced by factor of 2. A subsequent simulation using this new value for mold temperature resulted in a sigma level of 2.5, with less than 1% defects.

With this improved model the team addressed optimization, could process settings be configured so that a minimum quality goal could be reached while at the same time reducing the process cost per part? By analyzing the relationship between cost and process parameters, they found that the energy consumed by molding equipment was proportional to the product of cycle time and mold temperature ($\$ \infty$ Temp * Time) and that the labor cost to run molding equipment was proportional to cycle time ($\$ \infty$ Time). They built a process cost equation into the Excel model, where $PROCESS = K1^*Temp^*Time + K2^*Time$, and defined the cost response as a second Crystal Ball forecast. The per-part cost of the nonoptimized design was $2.03. The team assumed that they could control the nominal process settings but not the variation, so they defined the mean of each the three inputs as decision variables, where the range bounds were based on the low and high settings from the DOE (Table 23.5). In OptQuest, they chose to minimize process while improving the variation of part length to at least 4σ levels (in other words, the Z values of any forecast must have a lower bound of 4 to be considered a feasible solution).

After 800 OptQuest runs (each representing 1000 Crystal Ball simulations), the team had identified a new, robust design with a process cost of $1.16 per part (Fig 23.30) and improved quality to 4σ. For additional assurance, they reran the simulation with a greater number of trials to verify that the final design solution met performance requirements with a lower production cost than the cost of the first-cut design. For comparison purposes, the settings and results for each step of the project are listed in Table 23.6.

The continued successes of Monte Carlo simulation and stochastic optimization have made them increasingly popular tools. For project and financial managers, simulations explain cost,

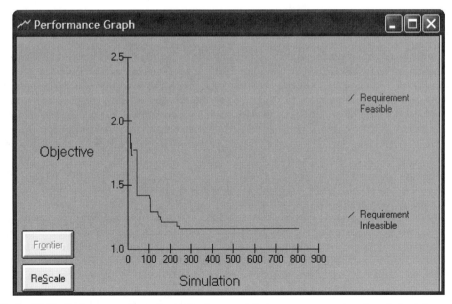

Figure 23.30 Performance chart of optimization to minimize process cost benefits of simulation and optimization.

TABLE 23.6   Iterations of Solutions from Original to Optimal States

Iteration #	Mold Temp Mean	Mold Temp SD	Cycle Time Mean	Cycle Time SD	Hold Press Mean	Hold Press SD	Sigma Level of Part Length	Process Cost ($)
1 (original settings)	160	10	100	10	130	5	1.94	2.03
2 (reduced mold temp SD)	160	5	100	10	130	5	2.53	2.03
3 (optimization)	150	5	61	10	140	5	4.01	1.16

schedule, and success uncertainty in initial stages of a project and estimate the impacts of a proposed change on customer satisfaction and profitability. The top five reasons for using simulation are as follows:

- *Avoid data paralysis*—analyze designs and processes in situations where data are partial, estimated, or nonexistent (Measure and Analyze phases).
- *Forecast performance*—predict performance of current processes and assess the associated risks and costs (measure phase).
- *Scope projects*—provide insight into which key factors have the highest impact on results and where to focus improvement efforts (define or analyze phase).
- *Improve decision quality*—communicate discoveries in a simple, highly visual manner that provides insights and helps to focus decisions (improve phase).
- *Perform optimization*—meet quality and cost objectives by determining optimal specifications and tolerances (improve phase).

In the end, any tool that harnesses the power of probability distributions to drive insights, to increase quality, and to improve customer satisfaction is one worth using. It is not enough to accurately describe variation or uncertainty; it's what you do with that knowledge that brings forth insights and solutions that drive successful projects.

Example 23.6: Aerospace R&D Project Portfolio   The business development division of a leading defense and aerospace corporation relayed the following story to Lawrence Goldman and Karl Luce of Oracle. Each year, this business division must decide which projects to fund from a possible portfolio of over 300 new marketing or R&D (research-and-development) projects. The goal is to fund the optimal portfolio of projects such that (1) returns are maximized, (2) budget and resource constraints respected, (3) risk is minimized, and (4) growth remains at or above market norms.

Prior to implementing the new project selection process several years ago, project selection decisions were often based on prior sales for the division requesting project funding. The project's opportunity, its associated risk, and correlation to the success of other projects were not an integrated part of the decisionmaking process. The division also found it difficult to collect and use metrics on past performance to improve future performance.

The division implemented a new process that included a Web-based application to capture project information, calculate its return, and pass the data to Microsoft Excel. Crystal Ball software was then used to analyze the projects and determine which projects met required returns, complied with budget constraints, and produced the necessary growth. The simulation-and-optimization model accounted for both positive and negative correlation between the projects. Although the business was already growing at a rate that exceeded the market norm, the new project selection process allowed this division to achieve a multi-million-dollar positive shift in performance in the first year, roughly equivalent to 10% of the overall budget. This money is now used to fund additional projects, further increasing the division's growth. Another important, yet less quantifiable, benefit was a substantial reduction in management resources dedicated to the project selection process, leading to overall management support for the new project selection process.   ■

23.17   DESIGN FOR SIX SIGMA PROJECT SELECTION EXAMPLE

Now let's see how this analytical selection technique works in practice. As a Six Sigma champion, you have been presented with eight possible projects for the upcoming year. For each project, your Six Sigma experts have computed (1) the expected change in revenue for each project; (2) the expected cost savings, or change in expenses; and (3) the initial investment required for each project. Using these figures, the finance manager has created a spreadsheet model to compute the gross profit and the economic value-added (EVA), or economic, profit for each project, as illustrated in Figure 23.31.

In the best of all worlds, you would run all eight projects; in reality, you have budget and labor limitations. If you select all eight projects, then you are $2,400,000 over budget and need 14 more team members than are available. Additionally, many of the variables, including project revenues, cost savings, investments, and staff requirements, are highly uncertain. Thus, the problem is to determine, on the basis of financial considerations, which Six Sigma projects should be selected to maximize the total EVA while staying within budget and labor limitations.

Complicating your job is that several of the eight projects appear to offer similar EVAs but different financial impacts. For some projects, the expected revenue is low or nonexistent, while the cost savings are high. These could be Six Sigma process improvement projects. Other projects show the reverse, with higher expected revenue and no cost savings. These could be design for Six Sigma (DFSS) projects.

Several of the projects fall between these extremes, showing both expected revenue and cost savings. How can you compare such dissimilar projects? One method is to compare each project's EVA and to select the subset of projects that leads to the highest, or optimal, EVA. This model was constructed such that, by entering a 1 or 0 for the investment decision column on the right, you can turn a project "on" or "off." If you enter a 0 for projects 3 and 8, then Excel recalculates, and you end up with a breakeven budget at an appropriate staffing level. You also see a relatively small drop in total EVA.

In a world without uncertainty, this would be an acceptable solution, but you know that many of the estimates in this model are uncertain, including the project revenues, cost savings, investments, and staff requirements. Thus, your goal is to incorporate risk analysis within the context of the project selection, and to select projects that best satisfy your constraints while optimizing your EVA. Next, you

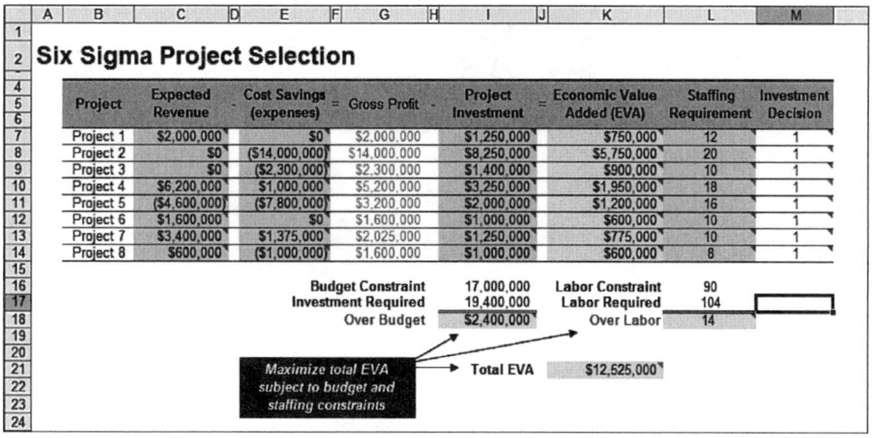

	Project	Expected Revenue	Cost Savings (expenses)	Gross Profit	Project Investment	Economic Value Added (EVA)	Staffing Requirement	Investment Decision
7	Project 1	$2,000,000	$0	$2,000,000	$1,250,000	$750,000	12	1
8	Project 2	$0	($14,000,000)	$14,000,000	$8,250,000	$5,750,000	20	1
9	Project 3	$0	($2,300,000)	$2,300,000	$1,400,000	$900,000	10	1
10	Project 4	$6,200,000	$1,000,000	$5,200,000	$3,250,000	$1,950,000	18	1
11	Project 5	($4,600,000)	($7,800,000)	$3,200,000	$2,000,000	$1,200,000	16	1
12	Project 6	$1,600,000	$0	$1,600,000	$1,000,000	$600,000	10	1
13	Project 7	$3,400,000	$1,375,000	$2,025,000	$1,250,000	$775,000	10	1
14	Project 8	$600,000	($1,000,000)	$1,600,000	$1,000,000	$600,000	8	1

Budget Constraint	17,000,000	Labor Constraint	90
Investment Required	19,400,000	Labor Required	104
Over Budget	$2,400,000	Over Labor	14

Maximize total EVA subject to budget and staffing constraints → Total EVA $12,525,000

Figure 23.31   Six Sigma project selection model for eight potential projects.

use the Crystal Ball software that includes a Monte Carlo simulation program and a stochastic optimizer; together, these software tools can optimize a project portfolio while accounting for uncertain inputs.

23.18   DEFINING SIMULATION INPUTS

As stated earlier, a Monte Carlo simulation requires that each uncertain variable in a model be represented by a fully specified distribution (normal, lognormal, poisson, and binomial). While it may initially seem to be a challenge to determine where to start, you can help yourself by asking: "Which of these inputs is just an estimate?" In the completed spreadsheet model in Figure 23.32, each value for project revenue, cost savings, project investment, and staff requirement is a single-point estimate or possibly an average value. These inputs are your uncertain factors. The next logical question is: "Which distribution do I use, and how do I determine the parameters?" The answer depends on how carefully the project sponsors did their homework. How did they determine the value for their inputs? Did they select the center of a range of values? Did they use historic data from similar projects? You must understand these details to help guide your selection of the proper probability distributions. For example, the given investment for project 1 is shown as $1,250,000. A discussion with the project sponsor reveals that this value is, indeed, the likeliest one, but that investment may be as low as $1,125,000 or as high as $1,562,500. You can use these three parameters to define a triangular distribution, as seen in Figure 23.33.

You also inquire about the revenues for project 3. The owner of this process held a strong opinion on the mean value for the revenues, $2,000,000, but when asked about uncertainty around the mean, she was less forthcoming. After discussion, they both agreed to implement the variable as a normal distribution where the range was defined using the 5th and 95th percentiles around the mean, illustrated in Figure 23.34.

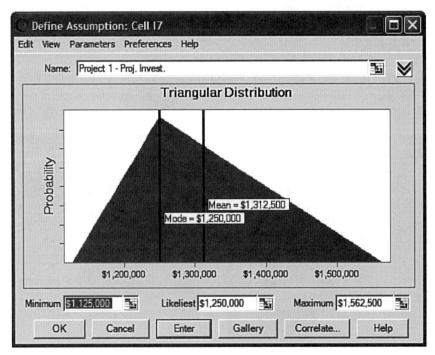

Figure 23.32   Triangular distribution representing the investment for project 3.

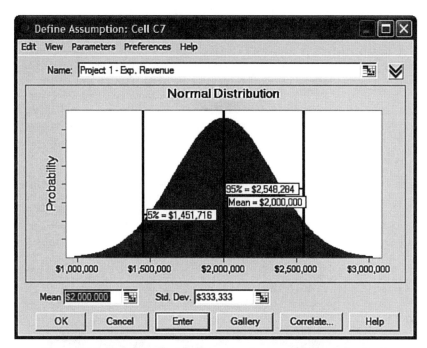

Figure 23.33   Normal distribution representing the expected revenue of project 3.

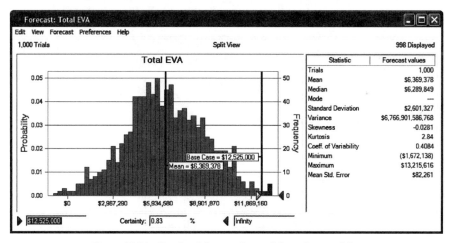

Figure 23.34   Simulated forecast for an eight-project portfolio.

You (the practitioner) should repeat this process until you have defined all of the uncertain inputs in your model. The input distribution types and parameters are not hardcoded; you can always adjust them as you simulate and analyze the model.

23.19   DEFINING OUTPUTS AND RUNNING A SIMULATION

After specifying distributions for all of the uncertain inputs in the model, you next define one or more output variables. These outputs, or forecasts, are the responses or effects of the system represented by the Y in $Y = f(x)$. The outputs in this model are the amount over budget, the number of staff over available labor, and the total EVA of the portfolio. Running a simulation is a simple task compared to creating the probabilistic model. In most simulation programs, you select a number of trials and then click on a run button. For each simulation trial, the software randomly selects a value from a defined distribution and enters that value into the spreadsheet, which then recalculates the formulas. The software then saves the forecast values from each simulation trial, and, when done with the simulation, calculates descriptive statistics for the entire group of trials. In Crystal Ball, the forecast values are shown in graphs and tables that help the user to interpret results of the simulation.

23.19.1   Analyzing a Simulation

The practitioner may first want to test run a single potential project portfolio. As shown earlier, one can deselect projects 3 and 8 to forecast an attractive total EVA of $11,025,000 that is within budget and within labor constraints. A 1000-trial simulation of this portfolio,[4] however, reveals a less rosy picture; the mean total EVA would be just less than $6,500,000, with a 1% certainty of exceeding the original estimate of $11,025,000, as shown in Figure 23.34. In the same project portfolio scenario, you are 52% certain to need more than the available staffing and 87% certain to be over budget.

With a forecast range of over $15,000,000, this simulated portfolio has become an unattractive, high-risk alternative. To determine what is driving the variation, you can use tools such as sensitivity analysis.

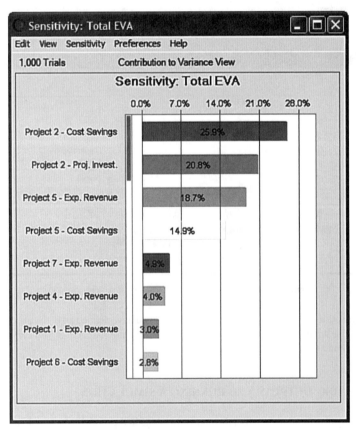

Figure 23.35   Sensitivity chart for the total EVA forecast.

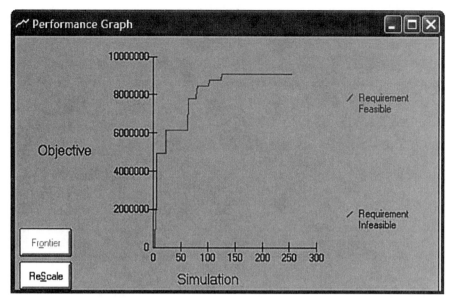

Figure 23.36 Simulation performance graph showing the highest possible total economic value added (EVA).

Crystal Ball software calculates sensitivity by computing rank correlation coefficients between every assumption and every forecast, and these normalized coefficients provide a meaningful measure of the degree to which assumptions and forecasts change together. This functionality helps modelers determine the relative influence that each distribution input has on a particular output. Sensitivity analysis ranks the assumptions according to their importance to each forecast cell.

Similar to a Pareto chart, the sensitivity chart, illustrated in Figure 23.35, displays these rankings and indicates which of the critical inputs in the analysis cause the predominance of variation in the output of interest. For the total EVA forecast, the cost savings and investment for project 2 and the expected revenue and cost savings for project 5 contributed most to the variation in the forecast. This insight can help improve the quality and accuracy of the model, help you understand what's driving

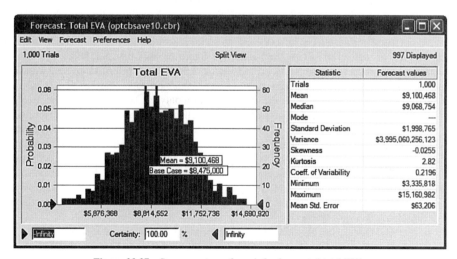

Figure 23.37 Screen capture of graph for forecasted total EVA.

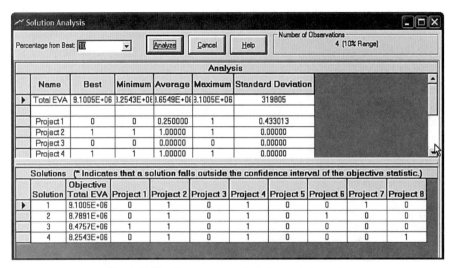

Figure 23.38   Screen capture of solution analysis graph.

cost and staff overruns, and perhaps guide you to an early elimination of certain projects from the overall portfolio.[5]

23.20   STOCHASTIC OPTIMIZATION: DISCOVERING THE BEST PORTFOLIO WITH THE LEAST RISK

The next step would be to run repeated simulations with different portfolios. But imagine how many consecutive simulations of project subsets you would have to run to eventually find the best portfolio, say, the one with minimal risk (lowest standard deviation) and the highest mean value. At most, you could have 2^n project portfolios, where $n = 8$ (the number of projects). Constraint equations would reduce this number of portfolios but still leave a daunting number of potential portfolios. With a manual approach, you would probably never find the best of all portfolios. Because simulation alone is insufficient to identify the best solution, your project portfolio will require a combination of simulation and optimization, referred to as *stochastic optimization*. Stochastic optimization helps modelers find the controllable variable settings that result in the best statistical parameters of forecast variables. For example, stochastic optimization enables the user to optimize an inventory system to minimize costs while ensuring enough inventory to meet uncertain future product demand. Other optimization examples include production scheduling, project/strategy selection and prioritization, and workforce planning. Using an optimizer in conjunction with a simulation tool, you can run consecutive simulations very quickly while having the optimizer return only the best portfolios. The optimizer has multiple search methods and intelligence features that allow it to train on good and bad results and then to learn what constitutes a good portfolio. The numerical methods used in OptQuest algorithms are collectively known as *metaheuristic optimization*.[6] In a Crystal Ball model, the controllable inputs are referred to as *decision variables*. Upper and lower bounds and value type define each discrete or continuous decision variable. A single OptQuest run comprises many Crystal Ball simulations, with perhaps 1000 trials per simulation, at different settings of the decision variables. The optimizer runs sequences of Monte Carlo simulations to find the right combination of decision variables for the best possible results.

In the portfolio model,[7] the decision variables are the investment decisions for each project. If the optimizer selects a "0," then the project is excluded from the portfolio. A "1" means that the project is included in the portfolio. Combinations of 1s and 0s define each portfolio that the optimizer simulates.

The objective of the optimization is to maximize the mean value of the forecasted total economic value added (EVA)[8] (see Figs. 23.36-23.38). Still, the highest total EVA may not be associated with a satisfactory portfolio given your budget and staffing restrictions. To meet these restrictions, you need to set requirements after each optimization to exclude portfolios that are not feasible. For a more reasonable result, you decide to include two requirements that describe an acceptable management: (1) that the over/underbudget forecast has 95% or more of the simulation trials below \$0 M (therefore accepting up to a 5% chance of requiring additional budget) and (2) that the over/understaffing forecast has 95% or more of the simulation trials below 0 personnel (thus accepting a 5% chance of requiring additional personnel). With the objective and two requirements, the optimizer will search for solutions that respect the budget and labor requirements but yield the highest possible total EVA.

23.21 CONCLUSIONS

The success of a Six Sigma implementation is highly vulnerable to factors such as changes in leadership, the quality of Belt training, the support of the company culture, and the success of projects implemented by the company. Although many project selection methods align projects to company strategy, shareholder value, and management metrics, these approaches generally lack assessment of the financial risks of individual projects. Monte Carlo simulation and stochastic optimization enable senior management a more granular and finance-based approach to portfolio analysis.

With the continuing development of faster personal computers and sophisticated software products, Monte Carlo simulation is quickly becoming a staple in the desktop analytic toolkit. When applied correctly, Monte Carlo simulation and stochastic optimization provide valuable insights not available through non-finance-based methods and can result in higher quality and more successful portfolios of Six Sigma projects. Crystal Ball, a Microsoft Excel–based suite of software tool, provides Monte Carlo simulation, optimization, and forecasting techniques that can help you predict capability, pinpoint critical-to-quality factors, and explore design alternatives. The Crystal Ball website (http://www.crystalball.com) offers Six Sigma papers, example models (including the project selection–EVA example described previously), and recorded Web seminars.

You can also download free trial versions of Crystal Ball and software tutorials. For specific inquiries, contact the software developers directly at sixsigma@crystalball.com. "Oracle" and "Crystal Ball" are registered trademarks of Oracle Corp. All rights reserved. Microsoft and Excel are registered trademarks of Microsoft Corporation in the United States and other countries. OptQuest is a registered trademark of OptTek Systems, Inc.

GLOSSARY

abscissa   Horizontal axis of a graph.

acceptance region alpha risk   Region of values for which the null hypothesis is accepted.

accuracy   Deviation of the measured or observed value from the true value.

alias   When the estimate of an effect includes the influence of one or more other effects—usually high-order interactions—the effects are said to be *aliased*.

alternate hypothesis   Probability of accepting the alternate hypothesis when the null hypothesis is true.

analysis of variance (ANOVA)   Mathematical process for separating the variability of a group of observations into assignable causes and setting up various significance tests.

assignable cause   Source of variation that is nonrandom.

assignable variations   Variations in a dataset that can be attributed to specific causes.

attribute   Characteristic that may take on only one value, for example, either 1 or 0.

attribute data   Numerical information at the nominal level. Data that represent the frequency of occurrence within some discrete category, for example, 42 solder defects in the X box.

average   Sum of values divided by the number of values.

awareness   Personal understanding of the interrelationship of quality and productivity, directing attention to the requirement for management commitment and statistical thinking to achieve *never-ending improvement*.

background variables   Variables that are not of experimental interest and are not held constant. Their effects are frequently assumed negligible or are randomized to ensure that contamination of the primary response does not occur.

balanced design   Experimental design where all cells (treatment combinations) have the same number of observations.

beta risk   Probability of accepting the null hypothesis when the alternate hypothesis is true.

blocking   Schedule for conducting treatment combinations in an experimental study such that any effects on the experimental results due to a known change in raw materials, operators, machines, etc. become concentrated in the levels of the blocking variable. The purpose of blocking is to isolate a systematic effect and prevent it from obscuring the main effects. Blocking is achieved by restricting randomization.

blocking variables   Relatively homogeneous set of conditions within which different conditions of the primary variable are compared.

capability   can be determined only after the process is in statistical control. When the process average $\pm 3\sigma$ spread of the distribution of individuals ($\bar{\bar{x}} + 3\sigma$) is contained within the specification tolerance (variables data), or when at least 99.73% of individuals are within specification (attributes data), a process is said to be capable.

causality   The principle that every change implies the operation of a cause.

causative   Effective as a cause.

cause   That which produces an effect or brings about a change.

cause–effect diagram   Graphical representation/diagram for individual or group problem solving that uses visual description of the various process elements to analyze potential sources of the variations of the process. Also called *fishbone diagram* or *Ishikawa diagram*.

centerpoints   Points at the center value of all factor ranges.

central tendency   Numerical average (mean, median, and mode). Centerline on a statistical process control chart.

centerline   Line on a statistical process control chart that represents the characteristic's central tendency.

champion   In Six Sigma, a member of senior management who is responsible for the logistics and business aspects of the organization program.

characteristic   Definable or measurable feature of a process, product, service, transaction, or variable.

classification   Differentiation of variables.

coding factor levels   Transforming the scale of measurement for a factor so that the high value becomes $+1$ and the low value becomes -1. After coding all factors in a two-level factorial experiment, the design matrix has all *orthogonal* columns.

coefficient of determination r^2 represents the adequacy of the regression model or the amount of variation explained by the regression equation.

common cause   Source of variation that affects all the individual values of the process output being studied in a control chart analysis. It appears as part of the random process variation.

common cause variation   Source of variation that is random. Inherent natural source of variation.

comparative experiments An experiment whose objective is to compare the treatments rather than to determine absolute values.

confidence interval Range in which a parameter is expected to perform on the basis of the amount of data collected and the desired confidence level.

consecutive Units of output produced in succession. A basis for selecting subgroup samples.

critical Y Effect that a series of sub-processes might have on a CTQ characteristic.

CTQ Critical-to-quality characteristic used to describe an element of a design, a characteristic of a part, a transaction, a product, or attribute of a service that is critical in the eyes of the customer.

decision variables Defined by upper and lower range bounds and value type (discrete or continuous) and increment.

Defect Failure to meet a Critical-to-Quality characteristic.

Design Set of experimental runs that allows fit to a particular model and estimation of estimate desired effects.

Design matrix Matrix description of an experiment that is useful for constructing and analyzing experiments.

detection Past-oriented strategy that attempts to identify unacceptable output after it has been produced and then separate it from the good output.

DMAVD An alternative/substitute abbreviation to DFSS (design for Six Sigma), and like DFSS, DMADV is central to Six Sigma initiatives. DMADV more specifically describes a method comprising linked steps—*define, measure, analyze, design, verify*—for ensuring that products and processes are designed at the outset to meet Six Sigma requirements.

DPMO Defects per million opportunities.

effect That which was produced by a cause.

effect of a factor Change in response produced by a change in the level of the factor.

effectiveness Focuses on results, for example; turning the right crank versus turning the crank right.

error Unexplained variation in a collection of observations. (*Note*: DOEs typically require understanding of both random error and lack of fit error.)

experiment Planned set of operations leading to a corresponding set of observations.

experimental unit One item to which a single treatment is applied in one replication of the basic experiment.

factor-independent variable Feature of the experimental conditions that may be varied from one observation to another. May be quantitative, qualitative, fixed, or random.

failure mode–effects analysis (FMEA) Process in which each potential failure mode in every subitem of an item is analyzed to determine its effect and impact on customer requirements or other aspects of the process.

individual Single unit, or single measurement of a characteristic.

interaction If the effect of one factor differs from effects at different levels of another factor, the two factors are said to *interact* or to have *interaction*.

hidden factory Those costs that cannot be broken down and traced to specific operations or are not made known to management.

histogram Graphical representation of the sample frequency distribution that describes the occurrence of grouped items.

levels of a factor Various values of a factor considered in the experiment are called *levels*.

location General concept for the typical values or central tendency of a distribution.

main effect Average effect of a factor is called the *main effect* of the factor.

mean Average of values in a group of measurements.

median   Middle value in a group of measurements when arranged from lowest to highest value. By convention, when the dataset is even, the median is the average of the two values in the center. When the dataset is odd, the median is the value in the center.

noncomparative experiment   An experiment whose objective is the determination of the properties or characteristics of a population.

nonconforming units   Units that do not conform to a specification or other inspection standard.

nonconformities   Specific occurrences of a condition that does not conform to specifications or other inspection standards. Sometimes called *discrepancies* of *defects*.

normal distribution   Continuous, symmetric, bell-shaped frequency distribution for variables data that underly the control charts for variables. When measurements have a normal distribution, about 68.26% of all individuals lie within $\pm 1\sigma$ unit of the mean. About 95.44% lie within $\pm 2\sigma$ of the mean, and about 99.73% lie within $\pm 3\sigma$ units of the mean. These percentages are the basis for control limits and control chart analysis.

operational definition   Means of clearly communicating quality expectations and performance.

optimization   Optimization examples include production scheduling, project/strategy selection and prioritization, workforce queuing, and inventory optimization.

outcome–response-dependent variable   Result of a trial with a given treatment.

Pareto diagram   Simple graphical representation (tool) for problem solving that involves ranking of all potential problem areas or sources of variation according to their contribution to cost or to total variation.

Pareto priority index (PPI)   There is an alternative approach for feasibility evaluation based on a known (PPI) calculation procedure. The feasibility estimate can be calculated as the geometric mean of values of all feasibility–related dimensions (cost, time consumption, chance of success, etc.).

Poisson distribution   Discrete probability distribution for attribute data that applies to nonconformities and underlies the c and U control charts.

precision   Measurement precision is related to its repeatability in terms of the deviation of a group of observations from a mean value.

prevention   Future-oriented strategy that improves quality and productivity by directing analysis and action toward correcting the *process* itself.

problem solving   Process of moving from symptoms to causes (special or/and common) to actions that improve performance.

process   Combination of people, equipment, materials, methods, and environment (mother nature) that produce an output—a given service, a product, a transaction, or other related element. A process can involve any aspect of a business. A key tool/method for managing processes is *statistical process control* (SPC).

process average   Location of the distribution of measured values of a particular process characteristic, usually designated as an overall average, $\bar{X}$

process control   See **statistical process control.**

project management   Application of knowledge, skills, tools, methods, and techniques to a rage of activities to meet objectives of a project.

process spread   Extent to which the distribution of individual values of the process characteristic vary. Often shown as the process average plus or minus some number of the standard deviation (e.g., in MINITAB and JMP found as a default as $\bar{\bar{X}} + 3\sigma$).

project management plan (output/input)   Formal approved document that defines how a project is to be executed, monitored, and controlled.

project management professional   Person certified as a PMP by the Project Management Institute.

quality functional deployment (QFD)   Structured method in which customer requirements are translated into the appropriate technical requirements for each stage of product development and production.

randomness   Condition in which individual values are not predictable, although they may come from a definable *distribution*.

replication   Performing the same treatment combination more than once.

response   Numerical result of a trial based on a given treatment combination.

response surface design   DOE that fully explores the process window and models the responses.

rotatability   Design that is rotatable if the variance of the predicted response at any point x depends only on the distance of x from the design centerpoint.

run   Consecutive number of points consistently increasing or decreasing, or above or below the *central line*.

run chart   Simple graphical representation of a characteristic of a process showing plotted values of some *statistic gathered* from the process.

scaling factor levels   Transforming factor levels so that the high value becomes $+1$ and low value becomes -1.

screening designs   DOE that identifies which of many factors have a significant effect on the response.

Shape   General concept for the overall pattern formed by a distribution of values.

sigma (σ)   Greek letter used to designate a *standard deviation*.

Simulation   Technique effectively applied to solve a wide variety of problems, including strategic analysis, value stream analysis, tolerance design, cost estimation, financial analysis, market forecasting, and resource allocation.

specification   Engineering requirement for "judging"/calculating acceptability of a particular product characteristic. A *specification* should *NEVER* be *confused* with a *control limit*.

spread   General concept for the extent by which values in a distribution differ from one another—*Dispersion.*

stability for control charts   Absence of special causes of variation.

standard deviation   Measure of the spread of the process output or the spread of a sampling statistic from the process of subgroup averages)—denoted by the Greek letter sigma (σ).

statistic   Value calculated from or based on sample data (subgroup average or range) used to make inferences about the process that produced the output from which the sample came.

Statistical process control (STC)   Use of statistical techniques to analyze a process to ensure that the process remains under control.

stochastic optimization   Combination of simulation and optimization.

strategic planning   Process used by organizations to envision its future and develop appropriate goals, objectives, and therefore action plans.

treatment   Treatment is a specific combination of factor levels whose effect is to be compared with other treatments.

treatment combination   Combination of the settings of several factors in a given experimental trial.

REFERENCES

1. Robert W. Galvin – Retired CEO and former chairman of Motorola Inc., founder of Galvin Electricity and recipient of the National Medal of Technology 1991 between others.
2. Mikel Harry, "Six Sigma Knowledge Design,"– Six Sigma Academy, 2001.
3. W.D. Kelton, and Law A., "Simulation Modeling & Analysis," McGraw-Hill, New York; 1991.

4. D.D. Faulder and Moseley F.L., "A Top Down Approach for Modern Portfolio Theory to Oil and Gas Property Investment", Proceedings of the 2004 Crystal Ball User Conference 2004; available online at: http://crystalball.com/cbuc/2004/papers/CBUC04-Moseley.pdf.

5. J. Rodriguez and Padua K., "An Application of Portfolio Optimization with Risk Assessment to E&P Projects," Proceedings of the 2005 Crystal Ball User Conference 2005; available online at: http://crystalball.com/cbuc/2004/papers/CBUC05-rodriguez.pdf.

6. M. Laguna, "Optimization of Complex Systems with OpQuest," 1997; available online at: http://www.crystalball.com/optquest/complexsystems.html.

7. T. de Lange, "Optimizing the Growth Portfolio of a Diversified Mining Company," 2005; Proceeding of the 2005 Crystal Ball User Conference; available online at: http://www.crystalball.com/cbuc/2005/papers/cbuc05-delange.pdf.

8. C. Hill, "portfolio Optimization Applied to Acquisition Evaluation," 2006; Proceedings of the 2006 Crystal Ball User Conference; available online at: http://www.crystalball.com/cbuc/2006/papers/cbuc06-hill.pdf.

24 Survey Methods and Sampling Techniques

24.1 OVERVIEW

We have previously introduced sample surveys and random samples. In this chapter, we explore these concepts in more detail and describe how to use a random-number table to select a sample. A sample statistic is a numerical summary measure calculated from sample data. The mean, median, mode, and standard deviation, when calculated from sample data, are called *sample statistics*. When the same numerical measures are calculated for the entire population, they are called *population parameters*. A population parameter is always a constant, whereas a sample statistic is always a random variable. Since every random variable possesses a probability distribution, each sample statistics possesses a probability distribution known as its *sampling distribution*. This chapter discusses sampling distributions for the sample mean and the sample proportion. These concepts form the foundation of inferential statistics discussed in previous chapters.

24.2 INTRODUCTION

Small companies know each of their customers personally. With success comes growth and soon both company employees and decision makers loss these insights. Much of this occurs because of increased numbers; however employees tend to have information based on recent experience while the customer related knowledge of decision makers might have been gained decades ago.

Use of carefully constructed surveys offers one way to improve knowledge and awareness of current and future customer requirements, needs, and desires.

24.3 THE SAMPLE SURVEY

For practical reasons such as time, cost, and resource constraints, we usually analyze a portion (sample) rather than the entire target population. Thus, a *sample survey* gathers information from a portion of the population. A variety of methods can be employed to conduct a sample survey; these include personal (face-to-face) interview, telephone, mail, the Internet, or direct observation. The personal interview offers high quality and high response rate. However, it is the most expensive and time-consuming technique currently in use. Telephone surveys also give high response rates, but are less expensive and require less time to administer. Nonetheless, there are problems with this technique as few people like to be called at home, and those without a telephone are excluded from the survey. A survey conducted by mail is the least expensive of the methods discussed here but has a poor the response rate because many of those queried do not return the questionnaires.

Practitioner's Guide for Statistics and Lean Six Sigma for Process Improvements. By Mikel J. Harry, Prem S. Mann, Ofelia C. de Hodgins, Christopher J. Lacke, and Richard Hulbert
Copyright © 2010 John Wiley & Sons, Inc.

Producing accurate survey results is a challenging task. The main questions facing the survey designer are defining the purpose of the survey, choosing the questions to be included in the survey, and determining appropriate future actions based on survey results.

Planning the Questionnaire

- Consider the advantages and disadvantages of using questionnaires.
- Prepare clearly written list of objectives for the study.
- Review the literature related to the survey goals.
- Determine the feasibility of administering the questionnaire to the population of interest.
- Prepare a timeline.
- Prepare the questions.

To quote Warren Mitosfsky, Director of Elections and Surveys for CBS News, any fool with 10 phones, a cell phone, and a word processor thinks he/she can conduct a poll. The preparation of a questionnaire is probably a difficult part of the survey process because the phrasing of the questions can affect the results of the entire survey.[1]

Even polls taken at the same time can produce dramatically different results depending on how the questions are phrased.

Example 24.1 *Is it a simple question?* Even a seemingly simple question can evoke complex responses. For example, "Do you own a car?" asked by Stanley Presser, a sociologist at the National Foundation Institute in Washington, DC. "That sounds like an awfully simple question"—but is it really? What does it mean?

Suppose that a wife is answering the poll, and the car is registered in her husband's name. How should she respond? What does "own" means? What if the car is leased? And what does "car" mean? Is it a van, a sport utility vehicle (SUV), or a family truck? It sounds like a simple question until we observe the many different factors involved in this simple question. In this way, the simplest question can quickly become complicated.

Assume that the question on car ownership is preceded by a series of related questions such as

1. Are you married?
2. Does your spouse drive an automotive vehicle?
3. Is it a regular car, a van, a SUV, or other vehicle?
4. Is it a lease, or does your spouse own the vehicle?
5. Now, how about your own car, do you own a car?

These above questions clarify the intended meaning of car ownership.

24.4 THE SURVEY SYSTEM

Knowing what the client wants is the key factor to success in any business. Many in the media, government agencies, business, and politics need to know what others think or feel about their performance, products, or services. Why does a carefully constructed marketing campaign fail? Why does a new product do well in one country and poorly in another? What can information in the company database tell us about our customers? How can an organization ensure that customers acquired through a merger will remain loyal to the merged organization?

Never before has so much attention been focused on understanding customers. It is no longer adequate to deliver what you think is best. Today's approach is to listen to the customer. In the world of Six Sigma, this is known as the *voice of customer* (VOC) (see Fig. 24.1).

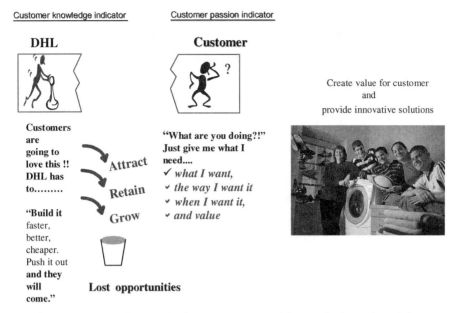

Figure 24.1 Figure showing what the customer wants and then creating innovative solutions.

There are many ways to learn what customers desire; one of them is a customer survey. This section offers suggestions and lists steps required to prepare a survey:

1. Clearly define project goals—what do you need to know about the customer [voice of customer (VOC)]?
2. Determine population—whom should be surveyed?
3. Select sample of respondents (identify the accessible population); avoid using samples of convenience.
4. Systematic sampling is a common sampling method.
5. Consider using random cluster sampling when each member of the population belongs to a subgroup.
6. Consider the need for precise results when choosing sample size and confidence interval.
7. Select survey method.
8. Create the questionnaire—what should be asked?
9. Pilot testing—test questions in a controlled environment.
10. Conduct interviews and collect data—ask the questions.
11. Analyze the data.
12. Prepare statistical tables and figures.
13. Consider using the mean to measure the centrality for equal-interval data.
14. If the median has been selected to measure centrality, use the interquartile range (see previous chapters) as the measure of variability.
15. Remember that the standard deviation (see previous chapters) has a special relationship to the normal curve that helps in its interpretation.
16. For moderately asymmetric distributions, the mode, median, and mean satisfy the formula $\text{mode} = 3 \times \text{median} - 2 \times \text{mean}$.
17. Estimate error margins.
18. Report results.

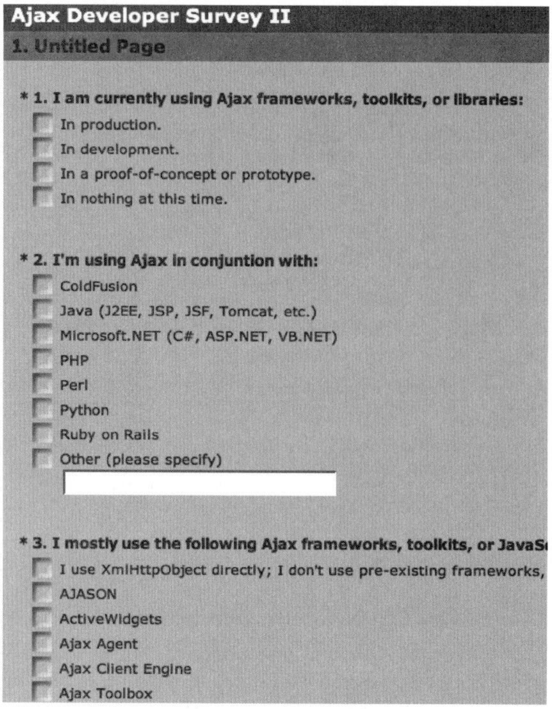

Ajax Developer Survey II

1. Untitled Page

*** 1. I am currently using Ajax frameworks, toolkits, or libraries:**

- In production.
- In development.
- In a proof-of-concept or prototype.
- In nothing at this time.

*** 2. I'm using Ajax in conjuntion with:**

- ColdFusion
- Java (J2EE, JSP, JSF, Tomcat, etc.)
- Microsoft.NET (C#, ASP.NET, VB.NET)
- PHP
- Perl
- Python
- Ruby on Rails
- Other (please specify)

*** 3. I mostly use the following Ajax frameworks, toolkits, or JavaS**

- I use XmlHttpObject directly; I don't use pre-existing frameworks,
- AJASON
- ActiveWidgets
- Ajax Agent
- Ajax Client Engine
- Ajax Toolbox

Figure 24.2 Typical survey questions.

Defining the goal of the survey determines whom should be surveyed and what questions to ask. If the goals are not clear, the results will be unclear. Example goals include potential market for a new book, rating of services, customer satisfaction levels, customer loyalty, television viewer demographics, and prediction of political contests. Remember that even the simplest questions can yield complex answers. If the goal is specific, it is easier to design the survey questionnaire (see example questionnaire in Fig. 24.2).

24.5 CLEAR GOALS

When the goals of a study can be expressed in a few clear and concise sentences, the design of the questionnaire becomes considerably easier. Questions are developed to directly address the goals of the study. One of the best ways to clarify your study goals is to decide how you intend to use the information. It is important to do this before you design the study. This sounds obvious, but many researchers neglect this task. Why do research if the results will not be used?

Commit the study goals to writing. Whenever you are unsure of a question, refer to the study goal, and a solution will become clear. Ask only questions that directly address the study goals. Avoid asking questions only because it would be interesting to know the answers. An example survey form based on clear goals is shown in Figure 24.3. Here, the objective is to understand how customers feel about the quality of the dining room service at a private golf club restaurant in Cornelius, NC.

24.6 TARGET POPULATION AND SAMPLE SIZE

There are two key components for determining whom should be interviewed: (1) we must identify the correct target population, and (2) we must choose an appropriate sample size. In previous chapters,

We encourage you to complete this card and leave it with the store manager, drop it in the comment card box on the wall, or mail it to our regional office. Our management routinely reviews this feedback and uses it to improve our operations. Thank you.

Quality	**Excellent**	**Good**	**Fair**	**Poor**
Quality of food	☐	☐	☐	☐
Value for the money	☐	☐	☐	☐

Did you receive exactly what you ordered? ____ Yes ____No

What did you order? _____

_____

Service				
Speed of service	☐	☐	☐	☐
Courtesy of employees	☐	☐	☐	☐
Employee appearance	☐	☐	☐	☐

Was your order taken promptly? ____Yes ____No

Cleanliness				
Inside store	☐	☐	☐	☐
Outside store	☐	☐	☐	☐
Restroom	☐	☐	☐	☐

Were any cleanliness improvements needed? ____ Yes ____ No

Dining room temperature: _____ too hot   _____ too cold or _____ OK

How many miles did you travel to get here?   ☐   ☐   ☐   ☐
                                            0–3 3–6 6–10   10 or more

Comments: _____

_____

_____

_____

_____

(Please print) Mr. Mrs.	Location
Name Ms. _____	

Address _____

City _____ State _____ Zip _____

Phone _____

Date of visit _____ Time _____ AM or PM

Did you visit our _____ Dining room _____ Pickup Window

Figure 24.3   Another example of a survey questionnaire.

we learned that larger samples more precisely describe the target population. In a survey, data are collected from members of a population or sample with little control over the factors that may affect the results of the survey. For example, to determine the amount that families spent on clothes last month, we would ask each family in the survey how much they spent on clothes and record the responses.

Previously we determined the sample size required for estimation of the mean μ for the population from which the sample was drawn. However, decisions on sample size should be based on the same factors used for design of experiments such as cost (budget), time available (cycle time), and confidence level (determining the sample size for the estimation of p). Many Websites provide "a sample size calculator" that assists the practitioner's choice of an appropriate survey sample size.

In chapters 16 and 17, we described two types of error and introduced the *significance level α*. Many survey experts become excited when they discover a "significant" finding. Simply, this means that we believe that the statistic is reliable; it does not mean that the finding is important. The logic underlying hypothesis testing of surveys is that *significance* is a statistical term used in measuring confidence that survey results reflect the attitudes or behavior of the entire target population.

Sample Size. What fraction of the target population do we need to sample? Two definitions are appropriate here:

> *Census*—a census includes every member of the target population. In practice, a census is rarely taken because it is very expensive and time-consuming. Furthermore, in many cases it is impossible to identify each member of the target population.
>
> *Sample survey*—usually, a portion of the target population is selected. This portion of the population is called a *sample*. Then the required information is collected from the elements included in the sample.

Until the sample becomes a sizable fraction of the population, accuracy is determined by sample size alone. For an item scored 0/1 for "No"/"Yes," the standard deviation is given by the following equation

$$SD = [p(1-p)]^{1/2}$$

where   SD = standard deviation
  p = proportion where score is 1
  n = sample size

The standard error of estimate (SE); the standard deviation of the possible p values based on the sample estimate), is given by

$$SE = \frac{SD}{n}$$

where SE = standard error. Therefore, SE is maximum when $p = .5$; thus, the worst-case scenario occurs when 50% agree and 50% disagree.

The sample size n can then be expressed as the largest integer $\leq [0.25/(SE \times SE)]$. For SE = 0.01 (1%), a sample size of 2500 would be required. Then

> For SE = 0.02 (2%), $N = 625$
> For SE = 0.03 (3%), $N = 278$
> For SE = 0.04 (4%), $N = 156$
> For SE = 0.05 (5%), $N = 100$

(*Note*: If the sample size is a small fraction of the total population, the actual size of the population is not relevant to this discussion.)

Sample Size—Binary Data. A general formula for determining sample size is

$$n = \frac{t^2 N \, p(1-p)}{t^2 p(1-p) + \alpha^2 (N-1)}$$

where N = size of total number of cases
  n = sample size
  α = expected error
  t = value taken from t distribution corresponding to a certain confidence interval
  p = probability of an event

There are several formulas for determining the sample size to satisfy a t test. A simple one is

$$n = \frac{2(Z\alpha + Z\beta)^2 \sigma^2}{D^2}$$

This formula underestimates the required sample size, but is reasonable for large sample sizes. A more accurate formula replaces the Z values with t values, and requires iteration since the df for the t distribution depends on sample size. Another formula, which offers greater accuracy, uses a noncentral t distribution and also requires iteration. The simplest approximation replaces the first Z value in the preceding formula with the value from the Student range statistic used to derive Tukey's follow-up test. If we do not have sufficiently detailed tables of the Studentized range, we can approximate the Tukey follow-up test using a Bonferroni correction by changing the first Z value for Z_α, where k is the number of comparisons.

Neither of these solutions is exact, but either approach is "probably" close enough, especially if the resulting sample size is larger than 30.

A better rule for Lean Sigma practitioners is as follows:

1. Test some minimum, predetermined, number of subjects.
2. Stop if the p value is $< .01$ or ≥ 0.36
3. Otherwise, increase the sample size (run more subjects)

(*Note*: This discussion assumes an $\alpha = 0.05$.)

An alternative option is to ***stop*** when the number of subjects becomes sufficiently large that the effect on significance can be ignored. We may evaluate *probability proportion to size* (PPS) sampling, stratification, ratio estimation, or any other form of *model-assisted estimation* using tools such as MINITAB and JMP that protect us from choosing a very unlucky sample. Which of these tools is preferable? Cost, availability of information guides, and potential payoff influence our choice of these tools; remember that none of these have much effect if the stratification/PPS ratio estimation variable is poorly correlated with the response (Y) of interest. (*Note:* Strictly speaking, the statistical parameters of the population are constant only at the time when the survey, experiment, or investigation is made. The population likely has an underlying trend that forces these parameters to be variables, just like the sample parameters.)

24.7 INTERVIEWING METHOD

Once the decision is made on what data should be collected, then it must be determined how these data should be collected. There are several methods:

Personal Interview. This requires the interviewer to ask the questions face-to-face. Personal interviews can take place almost anywhere such as on the street, outside a movie theater, or in a shopping mall. Personal interviews enable the surveyor to obtain in-depth and comprehensive information. The major disadvantage of personal interviews is cost. These interviews are expensive because of the one-to-one nature of the interview ($50 + per home interview including travel expense and time of the interviewer). Typically, the interviewer follows a written script and records the answers verbatim. Sometimes, the questionnaire simply lists topics that the researcher wishes to discuss with an industry expert. Personal interviews (because of their expense) are generally used only when subjects are not likely to respond to other survey methods. One advantage of personal interviews is the ability to find the target population. If, for example, we wish to know audience response to a recent movie, it is easier to find individuals who have seen the film outside the movie theater in which it is playing instead of randomly surveying by telephone.

Telephone Survey. This is the quickest way to gather information from a relatively large sample (100–400 respondents). The interviewer follows a prepared script that is essentially the same as

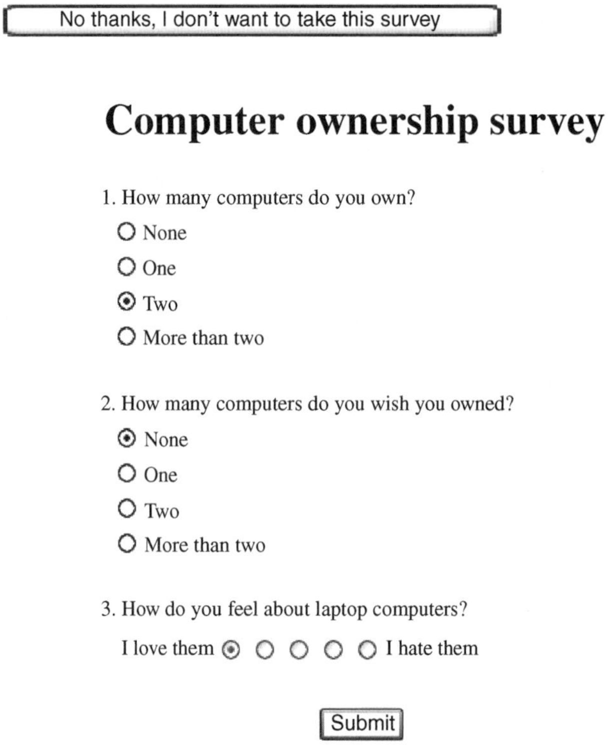

No thanks, I don't want to take this survey

Computer ownership survey

1. How many computers do you own?

○ None

○ One

◉ Two

○ More than two

2. How many computers do you wish you owned?

◉ None

○ One

○ Two

○ More than two

3. How do you feel about laptop computers?

I love them ◉ ○ ○ ○ ○ I hate them

Submit

Figure 24.4   A computer ownership questionnaire.

the written questionnaire. However, unlike a mail survey, the telephone survey allows opportunities for probing the respondent's opinions. Telephone surveys usually require less than 10 min for each interview. Typical survey costs are $4000–$6000 and can be completed over a period of a few weeks.

Mail Surveys. This is a cost-effective method for gathering information for large sample sizes, or when the sample comes from a wide geographic area. The cost is less than that of a survey conducted by telephone but requires at least twice as long to complete (8–12 weeks). Because there is no interviewer, that source of bias is eliminated. The main disadvantage of this method is the inability to probe respondents for more detailed information.

Computer Direct Interview. In this type of interview the respondent enters answers directly into the computer (see example questionnaire in Fig. 24.4). A disadvantage of this type of interview is that each respondent must have access to a computer.

Email Survey. This is economical and fast. Since most people have email, it is a better choice than a Webpage survey for many populations. Advantages of email surveys are speed, cost, and the ability to include pictures and sound. One disadvantage is that email programs have different capabilities ranging from being fully featured to supporting only plain text.

24.8 RESPONSE RATE, RESPONDENTS, AND NONRESPONDENTS

Response rate is the single most important indicator on how much confidence can be placed on survey results. It is the total number of respondents divided by the number sampled. The denominator includes the entire study population, even those who were selected but did not respond. Nonresponse reasons include refusal, lack of time, bad timing for the survey, language problems, illness, lack of interest, and being available.

A low response rate can seriously degrade the reliability of a study. Follow-ups and reminders are powerful tools for increasing the response rate. Traditionally, 10–60% of the sampled population respond without follow-up reminders. To encourage more responses, some companies have included cash, but many kept the cash, and still did not respond.

When designing a follow-up procedure, we must consider the unique characteristics of the members in the sample. The most successful follow-up approaches have used phone calls. Many studies indicate that a postcard follow-up slightly increased the response rate if a copy of the original survey was included. A meta-analysis revealed an aggregate gain of 3.5%. A postcard serves as a reminder for those who have forgotten or were not initially interested in participating in the survey.

Nonresponse Bias. Several researchers have studied the difference between respondents and nonrespondents. These studies have found that answers to questions given by late responders differed from those of early responders. These differences may have resulted from different levels of interest in the survey. One investigator surveyed an organization of volunteers and reported that active participants in the organization were likely to respond early. In one study, nonresponders tended to have lower education levels. Another study reported that demographic characteristics such as the age, education, and position in an organization were the same for respondents and nonrespondents. Another study found that most nonrespondents were single males.

In general, studies have shown that willingness to respond was a continuum ranging from fast to slow respondents; within this continuum, nonresponders were assumed to be similar to respondents at the slow end of the continuum. In fact, one study estimated the magnitude of the tendency to not respond by extrapolating the behavior of the slow responders. Another group concluded that response should not be considered to be a continuum, and that late respondents do not provide a convenient baseline for estimating the characteristics of nonrespondents.

24.9 SURVEY METHODS

The choice of survey method depends on several factors such as

Speed—email and Webpages are the quickest followed by telephone interviews. Mail surveys are the slowest.

Cost—personal interviews are the most costly method followed by telephone and then email.

Internet usage—Webpages and email surveys are fast. However, the survey results cannot be extended to reflect the entire population because the sample is essentially self-selected.

Literacy levels—less educated people rarely respond to surveys.

Sensitive questions—people are more willing to answer sensitive questions when the survey is placed on the Internet.

Video, sound, and graphics—when there is a need to obtain people's reactions to video, music, or pictures, a Webpage can display all of these media categories. Music can be played over the telephone and pictures can be placed in email surveys.

Keep It Short and Simple (the KISS Rule). A general consideration during questionnaire development is that it be both short and simple. A shock response occurs when an interviewee encounters a 20-page survey. Avoid the temptation to include too many questions. Place questions in groups and allow answers such as "don't know," "not applicable," "other," and "none." Be aware of cultural differences.

24.10   SOURCES OF INFORMATION AND DATA

Information can be obtained from many sources, including literature searches, Internet surveys, focus groups, personal interviews, telephone surveys, and mail surveys.

Literature search involves reviewing all readily available materials. These materials can include internal company information, relevant trade publications, newspapers, magazines, annual reports, company literature, online databases, and other published materials. Literature searches take several weeks and thus can be a very expensive source of information.

The *Internet* is a good way to get information during the initial stages of a research project. Often valuable, this type of information has questionable validity because it is highly subjective and might not be representative of the population. Before choosing an email or Internet survey, carefully consider how this potential bias might affect the results.

A *focus group* is frequently used to initially explore individual's ideas and attitudes and guide questionnaire design. This method is frequently used to discover customer concerns, and to test new marketing, innovative product, or advertising concepts. Suppose that a group of 6–20 people meet in a conference-room-like setting with a trained moderator. The room usually has a one-way mirror for viewing and recording responses from the participants. The moderator leads the group through meetings over several weeks at a cost of $2000–$3000. The disadvantages of this approach derive from the small size and that the group may not represent the attitudes of the target population.

24.11   ORDER OF THE QUESTIONS

Grouping similar questions simplifies the questionnaire and causes the respondent to feel more comfortable. Questions that use the same response formats or those that cover a specific topic should be grouped. Affinity diagrams (organizing ideas into common themes) can be used to create these groupings. Each question should comfortably follow the previous question, and transitions between questions should be smooth. Questionnaires that jump from one unrelated topic to another are disjointed and tend to reduce response rates.

Most investigators have found that the order of questions can affect responses. One study reported that questions in the latter half of a questionnaire were more likely to be omitted, and contained fewer extreme responses. Some researchers have suggested that asking general questions before specific ones avoids response contamination. Other researchers have reported that when specific questions were asked before general questions, respondents exhibited greater interest in the general questions.

In summary, it is not clear whether question order affects responses. A few researchers have reported that question order does not affect responses, while others have reported that it does. It is generally believed that question order effects exist in interviews but not in written surveys.

24.12   PILOT TESTING THE QUESTIONNAIRE

Test the questionnaire with a small number of interviews before conducting the entire survey. When questions are changed after the pilot test, do not combine the results from the pretest with those from the posttest as this dramatically decreases the quality of the results. Reducing measurement error through better questions is one approach to improve surveys. Clear understanding of what constitutes a good question, and how to use it is fundamental to good survey design. Therefore, it is always important to carefully design questions and use pilot testing before conducting the full survey.

24.13   BIASED SAMPLE OR RESPONSE ERROR

Response error occurs when a participant answers incorrectly. This may occur for many reasons; for instance, the respondent may have misunderstood the question, or the wording of the question may

1. Please rate your overall satisfaction
- Excellent
- Good
- Fair
- Poor
- Bad

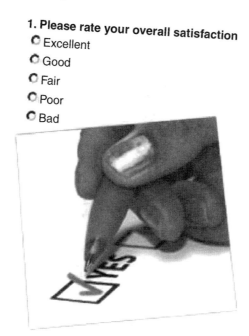

Figure 24.5   Illustration showing a voluntary five-choice response.

have led the respondent to answer incorrectly. It has been observed that participants respond differently when the wording of a question is altered. Usually this error is unintentional, but often, the interviewer's style may affect the respondent's answer. For surveys about race relations, the race of the surveyor may affect the answer.

Voluntary response (see Fig. 24.5) *error* occurs when a survey is conducted on a nonrandom sample. This occurs when a questionnaire is published in a magazine or newspaper and readers are invited to respond. Usually, only readers who have very strong opinions on the issues respond to such surveys. Surveys in which respondents are required to call a certain phone number and pay for the call often result in a self-selected sample. Consequently, the sample is neither random nor representative of the target population. To quote Larry King:[2]

> All over the board . . . the 900 telephone number is very popular these days, but viewers should be warned that in the case of political polling it has absolutely no basis in fact. Poor people in the audience can't contribute to the survey, so it's faulty to begin with . . . . So next time you see a poll based on 900 numbers, treat it as some sort of middle class amusement and then . . . forget about it!

24.14   SAMPLING—RANDOM AND NONRANDOM SAMPLES

A random sample is drawn such that each member of the population has some chance of being included in the sample. In a nonrandom sample, some members of the population may not have any chance of being selected.

Simple random sampling is the purest form of probability sampling. Each member of the population has an equal and known chance of being selected. A lottery is one way to select a simple random sample. For example, to select 10 students from a class of 55, we could write each of the names on separate pieces of paper. Then we place the pieces of paper in a container, mix them up thoroughly, and then draw a name from the container. We repeat this process 9 more times; the 10 names chosen are a simple random sample. A second selection method uses a table of random numbers. The following examples are all related to obtaining random sampling by the use of a different approach.

Example 24.2   Assume that we have a group of 600 people and need to randomly select 40 subjects from this group. To select a simple random sample, arrange the names of all 600 people in alphabetic order and assign a three-digit number, from 001 to 600, to each name. Next use the table of random numbers to select 40 subjects. Typically the random numbers in the table are recorded in blocks of five digits. To use this table, start anywhere. Start from any page, line and number. Then move in any desired direction. A rule is designed to derive a three-digit number from the list of five-digit numbers because a three-digit number is assigned to each of the 600 members of the population.

Solution   Using MINITAB

1. Calc > Random Data > Integer.
2. Enter 600 for the number of values.
3. Enter 1 for minimum.
4. Enter 600 for maximum.
5. Store in column C3.
6. Click OK.

After printing the results, cross-reference the original numbers with the MINITAB results, then select 40 from the cross-referenced 600 list.

Example 24.3   Using MINITAB to simulate the random selection of 100 companies.

1. Select Calc > Random Data > Integer.
2. Enter 100 for the number of values.
3. Enter 1 for minimum.
4. Enter 100 for maximum.
5. Click OK.

Solution   Write the names of the 100 companies on different pieces of paper and place them into a container. Mix them; draw 10 names. The results are a *random sample* of these 10 companies. However, if these 100 companies are arranged in alphabetical order and the first 10 names are selected, the result will be a *nonrandom sample* because 90 companies would have had no opportunity to be chosen.

   Note: For a random sample, each member of the population may not have the same chance of being included in the sample. Two types of non-random samples are known as: *convenience* and *judgment* sample. In a convenience sample, the most accessible members of the population are selected to quickly obtain results. In a judgment sample, the members are selected from the population based on the prior knowledge and judgment expert. Although these may be representative samples, the chances of this occurring are extremely small.

Example 24.4   A different method for selecting a random sample is to assign the same probability of selection to each member of the population. This is called a *simple random sample*.

Solution   One way to select a simple random sample is by lottery drawing. For example, if we need to select 5 students from a class of 50, we write each of the 50 names on separate pieces of paper. Then, we place all 50 names in a hat and mix them up thoroughly. Next, we draw one name randomly

TABLE 24.1   Example Data[a]

13049	85293	32747	17728	50495	34617	73707	33976	86177
86544	52703	74990	98288	61833	48803	75258	83382	79099
77295	70694	97326	35430	53881	94007	70471	66815	73042
54637	32831	59063	72353	87365	15322	33156	40331	93942
50938	12004	18585	23896	62559	44470	27701	66780	56157

[a]Commas have been omitted in these Five-digit numbers to demonstrate the "block of digits" point discussed above.

from the hat. We repeat this experiment four more times. The five drawn names constitute a simple random sample.

The second procedure uses a table of random numbers. A random process generates these numbers.

Example 24.5   Suppose that we have a group of 400 people and we need to randomly select 30 subjects from this group.

Solution   We arrange the names of all 400 people in alphabetic order and assign a three-digit number, from 001 to 400, to each person. Next, we use the table of random numbers to select 30 people. Frequently, the random numbers in these tables are recorded in blocks of five digits. To use this table, we can start anywhere. One way to do so is to close your eyes and point your finger anywhere on the table page and start at that point. From there, you can move in any direction. We need to pick three-digit numbers from the table because we have assigned three digit numbers to the 400 persons in the example population.

Let's start at the first block of the 31st row from the top of the table. Beginning with the first number, **130**49, of the selected row, we show five rows. If we use the first three digits of this block to select the first person from the population, then the first person selected will be the one we associated with the number **130**. If we move toward the right to make the next selection, the second block of "five digits" is **852**93. However, there are only 400 people in the population with assigned numbers of 001–400. Consequently we cannot use 852 to select a person. Thus we move to the next block of "five" digits without making a selection. The third block of digits is **327**47. The first three digits of the block are **327**, and thus we select the person. We continue with this process until 30 required persons are selected, giving us a simple random sample of 30 people. See Table 24.1.

Although a table of random numbers is included in many books on probability and statistics, we can easily construct a table of as many random numbers using most programming languages or an application such as MINITAB.

24.15   POPULATION DISTRIBUTION

Population distribution is the probability distribution derived from the information on all elements of a population.

Example 24.6   There are five employees working for a small company. Their annual salaries (in thousands of dollars) are listed below:

17	24	35	35	43

TABLE 24.2   Population Frequency Distribution

x	f
17	1
24	1
35	2
43	1
	$N = 5$

TABLE 24.3   Population Probability Distribution

x	$P(x)$
17	$\frac{1}{5} = 0.20$
24	$\frac{1}{5} = 0.20$
35	$\frac{2}{5} = 0.40$
43	$\frac{1}{5} = 0.20$
	Sum $= 1.0$

Let x denote the annual salary [in thousands of dollars] of an employee. Because there are only five data values, there is no need to group them; thus, we can we can create the frequency distribution of annual salaries shown in Table 24.2. Dividing each frequency class by the population size, we obtain the relative frequencies, which are the probabilities of those classes. Table 24.3 lists the probabilities of various x values in the population.

The values of the mean and standard deviation calculated from the probability distribution of Table 2.3 give the values of the population parameters, $\mu = 30.80$ and $\sigma = 9.174$.

24.16   SAMPLING DISTRIBUTION

As mentioned earlier in this chapter, the value of a population parameter is always constant. For any population dataset, there is only one value of the population mean μ. However, we would expect that different samples of the same size drawn from the same population would yield different values of the sample mean $\bar{x}$ as the value of the mean for each sample will depend on the elements included in that sample. Consequently, *the sample mean $\bar{x}$ is a random variable,* and like other random variables, the sample mean $\bar{x}$ possesses a probability distribution, called the *sampling distribution* of $\bar{x}$. Other sample statistics, such as the median, mode, and standard deviation, also possess sampling distributions.

Example 24.7   Reconsider the population of annual salaries for five employees given in Table 24.1 and list all possible samples of size 3 that can be selected without replacement from that population. The total number of possible samples is 10, as calculated by the combinatorial formula discussed in previous chapters.

$$\text{Total number of samples} = \frac{5!}{3!(5-3)} = \frac{5 \times 4 \times 3 \times 2 \times 1}{3 \times 2 \times 1 \times 2 \times 1} = 10$$

TABLE 24.4   All Possible Samples and Their Means for Sample Size 3

Sample	Salaries in the Sample	$\bar{x}$
ABC	17, 24, 35	25.33
ABD	17, 24, 35	25.33
ABE	17, 24, 43	28.00
ACD	17, 35, 35	29.00
ACE	17, 35, 43	31.67
ADE	17, 35, 43	31.67
BCD	24, 35, 35	31.33
BCE	24, 35, 43	34.00
BDE	24, 35, 43	34.00
CDE	35, 35, 43	37.67

If we assign letters A, B, C, D, and E to the salaries of the five employees so that

$$A = 17, B = 24, C = 35, D = 35, E = 43$$

then the possible samples (combinations) of three salaries each are

ABC, ABD, ABE, ACD, ACE, ADE, BCD, BCE, BDE, CDE

These 10 samples and their respective means are listed in Table 24.4. Note that the first two samples have the same three salaries. This is because the two of the employees, C and D, have the same annual salary. Therefore, the samples *ABC* and *ABD* have the same values. The mean of each sample is obtained by dividing the sum of the three salaries included in that sample by 3. For instance, the first sample mean is

$$\frac{17 + 24 + 35}{3} = 25.33$$

Note that the mean values in Table 24.4 are rounded two decimal places.

By dividing each value of $\bar{X}$ in Table 24.5 by the sum of all values, we obtain the relative frequency of each class, which can be used as the probabilities of that class. The resulting sampling distribution of $\bar{X}$ is shown in Table 24.6.

If we draw just one sample of three salaries from the population of five salaries, we may draw any of the 10 possible samples. Hence, the sample mean $\bar{X}$ can assume any value listed in Table 21.6 with the corresponding probability. For instance, the probability that the mean of a randomly drawn sample of

TABLE 24.5   Frequency Distribution of $\bar{x}$ for Sample Size 3

$\bar{X}$	f
25.33	2
28.00	1
29.00	1
31.33	1
31.67	2
34.00	2
37.67	1

TABLE 24.6   Sampling Distribution of $\bar{X}$ for Sample Size 3

$\bar{X}$	$P(\bar{X})$
25.33	$\frac{2}{10} = 0.20$
28.00	$\frac{1}{10} = 0.10$
29.00	$\frac{1}{10} = 0.10$
31.33	$\frac{1}{10} = 0.10$
31.67	$\frac{2}{10} = 0.20$
34.00	$\frac{2}{10} = 0.20$
37.67	$\frac{1}{10} = 0.10$
	$\sum P(X) = 1.0$

three salaries is 32.67 is 0.20:

$$P(\bar{X} = 31.67) = 0.20$$

24.17   SAMPLING AND NONSAMPLING ERRORS

The design of the survey sample selection, whether it involves stratification, clustering, or unequal probabilities of selection, affects the estimates of sampling error for a sample of a given size. Thus, the usual approach to describing sampling errors is to calculate what they would be for a simple random sample, followed by the calculation of the effects of deviations from a simple random sampling design. Although some sources of error in surveys are biasing and produce systematically distorted results, sampling error is a random result of sampling. When probability is used to select a sample, it is possible to calculate how much sample estimates will vary by chance because of sampling. Assuming that an infinite number of samples are drawn, the sample estimates of the means will form a normal distribution around the population value. The statistic more often used to describe sampling error is called the *standard error of a mean*. It is the standard deviation of the distribution of sample estimates of means that would be formed if an infinite number of samples of a given size are drawn. This is why it is safe to say that the larger the size of the sample and the less the variance of what is being measured, the more tightly the sample estimates will be around the true population value.

Usually, different samples selected from the same population will give different results because they contain different elements. The result obtained from any one sample will generally be different from the one obtained from the corresponding population. The difference between the value of a sample statistic obtained from a sample and the value of the corresponding population parameter obtained from the population is called the *sampling error*.

Sampling error is the difference between the value of a sample statistic and the value of the corresponding population parameter. In the case of the mean:

$$\text{Sampling error} = \bar{X} - \mu$$

assuming that the sample is random and nonsampling errors have not occurred. Sampling error captures errors caused by chance. *Nonsampling errors* are the errors that occur in the collection, recording, and tabulation of data. The most common error sources are listed below:

1. Poorly designed questionnaire
2. Use of an inadequate design
3. Recording and measurement errors
4. Nonresponse problems and related issues

Examples 24.6–24.8 describe some of the main causes of nonsampling errors. Nonsampling errors can be attributed to many sources, such as inability to obtain information about all cases in the sample, definitional difficulties, differences in the interpretation of questions, inability or unwillingness on the part of the respondents to provide correct information, inability to recall information, errors made in collection such as in recording or coding the data, errors made in processing the data, errors made in estimating values for missing data, biases resulting from the differing recall periods caused by the interviewing pattern used, and failure of all units in the universe to have some probability of being selected for the sample (undercoverage). The main reasons for the occurrence of nonsampling errors are listed in Example 24.8.

Example 24.8   If a sample is nonrandom (and hence, nonrepresentative), the sample results may differ from the census results. This example, from *US News & World Report*, describes how even a randomly selected sample can become nonrandom if some of the members included in the sample cannot be contacted.

A test poll conducted in the 1984 US presidential election found that if the poll were halted after interviewing only those subjects who could be reached on the first try, Reagan showed a 3% lead over Mondale. When the interviewers made a determined effort to reach everyone on their lists of randomly selected subjects, calling some as many as 30 times before finally reaching them, Reagan showed a 13% lead, much closer to the actual election result. As it turned out, people who were planning to vote Republican were simply less likely to be at home.[3]

Reasons for nonsampling error include the following:

1. The questions may be phrased in such a way that the members of the sample or population are confused. As a result, the answers obtained are not accurate. Finally, the respondents may intentionally give false information in response to some sensitive questions.

2. Individuals may not tell the truth about drinking habits, incomes, or opinions about minorities.

3. Asking two questions at one time (e.g., "Do you want to be rich and famous?")—the answers could be different.

4. Questions that may impose unwarranted assumptions (e.g., "Why the economy is in such a bad shape?" or "Do you think that in times like this it is a good idea to invest in the stockmarket?").

5. Question that include hidden contingencies (e.g., answers that may reflect what is to be measured by only a subset of the population—questions limited to analytical value and therefore not meaningful to the whole sample).

6. The polltaker may enter a wrong number in the records or make an error while entering the data on a computer as illustrated in Figure 24.6.

Observe that nonsampling errors can occur in both a sample survey and a census, whereas sampling errors occur only when a sample survey is conducted. Preparing the survey questionnaire carefully, and handling the data cautiously, can minimize nonsampling errors. However, it is almost impossible to avoid sampling error.

Example 24.9   Reconsider the population of five employees' salaries given in the example Tables 24.2 and 24.3. The salaries of the five employees in dollars ($) are: 17, 24, 35, 35 and 43 thousand. The population mean is

$$\mu = \frac{17 + 24 + 35 + 35 + 43}{5} = \$30.80 \text{ thousand}$$

Assume that we take a random sample of three salaries from this population that includes the salaries $16, $35, and $43 thousand. The mean for this second sample is

Figure 24.6   Figure illustrating how a polltaker may be making a mistake when entering the (wrong) number in the records.

$$\bar{x} = \frac{17 + 35 + 43}{3} = \$31.67 \text{ thousand}$$

Therefore

$$\text{Sampling error} = \bar{x} - \mu = 31.67 - 30.80 = \$0.87 \text{ thousand}$$

Thus, the mean salary estimated from the sample is \$0.87 thousand higher than the mean salary of the population. Note that this difference occurred by chance, that is, because we used a sample instead of the population. Assume that when we selected the above sample, by mistake we entered the second salary as \$37.00 thousand instead of \$35.00 thousand. As a consequence, we calculated the sample mean as:

$$\bar{x} = \frac{17 + 37 + 43}{3} = \$32.23 \text{ thousand}$$

Therefore, the difference between this sample mean and the population mean is:

$$\bar{x} - \mu = 32.33 - 30.80 = \$1.53 \text{ thousand}$$

The difference between the sample mean and the population mean *does not represent* the *sampling error*. As we calculated earlier, only \$0.87 thousand of this difference is due to sampling error; the second cause is human or nonsampling error. In other words, the remaining portion, which is equal to $1.53 - 0.87 = 0.66$ thousand dollars, represents a nonsampling error because it was caused by incorrectly recording the sample salaries. In this case we say that

- Sampling error = \$0.87 thousand
- Nonsampling error = \$0.66 thousand

Figure 24.7 Drawing showing sampling and nonsampling error of Example 21.5.

Note that in the real world, we do not know the mean of the population. Hence, we use the sample mean to estimate the population mean. Consequently, we will never know the size of the sampling error. Figure 24.7 illustrates sampling and nonsampling errors for the example given above.

GLOSSARY

activity symbol Rectangle used to designate an activity. A brief description of the activity is also included within this symbol in a process flow diagram.

affinity diagram Tool used to organize and present large amounts of data (ideas, issues, solutions) into logical categories on the basis of perception.

alpha risk Probability of accepting the alternative hypothesis (H_1), when, in reality the null hypothesis (H_0) is true

alternative hypothesis From hypothesis test—An alternative hypothesis is a claim about a population parameter that will be true if the null hypothesis is false.

attribute data Numerical information that is collected using a nominal measurement scale. Typical examples of attribute data are right/wrong, pass/fail and good/bad.

Bias Whenever convenience is used, either consciously or unconsciously, to select a sample, bias results. Examples include sampling based on friendship, appearance, cultural background, income level, and other similar approaches that underrepresent or omit parts of the population.

causality The principle that every change results from a cause.

cause That which produces an effect or brings about a change.

census Survey that includes every member of the population.

champion Member of senior management responsible for logistics and business aspects, including project selection and project tollgates, of a Six Sigma program.

characteristic Definable or measurable feature of a process, product, or variable.

centerline The line on a control chart that represents the central tendency of the characteristic distribution.

classification Allocation of variables to groups.

common cause Refers to the many causes of variation within a process that is in statitistical control.

complex level First level of the process tree. It is the last process performed to deliver the product/ service to the customer.

confidence level The probability that a random variable x lies within a defined interval.

confidence limits The two values that define the confidence level.

connections Arrows indicating the flow and sequence of the key activities in a process flow diagram.

connector symbol Circle used to indicate a break and its continuation elsewhere on the same page or another page on a process flow diagram.

continuous data Numerical information at the interval or ratio level, where subdivision of values is conceptually meaningful; elements can assume any number within an interval (e.g., 2.333 V).

critical requirements for quality, delivery, and cost The needs translated from the critical-to-satisfaction (CTS) characteristics.

critical-to-cost (CTC) characteristics   Product, service, and/or transactional characteristics that significantly influence one or more CTS in terms of cost.

critical-to-delivery (CTD) characteristic   A product, service, and/or transactional characteristic that significantly influences one or more CTS in terms of delivery.

critical-to-process (CTP) characteristic   A process parameter which significantly influences a CTQ, CTD, and/or CTC.

critical-to-quality (CTQ) characteristic   A product, service and/or transactional characteristic that significantly influences one or more CTS in terms of quality.

critical-to-satisfaction (CTS) characteristic   A characteristic required to meet customers' vital needs.

cycle time   Time required for each process step or subprocess.

data   Factual information used as a basis for reasoning, discussion, or calculation.

decision symbol   Diamond used to illustrate the point where a decision must be taken. Subsequent activities are dependent on the decision taken at this point in a process flow diagram.

delay symbol   Used to indicate that there is a delay or waiting period in a process flow diagram.

document symbol   Indicates a printed document pertinent to the process flow diagram.

effect   That which is produced by a cause.

element level   This level represents a breakdown of the processes listed at the subsystem level. Critical-to-process characteristics are identified directly below this level.

experiment   Test under defined conditions to determine an unknown effect.

factors   Independent variables.

flowline   Arrow that indicates the direction of the process and connects its elements in a process flow diagram.

independent variable   A controlled variable whose value is independent of the value of another variable.

interval scale   Measure used to express numerical information on a scale with equal distance between categories, but no absolute zero. Examples of interval scales include temperature ($°F$ and $°C$), a dial gage sitting on top of a gage block, and comparison of differences.

key process   A single step or subprocess of a process.

lower inner fence   Value in a dataset that is $1.5 \times IQR$ below the first quartile.

lower outer fence   Value in a dataset that is $3.0 \times IQR$ below the first quartile.

mean   Measure of central tendency calculated by dividing the sum of all values by the number of values in the dataset.

mean of $\bar{X}$ The mean of the sampling distribution, $\bar{X}$. μ_{bar} is equal to the population mean, μ.

measures of central tendency   Measures that describe the center of a distribution. The mean, median, and mode are three such measures.

measures of dispersion   Measures that give the spread of a distribution.

nominal scale   Measure used to classify elements into categories without considering any specific property. Examples of nominal scales include "causes" on fishbone diagrams, yes/no, and pass/fail.

nonrandom sample   In a nonrandom sample, some members of the population may not have any chance of being selected.

nonsampling errors   Errors that occur in the collection, recording, and tabulation of data.

null hypothesis   From hypothesis tests—a null hypothesis is a claim (or statement) about a population parameter that is assumed to be true until it is declared false.

ordinal scale   Measure used to rank nominal data based on a specific property. Examples of ordinal scales include relative height, Pareto charts, and customer satisfaction surveys.

outliers or extreme values   Values that are very small or very large relative to the majority of the values in a dataset.

parameter   Summary measure calculated for population data.

percentiles   Values that divide a ranked dataset into 100 equal parts.

perturbation   A nonrandom disturbance.

population   A group of similar items from which a sample is drawn. It is often referred to as the *universe*.

probability   The chance of something occurring.

probability of an event   Number of successful events divided by the total number of trials.

problem   Deviation from a specified standard.

problem solving   Process of solving problems.

project   A problem usually calling for planned action.

quartiles   Summary measures that divide a ranked dataset into four equal parts.

questionnaire   Group of questions that directly address study goals.

random   Selecting a sample so each item in the population has an equal chance of being selected; lack of predictability; without pattern.

random cause   Source of random variation.

randomness   Condition in which any individual event from a set of events has the same mathematical probability of occurrence as any other event within the specified set.

random sample   Sample chosen in such a way that each member of the population has some chance of being selected.

random variable   Variable that can assume any value from a set of possible values.

random variations   Variations in data, which result from causes that cannot be pinpointed or controlled.

range   Difference between the highest and the lowest values in a data set or values of a subgroup.

ratio   Numeric scale that has an absolute zero point and equal units of measure.

ratio scale   Measure used to express numerical information on a scale with equal difference between categories, but with an absolute zero in the range of measurement.

representative sample   Sample that accurately reflects a specific condition or set of conditions within the universe.

research   Critical and exhaustive investigation or experimentation aimed at the confirmation or revision of accepted conclusions because of newly discovered facts.

response rate   Indicator of the confidence level in the responses from a survey.

sample   Portion of the population selected for study.

sample survey   Technique for collecting information from a portion of the population.

sampling distribution of $\bar{X}$ The probability distribution of all values of $\bar{X}$ calculated from all possible samples of the same size selected from a population.

sampling error   The difference between the value of a sample statistic calculated from a random sample and the value of the corresponding population paramter. This type of error occurs as a result of chance.

scatter diagram   Graphical representation that displays the correlation or relationship between two variables.

simple random sample   Sample selected in such a way that each member of the population has the same chance of being included in the sample.

standard deviation   Statistical index of variability, which describes the spread.

statistic   Summary measure calculated for sample data.

survey   The collection of information from elements of a population or a sample.

survey interview   Face-to-face approach for conducting a survey with accurate and reliable results.

symbols   Terms that indicate the nature of each step (see Section 7.9.1 for more details).

terminal symbol   Rounded rectangle used to designate the beginning or end of a process flow. Usually identified as start, end, beginning, or stop in a process flow diagram.

variance   Measure of spread.

REFERENCES

1. R. Jaroslovsky, "What is in your mind America?" *Psychology Today* 54–59 (July/Aug. 1988).
2. "Larry King's people, news and views," *USA Today* (July 7, 1989).
3. "The numbers racket: How polls and statistics lie," *US News & World Report* (July 11, 1988).

APPENDIX A
Statistical Tables

The following tables are presented in this appendix:

Table I: Table of Binomial Probabilities
Table II: Standard Normal Distribution Table (see Figs. A.1 and A.2)
Table III: The t Distribution Table (see Fig. A.3)
Table IV: Chi-Square Distribution Table (see Fig. A.4)
Table V: The F Distribution Table (see Figs. A.5–A.12)
Table VI: Critical Values for the Mann–Whitney Test
Table VII: Critical Values for the Wilcoxon Signed-Rank Test
Table VIII: Sigma Conversion Table

Practitioner's Guide for Statistics and Lean Six Sigma for Process Improvements. By Mikel J. Harry, Prem S. Mann, Ofelia C. de Hodgins, Christopher J. Lacke, and Richard Hulbert
Copyright © 2010 John Wiley & Sons, Inc.

TABLE I   Table of Binomial Probabilities

n	x	.05	.1	.15	.2	.25	.3	.35	.4	.45	.5
2	0	.9025	.8100	.7225	.6400	.5625	.4900	.4225	.3600	.3025	.2500
	1	.0950	.1800	.2550	.3200	.3750	.4200	.4550	.4800	.4950	.5000
	2	.0025	.0100	.0225	.0400	.0625	.0900	.1225	.1600	.2025	.2500
3	0	.8574	.7290	.6141	.5120	.4219	.3430	.2746	.2160	.1664	.1250
	1	.1354	.2430	.3251	.3840	.4219	.4410	.4436	.4320	.4084	.3750
	2	.0071	.0270	.0574	.0960	.1406	.1890	.2389	.2880	.3341	.3750
	3	.0001	.0010	.0034	.0080	.0156	.0270	.0429	.0640	.0911	.1250
4	0	.8145	.6561	.5220	.4096	.3164	.2401	.1785	.1296	.0915	.0625
	1	.1715	.2916	.3685	.4096	.4219	.4116	.3845	.3456	.2995	.2500
	2	.0135	.0486	.0975	.1536	.2109	.2646	.3105	.3456	.3675	.3750
	3	.0005	.0036	.0115	.0256	.0469	.0756	.1115	.1536	.2005	.2500
	4	.0000	.0001	.0005	.0016	.0039	.0081	.0150	.0256	.0410	.0625
5	0	.7738	.5905	.4437	.3277	.2373	.1681	.1160	.0778	.0503	.0313
	1	.2036	.3281	.3915	.4096	.3955	.3602	.3124	.2592	.2059	.1563
	2	.0214	.0729	.1382	.2048	.2637	.3087	.3364	.3456	.3369	.3125
	3	.0011	.0081	.0244	.0512	.0879	.1323	.1811	.2304	.2757	.3125
	4	.0000	.0005	.0022	.0064	.0146	.0284	.0488	.0768	.1128	.1563
	5	.0000	.0000	.0001	.0003	.0010	.0024	.0053	.0102	.0185	.0313
6	0	.7351	.5314	.3771	.2621	.1780	.1176	.0754	.0467	.0277	.0156
	1	.2321	.3543	.3993	.3932	.3560	.3025	.2437	.1866	.1359	.0938
	2	.0305	.0984	.1762	.2458	.2966	.3241	.3280	.3110	.2780	.2344
	3	.0021	.0146	.0415	.0819	.1318	.1852	.2355	.2765	.3032	.3125
	4	.0001	.0012	.0055	.0154	.0330	.0595	.0951	.1382	.1861	.2344
	5	.0000	.0001	.0004	.0015	.0044	.0102	.0205	.0369	.0609	.0938
	6	.0000	.0000	.0000	.0001	.0002	.0007	.0018	.0041	.0083	.0156
7	0	.6983	.4783	.3206	.2097	.1335	.0824	.0490	.0280	.0152	.0078
	1	.2573	.3720	.3960	.3670	.3115	.2471	.1848	.1306	.0872	.0547
	2	.0406	.1240	.2097	.2753	.3115	.3177	.2985	.2613	.2140	.1641
	3	.0036	.0230	.0617	.1147	.1730	.2269	.2679	.2903	.2918	.2734
	4	.0002	.0026	.0109	.0287	.0577	.0972	.1442	.1935	.2388	.2734
	5	.0000	.0002	.0012	.0043	.0115	.0250	.0466	.0774	.1172	.1641
	6	.0000	.0000	.0001	.0004	.0013	.0036	.0084	.0172	.0320	.0547
	7	.0000	.0000	.0000	.0000	.0001	.0002	.0006	.0016	.0037	.0078

						p					
n	p	.5	.55	.6	.65	.7	.75	.8	.85	.9	.95
2	0	.2500	.2025	.1600	.1225	.0900	.0625	.0400	.0225	.0100	.0025
	1	.5000	.4950	.4800	.4550	.4200	.3750	.3200	.2550	.1800	.0950
	2	.2500	.3025	.3600	.4225	.4900	.5625	.6400	.7225	.8100	.9025
3	0	.1250	.0911	.0640	.0429	.0270	.0156	.0080	.0034	.0010	.0001
	1	.3750	.3341	.2880	.2389	.1890	.1406	.0960	.0574	.0270	.0071
	2	.3750	.4084	.4320	.4436	.4410	.4219	.3840	.3251	.2430	.1354
	3	.1250	.1664	.2160	.2746	.3430	.4219	.5120	.6141	.7290	.8574
4	0	.0625	.0410	.0256	.0150	.0081	.0039	.0016	.0005	.0001	.0000
	1	.2500	.2005	.1536	.1115	.0756	.0469	.0256	.0115	.0036	.0005
	2	.3750	.3675	.3456	.3105	.2646	.2109	.1536	.0975	.0486	.0135
	3	.2500	.2995	.3456	.3845	.4116	.4219	.4096	.3685	.2916	.1715
	4	.0625	.0915	.1296	.1785	.2401	.3164	.4096	.5220	.6561	.8145
5	0	.0313	.0185	.0102	.0053	.0024	.0010	.0003	.0001	.0000	.0000
	1	.1563	.1128	.0768	.0488	.0284	.0146	.0064	.0022	.0005	.0000
	2	.3125	.2757	.2304	.1811	.1323	.0879	.0512	.0244	.0081	.0011
	3	.3125	.3369	.3456	.3364	.3087	.2637	.2048	.1382	.0729	.0214
	4	.1563	.2059	.2592	.3124	.3602	.3955	.4096	.3915	.3281	.2036
	5	.0313	.0503	.0778	.1160	.1681	.2373	.3277	.4437	.5905	.7738
6	0	.0156	.0083	.0041	.0018	.0007	.0002	.0001	.0000	.0000	.0000
	1	.0938	.0609	.0369	.0205	.0102	.0044	.0015	.0004	.0001	.0000
	2	.2344	.1861	.1382	.0951	.0595	.0330	.0154	.0055	.0012	.0001
	3	.3125	.3032	.2765	.2355	.1852	.1318	.0819	.0415	.0146	.0021
	4	.2344	.2780	.3110	.3280	.3241	.2966	.2458	.1762	.0984	.0305
	5	.0938	.1359	.1866	.2437	.3025	.3560	.3932	.3993	.3543	.2321
	6	.0156	.0277	.0467	.0754	.1176	.1780	.2621	.3771	.5314	.7351
7	0	.0078	.0037	.0016	.0006	.0002	.0001	.0000	.0000	.0000	.0000
	1	.0547	.0320	.0172	.0084	.0036	.0013	.0004	.0001	.0000	.0000
	2	.1641	.1172	.0774	.0466	.0250	.0115	.0043	.0012	.0002	.0000
	3	.2734	.2388	.1935	.1442	.0972	.0577	.0287	.0109	.0026	.0002
	4	.2734	.2918	.2903	.2679	.2269	.1730	.1147	.0617	.0230	.0036
	5	.1641	.2140	.2613	.2985	.3177	.3115	.2753	.2097	.1240	.0406
	6	.0547	.0872	.1306	.1848	.2471	.3115	.3670	.3960	.3720	.2573
	7	.0078	.0152	.0280	.0490	.0824	.1335	.2097	.3206	.4783	.6983

							p				
n	x	**.05**	**.1**	**.15**	**.2**	**.25**	**.3**	**.35**	**.4**	**.45**	**.5**
8	**0**	.6634	.4305	.2725	.1678	.1001	.0576	.0319	.0168	.0084	.0039
	1	.2793	.3826	.3847	.3355	.2670	.1977	.1373	.0896	.0548	.0313
	2	.0515	.1488	.2376	.2936	.3115	.2965	.2587	.2090	.1569	.1094
	3	.0054	.0331	.0839	.1468	.2076	.2541	.2786	.2787	.2568	.2188
	4	.0004	.0046	.0185	.0459	.0865	.1361	.1875	.2322	.2627	.2734
	5	.0000	.0004	.0026	.0092	.0231	.0467	.0808	.1239	.1719	.2188
	6	.0000	.0000	.0002	.0011	.0038	.0100	.0217	.0413	.0703	.1094
	7	.0000	.0000	.0000	.0001	.0004	.0012	.0033	.0079	.0164	.0313
	8	.0000	.0000	.0000	.0000	.0000	.0001	.0002	.0007	.0017	.0039
9	**0**	.6302	.3874	.2316	.1342	.0751	.0404	.0207	.0101	.0046	.0020
	1	.2985	.3874	.3679	.3020	.2253	.1556	.1004	.0605	.0339	.0176
	2	.0629	.1722	.2597	.3020	.3003	.2668	.2162	.1612	.1110	.0703
	3	.0077	.0446	.1069	.1762	.2336	.2668	.2716	.2508	.2119	.1641
	4	.0006	.0074	.0283	.0661	.1168	.1715	.2194	.2508	.2600	.2461
	5	.0000	.0008	.0050	.0165	.0389	.0735	.1181	.1672	.2128	.2461
	6	.0000	.0001	.0006	.0028	.0087	.0210	.0424	.0743	.1160	.1641
	7	.0000	.0000	.0000	.0003	.0012	.0039	.0098	.0212	.0407	.0703
	8	.0000	.0000	.0000	.0000	.0001	.0004	.0013	.0035	.0083	.0176
	9	.0000	.0000	.0000	.0000	.0000	.0000	.0001	.0003	.0008	.0020
10	**0**	.5987	.3487	.1969	.1074	.0563	.0282	.0135	.0060	.0025	.0010
	1	.3151	.3874	.3474	.2684	.1877	.1211	.0725	.0403	.0207	.0098
	2	.0746	.1937	.2759	.3020	.2816	.2335	.1757	.1209	.0763	.0439
	3	.0105	.0574	.1298	.2013	.2503	.2668	.2522	.2150	.1665	.1172
	4	.0010	.0112	.0401	.0881	.1460	.2001	.2377	.2508	.2384	.2051
	5	.0001	.0015	.0085	.0264	.0584	.1029	.1536	.2007	.2340	.2461
	6	.0000	.0001	.0012	.0055	.0162	.0368	.0689	.1115	.1596	.2051
	7	.0000	.0000	.0001	.0008	.0031	.0090	.0212	.0425	.0746	.1172
	8	.0000	.0000	.0000	.0001	.0004	.0014	.0043	.0106	.0229	.0439
	9	.0000	.0000	.0000	.0000	.0000	.0001	.0005	.0016	.0042	.0098
	10	.0000	.0000	.0000	.0000	.0000	.0000	.0000	.0001	.0003	.0010
11	**0**	.5688	.3138	.1673	.0859	.0422	.0198	.0088	.0036	.0014	.0005
	1	.3293	.3835	.3248	.2362	.1549	.0932	.0518	.0266	.0125	.0054
	2	.0867	.2131	.2866	.2953	.2581	.1998	.1395	.0887	.0513	.0269
	3	.0137	.0710	.1517	.2215	.2581	.2568	.2254	.1774	.1259	.0806
	4	.0014	.0158	.0536	.1107	.1721	.2201	.2428	.2365	.2060	.1611
	5	.0001	.0025	.0132	.0388	.0803	.1321	.1830	.2207	.2360	.2256
	6	.0000	.0003	.0023	.0097	.0268	.0566	.0985	.1471	.1931	.2256
	7	.0000	.0000	.0003	.0017	.0064	.0173	.0379	.0701	.1128	.1611
	8	.0000	.0000	.0000	.0002	.0011	.0037	.0102	.0234	.0462	.0806
	9	.0000	.0000	.0000	.0000	.0001	.0005	.0018	.0052	.0126	.0269
	10	.0000	.0000	.0000	.0000	.0000	.0000	.0002	.0007	.0021	.0054
	11	.0000	.0000	.0000	.0000	.0000	.0000	.0000	.0000	.0002	.0005

						p					
n	x	.5	.55	.6	.65	.7	.75	.8	.85	.9	.95
8	0	.0039	.0017	.0007	.0002	.0001	.0000	.0000	.0000	.0000	.0000
	1	.0313	.0164	.0079	.0033	.0012	.0004	.0001	.0000	.0000	.0000
	2	.1094	.0703	.0413	.0217	.0100	.0038	.0011	.0002	.0000	.0000
	3	.2188	.1719	.1239	.0808	.0467	.0231	.0092	.0026	.0004	.0000
	4	.2734	.2627	.2322	.1875	.1361	.0865	.0459	.0185	.0046	.0004
	5	.2188	.2568	.2787	.2786	.2541	.2076	.1468	.0839	.0331	.0054
	6	.1094	.1569	.2090	.2587	.2965	.3115	.2936	.2376	.1488	.0515
	7	.0313	.0548	.0896	.1373	.1977	.2670	.3355	.3847	.3826	.2793
	8	.0039	.0084	.0168	.0319	.0576	.1001	.1678	.2725	.4305	.6634
9	0	.0020	.0008	.0003	.0001	.0000	.0000	.0000	.0000	.0000	.0000
	1	.0176	.0083	.0035	.0013	.0004	.0001	.0000	.0000	.0000	.0000
	2	.0703	.0407	.0212	.0098	.0039	.0012	.0003	.0000	.0000	.0000
	3	.1641	.1160	.0743	.0424	.0210	.0087	.0028	.0006	.0001	.0000
	4	.2461	.2128	.1672	.1181	.0735	.0389	.0165	.0050	.0008	.0000
	5	.2461	.2600	.2508	.2194	.1715	.1168	.0661	.0283	.0074	.0006
	6	.1641	.2119	.2508	.2716	.2668	.2336	.1762	.1069	.0446	.0077
	7	.0703	.1110	.1612	.2162	.2668	.3003	.3020	.2597	.1722	.0629
	8	.0176	.0339	.0605	.1004	.1556	.2253	.3020	.3679	.3874	.2985
	9	.0020	.0046	.0101	.0207	.0404	.0751	.1342	.2316	.3874	.6302
10	0	.0010	.0003	.0001	.0000	.0000	.0000	.0000	.0000	.0000	.0000
	1	.0098	.0042	.0016	.0005	.0001	.0000	.0000	.0000	.0000	.0000
	2	.0439	.0229	.0106	.0043	.0014	.0004	.0001	.0000	.0000	.0000
	3	.1172	.0746	.0425	.0212	.0090	.0031	.0008	.0001	.0000	.0000
	4	.2051	.1596	.1115	.0689	.0368	.0162	.0055	.0012	.0001	.0000
	5	.2461	.2340	.2007	.1536	.1029	.0584	.0264	.0085	.0015	.0001
	6	.2051	.2384	.2508	.2377	.2001	.1460	.0881	.0401	.0112	.0010
	7	.1172	.1665	.2150	.2522	.2668	.2503	.2013	.1298	.0574	.0105
	8	.0439	.0763	.1209	.1757	.2335	.2816	.3020	.2759	.1937	.0746
	9	.0098	.0207	.0403	.0725	.1211	.1877	.2684	.3474	.3874	.3151
	10	.0010	.0025	.0060	.0135	.0282	.0563	.1074	.1969	.3487	.5987
11	0	.0005	.0002	.0000	.0000	.0000	.0000	.0000	.0000	.0000	.0000
	1	.0054	.0021	.0007	.0002	.0000	.0000	.0000	.0000	.0000	.0000
	2	.0269	.0126	.0052	.0018	.0005	.0001	.0000	.0000	.0000	.0000
	3	.0806	.0462	.0234	.0102	.0037	.0011	.0002	.0000	.0000	.0000
	4	.1611	.1128	.0701	.0379	.0173	.0064	.0017	.0003	.0000	.0000
	5	.2256	.1931	.1471	.0985	.0566	.0268	.0097	.0023	.0003	.0000
	6	.2256	.2360	.2207	.1830	.1321	.0803	.0388	.0132	.0025	.0001
	7	.1611	.2060	.2365	.2428	.2201	.1721	.1107	.0536	.0158	.0014
	8	.0806	.1259	.1774	.2254	.2568	.2581	.2215	.1517	.0710	.0137
	9	.0269	.0513	.0887	.1395	.1998	.2581	.2953	.2866	.2131	.0867
	10	.0054	.0125	.0266	.0518	.0932	.1549	.2362	.3248	.3835	.3293
	11	.0005	.0014	.0036	.0088	.0198	.0422	.0859	.1673	.3138	.5688

							p				
n	x	.05	.1	.15	.2	.25	.3	.35	.4	.45	.5
12	0	.5404	.2824	.1422	.0687	.0317	.0138	.0057	.0022	.0008	.0002
	1	.3413	.3766	.3012	.2062	.1267	.0712	.0368	.0174	.0075	.0029
	2	.0988	.2301	.2924	.2835	.2323	.1678	.1088	.0639	.0339	.0161
	3	.0173	.0852	.1720	.2362	.2581	.2397	.1954	.1419	.0923	.0537
	4	.0021	.0213	.0683	.1329	.1936	.2311	.2367	.2128	.1700	.1208
	5	.0002	.0038	.0193	.0532	.1032	.1585	.2039	.2270	.2225	.1934
	6	.0000	.0005	.0040	.0155	.0401	.0792	.1281	.1766	.2124	.2256
	7	.0000	.0000	.0006	.0033	.0115	.0291	.0591	.1009	.1489	.1934
	8	.0000	.0000	.0001	.0005	.0024	.0078	.0199	.0420	.0762	.1208
	9	.0000	.0000	.0000	.0001	.0004	.0015	.0048	.0125	.0277	.0537
	10	.0000	.0000	.0000	.0000	.0000	.0002	.0008	.0025	.0068	.0161
	11	.0000	.0000	.0000	.0000	.0000	.0000	.0001	.0003	.0010	.0029
	12	.0000	.0000	.0000	.0000	.0000	.0000	.0000	.0000	.0001	.0002
13	0	.5133	.2542	.1209	.0550	.0238	.0097	.0037	.0013	.0004	.0001
	1	.3512	.3672	.2774	.1787	.1029	.0540	.0259	.0113	.0045	.0016
	2	.1109	.2448	.2937	.2680	.2059	.1388	.0836	.0453	.0220	.0095
	3	.0214	.0997	.1900	.2457	.2517	.2181	.1651	.1107	.0660	.0349
	4	.0028	.0277	.0838	.1535	.2097	.2337	.2222	.1845	.1350	.0873
	5	.0003	.0055	.0266	.0691	.1258	.1803	.2154	.2214	.1989	.1571
	6	.0000	.0008	.0063	.0230	.0559	.1030	.1546	.1968	.2169	.2095
	7	.0000	.0001	.0011	.0058	.0186	.0442	.0833	.1312	.1775	.2095
	8	.0000	.0000	.0001	.0011	.0047	.0142	.0336	.0656	.1089	.1571
	9	.0000	.0000	.0000	.0001	.0009	.0034	.0101	.0243	.0495	.0873
	10	.0000	.0000	.0000	.0000	.0001	.0006	.0022	.0065	.0162	.0349
	11	.0000	.0000	.0000	.0000	.0000	.0001	.0003	.0012	.0036	.0095
	12	.0000	.0000	.0000	.0000	.0000	.0000	.0000	.0001	.0005	.0016
	13	.0000	.0000	.0000	.0000	.0000	.0000	.0000	.0000	.0000	.0001
14	0	.4877	.2288	.1028	.0440	.0178	.0068	.0024	.0008	.0002	.0001
	1	.3593	.3559	.2539	.1539	.0832	.0407	.0181	.0073	.0027	.0009
	2	.1229	.2570	.2912	.2501	.1802	.1134	.0634	.0317	.0141	.0056
	3	.0259	.1142	.2056	.2501	.2402	.1943	.1366	.0845	.0462	.0222
	4	.0037	.0349	.0998	.1720	.2202	.2290	.2022	.1549	.1040	.0611
	5	.0004	.0078	.0352	.0860	.1468	.1963	.2178	.2066	.1701	.1222
	6	.0000	.0013	.0093	.0322	.0734	.1262	.1759	.2066	.2088	.1833
	7	.0000	.0002	.0019	.0092	.0280	.0618	.1082	.1574	.1952	.2095
	8	.0000	.0000	.0003	.0020	.0082	.0232	.0510	.0918	.1398	.1833
	9	.0000	.0000	.0000	.0003	.0018	.0066	.0183	.0408	.0762	.1222
	10	.0000	.0000	.0000	.0000	.0003	.0014	.0049	.0136	.0312	.0611
	11	.0000	.0000	.0000	.0000	.0000	.0002	.0010	.0033	.0093	.0222
	12	.0000	.0000	.0000	.0000	.0000	.0000	.0001	.0005	.0019	.0056
	13	.0000	.0000	.0000	.0000	.0000	.0000	.0000	.0001	.0002	.0009
	14	.0000	.0000	.0000	.0000	.0000	.0000	.0000	.0000	.0000	.0001

n	x	.5	.55	.6	.65	.7	.75	.8	.85	.9	.95
							p				
12	0	.0002	.0001	.0000	.0000	.0000	.0000	.0000	.0000	.0000	.0000
	1	.0029	.0010	.0003	.0001	.0000	.0000	.0000	.0000	.0000	.0000
	2	.0161	.0068	.0025	.0008	.0002	.0000	.0000	.0000	.0000	.0000
	3	.0537	.0277	.0125	.0048	.0015	.0004	.0001	.0000	.0000	.0000
	4	.1208	.0762	.0420	.0199	.0078	.0024	.0005	.0001	.0000	.0000
	5	.1934	.1489	.1009	.0591	.0291	.0115	.0033	.0006	.0000	.0000
	6	.2256	.2124	.1766	.1281	.0792	.0401	.0155	.0040	.0005	.0000
	7	.1934	.2225	.2270	.2039	.1585	.1032	.0532	.0193	.0038	.0002
	8	.1208	.1700	.2128	.2367	.2311	.1936	.1329	.0683	.0213	.0021
	9	.0537	.0923	.1419	.1954	.2397	.2581	.2362	.1720	.0852	.0173
	10	.0161	.0339	.0639	.1088	.1678	.2323	.2835	.2924	.2301	.0988
	11	.0029	.0075	.0174	.0368	.0712	.1267	.2062	.3012	.3766	.3413
	12	.0002	.0008	.0022	.0057	.0138	.0317	.0687	.1422	.2824	.5404
13	0	.0001	.0000	.0000	.0000	.0000	.0000	.0000	.0000	.0000	.0000
	1	.0016	.0005	.0001	.0000	.0000	.0000	.0000	.0000	.0000	.0000
	2	.0095	.0036	.0012	.0003	.0001	.0000	.0000	.0000	.0000	.0000
	3	.0349	.0162	.0065	.0022	.0006	.0001	.0000	.0000	.0000	.0000
	4	.0873	.0495	.0243	.0101	.0034	.0009	.0001	.0000	.0000	.0000
	5	.1571	.1089	.0656	.0336	.0142	.0047	.0011	.0001	.0000	.0000
	6	.2095	.1775	.1312	.0833	.0442	.0186	.0058	.0011	.0001	.0000
	7	.2095	.2169	.1968	.1546	.1030	.0559	.0230	.0063	.0008	.0000
	8	.1571	.1989	.2214	.2154	.1803	.1258	.0691	.0266	.0055	.0003
	9	.0873	.1350	.1845	.2222	.2337	.2097	.1535	.0838	.0277	.0028
	10	.0349	.0660	.1107	.1651	.2181	.2517	.2457	.1900	.0997	.0214
	11	.0095	.0220	.0453	.0836	.1388	.2059	.2680	.2937	.2448	.1109
	12	.0016	.0045	.0113	.0259	.0540	.1029	.1787	.2774	.3672	.3512
	13	.0001	.0004	.0013	.0037	.0097	.0238	.0550	.1209	.2542	.5133
14	0	.0001	.0000	.0000	.0000	.0000	.0000	.0000	.0000	.0000	.0000
	1	.0009	.0002	.0001	.0000	.0000	.0000	.0000	.0000	.0000	.0000
	2	.0056	.0019	.0005	.0001	.0000	.0000	.0000	.0000	.0000	.0000
	3	.0222	.0093	.0033	.0010	.0002	.0000	.0000	.0000	.0000	.0000
	4	.0611	.0312	.0136	.0049	.0014	.0003	.0000	.0000	.0000	.0000
	5	.1222	.0762	.0408	.0183	.0066	.0018	.0003	.0000	.0000	.0000
	6	.1833	.1398	.0918	.0510	.0232	.0082	.0020	.0003	.0000	.0000
	7	.2095	.1952	.1574	.1082	.0618	.0280	.0092	.0019	.0002	.0000
	8	.1833	.2088	.2066	.1759	.1262	.0734	.0322	.0093	.0013	.0000
	9	.1222	.1701	.2066	.2178	.1963	.1468	.0860	.0352	.0078	.0004
	10	.0611	.1040	.1549	.2022	.2290	.2202	.1720	.0998	.0349	.0037
	11	.0222	.0462	.0845	.1366	.1943	.2402	.2501	.2056	.1142	.0259
	12	.0056	.0141	.0317	.0634	.1134	.1802	.2501	.2912	.2570	.1229
	13	.0009	.0027	.0073	.0181	.0407	.0832	.1539	.2539	.3559	.3593
	14	.0001	.0002	.0008	.0024	.0068	.0178	.0440	.1028	.2288	.4877

						p					
n	*x*	.05	.1	.15	.2	.25	.3	.35	.4	.45	.5
15	0	.4633	.2059	.0874	.0352	.0134	.0047	.0016	.0005	.0001	.0000
	1	.3658	.3432	.2312	.1319	.0668	.0305	.0126	.0047	.0016	.0005
	2	.1348	.2669	.2856	.2309	.1559	.0916	.0476	.0219	.0090	.0032
	3	.0307	.1285	.2184	.2501	.2252	.1700	.1110	.0634	.0318	.0139
	4	.0049	.0428	.1156	.1876	.2252	.2186	.1792	.1268	.0780	.0417
	5	.0006	.0105	.0449	.1032	.1651	.2061	.2123	.1859	.1404	.0916
	6	.0000	.0019	.0132	.0430	.0917	.1472	.1906	.2066	.1914	.1527
	7	.0000	.0003	.0030	.0138	.0393	.0811	.1319	.1771	.2013	.1964
	8	.0000	.0000	.0005	.0035	.0131	.0348	.0710	.1181	.1647	.1964
	9	.0000	.0000	.0001	.0007	.0034	.0116	.0298	.0612	.1048	.1527
	10	.0000	.0000	.0000	.0001	.0007	.0030	.0096	.0245	.0515	.0916
	11	.0000	.0000	.0000	.0000	.0001	.0006	.0024	.0074	.0191	.0417
	12	.0000	.0000	.0000	.0000	.0000	.0001	.0004	.0016	.0052	.0139
	13	.0000	.0000	.0000	.0000	.0000	.0000	.0001	.0003	.0010	.0032
	14	.0000	.0000	.0000	.0000	.0000	.0000	.0000	.0000	.0001	.0005
	15	.0000	.0000	.0000	.0000	.0000	.0000	.0000	.0000	.0000	.0000
16	0	.4401	.1853	.0743	.0281	.0100	.0033	.0010	.0003	.0001	.0000
	1	.3706	.3294	.2097	.1126	.0535	.0228	.0087	.0030	.0009	.0002
	2	.1463	.2745	.2775	.2111	.1336	.0732	.0353	.0150	.0056	.0018
	3	.0359	.1423	.2285	.2463	.2079	.1465	.0888	.0468	.0215	.0085
	4	.0061	.0514	.1311	.2001	.2252	.2040	.1553	.1014	.0572	.0278
	5	.0008	.0137	.0555	.1201	.1802	.2099	.2008	.1623	.1123	.0667
	6	.0001	.0028	.0180	.0550	.1101	.1649	.1982	.1983	.1684	.1222
	7	.0000	.0004	.0045	.0197	.0524	.1010	.1524	.1889	.1969	.1746
	8	.0000	.0001	.0009	.0055	.0197	.0487	.0923	.1417	.1812	.1964
	9	.0000	.0000	.0001	.0012	.0058	.0185	.0442	.0840	.1318	.1746
	10	.0000	.0000	.0000	.0002	.0014	.0056	.0167	.0392	.0755	.1222
	11	.0000	.0000	.0000	.0000	.0002	.0013	.0049	.0142	.0337	.0667
	12	.0000	.0000	.0000	.0000	.0000	.0002	.0011	.0040	.0115	.0278
	13	.0000	.0000	.0000	.0000	.0000	.0000	.0002	.0008	.0029	.0085
	14	.0000	.0000	.0000	.0000	.0000	.0000	.0000	.0001	.0005	.0018
	15	.0000	.0000	.0000	.0000	.0000	.0000	.0000	.0000	.0001	.0002
	16	.0000	.0000	.0000	.0000	.0000	.0000	.0000	.0000	.0000	.0000

						p					
n	*x*	**.5**	**.55**	**.6**	**.65**	**.7**	**.75**	**.8**	**.85**	**.9**	**.95**
15	0	.0000	.0000	.0000	.0000	.0000	.0000	.0000	.0000	.0000	.0000
	1	.0005	.0001	.0000	.0000	.0000	.0000	.0000	.0000	.0000	.0000
	2	.0032	.0010	.0003	.0001	.0000	.0000	.0000	.0000	.0000	.0000
	3	.0139	.0052	.0016	.0004	.0001	.0000	.0000	.0000	.0000	.0000
	4	.0417	.0191	.0074	.0024	.0006	.0001	.0000	.0000	.0000	.0000
	5	.0916	.0515	.0245	.0096	.0030	.0007	.0001	.0000	.0000	.0000
	6	.1527	.1048	.0612	.0298	.0116	.0034	.0007	.0001	.0000	.0000
	7	.1964	.1647	.1181	.0710	.0348	.0131	.0035	.0005	.0000	.0000
	8	.1964	.2013	.1771	.1319	.0811	.0393	.0138	.0030	.0003	.0000
	9	.1527	.1914	.2066	.1906	.1472	.0917	.0430	.0132	.0019	.0000
	10	.0916	.1404	.1859	.2123	.2061	.1651	.1032	.0449	.0105	.0006
	11	.0417	.0780	.1268	.1792	.2186	.2252	.1876	.1156	.0428	.0049
	12	.0139	.0318	.0634	.1110	.1700	.2252	.2501	.2184	.1285	.0307
	13	.0032	.0090	.0219	.0476	.0916	.1559	.2309	.2856	.2669	.1348
	14	.0005	.0016	.0047	.0126	.0305	.0668	.1319	.2312	.3432	.3658
	15	.0000	.0001	.0005	.0016	.0047	.0134	.0352	.0874	.2059	.4633
16	0	.0000	.0000	.0000	.0000	.0000	.0000	.0000	.0000	.0000	.0000
	1	.0002	.0001	.0000	.0000	.0000	.0000	.0000	.0000	.0000	.0000
	2	.0018	.0005	.0001	.0000	.0000	.0000	.0000	.0000	.0000	.0000
	3	.0085	.0029	.0008	.0002	.0000	.0000	.0000	.0000	.0000	.0000
	4	.0278	.0115	.0040	.0011	.0002	.0000	.0000	.0000	.0000	.0000
	5	.0667	.0337	.0142	.0049	.0013	.0002	.0000	.0000	.0000	.0000
	6	.1222	.0755	.0392	.0167	.0056	.0014	.0002	.0000	.0000	.0000
	7	.1746	.1318	.0840	.0442	.0185	.0058	.0012	.0001	.0000	.0000
	8	.1964	.1812	.1417	.0923	.0487	.0197	.0055	.0009	.0001	.0000
	9	.1746	.1969	.1889	.1524	.1010	.0524	.0197	.0045	.0004	.0000
	10	.1222	.1684	.1983	.1982	.1649	.1101	.0550	.0180	.0028	.0001
	11	.0667	.1123	.1623	.2008	.2099	.1802	.1201	.0555	.0137	.0008
	12	.0278	.0572	.1014	.1553	.2040	.2252	.2001	.1311	.0514	.0061
	13	.0085	.0215	.0468	.0888	.1465	.2079	.2463	.2285	.1423	.0359
	14	.0018	.0056	.0150	.0353	.0732	.1336	.2111	.2775	.2745	.1463
	15	.0002	.0009	.0030	.0087	.0228	.0535	.1126	.2097	.3294	.3706
	16	.0000	.0001	.0003	.0010	.0033	.0100	.0281	.0743	.1853	.4401

							p				
n	x	.05	.1	.15	.2	.25	.3	.35	.4	.45	.5
17	0	.4181	.1668	.0631	.0225	.0075	.0023	.0007	.0002	.0000	.0000
	1	.3741	.3150	.1893	.0957	.0426	.0169	.0060	.0019	.0005	.0001
	2	.1575	.2800	.2673	.1914	.1136	.0581	.0260	.0102	.0035	.0010
	3	.0415	.1556	.2359	.2393	.1893	.1245	.0701	.0341	.0144	.0052
	4	.0076	.0605	.1457	.2093	.2209	.1868	.1320	.0796	.0411	.0182
	5	.0010	.0175	.0668	.1361	.1914	.2081	.1849	.1379	.0875	.0472
	6	.0001	.0039	.0236	.0680	.1276	.1784	.1991	.1839	.1432	.0944
	7	.0000	.0007	.0065	.0267	.0668	.1201	.1685	.1927	.1841	.1484
	8	.0000	.0001	.0014	.0084	.0279	.0644	.1134	.1606	.1883	.1855
	9	.0000	.0000	.0003	.0021	.0093	.0276	.0611	.1070	.1540	.1855
	10	.0000	.0000	.0000	.0004	.0025	.0095	.0263	.0571	.1008	.1484
	11	.0000	.0000	.0000	.0001	.0005	.0026	.0090	.0242	.0525	.0944
	12	.0000	.0000	.0000	.0000	.0001	.0006	.0024	.0081	.0215	.0472
	13	.0000	.0000	.0000	.0000	.0000	.0001	.0005	.0021	.0068	.0182
	14	.0000	.0000	.0000	.0000	.0000	.0000	.0001	.0004	.0016	.0052
	15	.0000	.0000	.0000	.0000	.0000	.0000	.0000	.0001	.0003	.0010
	16	.0000	.0000	.0000	.0000	.0000	.0000	.0000	.0000	.0000	.0001
	17	.0000	.0000	.0000	.0000	.0000	.0000	.0000	.0000	.0000	.0000
18	0	.3972	.1501	.0536	.0180	.0056	.0016	.0004	.0001	.0000	.0000
	1	.3763	.3002	.1704	.0811	.0338	.0126	.0042	.0012	.0003	.0001
	2	.1683	.2835	.2556	.1723	.0958	.0458	.0190	.0069	.0022	.0006
	3	.0473	.1680	.2406	.2297	.1704	.1046	.0547	.0246	.0095	.0031
	4	.0093	.0700	.1592	.2153	.2130	.1681	.1104	.0614	.0291	.0117
	5	.0014	.0218	.0787	.1507	.1988	.2017	.1664	.1146	.0666	.0327
	6	.0002	.0052	.0301	.0816	.1436	.1873	.1941	.1655	.1181	.0708
	7	.0000	.0010	.0091	.0350	.0820	.1376	.1792	.1892	.1657	.1214
	8	.0000	.0002	.0022	.0120	.0376	.0811	.1327	.1734	.1864	.1669
	9	.0000	.0000	.0004	.0033	.0139	.0386	.0794	.1284	.1694	.1855
	10	.0000	.0000	.0001	.0008	.0042	.0149	.0385	.0771	.1248	.1669
	11	.0000	.0000	.0000	.0001	.0010	.0046	.0151	.0374	.0742	.1214
	12	.0000	.0000	.0000	.0000	.0002	.0012	.0047	.0145	.0354	.0708
	13	.0000	.0000	.0000	.0000	.0000	.0002	.0012	.0045	.0134	.0327
	14	.0000	.0000	.0000	.0000	.0000	.0000	.0002	.0011	.0039	.0117
	15	.0000	.0000	.0000	.0000	.0000	.0000	.0000	.0002	.0009	.0031
	16	.0000	.0000	.0000	.0000	.0000	.0000	.0000	.0000	.0001	.0006
	17	.0000	.0000	.0000	.0000	.0000	.0000	.0000	.0000	.0000	.0001
	18	.0000	.0000	.0000	.0000	.0000	.0000	.0000	.0000	.0000	.0000

n	x	.5	.55	.6	.65	.7	.75	.8	.85	.9	.95
						p					
17	0	.0000	.0000	.0000	.0000	.0000	.0000	.0000	.0000	.0000	.0000
	1	.0001	.0000	.0000	.0000	.0000	.0000	.0000	.0000	.0000	.0000
	2	.0010	.0003	.0001	.0000	.0000	.0000	.0000	.0000	.0000	.0000
	3	.0052	.0016	.0004	.0001	.0000	.0000	.0000	.0000	.0000	.0000
	4	.0182	.0068	.0021	.0005	.0001	.0000	.0000	.0000	.0000	.0000
	5	.0472	.0215	.0081	.0024	.0006	.0001	.0000	.0000	.0000	.0000
	6	.0944	.0525	.0242	.0090	.0026	.0005	.0001	.0000	.0000	.0000
	7	.1484	.1008	.0571	.0263	.0095	.0025	.0004	.0000	.0000	.0000
	8	.1855	.1540	.1070	.0611	.0276	.0093	.0021	.0003	.0000	.0000
	9	.1855	.1883	.1606	.1134	.0644	.0279	.0084	.0014	.0001	.0000
	10	.1484	.1841	.1927	.1685	.1201	.0668	.0267	.0065	.0007	.0000
	11	.0944	.1432	.1839	.1991	.1784	.1276	.0680	.0236	.0039	.0001
	12	.0472	.0875	.1379	.1849	.2081	.1914	.1361	.0668	.0175	.0010
	13	.0182	.0411	.0796	.1320	.1868	.2209	.2093	.1457	.0605	.0076
	14	.0052	.0144	.0341	.0701	.1245	.1893	.2393	.2359	.1556	.0415
	15	.0010	.0035	.0102	.0260	.0581	.1136	.1914	.2673	.2800	.1575
	16	.0001	.0005	.0019	.0060	.0169	.0426	.0957	.1893	.3150	.3741
	17	.0000	.0000	.0002	.0007	.0023	.0075	.0225	.0631	.1668	.4181
18	0	.0000	.0000	.0000	.0000	.0000	.0000	.0000	.0000	.0000	.0000
	1	.0001	.0000	.0000	.0000	.0000	.0000	.0000	.0000	.0000	.0000
	2	.0006	.0001	.0000	.0000	.0000	.0000	.0000	.0000	.0000	.0000
	3	.0031	.0009	.0002	.0000	.0000	.0000	.0000	.0000	.0000	.0000
	4	.0117	.0039	.0011	.0002	.0000	.0000	.0000	.0000	.0000	.0000
	5	.0327	.0134	.0045	.0012	.0002	.0000	.0000	.0000	.0000	.0000
	6	.0708	.0354	.0145	.0047	.0012	.0002	.0000	.0000	.0000	.0000
	7	.1214	.0742	.0374	.0151	.0046	.0010	.0001	.0000	.0000	.0000
	8	.1669	.1248	.0771	.0385	.0149	.0042	.0008	.0001	.0000	.0000
	9	.1855	.1694	.1284	.0794	.0386	.0139	.0033	.0004	.0000	.0000
	10	.1669	.1864	.1734	.1327	.0811	.0376	.0120	.0022	.0002	.0000
	11	.1214	.1657	.1892	.1792	.1376	.0820	.0350	.0091	.0010	.0000
	12	.0708	.1181	.1655	.1941	.1873	.1436	.0816	.0301	.0052	.0002
	13	.0327	.0666	.1146	.1664	.2017	.1988	.1507	.0787	.0218	.0014
	14	.0117	.0291	.0614	.1104	.1681	.2130	.2153	.1592	.0700	.0093
	15	.0031	.0095	.0246	.0547	.1046	.1704	.2297	.2406	.1680	.0473
	16	.0006	.0022	.0069	.0190	.0458	.0958	.1723	.2556	.2835	.1683
	17	.0001	.0003	.0012	.0042	.0126	.0338	.0811	.1704	.3002	.3763
	18	.0000	.0000	.0001	.0004	.0016	.0056	.0180	.0536	.1501	.3972

						p					
n	x	.05	.1	.15	.2	.25	.3	.35	.4	.45	.5
19	0	.3774	.1351	.0456	.0144	.0042	.0011	.0003	.0001	.0000	.0000
	1	.3774	.2852	.1529	.0685	.0268	.0093	.0029	.0008	.0002	.0000
	2	.1787	.2852	.2428	.1540	.0803	.0358	.0138	.0046	.0013	.0003
	3	.0533	.1796	.2428	.2182	.1517	.0869	.0422	.0175	.0062	.0018
	4	.0112	.0798	.1714	.2182	.2023	.1491	.0909	.0467	.0203	.0074
	5	.0018	.0266	.0907	.1636	.2023	.1916	.1468	.0933	.0497	.0222
	6	.0002	.0069	.0374	.0955	.1574	.1916	.1844	.1451	.0949	.0518
	7	.0000	.0014	.0122	.0443	.0974	.1525	.1844	.1797	.1443	.0961
	8	.0000	.0002	.0032	.0166	.0487	.0981	.1489	.1797	.1771	.1442
	9	.0000	.0000	.0007	.0051	.0198	.0514	.0980	.1464	.1771	.1762
	10	.0000	.0000	.0001	.0013	.0066	.0220	.0528	.0976	.1449	.1762
	11	.0000	.0000	.0000	.0003	.0018	.0077	.0233	.0532	.0970	.1442
	12	.0000	.0000	.0000	.0000	.0004	.0022	.0083	.0237	.0529	.0961
	13	.0000	.0000	.0000	.0000	.0001	.0005	.0024	.0085	.0233	.0518
	14	.0000	.0000	.0000	.0000	.0000	.0001	.0006	.0024	.0082	.0222
	15	.0000	.0000	.0000	.0000	.0000	.0000	.0001	.0005	.0022	.0074
	16	.0000	.0000	.0000	.0000	.0000	.0000	.0000	.0001	.0005	.0018
	17	.0000	.0000	.0000	.0000	.0000	.0000	.0000	.0000	.0001	.0003
	18	.0000	.0000	.0000	.0000	.0000	.0000	.0000	.0000	.0000	.0000
	19	.0000	.0000	.0000	.0000	.0000	.0000	.0000	.0000	.0000	.0000
20	0	.3585	.1216	.0388	.0115	.0032	.0008	.0002	.0000	.0000	.0000
	1	.3774	.2702	.1368	.0576	.0211	.0068	.0020	.0005	.0001	.0000
	2	.1887	.2852	.2293	.1369	.0669	.0278	.0100	.0031	.0008	.0002
	3	.0596	.1901	.2428	.2054	.1339	.0716	.0323	.0123	.0040	.0011
	4	.0133	.0898	.1821	.2182	.1897	.1304	.0738	.0350	.0139	.0046
	5	.0022	.0319	.1028	.1746	.2023	.1789	.1272	.0746	.0365	.0148
	6	.0003	.0089	.0454	.1091	.1686	.1916	.1712	.1244	.0746	.0370
	7	.0000	.0020	.0160	.0545	.1124	.1643	.1844	.1659	.1221	.0739
	8	.0000	.0004	.0046	.0222	.0609	.1144	.1614	.1797	.1623	.1201
	9	.0000	.0001	.0011	.0074	.0271	.0654	.1158	.1597	.1771	.1602
	10	.0000	.0000	.0002	.0020	.0099	.0308	.0686	.1171	.1593	.1762
	11	.0000	.0000	.0000	.0005	.0030	.0120	.0336	.0710	.1185	.1602
	12	.0000	.0000	.0000	.0001	.0008	.0039	.0136	.0355	.0727	.1201
	13	.0000	.0000	.0000	.0000	.0002	.0010	.0045	.0146	.0366	.0739
	14	.0000	.0000	.0000	.0000	.0000	.0002	.0012	.0049	.0150	.0370
	15	.0000	.0000	.0000	.0000	.0000	.0000	.0003	.0013	.0049	.0148
	16	.0000	.0000	.0000	.0000	.0000	.0000	.0000	.0003	.0013	.0046
	17	.0000	.0000	.0000	.0000	.0000	.0000	.0000	.0000	.0002	.0011
	18	.0000	.0000	.0000	.0000	.0000	.0000	.0000	.0000	.0000	.0002
	19	.0000	.0000	.0000	.0000	.0000	.0000	.0000	.0000	.0000	.0000
	20	.0000	.0000	.0000	.0000	.0000	.0000	.0000	.0000	.0000	.0000

							p				
n	x	.5	.55	.6	.65	.7	.75	.8	.85	.9	.95
19	0	.0000	.0000	.0000	.0000	.0000	.0000	.0000	.0000	.0000	.0000
	1	.0000	.0000	.0000	.0000	.0000	.0000	.0000	.0000	.0000	.0000
	2	.0003	.0001	.0000	.0000	.0000	.0000	.0000	.0000	.0000	.0000
	3	.0018	.0005	.0001	.0000	.0000	.0000	.0000	.0000	.0000	.0000
	4	.0074	.0022	.0005	.0001	.0000	.0000	.0000	.0000	.0000	.0000
	5	.0222	.0082	.0024	.0006	.0001	.0000	.0000	.0000	.0000	.0000
	6	.0518	.0233	.0085	.0024	.0005	.0001	.0000	.0000	.0000	.0000
	7	.0961	.0529	.0237	.0083	.0022	.0004	.0000	.0000	.0000	.0000
	8	.1442	.0970	.0532	.0233	.0077	.0018	.0003	.0000	.0000	.0000
	9	.1762	.1449	.0976	.0528	.0220	.0066	.0013	.0001	.0000	.0000
	10	.1762	.1771	.1464	.0980	.0514	.0198	.0051	.0007	.0000	.0000
	11	.1442	.1771	.1797	.1489	.0981	.0487	.0166	.0032	.0002	.0000
	12	.0961	.1443	.1797	.1844	.1525	.0974	.0443	.0122	.0014	.0000
	13	.0518	.0949	.1451	.1844	.1916	.1574	.0955	.0374	.0069	.0002
	14	.0222	.0497	.0933	.1468	.1916	.2023	.1636	.0907	.0266	.0018
	15	.0074	.0203	.0467	.0909	.1491	.2023	.2182	.1714	.0798	.0112
	16	.0018	.0062	.0175	.0422	.0869	.1517	.2182	.2428	.1796	.0533
	17	.0003	.0013	.0046	.0138	.0358	.0803	.1540	.2428	.2852	.1787
	18	.0000	.0002	.0008	.0029	.0093	.0268	.0685	.1529	.2852	.3774
	19	.0000	.0000	.0001	.0003	.0011	.0042	.0144	.0456	.1351	.3774
20	0	.0000	.0000	.0000	.0000	.0000	.0000	.0000	.0000	.0000	.0000
	1	.0000	.0000	.0000	.0000	.0000	.0000	.0000	.0000	.0000	.0000
	2	.0002	.0000	.0000	.0000	.0000	.0000	.0000	.0000	.0000	.0000
	3	.0011	.0002	.0000	.0000	.0000	.0000	.0000	.0000	.0000	.0000
	4	.0046	.0013	.0003	.0000	.0000	.0000	.0000	.0000	.0000	.0000
	5	.0148	.0049	.0013	.0003	.0000	.0000	.0000	.0000	.0000	.0000
	6	.0370	.0150	.0049	.0012	.0002	.0000	.0000	.0000	.0000	.0000
	7	.0739	.0366	.0146	.0045	.0010	.0002	.0000	.0000	.0000	.0000
	8	.1201	.0727	.0355	.0136	.0039	.0008	.0001	.0000	.0000	.0000
	9	.1602	.1185	.0710	.0336	.0120	.0030	.0005	.0000	.0000	.0000
	10	.1762	.1593	.1171	.0686	.0308	.0099	.0020	.0002	.0000	.0000
	11	.1602	.1771	.1597	.1158	.0654	.0271	.0074	.0011	.0001	.0000
	12	.1201	.1623	.1797	.1614	.1144	.0609	.0222	.0046	.0004	.0000
	13	.0739	.1221	.1659	.1844	.1643	.1124	.0545	.0160	.0020	.0000
	14	.0370	.0746	.1244	.1712	.1916	.1686	.1091	.0454	.0089	.0003
	15	.0148	.0365	.0746	.1272	.1789	.2023	.1746	.1028	.0319	.0022
	16	.0046	.0139	.0350	.0738	.1304	.1897	.2182	.1821	.0898	.0133
	17	.0011	.0040	.0123	.0323	.0716	.1339	.2054	.2428	.1901	.0596
	18	.0002	.0008	.0031	.0100	.0278	.0669	.1369	.2293	.2852	.1887
	19	.0000	.0001	.0005	.0020	.0068	.0211	.0576	.1368	.2702	.3774
	20	.0000	.0000	.0000	.0002	.0008	.0032	.0115	.0388	.1216	.3585

TABLE II   Standard Normal Distribution Table

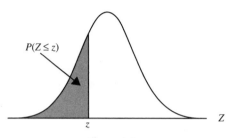

$P(Z \le z)$

Figure A.1

z	.09	.08	.07	.06	.05	.04	.03	.02	.01	.00
−6										.00000000099
−5.5										.00000001899
−5										.00000028665
−4.5										.00000339767
−4										.00003167124
−3.9	.000033	.000034	.000036	.000037	.000039	.000041	.000042	.000044	.000046	.000048
−3.8	.000050	.000052	.000054	.000057	.000059	.000062	.000064	.000067	.000069	.000072
−3.7	.000075	.000078	.000082	.000085	.000088	.000092	.000096	.000100	.000104	.000108
−3.6	.000112	.000117	.000121	.000126	.000131	.000136	.000142	.000147	.000153	.000159
−3.5	.000165	.000172	.000178	.000185	.000193	.000200	.000208	.000216	.000224	.000233
−3.4	.000242	.000251	.000260	.000270	.000280	.000291	.000302	.000313	.000325	.000337
−3.3	.000349	.000362	.000376	.000390	.000404	.000419	.000434	.000450	.000466	.000483
−3.2	.000501	.000519	.000538	.000557	.000577	.000598	.000619	.000641	.000664	.000687
−3.1	.000711	.000736	.000762	.000789	.000816	.000845	.000874	.000904	.000935	.000968
−3	.001001	.001035	.001070	.001107	.001144	.001183	.001223	.001264	.001306	.001350
−2.9	.001395	.001441	.001489	.001538	.001589	.001641	.001695	.001750	.001807	.001866
−2.8	.001926	.001988	.002052	.002118	.002186	.002256	.002327	.002401	.002477	.002555
−2.7	.002635	.002718	.002803	.002890	.002980	.003072	.003167	.003264	.003364	.003467
−2.6	.003573	.003681	.003793	.003907	.004025	.004145	.004269	.004396	.004527	.004661
−2.5	.004799	.004940	.005085	.005234	.005386	.005543	.005703	.005868	.006037	.006210
−2.4	.006387	.006569	.006756	.006947	.007143	.007344	.007549	.007760	.007976	.008198
−2.3	.008424	.008656	.008894	.009137	.009387	.009642	.009903	.010170	.010444	.010724
−2.2	.011011	.011304	.011604	.011911	.012224	.012545	.012874	.013209	.013553	.013903
−2.1	.014262	.014629	.015003	.015386	.015778	.016177	.016586	.017003	.017429	.017864
−2	.018309	.018763	.019226	.019699	.020182	.020675	.021178	.021692	.022216	.022750
−1.9	.023295	.023852	.024419	.024998	.025588	.026190	.026803	.027429	.028067	.028717
−1.8	.029379	.030054	.030742	.031443	.032157	.032884	.033625	.034380	.035148	.035930
−1.7	.036727	.037538	.038364	.039204	.040059	.040930	.041815	.042716	.043633	.044565
−1.6	.045514	.046479	.047460	.048457	.049471	.050503	.051551	.052616	.053699	.054799
−1.5	.055917	.057053	.058208	.059380	.060571	.061780	.063008	.064255	.065522	.066807
−1.4	.068112	.069437	.070781	.072145	.073529	.074934	.076359	.077804	.079270	.080757
−1.3	.082264	.083793	.085343	.086915	.088508	.090123	.091759	.093418	.095098	.096800
−1.2	.098525	.100273	.102042	.103835	.105650	.107488	.109349	.111232	.113139	.115070
−1.1	.117023	.119000	.121000	.123024	.125072	.127143	.129238	.131357	.133500	.135666
−1	.137857	.140071	.142310	.144572	.146859	.149170	.151505	.153864	.156248	.158655
−0.9	.161087	.163543	.166023	.168528	.171056	.173609	.176186	.178786	.181411	.184060
−0.8	.186733	.189430	.192150	.194895	.197663	.200454	.203269	.206108	.208970	.211855
−0.7	.214764	.217695	.220650	.223627	.226627	.229650	.232695	.235762	.238852	.241964
−0.6	.245097	.248252	.251429	.254627	.257846	.261086	.264347	.267629	.270931	.274253
−0.5	.277595	.280957	.284339	.287740	.291160	.294599	.298056	.301532	.305026	.308538
−0.4	.312067	.315614	.319178	.322758	.326355	.329969	.333598	.337243	.340903	.344578
−0.3	.348268	.351973	.355691	.359424	.363169	.366928	.370700	.374484	.378280	.382089
−0.2	.385908	.389739	.393580	.397432	.401294	.405165	.409046	.412936	.416834	.420740
−0.1	.424655	.428576	.432505	.436441	.440382	.444330	.448283	.452242	.456205	.460172
−0.0	.464144	.468119	.472097	.476078	.480061	.484047	.488034	.492022	.496011	.500000

TABLE II   Standard Normal Distribution Table (*Continued*)

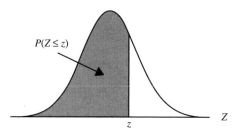

$P(Z \le z)$

Figure A.2

z	.00	.01	.02	.03	.04	.05	.06	.07	.08	.09
0.0	.500000	.503989	.507978	.511966	.515953	.519939	.523922	.527903	.531881	.535856
0.1	.539828	.543795	.547758	.551717	.555670	.559618	.563559	.567495	.571424	.575345
0.2	.579260	.583166	.587064	.590954	.594835	.598706	.602568	.606420	.610261	.614092
0.3	.617911	.621720	.625516	.629300	.633072	.636831	.640576	.644309	.648027	.651732
0.4	.655422	.659097	.662757	.666402	.670031	.673645	.677242	.680822	.684386	.687933
0.5	.691462	.694974	.698468	.701944	.705401	.708840	.712260	.715661	.719043	.722405
0.6	.725747	.729069	.732371	.735653	.738914	.742154	.745373	.748571	.751748	.754903
0.7	.758036	.761148	.764238	.767305	.770350	.773373	.776373	.779350	.782305	.785236
0.8	.788145	.791030	.793892	.796731	.799546	.802337	.805105	.807850	.810570	.813267
0.9	.815940	.818589	.821214	.823814	.826391	.828944	.831472	.833977	.836457	.838913
1.0	.841345	.843752	.846136	.848495	.850830	.853141	.855428	.857690	.859929	.862143
1.1	.864334	.866500	.868643	.870762	.872857	.874928	.876976	.879000	.881000	.882977
1.2	.884930	.886861	.888768	.890651	.892512	.894350	.896165	.897958	.899727	.901475
1.3	.903200	.904902	.906582	.908241	.909877	.911492	.913085	.914657	.916207	.917736
1.4	.919243	.920730	.922196	.923641	.925066	.926471	.927855	.929219	.930563	.931888
1.5	.933193	.934478	.935745	.936992	.938220	.939429	.940620	.941792	.942947	.944083
1.6	.945201	.946301	.947384	.948449	.949497	.950529	.951543	.952540	.953521	.954486
1.7	.955435	.956367	.957284	.958185	.959070	.959941	.960796	.961636	.962462	.963273
1.8	.964070	.964852	.965620	.966375	.967116	.967843	.968557	.969258	.969946	.970621
1.9	.971283	.971933	.972571	.973197	.973810	.974412	.975002	.975581	.976148	.976705
2.0	.977250	.977784	.978308	.978822	.979325	.979818	.980301	.980774	.981237	.981691
2.1	.982136	.982571	.982997	.983414	.983823	.984222	.984614	.984997	.985371	.985738
2.2	.986097	.986447	.986791	.987126	.987455	.987776	.988089	.988396	.988696	.988989
2.3	.989276	.989556	.989830	.990097	.990358	.990613	.990863	.991106	.991344	.991576
2.4	.991802	.992024	.992240	.992451	.992656	.992857	.993053	.993244	.993431	.993613
2.5	.993790	.993963	.994132	.994297	.994457	.994614	.994766	.994915	.995060	.995201
2.6	.995339	.995473	.995604	.995731	.995855	.995975	.996093	.996207	.996319	.996427
2.7	.996533	.996636	.996736	.996833	.996928	.997020	.997110	.997197	.997282	.997365
2.8	.997445	.997523	.997599	.997673	.997744	.997814	.997882	.997948	.998012	.998074
2.9	.998134	.998193	.998250	.998305	.998359	.998411	.998462	.998511	.998559	.998605
3.0	.998650	.998694	.998736	.998777	.998817	.998856	.998893	.998930	.998965	.998999
3.1	.999032	.999065	.999096	.999126	.999155	.999184	.999211	.999238	.999264	.999289
3.2	.999313	.999336	.999359	.999381	.999402	.999423	.999443	.999462	.999481	.999499
3.3	.999517	.999534	.999550	.999566	.999581	.999596	.999610	.999624	.999638	.999651
3.4	.999663	.999675	.999687	.999698	.999709	.999720	.999730	.999740	.999749	.999758
3.5	.999767	.999776	.999784	.999792	.999800	.999807	.999815	.999822	.999828	.999835
3.6	.999841	.999847	.999853	.999858	.999864	.999869	.999874	.999879	.999883	.999888
3.7	.999892	.999896	.999900	.999904	.999908	.999912	.999915	.999918	.999922	.999925
3.8	.999928	.999931	.999933	.999936	.999938	.999941	.999943	.999946	.999948	.999950
3.9	.999952	.999954	.999956	.999958	.999959	.999961	.999963	.999964	.999966	.999967
4.0	.99996832876									
4.5	.99999660233									
5	.99999971335									
5.5	.99999998101									
6	.99999999901									

TABLE III The *t* Distribution Table

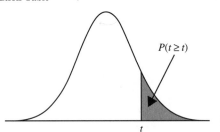

$P(t \geq t)$

Figure A.3

	Right-tail probability										
df	.20	.10	.05	.025	.02	.01	.005	.0025	.001	.0005	.0001
1	1.3764	3.0777	6.3138	12.706	15.895	31.821	63.657	127.32	318.31	636.62	3183.10
2	1.0607	1.8856	2.9200	4.3027	4.8487	6.9646	9.9248	14.089	22.327	31.599	70.7001
3	0.9785	1.6377	2.3534	3.1824	3.4819	4.5407	5.8409	7.4533	10.215	12.924	22.2037
4	0.9410	1.5332	2.1318	2.7764	2.9985	3.7469	4.6041	5.5976	7.1732	8.6103	13.0337
5	0.9195	1.4759	2.0150	2.5706	2.7565	3.3649	4.0321	4.7733	5.8934	6.8688	9.6776
6	0.9057	1.4398	1.9432	2.4469	2.6122	3.1427	3.7074	4.3168	5.2076	5.9588	8.0248
7	0.8960	1.4149	1.8946	2.3646	2.5168	2.9980	3.4995	4.0293	4.7853	5.4079	7.0634
8	0.8889	1.3968	1.8595	2.3060	2.4490	2.8965	3.3554	3.8325	4.5008	5.0413	6.4420
9	0.8834	1.3830	1.8331	2.2622	2.3984	2.8214	3.2498	3.6897	4.2968	4.7809	6.0101
10	0.8791	1.3722	1.8125	2.2281	2.3593	2.7638	3.1693	3.5814	4.1437	4.5869	5.6938
11	0.8755	1.3634	1.7959	2.2010	2.3281	2.7181	3.1058	3.4966	4.0247	4.4370	5.4528
12	0.8726	1.3562	1.7823	2.1788	2.3027	2.6810	3.0545	3.4284	3.9296	4.3178	5.2633
13	0.8702	1.3502	1.7709	2.1604	2.2816	2.6503	3.0123	3.3725	3.8520	4.2208	5.1106
14	0.8681	1.3450	1.7613	2.1448	2.2638	2.6245	2.9768	3.3257	3.7874	4.1405	4.9850
15	0.8662	1.3406	1.7531	2.1314	2.2485	2.6025	2.9467	3.2860	3.7328	4.0728	4.8800
16	0.8647	1.3368	1.7459	2.1199	2.2354	2.5835	2.9208	3.2520	3.6862	4.0150	4.7909
17	0.8633	1.3334	1.7396	2.1098	2.2238	2.5669	2.8982	3.2224	3.6458	3.9651	4.7144
18	0.8620	1.3304	1.7341	2.1009	2.2137	2.5524	2.8784	3.1966	3.6105	3.9216	4.6480
19	0.8610	1.3277	1.7291	2.0930	2.2047	2.5395	2.8609	3.1737	3.5794	3.8834	4.5899
20	0.8600	1.3253	1.7247	2.0860	2.1967	2.5280	2.8453	3.1534	3.5518	3.8495	4.5385
21	0.8591	1.3232	1.7207	2.0796	2.1894	2.5176	2.8314	3.1352	3.5272	3.8193	4.4929
22	0.8583	1.3212	1.7171	2.0739	2.1829	2.5083	2.8188	3.1188	3.5050	3.7921	4.4520
23	0.8575	1.3195	1.7139	2.0687	2.1770	2.4999	2.8073	3.1040	3.4850	3.7676	4.4152
24	0.8569	1.3178	1.7109	2.0639	2.1715	2.4922	2.7969	3.0905	3.4668	3.7454	4.3819
25	0.8562	1.3163	1.7081	2.0595	2.1666	2.4851	2.7874	3.0782	3.4502	3.7251	4.3517
26	0.8557	1.3150	1.7056	2.0555	2.1620	2.4786	2.7787	3.0669	3.4350	3.7066	4.3240
27	0.8551	1.3137	1.7033	2.0518	2.1578	2.4727	2.7707	3.0565	3.4210	3.6896	4.2987
28	0.8546	1.3125	1.7011	2.0484	2.1539	2.4671	2.7633	3.0469	3.4082	3.6739	4.2754
29	0.8542	1.3114	1.6991	2.0452	2.1503	2.4620	2.7564	3.0380	3.3962	3.6594	4.2539
30	0.8538	1.3104	1.6973	2.0423	2.1470	2.4573	2.7500	3.0298	3.3852	3.6460	4.2340
31	0.8534	1.3095	1.6955	2.0395	2.1438	2.4528	2.7440	3.0221	3.3749	3.6335	4.2155
32	0.8530	1.3086	1.6939	2.0369	2.1409	2.4487	2.7385	3.0149	3.3653	3.6218	4.1983
33	0.8526	1.3077	1.6924	2.0345	2.1382	2.4448	2.7333	3.0082	3.3563	3.6109	4.1822
34	0.8523	1.3070	1.6909	2.0322	2.1356	2.4411	2.7284	3.0020	3.3479	3.6007	4.1672
35	0.8520	1.3062	1.6896	2.0301	2.1332	2.4377	2.7238	2.9960	3.3400	3.5911	4.1531
36	0.8517	1.3055	1.6883	2.0281	2.1309	2.4345	2.7195	2.9905	3.3326	3.5821	4.1399
37	0.8514	1.3049	1.6871	2.0262	2.1287	2.4314	2.7154	2.9852	3.3256	3.5737	4.1275
38	0.8512	1.3042	1.6860	2.0244	2.1267	2.4286	2.7116	2.9803	3.3190	3.5657	4.1158
39	0.8509	1.3036	1.6849	2.0227	2.1247	2.4258	2.7079	2.9756	3.3128	3.5581	4.1047
40	0.8507	1.3031	1.6839	2.0211	2.1229	2.4233	2.7045	2.9712	3.3069	3.5510	4.0942
Conf. Level	60%	80%	90%	95%	96%	98%	99%	99.5%	99.8%	99.9%	99.98%

TABLE IV   Chi-Square Distribution Table

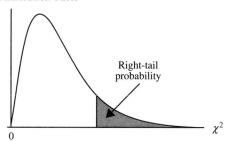

Figure A.4

				Right-tail probability					
df	0.25	0.20	0.10	0.05	0.025	0.02	0.01	0.005	0.001
1	1.323	1.642	2.706	3.841	5.024	5.412	6.635	7.879	10.828
2	2.773	3.219	4.605	5.991	7.378	7.824	9.210	10.597	13.816
3	4.108	4.642	6.251	7.815	9.348	9.837	11.345	12.838	16.266
4	5.385	5.989	7.779	9.488	11.143	11.668	13.277	14.860	18.467
5	6.626	7.289	9.236	11.070	12.833	13.388	15.086	16.750	20.515
6	7.841	8.558	10.645	12.592	14.449	15.033	16.812	18.548	22.458
7	9.037	9.803	12.017	14.067	16.013	16.622	18.475	20.278	24.322
8	10.219	11.030	13.362	15.507	17.535	18.168	20.090	21.955	26.124
9	11.389	12.242	14.684	16.919	19.023	19.679	21.666	23.589	27.877
10	12.549	13.442	15.987	18.307	20.483	21.161	23.209	25.188	29.588
11	13.701	14.631	17.275	19.675	21.920	22.618	24.725	26.757	31.264
12	14.845	15.812	18.549	21.026	23.337	24.054	26.217	28.300	32.909
13	15.984	16.985	19.812	22.362	24.736	25.472	27.688	29.819	34.528
14	17.117	18.151	21.064	23.685	26.119	26.873	29.141	31.319	36.123
15	18.245	19.311	22.307	24.996	27.488	28.259	30.578	32.801	37.697
16	19.369	20.465	23.542	26.296	28.845	29.633	32.000	34.267	39.252
17	20.489	21.615	24.769	27.587	30.191	30.995	33.409	35.718	40.790
18	21.605	22.760	25.989	28.869	31.526	32.346	34.805	37.156	42.312
19	22.718	23.900	27.204	30.144	32.852	33.687	36.191	38.582	43.820
20	23.828	25.038	28.412	31.410	34.170	35.020	37.566	39.997	45.315
21	24.935	26.171	29.615	32.671	35.479	36.343	38.932	41.401	46.797
22	26.039	27.301	30.813	33.924	36.781	37.659	40.289	42.796	48.268
23	27.141	28.429	32.007	35.172	38.076	38.968	41.638	44.181	49.728
24	28.241	29.553	33.196	36.415	39.364	40.270	42.980	45.559	51.179
25	29.339	30.675	34.382	37.652	40.646	41.566	44.314	46.928	52.620
26	30.435	31.795	35.563	38.885	41.923	42.856	45.642	48.290	54.052
27	31.528	32.912	36.741	40.113	43.195	44.140	46.963	49.645	55.476
28	32.620	34.027	37.916	41.337	44.461	45.419	48.278	50.993	56.892
29	33.711	35.139	39.087	42.557	45.722	46.693	49.588	52.336	58.301
30	34.800	36.250	40.256	43.773	46.979	47.962	50.892	53.672	59.703
40	45.616	47.269	51.805	55.758	59.342	60.436	63.691	66.766	73.402
50	56.334	58.164	63.167	67.505	71.420	72.613	76.154	79.490	86.661
60	66.981	68.972	74.397	79.082	83.298	84.580	88.379	91.952	99.607
70	77.577	79.715	85.527	90.531	95.023	96.388	100.425	104.215	112.317
80	88.130	90.405	96.578	101.879	106.629	108.069	112.329	116.321	124.839
90	98.650	101.054	107.565	113.145	118.136	119.648	124.116	128.299	137.208
100	109.141	111.667	118.498	124.342	129.561	131.142	135.807	140.169	149.449

TABLE V The *F* Distribution Table

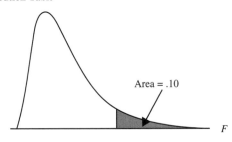

Area = .10

Figure A.5

	Right-tail probability = .10										
	num df										
den df	1	2	3	4	5	6	7	8	9	10	11
1	39.863	49.500	53.593	55.833	57.240	58.204	58.906	59.439	59.858	60.195	60.473
2	8.526	9.000	9.162	9.243	9.293	9.326	9.349	9.367	9.381	9.392	9.401
3	5.538	5.462	5.391	5.343	5.309	5.285	5.266	5.252	5.240	5.230	5.222
4	4.545	4.325	4.191	4.107	4.051	4.010	3.979	3.955	3.936	3.920	3.907
5	4.060	3.780	3.619	3.520	3.453	3.405	3.368	3.339	3.316	3.297	3.282
6	3.776	3.463	3.289	3.181	3.108	3.055	3.014	2.983	2.958	2.937	2.920
7	3.589	3.257	3.074	2.961	2.883	2.827	2.785	2.752	2.725	2.703	2.684
8	3.458	3.113	2.924	2.806	2.726	2.668	2.624	2.589	2.561	2.538	2.519
9	3.360	3.006	2.813	2.693	2.611	2.551	2.505	2.469	2.440	2.416	2.396
10	3.285	2.924	2.728	2.605	2.522	2.461	2.414	2.377	2.347	2.323	2.302
11	3.225	2.860	2.660	2.536	2.451	2.389	2.342	2.304	2.274	2.248	2.227
12	3.177	2.807	2.606	2.480	2.394	2.331	2.283	2.245	2.214	2.188	2.166
13	3.136	2.763	2.560	2.434	2.347	2.283	2.234	2.195	2.164	2.138	2.116
14	3.102	2.726	2.522	2.395	2.307	2.243	2.193	2.154	2.122	2.095	2.073
15	3.073	2.695	2.490	2.361	2.273	2.208	2.158	2.119	2.086	2.059	2.037
16	3.048	2.668	2.462	2.333	2.244	2.178	2.128	2.088	2.055	2.028	2.005
17	3.026	2.645	2.437	2.308	2.218	2.152	2.102	2.061	2.028	2.001	1.978
18	3.007	2.624	2.416	2.286	2.196	2.130	2.079	2.038	2.005	1.977	1.954
19	2.990	2.606	2.397	2.266	2.176	2.109	2.058	2.017	1.984	1.956	1.932
20	2.975	2.589	2.380	2.249	2.158	2.091	2.040	1.999	1.965	1.937	1.913
21	2.961	2.575	2.365	2.233	2.142	2.075	2.023	1.982	1.948	1.920	1.896
22	2.949	2.561	2.351	2.219	2.128	2.060	2.008	1.967	1.933	1.904	1.880
23	2.937	2.549	2.339	2.207	2.115	2.047	1.995	1.953	1.919	1.890	1.866
24	2.927	2.538	2.327	2.195	2.103	2.035	1.983	1.941	1.906	1.877	1.853
25	2.918	2.528	2.317	2.184	2.092	2.024	1.971	1.929	1.895	1.866	1.841
26	2.909	2.519	2.307	2.174	2.082	2.014	1.961	1.919	1.884	1.855	1.830
27	2.901	2.511	2.299	2.165	2.073	2.005	1.952	1.909	1.874	1.845	1.820
28	2.894	2.503	2.291	2.157	2.064	1.996	1.943	1.900	1.865	1.836	1.811
29	2.887	2.495	2.283	2.149	2.057	1.988	1.935	1.892	1.857	1.827	1.802
30	2.881	2.489	2.276	2.142	2.049	1.980	1.927	1.884	1.849	1.819	1.794
35	2.855	2.461	2.247	2.113	2.019	1.950	1.896	1.852	1.817	1.787	1.761
40	2.835	2.440	2.226	2.091	1.997	1.927	1.873	1.829	1.793	1.763	1.737
45	2.820	2.425	2.210	2.074	1.980	1.909	1.855	1.811	1.774	1.744	1.718
50	2.809	2.412	2.197	2.061	1.966	1.895	1.840	1.796	1.760	1.729	1.703
60	2.791	2.393	2.177	2.041	1.946	1.875	1.819	1.775	1.738	1.707	1.680
70	2.779	2.380	2.164	2.027	1.931	1.860	1.804	1.760	1.723	1.691	1.665
80	2.769	2.370	2.154	2.016	1.921	1.849	1.793	1.748	1.711	1.680	1.653
90	2.762	2.363	2.146	2.008	1.912	1.841	1.785	1.739	1.702	1.670	1.643
100	2.756	2.356	2.139	2.002	1.906	1.834	1.778	1.732	1.695	1.663	1.636
1000	2.711	2.308	2.089	1.950	1.853	1.780	1.723	1.676	1.638	1.605	1.577

TABLE V The *F* Distribution Table (*Continued*)

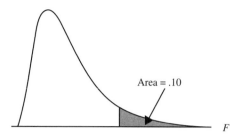

Figure A.6

					Right-tail probability = .10						
					num df						
den df	12	13	14	15	16	17	18	19	20	30	40
1	60.705	60.903	61.073	61.220	61.350	61.464	61.566	61.658	61.740	62.265	62.529
2	9.408	9.415	9.420	9.425	9.429	9.433	9.436	9.439	9.441	9.458	9.466
3	5.216	5.210	5.205	5.200	5.196	5.193	5.190	5.187	5.184	5.168	5.160
4	3.896	3.886	3.878	3.870	3.864	3.858	3.853	3.849	3.844	3.817	3.804
5	3.268	3.257	3.247	3.238	3.230	3.223	3.217	3.212	3.207	3.174	3.157
6	2.905	2.892	2.881	2.871	2.863	2.855	2.848	2.842	2.836	2.800	2.781
7	2.668	2.654	2.643	2.632	2.623	2.615	2.607	2.601	2.595	2.555	2.535
8	2.502	2.488	2.475	2.464	2.455	2.446	2.438	2.431	2.425	2.383	2.361
9	2.379	2.364	2.351	2.340	2.329	2.320	2.312	2.305	2.298	2.255	2.232
10	2.284	2.269	2.255	2.244	2.233	2.224	2.215	2.208	2.201	2.155	2.132
11	2.209	2.193	2.179	2.167	2.156	2.147	2.138	2.130	2.123	2.076	2.052
12	2.147	2.131	2.117	2.105	2.094	2.084	2.075	2.067	2.060	2.011	1.986
13	2.097	2.080	2.066	2.053	2.042	2.032	2.023	2.014	2.007	1.958	1.931
14	2.054	2.037	2.022	2.010	1.998	1.988	1.978	1.970	1.962	1.912	1.885
15	2.017	2.000	1.985	1.972	1.961	1.950	1.941	1.932	1.924	1.873	1.845
16	1.985	1.968	1.953	1.940	1.928	1.917	1.908	1.899	1.891	1.839	1.811
17	1.958	1.940	1.925	1.912	1.900	1.889	1.879	1.870	1.862	1.809	1.781
18	1.933	1.916	1.900	1.887	1.875	1.864	1.854	1.845	1.837	1.783	1.754
19	1.912	1.894	1.878	1.865	1.852	1.841	1.831	1.822	1.814	1.759	1.730
20	1.892	1.875	1.859	1.845	1.833	1.821	1.811	1.802	1.794	1.738	1.708
21	1.875	1.857	1.841	1.827	1.815	1.803	1.793	1.784	1.776	1.719	1.689
22	1.859	1.841	1.825	1.811	1.798	1.787	1.777	1.768	1.759	1.702	1.671
23	1.845	1.827	1.811	1.796	1.784	1.772	1.762	1.753	1.744	1.686	1.655
24	1.832	1.814	1.797	1.783	1.770	1.759	1.748	1.739	1.730	1.672	1.641
25	1.820	1.802	1.785	1.771	1.758	1.746	1.736	1.726	1.718	1.659	1.627
26	1.809	1.790	1.774	1.760	1.747	1.735	1.724	1.715	1.706	1.647	1.615
27	1.799	1.780	1.764	1.749	1.736	1.724	1.714	1.704	1.695	1.636	1.603
28	1.790	1.771	1.754	1.740	1.726	1.715	1.704	1.694	1.685	1.625	1.592
29	1.781	1.762	1.745	1.731	1.717	1.705	1.695	1.685	1.676	1.616	1.583
30	1.773	1.754	1.737	1.722	1.709	1.697	1.686	1.676	1.667	1.606	1.573
35	1.739	1.720	1.703	1.688	1.674	1.662	1.651	1.641	1.632	1.569	1.535
40	1.715	1.695	1.678	1.662	1.649	1.636	1.625	1.615	1.605	1.541	1.506
45	1.695	1.676	1.658	1.643	1.629	1.616	1.605	1.594	1.585	1.519	1.483
50	1.680	1.660	1.643	1.627	1.613	1.600	1.588	1.578	1.568	1.502	1.465
60	1.657	1.637	1.619	1.603	1.589	1.576	1.564	1.553	1.543	1.476	1.437
70	1.641	1.621	1.603	1.587	1.572	1.559	1.547	1.536	1.526	1.457	1.418
80	1.629	1.609	1.590	1.574	1.559	1.546	1.534	1.523	1.513	1.443	1.403
90	1.620	1.599	1.581	1.564	1.550	1.536	1.524	1.513	1.503	1.432	1.391
100	1.612	1.592	1.573	1.557	1.542	1.528	1.516	1.505	1.494	1.423	1.382
1000	1.552	1.531	1.511	1.494	1.478	1.464	1.451	1.439	1.428	1.350	1.304

TABLE V The *F* Distribution Table (*Continued*)

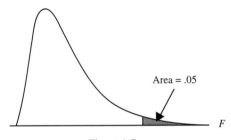

Area = .05

F

Figure A.7

den df	\multicolumn{11}{c}{Right-tail probability = .10}

| | \multicolumn{11}{c}{num df} |

den df	1	2	3	4	5	6	7	8	9	10	11
1	161.44	199.50	215.71	224.58	230.16	233.99	236.77	238.88	240.54	241.88	242.98
2	18.513	19.000	19.164	19.247	19.296	19.330	19.353	19.371	19.385	19.396	19.405
3	10.128	9.552	9.277	9.117	9.013	8.941	8.887	8.845	8.812	8.786	8.763
4	7.709	6.944	6.591	6.388	6.256	6.163	6.094	6.041	5.999	5.964	5.936
5	6.608	5.786	5.409	5.192	5.050	4.950	4.876	4.818	4.772	4.735	4.704
6	5.987	5.143	4.757	4.534	4.387	4.284	4.207	4.147	4.099	4.060	4.027
7	5.591	4.737	4.347	4.120	3.972	3.866	3.787	3.726	3.677	3.637	3.603
8	5.318	4.459	4.066	3.838	3.687	3.581	3.500	3.438	3.388	3.347	3.313
9	5.117	4.256	3.863	3.633	3.482	3.374	3.293	3.230	3.179	3.137	3.102
10	4.965	4.103	3.708	3.478	3.326	3.217	3.135	3.072	3.020	2.978	2.943
11	4.844	3.982	3.587	3.357	3.204	3.095	3.012	2.948	2.896	2.854	2.818
12	4.747	3.885	3.490	3.259	3.106	2.996	2.913	2.849	2.796	2.753	2.717
13	4.667	3.806	3.411	3.179	3.025	2.915	2.832	2.767	2.714	2.671	2.635
14	4.600	3.739	3.344	3.112	2.958	2.848	2.764	2.699	2.646	2.602	2.565
15	4.543	3.682	3.287	3.056	2.901	2.790	2.707	2.641	2.588	2.544	2.507
16	4.494	3.634	3.239	3.007	2.852	2.741	2.657	2.591	2.538	2.494	2.456
17	4.451	3.592	3.197	2.965	2.810	2.699	2.614	2.548	2.494	2.450	2.413
18	4.414	3.555	3.160	2.928	2.773	2.661	2.577	2.510	2.456	2.412	2.374
19	4.381	3.522	3.127	2.895	2.740	2.628	2.544	2.477	2.423	2.378	2.340
20	4.351	3.493	3.098	2.866	2.711	2.599	2.514	2.447	2.393	2.348	2.310
21	4.325	3.467	3.072	2.840	2.685	2.573	2.488	2.420	2.366	2.321	2.283
22	4.301	3.443	3.049	2.817	2.661	2.549	2.464	2.397	2.342	2.297	2.259
23	4.279	3.422	3.028	2.796	2.640	2.528	2.442	2.375	2.320	2.275	2.236
24	4.260	3.403	3.009	2.776	2.621	2.508	2.423	2.355	2.300	2.255	2.216
25	4.242	3.385	2.991	2.759	2.603	2.490	2.405	2.337	2.282	2.236	2.198
26	4.225	3.369	2.975	2.743	2.587	2.474	2.388	2.321	2.265	2.220	2.181
27	4.210	3.354	2.960	2.728	2.572	2.459	2.373	2.305	2.250	2.204	2.166
28	4.196	3.340	2.947	2.714	2.558	2.445	2.359	2.291	2.236	2.190	2.151
29	4.183	3.328	2.934	2.701	2.545	2.432	2.346	2.278	2.223	2.177	2.138
30	4.171	3.316	2.922	2.690	2.534	2.421	2.334	2.266	2.211	2.165	2.126
35	4.121	3.267	2.874	2.641	2.485	2.372	2.285	2.217	2.161	2.114	2.075
40	4.085	3.232	2.839	2.606	2.449	2.336	2.249	2.180	2.124	2.077	2.038
45	4.057	3.204	2.812	2.579	2.422	2.308	2.221	2.152	2.096	2.049	2.009
50	4.034	3.183	2.790	2.557	2.400	2.286	2.199	2.130	2.073	2.026	1.986
60	4.001	3.150	2.758	2.525	2.368	2.254	2.167	2.097	2.040	1.993	1.952
70	3.978	3.128	2.736	2.503	2.346	2.231	2.143	2.074	2.017	1.969	1.928
80	3.960	3.111	2.719	2.486	2.329	2.214	2.126	2.056	1.999	1.951	1.910
90	3.947	3.098	2.706	2.473	2.316	2.201	2.113	2.043	1.986	1.938	1.897
100	3.936	3.087	2.696	2.463	2.305	2.191	2.103	2.032	1.975	1.927	1.886
1000	3.851	3.005	2.614	2.381	2.223	2.108	2.019	1.948	1.889	1.840	1.798

TABLE V   The *F* Distribution Table (*Continued*)

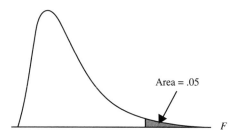

Area = .05

F

Figure A.8

den df	num df Right-tail probability = .05										
	12	13	14	15	16	17	18	19	20	30	40
1	243.91	244.69	245.36	245.95	246.46	246.92	247.32	247.69	248.01	250.10	251.14
2	19.413	19.419	19.424	19.429	19.433	19.437	19.440	19.443	19.446	19.462	19.471
3	8.745	8.729	8.715	8.703	8.692	8.683	8.675	8.667	8.660	8.617	8.594
4	5.912	5.891	5.873	5.858	5.844	5.832	5.821	5.811	5.803	5.746	5.717
5	4.678	4.655	4.636	4.619	4.604	4.590	4.579	4.568	4.558	4.496	4.464
6	4.000	3.976	3.956	3.938	3.922	3.908	3.896	3.884	3.874	3.808	3.774
7	3.575	3.550	3.529	3.511	3.494	3.480	3.467	3.455	3.445	3.376	3.340
8	3.284	3.259	3.237	3.218	3.202	3.187	3.173	3.161	3.150	3.079	3.043
9	3.073	3.048	3.025	3.006	2.989	2.974	2.960	2.948	2.936	2.864	2.826
10	2.913	2.887	2.865	2.845	2.828	2.812	2.798	2.785	2.774	2.700	2.661
11	2.788	2.761	2.739	2.719	2.701	2.685	2.671	2.658	2.646	2.570	2.531
12	2.687	2.660	2.637	2.617	2.599	2.583	2.568	2.555	2.544	2.466	2.426
13	2.604	2.577	2.554	2.533	2.515	2.499	2.484	2.471	2.459	2.380	2.339
14	2.534	2.507	2.484	2.463	2.445	2.428	2.413	2.400	2.388	2.308	2.266
15	2.475	2.448	2.424	2.403	2.385	2.368	2.353	2.340	2.328	2.247	2.204
16	2.425	2.397	2.373	2.352	2.333	2.317	2.302	2.288	2.276	2.194	2.151
17	2.381	2.353	2.329	2.308	2.289	2.272	2.257	2.243	2.230	2.148	2.104
18	2.342	2.314	2.290	2.269	2.250	2.233	2.217	2.203	2.191	2.107	2.063
19	2.308	2.280	2.256	2.234	2.215	2.198	2.182	2.168	2.155	2.071	2.026
20	2.278	2.250	2.225	2.203	2.184	2.167	2.151	2.137	2.124	2.039	1.994
21	2.250	2.222	2.197	2.176	2.156	2.139	2.123	2.109	2.096	2.010	1.965
22	2.226	2.198	2.173	2.151	2.131	2.114	2.098	2.084	2.071	1.984	1.938
23	2.204	2.175	2.150	2.128	2.109	2.091	2.075	2.061	2.048	1.961	1.914
24	2.183	2.155	2.130	2.108	2.088	2.070	2.054	2.040	2.027	1.939	1.892
25	2.165	2.136	2.111	2.089	2.069	2.051	2.035	2.021	2.007	1.919	1.872
26	2.148	2.119	2.094	2.072	2.052	2.034	2.018	2.003	1.990	1.901	1.853
27	2.132	2.103	2.078	2.056	2.036	2.018	2.002	1.987	1.974	1.884	1.836
28	2.118	2.089	2.064	2.041	2.021	2.003	1.987	1.972	1.959	1.869	1.820
29	2.104	2.075	2.050	2.027	2.007	1.989	1.973	1.958	1.945	1.854	1.806
30	2.092	2.063	2.037	2.015	1.995	1.976	1.960	1.945	1.932	1.841	1.792
35	2.041	2.012	1.986	1.963	1.942	1.924	1.907	1.892	1.878	1.786	1.735
40	2.003	1.974	1.948	1.924	1.904	1.885	1.868	1.853	1.839	1.744	1.693
45	1.974	1.945	1.918	1.895	1.874	1.855	1.838	1.823	1.808	1.713	1.660
50	1.952	1.921	1.895	1.871	1.850	1.831	1.814	1.798	1.784	1.687	1.634
60	1.917	1.887	1.860	1.836	1.815	1.796	1.778	1.763	1.748	1.649	1.594
70	1.893	1.863	1.836	1.812	1.790	1.771	1.753	1.737	1.722	1.622	1.566
80	1.875	1.845	1.817	1.793	1.772	1.752	1.734	1.718	1.703	1.602	1.545
90	1.861	1.830	1.803	1.779	1.757	1.737	1.720	1.703	1.688	1.586	1.528
100	1.850	1.819	1.792	1.768	1.746	1.726	1.708	1.691	1.676	1.573	1.515
1000	1.762	1.730	1.702	1.676	1.654	1.633	1.614	1.597	1.581	1.471	1.406

TABLE V   The *F* Distribution Table (*Continued*)

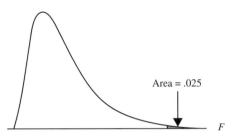

Figure A.9

	Right-tail probability $= .025$										
	num df										
den df	1	2	3	4	5	6	7	8	9	10	11
1	647.79	799.50	864.16	899.58	921.85	937.11	948.22	956.66	963.29	968.63	973.03
2	38.506	39.000	39.165	39.248	39.298	39.331	39.355	39.373	39.387	39.398	39.407
3	17.443	16.044	15.439	15.101	14.885	14.735	14.624	14.540	14.473	14.419	14.374
4	12.218	10.649	9.979	9.605	9.364	9.197	9.074	8.980	8.905	8.844	8.794
5	10.007	8.434	7.764	7.388	7.146	6.978	6.853	6.757	6.681	6.619	6.568
6	8.813	7.260	6.599	6.227	5.988	5.820	5.695	5.600	5.523	5.461	5.410
7	8.073	6.542	5.890	5.523	5.285	5.119	4.995	4.899	4.823	4.761	4.709
8	7.571	6.059	5.416	5.053	4.817	4.652	4.529	4.433	4.357	4.295	4.243
9	7.209	5.715	5.078	4.718	4.484	4.320	4.197	4.102	4.026	3.964	3.912
10	6.937	5.456	4.826	4.468	4.236	4.072	3.950	3.855	3.779	3.717	3.665
11	6.724	5.256	4.630	4.275	4.044	3.881	3.759	3.664	3.588	3.526	3.474
12	6.554	5.096	4.474	4.121	3.891	3.728	3.607	3.512	3.436	3.374	3.321
13	6.414	4.965	4.347	3.996	3.767	3.604	3.483	3.388	3.312	3.250	3.197
14	6.298	4.857	4.242	3.892	3.663	3.501	3.380	3.285	3.209	3.147	3.095
15	6.200	4.765	4.153	3.804	3.576	3.415	3.293	3.199	3.123	3.060	3.008
16	6.115	4.687	4.077	3.729	3.502	3.341	3.219	3.125	3.049	2.986	2.934
17	6.042	4.619	4.011	3.665	3.438	3.277	3.156	3.061	2.985	2.922	2.870
18	5.978	4.560	3.954	3.608	3.382	3.221	3.100	3.005	2.929	2.866	2.814
19	5.922	4.508	3.903	3.559	3.333	3.172	3.051	2.956	2.880	2.817	2.765
20	5.871	4.461	3.859	3.515	3.289	3.128	3.007	2.913	2.837	2.774	2.721
21	5.827	4.420	3.819	3.475	3.250	3.090	2.969	2.874	2.798	2.735	2.682
22	5.786	4.383	3.783	3.440	3.215	3.055	2.934	2.839	2.763	2.700	2.647
23	5.750	4.349	3.750	3.408	3.183	3.023	2.902	2.808	2.731	2.668	2.615
24	5.717	4.319	3.721	3.379	3.155	2.995	2.874	2.779	2.703	2.640	2.586
25	5.686	4.291	3.694	3.353	3.129	2.969	2.848	2.753	2.677	2.613	2.560
26	5.659	4.265	3.670	3.329	3.105	2.945	2.824	2.729	2.653	2.590	2.536
27	5.633	4.242	3.647	3.307	3.083	2.923	2.802	2.707	2.631	2.568	2.514
28	5.610	4.221	3.626	3.286	3.063	2.903	2.782	2.687	2.611	2.547	2.494
29	5.588	4.201	3.607	3.267	3.044	2.884	2.763	2.669	2.592	2.529	2.475
30	5.568	4.182	3.589	3.250	3.026	2.867	2.746	2.651	2.575	2.511	2.458
35	5.485	4.106	3.517	3.179	2.956	2.796	2.676	2.581	2.504	2.440	2.387
40	5.424	4.051	3.463	3.126	2.904	2.744	2.624	2.529	2.452	2.388	2.334
45	5.377	4.009	3.422	3.086	2.864	2.705	2.584	2.489	2.412	2.348	2.294
50	5.340	3.975	3.390	3.054	2.833	2.674	2.553	2.458	2.381	2.317	2.263
60	5.286	3.925	3.343	3.008	2.786	2.627	2.507	2.412	2.334	2.270	2.216
70	5.247	3.890	3.309	2.975	2.754	2.595	2.474	2.379	2.302	2.237	2.183
80	5.218	3.864	3.284	2.950	2.730	2.571	2.450	2.355	2.277	2.213	2.158
90	5.196	3.844	3.265	2.932	2.711	2.552	2.432	2.336	2.259	2.194	2.140
100	5.179	3.828	3.250	2.917	2.696	2.537	2.417	2.321	2.244	2.179	2.124
1000	5.039	3.703	3.129	2.799	2.579	2.421	2.300	2.204	2.126	2.061	2.006

TABLE V   The *F* Distribution Table (*Continued*)

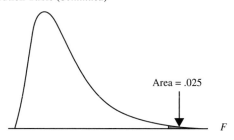

Area = .025

F

Figure A.10

					Right-tail probability = .025						
					num df						
den df	12	13	14	15	16	17	18	19	20	30	40
1	976.71	979.84	982.53	984.87	986.92	988.73	990.35	991.80	993.10	1001.4	1005.6
2	39.415	39.421	39.427	39.431	39.435	39.439	39.442	39.445	39.448	39.465	39.473
3	14.337	14.304	14.277	14.253	14.232	14.213	14.196	14.181	14.167	14.081	14.037
4	8.751	8.715	8.684	8.657	8.633	8.611	8.592	8.575	8.560	8.461	8.411
5	6.525	6.488	6.456	6.428	6.403	6.381	6.362	6.344	6.329	6.227	6.175
6	5.366	5.329	5.297	5.269	5.244	5.222	5.202	5.184	5.168	5.065	5.012
7	4.666	4.628	4.596	4.568	4.543	4.521	4.501	4.483	4.467	4.362	4.309
8	4.200	4.162	4.130	4.101	4.076	4.054	4.034	4.016	3.999	3.894	3.840
9	3.868	3.831	3.798	3.769	3.744	3.722	3.701	3.683	3.667	3.560	3.505
10	3.621	3.583	3.550	3.522	3.496	3.474	3.453	3.435	3.419	3.311	3.255
11	3.430	3.392	3.359	3.330	3.304	3.282	3.261	3.243	3.226	3.118	3.061
12	3.277	3.239	3.206	3.177	3.152	3.129	3.108	3.090	3.073	2.963	2.906
13	3.153	3.115	3.082	3.053	3.027	3.004	2.983	2.965	2.948	2.837	2.780
14	3.050	3.012	2.979	2.949	2.923	2.900	2.879	2.861	2.844	2.732	2.674
15	2.963	2.925	2.891	2.862	2.836	2.813	2.792	2.773	2.756	2.644	2.585
16	2.889	2.851	2.817	2.788	2.761	2.738	2.717	2.698	2.681	2.568	2.509
17	2.825	2.786	2.753	2.723	2.697	2.673	2.652	2.633	2.616	2.502	2.442
18	2.769	2.730	2.696	2.667	2.640	2.617	2.596	2.576	2.559	2.445	2.384
19	2.720	2.681	2.647	2.617	2.591	2.567	2.546	2.526	2.509	2.394	2.333
20	2.676	2.637	2.603	2.573	2.547	2.523	2.501	2.482	2.464	2.349	2.287
21	2.637	2.598	2.564	2.534	2.507	2.483	2.462	2.442	2.425	2.308	2.246
22	2.602	2.563	2.528	2.498	2.472	2.448	2.426	2.407	2.389	2.272	2.210
23	2.570	2.531	2.497	2.466	2.440	2.416	2.394	2.374	2.357	2.239	2.176
24	2.541	2.502	2.468	2.437	2.411	2.386	2.365	2.345	2.327	2.209	2.146
25	2.515	2.476	2.441	2.411	2.384	2.360	2.338	2.318	2.300	2.182	2.118
26	2.491	2.451	2.417	2.387	2.360	2.335	2.314	2.294	2.276	2.157	2.093
27	2.469	2.429	2.395	2.364	2.337	2.313	2.291	2.271	2.253	2.133	2.069
28	2.448	2.409	2.374	2.344	2.317	2.292	2.270	2.251	2.232	2.112	2.048
29	2.430	2.390	2.355	2.325	2.298	2.273	2.251	2.231	2.213	2.092	2.028
30	2.412	2.372	2.338	2.307	2.280	2.255	2.233	2.213	2.195	2.074	2.009
35	2.341	2.301	2.266	2.235	2.207	2.183	2.160	2.140	2.122	1.999	1.932
40	2.288	2.248	2.213	2.182	2.154	2.129	2.107	2.086	2.068	1.943	1.875
45	2.248	2.208	2.172	2.141	2.113	2.088	2.066	2.045	2.026	1.900	1.831
50	2.216	2.176	2.140	2.109	2.081	2.056	2.033	2.012	1.993	1.866	1.796
60	2.169	2.129	2.093	2.061	2.033	2.008	1.985	1.964	1.944	1.815	1.744
70	2.136	2.095	2.059	2.028	1.999	1.974	1.950	1.929	1.910	1.779	1.707
80	2.111	2.071	2.035	2.003	1.974	1.948	1.925	1.904	1.884	1.752	1.679
90	2.092	2.051	2.015	1.983	1.955	1.929	1.905	1.884	1.864	1.731	1.657
100	2.077	2.036	2.000	1.968	1.939	1.913	1.890	1.868	1.849	1.715	1.640
1000	1.958	1.916	1.879	1.846	1.816	1.789	1.765	1.743	1.722	1.581	1.499

TABLE V The *F* Distribution Table (*Continued*)

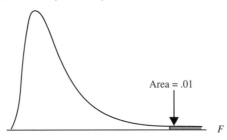

Figure A.11

					Right-tail probability = .01						
					num df						
den df	1	2	3	4	5	6	7	8	9	10	11
1	4052.2	4999.5	5403.4	5624.6	5763.7	5859.0	5928.4	5981.1	6022.5	6055.8	6083.3
2	98.503	99.000	99.166	99.249	99.299	99.333	99.356	99.374	99.388	99.399	99.408
3	34.116	30.817	29.457	28.710	28.237	27.911	27.672	27.489	27.345	27.229	27.133
4	21.198	18.000	16.694	15.977	15.522	15.207	14.976	14.799	14.659	14.546	14.452
5	16.258	13.274	12.060	11.392	10.967	10.672	10.456	10.289	10.158	10.051	9.963
6	13.745	10.925	9.780	9.148	8.746	8.466	8.260	8.102	7.976	7.874	7.790
7	12.246	9.547	8.451	7.847	7.460	7.191	6.993	6.840	6.719	6.620	6.538
8	11.259	8.649	7.591	7.006	6.632	6.371	6.178	6.029	5.911	5.814	5.734
9	10.561	8.022	6.992	6.422	6.057	5.802	5.613	5.467	5.351	5.257	5.178
10	10.044	7.559	6.552	5.994	5.636	5.386	5.200	5.057	4.942	4.849	4.772
11	9.646	7.206	6.217	5.668	5.316	5.069	4.886	4.744	4.632	4.539	4.462
12	9.330	6.927	5.953	5.412	5.064	4.821	4.640	4.499	4.388	4.296	4.220
13	9.074	6.701	5.739	5.205	4.862	4.620	4.441	4.302	4.191	4.100	4.025
14	8.862	6.515	5.564	5.035	4.695	4.456	4.278	4.140	4.030	3.939	3.864
15	8.683	6.359	5.417	4.893	4.556	4.318	4.142	4.004	3.895	3.805	3.730
16	8.531	6.226	5.292	4.773	4.437	4.202	4.026	3.890	3.780	3.691	3.616
17	8.400	6.112	5.185	4.669	4.336	4.102	3.927	3.791	3.682	3.593	3.519
18	8.285	6.013	5.092	4.579	4.248	4.015	3.841	3.705	3.597	3.508	3.434
19	8.185	5.926	5.010	4.500	4.171	3.939	3.765	3.631	3.523	3.434	3.360
20	8.096	5.849	4.938	4.431	4.103	3.871	3.699	3.564	3.457	3.368	3.294
21	8.017	5.780	4.874	4.369	4.042	3.812	3.640	3.506	3.398	3.310	3.236
22	7.945	5.719	4.817	4.313	3.988	3.758	3.587	3.453	3.346	3.258	3.184
23	7.881	5.664	4.765	4.264	3.939	3.710	3.539	3.406	3.299	3.211	3.137
24	7.823	5.614	4.718	4.218	3.895	3.667	3.496	3.363	3.256	3.168	3.094
25	7.770	5.568	4.675	4.177	3.855	3.627	3.457	3.324	3.217	3.129	3.056
26	7.721	5.526	4.637	4.140	3.818	3.591	3.421	3.288	3.182	3.094	3.021
27	7.677	5.488	4.601	4.106	3.785	3.558	3.388	3.256	3.149	3.062	2.988
28	7.636	5.453	4.568	4.074	3.754	3.528	3.358	3.226	3.120	3.032	2.959
29	7.598	5.420	4.538	4.045	3.725	3.499	3.330	3.198	3.092	3.005	2.931
30	7.562	5.390	4.510	4.018	3.699	3.473	3.304	3.173	3.067	2.979	2.906
35	7.419	5.268	4.396	3.908	3.592	3.368	3.200	3.069	2.963	2.876	2.803
40	7.314	5.179	4.313	3.828	3.514	3.291	3.124	2.993	2.888	2.801	2.727
45	7.234	5.110	4.249	3.767	3.454	3.232	3.066	2.935	2.830	2.743	2.670
50	7.171	5.057	4.199	3.720	3.408	3.186	3.020	2.890	2.785	2.698	2.625
60	7.077	4.977	4.126	3.649	3.339	3.119	2.953	2.823	2.718	2.632	2.559
70	7.011	4.922	4.074	3.600	3.291	3.071	2.906	2.777	2.672	2.585	2.512
80	6.963	4.881	4.036	3.563	3.255	3.036	2.871	2.742	2.637	2.551	2.478
90	6.925	4.849	4.007	3.535	3.228	3.009	2.845	2.715	2.611	2.524	2.451
100	6.895	4.824	3.984	3.513	3.206	2.988	2.823	2.694	2.590	2.503	2.430
1000	6.660	4.626	3.801	3.338	3.036	2.820	2.657	2.529	2.425	2.339	2.265

TABLE V   The *F* Distribution Table (*Continued*)

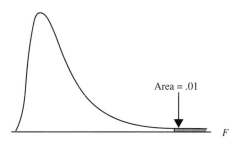

Area = .01

F

Figure A.12

den df	\multicolumn{10}{c	}{Right-tail probability = .01}

| | num df |||||||||||
den df	12	13	14	15	16	17	18	19	20	30	40
1	6106.3	6125.9	6142.7	6157.3	6170.1	6181.4	6191.5	6200.6	6208.7	6260.6	6286.8
2	99.416	99.422	99.428	99.433	99.437	99.440	99.444	99.447	99.449	99.466	99.474
3	27.052	26.983	26.924	26.872	26.827	26.787	26.751	26.719	26.690	26.505	26.411
4	14.374	14.307	14.249	14.198	14.154	14.115	14.080	14.048	14.020	13.838	13.745
5	9.888	9.825	9.770	9.722	9.680	9.643	9.610	9.580	9.553	9.379	9.291
6	7.718	7.657	7.605	7.559	7.519	7.483	7.451	7.422	7.396	7.229	7.143
7	6.469	6.410	6.359	6.314	6.275	6.240	6.209	6.181	6.155	5.992	5.908
8	5.667	5.609	5.559	5.515	5.477	5.442	5.412	5.384	5.359	5.198	5.116
9	5.111	5.055	5.005	4.962	4.924	4.890	4.860	4.833	4.808	4.649	4.567
10	4.706	4.650	4.601	4.558	4.520	4.487	4.457	4.430	4.405	4.247	4.165
11	4.397	4.342	4.293	4.251	4.213	4.180	4.150	4.123	4.099	3.941	3.860
12	4.155	4.100	4.052	4.010	3.972	3.939	3.909	3.883	3.858	3.701	3.619
13	3.960	3.905	3.857	3.815	3.778	3.745	3.716	3.689	3.665	3.507	3.425
14	3.800	3.745	3.698	3.656	3.619	3.586	3.556	3.529	3.505	3.348	3.266
15	3.666	3.612	3.564	3.522	3.485	3.452	3.423	3.396	3.372	3.214	3.132
16	3.553	3.498	3.451	3.409	3.372	3.339	3.310	3.283	3.259	3.101	3.018
17	3.455	3.401	3.353	3.312	3.275	3.242	3.212	3.186	3.162	3.003	2.920
18	3.371	3.316	3.269	3.227	3.190	3.158	3.128	3.101	3.077	2.919	2.835
19	3.297	3.242	3.195	3.153	3.116	3.084	3.054	3.027	3.003	2.844	2.761
20	3.231	3.177	3.130	3.088	3.051	3.018	2.989	2.962	2.938	2.778	2.695
21	3.173	3.119	3.072	3.030	2.993	2.960	2.931	2.904	2.880	2.720	2.636
22	3.121	3.067	3.019	2.978	2.941	2.908	2.879	2.852	2.827	2.667	2.583
23	3.074	3.020	2.973	2.931	2.894	2.861	2.832	2.805	2.781	2.620	2.535
24	3.032	2.977	2.930	2.889	2.852	2.819	2.789	2.762	2.738	2.577	2.492
25	2.993	2.939	2.892	2.850	2.813	2.780	2.751	2.724	2.699	2.538	2.453
26	2.958	2.904	2.857	2.815	2.778	2.745	2.715	2.688	2.664	2.503	2.417
27	2.926	2.871	2.824	2.783	2.746	2.713	2.683	2.656	2.632	2.470	2.384
28	2.896	2.842	2.795	2.753	2.716	2.683	2.653	2.626	2.602	2.440	2.354
29	2.868	2.814	2.767	2.726	2.689	2.656	2.626	2.599	2.574	2.412	2.325
30	2.843	2.789	2.742	2.700	2.663	2.630	2.600	2.573	2.549	2.386	2.299
35	2.740	2.686	2.639	2.597	2.560	2.527	2.497	2.470	2.445	2.281	2.193
40	2.665	2.611	2.563	2.522	2.484	2.451	2.421	2.394	2.369	2.203	2.114
45	2.608	2.553	2.506	2.464	2.427	2.393	2.363	2.336	2.311	2.144	2.054
50	2.562	2.508	2.461	2.419	2.382	2.348	2.318	2.290	2.265	2.098	2.007
60	2.496	2.442	2.394	2.352	2.315	2.281	2.251	2.223	2.198	2.028	1.936
70	2.450	2.395	2.348	2.306	2.268	2.234	2.204	2.176	2.150	1.980	1.886
80	2.415	2.361	2.313	2.271	2.233	2.199	2.169	2.141	2.115	1.944	1.849
90	2.389	2.334	2.286	2.244	2.206	2.172	2.142	2.114	2.088	1.916	1.820
100	2.368	2.313	2.265	2.223	2.185	2.151	2.120	2.092	2.067	1.893	1.797
1000	2.203	2.148	2.099	2.056	2.018	1.983	1.952	1.923	1.897	1.716	1.613

TABLE VI   Critical Values for the Mann–Whitney Test

n_1	probability	$n_2 = 4$	$n_2 = 5$	$n_2 = 6$	$n_2 = 7$	$n_2 = 8$	$n_2 = 9$	$n_2 = 10$
	.10	4	5	6	7	8	10	11
	.05	2	3	4	5	6	7	8
4	.025	1	2	3	4	5	5	6
	.01	0	1	2	2	3	4	4
	.005	0	0	1	1	2	2	3

n_1	probability	$n_2 = 4$	$n_2 = 5$	$n_2 = 6$	$n_2 = 7$	$n_2 = 8$	$n_2 = 9$	$n_2 = 10$
	.10	5	6	8	9	11	13	14
	.05	3	5	6	7	9	10	12
5	.025	2	3	4	6	7	8	9
	.01	1	2	3	4	5	6	7
	.005	0	1	2	2	3	4	5

n_1	probability	$n_2 = 4$	$n_2 = 5$	$n_2 = 6$	$n_2 = 7$	$n_2 = 8$	$n_2 = 9$	$n_2 = 10$
	.10	6	8	10	12	14	16	18
	.05	4	6	8	9	11	13	15
6	.025	3	4	6	7	9	11	12
	.01	2	3	4	5	7	8	9
	.005	1	2	3	4	5	6	7

n_1	probability	$n_2 = 4$	$n_2 = 5$	$n_2 = 6$	$n_2 = 7$	$n_2 = 8$	$n_2 = 9$	$n_2 = 10$
	.10	7	9	12	14	17	19	22
	.05	5	7	9	12	14	16	18
7	.025	4	6	7	9	11	13	15
	.01	2	4	5	7	8	10	12
	.005	1	3	4	5	7	8	10

n_1	probability	$n_2 = 4$	$n_2 = 5$	$n_2 = 6$	$n_2 = 7$	$n_2 = 8$	$n_2 = 9$	$n_2 = 10$
	.10	8	11	14	17	20	23	25
	.05	6	9	11	14	16	19	21
8	.025	5	7	9	11	14	16	18
	.01	3	5	7	8	10	12	14
	.005	2	3	5	7	8	10	12

n_1	probability	$n_2 = 4$	$n_2 = 5$	$n_2 = 6$	$n_2 = 7$	$n_2 = 8$	$n_2 = 9$	$n_2 = 10$
	.10	10	13	16	19	23	26	29
	.05	7	10	13	16	19	22	25
9	.025	5	8	11	13	16	18	21
	.01	4	6	8	10	12	15	17
	.005	2	4	6	8	10	12	14

n_1	probability	$n_2 = 4$	$n_2 = 5$	$n_2 = 6$	$n_2 = 7$	$n_2 = 8$	$n_2 = 9$	$n_2 = 10$
	.10	11	14	18	22	25	29	33
	.05	8	12	15	18	21	25	28
10	.025	6	9	12	15	18	21	24
	.01	4	7	9	12	14	17	20
	.005	3	5	7	10	12	14	17

Table created by Chris Lacke using Mathematica™

TABLE VI   Critical Values for the Wilcoxon Signed-Rank Test

Probability	$n = 5$	$n = 6$	$n = 7$	$n = 8$	$n = 9$	$n = 10$	$n = 11$	$n = 12$
.10	2	3	5	8	10	14	17	21
.05	0	2	3	5	8	10	13	17
.025	*	0	2	3	5	8	10	13
.01	*	*	0	1	3	5	7	10
.005	*	*	*	0	1	3	5	7

Probability	$n = 13$	$n = 14$	$n = 15$	$n = 16$	$n = 17$	$n = 18$	$n = 19$	$n = 20$
.10	26	31	36	42	48	55	62	69
.05	21	25	30	35	41	47	53	60
.025	18	21	25	29	34	40	46	52
.01	12	15	19	23	27	32	37	43
.005	9	12	15	19	23	27	32	37

Table created by Chris Lacke using Mathematica™

TABLE VIII   Sigma Conversion Table

Long-Term Yield	Long-Term Sigma	Short-Term Sigma	Defects Per Million	Defects Per 100
99.99966%	4.5	6.0	3.4	0.00034
99.9995%	4.4	5.9	5	0.0005
99.9992%	4.3	5.8	8	0.008
99.9990%	4.2	5.7	10	0.001
99.9980%	4.1	5.6	20	0.002
99.9970%	4.0	5.5	30	0.003
99.9960%	3.9	5.4	40	0.004
99.9930%	3.8	5.3	70	0.007
99.9900%	3.7	5.2	100	0.1
99.9850%	3.6	5.1	150	0.015
99.9770%	3.5	5.0	230	0.023
99.9670%	3.4	4.9	330	0.033
99.9520%	3.3	4.8	480	0.048
99.9320%	3.2	4.7	680	0.068
99.9040%	3.1	4.6	960	0.096
99.8650%	3.0	4.5	1,350	0.135
99.8140%	2.9	4.4	1,860	0.186
99.7450%	2.8	4.3	2,550	0.255
99.6540%	2.7	4.2	3,460	0.346
99.5340%	2.6	4.1	4,660	0.466
99.3790%	2.5	4.0	6,210	0.621
99.1810%	2.4	3.9	8,190	0.819
98.930%	2.3	3.8	10,700	1.07
98.610%	2.2	3.7	13,900	1.39
98.220%	2.1	3.6	17,800	1.78
97.730%	2.0	3.5	22,700	2.27
97.130%	1.9	3.4	28,700	2.87
96.410%	1.8	3.3	35,900	3.59
95.540%	1.7	3.2	44,600	4.46
94.520%	1.6	3.1	54,800	5.48
93.320%	1.5	3.0	66,800	6.68
91.920%	1.4	2.9	80,800	8.08

TABLE VIII   Sigma Conversion Table (*Continued*)

Long-Term Yield	Long-Term Sigma	Short-Term Sigma	Defects Per Million	Defects Per 100
90.320%	1.3	2.8	96,800	9.68
88.50%	1.2	2.7	115,000	11.5
86.50%	1.1	2.6	135,000	13.5
84.20%	1.0	2.5	158,000	15.8
81.60%	0.9	2.4	184,000	18.4
78.80%	0.8	2.3	212,000	21.2
75.80%	0.7	2.2	242,000	24.2
72.60%	0.6	2.1	274,000	27.4
69.20%	0.5	2.0	308,000	30.8
65.60%	0.4	1.9	344,000	34.4
61.80%	0.3	1.8	382,000	38.2
58.00%	0.2	1.7	420,000	42.0
54.00%	0.1	1.6	460,000	46.0

APPENDIX B
Answers to Selected Odd-Numbered Exercises

Chapter 8

8.1 **(a)**

Response	f
Y	21
N	15
D	4
Total	40

(b) Y: $\frac{21}{40}$, 52.5%; N: $\frac{15}{40}$, 37.5%; D: $\frac{4}{40}$, 10%; **(c)** 52.5%; **(d)** 47.5%.

8.3 **(a)**

Class	f
41–44	4
45–48	8
49–52	9
53–56	4
Total	25

(b) 41–44: $\frac{4}{25}$, 16%; $\frac{8}{25}$, 32%; $\frac{9}{25}$, 36%; 53–56, 16%.

(c) See Figure B.1.

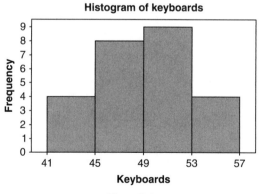

Figure B.1

Practitioner's Guide for Statistics and Lean Six Sigma for Process Improvements. By Mikel J. Harry, Prem S. Mann, Ofelia C. de Hodgins, Christopher J. Lacke, and Richard Hulbert
Copyright © 2010 John Wiley & Sons, Inc.

(d) 16%.

(e) See Figure B.2.

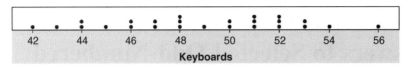

Figure B.2

(f)

4	23
4	445
4	6677
4	8889
5	00111
5	22233
5	4
5	6

8.5 (a)

Class	f
(b) 1–400	18
401–800	4
801–1200	2
1201–1600	2
1601–2000	2
2001–2400	2
Total	30

(b) 1–400: $\frac{18}{30}$, 60%; 401–800: $\frac{4}{30}$, 13.33%; 801–1200: $\frac{2}{30}$, 6.67%; 1201–1600: $\frac{2}{30}$, 6.67%; 1601–2000: $\frac{2}{30}$, 6.67%; 2001–2400: $\frac{2}{30}$, 6.67%.

(c) 20%.

(d) See Figure B.3.

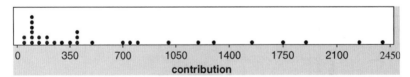

Figure B.3

8.7 (a)

Class	f
$70,000–$79,999	3
$80,000–$89,999	8
$90,000–$99,999	6
$100,000–$109,999	11
$110,000–$119,999	10
$120,000–$129,999	2
Total	40

(b) $70,000–$79,999: $\frac{3}{40}$, 7.5%; $80,000–$89,999: $\frac{8}{40}$, 20%; $90,000–$99,999: $\frac{6}{40}$, 15%; $100,000–$109,999: $\frac{11}{40}$, 27.5%; $110,000–$119,999: $\frac{10}{40}$, 25%; $120,000–$129,999: $\frac{2}{40}$, 5%.

(c) $10,000.

8.9 **(a)**

3	89
4	23456779
5	133556889
6	7

(b)

Mean	50.25
Median	50
Mode	Not uniquely defined (47, 53, 55, and 58 each occur twice)
Range	29
Variance	56.62
Standard deviation	7.52
Coefficient of variation	14.97
Q1	44.25
Q3	55.75
IQR	11.5

8.11 **(a)**

3	65 76 87
4	09 09 30 30 52 52 61 74 74 96
5	17 61 61 83 83 83
6	04 26 48 48 91 91
7	30 35 61 74 87
8	04 74
9	26 52
10	17

(b)

3–4	65 76 87 09 09 30 30 52 52 61 74 74 96
5–6	17 61 61 83 83 83 04 26 48 48 91 91
7–8	30 35 61 74 87 04 74
9–10	26 52 17

(c)

Mean	607.7
Median	583
Mode	583
Range	652
Variance	31,318.9
Standard deviation	177.0
Coefficient of variation	29.12%
Q1	452
Q3	735
IQR	283

8.13 **(a)**

	Mean	Median
With outlier	11189	2450
Without outlier	3838	2325

The mean changes by a larger amount (7351 vs. 125).

(b) The median.

8.15 **(a)**

Mean	24
Median	15.5
Mode	Not uniquely defined (7, 8, and 13 each occur twice)
Range	60
Variance	380.11
Standard deviation	19.50
Coefficient of variation	81.23%
Q1	8.5
Q3	36
IQR	27.5

(b) $P_{80} = 38$; 80% of the companies in this dataset have gross sales levels of $38 billion or less.

8.17 **(a)**

Mean	11.13
Median	10.5
Mode	6
Range	24
Variance	46.12
Standard deviation	6.79
Coefficient of variation	61.04%
Q1	6
Q3	16.75
IQR	10.75

(b) $P_{68} = 15$(by hand) or 14.64 (using software); 68% of the people in this dataset bought 15 or fewer books during the past year.

8.19 **(a)** See Figure B.4.

(b) $r = .973$, $r_s = .964$, $r_\tau = .905$.

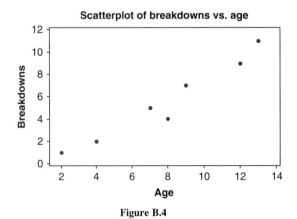

Figure B.4

8.21 **(a)** See Figure B.5.
(b) $r = .879$, $r_s = .929$, $r_\tau = .810$.

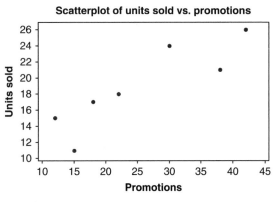

Figure B.5

Chapter 10

10.1 $S = \{AB, AC, BA, BC, CA, CB\}$.

10.3 Using C for correct and W for wrong: **(a)** WW, simple; **(b)** CC, WC, CW, compound; **(c)** CW, simple; **(d)** CW, WC, compound.

10.5 .6190; .3810; the values add up to 1 because a person either does or does not belong to a union, but not both. The events are complementary.

10.7 **(a)** yes; **(b)** no, **(c)** $A^c = \{1, 2, 5, 7, 8, 10\}$, $B^c = \{3, 4, 6, 7, 8, 9, 10\}$, $P(A^c) = .6$, $P(B^c) = .7$

10.9 **(a)** (i) .505, (ii) .35, (iii) .3, (iv) .1833; **(b)** no, yes, it is possible for a person to have a high school education and be better off, but it is impossible for a person to have both less than and more than a high school education; **(c)** no, P(worse off|more than high school) = .1167, P(worse off) = .285. The probability of being worse off changes when a person has more than a high school education.

10.11 **(a)** .4543; **(b)** .042

10.13 **(a)** (i) .18, (ii) .12; **(b)** the other events for which joint probability calculations can be made are "is single and doesn't own stocks or bonds" and "is married and doesn't own stocks or bonds"; .28, .42.

10.15 **(a)** .58; **(b)** .82.

10.17 .5862.

Chapter 11

11.1 **(a)**, **(c)**, **(d)**, and **(e)** are discrete, although **(c)** is often approximated by a continuous random variable; **(b)** and **(f)** are continuous

11.3 (i) .14; (ii) .25; (iii) .67; (iv) .33.

11.5   A consumer agency surveyed all 2500 families living in a small town to collect data on the number of television sets they owned. The following table lists the frequency distribution of the data collected by this agency.

Number of TV sets owned	0	1	2	3	4
Number of families	120	970	730	410	270

(a)

Number of TV sets owned	0	1	2	3	4
Probability	.048	.388	.292	.164	.108

(b) (i) .388; (ii) .272; (iii) .436; (iv) .834.

11.7   Despite all safety measures, accidents do happen at Brown's Manufacturing Corporation. Let x denote the number of accidents that occur during a month at this company. The following table, reproduced from Exercise 11.4, lists the probability distribution of x.

x	0	1	2	3	4
$P(x)$	.25	.30	.20	.15	.10

11.7 $\mu = 1.55$, $\sigma = 1.284$; on average, there are 1.55 accidents per month.

11.9   **(a)** Binomial. There are three trials (fixed), two events (red and blue), and replacing the ball each time results in independent trials with a constant probability of each color (.6 for red, .4 for blue).

 (b) Not binomial. Not replacing the ball causes the two assumptions in the last sentence to become invalid.

 (c) Binomial (approximately). The number of households in NYC is so large that the probability of owning stocks can be considered constant as long as the sample size is not too large.

11.11   **(a)** .2541; **(b)** .2553; **(c)** .1361.

11.13   **(a)** .5165; **(b)** .4835.

11.15   **(a)** .0733; **(b)** (i) .0183, (ii) .1465.

Chapter 12

12.1   **(a)** .7823; **(b)** .8553; **(c)** .4878; **(d)** .4996; **(e)** .0011; **(f)** .9884.

12.3   **(a)** 1.4; **(b)** -2.2; **(c)** -1.4; **(d)** 2.8.

12.5   **(a)** .7967; **(b)** .3372; **(c)** .0475; **(d)** .7734.

12.7   **(a)** 1.88%; **(b)** yes, there is no maximum value assumed, but it is highly unlikely.

12.9   64.658 oz.

12.11   **(a)** .0062; **(b)** .1525; **(c)** .8385; **(d)** yes, there is no maximum value assumed, but it is highly unlikely.

12.13   1.24%.

12.15   .2636.

12.17   **(a)** .0357; **(b)** .7364.

12.19   **(a)** .0821; **(b)** .3408; **(c)** .0986.

Chapter 13

13.1 **(a)**

x	6	8	12	20	25
$P(x)$	.2	.2	.2	.2	.2

(b)

Sample	$\bar{x}$	$P(\bar{x})$
6, 8, 12, 20	11.5	.2
6, 8, 12, 25	12.75	.2
6, 8, 20, 25	14.75	.2
6, 12, 20, 25	15.75	.2
8, 12, 20, 25	16.25	.2

(c) $\mu = 14.2$ Sampling error: $-2.7, -1.45, .55, 1.55, 2.05$.

13.3 The sampling distribution is normally distributed with a mean of 80 h. and a standard deviation of 1.5 h.

13.5 **(a)** .5910; **(b)** .1585; **(c)** .1820.

13.7 **(a)** The sampling distribution is approximately normally distributed with a mean of .65 and a standard deviation of .0570.

(b) .2152.

(c) .1469.

Chapter 14

14.1 **(a)** The sample size of 70 is large enough that the central-limit theorem is applicable, making the sampling distribution of the average hourly rate become approximately normal; **(b)** ($519.21, $580.79).

14.3 The confidence interval is (63.7594, 64.0106). This interval falls completely within the limits 63.7 and 64.3, so the machine will not need an adjustment.

14.5 **(a)** $\bar{x} = \$161.97$, $s = \$46.74$.

(b) The histogram in Figure B.6 is approximately symmetric and contains no outliers. Using the t procedures for calculating a confidence interval for a population mean can be done

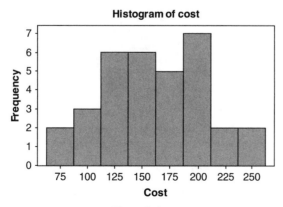

Figure B.6

with a small sample size under the aforementioned conditions, so a sample size of 33 is sufficiently large.

(c) ($142.08, $181.86).

14.7   (a) $\hat{p} = 1/3$; (b) (.2549, .4117); (c)$\frac{82}{244} = .3361$; (d) (.2582, .4140), the interval based on the Wilson estimator is .001 smaller than the interval based on the normal approximation; (e) yes, the respective products are 80 and 160, both of which are greater than or equal to 10.

14.9   Variance: (.00689, .03887); standard deviation: (.08303, .19715).

14.11   (a) ($79.47, $90.53); (b) you should discuss the tradeoffs of increasing the sample size and lowering the confidence level.

14.13   (a) (235.83, 260.17); (b) you should discuss the tradeoffs of increasing the sample size and lowering the confidence level.

14.15   The confidence interval is (24.0064, 24.0236). This interval falls completely within the limits 23.975 and 24.025, so the machine will not need an adjustment.

Chapter 16

16.1   (a) H_0: $\mu \leq \$2,103,000$, H_1 : $\mu > \$2,103,000$, right-tailed.

(b) H_0: $\mu = 15$, H_1: $\mu \neq 15$, two-tailed.

(c) H_0: $\mu \geq 45$, H_1: $\mu < 45$, left-tailed.

(d) H_0: $\mu \leq 35$, H_1: $\mu > 35$, right-tailed.

(e) H_0: $\mu = 50$, H_1: $\mu \neq 50$, two-tailed.

16.3   (a) H_0: $\mu = 16.6$, H_1: $\mu \neq 16.6$, $t = 2.176$, $.02 < p$ value $< .05$ (from MINITAB, p value .031), reject H_0, there is significant evidence that the average amount of leisure time is different than in 1988.

(b) The type I error in this case would be to conclude that the average amount of leisure time has changed when it really hasn't. The probability of making a type I error is .031.

16.5   (a) No, you will automatically fail to reject H_0, but this is not the same as accepting H_0.

(b) H_0: $\mu \geq 1200$, H_1: $\mu < 1200$, $t = 4.12$, p value $< .0005$, reject H_0, there is significant evidence that the average typing speed is less than 1200 words per hour, which contradicts the claim of the business school.

16.7 H_0: $\mu \leq 2$, H_1 : $\mu > 2$, $t = 1.68$, $.05 < p$ value $< .1$, fail to reject H_0, there is not significant evidence that the average learning time exceeds 2 hours, thus the company's claim is reasonable.

16.9 H_0: $p \geq .45$, H_1: $p < .45$, $z = 2.20$, p value $= .015$, reject H_0, there is significant evidence that the current percentage of workers who would change careers if possible is less than 45%.

Chapter 17

17.1   (a) $(0.5962, 2.4038)$; (b) H_0: $\mu_1 \leq \mu_2$, H_1: $\mu_1 > \mu_2$, (max s_i/min s_i) $= (1.8/1.35) = 1.33$ ≤ 2—use pooled t-test, $t = 4.38$, p value $< .0005$, reject H_0, there is significant evidence that the average number of policies sold per week is higher for salespersons with a business degree than for salespersons with a degree other than in business.

17.3   (a) $(-11.4895, -4.1105)$; (b) H_0: $\mu_1 \geq \mu_2$, H_1: $\mu_1 < \mu_2$, (max s_i/min s_i) $= (7.7/5.3) = 1.45 \leq 2$ —use pooled t-test, $t = -4.24$, p value $< .0005$, reject H_0, there is significant evidence that the average corn yield in Ohio is lower than that in Iowa.

17.5 H_0: $\eta_1 \leq \eta_2$, H_1: $\eta_1 > \eta_2$, $U = 81.5 - (8(8+1)/2) = 45.5$, $U_c = n_1 n_2 - U' = 8 \times 8 - 13 = 47$. Since $45.5 < 47$, fail to reject H_0, there is not significant evidence that the median waiting time with four tellers is shorter than the median waiting time with three tellers.

17.7   **(a)** $(-12.6293, 3.2007)$; **(b)** H_0: $\mu_d \geq 0$, H_1: $\mu_d < 0$, $t = -2.21$, p value $< .035$, fail to reject H_0, there is not significant evidence that there is, on average, an increase in the test score. **(c)** H_0: $\eta_d \geq 0$, H_1: $\eta_d < 0$, $W = 3.5$, p value $< .035$, $W_c = 1$, since $3.5 > 1$ we fail to reject H_0, there is not significant evidence that the median increase in the test score is positive.

17.9   **(a)** $(-0.1510, 0.0510)$; **(b)** H_0: $p_1 \geq p_2$, H_1: $p_1 < p_2$, $z = -1.07$, p value $= .141$, fail to reject H_0, there is not significant evidence that the proportion of all customers of the restaurant in area A who think that the food and service are excellent is less than the proportion of all customers of the restaurant in area B who think that the food and service are excellent.

17.11   **(a)** $(-0.281819, 0.128486)$; **(b)** H_0: $p_1 = p_2$, H_1: $p_1 \neq p_2$, $z = -0.97$, p value $= .332$, fail to reject H_0, there is not significant evidence of a difference between the airbag deployment rates of companies A and B.

Chapter 18

18.1   **(a)** 14.8603; **(b)** .7107.

18.3 H_0: $p_1 = .0994$, $p_2 = .2997$, $p_3 = .2716$, $p_4 = .3293$, H_1: at least two of the equalities do not hold, $\chi^2 = 8.7276$, df $= 3$, $.025 < p$ value $< .05$ (.033 from MINITAB), fail to reject H_0, there is not significant evidence that at least two of the percentages of workers in the US labor force by educational attainment have changed since June 2006.

18.5 H_0: $p_1 = p_2 = p_3 = p_4 = p_5 = .2$, H_1: at least two of the equalities do not hold, $\chi^2 = 6.75$, df $= 4$, $.1 < p$ value $< .2$ (.150 from MINITAB), fail to reject H_0, there is not significant evidence to conclude that the order distribution is not uniform over the 5 days of the week. In other words, the assumption that 20% of the orders come in on each of the five weekdays is reasonable.

18.7 H_0: punishment level and one's victim status are independent, H_1: punishment level and one's victim status are dependent, $\chi^2 = 19.216$, df $= 2$, p value $< .0005$, reject H_0, there is significant evidence that one's status as a victim and the punishment level handed out as a juror are dependent.

18.9 H_0: the distributions of support level for the farm subsidy program are the same for the Northeast, the Midwest, the South, and the West, H_1: the distributions of support level for the farm subsidy program are different for at least two of the aforementioned regions, $\chi^2 = 7.982$, df $= 6$, p value $> .2$ (.239 from MINITAB), fail to reject H_0, there is not significant evidence that the region one lives in has an impact on one's support level of a certain farm subsidy program.

18.11 H_0: the loan approval rates for all loans are the same for Thurow and Webber, H_1: the loan approval rates for all loans are different for Thurow and Webber, $\chi^2 = .699$, df $= 1$, p value $> .2$ (.403 from MINITAB), fail to reject H_0, there is not significant evidence that the loan approval rates are different for Thurow and Webber.

Chapter 19

19.1   **(a)** 3.11; **(b)** 2.29; **(c)** 6.47.

19.3 H_0: $\mu_1 = \mu_2 = \mu_3$, H_1: not all three population means are equal, $F = 7.86$, df $= (2,18)$, $.001 < p$ value $< .01$ (.004 from MINITAB), reject H_0, there is significant evidence to conclude that the average numbers of sales are not all the same for the three degrees. Post hoc comparison results (using Tukey's HSD): math–marketing: $(-8.653, -1.632)$, sociology – marketing: $(-7.653, -0.632)$, sociology–math: $(-2.510, 4.510)$. The confidence interval for the difference in the mean sales levels for sociology and math contains zero, so the average sales levels are not significantly different for sociology and math. Both intervals involving marketing do not contain zero, so the average sales level for marketing is considered to be significantly different from the average sales level for math and the average sales level for marketing.

19.5 $H_0: \mu_1 = \mu_2 = \mu_3$, H_1 : not all three population means are equal, $F = 26.48$, df $= (2,21)$, p value $< .001$ ($<.0005$ from MINITAB), reject H_0, there is significant evidence to conclude that the average wheat yields are not all the same for the three fertilizers. Post hoc comparison results (using Tukey's HSD): fertilizer 2 − fertilizer 1: $(-24.298, -11.702)$, fertilizer 3 − fertilizer 1: $(-12.923, -0.327)$, fertilizer 3 − fertilizer 2 $(5.077, 17.673)$. None of the intervals involving marketing contain zero, so the average wheat yields are significantly different for each of the three fertilizers.

Chapter 20

20.1   **(a)** $40; **(b)** each will pay the same amount because each person has to pay the $30 rental fee, they are charged the same rate per mile, and they all drive the same distance; **(c)** it is exact.

20.3   **(a)** $\hat{y} = 2.32 + 1.0935x$; **(b)** the estimated slope is 1.0935, which implies that as household income increases by $100, the monthly telephone bill increases, on average, by slightly more than $1.09—the estimated intercept **(a)** is 2.32, which would imply that a family with no income would have a monthly telephone bill of $2.32—however, this is an example of extrapolation because the smallest monthly income in the data set is $3000, so the intercept is meaningless in terms of the application; **(c)** $\hat{y} = 2.32 + 1.0935(50) = 56.995$, thus the predicted monthly phone bill is $57.00; **(d)** $s_e = 20.3132$, SST $= 9185.6$, SSE $= 3301.0$, SSR $= 5884.6$, $R^2 = .641$, $r = .801$; **(e)** $(.2545, 1.9325)$; **(f)** $H_0 : B \geq 1.5$, $H_1 : B < 1.5$, $t = -1.404$, $.05 < p$ value $< .1$, fail to reject H_0, there is not significant evidence that the slope is less than 1.5; **(g)** see figures B.7 and B.8.

As income increases, there is more variability in the amount of the phone bills and the residuals. The pattern of the residual plot is a fan, which implies that the constant variance assumption may be unreasonable. Therefore, a line may not be appropriate for modeling this relationship.

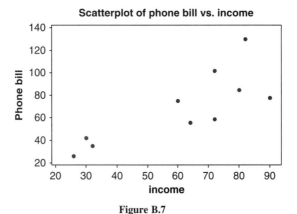

Figure B.7

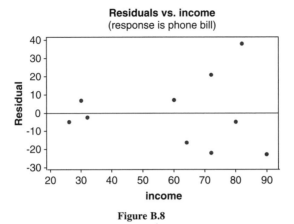

Figure B.8

20.5 (a) The production level is the explanatory variable, and the total cost is the response variable. The relationship should be positive because it costs more to produce more.

(b) $\hat{y} = 278.48 + 9.0104x$.

(c) The estimated slope is 9.0104, which implies that as the number of bats produced increases by 1000, the total cost increases, on average, by $9010.40. The estimated intercept (a) is 278.48, which would imply that a company that produces no bats would have a total cost of $278,400. However, this is an example of extrapolation because the smallest number of bats produced in the dataset is 40,000, so the intercept is meaningless in terms of the application.

(d) See Figures B.9 and B.10. On the basis of the two plots (in Figs. B.9 and B.10), using a line to model the relationship is appropriate. Note that the residual plot has no distinct pattern, which implies that the line was appropriate.

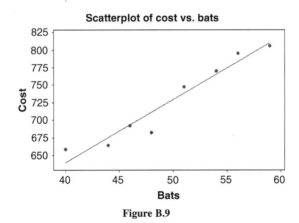

Figure B.9

(e) $R^2 = .937$, $r = .968$, the value of r is very close to 1, which implies that there is a strong linear association between the two variables.

(f) $\hat{y} = 278.48 + 9.0104(49) = 719.9896$; thus the predicted total cost is $719,989.60.

(g) $s_e = 16.2787$.

(h) (6.6694, 11.3514).

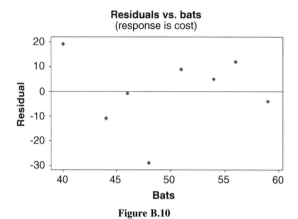

Figure B.10

(i) $H_0: B \leq 0$, $H_1: B > 0$, $t = 9.42$, p value $< .0005$, reject H_0; there is significant evidence that the slope is positive, implying a positive association between the number of bats made and total cost.

20.7 (a) $\hat{y} = -9.225 + .17332x_1 + .5177x_2$, where x_1 is population and x_2 is income. (b) the estimated coefficient of population is .17332, which implies that as the population within 5 mi increases by 1000 and there is no change in income, sales increase, on average, by $173.32; the estimated coefficient of income is .5177, which implies that as the mean annual income increases by $1000, with no change in population, sales increase, on average, by $517.70; the estimated intercept is meaningless because the implication is that there is no one living within 5 mi of the restaurant, which is not the way that chain restaurants operate; (c) $s_e = 4.1287$, $R^2 = .626$, $\bar{R}^2 = .533$; (d) SST $= 364.91$, SSR $= 228.54$, SSE $= 136.37$; (e) $\hat{y} = -9.225 + .17332(50) + .5177(52) = 26.3614$, thus, predicted weekly sales are $26,361.40; (f) $\hat{y} = -9.225 + .17332(45) + .5177(60) = 29.6364$, thus, the expected weekly sales are $29,636.40; (g) (.1891, .8463); (h) $H_0: B_2 = 0$, $H_1: B_2 \neq 0$, $t = 3.63$, p value $= .007$, reject H_0, there is significant evidence that the coefficient of income is significantly different from zero; (i) $H_0: B_1 = B_2 = 0$, H_1 : at least one $B_i \neq 0$, $F = 6.70$, p value $= .02$, reject H_0, there is significant evidence that at least one of the coefficients is significantly different from zero; (j) population and sales: $-.101$, income and sales: $.662$, population and income: $-.650$.

INDEX

Practitioner's Guide for Statistics and Lean Six Sigma for Process Improvements. Edited by Mikel J. Harry, Prem S. Mann, Ofelia C. de Hodgins, Christopher J. Lacke, and Richard Hulbert
Copyright © 2010 John Wiley & Sons, Inc.